The Electrical Engineering Handbook
Third Edition

Systems, Controls, Embedded Systems, Energy, and Machines

The Electrical Engineering Handbook Series

Series Editor
Richard C. Dorf
University of California, Davis

Titles Included in the Series

The Handbook of Ad Hoc Wireless Networks, Mohammad Ilyas
The Avionics Handbook, Cary R. Spitzer
The Biomedical Engineering Handbook, Third Edition, Joseph D. Bronzino
The Circuits and Filters Handbook, Second Edition, Wai-Kai Chen
The Communications Handbook, Second Edition, Jerry Gibson
The Computer Engineering Handbook, Vojin G. Oklobdzija
The Control Handbook, William S. Levine
The CRC Handbook of Engineering Tables, Richard C. Dorf
The Digital Signal Processing Handbook, Vijay K. Madisetti and Douglas Williams
The Electrical Engineering Handbook, Third Edition, Richard C. Dorf
The Electric Power Engineering Handbook, Leo L. Grigsby
The Electronics Handbook, Second Edition, Jerry C. Whitaker
The Engineering Handbook, Third Edition, Richard C. Dorf
The Handbook of Formulas and Tables for Signal Processing, Alexander D. Poularikas
The Handbook of Nanoscience, Engineering, and Technology, William A. Goddard, III, Donald W. Brenner, Sergey E. Lyshevski, and Gerald J. Iafrate
The Handbook of Optical Communication Networks, Mohammad Ilyas and Hussein T. Mouftah
The Industrial Electronics Handbook, J. David Irwin
The Measurement, Instrumentation, and Sensors Handbook, John G. Webster
The Mechanical Systems Design Handbook, Osita D.I. Nwokah and Yidirim Hurmuzlu
The Mechatronics Handbook, Robert H. Bishop
The Mobile Communications Handbook, Second Edition, Jerry D. Gibson
The Ocean Engineering Handbook, Ferial El-Hawary
The RF and Microwave Handbook, Mike Golio
The Technology Management Handbook, Richard C. Dorf
The Transforms and Applications Handbook, Second Edition, Alexander D. Poularikas
The VLSI Handbook, Wai-Kai Chen

The Electrical Engineering Handbook
Third Edition

Edited by
Richard C. Dorf

Circuits, Signals, and Speech and Image Processing

Electronics, Power Electronics, Optoelectronics, Microwaves, Electromagnetics, and Radar

Sensors, Nanoscience, Biomedical Engineering, and Instruments

Broadcasting and Optical Communication Technology

Computers, Software Engineering, and Digital Devices

Systems, Controls, Embedded Systems, Energy, and Machines

The Electrical Engineering Handbook
Third Edition

Systems, Controls, Embedded Systems, Energy, and Machines

Edited by
Richard C. Dorf
University of California
Davis, California, U.S.A.

Taylor & Francis
Taylor & Francis Group
Boca Raton London New York

A CRC title, part of the Taylor & Francis imprint, a member of the Taylor & Francis Group, the academic division of T&F Informa plc.

Published in 2006 by
CRC Press
Taylor & Francis Group
6000 Broken Sound Parkway NW, Suite 300
Boca Raton, FL 33487-2742

Printed in the United States of America on acid-free paper
10 9 8 7 6 5 4 3 2 1

International Standard Book Number-10: 0-8493-7347-6 (Hardcover)
International Standard Book Number-13: 978-0-8493-7347-3 (Hardcover)
Library of Congress Card Number 2005054347

Library of Congress Cataloging-in-Publication Data

Systems, controls, embedded systems, energy, and machines / edited by Richard C. Dorf.
p. cm.
Includes bibliographical references and index.
ISBN 0-8493-7347-6 (alk. paper)
1. Electric power systems--Control. 2. Electric power systems. 3. Systems engineering. I. Dorf, Richard C. II. Title.

TK1001.S93 2005
621.3--dc22 2005054347

informa
Taylor & Francis Group
is the Academic Division of Informa plc.

Visit the Taylor & Francis Web site at
http://www.taylorandfrancis.com

and the CRC Press Web site at
http://www.crcpress.com

Preface

Purpose

The purpose of *The Electrical Engineering Handbook, 3rd Edition* is to provide a ready reference for the practicing engineer in industry, government, and academia, as well as aid students of engineering. The third edition has a new look and comprises six volumes including:

Circuits, Signals, and Speech and Image Processing
Electronics, Power Electronics, Optoelectronics, Microwaves, Electromagnetics, and Radar
Sensors, Nanoscience, Biomedical Engineering, and Instruments
Broadcasting and Optical Communication Technology
Computers, Software Engineering, and Digital Devices
Systems, Controls, Embedded Systems, Energy, and Machines

Each volume is edited by Richard C. Dorf and is a comprehensive format that encompasses the many aspects of electrical engineering with articles from internationally recognized contributors. The goal is to provide the most up-to-date information in the classical fields of circuits, signal processing, electronics, electromagnetic fields, energy devices, systems, and electrical effects and devices, while covering the emerging fields of communications, nanotechnology, biometrics, digital devices, computer engineering, systems, and biomedical engineering. In addition, a complete compendium of information regarding physical, chemical, and materials data, as well as widely inclusive information on mathematics, is included in each volume. Many articles from this volume and the other five volumes have been completely revised or updated to fit the needs of today, and many new chapters have been added.

The purpose of *Systems, Controls, Embedded Systems, Energy, and Machines* is to provide a ready reference to subjects in the fields of energy devices, machines, and systems, as well as control systems and embedded systems. Here we provide the basic information for understanding these fields. We also provide information about the emerging fields of embedded systems.

Organization

The information is organized into three sections. The first two sections encompass 20 chapters, and the last section summarizes the applicable mathematics, symbols, and physical constants.

Most chapters include three important and useful categories: defining terms, references, and further information. *Defining terms* are key definitions and the first occurrence of each term defined is indicated in boldface in the text. The definitions of these terms are summarized as a list at the end most chapters or articles. The *references* provide a list of useful books and articles for follow-up reading. Finally, *further information* provides some general and useful sources of additional information on the topic.

Locating Your Topic

Numerous avenues of access to information are provided. A complete table of contents is presented at the front of the book. In addition, an individual table of contents precedes each of the sections. Finally, each chapter begins with its own table of contents. The reader should look over these tables of contents to become familiar with the structure, organization and content of the book. For example, see Section I: Energy, then

Chapter 2: Alternative Power Systems and Devices, and then Chapter 2.1: Distributed Power. This tree-and-branch table of contents enables the reader to move up the tree to locate information on the topic of interest.

Two indexes have been compiled to provide multiple means of accessing information: a subject index and an index of contributing authors. The subject index can also be used to locate key definitions. The page on which the definition appears for each key (defining) term is clearly identified in the subject index.

The Electrical Engineering Handbook, 3rd Edition is designed to provide answers to most inquiries and direct the inquirer to further sources and references. We hope that this volume will be referred to often and that informational requirements will be satisfied effectively.

Acknowledgments

This volume is testimony to the dedication of the Board of Advisors, the publishers, and my editorial associates. I particularly wish to acknowledge at Taylor and Francis Nora Konopka, Publisher; Helena Redshaw, Editorial Project Development Manager; and Mimi Williams, Project Editor. Finally, I am indebted to the support of Elizabeth Spangenberger, Editorial Assistant.

Richard C. Dorf
Editor-in-Chief

Editor-in-Chief

Richard C. Dorf, Professor of Electrical and Computer Engineering at the University of California, Davis, teaches graduate and undergraduate courses in electrical engineering in the fields of circuits and control systems. He earned a Ph.D. in electrical engineering from the U.S. Naval Postgraduate School, an M.S. from the University of Colorado, and a B.S. from Clarkson University. Highly concerned with the discipline of electrical engineering and its wide value to social and economic needs, he has written and lectured internationally on the contributions and advances in electrical engineering.

Professor Dorf has extensive experience with education and industry and is professionally active in the fields of robotics, automation, electric circuits, and communications. He has served as a visiting professor at the University of Edinburgh, Scotland; the Massachusetts Institute of Technology; Stanford University; and the University of California, Berkeley.

Professor Dorf is a Fellow of The Institute of Electrical and Electronics Engineers and a Fellow of the American Society for Engineering Education. Dr. Dorf is widely known to the profession for his *Modern Control Systems, 10th Edition* (Addison-Wesley, 2004) and *The International Encyclopedia of Robotics* (Wiley, 1988). Dr. Dorf is also the co-author of *Circuits, Devices and Systems* (with Ralph Smith), *5th Edition* (Wiley, 1992), and *Electric Circuits, 7th Edition* (Wiley, 2006). He is also the author of *Technology Ventures* (McGraw-Hill, 2005) and *The Engineering Handbook, 2nd Edition* (CRC Press, 2005).

Advisory Board

Contributors

Braden Allenby
AT&T Environment, Health & Safety
Bedminster, New Jersey

C.P. Arnold
University of Canterbury
Christchurch, New Zealand

Jos Arrillaga
University of Canterbury
Christchurch, New Zealand

Derek P. Atherton
University of Sussex
England, U.K.

Karen Blades
Lawrence Livermore National Laboratory
Livermore, California

Michele M. Blazek
AT&T
Pleasanton, California

Linda Sue Boehmer
LSB Technology
Clairton, Pennsylvania

Anjan Bose
Washington State University
Pullman, Washington

William L. Brogan
University of Nevada
Las Vegas, Nevada

George E. Cook
Vanderbilt University
Nashville, Tennessee

Reginald Crawford
Vanderbilt University
Nashville, Tennessee

David R. DeLapp
Vanderbilt University
Nashville, Tennessee

Mohamed E. El-Hawary
University of Nova Scotia
Halifax, Canada

Johan H.R. Enslin
KEMA T&D Consulting
Raleigh, North Carolina

Mehdi Ferdowsi
University of Missouri–Rolla
Rolla, Missouri

Jay C. Giri
AREVA T&D Corporation
Bellevue, Washington

J. Duncan Glover
Exponent
Natick, Massachusetts

Leo Grigsby
Auburn University
Jacksons Gap, Alabama

Charles A. Gross
Auburn University
Auburn, Alabama

R.B. Gungor
University of South Alabama
Mobile, Alabama

Anne R. Haake
Rochester Institute of Technology
Rochester, New York

Andrew Hanson
ABB, Inc.
Raleigh, North Carolina

Hans Hansson
Malardalen University
Vasteras, Sweden

Royce D. Harbor
University of West Florida
Eutaw, Alabama

Gregor Hoogers
Trier University of Applied Sciences
Birkenfled, Germany

Tien C. Hsia
University of California, Davis
Davis, California

Raymond G. Jacquot
University of Wyoming
Laramie, Wyoming

Hodge E. Jenkins
Mercer University
Macon, Georgia

George G. Karady
Arizona State University
Tempe, Arizona

Myron Kayton
Kayton Engineering Company
Santa Monica, California

Yong Deak Kim
Ajou University
Suwon, South Korea

Thomas R. Kurfess
Clemson University
Clemson, South Carolina

Ty A. Lasky
University of California, Davis
Davis, California

Luciano Lavagno
Cadence Berkeley Laboratory
Berkeley, California

Gordon K.F. Lee
San Diego State University
San Diego, California

Cornelius T. Leondes
University of California
San Diego, California

Thomas R. Mancini
Sandia National Laboratories
Albuquerque, New Mexico

Grant Martin
Tensilica Inc.
Santa Clara, California

Daniel A. Martinec
Aeronautical Radio, Inc.
Annapolis, Maryland

John E. McInroy
University of Wyoming
Laramie, Wyoming

Roger Messenger
Florida Atlantic University
Boca Raton, Florida

Mark L. Nagurka
Marquette University
Milwaukee, Wisconsin

Mikael Nolin
Malardalen University
Vasteras, Sweden

Thomas Nolte
Malardalen University
Vasteras, Sweden

Nicholas G. Odrey
Lehigh University
Bethlehem, Pennsylvania

Hitay Özbay
Bilkent University
Ankara, Turkey

Arun G. Phadke
Virginia Polytechnic Institute
Blacksburg, Virginia

Charles L. Phillips
Auburn University
Auburn, Alabama

Rama Ramakumar
Oklahoma State University
Stillwater, Oklahoma

Evelyn P. Rozanski
Rochester Institute of Technology
Rochester, New York

Andrew P. Sage
George Mason University
Fairfax, Virginia

Ioan Serban
University Politehnica of Timisoara
Timisoara, Romania

V.P. Shmerko
University of Calgary
Alberta, Canada

Tom Short
EPRI Solutions, Inc.
Schenectady, New York

Cary R. Spitzer
AvioniCon, Inc.
Williamsburg, Virginia

K. Neil Stanton
Stanton Associates
Bellevue, Washington

Elias G. Strangas
Michigan State University
East Lansing, Michigan

Alvin M. Strauss
Vanderbilt University
Nashville, Tennessee

Ronald J. Tallarida
Temple University
Philadelphia, Pennsylvania

Rao S. Thallam
Salt River Project
Phoenix, Arizona

Charles W. Therrien
Naval Postgraduate School
Monterey, California

Kleanthis Thramboulidis
University of Patras
Patras, Greece

Vyacheslav Tuzlukov
Ajou University
Suwon, South Korea

Darlusz Uciński
University of Zielona
Gora, Poland

Jerry Ventre
Florida Solar Energy Center
Cocoa, Florida

N.R. Watson
University of Canterbury
Christchurch, New Zealand

S.N. Yanushkevich
University of Calgary
Alberta, Canada

Won-Sik Yoon
Ajou University
Suwon, South Korea

Contents

SECTION I Energy

SECTION II Systems

SECTION III Mathematics, Symbols, and Physical Constants

Indexes

I

Energy

1

Conventional Power Generation

George G. Karady
Arizona State University

1.1 Introduction

The electric energy demand of the world is continuously increasing, and most of the energy is generated by conventional power plants, which remain the only cost-effective method for generating large quantities of energy.

Power plants utilize energy stored in the Earth and convert it to electrical energy that is distributed and used by customers. This process converts most of the energy into heat, thus increasing the entropy of the Earth. In this sense, power plants deplete the Earth's energy supply. Efficient operation becomes increasingly important to conserve energy.

Typical energy sources used by power plants include fossil fuel (gas, oil, and coal), nuclear fuel (uranium), geothermal energy (hot water, steam), and hydro energy (water falling through a head).

Around the turn of the century, the first fossil power plants used steam engines as the prime mover. These plants have 8- to 10-MW capacity, but increasing power demands resulted in the replacement by a more efficient steam boiler-turbine arrangement. The first commercial steam turbine was introduced by DeLaval in 1882. The boilers were developed from heating furnaces. Oil was the preferred and most widely used fuel in the beginning. The oil shortage promoted coal-fired plants, but the adverse environmental effects (sulfur dioxide generation, acid rain, dust pollution, etc.) curtailed their use in the late 1970s. Presently the most acceptable fuel is natural gas, which minimizes pollution and is available in large quantities. The increasing peak load demand led to the development of gas turbine power plants. These units can be started or stopped within a few minutes. The latest development is the combined-cycle power plant, which combines a gas turbine and a thermal unit. The hot exhaust gas from the gas turbine generates steam to drive a steam turbine, or the hot gas is used as the source of hot combustion air in a boiler, which provides steam to drive a turbine. During the next two decades, gas-fired power plants will dominate the electric industry.

The hydro plants' ancestors are water wheels used for pumping stations, mill driving, etc. Water-driven turbines were developed in the last century and have been used for generation of electricity since the beginning of their commercial use. However, most of the sites that can be developed economically are currently being utilized. No significant new development is expected in the United States in the near future.

Nuclear power plants appeared after the Second World War. The major development occurred during the 1960s; however, by the 1980s environmental considerations stopped plant development in the United States and slowed it down all over the world. Presently, the future of nuclear power generation is unclear, but the abundance of nuclear fuel and the expected energy shortage in the early part of the next century may rejuvenate nuclear development if safety issues can be resolved.

Geothermal power plants are the product of the clean energy concept, although the small-scale, local application of geothermal energy has a long history. Presently only a few plants are in operation. The potential for further development is limited because of the unavailability of geothermal energy sites that can be developed economically.

Typical technical data for different power plants is shown in Table 1.1.

1.2 Fossil Power Plants

The operational concept and major components of a fossil power plant are shown in Figure 1.1.

Fuel Handling

The most frequently used fuels are oil, natural gas, and coal. Oil and gas are transported by rail, on ships, or through pipelines. In the former case the gas is liquefied. Coal is transported by rail or ships if the plant is near a river or sea. The power plant requires several days of fuel reserve. Oil and gas are stored in large metal tanks, and coal is kept in open yards. The temperature of the coal layer must be monitored to avoid self-ignition.

Oil is pumped and gas is fed to the burners of the boiler. Coal is pulverized in large mills, and the powder is mixed with air and transported by air pressure, through pipes, to the burners. The coal transport from the yard to the mills requires automated transporter belts, hoppers, and sometimes manually operated bulldozers.

Boiler

Two types of boilers are used in modern power plants: the subcritical water-tube drum-type and the supercritical once-through type. The former operates around 2500 psi, which is under the water critical pressure of 3208.2 psi. The latter operates above that pressure, at around 3500 psi. The superheated steam temperature is about 1000°F (540°C) because of turbine temperature limitations.

A typical subcritical water-tube drum-type boiler has an inverted U shape. On the bottom of the rising part is the furnace where the fuel is burned. The walls of the furnace are covered by water pipes. The drum and the superheater are at the top of the boiler. The falling part of the U houses the reheaters, economizer (water heater), and air preheater, which is supplied by the forced-draft fan. The induced-draft fan forces the flue gases out of the system and sends them up the stack located behind the boiler. A flow diagram of the drum-type boiler is shown in Figure 1.2. The steam generator has three major systems: fuel, air-flue gas, and water-steam.

Fuel System

Fuel is mixed with air and injected into the furnace through burners. The burners are equipped with nozzles that are supplied by preheated air and carefully designed to assure the optimum air-fuel mix. The fuel mix is ignited by oil or gas torches. The furnace temperature is around 3000°F.

TABLE 1.1 Power Plant Technical Data

Generation Type	Typical MW Size	Capitalized Plant Cost, $/kW	Construction Lead Time, Years	Heat Rate, Btu/kWh	Fuel Cost, $/MBtu	Fuel Type	Equivalent Forced Outage Rate	Equivalent Scheduled Outage Rate	O & M Fixed, $/kW/year	Cost Variable, $/MWh
Nuclear	1200	2400	10	10,400	1.25	Uranium	20	15	25	8
Pulverized coal steam	500	1400	6	9,900	2.25	Coal	12	12	20	5
Atmospheric fluidized bed	400	1400	6	9,800	2.25	Coal	14	12	17	6
Gas turbine	100	350	2	11,200	4.00	Nat. gas	7	7	1	5
Combined-cycle	300	600	4	7,800	4.00	Nat. gas	8	8	9	3
Coal-gasification combined-cycle	300	1500	6	9,500	2.25	Coal	12	10	25	4
Pumped storage hydro	300	1200	6	—	—	—	5	5	5	2
Conventional hydro	300	1700	6	—	—	—	3	4	5	2

Source: H.G. Stoll, *Least-Cost Electric Utility Planning*, © 1989 John Wiley & Sons. Reprinted by permission of John Wiley & Sons, Inc.

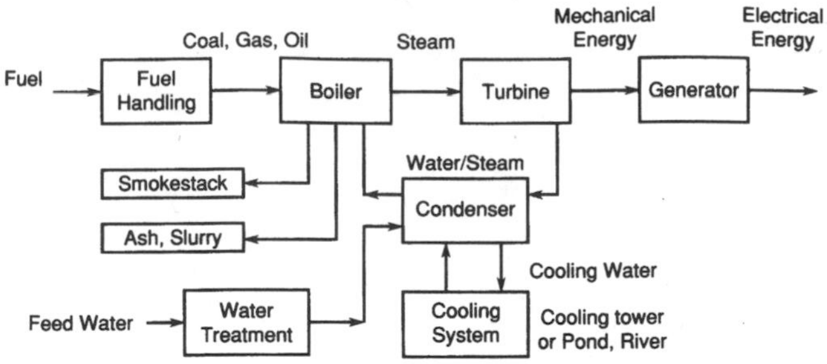

FIGURE 1.1 Major components of a fossil power plant.

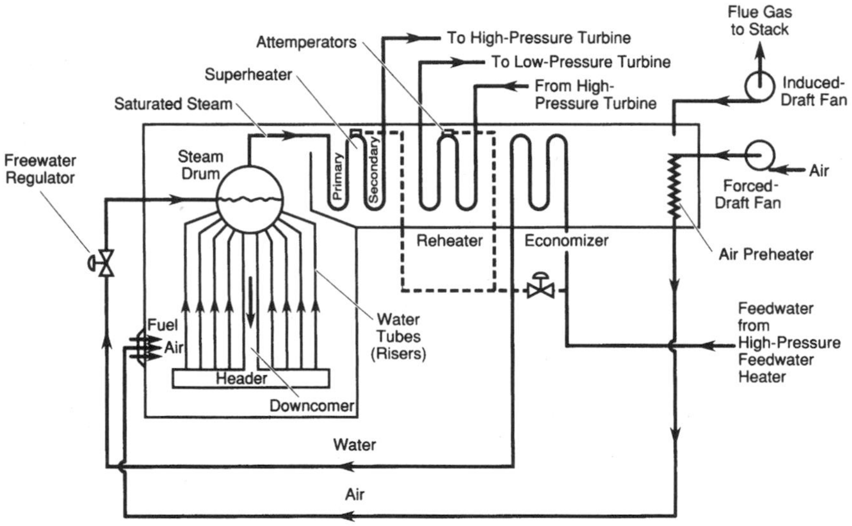

FIGURE 1.2 Flow diagram of a typical drum-type steam boiler. (*Source*: M.M. El-Wakil, *Power Plant Technology*, New York: McGraw-Hill, 1984, p. 210. With permission.)

Air-Flue Gas System

Ambient air is driven by the forced-draft fan through the air preheater which is heated by the high-temperature (600°F) flue gases. The air is mixed with fuel in the burners and enters the furnace, where it supports the fuel burning. The hot combustion flue gas generates steam and flows through the boiler to heat the superheater, reheaters, economizer, etc. Induced-draft fans, located between the boiler and the stack, increase the flow and send the 300°F flue gases to the atmosphere through the stack.

Water-Steam System

Large pumps drive the feedwater through the high-pressure heaters and the economizer, which further increases the water temperature (400 to 500°F). The former is heated by steam removed from the turbine; the latter is heated by the flue gases. The preheated water is fed to the steam drum. Insulated tubes, called downcomers, are located outside the furnace and lead the water to a header. The header distributes the hot water among the risers. These are water tubes that line the furnace walls. The water tubes are heated by the combustion gases through both convection and radiation. The steam generated in these tubes flows to the drum, where it is separated from the water. Circulation is maintained by the density difference between the water in the downcomer and the water tubes. Saturated steam, collected in the drum, flows through the superheater. The superheater increases the steam temperature to about 1000°F. Dry superheated steam drives the high-pressure turbine. The exhaust from the high-pressure turbine goes to the reheater, which again increases the steam temperature. The reheated steam drives the low-pressure turbine.

The typical supercritical once-through-type boiler concept is shown in Figure 1.3.

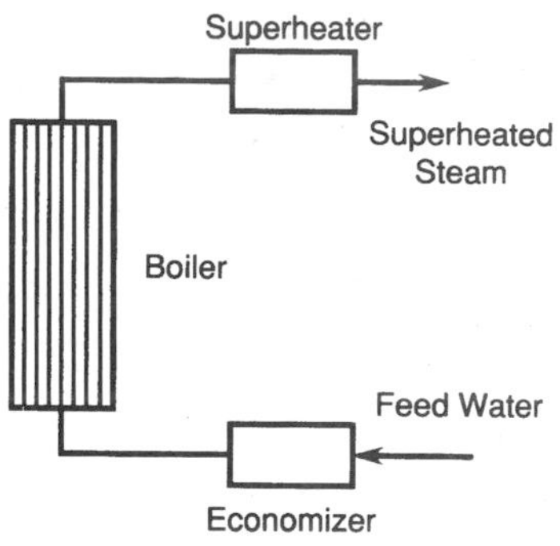

FIGURE 1.3 Concept of once-through-type steam generator.

The feedwater enters through the economizer to the boiler consisting of riser tubes that line the furnace wall. All the water is converted to steam and fed directly to the superheater. The latter increases the steam temperature above the critical temperature of the water and drives the turbine. The construction of these steam generators is more expensive than the drum-type units but has a higher operating efficiency.

Figure 1.4 shows an approximate layout and components of a coal-fired power plant. The solid arrows show the flow of flue gas. The dotted arrows show the airflow through the preheating system. The figure shows the approximate location of the components

Turbine

The turbine converts the heat energy of the steam into mechanical energy. Modern power plants usually use one high-pressure and one or two lower-pressure turbines. A typical turbine arrangement is shown in Figure 1.5.

The figure shows that only one bearing is between each of the machines. The shafts are connected to form a tandem compound steam-turbine unit. High-pressure steam enters the high-pressure turbine to flow through and drive the turbine. The exhaust is reheated in the boiler and returned to the lower-pressure units. Both the rotor and the stationary part of the turbine have blades. The length of the blades increases from the steam entrance to the exhaust.

Figure 1.6 shows the blade arrangement of an impulse-type turbine. Steam enters through nozzles and flows through the first set of moving rotor blades. The following stationary blades change the direction of the flow and direct the steam into the next set of moving blades. The nozzles increase the steam speed and reduce pressure, as shown in the figure. The impact of the high-speed steam, generated by the change of direction and speed in the moving blades, drives the turbine.

The reaction-type turbine has nonsymmetrical blades, which assure that the pressure continually drops through all rows of blades but that steam velocity decreases in the moving blades and increases in the stationary blades.

Figure 1.7 shows the rotor of a large steam turbine. The figure shows that the diameters of the blades vary. The high-pressure steam enters at the middle (the low blade diameter side) of the turbine. As the high-pressure steam passes through the blades its pressure decreases. In order to maintain approximately constant driving force, the blades' diameter is increased towards the end of the turbine.

Generator

The generator converts mechanical energy from the turbines into electrical energy. The major components of the generator are the frame, stator core and winding, rotor and winding, bearings, and cooling system. Figure 1.8 shows the cross section of a modern hydrogen-cooled generator.

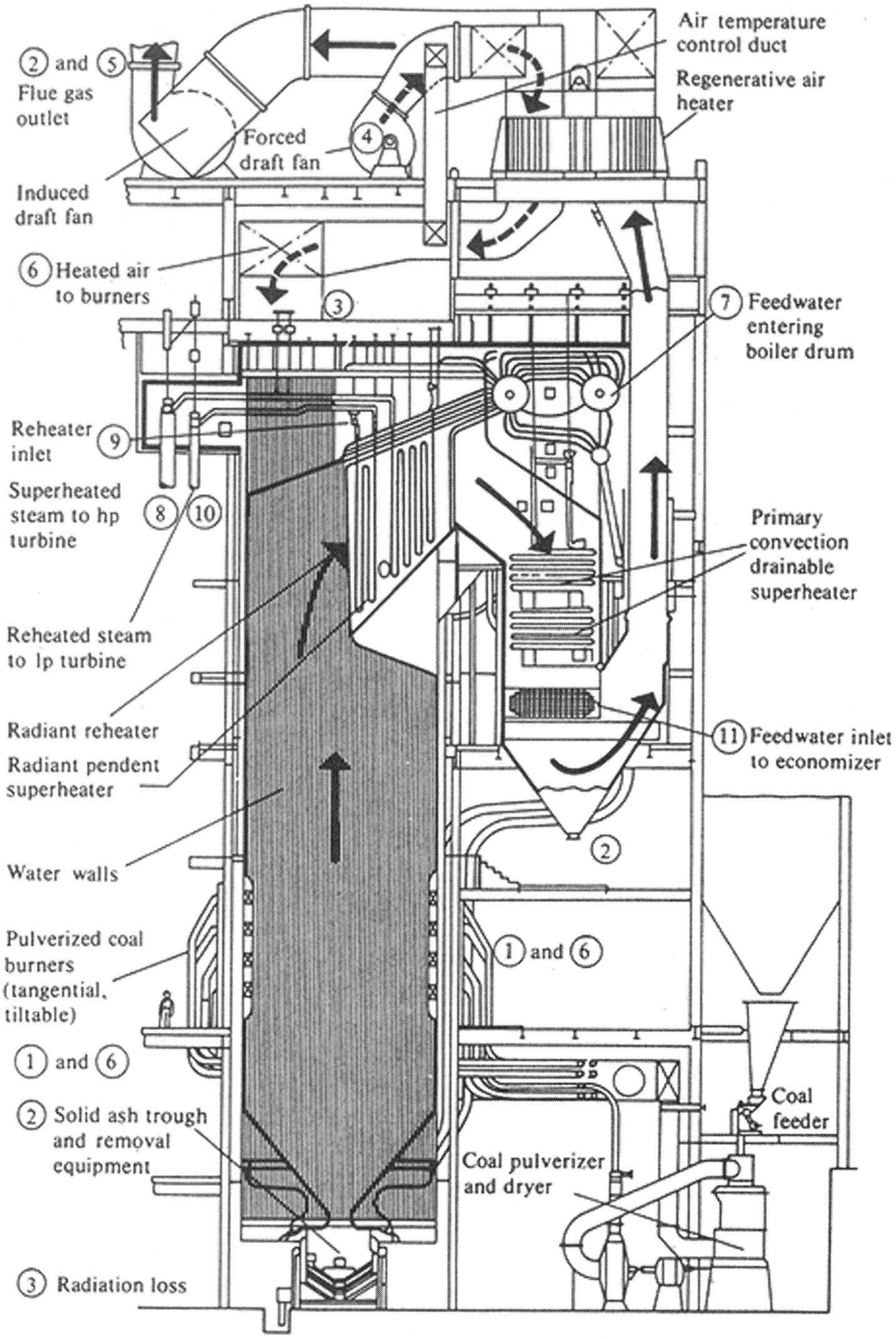

FIGURE 1.4 Major components and physical layout of a coal-fired fossil power plant. (*Source*: A.W. Culp, *Principles of Energy Conversion*, 2nd ed., New York: McGraw Hill, 1991, p. 220. With permission.)

The stator has a laminated and slotted silicon steel iron core. The stacked core is clamped and held together by insulated axial through bolts. The stator winding is placed in the slots and consists of a copper-strand configuration with woven glass insulation between the strands and mica flakes, mica

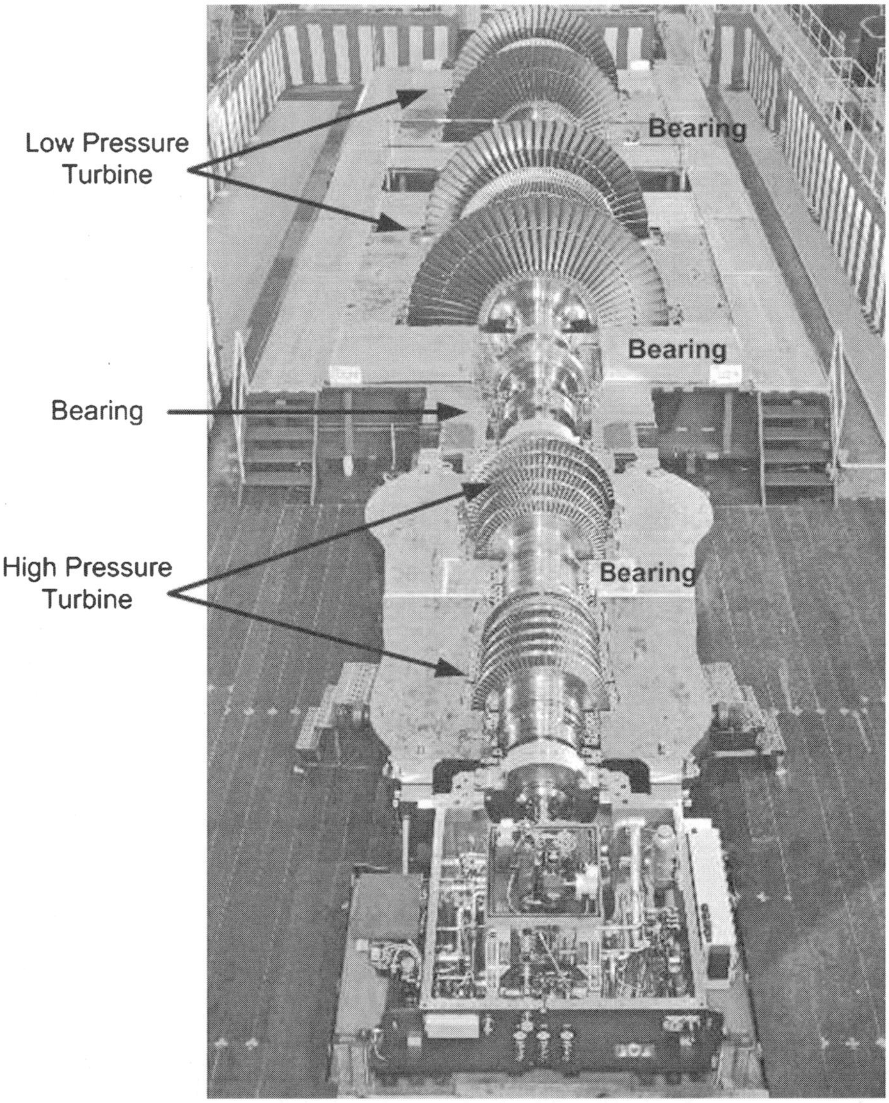

FIGURE 1.5 Open large tandem compound steam turbine. (Courtesy of Toshiba.)

mat, or mica paper ground-wall insulation. To avoid insulation damage caused by vibration, the ground-wall insulation is reinforced by asphalt, epoxy-impregnated fiberglass, or Dacron. The largest machine stator is Y-connected and has two coils per phase, connected in parallel. Most frequently, the stator is hydrogen cooled; however, small units may be air cooled and very large units may be water cooled.

The solid steel rotor has slots milled along the axis. The multiturn, copper rotor winding is placed in the slots and cooled by hydrogen. Cooling is enhanced by subslots and axial cooling passages. The rotor winding is restrained by wedges inserted in the slots.

The rotor winding is supplied by dc current, either directly by a brushless excitation system or through collector rings. The rotor is supported by bearings at both ends. The nondrive-end bearing is insulated to avoid shaft current generated by stray magnetic fields. The hydrogen is cooled by a hydrogen-to-water heat exchanger mounted on the generator or installed in a closed-loop cooling system.

The dc current of the rotor generates a rotating magnetic field that induces an ac voltage in the stator winding. This voltage drives current through the load and supplies the electrical energy.

Figure 1.9 shows the typical arrangement of a turbine, generator, and exciter installed in a power plant. The figure shows that the three units are connected in series. The turbine has a high- and low-pressure stage and drives the generator. The generator drives the exciter that produces the dc current for the generator rotor.

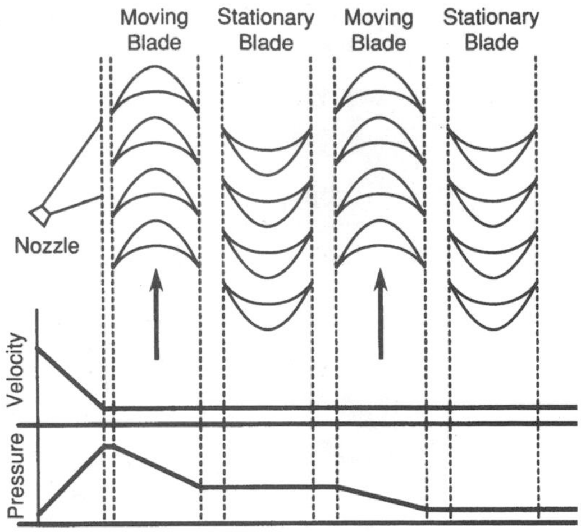

FIGURE 1.6 Velocity and pressure variation in an impulse turbine.

Electric System

Energy generated by the power plant supplies the electric network through transmission lines. The power plant operation requires auxiliary power to operate mills, pumps, etc. The auxiliary power requirement is approximately 10 to 15%.

Smaller generators are directly connected in parallel using a busbar. Each generator is protected by a circuit breaker. The power plant auxiliary system is supplied from the same busbar. The transmission lines are connected to the generator bus, either directly or through a transformer.

The larger generators are unit-connected. In this arrangement the generator is directly connected, without a circuit breaker, to the main transformer. A conceptual one-line diagram is shown in Figure 1.10. The generator supplies main and auxiliary transformers without circuit breakers. The units are connected in parallel at the high-voltage side of the main transformers by a busbar. The transmission lines are also supplied from this bus. Circuit breakers are installed at the secondary side of the main and auxiliary transformer. The application of a generator circuit breaker is not economical in the case of large generators. Because of the generator's large short-circuit current, special expensive circuit breakers are required. However, the transformers reduce the short-circuit current and permit the use of standard circuit breakers

FIGURE 1.7 Rotor of a large steam turbine. (Courtesy of Siemens.)

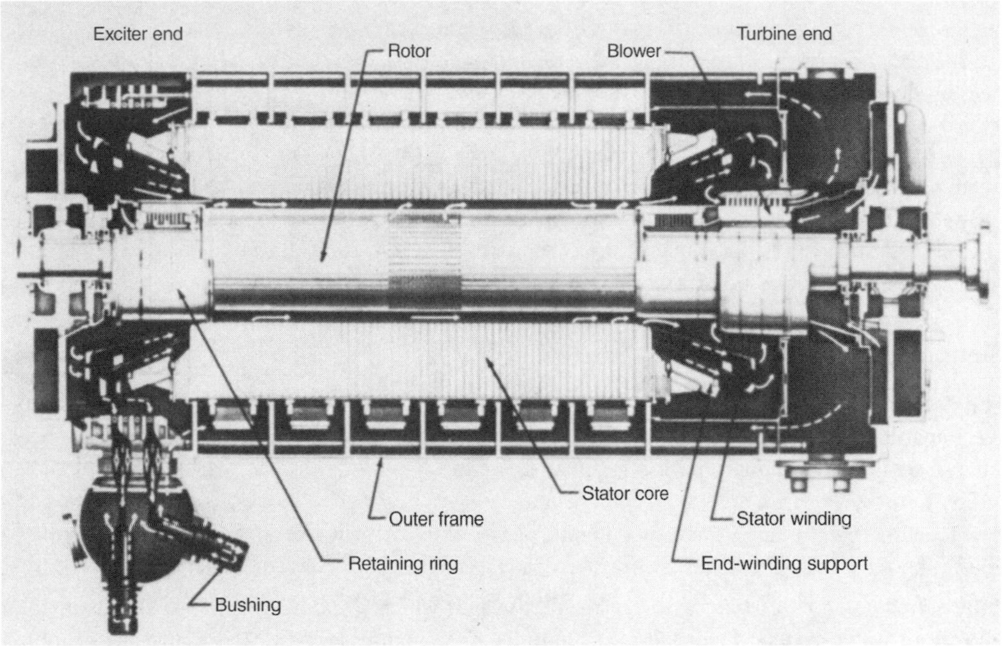

FIGURE 1.8 Direct hydrogen-inner-cooled generator. (*Source*: R.W. Beckwith, Westinghouse Power Systems Marketing Training Guide on Large Electric Generators, Pittsburgh: Westinghouse Electric Corp., 1979, p. 54. With permission.)

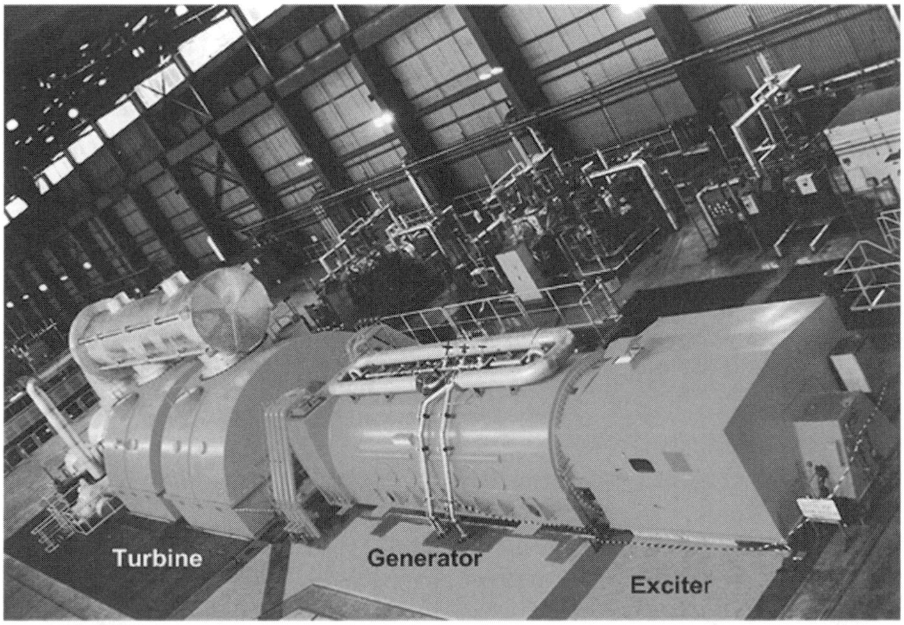

FIGURE 1.9 Turbine, generator, and exciter. (Courtesy of Siemens.)

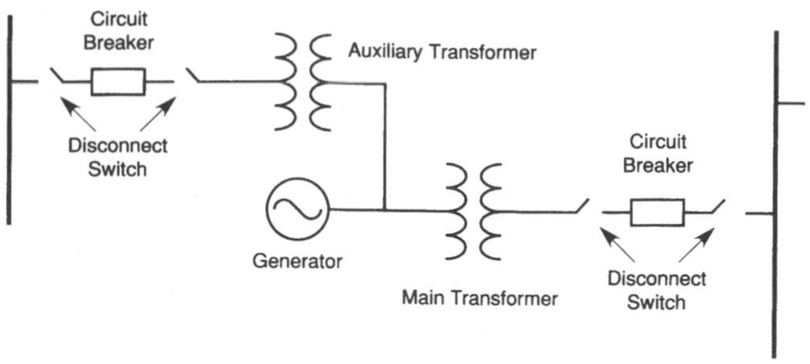

FIGURE 1.10 Conceptual one-line diagram for a unit-connected generator.

at the secondary side. The disconnect switches permit visual observation of the off state and are needed for maintenance of the circuit breakers.

Condenser

The condenser condenses turbine exhaust steam to water, which is pumped back to the steam generator through various water heaters. The condensation produces a vacuum necessary to exhaust the steam from the turbine. The condenser is a shell-and-tube heat exchanger, where steam condenses on water-cooled tubes. Cold water is obtained from the cooling towers or other cooling systems. The condensed water is fed through a deaerator, which removes absorbed gases from the water. Next, the gas-free water is mixed with the feedwater and returned to the boiler. The gases absorbed in the water may cause corrosion (oxygen) and increase condenser pressure, adversely affecting efficiency. Older plants use a separate deaerator heater, while deaerators in modern plants are usually integrated in the condenser, where injected steam jets produce pressure drop and remove absorbed gases.

Stack and Ash Handling

The stack is designed to disperse gases into the atmosphere without disturbing the environment. This requires sufficient stack height to assist the fans in removing gases from the boiler through natural convection. The gases contain both solid particles and harmful chemicals. Solid particles, like dust, are removed from the flue gas by electrostatic precipitators or baghouse filters. Harmful sulfur dioxide is eliminated by scrubbers. The most common is the lime/limestone scrubbing process.

Coal-fired power plants generate a significant amount of ash. The disposition of the ash causes environmental problems. Several systems have been developed in past decades. Large ash particles are collected by a water-filled ash hopper located at the bottom of the furnace. Fly ash is removed by filters, then mixed with water. Both systems produce sludge that is pumped to a clay-lined pond where water evaporates and the ash fills disposal sites. The clay lining prevents intrusion of groundwater into the pond.

Cooling and Feedwater System

The condenser is cooled by cold water. The open-loop system obtains the water from a river or sea, if the power plant location permits it. The closed-loop system utilizes cooling towers, spray ponds, or spray canals. In the case of spray ponds or canals, the water is pumped through nozzles which generate fine sprays. Evaporation cools the water sprays as they fall back into the pond. Several different types of cooling towers have been developed. The most frequently used is the wet cooling tower, where the hot water is sprayed on top of a latticework of horizontal bars. The water drifts downward and is cooled through evaporation by the air which is forced through by fans or natural updraft.

The power plant loses a small fraction of the water through leakage. The feedwater system replaces this lost water. Replacement water has to be free from absorbed gases, chemicals, etc. because the impurities cause severe corrosion in the turbines and boiler. The water-treatment system purifies replacement water by pretreatment, which includes filtering, chlorination, demineralization, condensation, and polishing. These complicated chemical processes result in a corrosion-free high-quality feedwater.

1.3 Gas-Turbine and Combined-Cycle Power Plants

Gas Turbine

The increase of peak demand resulted in the development of gas-turbine power plants. Figure 1.11 shows the main components of a gas-turbine power plant, which are:

- Compressor
- Combustion chamber
- Turbine
- Exhaust chute

The air is drawn by the compressor, which increases the air pressure. This high-pressure air is injected into the combustion chamber. The natural gas or the vaporized fossil fuel is injected in the combustion chamber to form an air-fuel mixture. The air-fuel mixture is ignited. The combustion of the fuel mixture produces a high-pressure, high-temperature gas which is injected into the turbine through nozzles. This gas mixture drives the rotor of the turbine. The pressure drops and the spent fuel mixture is exhausted into the air through a short tube. The temperature of the exhaust is rather high, in the range of 800 to 1000°C. The turbine drives the generator. The gas turbine requires a starting motor which drives the unit and generates a sufficient amount of compressed air to start the combustion process. One solution is the use of a generator as a motor. This can be achieved by supplying the generator with variable-frequency voltage to work as a motor. In some cases a gearbox is connected between the turbine and generator to reduce the turbine speed. The generator synchronous speed is 3600 rpm, and the optimum turbine speed is above 10,000 rpm. The capacity of this plant is in the region of 50 to 100 MW.

Combined-Cycle Plant

The recognition that the exhaust of a gas turbine has high temperature led to the development of combined-cycle power plants. Figure 1.12 shows the major components of a combined-cycle power plant.

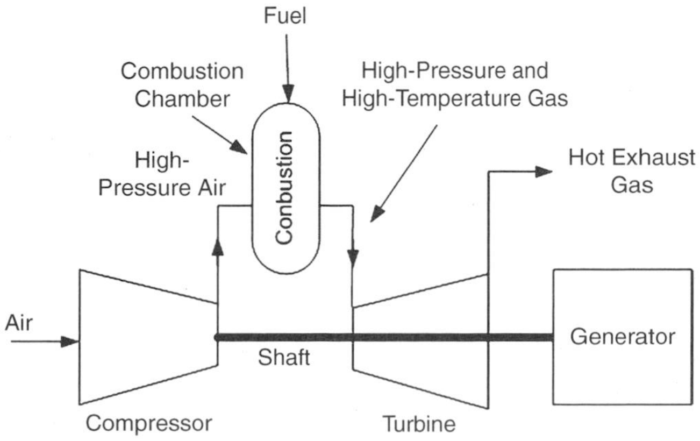

FIGURE 1.11 Gas-turbine power plant.

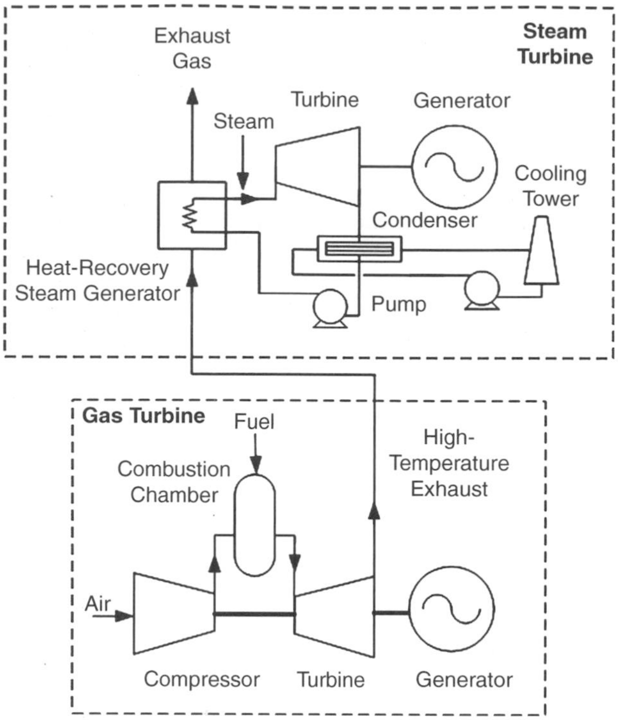

FIGURE 1.12 Combined-cycle power plant.

The plant contains a gas turbine and a conventional steam turbine. The hot exhaust gas from the gas turbine supplies a heat exchanger to generate steam. The steam drives a convectional turbine and generator. The steam condenses in the condenser, which is cooled by water from a cooling tower or fresh water from a lake or river. The combined-cycle power plant has higher efficiency (around 42%) than a conventional plant (27 to 32%) or a gas-turbine power plant. The capacity of this plant is 100 to 500 MW.

1.4 Nuclear Power Plants

More than 500 nuclear power plants operate around the world. Close to 300 operate pressurized water reactors (PWRs); more than 100 are built with boiling-water reactors (BWRs); about 50 use gas-cooled reactors; and the rest are heavy-water reactors. In addition a few fast-breeder reactors are in operation. These reactors are built for better utilization of uranium fuel. The modern nuclear plant size varies from 100 to 1300 MW. Figure 1.13 shows the large Paulo Verde nuclear power plant in Arizona. The 4000-MW-capacity power plant has three 1300-MW pressurized-water reactors, The figure shows the large concrete domes housing the nuclear reactors, the cooling towers, generator building, and the large switchyard.

Pressurized-Water Reactor

The general arrangement of a power plant with a PWR is shown in Figure 1.14(a).

The reactor heats the water from about 550 to about 650°F. High pressure, at about 2235 psi, prevents boiling. Pressure is maintained by a pressurizer, and the water is circulated by a pump through a heat exchanger. The heat exchanger evaporates the feedwater and generates steam that supplies a system similar to a

FIGURE 1.13 View of Palo Verde nuclear power plant. (Courtesy of Salt River Project.)

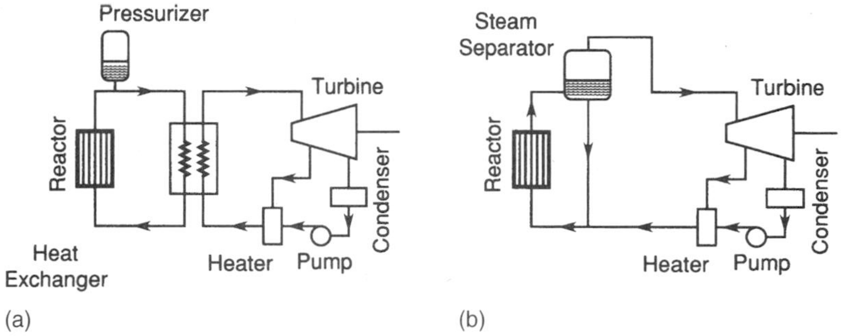

FIGURE 1.14 (a) Power plant with PWR; (b) power plant with BWR.

conventional power plant. The advantage of this two-loop system is the separation of the potentially radioactive reactor cooling fluid from the water-steam system.

The reactor core consists of fuel and control rods. Grids hold both the control and fuel rods. The fuel rods are inserted in the grid following a predetermined pattern. The fuel elements are zircaloy-clad rods filled with UO2 pellets. The control rods are made of a silver (80%), cadmium (5%), and indium (15%) alloy protected by stainless steel. The reactor operation is controlled by the position of the rods. In addition, control rods are used to shut down the reactor. The rods are released and fall into the core when emergency shutdown is required. Cooling water enters the reactor from the bottom, flows through the core, and is heated by nuclear fission.

Boiling-Water Reactor

In the BWR shown in Figure 1.14(b), the pressure is low, about 1000 psi. The nuclear reaction heats the water directly to evaporate it and produce wet steam at about 545°F. The remaining water is recirculated and mixed with feedwater. The steam drives a turbine that typically rotates at 1800 rpm. The rest of the plant is similar to a conventional power plant. A typical reactor arrangement is shown in Figure 1.15. The figure shows all the major components of a reactor. The fuel and control rod assembly is located in the lower part. The steam separators are above the core, and the steam dryers are at the top of the reactor. The reactor is enclosed by a concrete dome.

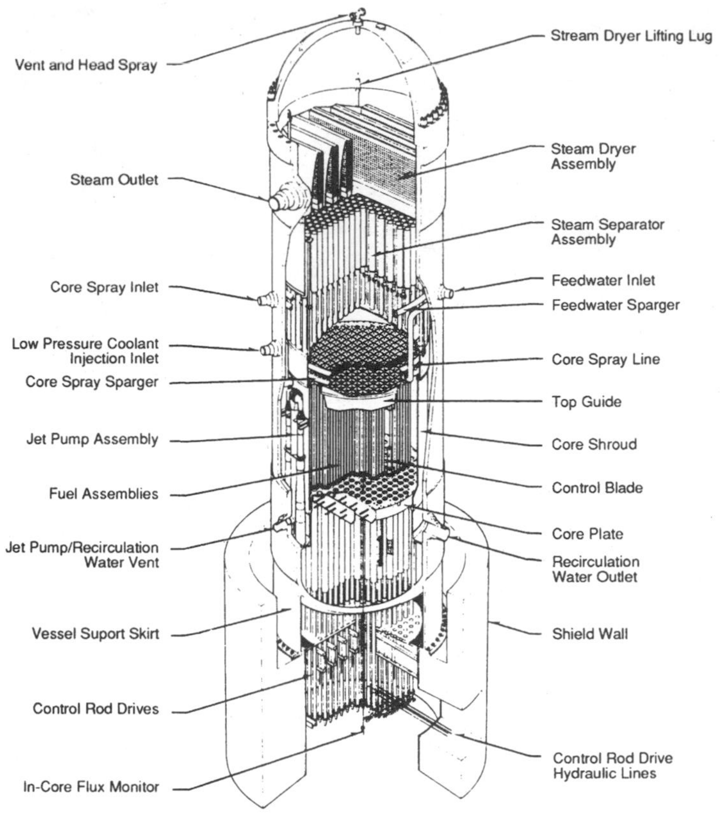

FIGURE 1.15 Typical BWR reactor arrangement. (*Source*: Courtesy of General Electric Company.)

1.5 Geothermal Power Plants

The solid crust of the Earth is an average of 20 miles (32 km) deep. Under the solid crust is a molten mass, the magma. The heat stored in the magma is the source of geothermal energy. The hot molten magma comes close to the surface at certain points in the Earth and produces volcanoes, hot springs, and geysers. These are the signs of a possible geothermal site. Three forms of geothermal energy are considered for development.

Hydrothermal Source

This is the most developed source. Power plants, up to a capacity of 2000 MW, are in operation worldwide. Heat from the magma is conducted upward by the rocks. The groundwater drifts down through the cracks and fissures to form reservoirs when water-impermeable solid rockbed is present. The water in this reservoir is heated by the heat from the magma. Depending on the distance from the magma and rock configuration, steam, hot pressurized water, or a mixture of the two is generated. Signs of these underwater reservoirs include hot springs and geysers. The reservoir is tapped by a well, which brings the steam-water mixture to the surface to produce energy. The geothermal power plant concept is illustrated in Figure 1.16

The hot water and steam mixture is fed into a separator. If the steam content is high, a centrifugal separator is used to remove the water and other particles. The obtained steam drives a turbine. The typical pressure is around 100 psi, and the temperature is around 400°F (200°C). If the water content is high, the

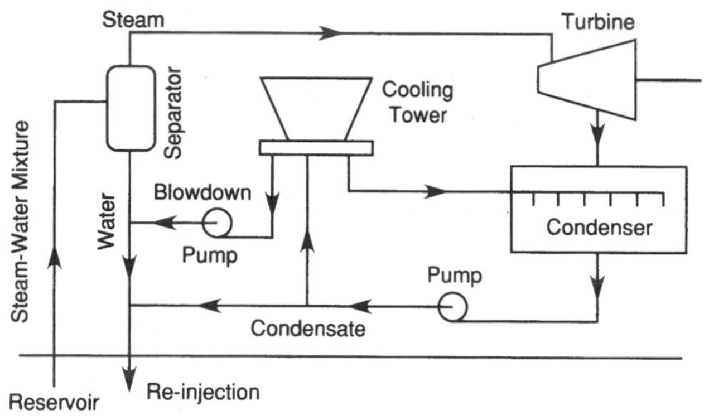

FIGURE 1.16 Concept of a geothermal power plant.

water-steam mixture is led through a flashed-steam system where the expansion generates a better quality of steam and separates the steam from the water. The water is returned to the ground, and the steam drives the turbine. Typically the steam entering the turbine has a temperature of 120 to 150°C and a pressure of 30 to 40 psi.

The turbine drives a conventional generator. The typical rating is in the 20- to 100-MW range. The exhaust steam is condensed in a direct-contact condenser. A part of the obtained water is reinjected into the ground. The rest of the water is fed into a cooling tower to provide cold water to the condenser.

Major problems with geothermal power plants are the minerals and noncondensable gases in the water. The minerals make the water highly corrosive, and the separated gases cause air pollution. An additional problem is noise pollution. The centrifugal separator and blowdowns require noise dampers and silencers.

Petrothermal Source

Some fields have only hot rocks under the surface. Utilization of this petrothermal source requires pumping surface water through a well in a constructed hole to a reservoir. The hot water is then recovered through another well. The problem is the formation of a reservoir. The U.S. government is studying practical uses of petrothermal sources.

Geopressured Source

In deep underground holes (8,000 to 30,000 ft) a mixture of pressurized water and natural gas, like methane, may sometimes be found. These geopressured sources promise power generation through the combustion of methane and the direct recovery of heat from the water. The geopressured method is currently in an experimental stage, with operating pilot plants.

1.6 Hydroelectric Power Plants

Hydroelectric power plants convert energy produced by a water head into electric energy. A typical hydroelectric power plant arrangement is shown in Figure 1.17.

The head is produced by building a dam across a river, which forms the upper-level reservoir. In the case of low head, the water forming the reservoir is fed to the turbine through the intake channel or the turbine is integrated in the dam. The latter arrangement is shown in Figure 1.17(a). Penstock tubes or tunnels are used

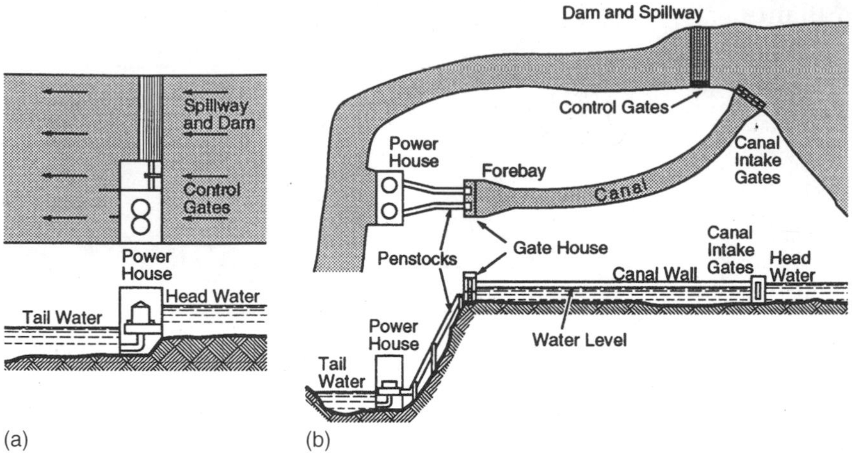

FIGURE 1.17 Hydroelectric power plant arrangement. (a) Low-head plant, (b) medium-head plant. (*Source*: D.G. Fink, *Standard Handbook for Electrical Engineers*, New York: McGraw-Hill, 1978. With permission.)

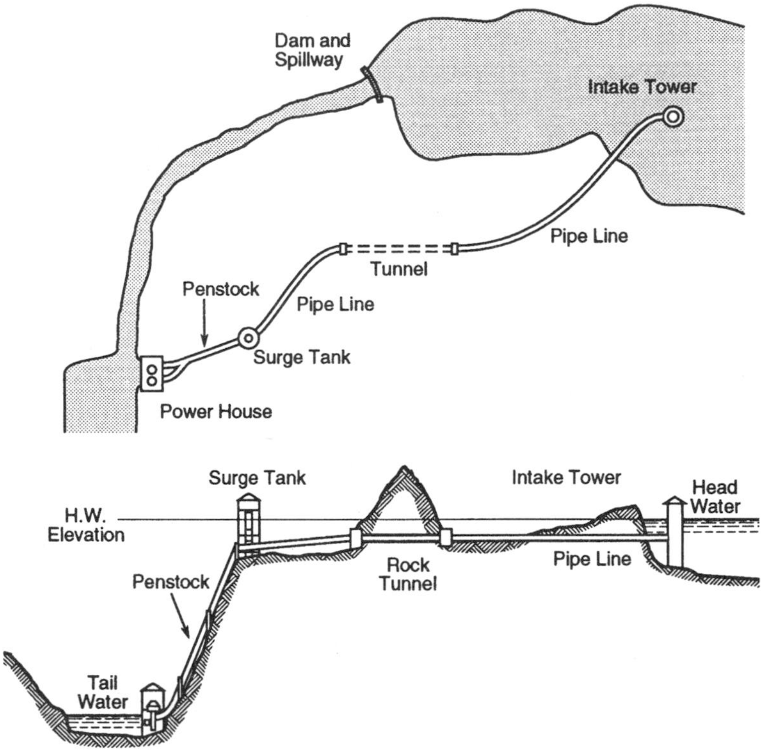

FIGURE 1.18 Hydroelectric power plant arrangement, high-head plant. (*Source*: D.G. Fink, *Standard Handbook for Electrical Engineers*, New York: McGraw-Hill, 1978. With permission.)

for medium- (Figure 1.17(b)) and high-head plants (Figure 1.18). The spillway regulates the excess water by opening gates at the bottom of the dam or permitting overflow on the spillway section of the dam. The water discharged from the turbine flows to the lower or tail-water reservoir, which is usually a continuation of the original water channel.

High-Head Plants

High-head plants (Figure 1.18) are built with impulse turbines, where the head-generated water pressure is converted into velocity by nozzles and the high-velocity water jets drive the turbine runner.

Figure 1.19 shows an aerial view of a hydropower plant in Arizona. The figure shows the large dam and the relatively small power plant. The dam blocks the Salt River and forms a lake as an upper reservoir; the lower reservoir is also a lake as shown in the picture.

Low- and Medium-Head Plants

Low- and medium-head installations (Figure 1.17) are built with reaction-type turbines, where the water pressure is mostly converted to velocity in the turbine. The two basic classes of reaction turbines are the propeller or Kaplan type, mostly used for low-head plants, and the Francis type, mostly used for medium-head plants. The cross section of a typical low-head Kaplan turbine is shown in Figure 1.20.

The vertical-shaft turbine and generator are supported by a thrust bearing immersed in oil. The generator is in the upper, watertight chamber. The turbine runner has four to ten propeller types and adjustable pitch blades. The blades are regulated from 5 to 35° by an oil pressure operated servomechanism. The water is evenly distributed along the periphery of the runner by a concrete spiral case and regulated by adjustable wicket blades. The water is discharged from the turbine through an elbow-shaped draft tube. The conical profile of the tube reduces the water speed from the discharge speed of 10 to 30 ft/s to 1 ft/s to increase turbine efficiency.

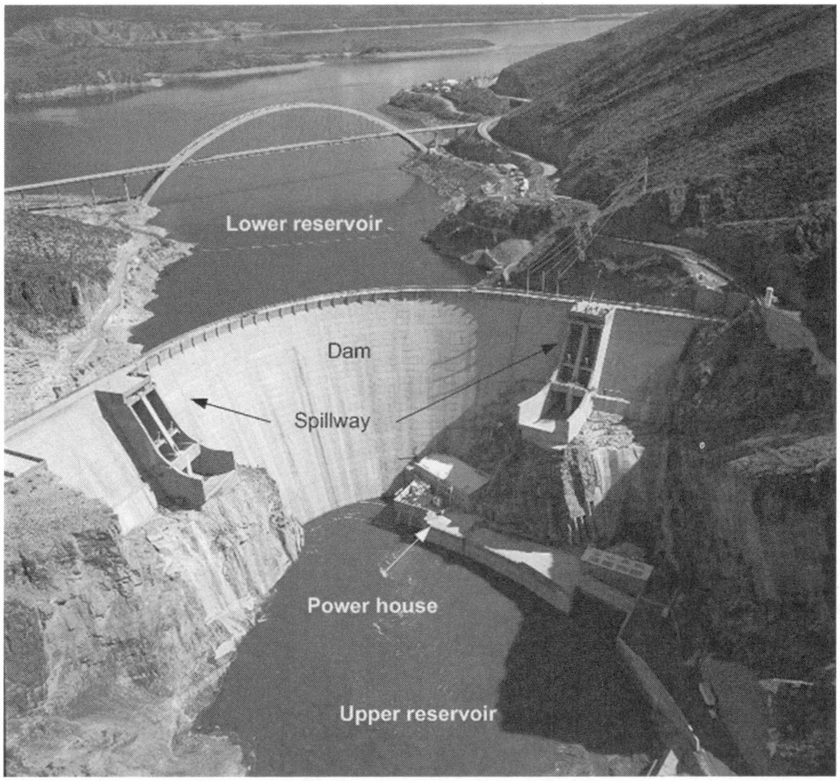

FIGURE 1.19 Aerial view of Roosevelt Dam and hydropower plant (36 MW) in Arizona. (Courtesy of Salt River Project.)

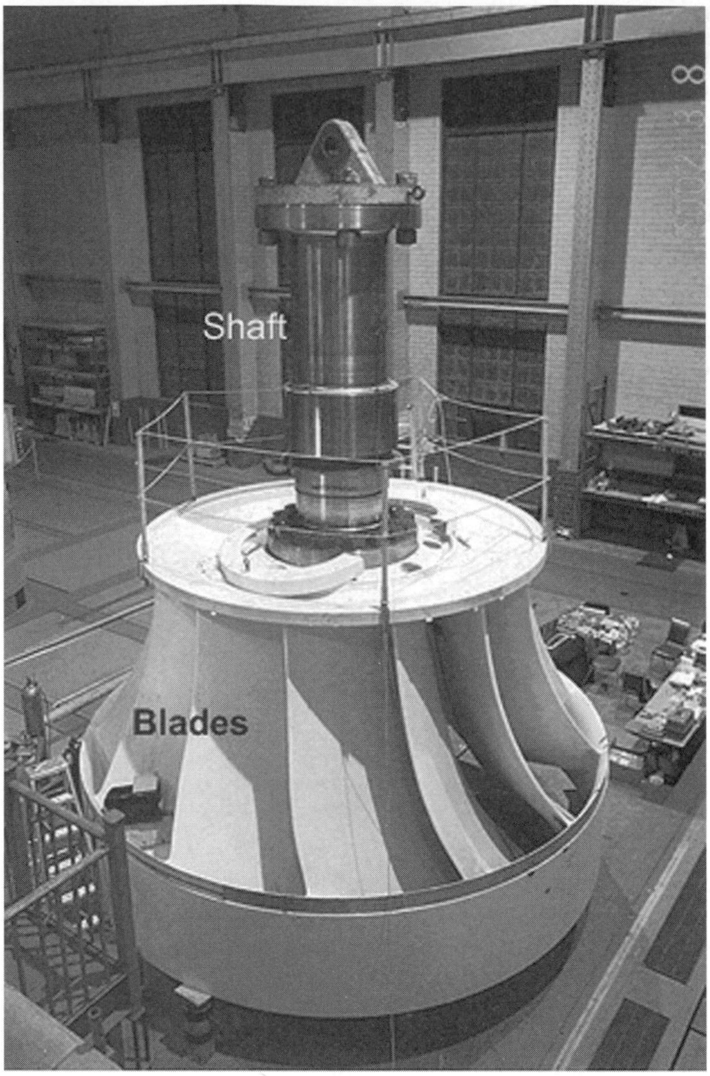

FIGURE 1.20 Typical low-head hydroplant with Kaplan turbine. (Courtesy of Hydro-Quèbec.)

Hydrogenerators

The hydrogenerator is a low-speed (100 to 360 rpm) salient-pole machine with a vertical shaft. A typical number of poles is from 20 to 72. They are mounted on a pole spider, which is a welded, spoked wheel. The spider is mounted on the forged steel shaft. The poles are built with a laminated iron core and stranded copper winding. Damper bars are built in the pole faces. The stator is built with a slotted, laminated iron core that is supported by a welded steel frame. Windings are made of stranded conductors insulated between the turns by glass fiber or Dacron glass. The ground insulation is multiple layers of mica tape impregnated by epoxy or polyester resins. The older machines use asphalt and mica-tape insulation, which is sensitive to corona-discharge-caused insulation deterioration. Direct water cooling is used for very large machines, while the smaller ones are air- or hydrogen-cooled. Some machines use forced-air cooling with an air-to-water heat exchanger. A braking system is installed in larger machines to stop the generator rapidly and to avoid damage to the thrust.

Defining Terms

Boiler: A steam generator that converts the chemical energy stored in the fuel (coal, gas, etc.) to thermal energy by burning. The heat evaporates the feedwater and generates high-pressure steam.

Economizer: A heat exchanger that increases the feedwater temperature. The flue gases heat the heat exchanger.

Fuel: Thermal power plants use coal, natural gas, and oil as a fuel, which is burned in the boiler. Nuclear power plants use uranium as a fuel.

Penstock: A water tube that feeds the turbine. It is used when the slope is too steep to use an open canal.

Reactor: A container where the nuclear reaction takes place. The reactor converts the nuclear energy to heat.

Superheater: A heat exchanger that increases the steam temperature to about 1000°F. It is heated by the flue gases.

Surge tank: An empty vessel that is located at the top of the penstock. It is used to store water surge when the turbine valve is suddenly closed.

References

A.J. Ellis, "Using geothermal energy for power," *Power*, vol. 123, no. 10, 1979.
M.M. El-Wakil, *Power Plant Technology*, New York: McGraw-Hill, 1984.
A.V. Nero, *A Guidebook to Nuclear Reactors*, Berkeley: University of California Press Ltd., 1979.
J. Weisman and L.E. Eckart, *Modern Power Plant Engineering*, Englewood Cliffs, N.J.: Prentice-Hall, 1985.

Further Information

Other recommended publications include the "Power Plant Electrical References Series," published by EPRI, which consists of several books dealing with power plant electrical system design. A good source of information on the latest developments is *Power* magazine, which regularly publishes articles on power plants.

Additional books include the following:

S. Glasstone and M.C. Edlund, *The Elements of Nuclear Reactor Theory*, New York: Van Nostrand, 1952, p. 416.
G. Murphy, *Elements of Nuclear Engineering*, New York: Wiley, 1961.
M.A. Schultz, *Control of Nuclear Reactors and Power Plants*, New York: McGraw-Hill, 1955.
R.H. Shannon, *Handbook of Coal-Based Electric Power Generation*, Park Ridge, N.J.: Noyes, 1982, p. 372.
E.J.G. Singer, *Combustion: Fossil Power Systems*, Windsor, Conn.: Combustion Engineering, Inc., 1981.
B.G.A. Skrotzki and W.A. Vopat, *Power Station Engineering and Economy*, New York: McGraw-Hill, 1960.
M.J. Steinberg and T.H. Smith, *Economy Loading of Power Plants and Electric Systems*, New York: Wiley, 1943, p. 203.
Various, *Electric Generation: Steam Stations*, B.G.A. Skrotzki, Ed., New York: McGraw-Hill, 1970, p. 403.
Various, *Steam*, New York: Babcock & Wilcox, 1972.
Various, *Steam: Its Generation and Use*, New York: Babcock & Wilcox, 1978.

2

Alternative Power Systems and Devices

Johan H.R. Enslin
KEMA T&D Consulting

Rama Ramakumar
Oklahoma State University

Thomas R. Mancini
Sandia National Laboratories

Roger Messenger
Florida Atlantic University

Jerry Ventre
Florida Solar Energy Center

Gregor Hoogers
Trier University of Applied Sciences

2.1 Distributed Power

Johan H.R. Enslin and Rama Ramakumar

Introduction

Distributed power (DP) refers to the process and concepts in which small to medium (a few kW up to 50 MW or more) power generation facilities, energy storage facilities (thermal, flywheel, hydro, flow, and regular batteries), and other strategies are located at or near the customer's loads and premises. DP technologies installed near customers' loads operate as grid-connected or islanded resources at the distribution or subtransmission level and are geographically scattered throughout the service area. DP generation harnesses renewable and nonrenewable energy sources, such as solar insolation, wind, biomass, tides, hydro, waves, geothermal, biogas, natural gas, hydrogen, and diesel, in a distributed manner. DP also includes several nonutility sources of electricity, including facilities for self-generation, energy storage, and combined heat and power (CHP) or cogeneration systems.

Interest in DP has been growing steadily, due to inherently high reliability and possible efficiency improvements over conventional generation. In addition to the obvious advantages realized by the development of renewable energy sources, DP is ideally suited to power sensitive loads, small remote loads located far from the grid, and integrated renewable energy sources into the grid.

To interface these different power concepts, electrical networks are currently evolving into hybrid networks, including ac and dc, with energy storage for individual households, office buildings, industry, and utility feeders.

DP technologies can be depicted as an energy web and are also the structure for a possible future hydrogen-integrated society. This energy-web concept combines the DP technologies with the energy market, information datalinks, and hybrid ac and dc electrical networks including gas and hydrogen infrastructures, as shown in Figure 2.1.

Emerging DP technologies are available or have matured as the preferred technology for site-specific applications of power quality and reliability improvement, integrated CHP, and voltage support on transmission and distribution networks.

Several governments and utilities worldwide promote the use of DP technologies using renewable energy sources with subsidies and customer participation programs. Some examples include offshore wind farms and several "green" suburbs where roof-mounted photovoltaic (PV) arrays are installed on most of the roofs of individual homes, apartments, and communal buildings.

General Features

DP will have one or more of the following features:

- Small to medium size, geographically distributed
- Intermittent input resource, e.g., wind and insolation power
- Stand-alone or interface at the distribution or subtransmission level
- Utilize site-specific energy sources
- Located near the loads
- Integration of energy storage and control with power generation.

Potential and Future

Globally, the potential for DP is vast. Concerns over the unrestricted use of conventional energy resources and the associated environmental problems, such as the greenhouse effect and global warming, are providing the impetus necessary for the continued development of technologies for DP.

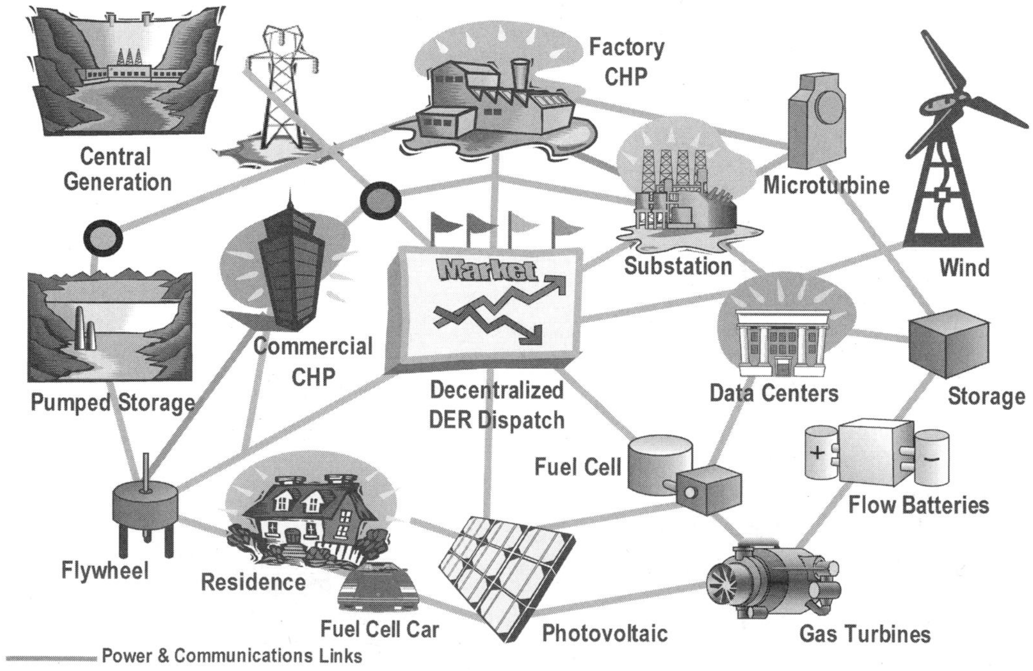

FIGURE 2.1 Presentation of the energy-web concept.

Motivation

Some of the main motivations for the use of DP are:

- Low incremental capital investments
- Long lead-times to site DP plants
- Likely to result in improved reliability, availability, and power quality
- Location near load centers lowers the energy transport losses and decreases transmission and distribution costs
- Better utilization of distribution network infrastructure
- Integration of control and protection with energy flow stream
- Preferred technology for digital society, requiring high-reliability figures.

DP Technologies

Many technologies have been proposed and employed for DP. Power ratings of DP systems vary from kilowatts to megawatts, depending on the application. A listing of some of the DP technologies is given below.

- Photovoltaic
- Wind energy conversion systems
- Mini and micro hydro
- Geothermal plants
- Tidal and wave energy conversion
- Fuel cell
- Solar-thermal-electric conversion
- Biomass utilization
- Thermoelectrics and thermionics
- Micro and mini turbines as cogeneration plants, powered by natural gas and supplying electricity and heat
- Energy storage technologies, including flow and regular batteries, pump-storage hydro, flywheels, and thermal energy storage concepts.
- Small-scale nuclear reactors are also in the development phase as pebble-bed modular nuclear reactors (PBMR).

These technologies will not all be described here. Mainly an emphasis will be provided on PV, wind, micro turbine, and storage technologies. Distributed wind energy generation and solar PV power generation are the best-established renewable distributed power options and will be discussed in more detail.

Photovoltaic (PV) Power Generation

PV refers to the direct conversion of insolation (incident solar radiation) to electricity. A PV cell (also known as a solar cell) is simply a large-area semiconductor pn junction diode, with the junction positioned very close to the top surface. Typically, a metallic grid structure on the top and a sheet structure in the bottom collect the minority carriers crossing the junction and serve as terminals. The minority carriers are generated by the incident photons with energies greater than or equal to the energy gap of the semiconductor material.

Since the output of an individual cell is rather low (1 or 2 W at a fraction of a volt), several (30 to 60) cells are combined to form a module. Typical module ratings range from 40 to 100 W at open-circuit voltages of 15 to 17 V. PV modules are progressively put together to form panels, arrays, and ultimately a PV plant consisting of several array segments. Plants rated at several MW have been built and operated successfully. Currently new houses and buildings are built with integrated PV panels on the roofs, forming complete subdivision communities with integrated PV roofs.

Advantages of PV include demonstrated low operation and maintenance costs, no moving parts, silent and simple operation, long lifetime if properly cared for, no recurring fuel costs, modularity, and minimal environmental effects. The disadvantages are the cost, energy requirements during manufacturing, need for large collector areas due to the diluteness of insolation, and the diurnal and seasonal variability of the output.

PV systems can be flat-plate or concentrating types. While flat-plate systems utilize the global (direct and diffuse) radiation, concentrator systems harness only the direct or beam radiation. As such, concentrating systems must track (one axis or two axes) the sun. Flat-plate systems may or may not be mounted on trackers.

By 1990, efficiencies of flat-plate crystalline and thin-film cells had reached 23% and 15%, respectively. Efficiencies as high as 34% were recorded for concentrator cells. Single-crystal and amorphous PV module efficiencies of 12% and 5%, respectively, were achieved by the early 1990s. For an average module efficiency of 10% and an insolation of 1 kW/m^2 on a clear afternoon, 10 m^2 of collector area is required for each of output.

The output of a PV system is dc, and inversion is required for supplying ac loads or for utility-interactive operation. While the required fuel input to a conventional power plant depends on its output, the input to a PV system is determined by external factors, such as cloud cover, time of day, season of the year, geographic location, orientation, and geometry of the collector. Therefore, PV systems are operated, as far as possible, at or near their maximum outputs. In addition, PV plants have no energy storage capabilities, and their power output is subject to rapid changes due to moving clouds.

The current-voltage (*IV*) characteristic of an illuminated solar cell is shown in Figure 2.2. It is given as

$$I = I_s - I_0\left[\exp\left(\frac{eV}{kT}\right) - 1\right].$$

where: I_0 and I_s are the dark and source currents, respectively, k is the Boltzmann constant (1.38×10^{-23} J/K), T is the temperature in K, and e is the electron charge. Under ideal conditions (identical cells), for a PV

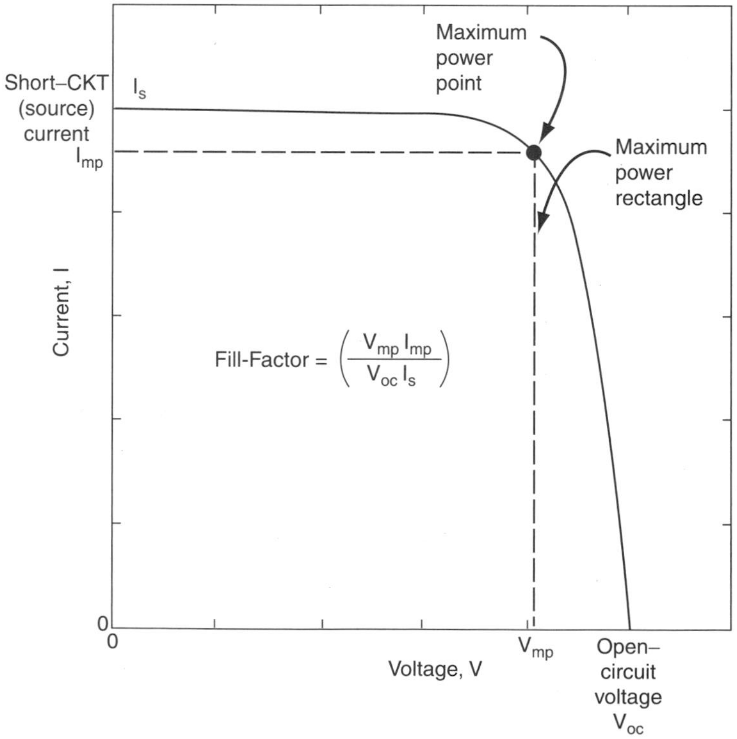

FIGURE 2.2 Typical current-voltage characteristic of an illuminated solar cell.

module with a series-parallel arrangement of cells, the *IV* characteristic will be similar, except that the current scale should be multiplied by the number of parallel branches and the voltage scale by the number of cells in series in the module. The source current varies linearly with insolation. The dark current increases as the cell operating temperature increases. Also, the larger the energy gap of the material, the smaller is the dark current. For improved operation, the ratio of source current to dark current should be made as large as possible.

Single-crystal silicon is still the dominant technology for fabricating PV devices. Polycrystalline, semicrystalline, and amorphous silicon technologies are developing rapidly to challenge this. Highly innovative technologies such as spheral cells are being introduced to reduce costs. Concentrator systems typically employ gallium arsenide or multiple junction cells. Many other materials and thin-film technologies are under investigation as potential candidates.

PV applications range from milliwatts (consumer electronics) to megawatts (central station plants). They are suitable for portable, remote, stand-alone, and network interactive applications. PV systems should be considered as energy sources and their design should maximize the conversion of insolation into useable electrical form. Power requirements of practical loads are met using energy storage, maximum power point tracking (MPPT) choppers, and inverters for network interconnection. Concentrating systems have been designed and operated to provide both electrical and low-grade thermal outputs with combined peak utilization efficiencies approaching 60%.

The vigorous growth of PV technology is manifested by the large growth of PV installations. At the end of 2003, cumulative installed PV systems in Japan, Germany, and California totaled 640 peak MW, 375 peak MW, and 60 peak MW, respectively. In 2003, about 790 MW of PVs were manufactured worldwide, up from 560 MW delivered in 2002.

An example of a Dutch rooftop project is shown in Figure 2.3, where all the homes in a subdivision are installed with solar PV panels. In this Dutch Nieuwland project near Amersfoort, a total of 12,000 m^2 of PV arrays have been installed on 500 homes. A total of 1 GWh of renewable energy is generated annually by this project.

Over the years prices of PV installations have steadily dropped from an original capital cost of $7000 per kWp (kilowatt peak) in 1988 to an average cost in 2003 of $5000 per kWp. Under the 100,000 Roof Program in Germany, which ended in June 2003, 345 MW peak were installed at an average cost in 2003 of €4200 per kWp. In 2004 the well-known feed-in tariff for renewable energy plants in Germany was increased from € 0.45 to 0.57 per kWh delivered.

Wind Energy Conversion

Wind energy is intermittent, highly variable, site-specific, and the least dependent upon latitude among all renewable resources. The power density (in W/unit area) in moving air (wind) is a cubic function of wind speed, and therefore even small increases in average wind speeds can lead to significant increases in the capturable energy. Wind sites are typically classified as good, excellent, or outstanding, with associated mean wind speeds of 13, 16, and 19 mph, respectively.

Wind turbines employ lift and drag forces to convert wind energy to rotary mechanical energy, which is then converted to electrical energy by coupling a suitable generator. The power coefficient C_p of a wind turbine is the fraction of the incident power converted to mechanical shaft power, and it is a function of the tip speed ratio λ, as shown in Figure 2.4. For a given propeller configuration, at any given wind speed, there is an optimum tip speed that maximizes C_p.

Several types of wind turbines are available. They can have horizontal or vertical axes, number of blades ranging from one to several, mounted upwind or downwind, and fixed- or variable-pitch blades with full blade or tip control. Vertical-axis (Darrieus) turbines are not self-starting and require a starting mechanism. Today, horizontal-axis turbines with two or three blades are the most prevalent at power levels of 3 to 5 MW each. These turbines propel electrical induction or synchronous generators through a gearbox or directly using a large ring gear.

FIGURE 2.3 Roof-mounted Dutch PV subdivision, Nieuwland, in Amersfoort.

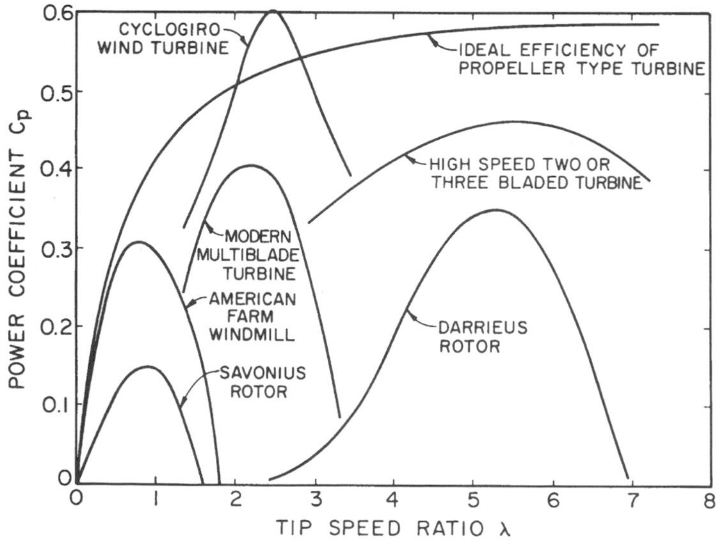

FIGURE 2.4 Typical wind turbine characteristics.

The electrical output P_e of a wind energy conversion system is given as

$$P_e = \eta_g \eta_m A C_p K v^3$$

where: η_g and η_m are the efficiencies of the electrical generator and mechanical interface, respectively, A is the swept area, K is a constant tip-speed ratio C_p, and v is the wind speed incident on the turbine.

There are two basic options for wind energy conversion. With varying wind speeds, the wind turbine can be operated at a constant speed, controlled by blade-pitch control. A conventional synchronous or induction machine is then employed to generate constant-frequency ac. Alternatively, the wind turbine rotational speed can be allowed to vary together with the wind speed to maintain a constant and optimum tip-speed ratio. An induction or synchronous generator is used in combination with a back-to-back three-phase power electronic converter, interfacing the wind generator with the electrical power network. This converter can also provide reactive power at the interconnection point. The variable-speed option allows optimum efficiency operation of the turbine over a wide range of wind speeds, resulting in increased outputs with lower structural loads and stresses. The power fluctuations into the electrical power network are also reduced with a variable-speed generating system. Most future advanced turbines are expected to operate in the variable-speed mode and to use power electronics to convert the variable-frequency output to constant frequency with minimal harmonic distortion.

Individual or small groups of wind-power generators in the 0.5- to 5-MW power levels are interfaced to the electrical distribution network on a distributed level. Large-scale harnessing of wind energy (in the 50- to 1000-MW power range) requires ten to hundreds of wind generators arranged in a wind farm with spaces of about two to three diameters crosswind and about ten diameters apart downwind. The power output of an individual wind generator will fluctuate over a wide range, and its statistics strongly depend on the wind statistics. When many wind generators are used in a wind farm, some smoothing of the total power output will result, depending on the statistical independence of the outputs of the individual generators. This is desirable, especially with high (>20%) penetration of wind power in the generation mix of a network section. One of the major disadvantages of wind generators is that they are not dispatchable, having an intermittent power output.

Although wind energy conversion has overall minimum environmental impacts, the large rotating structures involved do generate some noise and introduce visual aesthetics problems. By locating wind energy systems sufficiently far from centers of population, these effects can be minimized. The envisaged potential for bird kills turned out not to be a serious problem. Wind energy systems occupy only a very small fraction of the land. However, the area surrounding them can be used only for activities such as farming and livestock grazing. Thus, there is some negative impact on land use. To minimize these environmental problems, utilize constant and higher wind speeds, and improve the cost-effectiveness of wind power generation, 1000-MW and even larger wind farms are proposed to be installed offshore. Some offshore wind farms have already been installed off the coast of Denmark and the U.K. Large offshore programs are currently being investigated in the North Sea, off the Dutch coast, off the German northeast coast, off the west coast of Spain, and off the coast of New York.

Hydro

Hydropower is a mature and one of the most promising renewable energy technologies. In the context of DP, small (less than 100 MW), mini (less than 1 MW), and micro (less than 100 kW) hydroelectric plants are of interest. The source of hydropower is the hydrologic cycle driven by the energy from the sun. Most of the sites for DP hydro are either low-head (2 to 20 m) or medium-head (20 to 150 m). The global hydroelectric potential is vast; 900 MW of new small hydro capacity was added worldwide in 2003. One estimate puts small-scale hydro at 31 GW for Indonesia alone. The installed capacity of small hydro in the People's Republic of China was exceeding 7 GW, even by 1980. The Three Gorges large-scale hydro dam project in China will be completed by 2009 with 26 generators generating 18.2 GW of power.

Both impulse and reaction turbines have been employed for small-scale hydro for DP. Several standardized units are available in the market. Most of the units are operated at constant speed with governor control and are coupled to synchronous machines to generate ac power. If the water source is highly variable, it may be necessary to employ variable-speed generation. Similarly to variable wind generation, variable-speed hydro generation utilizes back-to-back power electronic converters to convert the variable generator output to constant-frequency ac power. In a very small system of a few, it may be required to install energy storage on the dc link of the power electronic converter.

Geothermal

Geothermal plants exploit the heat stored in the form of hot water and steam in the Earth's crust at depths of 800 to 2000 m. By nature, these resources are extremely site-specific and slowly run down (depletable) over a period of years. For electric power generation, the resource should be at least around 200°C. Depending on the temperature and makeup, dry steam, flash steam, or binary technology can be employed. Of these, dry natural steam is the best, since it eliminates the need for a boiler.

The three basic components of a geothermal plant are: (1) a production well to bring the resource to the surface, (2) a turbine generator system for energy conversion, and (3) an injection well to recycle the spent geothermal fluids back into the reservoir.

The typical size of back-pressure plants ranges from 1 to 10 MW. Indirect and binary cycle plants are usually in the 10- to 60-MW range, while condensing units are on the order of 15 to 110 MW. The variable size and nature of geothermal generation plants has the advantage of enabling units to be installed in modules. About 6800 MW (38 TWh/year) of electricity generating capacity is currently installed around the world in places where both steam and water are produced at temperatures over 200°C. Half of this capacity is located in the U.S.A. The Geysers plant, north of San Francisco, is the largest in the world, with an installed capacity of 516 MW. In some developing countries, the Philippines, for example, geothermal plants supply nearly 20% of their electrical needs.

Tidal Energy

The origin of tidal energy is the upward-acting gravitational force of the moon, which results in a cyclic variation in the potential energy of water at a point on the Earth's surface. Topographical features, such as the shape and size of estuaries, amplify these variations. The ratio between maximum spring tide and minimum at neap can be as much as three to one. In estuaries, the tidal range can be as large as 10 to 15 m.

Power can be generated from a tidal estuary in two basic ways. A single basin can be used with a barrage at a strategic point along the estuary. By installing turbines at this point, electricity can be generated either when the tide is ebbing or flooding. In the two-basin scheme, generation can be time-shifted to coincide with hours of peak demand by using the basins alternately.

As can be expected, tidal energy conversion is very site-specific. The La Rance tidal power plant in Brittany, France, generating 240 MW with 24-vane-type horizontal turbines and alternator motors, each rated at 10 MVA, is the largest operating tidal facility in the world. It is far bigger than the next largest, the Annapolis facility (20 MW) in Canada. The Rance tidal plant has been in operation since 1961 (refurbished in 1995), with good technical and economic results. It has generated, on the average, 500 GWh of net energy per year.

Some very ambitious projects are currently being studied: Severn and Mersey, England (8600 MW and 700 MW, respectively), Bay of Fundy, Canada (5300 MW) with the highest known tidal range of 17 m, San José, Argentina (5000 MW), Kutch, India (900 MW), and Garolim, Korea (480 MW).

Fuel Cells

A fuel cell is a simple static device that converts the chemical energy in fuel directly, isothermally, and continuously into electrical energy. Fuel and oxidant (typically oxygen in air) are fed to the device in which an electrochemical reaction takes place that oxidizes the fuel, reduces the oxidant, and releases energy. The energy released is in both electrical and thermal forms. The electrical part provides the required output. Since a fuel cell completely bypasses the thermal-to-mechanical conversion involved in a conventional power plant and its operation is isothermal, fuel cells are not Carnot-limited. Electrical efficiencies up to 80% are theoretically possible, and practically, the efficiency can be in the range of 43 to 55% for modular dispersed generators featuring fuel cells.

Although their structure is somewhat like that of a battery, fuel cells never need recharging or replacing and can consistently produce electricity as long as they are supplied with hydrogen and oxygen. Fossil fuels (coal, oil, and natural gas), biomass (plant material), or pure hydrogen can be used as the source of fuel. If pure hydrogen is used, the emissions from a fuel cell are only electricity and water. Fuel cells are small and modular in nature, and therefore fuel cell power plants can be used to provide electricity in many different applications, from small mobile devices, like electric vehicles, to large, grid-connected utility power plants. First used in the U.S. space program in the 1950s, fuel cells are only a developing technology with some commercial uses today, but may emerge as a significant source of electricity in the near future.

The low (< 0.02 kg/MWh) airborne emissions of fuel cell plants make them prime candidates for siting in urban areas. The possibility of using fuel cells in combined heat and power (CHP) units provides the cleanest and most efficient (> 90%) energy system option utilizing valuable (or imported) natural gas resources.

Hydrocarbon fuel (natural gas or LNG) or gasified coal is reformed first to produce hydrogen-rich (and sulfur-free) gas that enters the fuel cell stack where it is electrochemically "burned" to produce electrical and thermal outputs. The electrical output of a fuel cell is low-voltage high-current dc. Most commercially successful fuel cells are currently based on a PEM (proton exchange membrane, also called polymer electrolyte membrane) fuel cell. At the anode of a PEM fuel cell, hydrogen molecules give up electrons, forming hydrogen ions. This process is made possible by the platinum catalyst. The proton exchange membrane now allows the protons to flow through, but not electrons. As a result, the hydrogen ions flow directly through the proton exchange membrane to the cathode, while the electrons flow through an external electrical circuit. By utilizing a properly organized stack of cells, adding some dc energy storage to provide some buffering and a power electronic inverter, utility-grade ac output is obtained.

Solar-Thermal Electric Conversion

A solar-thermal electric power plant generates heat by using concentrated sunlight through lenses and reflectors. Because the heat can be stored, these plants are unique as they can generate power when it is needed. Parabolic troughs, parabolic dishes, and central receivers are used to generate temperatures in the range of 400 to 500, 800 to 900, and > 500°C, respectively.

The technical feasibility of the central receiver system was demonstrated in the early 1980s by the 10-MWe Solar One system in Barstow, California. Over a six-year period, this system delivered 37 GWh of net energy to the Southern California Edison Grid, with an overall system efficiency in the range of 7 to 8%. With improvements in heliostat and receiver technologies, annual system efficiencies of 14 to 15% and generation costs of 8 to 12¢/kWh have been projected.

In the late 1970s and early 1980s, parabolic-dish electric-transport technology for DP was under active development at the Jet Propulsion Laboratory (JPL) in Pasadena, California. Prototype modules with Stirling engines reached a record 29% overall efficiency of conversion from insolation to electrical output. Earlier parabolic-dish designs collected and transported thermal energy to a central location for conversion to electricity. Advanced designs, such as the one developed at JPL, employed engine-driven generators at the focal points of the dishes, and energy was collected and transported in electrical form.

By far the largest installed capacity (nearly 400 MW) of solar-thermal electric DP employs parabolic-trough collectors and oil to transport the thermal energy to a central location for conversion to electricity via a steam-Rankine cycle. With the addition of a natural gas burner for hybrid operation, this technology, developed by LUZ under the code name SEGS (solar-electric generating system), accounts for more than 90% of the world's solar electric capacity, all located in Daggett, Kramer Junction, and Harper Lake in California. Generation costs of around 8 to 9¢/kWh have been realized with SEGS. This hybrid technology burns the natural gas to compensate for the temporary variations of insolation and balance the power delivered by the system.

This compensation may come from 7 to 11 p.m. in summer and 8 a.m. to 5 p.m. in winter. SEGS will require about 5 acres/MW or can deliver 130 MW/mi^2 of land area.

Solar Two, a "power tower" electricity generating plant in California, is a 10-MW prototype for large-scale commercial power plants. It stores the solar-energy-molten salt at 560°C, which allows the plant to generate continuous power. Construction was completed in March 1996. Over 700 MW of solar-thermal electric systems was deployed by 2003 globally. The market for these systems may exceed 5000 MW by 2010.

Biomass Energy

Biological sources provide a wide array of materials that have been and continue to be used as energy sources. Wood, wood waste, and residue from wood processing industries, sewage or municipal solid waste, cultivated herbaceous and other energy crops, waste from food processing industries, and animal wastes are lumped together by the term biomass. The most compelling argument for the use of biomass technologies is the inherent recycling of the carbon by photosynthesis. In addition to the obvious method of burning biomass, conversion to liquid and gaseous fuels is possible, thus expanding the application possibilities.

In the context of electric power generation, the role of biomass is expected to be for repowering old units and for use in small (20 to 50 MW) new plants. Several new high-efficiency conversion technologies are either already available or under development for the utilization of biomass. The technologies and their overall conversion efficiencies are listed below.

- FBC (fluidized-bed combustor), 36 to 38%
- EPS (energy performance system) combustor, 34 to 36%
- BIG/STIG (biomass-integrated gasifier/steam-injected gas turbine), 38 to 47%.

Acid or enzymatic hydrolysis, gasification, and aqueous pyrolysis are some of the other technology options available for biomass utilization.

Anaerobic digestion of animal wastes is being used extensively in developing countries to produce biogas, which is utilized directly as a fuel in burners and for lighting. An 80:20 mixture of biogas and diesel has been used effectively in biogas engines to generate electricity in small quantities.

Biomass-fueled power plants are best suited in small (<100 MW) sizes for DP to serve base load and intermediate loads in the eastern United States and in many other parts of the world. This contribution is clean, renewable, and reduces CO_2 emissions. Since biomass fuels are sulfur-free, these plants can be used to offset CO_2 and SO_2 emissions from new fossil power plants. Ash from biomass plants can be recycled and used as fertilizer. A carefully planned and well-managed short-rotation woody crop (SRWC) plantation program with yields in the range of 6 to 12 dry tons/acre/year can be effectively used to mitigate greenhouse gases and contribute thousands of megawatts of DP to the U.S. grid by the turn of the century.

Thermoelectrics

Thermal energy can be directly converted to electrical energy by using the thermoelectric effects in materials. Semiconductors offer the best option as thermocouples, since thermojunctions can be constructed using a *p*-type and an *n*-type material to cumulate the effects around a thermoelectric circuit. Moreover, by using solid solutions of tellurides and selenides doped to result in a low density of charge carriers, relatively moderate thermal conductivities and reasonably good electrical conductivities can be achieved.

In a thermoelectric generator, the Seebeck voltage generated under a temperature difference drives a dc current through the load circuit. Even though there is no mechanical conversion, the process is still Carnot-limited, since it operates over a temperature difference. In practice, several couples are assembled in a series-parallel configuration to provide dc output power at the required voltage.

Typical thermoelectric generators employ radioisotope or nuclear reactor or hydrocarbon burner as the heat source. They are custom-made for space missions, as exemplified by the systems for nuclear auxiliary power (SNAP) series and the radioisotope thermoelectric generator (RTG) used by the Apollo astronauts. Maximum performance over a large temperature range is achieved by cascading stages. Each stage consists of thermocouples electrically in series and thermally in parallel. The stages themselves are thermally in series and electrically in parallel.

Tellurides and selenides are used for power generation up to 600°C. Silicon germanium alloys turn out better performance above this up to 1000°C. With the materials available at present, conversion efficiencies in the 5 to 10% range can be expected. Whenever small amounts of silent reliable power are needed for long periods, thermoelectrics offer a viable option. Space, underwater, biomedical, and remote terrestrial power, such as cathodic protection of pipelines, fall into this category.

Thermionics

Direct conversion of thermal energy into electrical energy can be achieved by employing the Edison effect—the release of electrons from a hot body, also known as thermionic emission. The thermal input imparts sufficient energy (≧ work function) to a few electrons in the emitter (cathode), which helps them escape. If these electrons are collected, using a collector (anode), and a closed path through a load is established for them to complete the circuit back to the cathode, then electrical output is obtained. Thermionic converters are heat engines with electrons as the working fluid and, as such, are subject to Carnot limitations.

Converters filled with ionizable gasses, such as cesium vapor in the interelectrode space, yield higher power densities due to space charge neutralization. Barrier index is a parameter that signifies the closeness to ideal performance with no space charge effects. As this index is reduced, more applications become feasible.

A typical example of developments in thermionics is the TFE (thermionic fuel element) that integrates the converter and nuclear fuel for space nuclear power in the kW to MW level for very long (7 to 10 years) duration missions. Another niche is the thermionic cogeneration burner module, a high-temperature burner equipped with thermionic converters. Electrical outputs of 50 kW/MW of thermal output have been achieved. High (600 to 650°C) heat rejection temperatures of thermionic converters are ideally suited for producing flue gas in the 500 to 550°C range for industrial processes. A long-range goal is to use thermionic converters as toppers for conventional power plants. Such concepts are not economical at present.

Energy Storage

An energy storage system, integrated with power electronics, forms the ideal mitigation option to solve the technical interconnection and remote power supply problems associated with intermittent distributed power resources. For a DP with network connection, remote area power supply, and interconnection of hybrid networks, energy storage plays a crucial role. An energy storage system should be considered to compensate the fluctuating output of wind farms and to improve power quality at the interconnection point between the DP and the network. The total intermittent power output of a wind farm can, for example, be balanced using a large-scale energy storage plant, providing dispatchable and controllable power at the output of the plant. Hydro pump storage has been used for decades to do power balancing between generation and load demand. For DP systems, medium to small energy storage options are considered. Several storage technologies exist and are continuously improved for these applications.

Hydro pump storage uses two reservoirs at different elevations, bidirectional pumps, and synchronous machines connected to the utility network. Power rating and energy storage capacity are in the 100 to 2000 MW and 500 to 40000 MWh range, respectively. High environmental impacts and limited and site-specific locations make this energy storage option only applicable for large applications. Underground pumped storage, using flooded mine shafts or other cavities, is technically also possible. The open sea can also be used as the lower reservoir. A seawater-pumped hydro plant was first built in Japan in 1999 (Yanbaru, 30 MW).

Compressed-air energy storage (CAES) is a peaking gas turbine power plant that consumes less than 40% of the gas used in conventional gas turbine to produce the same amount of electric output power by precompressing air using the low-cost electricity from the power grid at off-peak times. This energy is then utilized later along with some gas fuel to generate electricity as needed. The compressed air is often stored in appropriate underground mines or caverns created inside salt rocks. The first commercial CAES was a 290-MW unit built in Hundorf, Germany, in 1978. The second commercial CAES was a 110-MW unit built in

McIntosh, Alabama, in 1991. The construction took 30 months and cost $65 million (about $591/kW). This unit comes on line within 14 minutes. Some storage concepts also combine compressed air, natural gas, or hydrogen with pumped hydro storage into one design, using underground cavities that may be interesting in the future with higher utilization of hydrogen.

Traditional and flow batteries are still considered the most important storage technology for DP applications. Lead-acid batteries are used in several mobile and stationary applications. It is a low-cost and popular storage choice for remote area power supply options, power quality, uninterruptible power supplies, and some spinning reserve applications. Its application for energy management, however, has been very limited due to its short cycle life. The lead-acid battery plant is a 40-MWh system in Chino, California, built in 1988.

Flow batteries are regenerative fuel cell technology with membranes that provide a reversible electrochemical reaction between two electrolytes, stored in separate storage tanks. For flow batteries the power and energy ratings are independent of each other. The power rating is associated with the size and number of membrane stacks, while the energy storage capacity is associated with the size of the storage electrolyte tanks. Flow batteries are now industrialized up to 15 MW, 120 MWh using polysulfide bromide (PSB) technologies. Vanadium redox flow batteries (VRB) employ vanadium redox couples stored in mild sulfuric acid solutions. VRB has been commercialized up to 500 kW, 10 hrs (5 MWh) and installed in Japan and the U.S. Other battery technologies using zn-br, metal-air, and NaS are continuously developed and improved. Flywheel, ultracapacitors, and superconducting magnetic energy storage (SMES) systems have been commercialized and installed for power quality and network stability mitigation solutions. The net efficiency of these batteries is in the 60 to 80% range, with capital prices of $800 to 3000/kW and $50 to 1000/kWh.

Network Impacts

The response of distribution systems to high penetrations of DP is not yet fully investigated. Also, the nature of the response will depend on the DP technology involved. However, there are some general areas of potential impacts common to most of the technologies: (1) voltage flicker, imbalance, regulation, etc.; (2) power quality; (3) real and reactive power flow modifications; (4) islanding; (5) synchronization during system restoration; (6) transients; (7) protection issues; (8) load-following capability; and (9) dynamic interaction with the rest of the system. Since there are very few systems with high penetration of DP, studies based on detailed models should be undertaken to forecast potential problems and arrive at suitable solutions.

Defining Terms

Biomass: General term used for wood, wood wastes, sewage, cultivated herbaceous and other energy crops, and animal wastes.

Distributed generation (DG): Medium to small power plants at or near loads and scattered throughout the service area.

Distributed power (DP): System concepts in which distributed generation facilities, energy storage facilities, control, and protection strategies are located near loads and scattered throughout the service area.

Fuel cell: Device that converts the chemical energy in a fuel directly and isothermally into electrical energy.

Geothermal energy: Thermal energy in the form of hot water and steam in the Earth's crust.

Hydropower: Conversion of potential energy of water into electricity using generators coupled to impulse or reaction water turbines.

Insolation: Incident solar radiation.

Photovoltaics: Conversion of insolation into dc power by means of solid-state pn-junction diodes.

Solar-thermal-electric conversion: Collection of solar energy in thermal form using flat-plate or concentrating collectors and its conversion to electrical form.

Thermionics: Direct conversion of thermal energy into electrical energy by using the Edison effect (thermionic emission).

Thermoelectrics: Direct conversion of thermal energy into electrical energy using the thermoelectric effects in materials, typically semiconductors.

Tidal energy: The energy contained in the varying water level in oceans and estuaries, originated by lunar gravitational force.

Wind energy conversion: The generation of electrical energy using electromechanical energy converters driven by wind turbines.

References

A-M. Borbely and J.F. Kreider, Eds., *Distributed Generation: The Power Paradigm for the New Millennium*, Boca Raton, FL: CRC Press, 2001.

R.C. Dorf, *Energy, Resources, and Policy*, Reading, MA: Addison-Wesley, 1978.

J.H.R. Enslin, "In store for the future? Interconnection and energy storage for offshore wind farms," *Proceedings of Renewable Energy World*. James & James Ltd., 2004, pp. 104–113.

J.H.R. Enslin, W.T.J. Hulshorst, A.M.S. Atmadji, P.J.M. Heskes, A. Kotsopoulos, J.F.G. Cobben, and P. Van der Sluijs, "Harmonic interaction between large numbers of photovoltaic inverters and the distribution network," *Proceedings of Renewable Energy 2003*, London: Sovereign Publications Ltd., 2003, pp. 109–113.

EPRI and DOE, *Handbook of Energy Storage for Transmission and Distribution Applications*, EPRI Product No. 1001834, 2003.

J.J. Fritz, *Small and Mini Hydropower Systems*, New York: McGraw-Hill, 1984.

S. Heier, *Grid Integration of Wind Energy Conversion Systems*, New York: Wiley, 1998.

J.F. Kreider and F. Kreith, Eds., *Solar Energy Handbook*, New York: McGraw-Hill, 1981.

R. Ramakumar and J.E. Bigger, "Photovoltaic systems," *Proc. IEEE*, vol. 81, no. 3, pp. 365–377, 1993.

R. Ramakumar, I. Abouzahr, and K. Ashenayi, "A knowledge-based approach to the design of integrated renewable energy systems," *IEEE Trans. Energy Convers.*, vol. EC-7, no. 4, pp. 648–659, 1992.

2.2 Solar Electric Systems[1]

Thomas R. Mancini, Roger Messenger, and Jerry Ventre

Solar Thermal Electric Systems

Solar thermal power systems, which are also referred to as concentrating solar power systems, use the heat generated from the concentration and absorption of solar energy to drive heat engine/generators and, thereby, produce electrical power. Three generic solar thermal systems, trough, power tower, and dish-engine systems, are used in this capacity. Trough systems use linear parabolic concentrators to focus sunlight along the focal lines of the collectors. In a power tower system, a field of two-axis tracking mirrors, called heliostats, reflects the solar energy onto a receiver that is mounted on top of a centrally located tower. Dish-engine systems, the third type of solar thermal system, continuously track the sun, providing concentrated sunlight to a thermal receiver and heat engine/generator located at the focus of the dish.

Trough Systems

Of the three solar thermal technologies, trough-electric systems are the most mature with 354 MW installed in the Mojave Desert of Southern California. Trough systems produce about 75 suns concentration and operate at temperatures of up to 400°C at annual efficiencies of about 12%. These systems use linear-parabolic concentrators to focus the sunlight on a glass-encapsulated tube that runs along the focal line of the collector,

[1]The Concentrating Solar Power activities presented herein are funded by the U.S. Department of Energy through Sandia National Laboratories, a multi-program laboratory operated by Sandia Corporation, a Lockheed Martin Company, for the U.S. Department of Energy under Contract DE-AC04–94-AL85000.

FIGURE 2.5 Solar collector field at a SEGS plant located at Kramer Junction, CA. (Photograph courtesy of Sandia National Laboratories.)

shown in Figure 2.5. Troughs are usually oriented with their long axis north–south, tracking the sun from east to west, to have the highest collection efficiencies. The oil working fluid is heated as it circulates through the receiver tubes and before passing through a steam-generator heat exchanger. In the heat exchanger, water boils producing the steam that is used to drive a conventional Rankine-cycle turbine generator. The optimal size for a trough-electric system is thought to be about 200 megawatts, limited mainly by the size of the collector field.

A major challenge facing these plants has been to reduce the operating and maintenance costs, which represent about a quarter of the cost of the electricity they produce. The newer plants are designed to operate at 10 to 14% annual efficiency and to produce electricity for $.08 to $.14 a kilowatt-hour, depending on interest rates and tax incentives with maintenance costs estimated to be about $.02 a kilowatt hour. New systems are currently being designed for use in the Southwest U.S. that will utilize ongoing research and development addressing issues such as improved receiver tubes, advanced working/storage fluids, and molten-salt storage for trough plants.

Power Towers

In the Spring of 2003, the first commercial power towers are being designed for installation in Spain in response to a very attractive solar incentive program. Power towers, shown in Figure 2.6, are also known as central receivers and, while close, they are not as commercially mature technologically as trough systems. In a power tower, two-axis tracking mirrors called heliostats reflect solar energy onto a thermal receiver that is located at the top of a tower in the center of the heliostat field. To maintain the sun's image on the centrally located receiver, each heliostat must track a position in the sky that is midway between the receiver and the sun. Power towers have been designed for different working fluids, including water/steam, sodium nitrate salts, and air. Each working fluid brings an associated set of design and operational issues; the two active tower design concepts in Europe and the U.S. utilize air and molten-nitrate salt as the working fluids. Both system designs incorporate thermal storage to increase the operating time of the plant, thereby allowing the solar energy to be collected when the sun shines, stored, and used to produce power when the sun is not shining. This feature, which allows power to be dispatched when needed, increases the "value" of the electricity generated with solar energy.

In the U. S., the choice of a molten-nitrate salt as the working fluid provides for high-temperature, low-pressure operation and thermal storage in the hot salt. The molten-salt approach uses cold salt (290°C) that is pumped from a cold storage tank, through the receiver where it is heated with 800 suns to 560°C. The hot salt

FIGURE 2.6 Warming the receiver at Solar 2 in Dagget, CA. The "wings" one either side of the receiver result from atmospheric scattering of the sunlight reflected from the heliostat field. (Photograph courtesy of Sandia National Laboratories.)

is delivered to a hot salt storage tank. Electrical power is produced when hot salt is pumped from the hot tank and through a steam generator; the steam is then used to power a conventional Rankine turbine/generator. The salt, which has cooled, is returned to the cold tank. The optimal size for power towers is in the 100- to 300-megawatt range. Studies have estimated that power towers could operate at annual efficiencies of 15 to 18% and produce electrical power at a cost of $.06 to $.11 per kilowatt-hour. Advanced systems are focusing on reducing the cost of heliostats and on operational issues associated with plant operation and thermal storage.

Dish–Stirling Systems

Dish–Stirling systems track the sun and focus solar energy into a cavity receiver where it is absorbed and transferred to a heat engine/generator. Figure 2.7 is a picture of a Dish–Stirling system. Although a Brayton engine has been tested on a dish and some companies are considering adapting microturbine technology to dish engine systems, kinematic Stirling engines are currently being used in all four Dish–Stirling systems under development today. Stirling engines are preferred for these systems because of their high efficiencies (thermal-to-mechanical efficiencies in excess of 40% have been reported), high power density (40 to 70 kW/liter for solar engines), and their potential for long-term, low-maintenance operation. Dish–Stirling systems have demonstrated the highest efficiency of any large solar power technology, producing more than 3000 suns concentration, operating at temperatures in excess of 750°C, and at annual efficiencies of 23% solar-to-electric conversion. Dish–Stirling systems are modular, i.e., each system is a self-contained power generator, allowing their assembly into plants ranging in size from a few kilowatts to tens of megawatts. The near-term markets identified by the developers of these systems include remote power, water pumping, grid-connected power in developing countries, and end-of-line power conditioning applications.

The major issues being addressed by system developers are reliability, due to high-temperature, high-heat flux conditions, system availability, and cost. Current systems range in size from 10 to 25 kW, primarily because of the available engines. These systems are the least developed of the three concentrating solar power systems. Research and development is focusing on reliability improvement and solar concentrator cost reductions.

FIGURE 2.7 Advanced dish development system (ADDS) on test at Sandia National Laboratories' National Solar Thermal Test Facility. (Photograph courtesy of Sandia National Laboratories.)

Photovoltaic Power Systems

In 2002, worldwide shipments of photovoltaic (PV) modules passed the 500-megawatt mark as PV module shipments continued to grow at an annual rate near 25% [Maycock, 2002]. Another important milestone in the PV industry occurred in 2001. Of the 395 MW of PV installed during 2001, the capacity of grid-connected PV systems installed (204 MW) surpassed the capacity of stand-alone systems installed (191 MW). Improvements in all phases of deployment of PV power systems have contributed to the rapid acceptance of this clean electric power-producing technology.

The photovoltaic effect occurs when a photodiode is operated in the presence of incident light when connected to a load. Figure 2.8a shows the *I–V* characteristic of the photodiode. Note that the characteristic passes through the first, third, and fourth quadrants of the *I–V* plane. In the first and third quadrants,

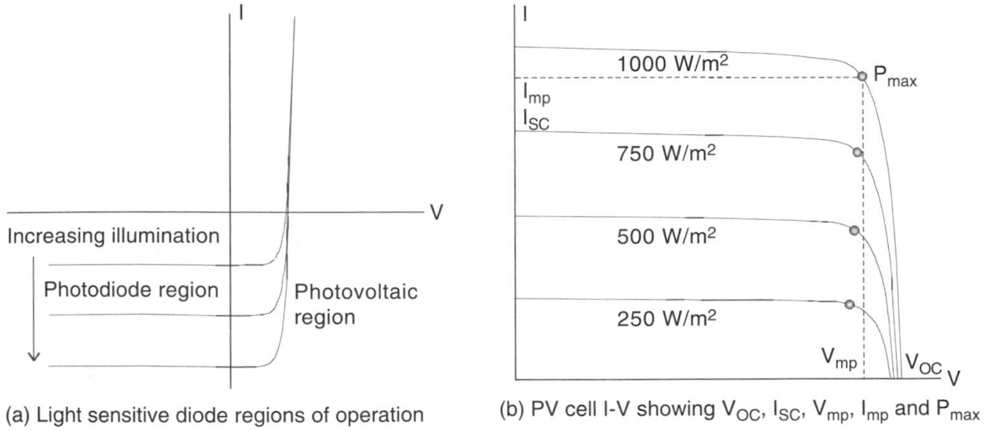

FIGURE 2.8 *I–V* characteristics of light sensitive diode.

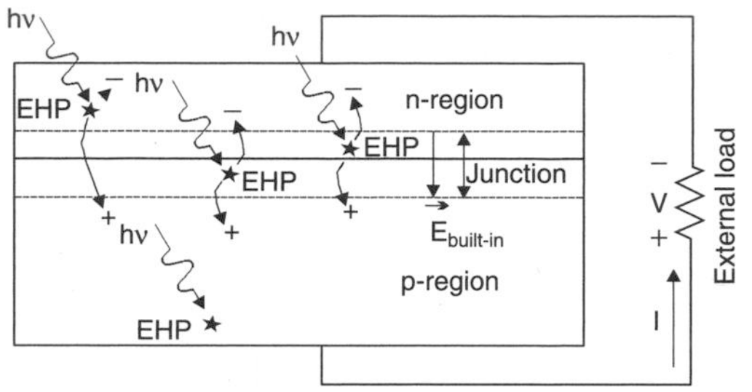

FIGURE 2.9 The illuminated pn junction showing the creation of electron-hole pairs.

the device dissipates power supplied to the device from an external source. But in the fourth quadrant, the device generates power, since current leaves the positive terminal of the device. Photo-voltaic cells are characterized by an open circuit voltage and a short circuit current, as shown in Figure 2.8b. Note that in Figure 2.8b, the current axis is inverted such that a positive current represents current leaving the device positive terminal. The short circuit current is directly proportional to the incident light intensity, while the open circuit voltage is proportional to the logarithm of the incident light intensity. Figure 2.8b shows that the cell also has a single point of operation where its output power is maximum. This point, labeled P_{max}, occurs at current, I_{mp} and voltage, V_{mp}. Since cells are relatively expensive, good system designs generally ensure that the cells operate as close as possible to their maximum power points.

The power generation mechanism that applies to conventional PV cells is shown in Figure 2.9. Photovoltaic power generation requires the presence of a pn junction. The pn junction is formed from a material, such as crystalline silicon, that has holes as the dominant charge carriers on the p-side and electrons as dominant carriers on the n-side. In simple terms, incident photons with sufficient energy interact with crystal lattice atoms in the PV cell to generate electron-hole pairs (EHP). If the EHP is generated sufficiently close to the pn junction, the negatively charged electron will be swept to the n-side of the junction and the positively charged hole will be swept to the p-side of the junction by the built-in electric field of the junction. The separation of the photon-generated charges creates a voltage across the cell, and if an electrical load is connected to the cell, the photon-generated current will flow through the load as shown in Figure 2.9.

Individual PV cells, with cell areas ranging from small to approximately 225 cm^2, generally produce only a few watts or less at about half a volt. Thus, to produce large amounts of power, cells must be connected in series–parallel configurations. Such a configuration is called a module. A typical crystalline silicon module will contain 36 cells and be capable of producing about 100 watts when operated at approximately 17 V and a current of approximately 6 A. Presently, the largest available module is a 300-watt unit that measures approximately 4 × 6 ft (see http://www.asepv.com/aseprod.html for more information). Although higher conversion efficiencies have been obtained, typical conversion efficiencies of commercially available crystalline silicon PV modules are in the 14% range, which means the modules are capable of generating approximately 140 W/m^2 of module surface area under standard test conditions. However, under some operating conditions where the module temperatures may reach 60°C, the module output power may be degraded by as much as 15%.

If the power of a module is still inadequate for the needs of the system, additional modules may be connected in series or in parallel until the desired power is obtained.

Stand-Alone PV Systems

Figure 2.10 shows a hierarchy of stand-alone PV systems. Stand-alone PV systems may be as simple as a module connected to an electrical load. The next level of complexity is to use a maximum power tracker or a linear current booster between module and load as a power matching device to ensure that the module delivers maximum power to the load at all times. These systems are common for pumping water and are the equivalent of a dc-dc matching transformer that matches source resistance to load resistance.

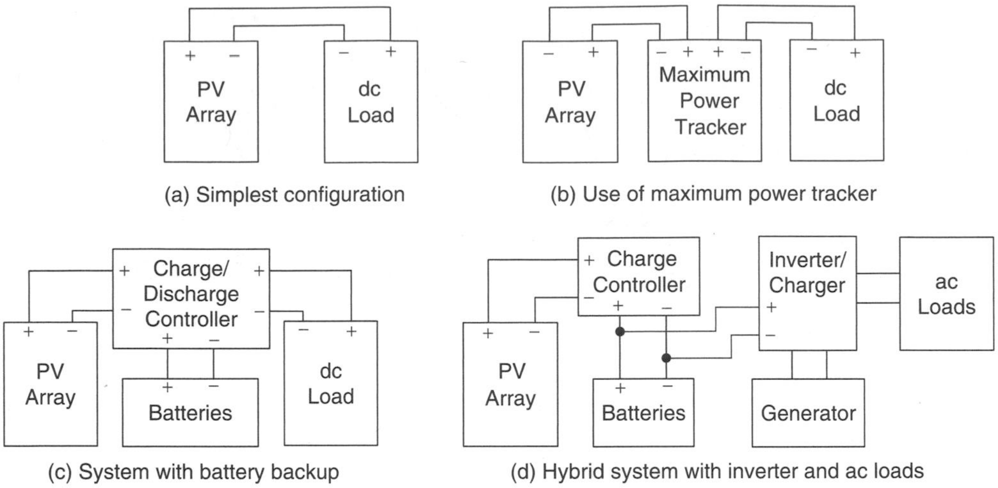

FIGURE 2.10 Four examples of stand-alone PV systems.

The next level of complexity involves the use of storage batteries to allow for the use of the PV-generated electricity when the sun is not shining. These systems also generally include an electronic controller to prevent battery overcharge and another controller to prevent overdischarge of the batteries. Many charge controllers perform both functions.

If loads are ac, an inverter can be incorporated into the system to convert the dc from the PV array to ac. A wide range of inverter designs are available with a wide range of output waveforms ranging from square waves to relatively well-approximated sine waves with minimal harmonic distortion. Most good inverters are capable of operating at conversion efficiencies greater than 90% over most of their output power range.

Available sunlight is measured in $kWh/m^2/day$. However, since peak solar power, or "peak sun" is defined as $1\ kW/m^2$, the term peak sun hours (psh) is often used as an equivalent expression for available solar energy (i.e., $1\ psh = 1\ kWh/m^2/day$). A number of tabulations of psh are available [Sandia National Laboratories, 1995; see also www.nrel.gov and http://solstice.crest.org/renewables/solrad/]. In some regions, there is a large difference in seasonal available sunlight. In these regions with large seasonal differences in psh, it often makes sense to back up the solar electrical production with a fossil-fueled or wind generator. When more than one source of electricity is incorporated in a stand-alone system, the system is called a hybrid system. Elegant inverter designs incorporate circuitry for starting a generator if the PV system battery voltage drops too low.

Grid-Connected PV Systems

Grid-connected, or utility interactive, PV systems connect the output of the PV inverter directly to the utility grid and are thus capable of supplying power to the utility grid. An inverter operating in the utility interactive mode must be capable of disconnecting from the utility in the event of a utility failure. Elegant control mechanisms have been developed that enable the inverter to detect either an out-of-frequency-range or an out-of-voltage-range utility event. In essence, the output of the inverter monitors the utility voltage and delivers power to the grid as a current source rather than as a voltage source. Since the inverters are microprocessor controlled, they are capable of monitoring the grid in intervals separated by less than a millisecond, so it is possible for the inverter to keep the phase of the injected current very close to the phase of the utility voltage. Most inverter operating power factors are close to unity.

The simplest utility-interactive PV system connects the PV array to the inverter and connects the inverter output to the utility grid. It is also possible to incorporate battery backup into a utility-interactive PV system so that in the event of loss of the grid, the inverter will be able to power emergency loads. The emergency loads are connected to a separate emergency distribution panel so the inverter can still disconnect from the utility if

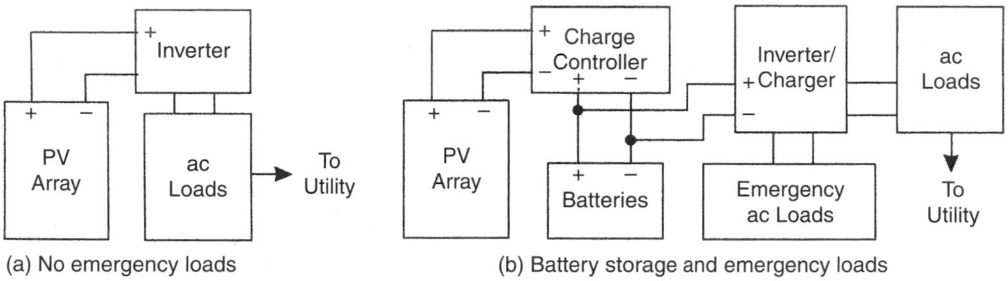

FIGURE 2.11 Utility interactive PV systems.

the utility fails but then instantaneously switch from current source mode to voltage source mode to supply the emergency loads. Figure 2.11 shows the two types of utility interactive systems.

Defining Terms

Grid-connected PV systems: Photovoltaic systems that interact with the utility grid, using the utility grid as a destination for excess electricity generated by the PV system or as a source of additional electricity needed to supply loads that exceed the PV system capacity.

Heliostats: A system of mirrors that are controlled to reflect sunlight onto a power tower.

Photovoltaic cells: Cells made of various materials that are capable of converting light directly into electricity.

Photovoltaic systems: Systems that incorporate series and parallel combinations of photovoltaic cells along with power conditioning equipment to produce high levels of dc or ac power.

Power tower: A central receiving station upon which the sun is reflected by a system of mirrors to produce solar thermal electricity.

Solar thermal electric: The process of using the sun to heat a material that can be used to generate steam to drive a turbine to generate electricity.

Stand-alone PV systems: Photovoltaic systems that operate independently of a utility grid.

Trough electric systems: A system of parabolic troughs in which a liquid is passed through the focal point for superheating to produce solar thermal electricity.

References

G.E. Cohen, D.W. Kearney, and G.J. Kolb. Final report on the operation and maintenance improvement program for concentrating solar power plants, *Sandia National Laboratories Report*, 1999, SAND 99–1290, (June).

R. Davenport, J. Mayette, and R. Forristall. The Salt River Project SunDish Dish/Stirling System, *ASME Solar Forum 2001*, Washington, D.C., April 21–25, 2001.

R. Diver and C. Andraka. Integration of the advanced dish development system, paper no. ISEC2003–44238, *Proceedings of the ASME International Solar Energy Conference*, Kohala Coast, Hawaii, March 15–18, 2003.

R.B. Diver, C.E. Andraka, K.S. Rawlinson, V. Goldberg and G. Thomas. The advanced dish development system project, Proceedings of Solar Forum 2001, Solar Energy, The Power to Choose, Washington, D.C. April 21–25, 2001.

R. Diver, C. Andraka, K. Rawlinson, T. Moss, V. Goldberg, and G. Thomas. Status of the advanced dish development system project, paper no. ISEC2003–44237, Proceedings of the ASME International Solar Energy Conference, Kohala Coast, Hawaii, March 15–18, 2003.

P. Falcone. *A Handbook for Solar Central Receiver Design*, Sand 86–8009, Albuquerque, NM: Sandia National Laboratories, 1986.

M. Haeger, L. Keller, R. Monterreal, and A. Valverde. PHOEBUS Technology Program Solar Air Receiver (TSA): Experimental Set-up for TSA at the CESA Test Facility of the Plataforma Solar de Almeria (PSA), Solar Engineering 1994, proceedings of ASME/JSME/JSES International Solar Energy Conference, ISBN 0–7918–1192–1. 1994, pp. 643–647.

P. Heller, A. Baumüller, and W. Schiel. EuroDish — The Next Milestone to Decrease the Costs of Dish/Stirling System towards Competitiveness, 10th International Symposium on Solar Thermal Concentrating Technologies, Sydney, 2000.

M. Lotker. Barriers to Commercialization of Large-Scale Solar Electricity: Lessons Learned form the LUZ Experience. Albuquerque, NM: Sandia National Laboratories, 1991, SAND91–7014.

T. Mancini, P. Heller, et al. Dish Stirling systems: an overview of development and status, *ASME J. of Solar Energy Eng.*, 2003.

P. Maycock, The world PV market, production increases 36%, *Renewable Energy World,* pp. 147–161, 2002, (July–August).

R. Messenger and J. Ventre. *Photovoltaic Systems Engineering,* Boca Raton, FL: CRC Press, 2000.

J. Pacheco, et al. *Final Test and Evaluation Results form the Solar Two Project*, Sand 2002–0120, Albuquerque, NM: Sandia National Laboratories, 2002.

H. Price. A parabolic trough solar power plant simulation model. 2003. paper no. ISEC2003–44241, Proceedings of the ASME International Solar Energy Conference, Kohala Coast, Hawaii, March 15–18, 2003.

H. Price and D. Kearney. Reducing the cost of energy from parabolic trough solar power plants, paper no. ISEC2003–44069, Proceedings of the ASME International Solar Energy Conference, Hawaii, March 15–18, 2003.

H. Reilly and Ang G. Kolb. *An Evaluation of Molten-Salt Power Towers Including Results of the Solar Two Project, Sand* 2001–3674, Albuquerque, NM: Sandia National Laboratories, 2001, November.

Sandia National Laboratories. *Stand-Alone Photovoltaic Systems: A Handbook of Recommended Design Practices,* Albuquerque, NM: Sandia National Laboratories, 1995.

K. Stone, E. Leingang, R. Liden, E. Ellis, T. Sattar, T.R. Mancini, and H. Nelving. SES/Boeing Dish Stirling System Operation, Solar Energy: The Power to Choose, Washington, DC, April 21–25, 2001, (ASME/ASES/AIA/ASHRAE/SEIA).

K. Stone, G. Rodriguez, J. Paisley, J.P. Nguyen, T.R. Mancini, and H. Nelving. Performance of the SES/Boeing Dish Stirling System, Solar Energy: The Power to Choose, Washington, D.C., April 21–25, 2001, (ASME/ASES/AIA/ASHRAE/SEIA)

R. Tamme, D. Laing, and W. Steinmann. Advanced thermal energy storage technology for parabolic trough, paper no. ISEC2003–44033, Proceedings of the ASME International Solar Energy Conference, Kohala Coast, Hawaii, March 15–18, 2003.

C. Tyner, G. Kolb, M. Geyer, and M. Romero, Concentrating Solar Power in 2001: An IEA/SolarPACES. SolarPACES Task I: Electric Power Systems, Solar Paces Task 1, Electric Power Systems.

2.3 Fuel Cells

Gregor Hoogers

As early as 1838/1839 Friedrich Wilhelm Schönbein and William Grove discovered the basic operating principle of fuel cells by reversing water electrolysis to generate electricity from hydrogen and oxygen [Bossel, 2000]. This has not changed since:

A fuel cell is an electrochemical device that continuously converts chemical energy into electric energy — and heat — for as long as fuel and oxidant are supplied.

Fuel cells therefore bear similarities both to batteries, with which they share the electrochemical nature of the power generation process, and with engines which, unlike batteries, will work continuously consuming a fuel of some sort. When hydrogen is used as fuel, they generate only power and pure water: they are therefore called zero-emission engines. Thermodynamically, the most striking difference is that thermal engines are limited by the well-known Carnot efficiency, while fuel cells are not. This is an advantage at low-temperature operation — if one considers, for example, the speculative Carnot efficiency of a biological system working, instead, as a heat engine. At high temperatures, the Carnot efficiency theoretically surpasses the electrochemical process efficiency. Therefore, one has to look in detail at the actual efficiencies achieved in working thermal and fuel cell systems where, in both cases, losses rather than thermodynamics dominate the actual performance.

Fuel cells are currently being developed for three main markets: automotive propulsion, electric power generation, and portable systems. Each main application is dominated by specific systems requirements and even different types of fuel cells, while the underlying operating principle remains the same.

Fundamentals

Operating Principle

The underlying principle of fuel cell operation is the same for all types that have emerged over the past 160 years: a so-called redox reaction is carried out in two half-reactions located at two electrodes separated by an electrolyte. The purpose of the electrolyte is electronic separation of and ionic connection between the two electrodes, which will assume different electrochemical potentials. The advantage of carrying out the reaction in two parts is direct electrochemical conversion of chemical energy into electric energy by employing the resulting potential difference between the two electrodes, anode and cathode. The simplest and most relevant reaction in this context is the formation of water from the elements,

$$H_2 + \int O_2 = H_2O \tag{2.1}$$

Rather than carrying out reaction (Equation (2.1)) as a gas phase reaction, i.e., thermally after supplying activation energy (ignition) or, more gently, by passing the two reactants over an oxidizing catalyst such as platinum and generating merely heat, the electrochemical reaction requires the half-reactions

$$H_2 = 2H^+ + 2e^- \tag{2.2}$$

and

$$\int O_2 + 2H^+ + 2e^- = H_2O \tag{2.3}$$

to take place at an anode (Equation (2.2)) and cathode (Equation (2.3)), respectively. Therefore, the two electrodes assume their electrochemical potentials at — theoretically — 0 and 1.23 V, respectively, and an electronic current will flow when the electrodes are connected through an external circuit. The anode reaction is the hydrogen oxidation reaction (HOR); the cathode reaction is the oxygen reduction reaction (ORR). In this specific case, the electrolyte is an acidic medium which only allows the passage of protons, H^+, from anode to cathode. Water is formed at the cathode.

Fuel Cell Types

The different fuel cells differ from each other by and are named after the choice of electrolyte. The electrolyte will also determine the nature of the ionic charge carrier and whether it flows from anode to cathode or cathode to anode. Electrode reactions and electrolytes are listed in Table 2.1.

Grove's fuel cell operating in dilute sulfuric acid electrolyte, but also the phosphoric acid fuel cell (PAFC, operating at around 200°C) and the proton exchange membrane fuel cell (PEMFC or PEFC, operating at

TABLE 2.1 Fuel Cell Types, Electrolytes, and Electrode Reactions

Fuel Cell Type	Electrolyte	Charge Carrier	Anode Reaction	Cathode Reaction	Operating Temperature
Alkaline FC (AFC)	KOH	OH^-	$H_2 + 2OH^- = 2H_2O + 2e^-$	$1/2\ O_2 + H_2O + 2e^- = 2OH^-$	60–120°C
Proton exchange membrane FC (PEMFC, SPFC)	Solid polymer (such as Nafion)	H^+	$H_2 = 2H^+ + 2e^-$	$1/2\ O_2 + 2H^+ + 2e^- = H_2O$	50–100°C
Phosphoric acid FC (PAFC)	Phosphoric acid	H^+	$H_2 = 2H^+ + 2e^-$	$1/2\ O_2 + 2H^+ + 2e^- = H_2O$	~220°C
Molten carbonate FC (MCFC)	Lithium and potassium carbonate	CO_3^{2-}	$H_2 + CO_3^{2-} = H_2O + CO_2 + 2e^-$	$1/2O_2 + CO_2 + 2e^- = CO_3^{2-}$	~650°C
Solid oxide FC (SOFC)	Solid oxide electrolyte (yttria stabilized zirconia, YSZ)	O^{2-}	$H_2 + O^{2-} = H_2O + 2e^-$	$1/2\ O_2 + 2e^- = O^{2-}$	~1000°C

80°C, also called solid polymer fuel cell, SPFC) are examples for cells with acidic, proton-conducting electrolytes as in reactions (Equation (2.2)) and (Equation (2.3)).

In contrast, when using an alkaline electrolyte, the primary charge carrier is the OH^- ion, which flows in the opposite direction, so water is formed at the anode. The overall reaction (Equation (2.1)) does not change. The resulting fuel cell is called an alkaline fuel cell or AFC and leads a very successful niche existence in supplying electric power to space craft such as Apollo and the Shuttle.

Phosphoric acid fuel cells (PAFCs) use molten H_3PO_4 as an electrolyte. The PAFC has been mainly developed for the medium-scale power generation market, and 200 kW demonstration units have now clocked up many thousands of hours of operation. However, in comparison with the two low-temperature fuel cells, alkaline and proton exchange membrane fuel cells (AFCs, PEMFCs), PAFCs achieve only moderate current densities.

The proton exchange membrane fuel cell, PEMFC, takes its name from the special plastic membrane that it uses as its electrolyte. Robust cation exchange membranes were originally developed for the chlor-alkali industry by Du Pont and have proven instrumental in combining all the key parts of a fuel cell, anode and cathode electrodes and the electrolyte, in a very compact unit. This membrane electrode assembly (MEA), not thicker than a few hundred microns, is the heart of a PEMFC and, when supplied with fuel and air, generates electric power at cell voltages up to 1 V and power densities of up to about 1 Wcm^{-2}.

The membrane relies on the presence of liquid water to be able to conduct protons effectively, and this limits the temperature up to which a PEMFC can be operated. Even when operated under pressure, operating temperatures are limited to below 100°C. Therefore, to achieve good performance, effective electrocatalyst technology is required. The catalysts form thin (several microns to several tens of microns) gas-porous electrode layers on either side of the membrane. Ionic contact with the membrane is often enhanced by coating the electrode layers using a liquid form of the membrane ionomer.

The MEA is typically located between a pair of current collector plates with machined flow fields for distributing fuel and oxidant to anode and cathode, respectively (compare with Figure 2.12). A water jacket for cooling may be inserted at the back of each reactant flow field followed by a metallic current collector plate. The cell can also contain a humidification section for the reactant gases, which helps to keep the membrane electrolyte in a hydrated, proton-conduction form. The technology is given a more thorough discussion in the following chapter.

Another type of fuel cell is already built into most cars today: the lambda sensor measuring the oxygen concentration in the exhaust of four-stroke spark ignition engines is based on high temperature oxygen ion, O^{2-}, conductors. Typically, this is yttria stabilized zirconia or YSZ, which is also used as solid electrolyte in high temperature (up to 1100°C or 1850°F) solid oxide fuel cells (SOFC). In fact, a lambda sensor works rather well as a high-temperature fuel cell when a hydrogen or a hydrocarbon is used instead of exhaust gas. Table 2.1 shows the electrode reactions.

The second high-temperature fuel cell uses molten carbonate salts at 650°C (1200°F) as electrolyte which conducts carbonate ions, CO_3^{2-}. It is unique that in this type of fuel cell, the ions are formed not from the reactants but from CO_2 that is injected into the cathode gas stream, and usually recycled from the anode exhaust — compare Table 2.1.

The two high-temperature fuel cells, solid oxide and molten carbonate (SOFC and MCFC), have mainly been considered for large scale (MW) stationary power generation. In these systems, the electrolytes consist of anionic transport materials, as O^{2-} and CO_3^- are the charge carriers. These two fuel cells have two major advantages over low-temperature types. They can achieve high electric efficiencies — prototypes have achieved over 45% — with over 60% currently targeted. This makes them particularly attractive for fuel-efficient stationary power generation.

The high operating temperatures also allow direct, internal processing of fuels such as natural gas. This reduces the system complexity compared with low-temperature power plants, which require hydrogen generation in an additional process step. The fact that high-temperature fuel cells cannot easily be turned off is acceptable in the stationary sector, but most likely only there.

Thermodynamics and Efficiency of Fuel Cells

Thermodynamics [Atkins, 1994] teaches that the right thermodynamic potential to use for processes with nonmechanical work is ΔG, the Gibbs free energy. The Gibbs free energy determines the maximum of the nonmechanical work that can be expected from such a reaction, for example, a biological or electrochemical reaction. Another useful quantity in this context is the standard potential of an electrochemical reaction, E_0. E_0 is related to the Gibbs free energy of the reaction by

$$\Delta G = nFE_0 \tag{2.4}$$

where n is the number of charges in the reaction, and $F = 96485\ \text{Cmol}^{-1}$ is the Faraday constant. ΔG essentially determines the upper limit of the electric work coming out of a fuel cell. Both Gibbs free energies and standard reaction potentials are tabulated. So are standard reaction enthalpies, ΔH.

In order to derive the highest possible cell voltage and the cell efficiency, let us consider reaction (Equation (2.1)). The reaction Gibbs free energy is determined in the usual way from the Gibbs free energies of formation, ΔG_f, of the product side minus ΔG_f on the reactant side.

$$\Delta G = \Delta G_f(H_2O) - \Delta G_f(H_2) - 1/2\,\Delta G_f(O_2) = \Delta G_f(H_2O) = -237.13\ \text{kJmol}^{-1} \tag{2.5}$$

(The Gibbs free energies of elements in their standard states are zero.)

From (Equation (2.4)) we can now calculate the maximum cell voltage for a cell based on reaction (Equation (2.1)):

$$E_0 = \Delta G(H_2O)/nF = 237.13\ \text{kJmol}^{-1}/(2.96485\ \text{C mol}^{-1}) = 1.23\ \text{V} \tag{2.6}$$

Here, $n = 2$ because reaction (Equation (2.3)) shows that two electrons are exchanged for the formation of one mol of water.

The higher heating value (HHV) of the same reaction is simply ΔH, the reaction enthalpy for reaction (Equation (2.1)) with liquid product water. In analogy with reaction (Equation (2.5)), we can calculate the reaction enthalpy from the tabulated enthalpies of formation. Again:

$$\Delta H = \Delta H_f(H_2O, l) = -285.83\ \text{kJmol}^{-1} \tag{2.7}$$

Comparison with (Equation (2.5)) shows that an electrochemical cell based on reaction (Equation (2.1)) can at best achieve an electric efficiency of

$$\eta_{el}^{max} = \Delta G/\Delta H = 0.83 = 83\%$$

based on the HHV of the fuel.

Measuring the actual electric efficiency of a fuel cell is extremely simple: A measurement of the operating cell voltage E is sufficient:

$$\eta_{el} = E/E_0 \bullet \Delta G/\Delta H = (nF/\Delta H) \bullet E = 0.68\,\mathrm{V}^{-1} E \tag{2.8}$$

For example, a hydrogen/air fuel cell achieving a cell voltage of 0.7 V under operation converts $0.68\,\mathrm{V}^{-1} \bullet 0.7\,\mathrm{V} = 0.47 = 47\%$ of the chemical energy supplied into electric energy.

Effects of Pressure and Temperature

Everything that has been said so far was based on standard values for pressure and temperature, i.e., $T = 298$ K and $P = 1$ bar. In practice, fuel cells usually operate at elevated temperatures and, often, above ambient pressure.

Generally, for a chemical reaction of the type

$$a\mathrm{A} + b\mathrm{B} \quad m\mathrm{M} + n\mathrm{N} \tag{2.9}$$

the effect of the partial pressures (strictly speaking: activities) of the reactants and products on the change in Gibbs free energy is given as

$$\Delta G = \Delta G^\circ + RT \ln \frac{P_\mathrm{M}^m P_\mathrm{N}^n}{P_\mathrm{A}^a P_\mathrm{B}^b} \tag{2.10}$$

When converted to potentials, using (Equation (2.4)), this turns into the well-known Nernst equation. It is clear from (Equation (2.1)), (Equation (2.9)), and (Equation (2.10)) that an increase in either $P(H_2)$ or $P(O_2)$ will lead to a higher cell potential.

The temperature dependence is given by the temperature dependence of the Gibbs free energy, $\Delta G(T)$.

Since $\Delta G = \Delta H - T\Delta S$, and ΔH depends only weakly on T, the temperature derivative of ΔG is given by: $\mathrm{d}G/\mathrm{d}T = -\Delta S$ or using (Equation (2.4))

$$\mathrm{d}E_0/\mathrm{d}T = -\Delta S/nF \tag{2.11}$$

System Efficiency

The *system efficiency* in this case will, of course, be lower due to the consumption of the power system itself and due to wastage of hydrogen fuel or air. The reason for this is often system related, i.e., a certain amount of hydrogen is passing through the cell unused in order to remove continuously water or contaminants from the anode. The hydrogen leaving the cell is either recycled or burned. The hydrogen excess is quantified by two terms used by different authors, stoichiometry and the fuel utilization. A 50% excess of hydrogen feed is expressed in a fuel stoichiometry of 1.5 or in a fuel utilization of 1/1.5 = 2/3. An air surplus is also used routinely. This will affect the system efficiency through the energy lost on compressor power.

Another consideration is linked to the method of providing hydrogen. When hydrogen is made from another chemical that is more apt for storage (for example, on board a vehicle) or for supply from the natural gas grid, the fuel processor or *reformer* will also have a certain efficiency.

Finally, the DC power generated by a fuel cell is often converted into AC, either for an electric motor or for supplying mains current. The alternator, again, affects the overall system efficiency by a load-dependent loss factor.

Kinetics of Fuel Cell Reactions

As may be expected, the treatment of the kinetics of fuel cell reactions is harder than the thermodynamics but nonetheless much simpler than the details of the energy conversion in thermal engines.

A full treatment of the electrochemical kinetics (Butler–Volmer Equation) is beyond the scope of this chapter and can be found in the general electrochemical [Hamann, 1998] or dedicated fuel cell literature [Hoogers, 2003]. It is a striking result of the kinetic work on fuel cells that anode reaction kinetics are far superior to cathode kinetics and that the anode potential — with the cell operating on clean hydrogen — is close to the reversible hydrogen potential at 0 V. The cell performance is therefore dominated by the cathode potential, i.e., $E_{cell} = E_c$.

Essentially, the kinetic description of electrode kinetics predicts an exponential dependence of the cell current on the cell voltage. For practical purposes, it is often more desirable to express the cell potential as a function of current drawn from the cell, i.e.,

$$E_c = E_r - b \log_{10}(i/i_o) - ir \tag{2.12}$$

where E_r = reversible potential for the cell; i_o = exchange current density for oxygen reduction; b = Tafel slope for oxygen reduction; r = (area specific) ohmic resistance; and i = current density, in which ohmic losses have been included by adding the term $(-ir)$.

This is the so called *Tafel equation* (with addition of the *ir*-dependence). The *Tafel slope b* is determined by the nature of the electrochemical process. b can be expressed as

$$b = RT/\beta F \tag{2.13}$$

where $R = 8.314\ \mathrm{Jmol^{-1}K^{-1}}$ denotes the universal gas constant, F is again the Faraday constant, T is the temperature (in K), and β is the transfer coefficient, a parameter related to the symmetry of the chemical transition state, usually taken to be 0.5, when no first principle information is available. For the oxygen reduction reaction in practical proton exchange membrane fuel cells, b is usually experimentally determined with values between 40 and 80 mV.

The main factor controlling the activation overpotential and hence the cell potential, $E_{cell} = E_c$, is the (apparent) exchange current density i_0. (Equation (2.12)) demonstrates that, due to the logarithm, a tenfold increase in i_0 leads to an increase in cell potential at the given current by one unit of b, or typically 60 mV. It is important to dwell on this point. While the reversible potential E_r is given by thermodynamics and the Tafel slope b is dictated by the chemical reaction (and the temperature), the value for i_0 depends entirely on reaction kinetics. Ultimately, it depends on the skill of the MEA and electrocatalyst producer to increase this figure.

For doing so there are, in principle, the following possible approaches:

- The magnitude of i_0 can be increased (within limits) by adding more electrocatalyst to the cathode. As today's electrocatalysts (used in low temperature fuel cells) contain platinum, there are economic reasons why MEA makers do not just put more platinum inside their products.
- Clearly, there have been many attempts to do away with platinum as the leading cathode catalyst for low-temperature fuel cells altogether. Unfortunately, to date, there appears to be no convincing alternative to platinum or related noble metals. This is not merely due to the lack of catalytic activity of other catalyst systems but often is a result of insufficient chemical stability of the materials considered.
- A logical and very successful approach is the more effective use of platinum in fuel cell electrodes. A technique borrowed from gas phase catalysis is the use of supported catalysts with small, highly dispersed platinum particles. Of course, electrocatalysts have to use electrically conducting substrate materials, usually specialized carbons.
- Of course, it is not sufficient to improve the surface area of the catalyst employed; there has to be good electrochemical contact between the membrane and the catalyst layer. *In-situ* measurement of the effective platinum surface area (EPSA) is a critical test for the quality of an electrode structure. The

EPSA may be measured by electronic methods or, more commonly, by carbon monoxide adsorption and subsequent electrooxidation to carbon dioxide with charge measurement. For more information see Chapter 6 in [Hoogers, 2003]. Impedance measurements under practical loads can give valuable information on the catalyst utilization under operating conditions.

Fueling and Fuel Cell Systems

Not all fuel cells require these two reactants, but all fuel cells will function well with hydrogen and oxygen or air. In stationary power systems, the most convenient fuel is usually natural gas from the national grid. This has to be converted into the right fuel for the fuel cell — see below. In automotive systems, but also for portable power fuel, storage is an issue.

Hydrogen Storage

Pressure Cylinders. The need for lighter gas storage has led to the development of lightweight composite rather than steel cylinders. Carbon-wrapped aluminum cylinders can store hydrogen at pressures of up to 55 MPa (550 bar/8000 PSI). In most countries, gas cylinders are typically filled up to a maximum of 24.8 or 30 MPa (248 bar/3600 PSI and 300 bar/4350 PSI, respectively). At the higher pressure, a modern composite tank reaches a hydrogen mass fraction of approximately 3%, i.e., only 3% of the weight of the full cylinder consists of hydrogen. In a further development, so called "conformable" tanks have been produced in order to give a better space filling than packed cylinders.

General Motors' current compressed hydrogen gas storage systems typically hold 2.1 kg of hydrogen in a 140 liter/65 kg tank at 350 bar, which is good for 170 km (106 miles). The target here is a 230 liter/110 kg tank that would hold 7 kg of hydrogen at 700 bar, giving the same range as liquid hydrogen (see below), 700 km (438 miles) (H&FC, 2001).

Californian Quantum Technologies WorldWide has demonstrated a composite hydrogen pressure storage tank with a nominal operating pressure of almost 700 bar (10,000 PSI) giving an 80% capacity increase over tanks operating at 350 bar. The new tank underwent a hydrostatic burst test at which it failed under 1620 bar (23,500 psi). This test was done along the lines given in the draft regulations by the European Integrated Hydrogen Project (EIHP). The tank has an in-tank regulator that provides a gas supply under no more than 10 bar (150 PSI).

Liquid Hydrogen. It is unfortunate that the critical temperature of hydrogen, i.e., the temperature below which the gas can be liquefied, is at 33 K. Storage in cryogenic tanks at the boiling point of hydrogen, 20.39 K (−252.76°C/−422.97°F) at 1 atm (981 hPa), allows higher storage densities at the expense of the energy required for the liquefaction process. Lowering the temperature of hydrogen to its boiling point at 20.39 K (−252.76°C/−422.97°F) at atmospheric pressure requires approximately 39.1 kJ/g or 79 kJ mol^{-1}. To put this figure into perspective, this energy amounts to over a quarter of the higher heating value (286 kJ mol^{-1}) of hydrogen.

Another problem with cryogenic storage is hydrogen boil-off. Despite good thermal insulation, the heat influx into the cryogenic tank is continuously compensated by boiling off quantities of the liquid (heat of evaporation). In cryogenic storage systems on-board cars, the boil-off rate is estimated by most developers at approximately 1% per day, which results in further efficiency losses.

Cryogenic tanks consist of a multi-layered aluminum foil insulation. A typical tank stores 120 l of cryogenic hydrogen or 8.5 kg, which corresponds to an extremely low (liquid) density of 0.071 kg dm^{-3}. The empty tank has a volume of approximately 200 l and weighs 51.5 kg [Larminie, 2000]. This corresponds to a hydrogen mass fraction of 14.2%.

In General Motors' HydroGen1, 5 kg of hydrogen are stored in a 130 l/50 kg tank that gives the vehicle a 400 km (250 mile) drive range. The future target is a 150-l tank that is lighter yet, holding 7 kg for a range of 700 km (438 miles), as well as reduced boil-off time via an additional liquefied/dried air cooling shield developed by Linde.

The actual handling of cryogenic hydrogen poses a problem to the filling station, requiring special procedures such as fully automated, robotic filling stations.

Metal Hydrides. Most elements form ionic, metallic, covalent, or polymeric hydrides or mixtures thereof [Greenwood, 1984]. Ionic and metallic types are of particular interest, as they allow reversible storage of hydrogen [Sandrock, 1994].

The formation of the hydride is an exothermal process. Important parameters in this context are the enthalpy of formation of the hydride, which may range between several kJ and several hundred kJ per mol of hydrogen stored, and the resulting temperature and pressure to release the hydrogen from the hydride. In order to adjust these figures to technically acceptable levels, intermetallic compounds have been developed. Depending on the hydride used, the mass fraction of hydrogen ranges between 1.4 and 7.7% of total mass. Many hydrides actually store more hydrogen by volume than liquid hydrogen does. However, the storage process itself is exothermal and may amount to 25 to 45% of the hydrogen HHV (for full details, see Chapter 5 in [Hoogers, 2003]). Clearly, for large storage capacities such as the typical automotive tank, this is entirely unacceptable. Yet, for small, hydrogen-based portable systems, metal hydrides may form a convenient method of energy storage if cartridges are made universally available for purchase and recycling.

Sodium Borohydride. Sodium borohydride, $NaBH_4$, has recently received much attention through the work of Millenium Cell and DaimlerChrysler. Millennium Cell has patented a process that releases hydrogen from an aqueous solution of sodium borohydride, $NaBH_4$, in an exothermal reaction. (Sodium borohydride is usually made from borax using diborane, a highly reactive, highly toxic gas.) Hydrogen is only produced when the liquid fuel is in direct contact with a catalyst. The only other reaction product, sodium metaborate (analogous to borax), is water soluble and environmentally benign. The 35 wt% solution (35 wt% $NaBH_4$, 3 wt% NaOH, 62 wt% H_2O) will store 7.7 wt% of hydrogen or 77 g/921 standard liters of hydrogen in one liter of solution.

DaimlerChrysler has presented a fuel cell vehicle, the *Natrium*, that incorporates a sodium borohydride tank of about the size of a regular gas tank, which can power the concept vehicle about 300 miles.

There are numerous questions regarding production, energy efficiency, infrastructure, and stability of the solutions (Millenium quotes a half-life of 450 days, equivalent to 0.15% decomposition per day — this would be far less than the liquid hydrogen boil-off). The technique may well have its merits for special applications, particularly if an environmentally benign production process can be developed.

Fuel Reforming. Hydrogen is currently produced in large quantities for mainly two applications. Roughly 50% of the world hydrogen production is consumed for the hydro-formulation of oil in refineries producing mainly automotive fuels. Approximately 40% is produced for subsequent reaction with nitrogen to ammonia, the only industrial process known to bind atmospheric nitrogen. Ammonia is used in a number of applications, with fertilizer production playing a key role.

Storage of some hydrocarbon-derived liquid fuel followed by hydrogen generation on-board vehicles has been one option for chemical storage of hydrogen. So far, practical prototype cars have only been produced with methanol reformers (DaimlerChrysler). In contrast, the work on gasoline reformers by GM/Toyota/Exxon on the one hand and HydrogenSource (Shell Hydrogen and UTC Fuel Cells) on the other has not yet led to working fuel cell prototype cars.

Fortunately, there is no dispute about natural gas reforming for stationary power generation. An exhaustive discussion of the catalysis involved in fuel processing is presented in [Trimm, 2001].

The main techniques are briefly described in the following.

Steam Reforming (SR). Steam reforming, SR, of methanol is given by the following chemical reaction equation:

$$CH_3OH + H_2O \rightarrow CO_2 + 3\,H_2 \qquad \Delta H = 49\,\text{kJmol}^{-1} \tag{2.14}$$

Methanol and water are evaporated and react in a catalytic reactor to carbon dioxide and hydrogen, the desired product. Methanol steam reforming is currently performed at temperatures between 200 and 300C (390 and 570°F) over copper catalysts supported by zinc oxide [Emonts, 1998]. One mole of methanol reacts to three moles of dihydrogen. This means that an extra mole of hydrogen originates from the added water.

In practice, reaction (Equation (2.14)) is only one of a whole series, and the raw reformer output consists of hydrogen, carbon dioxide, and carbon monoxide. Carbon monoxide is converted to carbon dioxide and more hydrogen in a high temperature shift, HTS, stage followed by a low temperature shift, LTS, stage. In both stages, the ***water–gas shift reaction***

$$CO + H_2O \rightarrow CO_2 + H_2 \qquad \Delta H = -41\,\mathrm{kJmol^{-1}} \tag{2.15}$$

takes place.

As water–gas shift is an exothermal reaction, if too much heat is generated, it will eventually drive the reaction towards the reactant side (Le Chatelier's principle). Therefore, multiple stages with interstage cooling are used in practice. The best catalyst for the HTS reaction is a mixture of iron and chromium oxides (Fe_3O_4 and Cr_2O_3) with good activity between 400 and 550°C (750 and 1020°F). LTS uses copper catalysts similar to and under similar operating conditions to those used in methanol steam reforming (Equation (2.14)).

Steam reforming of methane from natural gas is the standard way of producing hydrogen on an industrial scale. It is therefore of general importance to a hydrogen economy. In addition, smaller-scale methane steam reformers have been developed to provide hydrogen for stationary power systems based on low-temperature fuel cells, PEMFC and PAFC.

The methane steam reforming reaction is described by

$$CH_4 + H_2O \rightarrow CO + 3\,H_2 \qquad \Delta H = 206\,\mathrm{kJmol^{-1}} \tag{2.16}$$

It is again followed by the shift reactions (Equation (2.15)).

Methane steam reforming is usually catalyzed by nickel [Ridler, 1996] at temperatures between 750 and 1000°C (1380 and 1830°F), with excess steam to prevent carbon deposition ("coking") on the nickel catalyst [Trimm, 2001].

Partial Oxidation (POX). The second important reaction for generating hydrogen on an industrial scale is partial oxidation. It is generally employed with heavier hydrocarbons or when there are special preferences because certain reactants (for example, pure oxygen) are available within a plant.

It can be seen as oxidation with less than the stoichiometric amount of oxygen for full oxidation to the stable end products, carbon dioxide and water. For example, for methane

$$CH_4 + \int O_2 \rightarrow CO + 2H_2 \qquad \Delta H = -36\,\mathrm{kJmol^{-1}} \tag{2.17a}$$

and/or

$$CH_4 + O_2 \rightarrow CO_2 + 2H_2 \qquad \Delta H = -319\,\mathrm{kJmol^{-1}} \tag{2.17b}$$

While the methanol reformers used in fuel cell vehicles presented by DaimlerChrysler are based on steam reforming, Epyx (a subsidary of Arthur D. Little, now part of Nuvera) and Shell (now incorporated into HydrogenSource) are developing partial oxidation reactors for processing gasoline.

Autothermal Reforming (ATR). Attempts have been made to combine the advantages of both concepts, steam reforming and partial oxidation. Ideally, the exothermal reaction (Equation (2.17)) would be used for startup and for providing heat to the endothermal process (Equation (2.16)) during steady-state operation. The reactions can either be run in separate reactors that are in good thermal contact or in a single catalytic reactor.

Comparison of Reforming Technologies. So, when does one use which technique for reforming? The first consideration is the ease by which the chosen fuel can be reformed using the respective method. Generally speaking, methanol is most readily reformed at low temperatures and can be treated well in any type of reformer. Methane and, similarly, LPG (liquefied petroleum gas) require much higher temperatures but again can be processed by any of the methods discussed. With higher hydrocarbons, the current standard fuels used in the automotive sector, one usually resorts to POX reactors.

Table 2.2 compares the theoretical (thermodynamic) efficiencies for proton exchange membrane fuel cells operating directly (DMFC) or indirectly on methanol (using one of the reformers discussed). The efficiencies currently achieved are labeled η_{el}(tech) – typ, while the technical limits are estimated by the author at η_{el}(tech) – max.

Table 2.3 and Table 2.4 show gas compositions from methanol reformers calculated as upper limit (by the author) and determined experimentally, respectively. Table 2.3 also includes information on the feed mix, i.e., X, the molar percentage, and wt%, the weight percentage of methanol in the methanol/water feed. A disadvantage of the POX reactors is the presence of nitrogen (from partial oxidation with ambient air) in the reformer output. This leads to lower hydrogen concentrations.

Steam reforming gives the highest hydrogen concentration. At the same time, a system relying entirely on steam reforming operates best under steady-state conditions because it does not lend itself to rapid dynamic response. This also applies to startup. Partial oxidation, in contrast, offers compactness, fast startup, and rapid dynamic response, while producing lower concentrations of hydrogen. In addition to differences in product stoichiometries between SR and POX reformers, the output of a POX reformer is necessarily further diluted by nitrogen. Nitrogen is introduced into the system from air, which is usually the only economical source of oxygen, and carried through as an inert. Autothermal reforming offers a compromise.

Yet, the fuel processor cannot be seen on its own. Steam reforming is highly endothermic. Heat is usually supplied to the reactor, for example, by burning extra fuel. In a fuel cell system, (catalytic) oxidation of excess hydrogen exiting from the anode provides a convenient way of generating the required thermal energy. In stationary power generation, it is worth considering that the PAFC fuel cell stack operates at a high enough temperature to allow generating steam and feeding it to the fuel processor. Steam reforming may be appropriate here whereas autothermal reforming could be considered in a PEMFC system, which has only low-grade heat available.

Fuel efficiency also deserves careful attention. Though always important, the cost of fuel is the most important factor in stationary power generation (on a par with plant availability). Hence, the method offering the highest overall hydrogen output from the chosen fuel, usually natural gas, is selected. Steam reforming delivers the highest hydrogen concentrations. Therefore, the fuel cell stack efficiency at the higher hydrogen content may offset the higher fuel demand for steam generation. This is probably the reason why steam reforming is currently also the preferred method for reforming natural gas in stationary power plants based on PEM fuel cells.

TABLE 2.2 Thermodynamics of PEM Fuel Cell Systems Operating on Methanol

	ΔH [kJ/mol]	ΔG [kJ/mol]	η_{el} (theor.)	η_{el} (tech) – typ	η_{el} (tech) – max
DMFC		–702,35	0,97	0,27	0,40
SR + H_2 – PEM	130,98	–711,39	0,83	0,32	0,54
POX + H_2 – PEM	–154,85	–474,26	0,65	0,25	0,42
ATR + H_2 – PEM	0	–602,73	0,83	0,32	0,54

Note: The DMFC is a fuel cell that can electro-oxidize methanol directly while the other systems are based on hydrogen fuel cells with reformer: Steam-reformer (SR), partial oxidation reformer (POX), and autothermal (ATR) reformer. The efficiencies have been determined by thermodynamics (theo.) or estimated by the author (typ./max.).

TABLE 2.3 Stoichiometric Input Feed and Output Gas Compositions for Different Concepts of Methanol Reforming

	X (MeOH)	wt% MeOH	H_2	CO_2	N_2
DMFC	0,5	64			
$SR + H_2 - PEM$	0,5	64	75%	25%	0%
$POX + H_2 - PEM$	1,0	100	41%	20%	39%
$ATR + H_2 - PEM$	0,65	77	58%	23%	20%

TABLE 2.4 Experimental Gas Compositions Obtained as Reformer Outputs

	H_2	CO_2	N_2	CO
SR [PASE 00]	67%	22%	–	
POX [PASE 00]	45%	20%	22%	
ATR [GOLU 98]	55%	22%	21%	2%

In automotive applications, the dynamic behavior of the reformer system may control the whole drive train, depending on whether backup batteries, super-capacitors or other techniques are used for providing peak power. A POX reformer offers the required dynamic behavior and fast startup and is likely to be the best choice with higher hydrocarbons. For other fuels, in particular, methanol, an ATR should work best. Yet, the reformers used by DaimlerChrysler in their NECAR 3 and 5 vehicles are steam reformers. This perhaps surprising choice can be reconciled when one considers that during startup, additional air is supplied to the reformer system to do a certain degree of partial oxidation. During steady-state operation, the reformer operates solely as SR with heat supplied from excess hydrogen. Clearly, it is not always possible to draw clear borderlines between different types of reformers.

Striking advantages of liquid fuels are their high energy storage densities and their ease of transport and handling. Liquid fuel tanks are readily available, and their weight and volume are essentially dominated by the fuel itself. LPG is widely applied in transportation in some countries, and storage is comparably straightforward since LPG is readily liquefied under moderate pressures (several bar).

CO Removal/Pd-Membrane Technology. As was noted in the introduction to this chapter, different fuel cells put different demands on gas purity. CO removal is of particular concern to the operation of the PEMFC, less so with fuel cells operating at higher temperatures.

After reforming and water gas shift, the CO concentration in the reformer gas is usually reduced to 1 to 2%. A PEMFC will require further cleanup down to levels in the lower ppm range. Another reason for having further CO cleanup stages is the risk of CO spikes, which may result from rapid load changes of the reformer system as can be expected from automotive applications.

There are a number of ways to clean the raw reformer gas of CO. Alternatively, ultra-pure hydrogen void of any contaminant can be produced using Pd membrane technology.

We will discuss these methods in turn.

Preferential Oxidation. Oxidative removal of CO is one option often applied in fuel cell systems working with hydrocarbon reformers. Unfortunately, this increases the system complexity because well-measured concentrations of air have to be added to the fuel stream.

The reaction

$$CO + \int O_2 \rightarrow CO_2 \qquad \Delta H = -260\,\text{kJmol}^{-1} \tag{2.18}$$

works surprisingly well despite the presence of CO_2 and H_2 in the fuel gas. This is due to the choice of catalyst, which is typically a noble metal such as platinum, ruthenium, or rhodium supported on alumina. Gold catalysts supported on reducible metal oxides have also shown some benefit, particularly at temperatures below 100°C [Plzak, 1999]. CO bonds very strongly to noble metal surfaces at low to moderate temperatures. So, the addition reaction (Equation (2.18)) takes place on the catalytic surface, in preference to the undesirable direct catalytic oxidation of hydrogen. Therefore, this technique is referred to as preferential oxidation, or PROX. The selectivity of the process has been defined as the ratio of oxygen consumed for oxidizing CO divided by the total consumption of oxygen.

The term selective oxidation, or SELOX, is also used. But this should better be reserved for the case where CO removal takes place *within* the fuel cell.

An elegant way of making the fuel cell more carbon monoxide-tolerant is the development of CO-tolerant anode catalysts and electrodes [Cooper, 1997]. A standard technique is the use of alloys of platinum and ruthenium. Another, rather crude, way of overcoming anode poisoning by CO is the direct oxidation of CO by air in the anode itself [Gottesfeld, 1988]. One may see this as an internal form of the preferential oxidation discussed above. In order to discriminate the terminology, this method is often referred to as selective oxidation or SELOX. The air for oxidizing CO is "bled" into the fuel gas stream at concentrations of around 1%. Therefore, this technique has been termed *air bleed.* It is a widely accepted way of operating fuel cells on reformer gases.

Bauman et al. have also shown that anode performance after degradation due to CO "spikes," which are likely to appear in a reformer-based fuel cell system upon rapid load changes, recovers much more rapidly when an air bleed is applied [Bauman, 1999].

Pd-Membranes. In some industries, such as the semiconductor industry, there is a demand for ultra-pure hydrogen. Since purchase of higher-grade gases multiplies the cost, hydrogen is either generated on site or low-grade hydrogen is further purified.

A well-established method for hydrogen purification (and only applicable to hydrogen) is permeation through palladium membranes [McCabe, 1997], and plug-and-play hydrogen purifiers are commercially available. Palladium only allows hydrogen to permeate and retains any other gas components, such as nitrogen, carbon dioxide, carbon monoxide, and any trace impurities, on the upstream side. As carbon monoxide adsorbs strongly onto the noble metal, concentrations in the lower percent range may hamper hydrogen permeation through the membrane, unless membrane operating temperatures high enough to oxidize carbon monoxide (in the presence of some added air) to carbon dioxide are employed. In a practical test, operating temperatures in excess of 350°C and operating pressures above 20 hPa (20 bar or 290 PSI) had to be used [Emonts, 1998].

For economic reasons, thin film membranes consisting of palladium/silver layers deposited on a ceramic support are being developed. Thin film membranes allow reducing the amount of palladium employed and improve the permeation rate. Silver serves to stabilize the desired metallic phase of palladium under the operation conditions. Yet, thermal cycling and hydrogen embrittlement pose potential risks to the integrity of membranes no thicker than a few microns [Emonts, 1998].

The main problems with palladium membranes in automotive systems appear to be the required high pressure differential, which takes its toll from overall systems efficiency, the cost of the noble metal, and/or membrane lifetime. Other applications, in particular compact power generators, may benefit from reduced system complexity.

A number of companies, including Misubishi Heavy Industries [Kuroda, 1996] and IdaTech [Edlund, 2000], have developed reformers based on Pd membranes. Here, the reforming process takes place inside a membrane tube or in close contact with the Pd membrane. IdaTech achieved CO and CO_2 levels of less than 1 ppm with a SR operating inside the actual cleanup membrane unit using a variety of fuels.

Methanation. Methanation,

$$CO + 3H_2 \rightarrow CH_4 + H_2O \qquad \Delta H = -206\,\text{kJmol}^{-1} \tag{2.19}$$

is another option for removing CO in the presence of large quantities of H_2.

Unfortunately, due to current lack of selective catalysts, the methanation of CO_2 is usually also catalyzed. Therefore, for cleaning up small concentrations of CO in the presence of large concentrations of CO_2, methanation is not possible. IdaTech uses methanation in conjunction with Pd-membranes as second cleanup stage. This is possible in this particular instance because the membrane removes both CO and CO_2 down to ppm levels.

Other System Components

Fuel Cell Stack. Figure 2.12 demonstrates how MEAs are supplied with reactant gases and put together to form a fuel cell stack. The gas supply is a compromise between the flat design necessary for reducing ohmic losses and sufficient access of reactants. Therefore, so-called flow field plates are employed to feed hydrogen to the anodes and air/oxygen to the cathodes present in a fuel cell stack. Other stack components include cooling elements, current collector plates for attaching power cables, end plates, and, possibly, humidifiers. End plates give the fuel cell stack mechanical stability and enable sealing of the different components by compression. A number of designs have been presented. Ballard Power Systems has used both threaded rods running along the whole length of the stack and metal bands tied around the central section of the stack for compression.

Bipolar Plates. Flow field plates in early fuel cell designs — and still in use in the laboratory — were usually made of graphite into which flow channels were conveniently machined. These plates have high electronic and good thermal conductivity and are stable in the chemical environment inside a fuel cell. Raw bulk graphite is made in a high-temperature sintering process that takes several weeks and leads to shape distortions and the introduction of some porosity in the plates. Hence, making flow field plates is a lengthy and labor-intensive process, involving sawing blocks of raw material into slabs of the required thickness, vacuum-impregnating the blocks or the cut slabs with some resin filler for gas-tightness, and grinding and polishing to the desired surface finish. Only then can the gas flow fields be machined into the blank plates by a standard milling and engraving process. The material is easily machined but abrasive. Flow field plates made in this way are usually several millimeters (1 mm = 0.04 inch) thick, mainly to give them mechanical strength and allow the engraving of flow channels. This approach allows the greatest possible flexibility with respect to designing and optimizing the flow field.

When building stacks, flow fields can be machined on either side of the flow field plate such that it forms the cathode plate on one and the anode plate on the other side. Therefore, the term *bipolar plate* in often used

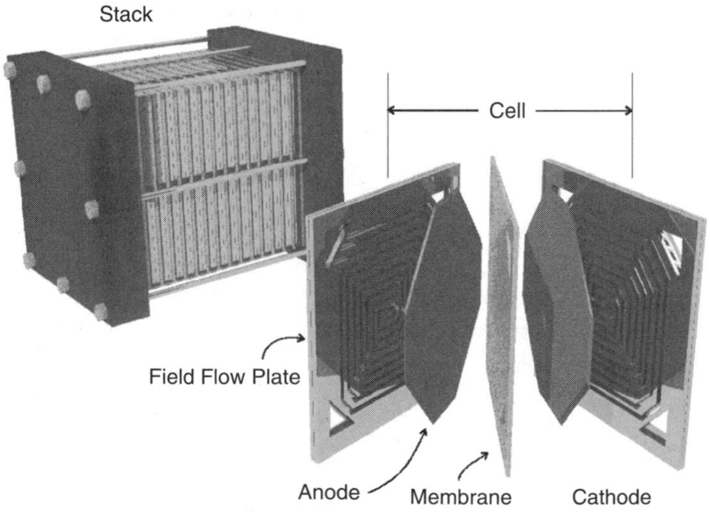

FIGURE 2.12 Fuel cell stack made up of flow field plates (or bipolar plates) and MEAs (shown in the insert).

in this context. The reactant gases are then passed through sections of the plates and essentially the whole fuel cell stack — compare Figure 2.1.

Graphite-Based Materials. The choice of materials for producing bipolar plates in commercial fuel cell stacks is not only dictated by performance considerations but also cost. Present blank graphite plates cost between US$20 and US$50 apiece in small quantities, i.e., up to US$1000/m^2, or perhaps over US$100 per kW assuming one plate per MEA plus cooling plates at an MEA power density of 1 Wcm^{-2}. Again, automotive cost targets are well beyond reach, even ignoring additional machining and tooling time.

This dilemma has sparked off several alternative approaches. Ballard Power Systems has developed plates based on (laminated) graphite foil, which can be cut, molded, or carved in relief in order to generate a flow field pattern. This may open up a route to low-cost volume production of bipolar plates. Potential concerns are perhaps the uncertain cost and the availability of the graphite sheet material in large volumes.

Another cost-effective volume production technique is injection or compression molding. Difficulties with molded plates lie in finding the right composition of the material, which is usually a composite of graphite powder in a polymer matrix. While good electronic conductivity requires a high graphite fill, this hampers the flow and hence the moldability of the composite. Thermal stability and resistance towards chemical attack of the polymers limit the choice of materials.

Plug Power has patented flow field plates that consist of conducting parts framed by nonconducting material, which may form part of the flow field systems.

Metallic Bipolar Plates. Metals are very good electronic and thermal conductors and exhibit excellent mechanical properties. Undesired properties include their limited corrosion resistance and the difficulty and cost of machining.

The metals contained in the plates bear the risk of leaching in the harsh electrochemical environment inside a fuel cell stack; leached metal may form damaging deposits on the electrocatalyst layers or could be ion-exchanged into the membrane or the ionomer, thereby decreasing the conductivity. Corrosion is believed to be more serious at the anode, probably due to weakening of the protective oxide layer in the hydrogen atmosphere.

Several grades of stainless steel (310, 316, 904L) have been reported to survive the highly corroding environment inside a fuel cell stack for 3000 h without significant degradation [Davies, 2000] by forming a protective passivation layer.

Clearly, the formation of oxide layers reduces the conductivity of the materials employed. Therefore, coatings have been applied in some cases. In the simplest case, this may be a thin layer of gold or titanium. Titanium nitride layers are another possibility and have been applied to lightweight plates made of aluminum or titanium cores with corrosion-resistant spacer layers. Whether these approaches are commercially viable depends on the balance between materials and processing costs.

Meanwhile, mechanical machining of flow fields into solid stainless steel plates is difficult. A number of companies such as Microponents (Birmingham, U.K.) and PEM (Germany) are attempting to achieve volume production of flow field plates by employing chemical etching techniques. Yet, etching is a slow process and generates slurries containing heavy metals, and it is hence of limited use for mass production.

Another solution to the problem of creating a (serpentine) flow field is using well-known metal stamping techniques. To date, there appear to be no published data on flow fields successfully produced in this fashion.

Humidifiers and Cooling Plates. A fuel cell stack may contain other components. The most prominent ones are cooling plates or other devices and techniques for removing reaction heat and, possibly, humidifiers.

Cooling is vital to maintain the required point of operation for a given fuel cell stack. This may either be an isothermal condition, or perhaps a temperature gradient may be deliberately superimposed in order to help water removal. Relatively simple calculations show that for very high power densities such as those attained in automotive stacks, liquid cooling is mandatory. This is traditionally done by introducing dedicated cooling plates into the stack, through which water is circulated.

In less demanding applications, such as portable systems, where the system has to be reduced to a bare minimum of components, air cooling is sometimes applied. In the simplest case, the cathode flow fields are open to ambient, and reactant air is supplied by a fan, at the same time providing cooling. No long-term performance data have been reported for this type of air-cooled stack.

A second function sometimes integrated into the stack is reactant humidification. It is currently unknown whether innovative membrane concepts will ever allow unhumidified operation in high-performance fuel cell stacks. In most automotive stacks to date, both fuel gas and air are most likely humidified because maximum power is required, which is only achieved with the lowest possible membrane resistance.

The literature knows several types of humidifiers, bubblers, membrane, or fiber-bundle humidifiers and water evaporators. The simplest humidifier is the well-known "bubbler," essentially the wash bottle design with gas directly passing through the liquid. Clearly, this approach allows only poor control of humidification, is less suited within a complex fuel cell system, and may cause a potential safety hazard due to the direct contact of the fluids. Another approach is using a membrane humidifier. A semipermeable membrane separates a compartment filled with water from a compartment with the reactant gas. Ideally, the gas is conducted along the membrane and continually increases its humidity up to or close to saturation as it passes from the gas inlet to the gas outlet. Some concepts combine humidification and cooling [Vitale, 2000].

This concludes the list of the most important functional components of a fuel cell stack. The fuel cell system contains a large number of other components for fuel generation, pumping, compression, etc., which are usually just summarized under the term balance-of-plant, BOP.

Direct Methanol Fuel Cells

No doubt one of the most elegant solutions would be to make fuel cells operate on a liquid fuel. This is particularly so for the transportation and portable sectors. The direct methanol fuel cell (DMFC), a liquid- or vapor-fed PEM fuel cell operating on a methanol/water mix and air, therefore deserves careful consideration. The main technological challenges are the formulation of better anode catalysts, which lower the anode overpotentials (currently several hundred millivolts at practical current densities), and the improvement of membranes and cathode catalysts in order to overcome cathode poisoning and fuel losses by migration of methanol from anode to cathode. Current prototype DMFCs generate up to 0.2 Wcm^{-2} (based on the MEA area) of electric power, but not yet under practical operating conditions or with acceptable platinum loadings.

Therefore, there is currently little hope that DMFCs will ever be able to power a commercially viable car. The strength of the DMFC lies in the inherent simplicity of the entire system and the fact that the liquid fuel holds considerably more energy than do conventional electrochemical storage devices such as primary and secondary batteries. Therefore, DMFCs are likely to find a range of applications in supplying electric power to portable or grid-independent systems with small power consumption but long, service-free operation.

Applications

It now looks as though fuel cells are eventually coming into widespread commercial use. Before starting a fuel cell development project, one should consider the benefits to be expected in the respective application.

Table 2.5 lists a range of applications for each type of fuel cell.

Transportation. In the transportation sector, fuel cells are probably the most serious contenders as competitors to internal combustion engines. They are highly efficient as they are electrochemical rather than thermal engines. Hence they can help to reduce the consumption of primary energy and the emission of CO_2.

What makes them most attractive for transport applications, though, is the fact that they emit zero or ultra-low emissions. And this is what mainly inspired automotive companies and other fuel cell developers in the 1980s and 1990s to start developing fuel cell-powered cars and buses. Leading developers realized that although the introduction of the three-way catalytic converter had been a milestone, keeping up the pace in cleaning up car emissions further was going to be very tough indeed. Legislation such as California's Zero

TABLE 2.5 Currently Developed Types of Fuel Cells and Their Characteristics and Applications

Fuel Cell Type	Electrolyte	Charge Carrier	Operating Temperature	Fuel	Electric Efficiency (System)	Power Range/ Application
Alkaline FC (AFC)	KOH	OH	60–120°C	Pure H_2	35–55%	<5 kW, niche markets (military, space)
Proton exchange membrane FC (PEMFC)	Solid polymer (such as Nafion)	H^+	50–100°C	Pure H_2 (tolerates CO_2)	35–45%	Automotive, CHP (5–250 kW), portable
Phosphoric acid FC (PAFC)	Phosphoric acid	H^+	~220°C	Pure H_2 (tolerates CO_2, approx. 1% CO)	40%	CHP (200 kW)
Molten carbonate FC (MCFC)	Lithium and potassium carbonate	CO_3^{2-}	~650°C	H_2, CO, CH_4, other hydrocarbons tolerates CO_2	>50%	200 kW-MW range, CHP and standalone
Solid oxide FC (SOFC)	Solid oxide electrolyte (yttria, zirconia)	O^{2-}	~1000°C	H_2, CO, CH_4, other hydrocarbons tolerates CO_2	>50%	2 kW-MW range, CHP and standalone

Emission Mandate has come in and initially, only battery-powered vehicles were seen as a solution to the problem of building zero-emission vehicles. Meanwhile, it has turned out that the storage capacity of batteries is unacceptable for practical use because customers ask for the same drive range they are used to from internal combustion engines (ICEs). In addition, the battery solution is unsatisfactory for yet another reason: with battery-powered cars the point where air pollution takes place is only shifted back to the electric power plant providing the electricity for charging. This is the point where fuel cells were first seen as the only viable technical solution to the problem of car-related pollution.

It has now become clear that buses will make the fastest entry into the market because the hydrogen storage problem has been solved with roof-top pressure tanks. Also, fueling is not an issue due to the fleet nature of buses. For individual passenger cars, the future remains unclear. Most developers have now moved away from on-board reforming and promote direct hydrogen storage. Yet, this will require a new fuel infrastructure, which is not easy to establish.

Another automotive application is auxiliary power units (APUs). APUs are designed not to drive the main power train but to supply electric power to all devices on-board conventional internal combustion engines, even when the main engine is not operating. A typical example for a consumer requiring large amounts of power (several kW) is an air-conditioning system. Currently, due to the difficulties with reforming gasoline, high-temperature fuel cells are considered a good option.

Stationary Power. Cost targets were first seen as an opportunity: The reasoning went that, when fuel cells were going to meet automotive cost targets, other applications, including stationary power, would benefit from this development, and a cheap multipurpose power source would become available.

Meanwhile, stationary power generation, in addition to buses, is viewed as the leading market for fuel cell technology. The reduction of CO_2 emissions is an important argument for the use of fuel cells in small stationary power systems, particularly in combined heat and power generation. In fact, fuel cells are currently the only practical engines for micro-CHP systems in the domestic environment at less than, say, 5 kW of electric power output. The higher capital investment for a CHP system would be offset against savings in domestic energy supplies and — in more remote locations — against power distribution cost and complexity. It is important to note that due to the use of CHP, the electrical efficiency is less critical than in other applications. Therefore, in principle all fuel cell types are applicable. In particular, PEMFC systems and SOFC systems are currently being developed in this market segment. PEMFC systems offer the advantage that they can be easily turned off when not required. SOFC systems, with their higher operating temperature, are usually operated continuously. They offer more high-grade heat and provide simpler technology for turning the main feed — natural gas, sometimes propane — into electric power than the more poison-sensitive PEMFC technology.

In the 50 to 500 kW range there will be competition with spark or compression ignition engines modified to run on natural gas. So far, several hundred 200-kW phosphoric acid fuel cell plants manufactured by ONSI (IFC) have been installed worldwide. Yet, in this larger power range, high-temperature fuel cells offer distinct advantages. They allow simpler gas pretreatment than low-temperature fuel cells do and achieve higher electrical efficiencies. This is important when one considers that a standard motor CHP unit at several hundred kilowatts of electric power output will generally achieve close to 40% electrical efficiency — at a fraction of the cost of a fuel cell. High-temperature fuel cells also offer process heat in the form of steam rather than just hot water. Both MCFC and SOFC are currently being developed for industrial CHP systems.

Portable Power. The portable market is less well defined, but a potential for quiet fuel cell power generation is seen in the sub-1kW portable range. The term "portable fuel cells" often includes grid-independent applications such as camping, yachting, and traffic monitoring. Fuels considered vary from one application to another. And fuels are not the only aspect that varies. Different fuel cells may be needed for each subsector in the portable market.

It is currently not clear whether DMFC systems will ever be able to compete with lithium ion batteries in mass markets such as cellular phones or portable computers. But in applications where size does not matter and long, unattended service at low power levels is required, DMFCs are the ideal power source.

There may also be room for hydrogen-fueled PEMFC systems where higher peak power demand is needed for short durations. Storage of hydrogen and the implementation of a supply infrastructure remain problems to be solved.

At more than several hundred watts of electric power, reformer-based systems remain the only option to date. This is the range of the APU (see Transportation, above), and it is likely that similar systems will provide power to yachts and motor homes, mountaineering huts, etc. They may be fueled by propane or perhaps some liquid fuel — compare the following section.

Choice of Fuel

Before even deciding which fuel cell type to use, one needs to identify the best fuel base and the allowable size and complexity of the resulting fuel cell system.

Of the three key applications for fuel cells — automotive propulsion, stationary power generation and portable power — the automotive case is most readily dealt with. For generating the propulsion power of a car, bus, or truck, only PEMFCs are currently being considered for their superior volumetric power density with operation on hydrogen. Leading developers originally favored hydrogen generation from methanol on board and, more recently, either hydrogen generation from reformulated gasoline on board or hydrogen storage on board. As hydrogen generation from gasoline is technically difficult, questionable for energy efficiency reasons, and methanol as a fuel base has been demonstrated but is unwanted by most developers, hydrogen storage on board appears to emerge as the overall compromise. Moreover, automotive propulsion is still at least 10 years away from widespread market introduction; it is a highly challenging, specialist job that is well underway at the leading developers' laboratories — and it should be left there.

The other extreme is portable power, where again the PEMFC is the only fuel cell seriously considered. In this case, there are two options for fuel, hydrogen or direct methanol, i.e., without prior conversion to hydrogen. The latter type of PEMFC is usually referred to as DMFC. For systems larger than, say, 1 kW that operate continuously, direct methanol is not an option because of the expense of the fuel cell. Direct hydrogen is not an option because of the lack of storage capacity. Therefore, one may consider converting a chemical such as LPG, propane, butane, methanol, gasoline, or diesel into hydrogen. Such larger systems will be considered among the stationary power systems.

Stationary power generation, at the first glance, is the least clear application in terms of the technology base: four types of fuel cell are currently being developed for stationary power generation.

In contrast, the fuel base is rather clear in this particular case: for residential and small combined heat and power generation, natural gas is the usual source of energy, and fuel cell systems have to adapt to this fuel. In areas without a gas — and electricity — grid, LPG may be an option and is being actively pursued by a number of developers.

Table 2.6 summarizes the most likely applications and fueling options for all types of fuel cells.

TABLE 2.6 Fuel Cell Applications and Likely Fueling Options

	Automotive		Portable		Stationary		
Fuel Cell	Main Drive Train	Auxiliary Power	Battery Replacement Long Operation, Low Power	Battery Replacement Short Operation, High Power	Residential Remote Power or Residential Heat and Power (few kW range)	Small-Scale Heat and Power (100 kW to several MW)	Large Central Power Station (many MW)
PEMFC	50 to 100 kW hydrogen powered (liquid or compressed) other fuels less likely	Less likely	0.1 to 500 W direct methanol powered = DMFC	100 W to few kW hydrogen (compressed or metal hydride)	1 to 10 kW natural gas or LPG powered (possibly methanol or fuel oil)	100 kW to 10 MW natural gas powered less likely	
MCFC		Less likely			Less likely	100 kW to 10 MW natural gas powered	Perhaps, in combination with gas turbine coal gas fired
SOFC		Several kW gasoline or diesel powered			1 to 10 kW natural gas or LPG powered	100 kW to 10 MW natural gas powered	Perhaps, in combination with gas turbine coal gas fired
PAFC						100 kW to 10 MW natural gas powered less likely	

Defining Terms

AFC: Alkaline fuel cell.

APU: Auxiliary power unit — supplying electric power to cars independent of the main engine.

Bipolar plate: Metal, graphite or composite plate separating two adjacent cells in planar PEMFC, AFC, PAFC, and MCFC stacks (possibly also planar SOFC). One side consequently acts as positive electrode (cathode), while the opposite side acts as negative electrode (anode) to the following cell. The two sides contain channels to distribute the reactants, usually hydrogen-rich gas (anode) and air (cathode).

CHP: Combined heat and power generation.

DMFC: Direct methanol fuel cell. Liquid or vaporized methanol/water fueled version of PEMFC.

Humidifier: Unit required for providing high water vapor concentration in inlet gas streams of PEMFC.

MCFC: Molten carbonate fuel cell.

MEA: Membrane electrode assembly — central electrochemical part of a PEMFC. Consists of anode, cathode and membrane electrolyte.

PAFC: Phosphoric acid fuel cell.

PEMFC: Proton exchange membrane fuel cell — also called SPFC.

Reformer: Gas or liquid processing unit providing hydrogen-rich gas to run fuel cells off readily available hydrocarbon fuels such as methane, methanol, propane, gasoline, or diesel.

SPFC: Solid polymer fuel cell — see PEMFC.

SOFC: Solid oxide fuel cell.

Stack: Serial combination (pile) of alternating planar cells and bipolar plates in order to increase the overall voltage.

References

P.W. Atkins. *Physical Chemistry,* Oxford: Oxford University Press, 2000.

J.W. Bauman, T.A. Zawodzinski, Jr., and S. Gottesfeld. in *Proceedings of the Second International Symposium on Proton Conducting Membrane Fuel Cells II*, S. Gottesfeld and T.F. Fuller (Eds.), Pennington, NJ: Electrochem. Soc, 1999.

U. Bossel. *The Birth of the Fuel Cell 1835–1845,* Oberrohrdorf, Switzerland: European Fuel Cell Forum, 2000.

S.J. Cooper, A.G. Gunner, G. Hoogers, and D. Thompsett. Reformate tolerance in proton exchange membrane fuel cells: electrocatalyst solutions. in *New Materials for Fuel Cells and Modern Battery Systems II*, O. Savadogo and P. R. Roberge (Eds.), 1997.

D.P. Davies, P.L. Adcock, M. Turpin, and S.J. Rowen. *J. Power Sources*, vol. 86, 237, 2000.

D. Edlund. A versatile, low-cost, and compact fuel processor for low-temperature fuel cells, *Fuel Cells Bulletin No. 14*, 2000.

B. Emonts, J. Bøgild Hansen, S. Lœgsgaard Jørgensen, B. Höhlein, and R. Peters. Compact methanol reformer test for fuel-cell powered light-duty vehicles, *J. Power Sources*, vol. 71, 288, 1998.

S. Gottesfeld and J. Pafford, A new approach to the problem of carbon monoxide poisoning in fuel cells operating at low temperatures, *J. Electrochem Soc.*, vol. 135, 2651, 1988.

N.N. Greenwood and A. Earnshaw, *Chemistry of the Elements*, Oxford: Pergamon Press, 1984.

C. Hamann, A. Hamnett, and W. Vielstich. *Electrochemistry.* Weinheim: Wiley-VCH, 1998.

H&FC. *Hydrogen & Fuel Cell Letter,* 2001, (June).

K. Kuroda, K. Kobayashi, N. O'Uchida, Y. Ohta, and Y. Shirasaki. Study on performance of hydrogen production from city gas equipped with palladium membranes, *Mitsubishi Juko Giho*, vol. 33, no. 5, 1996.

R.W. McCabe, and P.J. Mitchell, *J. Catal.*, vol. 103, 419, 1987.

V. Plzak, B. Rohland, and L. Jörissen. Preparation and screening of Au/MeO catalysts for the preferential oxidation of CO in H2-containing gases, Poster, 50th ISE Meeting, Pavia, 1999.

D.E. Ridler and M.V. Twigg. Steam reforming, in *Catalyst Handbook*, M.V. Twigg (Ed.) London: Manson Publishing, 1996, p. 225.

G. Sandrock. Intermetallic hydrides: history and applications, in *Proceedings of the Symposium on Hydrogen and Metal Hydride Batteries*, P.D. Bennett and T. Sakai (Eds.), 1994.

D.L. Trimm and Z. Önsan. On-board fuel conversion for hydrogen-fuel-cell-driven vehicles, *Catalysis Reviews*, vol. 43, 31–84, 2001.

N.G. Vitale, and D.O. Jones. (Plug Power Inc). U.S. Patent US6,066,408, 2000.

Further Reading

G. Hoogers, (Ed.) *Fuel Cell Technology Handbook,* Boca Raton, FL: CRC Press, 2003.

J. Larminie and A. Dicks, *Fuel Cell Systems Explained,* Wiley-VCH, 2000.

W. Vielstich, A. Lamm, and H. Gasteiger (Eds.) *Handbook of Fuel Cells — Fundamentals, Technology, Applications* (4 Volumes). Wiley-VCH, 2003.

3

Transmissions

Rao S. Thallam
Salt River Project

Mohamed E. El-Hawary
University of Nova Scotia

Charles A. Gross
Auburn University

Arun G. Phadke
Virginia Polytechnic Institute

R.B. Gungor
University of South Alabama

J. Duncan Glover
Exponent Natick

3.1 High-Voltage Direct-Current Transmission

Rao S. Thallam

The first commercial high-voltage direct-current (HVDC) power transmission system was commissioned in 1954, with an interconnection between the island of Gotland and the Swedish mainland. It was an undersea cable, 96 km long, with ratings of 100 kV and 20 MW. There are now more than 50 systems operating throughout the world, and several more are in the planning, design, and construction stages. HVDC transmission has become acceptable as an economical and reliable method of power transmission and interconnection. It offers advantages over alternating current (ac) for long-distance power transmission and as asynchronous interconnection between two ac systems and offers the ability to precisely control the power flow without inadvertent loop flows in an interconnected ac system. Table 3.1 lists the HVDC projects to date (1995), their ratings, year commissioned (or the expected year of commissioning), and other details. The largest system in operation, Itaipu HVDC transmission, consists of two ±600-kV, 3150-MW-rated bipoles, transmitting a total of 6300 MW power from the Itaipu generating station to the Ibiuna (formerly Sao Roque) converter station in southeastern Brazil over a distance of 800 km.

Table 3.1 HVDC Projects Data

	HVDC Supplier†	Year Commissioned	Power Rating, MW	DC Volts, kV	Line/Cable, km	Location
			Mercury Arc Valves			
Moscow-Kashira[a]	F	1951	30	±100	100	Russia
Gotland I[a]	A	1954	20	±100	96	Sweden
English Channel	A	1961	160	±100	64	England-France
Volgograd-Donbass[b]	F	1965	720	±400	470	Russia
Inter-Island	A	1965	600	±250	609	New Zealand
Konti-Skan I	A	1965	250	250	180	Denmark-Sweden
Sakuma	A	1965	300	2125	B-B[f]	Japan
Sardinia	I	1967	200	200	413	Italy
Vancouver I	A	1968	312	260	69	Canada
Pacific Intertie	JV	1970	1440	±400	1362	USA
		1982	1600			
Nelson River I[c]	I	1972	1620	±450	892	Canada
Kingsnorth	I	1975	640	±266	82	England
			Thyristor Valves			
Gotland Extension	A	1970	30	±150	96	Sweden
Eel River	C	1972	320	2×80	B-B	Canada
Skagerrak I	A	1976	250	250	240	Norway-Denmark
Skagerrak II	A	1977	500	±250	240	Norway-Denmark
Skagerrak III	A	1993	440	±350	240	Norway-Denmark
Vancouver II	C	1977	370	–280	77	Canada
Shin-Shinano	D	1977	300	2×125	B-B	Japan
		1992	600	3×125		
Square Butte	C	1977	500	±250	749	USA
David A. Hamil	C	1977	100	50	B-B	USA
Cahora Bassa	J	1978	1920	±533	1360	Mozambique-S. Africa
Nelson River II	J	1978	900	±250	930	Canada
		1985	1800	±500		
C-U	A	1979	1000	±400	710	USA
Hokkaido-Honshu	E	1979	150	125	168	Japan
	E	1980	300	250		
		1993	600	±250		
Acaray	G	1981	50	25.6	B-B	Paraguay
Vyborg	F	1981	355	1×170 (±85)	B-B	Russia (tie with Finland)
	F	1982	710	2×170		
			1065	3×170		
Duernrohr	J	1983	550	145	B-B	Austria
Gotland II	A	1983	130	150	100	Sweden
Gotland III	A	1987	260	±150	103	Sweden
Eddy County	C	1983	200	82	B-B	USA
Chateauguay	J	1984	1000	2×140	B-B	Canada
Oklaunion	C	1984	200	82	B-B	USA
Itaipu	A	1984	1575	±300	785	Brazil
	A	1985	2383			
	A	1986	3150	±600		
	A	1987	6300	2×±600		
Inga-Shaba	A	1982	560	±500	1700	Zaire
Pac Intertie Upgrade	A	1984	2000	±500	1362	USA
Blackwater	B	1985	200	57	B-B	USA
Highgate	A	1985	200	±56	B-B	USA
Madawaska	C	1985	350	140	B-B	Canada
Miles City	C	1985	200	±82	B-B	USA
Broken Hill	A	1986	40	2×17(±8.33)	B-B	Australia

(Continued)

Table 3.1 Continued

	HVDC Supplier†	Year Commissioned	Power Rating, MW	DC Volts, kV	Line/Cable, km	Location
Intermountain	A	1986	1920	±500	784	USA
Cross-Channel						
Les Mandarins	H	1986	2000	±270	72	France
Sellindge	I	1986	2000	±270	72	England
Descantons-Comerford	C	1986	690	±450	172	Canada-USA
SACOI[d]	H	1986	200	200	415	Corsica Island
SACOI[e]		1992	300			Italy
Urguaiana Freq. Conv.	D	1987	53.7	17.9	B-B	Brazil (tie with Uruguay)
Virginia Smith (Sidney)	G	1988	200	55.5	B-B	USA
Gezhouba-Shanghai	B + G	1989	600	500	1000	China
		1990	1200	±500		
Konti-Skan II	A	1988	300	285	150	Sweden-Denmark
Vindhyachal	A	1989	500	2×69.7	B-B	India
Pac Intertie Expansion	B	1989	1100	±500	1362	USA
McNeill	I	1989	150	42	B-B	Canada
Fenno-Skan	A	1989	500	400	200	Finland-Sweden
Sileru-Barsoor	K	1989	100	+100	196	India
			200	+200		
			400	±200		
Rihand-Delhi	A	1991	750	+500	910	India
		1991	1500	±500		
Hydro Quebec-New Eng.	A	1990	2000[g]	±450	1500	Canada-USA
Welch-Monticello		1995	300		B-B	USA
		1998	600			
Etzenricht		1993	600	160	B-B	Germany (tie with Czech)
Vienna South-East	G	1993	600	160	B-B	Austria (tie with Hungary)
DC Hybrid Link	AB	1993	992	+270/−350	617	New Zealand
Chandrapur-Padghe		1997	1500	±500	900	India
Chandrapur-Ramagundam		1996	1000	2×205	B-B	India
Leyte-Luzun		1997	1000	350	440	Philippines
Haenam-Cheju I		1997	300	±180	100	South Korea
Baltic Cable Project		1994	600	450	250	Sweden-Germany
Victoria-Tasmania		1995	300	300		Australia
Kontek HVDC Intercon		1995	600	600		Denmark
Scotland-N. Ireland		1998	250	150	60	United Kingdom
Greece-Italy		1998	500			Italy
Tiang-Guang		1998	1800	500	903	China
Visakhapatnam		1998	500	205	B-B	India
Thailand-Malaysia		1998	300	300	110	Malaysia-Thailand
Rivera		1998	70		B-B	Urguay

†A–ASEA; B–Brown Boveri; C–General Electric; D–Toshiba; E–Hitachi; F–Russian; G–Siemens; H–CGEE Alsthom; I–GEC (formerly Eng. Elec.); J–HVDC W.G. (AEG, BBC, Siemens); K–(Independent); AB–ABB Brown Boveri; JV–Joint Venture (GE and ASEA).

[a]Retired from service.
[b]Two valve groups replaced with thyristors in 1977.
[c]Two valve groups in Pole 1 replaced with thyristors by GEC in 1991.
[d]50-MW thyristor tap.
[e]Uprate with thyristor valves.
[f]Back-to-back HVDC system.
[g]Multiterminal system. Largest terminal is rated 2250 MW.
Source: Data compiled by D.J. Melvold, Los Angeles Department of Water and Power.

Configurations of DC Transmission

HVDC transmission systems can be classified into three categories:

1. Back-to-back systems
2. Two-terminal, or point-to-point, systems
3. Multiterminal systems

These will be briefly described here.

Back-to-Back DC System

In a back-to-back dc system (Figure 3.1), both rectifier and inverter are located in the same station, usually in the same building. The rectifier and inverter are usually tied with a reactor, which is generally of outdoor, air-core design. A back-to-back dc system is used to tie two **asynchronous ac systems** (systems that are not in synchronism). The two ac systems can be of different operating frequencies, for example, one 50 Hz and the other 60 Hz. Examples are the Sakuma and Shin-Shinano converter stations in Japan. Both are used to link the 50- and 60-Hz ac systems. The Acaray station in Paraguay links the Paraguay system (50 Hz) with the Brazilian system, which is 60 Hz. Back-to-back dc links are also used to interconnect two ac systems that are of the same frequency but are not operating in synchronism. In North America, eastern and western systems are not synchronized, and Quebec and Texas are not synchronized with their neighboring systems. A dc link offers a practical solution as a tie between nonsynchronous systems. Thus to date, there are ten back-to-back dc links in operation interconnecting such systems in North America. Similarly, in Europe, eastern and western systems are not synchronized, and dc offers the practical choice for interconnection between them.

Two-Terminal, or Point-to-Point, DC Transmission

Two-terminal dc systems can be either **bipolar** or **monopolar**. Bipolar configuration, shown in Figure 3.2, is the commonly used arrangement for systems with overhead lines. In this system, there will be two conductors, one for each polarity (positive and negative) carrying nearly equal currents. Only the difference of these currents, which is usually small, flows through ground return.

A monopolar system will have one conductor, either positive or negative polarity, with current returning through either ground or another metallic return conductor. The monopolar ground return current configuration, shown in Figure 3.3, has been used for undersea cable systems, where current returns through the sea. This configuration can also be used for short-term emergency operation for a two-terminal dc line system in the event of a pole outage. However, concerns for corrosion of underground metallic structures and interference with telephone and other utilities will restrict the duration of such operation. The total ampere-hour operation per year is usually the restricting criterion.

In a monopolar metallic return system, shown in Figure 3.4, return current flows through a conductor, thus avoiding problems associated with ground return current. This method is generally used as a contingency mode of operation for a normal bipolar transmission system in the event of a partial converter (one-pole

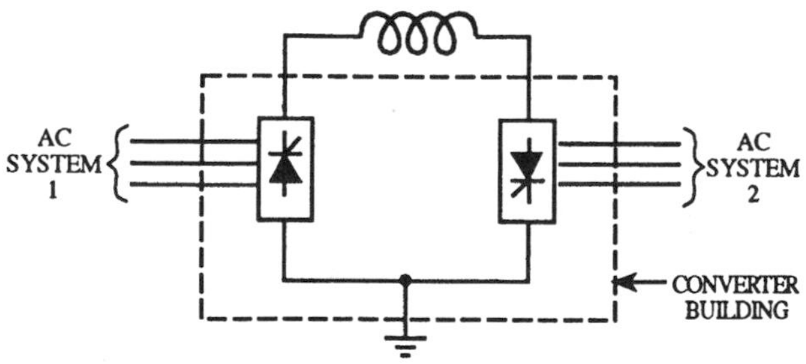

FIGURE 3.1 Back-to-back dc system.

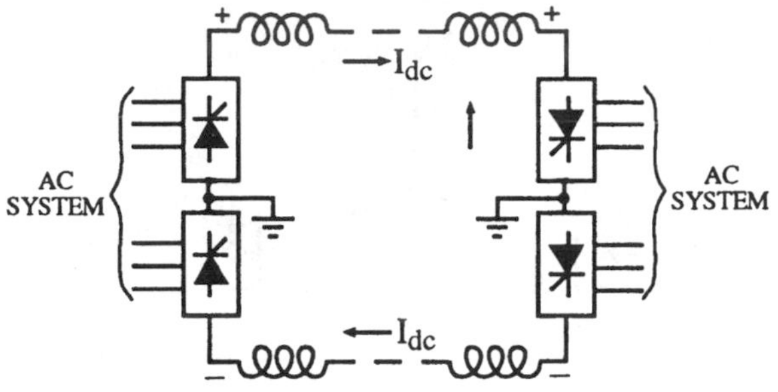

FIGURE 3.2 Bipolar dc system.

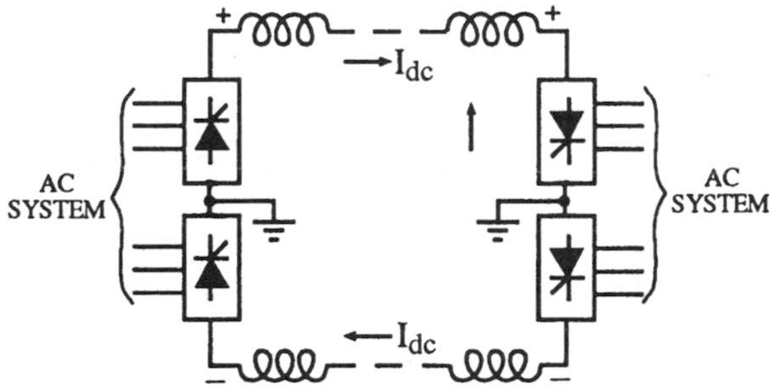

FIGURE 3.3 Monopolar ground return dc system.

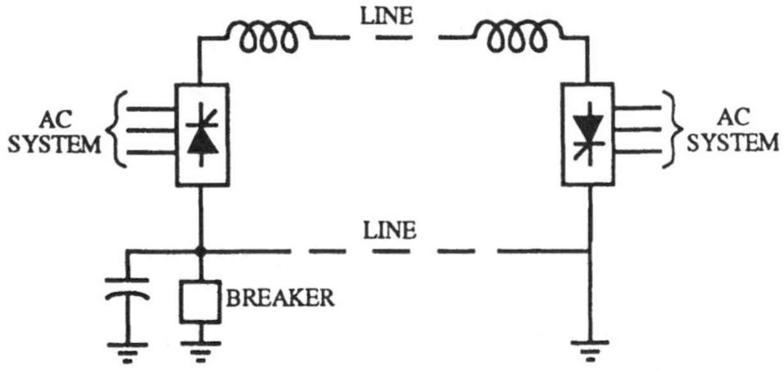

FIGURE 3.4 Monopolar metallic return dc system.

equipment) outage. In the case of outage of a one-pole converter, the conductor of the affected pole will be used as the returning conductor. A metallic return transfer breaker will be opened, diverting the return current from the ground path and into the pole conductor. This conductor will be grounded at one end and will be insulated at the other end. This system can transmit half the power of the normal bipolar system capacity (and can be increased if overload capacity is available). However, the line losses will be doubled compared to the normal bipolar operation for the same power transmitted.

Multiterminal DC Systems

There are two basic configurations in which the dc systems can be operated as multiterminal systems (MTDC):

1. Parallel configuration
2. Series configuration

Parallel configuration can be either radial-connected (Figure 3.5(a)) or mesh-connected (Figure 3.5(b)). In a parallel-connected multiterminal dc system, all converters operate at the same nominal dc voltage, similar to

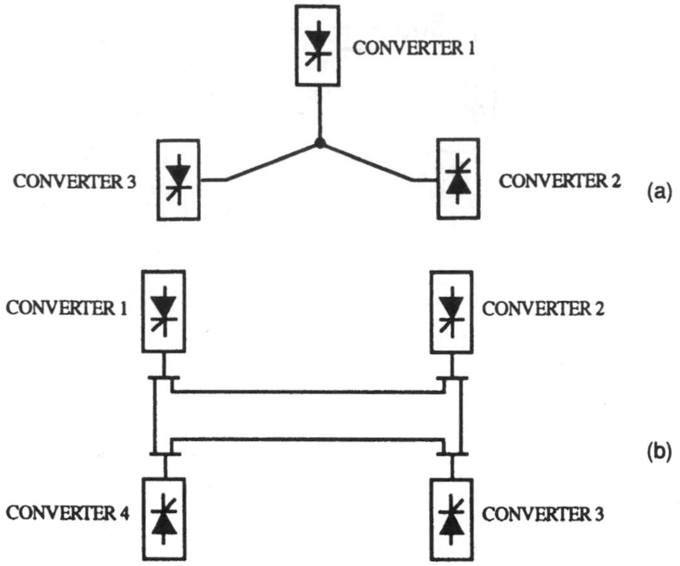

FIGURE 3.5 (a) Parallel-controlled radial MTDC system; (b) parallel-connected mesh-type MTDC system.

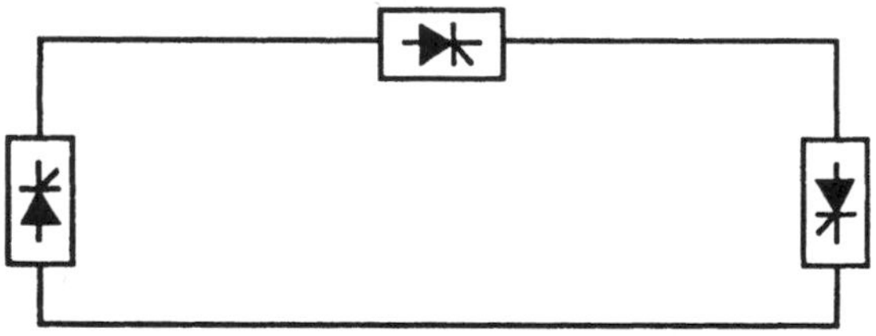

FIGURE 3.6 Series-connected MTDC system.

ac system interconnections. In this operation, one converter determines the operating voltage, and all other terminals operate in a current-controlling mode.

In a series-connected multiterminal dc system (Figure 3.6), all converters operate at the same current. One converter sets the current that will be common to all converters in the system. Except for the converter that sets the current, the remaining converters operate in voltage control mode (constant **firing angle** or constant **extinction angle**). The converters operate almost independently without requirement for high-speed communication between them. The power output of a non-current-controlling converter is varied by varying its voltage. At all times, the sum of the voltages across the rectifier stations must be larger than the sum of voltages across the inverter stations. Disadvantages of a series-connected system are (1) reduced efficiency because full line insulation is not used at all times, and (2) operation at higher firing angles will lead to high converter losses and higher reactive power requirements from the ac system.

There are now two truly multiterminal dc systems in operation. The Sardinia–Corsica–Italy three-terminal dc system was originally commissioned as a two-terminal (Sardinia–Italy) system in 1967 with a 200-MW rating. In 1986, the Corsica tap was added and the system was upgraded to a 300-MW rating. The two-terminal Hydro Quebec–New England HVDC interconnection (commissioned in 1985) was extended to a five-terminal system and commissioned in 1990 (see Table 3.1). The largest terminal of this system at Radisson station in Quebec is rated at 2250 MW. Two more systems, the Nelson River system in Canada and the Pacific NW-SW Intertie in the United States, also operate as multiterminal systems. Each of these systems has two converters at each end of the line, but the converters at each end are constrained to operate in the same mode, either rectifier or inverter.

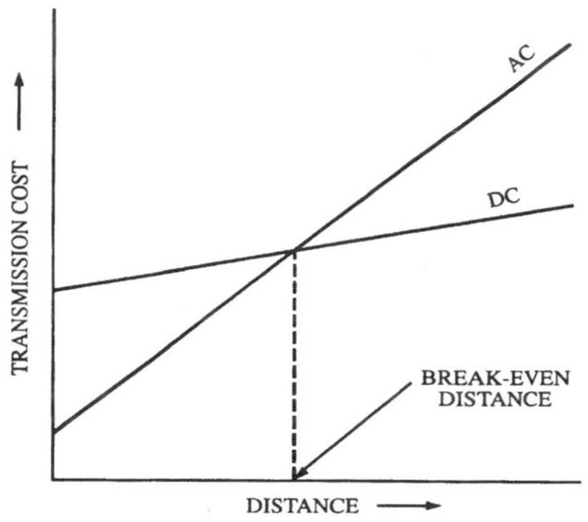

FIGURE 3.7 Transmission cost as function of line length.

Economic Comparison of AC and DC Transmission

In cases where HVDC is selected on technical considerations, it may be the only practical option, as in the case of an asynchronous interconnection. However, for long-distance power transmission, where both ac and HVDC are practical, the final decision is dependent on total costs of each alternative. Total cost of a transmission system includes the line costs (conductors, insulators, and towers) plus the right-of-way (R-o-W) costs. A dc line with two conductors can carry almost the same amount of power as the three-phase ac line with the same size of line conductors. However, dc towers with only two conductors are simpler and cheaper than three-phase ac towers. Hence the per-mile costs of line and R-o-W will be lower for a dc line. Power losses in the dc line are also lower than for ac for the same power transmitted. However, the HVDC system requires converters at the two ends of the line; hence the terminal costs for dc are higher than for ac. Variation of total costs for ac and dc as a function of line length are shown in Figure 3.7. There is a break-even distance above which the total costs of dc option will be lower than the ac transmission option. This is in the range of 500 to 800 km for overhead lines but much shorter for cables. It is between 20 and 50 km for submarine cables and twice as far for underground cables.

Principles of Converter Operation

Converter Circuit

Since the generation and most of the transmission and utilization is alternating current, HVDC transmission requires conversion from ac to dc (called rectification) at the sending end and conversion back from dc to ac (called inversion) at the receiving end. In HVDC transmission, the basic device used for conversion from ac to dc and from dc to ac is a three-phase full-wave bridge converter, which is also known as a Graetz circuit. This is a three-phase six-pulse converter. A three-phase twelve-pulse converter will be composed of two three-phase six-pulse converters, supplied with voltages differing in phase by 30 degrees (Figure 3.8). The phase difference of 30 degrees is obtained by supplying one six-pulse bridge with a Y/Y transformer and the other by Y/Δ transformer.

Relationships between AC and DC Quantities

Voltages and currents on ac and dc sides of the converter are related and are functions of several converter parameters, including the converter transformer. The following equations are provided here for easy reference. Detailed derivations are given in Kimbark [1971].

E_{LL} = rms line-to-line voltage of the converter ac bus
I_1 = rms value of fundamental frequency component of the converter ac current
h = harmonic number

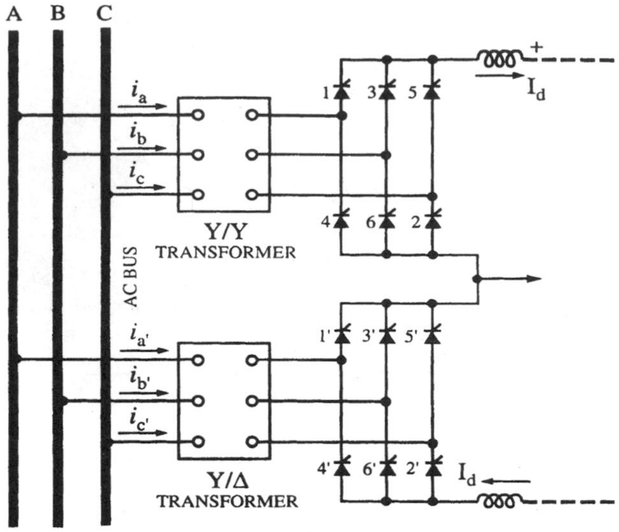

FIGURE 3.8 Basic circuit of a 12-pulse HVDC converter.

α = valve firing delay angle (from the instant the valve voltage is positive)
u = overlap angle (also called commutation angle)
ϕ = phase angle between voltage and current
$\cos\phi$ = displacement power factor
V_{d0} = ideal no-load dc voltage (at $\alpha = 0$ and $u = 0$)
L_c = commutating circuit inductance
$\beta = 180-\alpha$ = angle of advance for inverter
$\gamma = 180-(\alpha+u)$ = margin angle for inverter

with $\alpha = 0$, $u = 0$,

$$V_{d0} = \frac{3\sqrt{2}}{\pi}E_{LL} = 1.35E_{LL} \tag{3.1}$$

With $\alpha > 0$, and $u = 0$

$$V_d = V_{d0}\cos\alpha \tag{3.2}$$

Theoretically α can vary from 0 to 180 degrees (with $u = 0$); hence V_d can vary from $+V_{d0}$ to $-V_{d0}$. Since the valves conduct current in only one direction, variation of dc voltage from V_{d0} to $-V_{d0}$ means reversal of power flow direction and the converter mode of operation changing from rectifier to inverter.

$$I_1 = \frac{\sqrt{6}}{\pi}I_d = 0.78I_d \tag{3.3}$$

$$\cos\phi = \cos\alpha = \frac{V_d}{V_{d0}} \tag{3.4}$$

With $\alpha > 0$ and $0 < u > 60°$,

$$V_d = V_{d0}\frac{\cos\alpha + \cos(\alpha+u)}{2} \tag{3.5}$$

$$V_d = \frac{3\sqrt{2}}{\pi}E_{LL}\frac{\cos\alpha + \cos(\alpha + u)}{2} \tag{3.6}$$

$$I_1 \approx \frac{\sqrt{6}}{\pi}I_d \tag{3.7}$$

The error in Equation (3.7) is only 4.3% at $u = 60$ degrees (maximum overlap angle for normal steady-state operation), and it will be even lower (1.1%) for most practical cases when u is 30 degrees or less. It can be seen from Equations (3.6) and (3.7) that the ratio between ac and dc currents is almost fixed, but the ratio between ac and dc voltages varies as a function of α and u. Hence the HVDC converter can be viewed as a variable-ratio voltage transformer, with almost fixed current ratio.

$$P_{dc} = V_d I_d \tag{3.8}$$

$$P_{ac} = \sqrt{-3E_{LL}I_1\cos\,\phi} \tag{3.9}$$

Substituting for V_d and I_d in Equation (3.8) and comparing with Equation (3.9),

$$\cos\,\phi \approx \frac{\cos\,\alpha + \cos(\alpha + u)}{2} \tag{3.10}$$

From Equations (3.5) and (3.10),

$$\cos\,\phi \approx \frac{V_d}{V_{d0}} \tag{3.11}$$

From Equations (3.1), (3.5), and (3.10),

$$V_d \approx 1.35E_{LL}\cos\,\phi \tag{3.12}$$

AC Current Harmonics

The HVDC converter is a harmonic current source on the ac side. Fourier analysis of an ac current waveform, shown in Figure 3.9, shows that it contains the fundamental and harmonics of the order 5, 7, 11, 13, 17, 19, etc. The current for zero degree overlap angle can be expressed as

$$i(t) = \frac{2\sqrt{3}}{\pi}I_d\left(\begin{array}{c}\cos\omega t - \frac{1}{5}\cos 5\omega t + \frac{1}{7}\cos 7\omega t \\ -\frac{1}{11}\cos 11\omega t + \frac{1}{13}\cos 13\omega t + \cdots\end{array}\right) \tag{3.13}$$

and

$$I_{h0} = \frac{I_{10}}{h} \tag{3.14}$$

where I_{10} and I_{h0} are the fundamental and harmonic currents, respectively, at $\alpha = 0$ and $u = 0$.

Equation (3.14) indicates that the magnitudes of harmonics are inversely proportional to their order.

Converter ac current waveform $i_{a'}$ for phase a with a Y/Δ transformer is also shown in Figure 3.9. Fourier analysis of this current shows that the fundamental and harmonic components will have the same magnitude as in the case of the Y/Y transformer. However, harmonics of order 5, 7, 17, 19, etc. are in phase opposition,

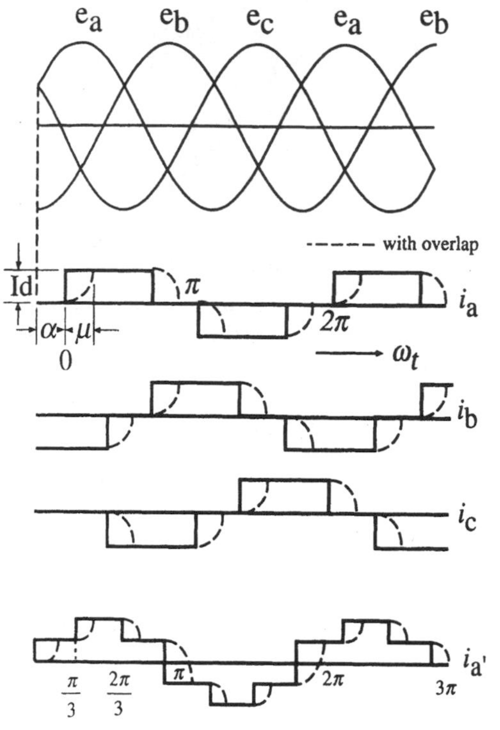

FIGURE 3.9 AC line current waveforms, i_a, i_b, i_c with Y/Y transformer and $i_{a'}$ with Y/Δ transformer.

whereas harmonics of order 11, 13, 23, 25, etc. are in phase with the Y/Y transformer case. Hence harmonics of order 5, 7, 17, 19, etc. will be canceled in a 12-pulse converter and do not appear in the ac system. In practice they will not be canceled completely because of imbalances in converter and transformer parameters.

Effect of Overlap. The effect of overlap due to commutation angle is to decrease the amplitude of harmonics from the case with zero overlap. Magnitudes of harmonics for a general case with finite firing angle (α) and overlap angle (u) are given by

$$\frac{I_h}{I_{h0}} = \frac{1}{x}\left[A^2 + B^2 - 2AB\cos(2\alpha + u)\right]^{1/2} \tag{3.15}$$

where

$$A = \frac{\sin(h+1)\frac{u}{2}}{h+1} \qquad B = \frac{\sin(h-1)\frac{u}{2}}{h-1}$$

$$x = \cos\alpha - \cos(\alpha + u)$$

Noncharacteristic Harmonics. In addition to the harmonics described above, converters also generate other harmonics due to "nonideal" conditions of converter operation. Examples of the nonideal conditions are converter ac bus voltage imbalance, perturbation of valve firing pulses, distortion of ac bus voltages, and unbalanced converter transformer impedances. Harmonics generated due to these causes are called *noncharacteristic* harmonics. These are usually smaller in magnitude compared to characteristic harmonics but can create problems if resonances exist in the ac system at these frequencies. In several instances additional filters were installed at the converter ac bus to reduce levels of these harmonics flowing into the ac system.

Converter Control

The static characteristic of a HVDC converter is shown in Figure 3.10. There are three distinct features of this characteristic.

Constant Firing Angle Characteristic (A–B). If the converter is operating under constant firing angle control, the converter characteristic can be described by the equation

$$V_d = V_{d0}\cos\alpha - \frac{3\omega L_c}{\pi} I_d \tag{3.16}$$

When the ordered current is too high for the converter to deliver, it will operate at the minimum firing angle (usually 5 degrees). Then the current will be determined by the voltage V_d and the load. This is also referred to as the natural voltage characteristic. The converter in this mode is equivalent to a dc voltage source with internal resistance R_c, where

$$R_c = \frac{3\omega L_c}{\pi} \tag{3.17}$$

Constant Current Control. This is the usual mode of operation of the rectifier. When the converter is operating in constant current control mode, the firing angle is adjusted to maintain dc current at the ordered value. If the load current goes higher than the ordered current for any reason, control increases the firing angle to reduce dc voltage and the converter operation moves in the direction from B to C. At point C, the firing angle reaches 90 degrees (neglecting overlap angle), the voltage changes polarity, and the converter becomes an inverter. From C to D, the converter works as an inverter.

Constant Extinction Angle Control. At point D, the inverter firing angle has increased to a point where further increase can cause commutation failure. The inverter, for its safe operation, must be operated with sufficient angle of advance β, such that under all operating conditions the extinction angle γ is greater than the valve deionization angle. The deionization angle is defined as the time in electrical degrees from the instant current reaches zero in a particular valve to the time the valve can withstand the application of positive voltage. Typical minimum values of γ are 15 to 20 degrees for mercury arc valves and slightly less for

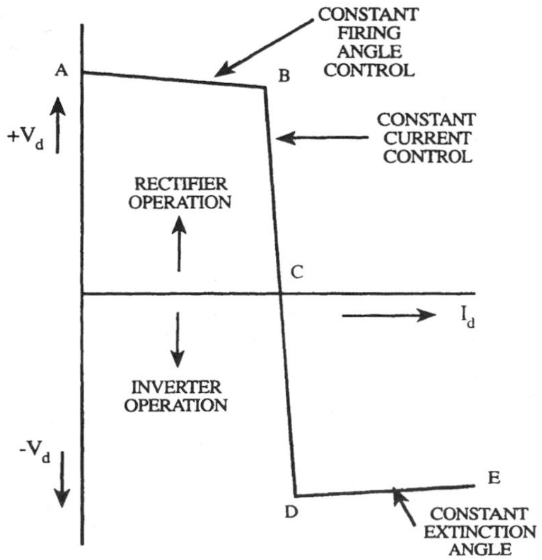

FIGURE 3.10 HVDC converter static characteristic.

thyristor valves. During the range D to E, the increase of load current increases the overlap angle, which reduces the dc voltage. This is the negative resistance characteristic of the inverter.

The functional requirements for HVDC converter control are:

1. Minimize the generation of noncharacteristic harmonics.
2. Safe inverter operation with fewest possible commutation failures even with distorted ac voltages.
3. Lowest possible consumption of reactive power. This requires operation with smallest possible delay angle α and extinction angle γ without increased risk of commutation failures.
4. Smooth transition from current control to extinction angle control.
5. Sufficient stability margins and response time when the ratio of the ac system short-circuit strength and the rated dc power (short-circuit ratio) is low.

Individual Phase Control

In the early HVDC systems individual phase control systems were used. The firing angle of each valve is calculated individually and operated either as constant α or constant γ control.

A schematic of the individual phase control system is shown in Figure 3.11. Six timing voltages are derived from the ac bus voltage, and the six grid pulses are generated at nominally identical delay times subsequent to the respective voltage-zero crossings. The delays are produced by independent delay circuits and controlled by a common direct voltage V_c, which is derived through a feedback loop to control constant dc line current or constant power. Several variations of this control were used until the late 1960s.

Disadvantages of Individual Phase Control. With distorted ac bus voltages, the firing pulses will be unequally spaced, thus generating noncharacteristic harmonics in ac current. This in turn will further distort the ac bus voltage. This process could lead to harmonic instability, particularly with ac systems of low short-circuit capacity (high-impedance system). Control system filters were tried to solve this problem. However, the filters could increase the potential for commutation failures and also reduce the speed of control system response for faults or disturbances in the ac system.

Equal Pulse Spacing Control

A control system based on the principle of equal spacing of firing pulses at intervals of 60 degrees (electrical) independent of ac bus voltages was developed in the late 1960s. The basic components of this system, shown in Figure 3.12, consist of a voltage-controlled oscillator and ring counter. The frequency of the oscillator is directly proportional to the dc control voltage V_c. Under steady-state conditions, pulse frequency is precisely $6f$, where f is the ac system frequency. The phase of each grid pulse will have some arbitrary value relative to the ac bus voltage. If the three-phase ac bus voltages are symmetrical sine waves with no distortion, then α is the same for all valves. The oscillator will be phase-locked with the ac system frequency to avoid drifting. The dc control voltage V_c is derived from a feedback loop for constant current, constant power, or constant extinction angle γ.

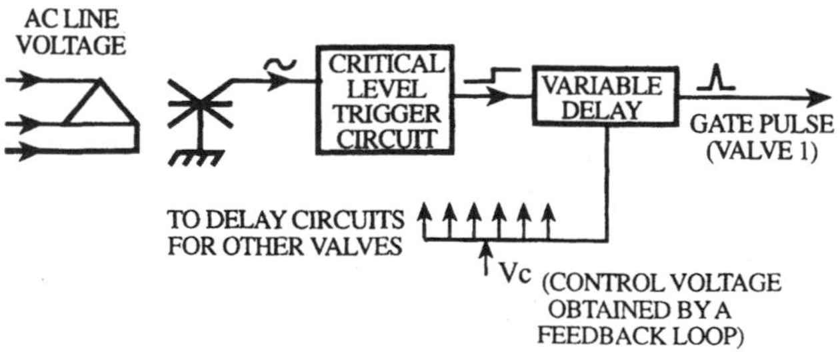

FIGURE 3.11 Constant α control.

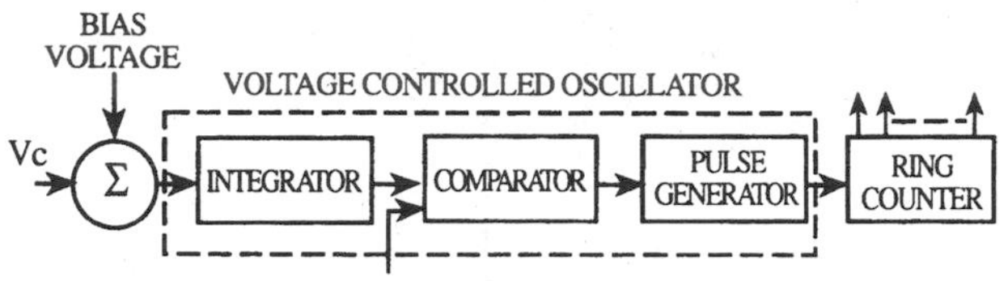

FIGURE 3.12 Equal pulse spacing control.

The control systems used in recent projects are digital-based and much more sophisticated than the earlier versions.

Developments

During the last two or three decades, several developments in HVDC technology have taken place that improved viability of the HVDC transmission. Prior to 1970 mercury arc valves were used for converting from ac to dc and dc to ac. They had several operational problems including frequent arcbacks. Arcback is a random phenomenon that results in failure of a valve to block conduction in the reverse direction. This is most common in the rectifier mode of operation. In rectifier operation, the valve is exposed to inverse voltage for approximately two-thirds of each cycle. Arcbacks result in line-to-line short circuits, and sometimes in three-phase short circuits, which subject the converter transformer and valves to severe stresses.

Thyristors

Thyristor valves were first used for HVDC transmission in the early 1970s, and since then have completely replaced mercury arc valves. The term thyristor valve, carried over from mercury arc valve, is used to refer to an assembly of series and parallel connection of several thyristors to make up the required voltage and current ratings of one arm of the converter. The first test thyristor valve in a HVDC converter station was installed in 1967, replacing a mercury arc valve in the Ygne converter station on the island of Gotland (see Gotland I in Table 3.1). The Eel River back-to-back station in New Brunswick, Canada, commissioned in 1972, was the first all-thyristor HVDC converter station. The voltage and current ratings of thyristors have increased steadily over the last two decades. Figure 3.13 shows the maximum blocking voltage of thyristors from the late 1960s to date. The current

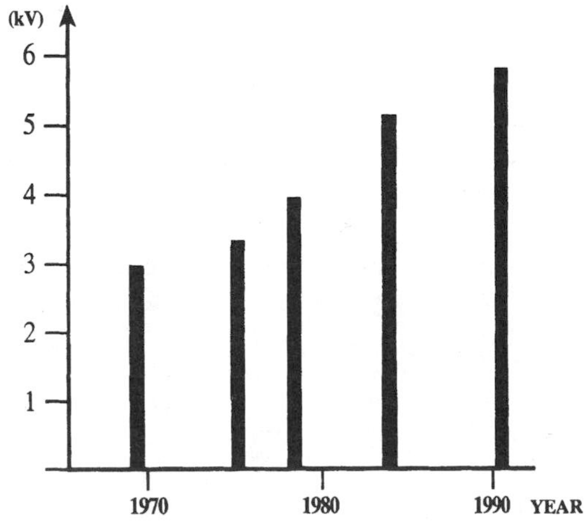

FIGURE 3.13 Maximum blocking voltage development.

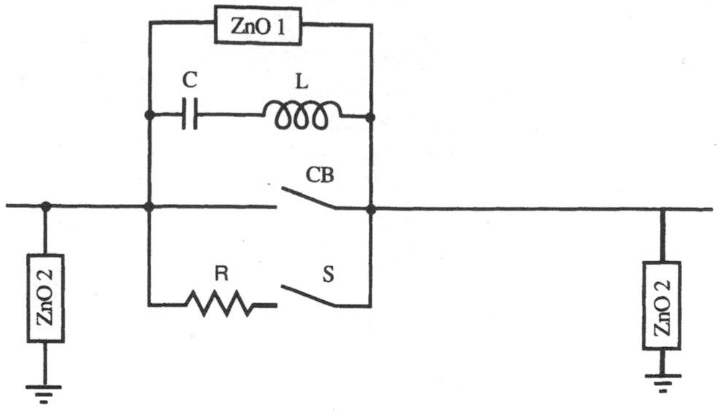

FIGURE 3.14 DC circuit breaker (one module).

ratings have also increased in this period from 1 to 4 kA. Some of the increased current ratings were achieved with large-diameter silicon wafers (presently 100-mm diameter) and with improved cooling systems. Earlier projects used air-cooled thyristors. Water-cooled thyristors are used for all the recent projects.

Other recent developments include direct light-triggered thyristors and gate turn-off (GTO) thyristors. GTO thyristors have some advantages for HVDC converters connected to weak ac systems. They are now available in ratings up to 4.5 kV and 4 kA but have not yet been applied in HVDC systems.

DC Circuit Breakers

Interrupting the current in ac systems is aided by the fact that ac current goes through zero every half-cycle or approximately every 8 ms in a 60-Hz system. The absence of natural current zero in dc makes it difficult to develop a dc circuit breaker. There are three principal problems in designing a dc circuit breaker:

1. Forcing current zero in the interrupting element
2. Controlling the overvoltages caused by large di/dt in a highly inductive circuit
3. Dissipating large amounts of energy (tens of megajoules)

The second and third problems are solved by the application of zinc oxide varistors connected line to ground and across the breaking element. The first is the major problem, and several different solutions are adopted by different manufacturers. Basically, current zero is achieved by inserting a counter voltage into the circuit.

In the circuit shown in Figure 3.14, opening CB (air-blast circuit breaker) causes current to be commutated to the parallel LC circuit. The commutating circuit will be oscillatory, which creates current zero in the circuit breaker. The opening of CB increases the voltage across the commutating circuit, which will be limited by the zinc oxide varistor ZnO_1 by entering into conduction. The resistance R is the closing resistor in series with switch S.

It should be noted that a two-terminal dc system does not need a dc breaker, since the fast converter control response can bring the current quickly to zero. In multiterminal systems, dc breakers can provide additional flexibility of operation. The multiterminal dc systems commissioned so far have not employed dc breakers.

HVDC Light

HVDC converter technology has undergone several stages of evolution. In the 1950s and 1960s, mercury arc rectifiers were used for conversion from ac to dc and *vice versa*. In the 1970s, thyristors of sufficient voltage rating became available to build HVDC converters. Thyristors greatly improved the performance of converters over the mercury arc rectifiers, which suffered from frequent arcbacks. Although the use of thyristors has greatly improved the performance, HVDC transmission with thyristor converters has some disadvantages. The latest development, in the 1990s, is the application of insulated gate bipolar transistor (IGBT) valves and voltage source converters (VSC). This technology is called "HVDC light."

Voltage source converters when applied for HVDC transmission have some advantages over the traditional HVDC transmission. Traditional HVDC converters with thyristors are basically current source converters and require an ac voltage source and reactive power supply for satisfactory operation. The weaker the ac source, the more reactive the power supply required, and reactive power must be regulated over a strict close range. In most cases, shunt capacitor banks/ harmonic filters connected to the ac bus provide a reactive power supply. In some cases when the ac system is very weak, where the short-circuit ratio is less than 2.0, static var controllers (SVC) or synchronous condensers have to be installed to improve the performance.

HVDC light using voltage source converters will not require a reactive power supply and can even operate without an ac source. VSCs with IGBTs use pulse width modulation (PWM) to synthesize an ac voltage source. IGBTs are fired at a frequency of 1 to 3 kHz, which means less harmonic filtering, and smaller inductors and capacitors. HVDC light with VSC converters can also provide voltage support to the ac system. The VSC design is capable of operating in four-quadrant mode, which means polarity reversal is not needed to reverse the direction of power.

VSC Converter Design

Voltage-sourced converter design comprises valves with series-connected IGBTs. Each IGBT is rated 2.5 kV and 1500 A. The converter station is built with modular valve housings, which are electromagnetically shielded. A three-level voltage-sourced converter schematic is shown in Figure 3.15. The midpoint of the VSC is effectively grounded, and is connected to the ac system through series phase reactors and a power transformer. Shunt ac filters are connected to filter the harmonics. The filter size required for HVDC light will be much smaller than a traditional HVDC system.

Three-phase, three-level VSC converter connection and the PWM signal are shown in Figure 3.16.

The converter design also uses compact, dry, high-voltage dc capacitors on the dc side of the VSC for a back-to-back dc station. For transmission applications, dc filters may be required.

VSC-based HVDC light projects to date (2003) are listed in Table 3.2.

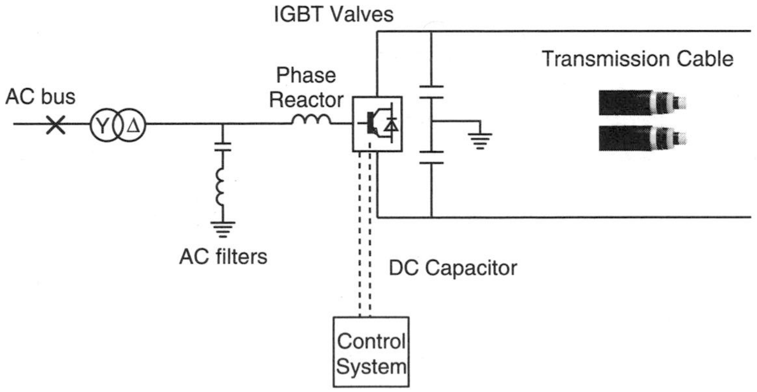

FIGURE 3.15 VSC-based dc transmission (HVDC light).

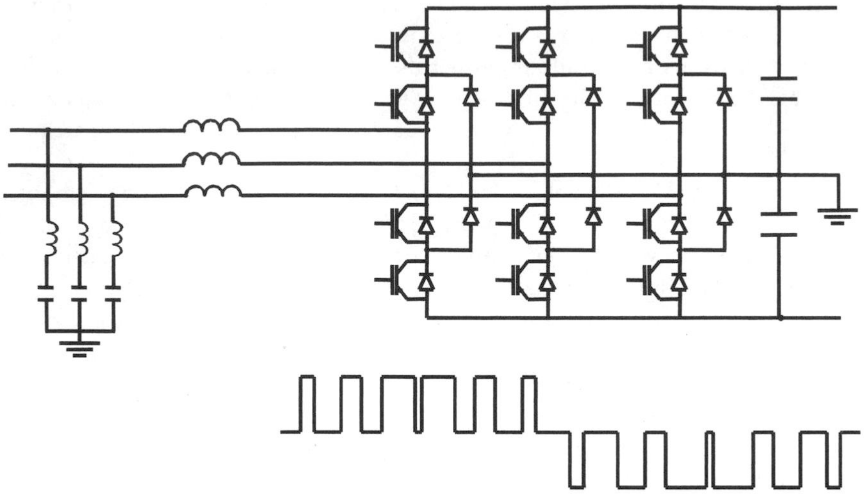

FIGURE 3.16 PWM signal for three-level VSC.

Table 3.2 HVDC Light Projects (2003)

Name of Project	Rating	Reasons for Using VSC
Helljon	3 MW, ± 10 kV, 10 km	Development
Gotland	50 MW, ± 80 kV, 70 km underground cable	Wind power, Environmental, Voltage support, Stabilize ac lines
Tjaereborg	7 MW, ±10 kV, 4 km underground	Wind power testing, Variable frequency, Voltage support
Directlink	3 x 60 MW, +80 kV, 65 km underground	Asynchronous tie, Weak systems, Environmental, Permitting
Eagle Pass	36 MW, back-to-back	Weak system, Voltage support, Asynchronous tie, Black start capability
Cross Sound Cable Interconnection	330 MW, +150 KV, 180 km undersea cable	Asynchronous tie, Weak systems, Environmental, Permitting
Troll A	2×40 MW, +60 kV, 70 km submarine cable	Offshore platform, Var. speed drive, Environmental, Black start

Defining Terms

Asynchronous ac systems: AC systems with either different operating frequencies or that are not in synchronism.

Bipole: DC system with two conductors, one positive and the other negative polarity. Rated voltage of a bipole is expressed as ±100 kV, for example.

Commutation: Process of transferring current from one valve to another.

Commutation angle (overlap angle): Time in electrical degrees from the start to the completion of the commutation process.

Extinction angle: Time in electrical degrees from the instant the current in a valve reaches zero (end of conduction) to the time the valve voltage changes sign and becomes positive.

Firing angle (delay angle): Time in electrical degrees from the instant the valve voltage is positive to the application of firing pulse to the valve (start of conduction).

Pulse number of a converter: Number of ripples in dc voltage per cycle of ac voltage. A three-phase two-way bridge is a six-pulse converter.

Thyristor valve: Assembly of series and parallel connection of several thyristors to make up the required current and voltage ratings of one arm of the converter.

References

J.D. Ainsworth, “The phase locked oscillator: a new control system for controlled static converters,” *IEEE Trans. Power Appar. Syst.*, vol. PAS-87, pp. 859–865, 1968.

M.P. Bahrman, J.G. Johansson, and B.A. Nilsson, “Voltage source converter transmission technologies — the right fit for the application, Panel Session on VSC Technology,” *IEEE Summer Meeting, Toronto*, 2003 (June).

A. Ekstrom and L. Eklund, “HVDC thyristor valve development,” in *Proc. Int. Conf. DC Power Transm. Montreal*, pp. 220–227, 1984.

A. Ekstrom and G. Liss, “A refined HVDC control system,” *IEEE Trans. Power Appar. Syst.*, vol. PAS-89, no. 5, pp. 723–732, 1970.

E. W. Kimbark, *Direct Current Transmission*, Vol. 1, New York: Wiley-Interscience, 1971.

W.F. Long et al., “Considerations for implementing multiterminal dc systems,” *IEEE Trans. Power Appar. Syst.*, pp. 2521–2530, 1985.

K.R. Padiyar, *HVDC Power Transmission Systems — Technology and System Interactions*, New Delhi: Wiley Eastern Limited, 1990.

J. Reeve and P.C.S. Krishnayya, “Unusual current harmonics arising from high-voltage dc transmission,” *IEEE Trans. Power Appar. Syst.*, vol. PAS-87, no. 3, pp. 883–893, 1968.

R.S. Thallam and J. Reeve, “Dynamic analysis of harmonic interaction between ac and dc power systems,” *IEEE Trans. Power Appar. Syst.*, vol. PAS-93, no. 2, pp. 640–646, 1974.

E. Uhlmann, *Power Transmission by Direct Current*, Berlin: Springer-Verlag, 1975.

Further Information

The three textbooks cited under References are excellent for further reading. The IEEE (U.S.) and IEE (U.K.) periodically hold conferences on “dc transmission.” The last IEEE conference was held in 1984 in Montreal, and the IEE conference was held in 1991 (conf. publ. no. 345) in London. Proceedings can be ordered from these organizations.

3.2 Compensation

Mohamed E. El-Hawary

The term *compensation* is used to describe the intentional insertion of reactive power-producing devices, either capacitive or inductive, to achieve a desired effect in the electric power system. The effects include improved voltage profiles, enhanced stability performance, and improved transmission capacity. The devices are connected either in series or in shunt (parallel) at a particular point in the power circuit.

For illustration purposes, we consider the circuit of Figure 3.17, where the link has an impedance of $R + jX$, and it is assumed that $V_1 > V_2$ and V_1 leads V_2. The corresponding phasor diagram for zero R and lagging load current I is shown in Figure 3.18. The approximate relationship between the scalar voltage difference between two nodes in a network and the flow of reactive power Q can be shown to be [Weedy, 1972]

$$\Delta V = \frac{RP_2 + XQ_2}{V_2} \tag{3.18}$$

In most power circuits, $X >> R$ and the voltage difference ΔV determines Q.

The flow of power and reactive power is from A to B when $V_1 > V_2$ and V_1 leads V_2. Q is determined mainly by $V_1 - V_2$. The direction of reactive power can be reversed by making $V_2 > V_1$. It can thus be seen that if a scalar voltage difference exists across a largely reactive link, the reactive power flows toward the node of lower voltage. Looked at from an alternative point of view, if there is a reactive power deficit at a point in an electric network, this deficit has to be supplied from the rest of the circuit and hence the voltage at that point falls.

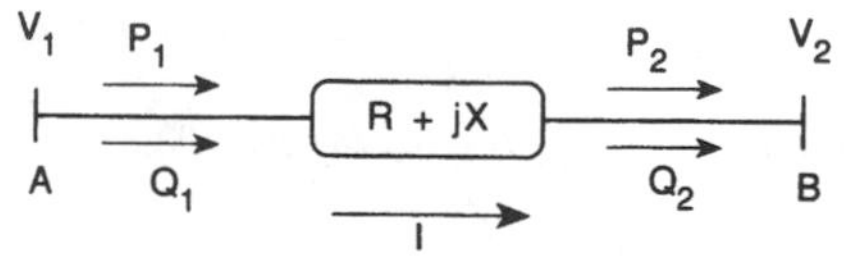

FIGURE 3.17 Two nodes connected by a link.

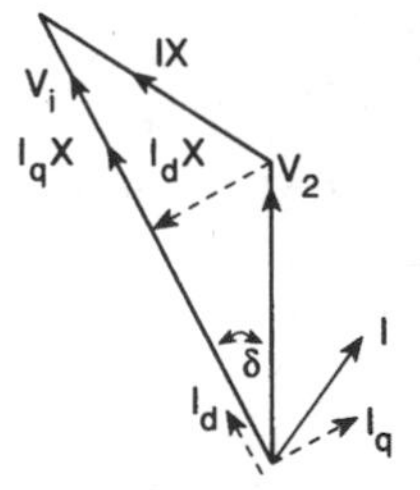

FIGURE 3.18 Phasor diagram for system shown in Figure 3.20.

Of course, a surplus of reactive power generated will cause a voltage rise. This can be interpreted as providing voltage support by supplying reactive power at that point.

Assuming that the link is reactive, i.e., with $R = 0$, then $P_1 = P_2 = P$. In this case, the active power transferred from point A to point B can be shown to be given by [El-Hawary, 1995]

$$P = P_{\max} \sin \delta \tag{3.19}$$

The maximum power transfer $P_{\max}$ is given by

$$P_{\max} = \frac{V_1 V_2}{X} \tag{3.20}$$

It is clear that the power transfer capacity defined by Equation (3.20) is improved if V_2 is increased.

Series Capacitors

Series capacitors are employed to neutralize part of the inductive reactance of a power circuit, as shown in Figure 3.19. From the phasor diagram of Figure 3.20 we see that the load voltage is higher with the capacitor inserted than without the capacitor.

Introducing series capacitors is associated with an increase in the circuit's transmission capacity [from Equation (3.20) with a net reduction in X] and enhanced stability performance as well as improved voltage conditions on the circuit. They are also valuable in other aspects such as:

- Controlling reactive power balance
- Load distribution and control of overall transmission losses

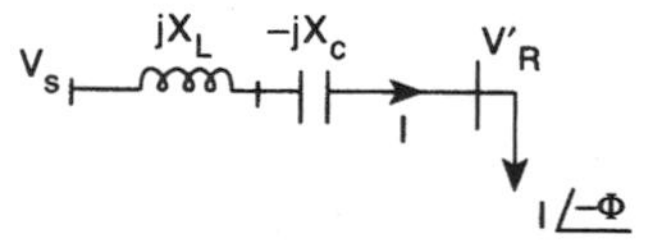

FIGURE 3.19 Line with series capacitor.

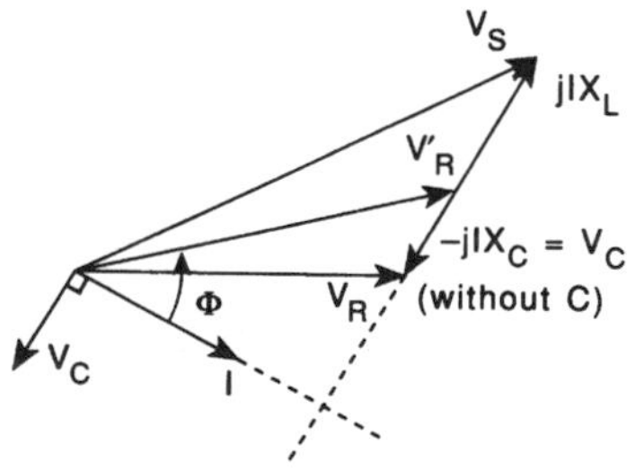

FIGURE 3.20 Phasor diagram corresponding to Figure 3.22.

Series-capacitor compensation delays investments in additional overhead lines for added transmission capacity, which is advantageous from an environmental point of view.

The first worldwide series-capacitor installation was a 33-kV 1.25-MVAR bank on the New York Power & Light system, which was put in service in 1928. Since then, many higher-capacity, higher-voltage installations have been installed in the United States, Canada, Sweden, Brazil, and other countries.

The reduction in a circuit's inductive reactance increases the short-circuit current levels over those for the noncompensated circuit. Care must be taken to avoid exposing series capacitors to such large short-circuit currents, since this causes excessive voltage rise as well as heating that can damage the capacitors. Specially calibrated spark gaps and short-circuiting switches are deployed within a predetermined time interval to avoid damage to the capacitors.

The interaction between a series-capacitor-compensated ac transmission system in electrical **resonance** and a turbine-generator mechanical system in torsional mechanical resonance results in the phenomenon of **subsynchronous resonance** (SSR). Energy is exchanged between the electrical and mechanical systems at one or more natural frequencies of the combined system below the synchronous frequency of the system. The resulting mechanical oscillations can increase until mechanical failure takes place.

Techniques to counteract SSR include the following:

- *Supplementary excitation control:* The subsynchronous current and/or voltage is detected and the excitation current is modulated using high-gain feedback to vary the generator output voltage, which counters the subsynchronous oscillations [see El-Serafi and Shaltout, 1979].
- *Static filters:* These are connected in series with each phase of each main generator. Step-up transformers are employed. The filters are tuned to frequencies that correspond to the power system frequency and the troublesome machine natural modes of oscillations [see Tice and Bowler, 1975].
- *Dynamic filters:* In a manner similar to that of excitation control, the subsynchronous oscillation is detected, and a counter emf is generated by a thyristor cycloconverter or a similar device and injected in the power line through a series transformer [see Kilgore et al., 1975].
- *Bypassing series capacitors:* To limit transient torque buildup, complete or partial bypass with the aid of low set gaps.
- Amortisseur windings on the pole faces of the generator rotors can be employed to improve damping.
- A more recent damping scheme [see Hingorani, 1981] is based on measuring the half-cycle period of the series-capacitor voltage, and if this period exceeds a preset value, the capacitor's charge is dissipated into a resistor shunting the capacitor through two antiparallel thyristors.
- A passive SSR countermeasure scheme [see Edris, 1990] involves using three different combinations of inductive and capacitive elements on the three phases. The combinations will exhibit the required equal degree of capacitive compensation in the three phases at power frequency. At any other frequency, the three combinations will appear as unequal reactances in the three phases. In this manner, asynchronous oscillations will drive unsymmetrical three-phase currents in the generator's armature windings. This creates an mmf with a circular component of a lower magnitude, compared with the corresponding component if the currents were symmetrical. The developed interacting electromagnetic torque will be lower.

Synchronous Compensators

A synchronous compensator is a synchronous motor running without a mechanical load. Depending on the value of excitation, it can absorb or generate reactive power. The losses are considerable compared with static capacitors. When used with a voltage regulator, the compensator can run automatically overexcited at high-load current and underexcited at low-load current. The cost of installation of synchronous compensators is high relative to capacitors.

Shunt Capacitors

Shunt capacitors are used to supply capacitive kVAR to the system at the point where they are connected, with the same effect as an overexcited synchronous condenser, generator, or motor. Shunt capacitors supply reactive power to counteract the out-of-phase component of current required by an inductive load. They are either energized continuously or switched on and off during load cycles.

Figure 3.21(a) displays a simple circuit with shunt capacitor compensation applied at the load side. The line current I_L is the sum of the motor load current I_M and the capacitor current I_c. From the current phasor diagram of Figure 3.21(b), it is clear that the line current is decreased with the insertion of the shunt capacitor. Figure 3.21(c) displays the corresponding voltage phasors. The effect of the shunt capacitor is to reduce the source voltage to V_{s1} from V_{s0}.

From the above considerations, it is clear that shunt capacitors applied at a point in a circuit supplying a load of lagging power factor have the following effects:

- Increase voltage level at the load
- Improve voltage regulation if the capacitor units are properly switched
- Reduce I^2R power loss and I^2X kVAR loss in the system because of reduction in current
- Increase power factor of the source generators
- Decrease kVA loading on the source generators and circuits to relieve an overloaded condition or release capacity for additional load growth
- By reducing kVA load on the source generators, additional active power loading may be placed on the generators if turbine capacity is available
- Reduce demand kVA where power is purchased
- Reduce investment in system facilities per kW of load supplied

To reduce high inrush currents in starting large motors, a capacitor starting system is employed. This maintains acceptable voltage levels throughout the system. The high inductive component of normal reactive starting current is offset by the addition, during the starting period only, of capacitors to the motor bus. This differs from applying capacitors for motor power factor correction.

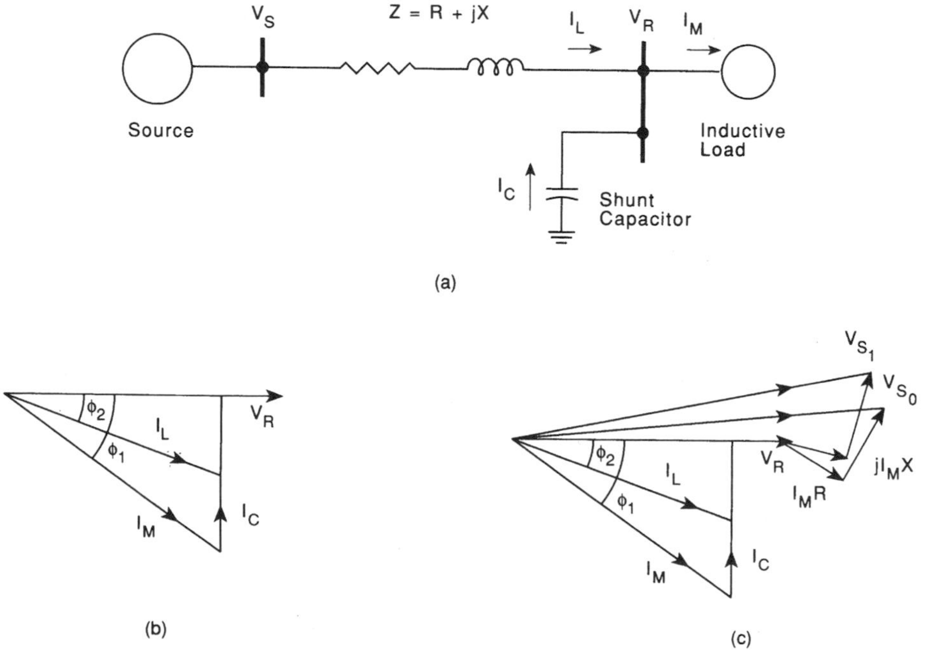

FIGURE 3.21 (a) Shunt-capacitor-compensated load; (b) current phasor diagram; (c) voltage phasor diagram.

When used for voltage control, the action of shunt capacitors is different from that of synchronous condensers, since their reactive power varies as the square of the voltage, whereas the synchronous machine maintains approximately constant kVA for sudden voltage changes. The synchronous condenser has a greater stabilizing effect upon system voltages. The losses of the synchronous condenser are much greater than those of capacitors.

Note that in determining the amount of shunt capacitor kVAR required, since a voltage rise increases the lagging kVAR in the exciting currents of transformer and motors, some additional capacitor kVAR above that based on initial conditions without capacitors may be required to get the desired correction. If the load includes synchronous motors, it may be desirable, if possible, to increase the field currents to these motors.

The following are the relative merits of shunt and series capacitors:

- If the total line reactance is high, series capacitors are very effective.
- If the voltage drop is the limiting factor, series capacitors are effective; also, voltage fluctuations are evened out.
- If the reactive power requirements of the load are small, the series capacitor is of little value.
- If thermal considerations limit the current, then series capacitors are of little value since the reduction in line current associated with them is small.

Applying capacitors with harmonic-generating apparatus on a power system requires considering the potential of an excited harmonic resonance condition. Either a series or a shunt resonance condition may take place. In actual electrical systems utilizing compensating capacitors, either type of resonance or a combination of both can occur if the resonant point happens to be close to one of the frequencies generated by harmonic sources in the system. The outcome can be the flow of excessive amounts of harmonic current or the appearance of excessive harmonic overvoltages, or both. Possible effects of this are excessive capacitor fuse operation, capacitor failure, overheating of other electrical equipment, or telephone interference.

Shunt Reactors

Shunt reactor compensation is usually required under conditions that are the opposite of those requiring shunt capacitor compensation (see Figure 3.22). Shunt reactors are installed to remedy the following situations:

- Overvoltages that occur during low load periods at stations served by long lines as a result of the line's capacitance (Ferranti effect).
- Leading power factors at generating plants resulting in lower transient and steady-state stability limits, caused by reduced field current and the machine's internal voltage. In this case, shunt reactors are usually installed at either side of the generator's step-up transformers.
- Open-circuit line charging kVA requirements in extra-high-voltage systems that exceed the available generation capabilities.

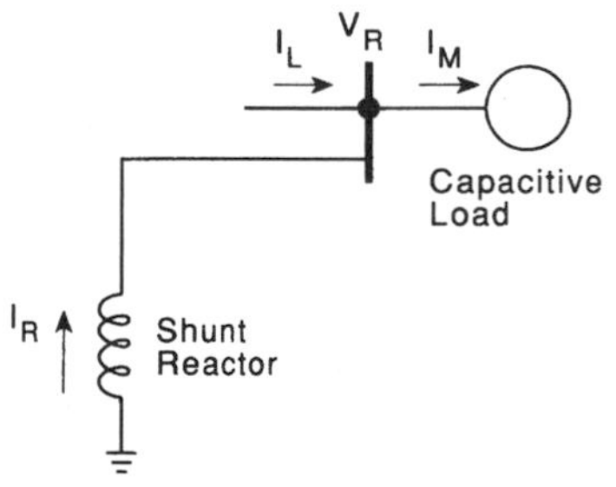

FIGURE 3.22 Shunt-reactor-compensated load.

Coupling from nearby energized lines can cause severe resonant overvoltages across the shunt reactors of unenergized compensated lines.

Static VAR Compensators (SVC)

Advances in **thyristor** technology for power systems applications have lead to the development of the static VAR compensators (SVC). These devices contain standard shunt elements (**reactors,** capacitors) but are controlled by thyristors [El-Hawary, 1995].

Static VAR compensators provide solutions to two types of compensation problems normally encountered in practical power systems [Gyugyi et al., 1978]. The first is load compensation, where the requirements are usually to reduce or cancel the reactive power demand of large and fluctuating industrial loads, such as electric arc furnaces and rolling mills, and to balance the real power drawn from the ac supply lines. These types of heavy industrial loads are normally concentrated in one plant and served from one network terminal, and thus can be handled by a local compensator connected to the same terminal. The second type of compensation is related to voltage support of transmission lines at a given terminal in response to disturbances of both load and generation. The voltage support is achieved by rapid control of the SVC reactance and thus its reactive power output. The main objectives of dynamic VAR compensation are to increase the stability limit of the ac power system, to decrease terminal voltage fluctuations during load variations, and to limit overvoltages subsequent to large disturbances. SVCs are essentially thyristor-controlled reactive power devices.

The two fundamental thyristor-controlled reactive power device configurations are [Olwegard et al., 1981]:

- Thyristor-switched shunt capacitors (TSC): The idea is to split a **capacitor bank** into sufficiently small capacitor steps and switch those steps on and off individually. Figure 3.23(a) shows the concept of the TSC. It offers stepwise control, virtually no transients, and no harmonic generation. The average delay for executing a command from the regulator is half a cycle.
- Thyristor-switched shunt reactors (TCR): In this scheme the fundamental frequency current component through the reactor is controlled by delaying the closing of the thyristor switch with respect to the natural zero crossings of the current. Figure 3.23(b) shows the concept of the TCR. Harmonic currents are generated from the phase-angle-controlled reactor.

The magnitude of the harmonics can be reduced using two methods. In the first, the reactor is split into smaller steps, while only one step is phase-angle controlled. The other reactor steps are either on or off. This decreases the magnitude of all harmonics. The second method involves the 12-pulse arrangement, where two identical connected thyristor-controlled reactors are used, one operated from a wye-connected secondary winding, the other from a delta-connected winding of a step-up transformer. TCR units are characterized by continuous control, and there is a maximum of one half-cycle delay for executing a command from the regulator.

In many applications, the arrangement of an SVC consists of a few large steps of thyristor-switched capacitor and one or two thyristor-controlled reactors, as shown in Figure 3.23(c). The following are some practical schemes.

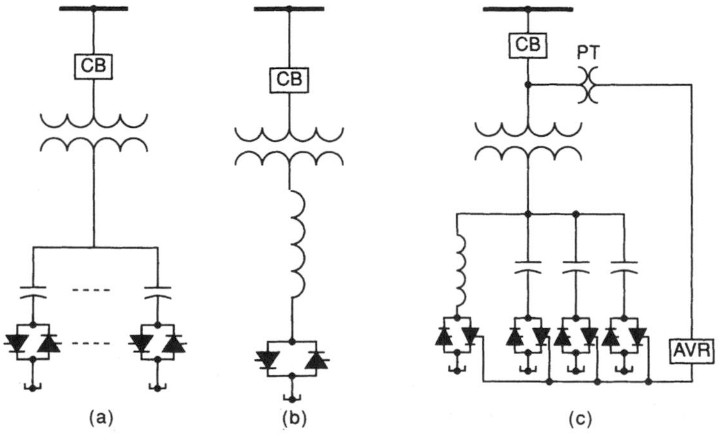

FIGURE 3.23 Basic static VAR compensator configurations. (a) Thyristor-switched shunt capacitors (TSC); (b) thyristor-switched shunt reactors (TCR); (c) combined TSC/TCR.

Fixed-Capacitor, Thyristor-Controlled Reactor (FC-TCR) Scheme

This scheme was originally developed for industrial applications, such as arc furnace "flicker" control [Gyugyi and Taylor, 1980]. It is essentially a TCR (controlled by a delay angle α) in parallel with a fixed capacitor. Figure 3.24 shows a basic fixed-capacitor, thyristor-controlled reactor-type compensator and associated waveforms. Figure 3.25 displays the steady-state reactive power versus terminal voltage characteristics of a static VAR compensator. In the figure, B_C is the imaginary part of the admittance of the capacitor C, and B_L is the imaginary part of the equivalent admittance of the reactor L at delay angle α. The relation between the output VARs and the applied voltage is linear over the voltage band of regulation. In practice, the fixed capacitor is usually replaced by a filter network that has the required capacitive reactance at the power system frequency but exhibits a low impedance at selected frequencies to absorb troublesome harmonics.

The behavior and response of the FC-TCR type of compensator under large disturbances is uncontrollable, at least during the first few cycles following the disturbance. The resulting voltage transients are essentially determined by the fixed capacitor and the power system impedance. This can lead to overvoltage and resonance problems.

At zero VAR demand, the capacitive and reactive VARs cancel out, but the capacitor bank's current is circulated through the reactor bank via the thyristor switch. As a result, this configuration suffers from no load (standby) losses. The losses decrease with increasing the capacitive VAR output and, conversely, increase with increasing the inductive VAR output.

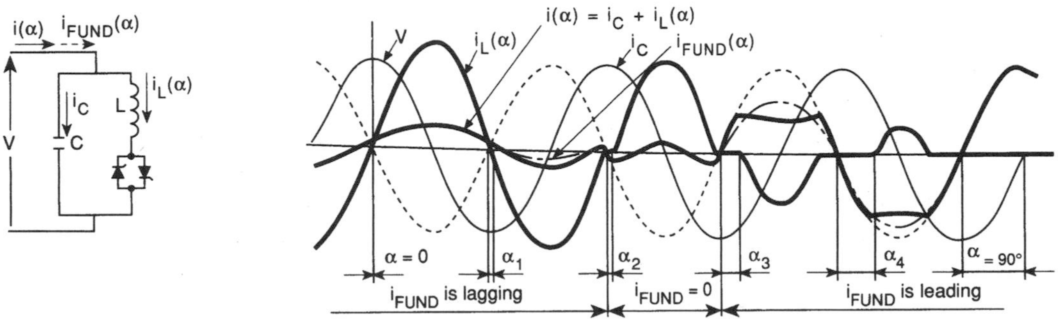

FIGURE 3.24 Basic fixed-capacitor, thyristor-controlled reactor-type compensator and associated waveforms.

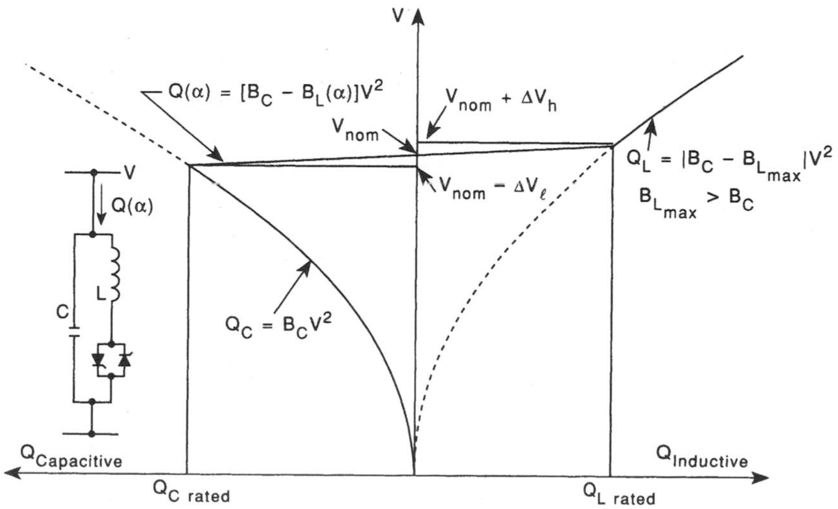

FIGURE 3.25 The steady-state reactive power versus terminal voltage characteristics of a static VAR compensator.

Thyristor-Switched Capacitor, Thyristor-Controlled Reactor (TSC-TCR) Scheme

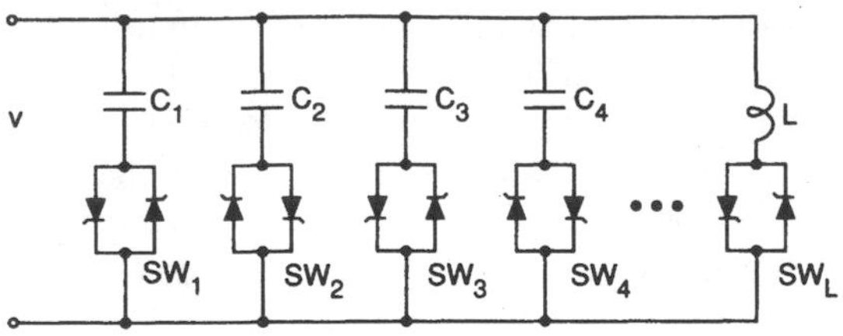

FIGURE 3.26 Basic thyristor-switched capacitor, thyristor-controlled reactor-type compensator

This hybrid compensator was developed specifically for utility applications to overcome the disadvantages of the FC-TCR compensators (behavior under large disturbances and loss characteristic). Figure 3.26 shows a basic circuit of this compensator. It consists in general of a thyristor-controlled reactor bank (or banks) and a number of capacitor banks, each in series with a solid-state switch, which is composed of either a reverse-parallel-connected thyristor pair or a thyristor in reverse parallel with a diode. The reactor's switch is composed of a reverse-parallel-connected thyristor pair that is capable of continuously controlling the current in the reactor from zero to maximum rated current.

The total capacitive range is divided into n operating intervals, where n is the number of capacitor banks in the compensator. In the first interval one capacitor bank is switched in, and at the same time the current in the TCR bank is adjusted so that the resultant VAR output from capacitor and reactor matches the VAR demand. In the ith interval the output is controllable in the range $[(i-1)\text{VAR}_{\max}/n]$ to $(i\text{VAR}_{\max}/n)$ by switching in the ith capacitor bank and using the TCR bank to absorb the surplus capacitive VARs. This scheme can be considered as a conventional FC-TCR, where the rating of the reactor bank is kept relatively small ($1/n$ times the maximum VAR output) and the value of the capacitor bank is changed in discrete steps so as to keep the operation of the reactor bank within its normal control range.

The losses of the TSC-TCR compensator at zero VARs output are inherently low, and they increase in proportion to the VAR output.

The mechanism by which SVCs introduce damping into the system can be explained as a result of the change in system voltage due to switching of a capacitor/reactor. The electrical power output of the generators is changed immediately due to the change in power transfer capability and the change in load power requirements.

Among the early applications of SVC for power system damping is the application to the Scandinavian system as discussed in Olwegard et al. [1981]. More recently, SVC control for damping of system oscillations based on local measurements has been proposed. The scheme uses phase-angle estimates based on voltage and power measurements at the SVC location as the control signal [see Lerch et al., 1991].

For a general mathematical model of an SVC and an analysis of its stabilizing effects, see Hammad [1986]. Representing the SVC in transient analysis programs is an important consideration [see Gole and Sood, 1990; Lefebvre and Gerin-Lajoie, 1992].

It is important to recognize that applying static VAR compensators to series-clompensated ac transmission lines results in three distinct resonant modes [Larsen et al., 1990]:

- Shunt-capacitance resonance involves energy exchange between the shunt capacitance (line charging plus any power factor correction or SVCs) and the series inductance of the lines and the generator.
- Series-line resonance involves energy exchange between the series capacitor and the series inductance of the lines, transformers, and generators. The resonant frequency will depend on the level of series compensation.
- Shunt-reactor resonance involves energy exchange between shunt reactors at the intermediate substations of the line and the series capacitors.

The applications of SVCs are part of the broader area of flexible ac transmission systems (FACTS) [Hingorani, 1993]

Defining Terms

Capacitor bank: An assembly at one location of capacitors and all necessary accessories, such as switching equipment, protective equipment, and controls, required for a complete operating installation.

Reactor: A device whose primary purpose is to introduce reactance into a circuit. Inductive reactance is frequently abbreviated inductor.

Resonance: The enhancement of the response of a physical system to a periodic excitation when the excitation frequency is equal to a natural frequency of the system.

Shunt: A device having appreciable impedance connected in parallel across other devices or apparatus and diverting some of the current from it. Appreciable voltage exists across the shunted device or apparatus, and an appreciable current may exist in it.

Shunt reactor: A reactor intended for connection in shunt to an electric system to draw inductive current.

Subsynchronous resonance: An electric power system condition where the electric network exchanges energy with a turbine generator at one or more of the natural frequencies of the combined system below the synchronous frequency of the system.

Thyristor: A bistable semiconductor device comprising three or more junctions that can be switched from the off state to the on state, or vice versa, such switching occurring within at least one quadrant of the principal voltage-current characteristic.

References

I.S. Benko, B. Bhargava, and W.N. Rothenbuhler, "Prototype NGH subsynchronous resonance damping scheme, part II—Switching and short circuit tests," *IEEE Trans. Power Syst.*, vol. 2, pp. 1040–1049, 1987.

L.E. Bock and G.R. Mitchell, "Higher line loadings with series capacitors," *Transmission Magazine*, March 1973.

E.W. Bogins and H.T. Trojan, "Application and design of EHV shunt reactors," *Transmission Magazine*, March 1973.

C.E. Bowler, D.N. Ewart, and C. Concordia, "Self excited torsional frequency oscillations with series capacitors," *IEEE Trans. Power Appar. Syst.*, vol. 93, pp. 1688–1695, 1973.

G.D. Brewer, H.M. Rustebakke, R.A. Gibley, and H.O. Simmons, "The use of series capacitors to obtain maximum EHV transmission capability," *IEEE Trans. Power Appar. Syst.*, vol. 83, pp. 1090–1102, 1964.

C. Concordia, "System compensation, an overview," *Transmission Magazine*, March 1973.

S.E.M. de Oliveira, I. Gardos, and E.P. Fonseca, "Representation of series capacitors in electric power system stability studies," *IEEE Trans. Power Syst.*, vol. 6, no. 3, pp. 1119–1125, 1991.

A.A. Edris, "Series compensation schemes reducing the potential of subsynchronous resonance," *IEEE Trans. Power Syst.*, vol. 5, no. 1, pp. 219–226, 1990.

M.E. El-Hawary, *Electrical Power Systems: Design and Analysis*, Piscataway, N.J.: IEEE Press, 1995.

A.M. El-Serafi and A. A. Shaltout, "Damping of SSR Oscillations by Excitation Control," *IEEE PES Summer Meeting*, Vancouver, 1979.

A.M. Gole and V.K. Sood, "A static compensator model for use with electromagnetic transients simulation programs," *IEEE Trans. Power Delivery*, vol. 5, pp. 1398–1407, 1990.

L. Gyugyi, R.A. Otto, and T.H. Putman, "Principles and applications of static thyristor-controlled shunt compensators," *IEEE Trans. Power Appar. Syst.*, vol. PAS-97, pp. 1935–1945, 1978.

L. Gyugyi and E.R. Taylor, Jr., "Characteristics of static thyristor-controlled shunt compensators for power transmission system applications," *IEEE Trans. Power Appar. Syst.*, vol. PAS-99, pp. 1795–1804, 1980.

A.E. Hammad, "Analysis of power system stability enhancement by static VAR compensators," *IEEE Trans. Power Syst.*, vol. 1, pp. 222–227, 1986.

J.F. Hauer, "Robust damping controls for large power systems," *IEEE Control Systems Magazine*, pp. 12–18, January 1989.

R.A. Hedin, K.B. Stump, and N.G. Hingorani, "A new scheme for subsynchronous resonance damping of torsional oscillations and transient torque—Part II," *IEEE Trans. Power Appar. Syst.*, vol. PAS-100, pp. 1856–1863, 1981.

N.G. Hingorani, "A new scheme for subsynchronous resonance damping of torsional oscillations and transient torque—Part I," *IEEE Trans. Power Appar. Syst.*, vol. PAS-100, pp. 1852–1855, 1981.

N.G. Hingorani, B. Bhargava, G.F. Garrigue, and G.D. Rodriguez, "Prototype NGH subsynchronous resonance damping scheme, part I—Field installation and operating experience," *IEEE Trans. Power Syst.*, vol. 2, pp. 1034–1039, 1987.

N.G. Hingorani, "Flexible AC transmission," *IEEE Spectrum*. vol. 30, no. 4, pp. 40–45, 1993.

IEEE Subsynchronous Resonance Working Group, "Proposed terms and definitions for subsynchronous oscillations," *IEEE Trans. Power Appar. Syst.*, vol. PAS-99, pp. 506–511, 1980.

IEEE Subsynchronous Resonance Working Group, "Countermeasures to subsynchronous resonance problems," *IEEE Trans. Power Appar. Syst.*, vol. PAS-99, pp. 1810–1818, 1980.

IEEE Subsynchronous Resonance Working Group, "Series capacitor controls and settings as countermeasures to subsynchronous resonance," *IEEE Trans. Power Appar. Syst.*, vol. PAS-101, pp. 1281–1287, June 1982.

G. Jancke, N. Fahlen, and O. Nerf, "Series capacitors in power systems," *IEEE Transactions on Power Appar. Syst.*, vol. PAS-94, pp. 915–925, May/June 1975.

L.A. Kilgore, D.G. Ramey, and W.H. South, "Dynamic filter and other solutions to the subsynchronous resonance problem," *Proceedings of the American Power Conference*, vol. 37, p. 923, 1975.

E.W. Kimbark, *Power System Stability*, vol. I, *Elements of Stability Calculations*, New York: Wiley, 1948.

E.W. Kimbark, "Improvement of system stability by switched series capacitors," *IEEE Trans. Power Appar. Syst.*, vol. 85, pp. 180–188, February 1966.

J.J. LaForest, K.W. Priest, Ramirez, and H. Nowak, "Resonant voltages on reactor compensated extra-high-voltage lines," *IEEE Trans. Power Appar. Syst.*, vol. PAS-91, pp. 2528–2536, November/December 1972.

E.V. Larsen, D.H. Baker, A.F. Imece, L. Gerin-Lajoie, and G. Scott, "Basic aspects of applying SVC's to series-compensated ac transmission lines," *IEEE Trans. Power Delivery*, vol. 5, pp. 1466–1472, July 1990.

S. Lefebvre and L. Gerin-Lajoie, "A static compensator model for the EMTP," *IEEE Trans. Power Systems*, vol. 7, no. 2, pp. 477–486, May 1992.

E. Lerch, D. Povh, and L. Xu, "Advanced SVC control for damping power system oscillations," *IEEE Trans. Power Syst.*, vol. 6, pp. 524–531, May 1991.

S.M. Merry and E.R. Taylor, "Overvoltages and harmonics on EHV systems," *IEEE Trans. Power Appar. Syst.*, vol. PAS-91, pp. 2537–2544, November/December 1972.

A. Olwegard, K. Walve, G. Waglund, H. Frank, and S. Torseng, "Improvement of transmission capacity by thyristor controlled reactive power," *IEEE Trans. Power Appar. Syst.*, vol. PAS-100, pp. 3930–3939, 1981.

J.B. Tice and C.E.J. Bowler, "Control of phenomenon of subsynchronous resonance," *Proceedings of the American Power Conference*, vol. 37, pp. 916–922, 1975.

B.M. Weedy, *Electric Power Systems*, London: Wiley, 1972.

Further Information

An excellent source of information on the application of capacitors on power systems is the Westinghouse *Transmission and Distribution* book, published in 1964. A most readable treatment of improving system stability by series capacitors is given by Kimbark's paper [1966]. Jancke et al. [1975] give a detailed discussion of experience with the 400-kV series-capacitor compensation installations on the Swedish system and aspects of the protection system. Hauer [1989] presents a discussion of practical stability controllers that manipulate series and/or shunt reactance.

An excellent summary of the state of the art in static VAR compensators is the record of the IEEE Working Group symposium conducted in 1987 on the subject (see IEEE Publication 87TH0187-5-PWR, Application of Static VAR Systems for System Dynamic Performance).

For state-of-the-art coverage of subsynchronous resonance and countermeasures, two symposia are available: IEEE Publication 79TH0059–6-PWR, State-of-the-Art Symposium—Turbine Generator Shaft Torsionals, and IEEE Publication 81TH0086–9-PWR, Symposium on Countermeasures for Subsynchronous Resonance.

3.3 Fault Analysis in Power Systems

Charles A. Gross

A **fault** in an electrical power system is the unintentional and undesirable creation of a conducting path (a *short circuit*) or a blockage of current (an *open circuit*). The short-circuit fault is typically the most common and is usually implied when most people use the term *fault*. We restrict our comments to the short-circuit fault.

The causes of faults include lightning, wind damage, trees falling across lines, vehicles colliding with towers or poles, birds shorting out lines, aircraft colliding with lines, vandalism, small animals entering switchgear, and line breaks due to excessive ice loading. Power system faults may be categorized as one of four types: single line-to-ground, line-to-line, double line-to-ground, and balanced three-phase. The first three types constitute severe unbalanced operating conditions.

It is important to determine the values of system voltages and currents during faulted conditions so that protective devices may be set to detect and minimize their harmful effects. The time constants of the associated transients are such that sinusoidal steady-state methods may still be used. The method of symmetrical components is particularly suited to fault analysis.

Our objective is to understand how symmetrical components may be applied specifically to the four general fault types mentioned and how the method can be extended to any unbalanced three-phase system problem.

Note that phase values are indicated by subscripts, *a, b, c;* sequence (symmetrical component) values are indicated by subscripts 0, 1, 2. The transformation is defined by

$$\begin{bmatrix} \bar{V}_a \\ \bar{V}_b \\ \bar{V}_c \end{bmatrix} = \begin{bmatrix} 1 & 1 & 1 \\ 1 & a^2 & a \\ 1 & a & a^2 \end{bmatrix} \begin{bmatrix} \bar{V}_0 \\ \bar{V}_1 \\ \bar{V}_2 \end{bmatrix} = [T] \begin{bmatrix} \bar{V}_0 \\ \bar{V}_1 \\ \bar{V}_2 \end{bmatrix}$$

$$\begin{bmatrix} \bar{V}_0 \\ \bar{V}_1 \\ \bar{V}_2 \end{bmatrix} = \frac{1}{3} \begin{bmatrix} 1 & 1 & 1 \\ 1 & a & a^2 \\ 1 & a^2 & a \end{bmatrix} \begin{bmatrix} \bar{V}_0 \\ \bar{V}_1 \\ \bar{V}_2 \end{bmatrix} = [T]^{-1} = \begin{bmatrix} \bar{V}_a \\ \bar{V}_b \\ \bar{V}_c \end{bmatrix}$$

Simplifications in the System Model

Certain simplifications are possible and usually employed in fault analysis.

- Transformer magnetizing current and core loss will be neglected.
- Line shunt capacitance is neglected.
- Sinusoidal steady-state circuit analysis techniques are used. The so-called dc offset is accounted for by using correction factors.
- Prefault voltage is assumed to be $1\angle 0°$ per-unit. One per-unit voltage is at its nominal value prior to the application of a fault, which is reasonable. The selection of zero phase is arbitrary and convenient. Prefault load current is neglected.

For hand calculations, neglect series resistance is usually neglected (this approximation will not be necessary for a computer solution). Also, the only difference in the positive and negative sequence networks is introduced by the machine impedances. If we select the subtransient reactance X''_d for the positive sequence reactance, the difference is slight (in fact, the two are identical for nonsalient machines). The simplification is important, since it reduces computer storage requirements by roughly one-third. Circuit models for generators, lines, and transformers are shown in Figures 3.27, 3.28, and 3.29, respectively.

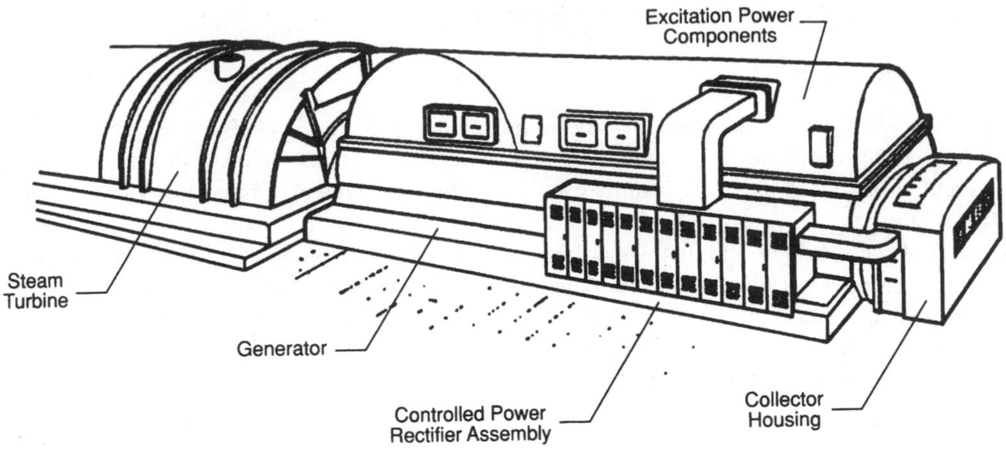

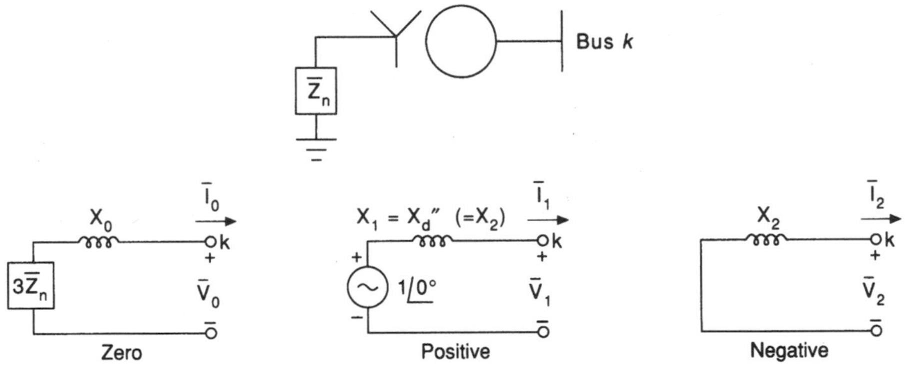

FIGURE 3.27 Generator sequence circuit models.

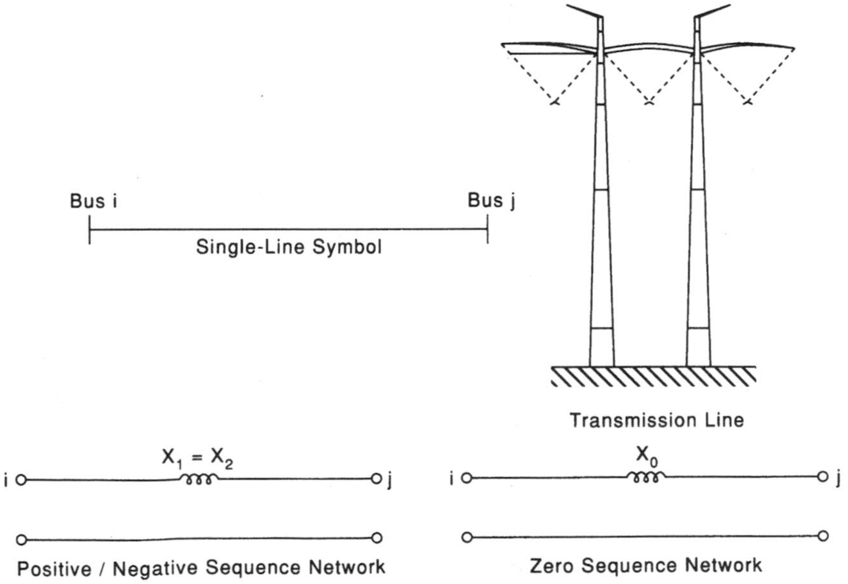

FIGURE 3.28 Line sequence circuit models.

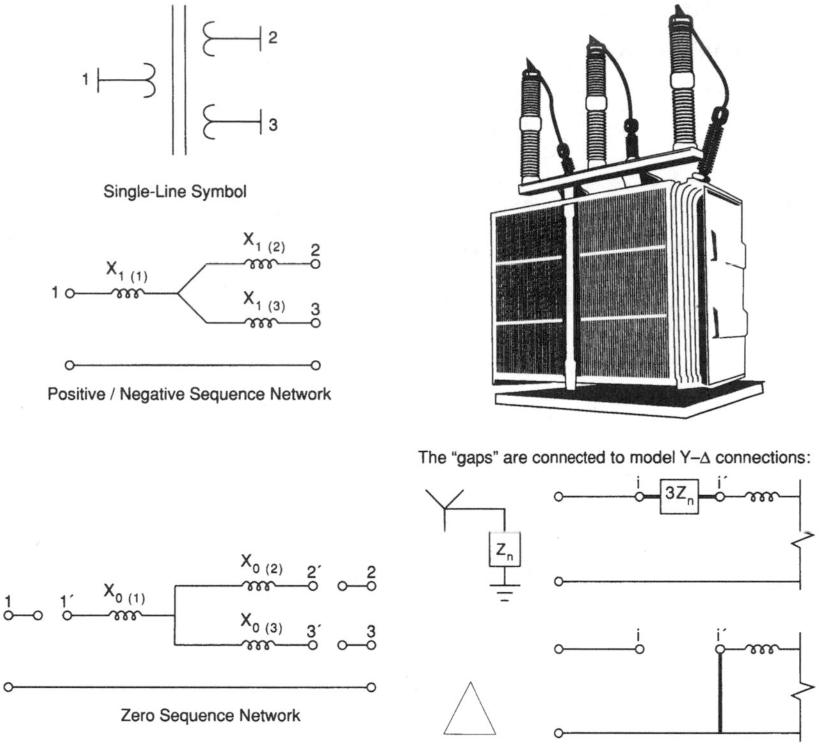

FIGURE 3.29 Transformer sequence circuit models.

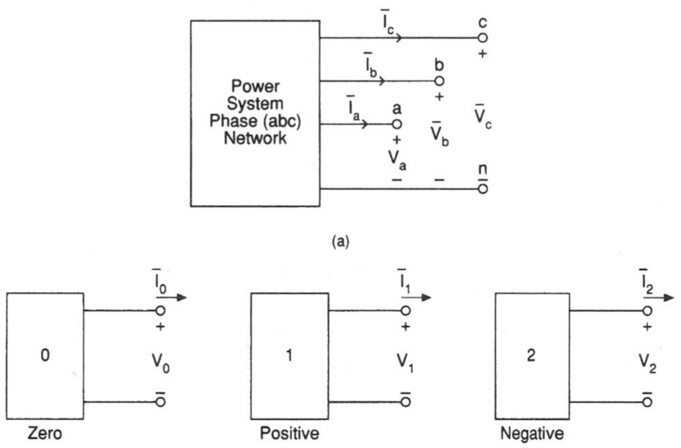

FIGURE 3.30 General fault port in an electric power system. (a) General fault port in phase (*abc*) coordinates; (b) corresponding fault ports in sequence (012) coordinates.

Our basic approach to the problem is to consider the general situation suggested in Figure 3.30(a). The general terminals brought out are for purposes of external connections that will simulate faults. Note carefully the positive assignments of phase quantities. Particularly note that the currents flow *out of* the system. We can construct general *sequence* equivalent circuits for the system, and such circuits are indicated in Figure 3.30(b). The ports indicated correspond to the general three-phase entry port of Figure 3.30(a). The positive sense of sequence values is compatible with that used for phase values.

The Four Basic Fault Types

The Balanced Three-Phase Fault

Imagine the general three-phase access port terminated in a fault impedance ($\bar{Z}_f$) as shown in Figure 3.31(a). The terminal conditions are

$$\begin{bmatrix} \bar{V}_a \\ \bar{V}_b \\ \bar{V}_c \end{bmatrix} = \begin{bmatrix} \bar{Z}_f & 0 & 0 \\ 0 & \bar{Z}_f & 0 \\ 0 & 0 & \bar{Z}_f \end{bmatrix} \begin{bmatrix} \bar{I}_a \\ \bar{I}_b \\ \bar{I}_c \end{bmatrix}$$

Transforming to $[Z_{012}]$,

$$[Z_{012}] = [T]^{-1} \begin{bmatrix} \bar{Z}_f & 0 & 0 \\ 0 & \bar{Z}_f & 0 \\ 0 & 0 & \bar{Z}_f \end{bmatrix} [T] = \begin{bmatrix} \bar{Z}_f & 0 & 0 \\ 0 & \bar{Z}_f & 0 \\ 0 & 0 & \bar{Z}_f \end{bmatrix}$$

The corresponding network connections are given in Figure 3.32(a). Since the zero and negative sequence networks are passive, only the positive sequence network is nontrivial.

$$\bar{V}_0 = \bar{V}_2 = 0 \tag{3.21}$$

$$\bar{I}_0 = \bar{I}_2 = 0 \tag{3.22}$$

$$\bar{V}_1 = \bar{Z}_f \bar{I}_1 \tag{3.23}$$

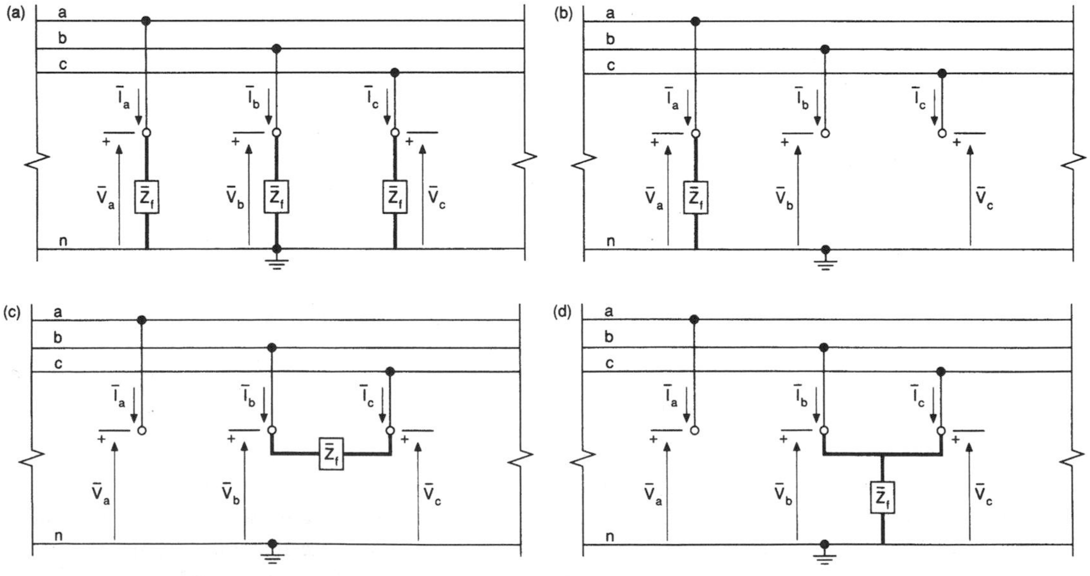

FIGURE 3.31 Fault types. (a) Three-phase fault; (b) single phase-to-ground fault; (c) phase-to-phase fault; (d) double phase-to-ground fault.

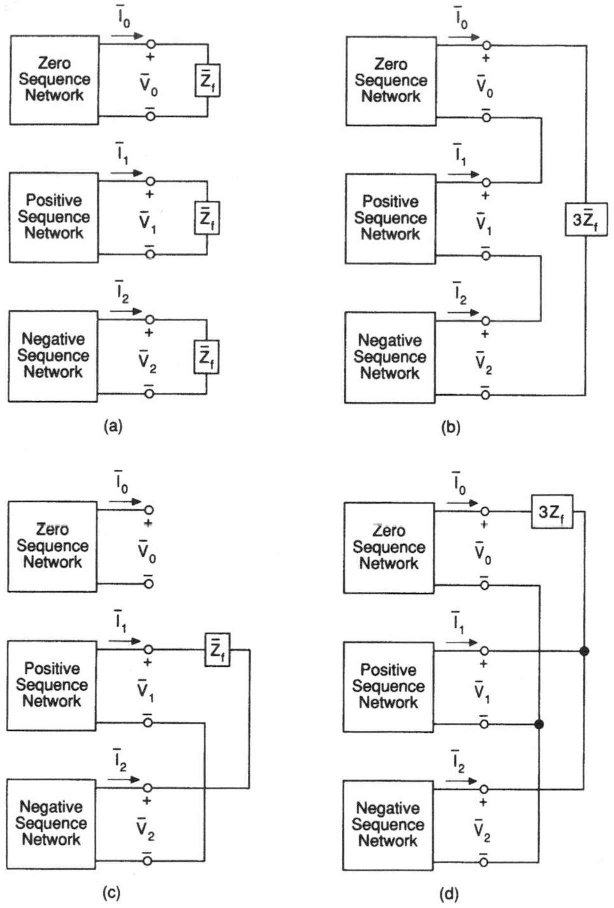

FIGURE 3.32 Sequence network terminations for fault types. (a) Balanced three-phase fault; (b) single phase-to-ground fault; (c) phase-to-phase fault; (d) double phase-to-ground fault.

The Single Phase-to-Ground Fault

Imagine the general three-phase access port terminated as shown in Figure 3.31(b). The terminal conditions are

$$\bar{I}_b = 0 \quad \bar{I}_c = 0 \quad \bar{V}_a = \bar{I}_a \bar{Z}_f$$

Therefore

$$\bar{I}_0 + a^2\bar{I}_1 + a\bar{I}_2 = \bar{I}_0 + a\bar{I}_1 + a^2\bar{I}_2 = 0$$

or

$$\bar{I}_1 = \bar{I}_2$$

Also

$$\bar{I}_b = \bar{I}_0 + a^2\bar{I}_1 + a\bar{I}_2 = \bar{I}_0 + (a^2 + a)\bar{I}_1 = 0$$

or

$$\bar{I}_0 = \bar{I}_1 = \bar{I}_2 \tag{3.24}$$

Furthermore it is required that

$$\begin{aligned} \bar{V}_a &= \bar{Z}_f \bar{I}_a \\ \bar{V}_0 + \bar{V}_1 + \bar{V}_2 &= 3\bar{Z}_f \bar{I}_1 \end{aligned} \tag{3.25}$$

In general then, Equations (3.24) and (3.25) must be simultaneously satisfied. These conditions can be met by interconnecting the sequence networks as shown in Figure 3.32(b).

The Phase-to-Phase Fault

Imagine the general three-phase access port terminated as shown in Figure 3.31(c). The terminal conditions are such that we may write

$$\bar{I}_0 = 0 \quad \bar{I}_b = -\bar{I}_c \quad \bar{V}_b = \bar{Z}_f \bar{I}_b + \bar{V}_c$$

It follows that

$$\bar{I}_0 = \bar{I}_1 = \bar{I}_2 = 0 \tag{3.26}$$

$$\bar{I}_0 = 0 \tag{3.27}$$

$$\bar{I}_1 = -\bar{I}_2 \tag{3.28}$$

In general then, Equations (3.26), (3.27), and (3.28) must be simultaneously satisfied. The proper interconnection between sequence networks appears in Figure 3.32(c).

The Double Phase-to-Ground Fault

Consider the general three-phase access port terminated as shown in Figure 3.31(d). The terminal conditions indicate

$$\bar{I}_a = 0 \quad \bar{V}_b = \bar{V}_c \quad \bar{V}_b = (\bar{I}_b + \bar{I}_c)\bar{Z}_f$$

It follows that

$$\bar{I}_0 = \bar{I}_1 = \bar{I}_2 = \bar{0} \tag{3.29}$$

$$\bar{V}_1 = \bar{V}_2 \tag{3.30}$$

and

$$\bar{V}_0 = \bar{V}_1 = 3\bar{Z}_f \bar{I}_0 \tag{3.31}$$

For the general double phase-to-ground fault, Equations (3.29), (3.30), and (3.31) must be simultaneously satisfied. The sequence network interconnections appear in Figure 3.32(d).

An Example Fault Study

Case: EXAMPLE SYSTEM
Run:
System has data for 2 Line(s); 2 Transformer(s);
4 Bus(es); and 2 Generator(s)

Transmission Line Data

Line	Bus	Bus	Seq	R	X	B	Srat
1	2	3	pos	0.00000	0.16000	0.00000	1.0000
			zero	0.00000	0.50000	0.00000	
2	2	3	pos	0.00000	0.16000	0.00000	1.0000
			zero	0.00000	0.50000	0.00000	

Transformer Data

Transformer	HV Bus	LV Bu	Seq	R	X	C	Srat
1	2	1	pos	0.00000	0.05000	1.00000	1.0000
	Y	Y	zero	0.00000	0.05000		
2	3	4	pos	0.00000	0.05000	1.00000	1.0000
	Y	D	zero	0.00000	0.05000		

Generator Data

No.	Bus	Srated	Ra	Xd''	Xo	Rn	Xn	Con
1	1	1.0000	0.0000	0.200	0.0500	0.0000	0.0400	Y
2	4	1.0000	0.0000	0.200	0.0500	0.0000	0.0400	Y

Zero Sequence [Z] Matrix

0.0 + j(0.1144)	0.0 + j(0.0981)	0.0 + j(0.0163)	0.0 + j(0.0000)
0.0 + j(0.0981)	0.0 + j(0.1269)	0.0 + j(0.0212)	0.0 + j(0.0000)
0.0 + j(0.0163)	0.0 + j(0.0212)	0.0 + j(0.0452)	0.0 + j(0.0000)
0.0 + j(0.0000)	0.0 + j(0.0000)	0.0 + j(0.0000)	0.0 + j(0.1700)

Positive Sequence [Z] Matrix

0.0 + j(0.1310)	0.0 + j(0.1138)	0.0 + j(0.0862)	0.0 + j(0.0690)
0.0 + j(0.1138)	0.0 + j(0.1422)	0.0 + j(0.1078)	0.0 + j(0.0862)
0.0 + j(0.0862)	0.0 + j(0.1078)	0.0 + j(0.1422)	0.0 + j(0.1138)
0.0 + j(0.0690)	0.0 + j(0.0862)	0.0 + j(0.1138)	0.0 + j(0.1310)

The single-line diagram and sequence networks are presented in Figure 3.33.

Suppose bus 3 in the example system represents the fault location and $\bar{Z}_f = 0$. The positive sequence circuit can be reduced to its Thévenin equivalent at bus 3:

$$E_{\mathrm{T1}} = 1.0\angle 0^\circ \qquad \bar{Z}_{\mathrm{T1}} = j0.1422$$

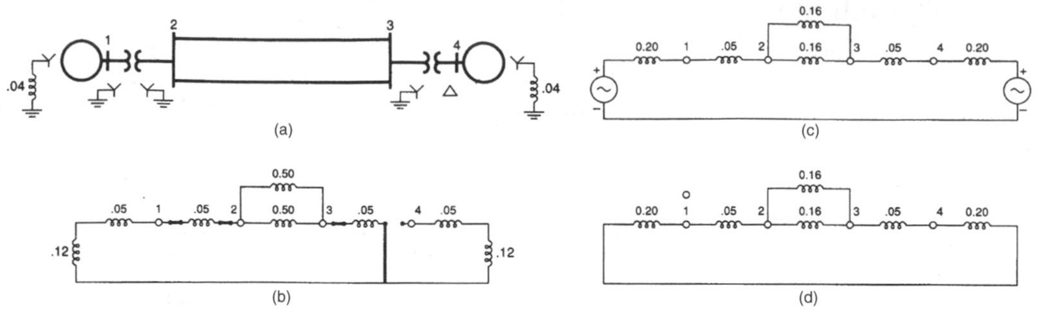

FIGURE 3.33 Example system. (a) Single-line diagram; (b) zero sequence network; (c) positive sequence network; (d) negative sequence network.

Similarly, the negative and zero sequence Thévenin elements are

$$\bar{E}_{T2} = 0 \quad \bar{Z}_{T2} = j0.1422$$
$$\bar{E}_{T0} = 0 \quad Z_{T0} = j0.0452$$

The network interconnections for the four fault types are shown in Figure 3.34. For each of the fault types, compute the currents and voltages at the faulted bus.

Balanced Three-Phase Fault

The sequence networks are shown in Figure 3.34(a). Obviously,

$$\bar{V}_0 = \bar{I}_0 = \bar{V}_2 = \bar{I}_2 = 0$$

$$\bar{I}_1 = \frac{1\angle 0^\circ}{j0.1422} = -j7.032; \quad \text{also } \bar{V}_1 = 0$$

To compute the phase values,

$$\begin{bmatrix} \bar{I}_a \\ \bar{I}_b \\ \bar{I}_c \end{bmatrix} = [T]\begin{bmatrix} \bar{I}_0 \\ \bar{I}_1 \\ \bar{I}_2 \end{bmatrix} = \begin{bmatrix} 1 & 1 & 1 \\ 1 & a^2 & a \\ 1 & a & a^2 \end{bmatrix}\begin{bmatrix} 0 \\ -j7.032 \\ 0 \end{bmatrix} = \begin{bmatrix} 7.032\angle -90^\circ \\ 7.032\angle 150^\circ \\ 7.032\angle 30^\circ \end{bmatrix} \begin{bmatrix} \bar{V}_a \\ \bar{V}_b \\ \bar{V}_c \end{bmatrix} = [T]\begin{bmatrix} 0 \\ 0 \\ 0 \end{bmatrix} = \begin{bmatrix} 0 \\ 0 \\ 0 \end{bmatrix}$$

Single Phase-to-Ground Fault

The sequence networks are interconnected as shown in Figure 3.34(b).

$$\bar{I}_0 = \bar{I}_1 = \bar{I}_2 = \frac{1\angle 0^\circ}{j0.0452 + j0.1422 + j0.1422} = -j3.034$$

$$\begin{bmatrix} \bar{I}_a \\ \bar{I}_b \\ \bar{I}_c \end{bmatrix} = \begin{bmatrix} 1 & 1 & 1 \\ 1 & a^2 & a \\ 1 & a & a^2 \end{bmatrix}\begin{bmatrix} -j3.034 \\ -j3.034 \\ -j3.034 \end{bmatrix} = \begin{bmatrix} -j9.102 \\ 0 \\ 0 \end{bmatrix}$$

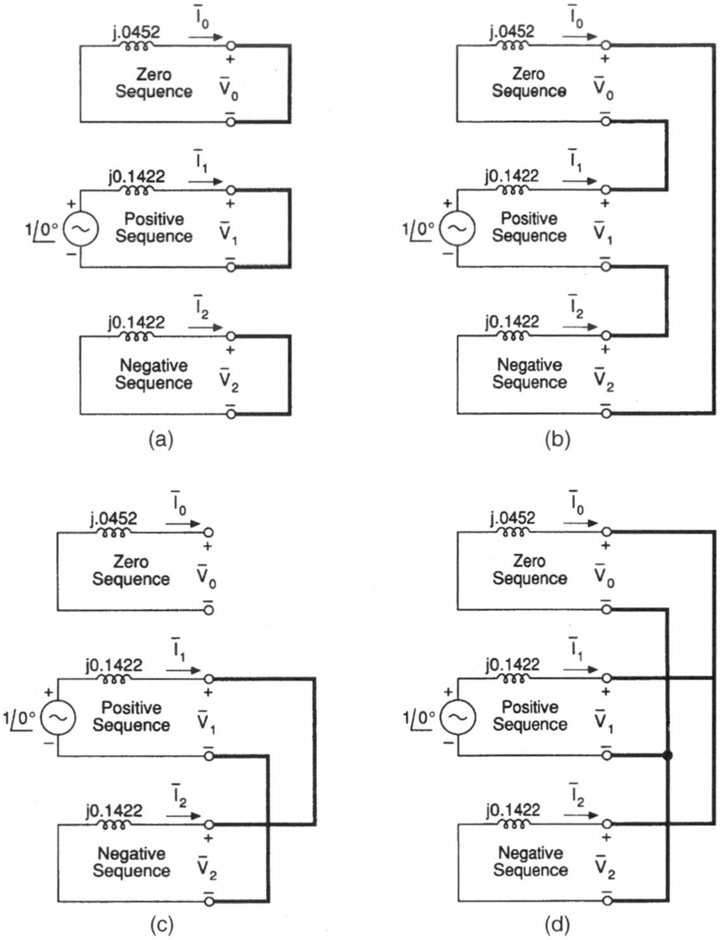

FIGURE 3.34 Example system faults at bus 3. (a) Balanced three-phase; (b) single phase-to-ground; (c) phase-to-phase; (d) double phase-to-ground.

The sequence voltages are

$$\bar{V}_0 = -j0.0452(-j3.034) = -1371$$

$$\bar{V}_1 = 1.0 - j0.1422(-j3.034) = 0.5685$$

$$\bar{V}_2 = -j0.0422(-j3.034) = -0.4314$$

The phase voltages are

$$\begin{bmatrix} \bar{V}_a \\ \bar{V}_b \\ \bar{V}_c \end{bmatrix} = \begin{bmatrix} 1 & 1 & 1 \\ 1 & a^2 & a \\ 1 & a & a^2 \end{bmatrix} \begin{bmatrix} -0.1371 \\ 0.5685 \\ -0.4314 \end{bmatrix} = \begin{bmatrix} 0 \\ 0.8901\angle - 103.4^\circ \\ 0.8901\angle - 103.4^\circ \end{bmatrix}$$

Phase-to-phase and double phase-to-ground fault values are calculated from the appropriate networks [Figures 3.34(c) and (d)]. Complete results are provided.

FaultedBus	Phase a	Phase b	Phase c
3	G	G	G

Sequence Voltages

Bus	V0		V1		V2	
1	0.0000/	0.0	0.3939/	0.0	0.0000/	0.0
2	0.0000/	0.0	0.2424/	0.0	0.0000/	0.0
3	0.0000/	0.0	0.0000/	0.0	0.0000/	0.0
4	0.0000/	0.0	0.2000/	−30.0	0.0000/	30.0

Phase Voltages

Bus	Va		Vb		Vc	
1	0.3939/	0.0	0.3939/	−120.0	0.3939/	120.0
2	0.2424/	0.0	0.2424/	−120.0	0.2424/	120.0
3	0.0000/	6.5	0.0000/	−151.2	0.0000/	133.8
4	0.2000/	−30.0	0.2000/	−150.0	0.2000/	90.0

Sequence Currents

Bus to Bus		I0		I1		I2	
1	2	0.0000/	167.8	3.0303/	−90.0	0.0000/	90.0
1	0	0.0000/	−12.2	3.0303/	90.0	0.0000/	−90.0
2	3	0.0000/	167.8	1.5152/	−90.0	0.0000/	90.0
2	3	0.0000/	167.8	1.5152/	−90.0	0.0000/	90.0
2	1	0.0000/	−12.2	3.0303/	90.0	0.0000/	−90.0
3	2	0.0000/	−12.2	1.5152/	90.0	0.0000/	−90.0
3	2	0.0000/	−12.2	1.5152/	90.0	0.0000/	−90.0
3	4	0.0000/	−12.2	4.0000/	90.0	0.0000/	−90.0
4	3	0.0000/	0.0	4.0000/	−120.0	0.0000/	120.0
4	0	0.0000/	0.0	4.0000/	60.0	0.0000/	−60.0

Faulted Bus	Phase a	Phase b	Phase c
3	G	G	G

Phase Currents

Bus to Bus		Ia		Ib		Ic	
1	2	3.0303/	−90.0	3.0303/	150.0	3.0303/	30.0
1	0	3.0303/	90.0	3.0303/	−30.0	3.0303/	−150.0
2	3	1.5151/	−90.0	1.5151/	150.0	1.5151/	30.0
2	3	1.5151/	−90.0	1.5151/	150.0	1.5151/	30.0
2	1	3.0303/	90.0	3.0303/	−30.0	3.0303/	−150.0
3	2	1.5151/	90.0	1.5151/	−30.0	1.5151/	−150.0
3	2	1.5151/	90.0	1.5151/	−30.0	1.5151/	−150.0
3	4	4.0000/	90.0	4.0000/	−30.0	4.0000/	−150.0
4	3	4.0000/	−120.0	4.0000/	120.0	4.0000/	−0.0
4	0	4.0000/	60.0	4.0000/	−60.0	4.0000/	−180.0

Faulted Bus	Phase a	Phase b	Phase c
3	G	0	0

Sequence Voltages

Bus	V0		V1		V2	
1	0.0496/	180.0	0.7385/	0.0	0.2615/	180.0
2	0.0642/	180.0	0.6731/	0.0	0.3269/	180.0
3	0.1371/	180.0	0.5685/	0.0	0.4315/	180.0
4	0.0000/	0.0	0.6548/	−30.0	0.3452/	210.0

Phase Voltages

Bus	Va		Vb		Vc	
1	0.4274/	0.0	0.9127/	−108.4	0.9127/	108.4
2	0.2821/	0.0	0.8979/	−105.3	0.8979/	105.3
3	0.0000/	89.2	0.8901/	−103.4	0.8901/	103.4
4	0.5674/	−61.8	0.5674/	−118.2	1.0000/	90.0

Sequence Currents

Bus to Bus		I0		I1		I2	
1	2	0.2917/	−90.0	1.3075/	−90.0	1.3075/	−90.0
1	0	0.2917/	90.0	1.3075/	90.0	1.3075/	90.0
2	3	0.1458/	−90.0	0.6537/	−90.0	0.6537/	−90.0
2	3	0.1458/	−90.0	0.6537/	−90.0	0.6537/	−90.0
2	1	0.2917/	90.0	1.3075/	90.0	1.3075/	90.0
3	2	0.1458/	90.0	0.6537/	90.0	0.6537/	90.0
3	2	0.1458/	90.0	0.6537/	90.0	0.6537/	90.0
3	4	2.7416/	90.0	1.7258/	90.0	1.7258/	90.0
4	3	0.0000/	0.0	1.7258/	−120.0	1.7258/	−60.0
4	0	0.0000/	90.0	1.7258/	60.0	1.7258/	120.0

Faulted Bus	Phase a	Phase b	Phase c
3	G	0	0

Phase Currents

Bus to Bus		Ia		Ib		Ic	
1	2	2.9066/	−90.0	1.0158/	90.0	1.0158/	90.0
1	0	2.9066/	90.0	1.0158/	−90.0	1.0158/	−90.0
2	3	1.4533/	−90.0	0.5079/	90.0	0.5079/	90.0
2	3	1.4533/	−90.0	0.5079/	90.0	0.5079/	90.0
2	1	2.9066/	90.0	1.0158/	−90.0	1.0158/	−90.0
3	2	1.4533/	90.0	0.5079/	−90.0	0.5079/	−90 0
3	2	1.4533/	90.0	0.5079/	−90.0	0.5079/	−90 0
3	4	6.1933/	90.0	1.0158/	90.0	1.0158/	90.0
4	3	2.9892/	−90.0	2.9892/	90.0	0.0000/	−90.0
4	0	2.9892/	90.0	2.9892/	−90.0	0.0000/	90.0

Faulted Bus	Phase a	Phase b	Phase c
3	0	C	B

Sequence Voltages

Bus	V0		V1		V2	
1	0.0000/	0.0	0.6970/	0.0	0.3030/	0.0
2	0.0000/	0.0	0.6212/	0.0	0.3788/	0.0
3	0.0000/	0.0	0.5000/	0.0	0.5000/	0.0
4	0.0000/	0.0	0.6000/	−30.0	0.4000/	30.0

Phase Voltages

Bus	Va		Vb		Vc	
1	1.0000/	0.0	0.6053/	−145.7	0.6053/	145.7
2	1.0000/	0.0	0.5423/	−157.2	0.5423/	157.2
3	1.0000/	0.0	0.5000/	−180.0	0.5000/	−180.0
4	0.8718/	−6.6	0.8718/	−173.4	0.2000/	90.0

Sequence Currents

Bus to Bus		I0		I1		I2	
1	2	0.0000/	−61.0	1.5152/	−90.0	1.5152/	90.0
1	0	0.0000/	119.0	1.5152/	90.0	1.5152/	−90.0
2	3	0.0000/	−61.0	0.7576/	−90.0	0.7576/	90.0
2	3	0.0000/	−61.0	0.7576/	−90.0	0.7576/	90.0
2	1	0.0000/	119.0	1.5152/	90.0	1.5152/	−90.0
3	2	0.0000/	119.0	0.7576/	90.0	0.7576/	−90.0
3	2	0.0000/	119.0	0.7576/	90.0	0.7576/	−90.0
3	4	0.0000/	119.0	2.0000/	90.0	2.0000/	−90.0
4	3	0.0000/	0.0	2.0000/	−120.0	2.0000/	120.0
4	0	0.0000/	90.0	2.0000/	60.0	2.0000/	−60.0

Faulted Bus	Phase a	Phase b	Phase c
3	0	C	B

Phase Currents

Bus to Bus		Ia		Ib		Ic	
1	2	0.0000/	180.0	2.6243/	180.0	2.6243/	0.0
1	0	0.0000/	180.0	2.6243/	0.0	2.6243/	180.0
2	3	0.0000/	−180.0	1.3122/	180.0	1.3122/	0.0
2	3	0.0000/	−180.0	1.3122/	180.0	1.3122/	0.0
2	1	0.0000/	180.0	2.6243/	0.0	2.6243/	180.0
3	2	0.0000/	−180.0	1.3122/	0.0	1.3122/	180.0
3	2	0.0000/	−180.0	1.3122/	0.0	1.3122/	180.0
3	4	0.0000/	−180.0	3.4641/	0.0	3.4641/	180.0
4	3	2.0000/	−180.0	2.0000/	180.0	4.0000/	0.0
4	0	2.0000/	0.0	2.0000/	0.0	4.0000/	− 180.0

Faulted Bus	Phase a	Phase b	Phase c
3	0	G	G

Sequence Voltages

Bus	V0		V1		V2	
1	0.0703/	0.0	0.5117/	0.0	0.1177/	0.0
2	0.0909/	0.0	0.3896/	0.0	0.1472/	0.0
3	0.1943/	−0.0	0.1943/	0.0	0.1943/	0.0
4	0.0000/	0.0	0.3554/	−30.0	0.1554/	30.0

Phase Voltages

Bus	Va		Vb		Vc	
1	0.6997/	0.0	0.4197/	−125.6	0.4197/	125.6
2	0.6277/	0.0	0.2749/	−130.2	0.2749/	130.2
3	0.5828/	0.0	0.0000/	−30.7	0.0000/	−139.6
4	0.4536/	−12.7	0.4536/	−167.3	0.2000/	90.0

Sequence Currents

Bus to Bus		I0		I1		I2	
1	2	0.4133/	90.0	2.4416/	−90.0	0.5887/	90.0
1	0	0.4133/	−90.0	2.4416/	90.0	0.5887/	−90.0
2	3	0.2067/	90.0	1.2208/	−90.0	0.2943/	90.0
2	3	0.2067/	90.0	1.2208/	−90.0	0.2943/	90.0
2	1	0.4133/	−90.0	2.4416/	90.0	0.5887/	−90.0
3	2	0.2067/	−90.0	1.2208/	90.0	0.2943/	−90.0
3	2	0.2067/	−90.0	1.2208/	90.0	0.2943/	−90.0
3	4	3.8854/	−90.0	3.2229/	90.0	0.7771/	−90.0
4	3	0.0000/	0.0	3.2229/	−120.0	0.7771/	120.0
4	0	0.0000/	−90.0	3.2229/	60.0	0.7771/	−60.0

Faulted Bus	Phase a	Phase b	Phase c
3	0	G	G

Phase Currents

Bus to Bus		Ia		Ib		Ic	
1	2	1.4396/	−90.0	2.9465/	153.0	2.9465/	27.0
1	0	1.4396/	90.0	2.9465/	−27.0	2.9465/	−153.0
2	3	0.7198/	−90.0	1.4733/	153.0	1.4733/	27.0
2	3	0.7198/	−90.0	1.4733/	153.0	1.4733/	27.0
2	1	1.4396/	90.0	2.9465/	−27.0	2.9465/	−153.0
3	2	0.7198/	90.0	1.4733/	−27.0	1.4733/	−153.0
3	2	0.7198/	90.0	1.4733/	−27.0	1.4733/	−153.0
3	4	1.4396/	−90.0	6.1721/	−55.9	6.1721/	−124.1
4	3	2.9132/	−133.4	2.9132/	133.4	4.0000/	−0.0
4	0	2.9132/	46.6	2.9132/	−46.6	4.0000/	−180.0

Further Considerations

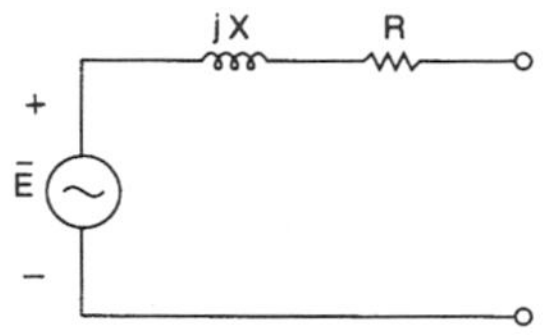

FIGURE 3.35 Positive sequence circuit locking back into faulted bus.

Generators are not the only sources in the system. All rotating machines are capable of contributing to fault current, at least momentarily. Synchronous and induction motors will continue to rotate due to inertia and function as sources of fault current. The impedance used for such machines is usually the transient reactance X'_d or the subtransient X''_d, depending on protective equipment and speed of response. Frequently motors smaller than 50 hp are neglected. Connecting systems are modeled with their Thévenin equivalents.

Although we have used ac circuit techniques to calculate faults, the problem is fundamentally transient since it involves sudden switching actions. Consider the so-called dc offset current. We model the system by determining its positive sequence Thévenin equivalent circuit, looking back into the positive sequence network at the fault, as shown in Figure 3.35. The transient fault current is

$$i(t) = I_{ac}\sqrt{2}\cos(\omega t - \beta) + I_{dc}e^{-t/\tau}$$

This is a first-order approximation and strictly applies only to the three-phase or phase-to-phase fault. Ground faults would involve the zero sequence network also.

$$I_{ac} = \frac{E}{\sqrt{R^2 + X^2}} = \text{rms ac current}$$

$$I_{dc}(t) = I_{dc}e^{-t/\tau} = \text{dc offset current}$$

The maximum initial dc offset possible would be

$$\text{Max}\, I_{dc} = I_{max} = \sqrt{2}I_{ac}$$

The dc offset will exponentially decay with time constant τ, where

$$\tau = \frac{L}{R} = \frac{X}{\omega R}$$

The maximum dc offset current would be $I_{dc}(t)$

$$I_{dc}(t) = I_{dc}e^{-t/\tau} = \sqrt{2}I_{ac}e^{-t/\tau}$$

The *transient rms* current $I(t)$, accounting for both the ac and dc terms, would be

$$I(t) = \sqrt{I_{ac}^2 + I_{dc}^2(t)} = I_{ac}\sqrt{1 + 2e^{-2t/\tau}}$$

Define a multiplying factor k_i such that I_{ac} is to be multiplied by k_i to estimate the interrupting capacity of a breaker which operates in time T_{op}. Therefore,

$$k_i = \frac{I(T_{op})}{I_{ac}} = \sqrt{1 + 2e^{-2T_{op}/\tau}}$$

Observe that the maximum possible value for k_i is $\sqrt{3}$.

Example

In the circuit of Figure 3.35, $E = 2400$ V, $X = 2\,\Omega$, $R = 0.1\,\Omega$, and $f = 60$ Hz. Compute k_i and determine the interrupting capacity for the circuit breaker if it is designed to operate in two cycles. The fault is applied at $t = 0$.

Solution

$$I_{ac} \cong \frac{2400}{2} = 1200\text{ A}$$

$$T_{op} = \frac{2}{60} = 0.0333\text{ s}$$

$$\tau = \frac{X}{\omega R} = \frac{2}{37.7} = 0.053$$

$$k_i = \sqrt{1 + 2e^{-2T_{op}/\tau}} = \sqrt{1 + 2e^{-0.0067/0.053}} = 1.252$$

Therefore

$$I = k_i I_{ac} = 1.252(1200) = 1503\text{ A}$$

The Thévenin equivalent at the fault point is determined by normal sinusoidal steady-state methods, resulting in a first-order circuit as shown in Figure 3.35. While this provides satisfactory results for the steady-state component I_{ac}, the X/R value so obtained can be in serious error when compared with the rate of decay of $I(t)$ as measured by oscillographs on an actual faulted system. The major reasons for the discrepancy are, first of all, that the system, for transient analysis purposes, is actually high-order, and second, the generators do not hold constant impedance as the transient decays.

Summary

Computation of fault currents in power systems is best done by computer. The major steps are summarized below:

- Collect, read in, and store machine, transformer, and line data in per-unit on common bases.
- Formulate the sequence impedance matrices.
- Define the faulted bus and Z_f. Specify type of fault to be analyzed.
- Compute the sequence voltages.
- Compute the sequence currents.
- Correct for wye-delta connections.
- Transform to phase currents and voltages.

For large systems, computer formulation of the sequence impedance matrices is required. Refer to Further Information for more detail. Zero sequence networks for lines in close proximity to each other (on a common right-of-way) will be mutually coupled. If we are willing to use the same values for positive and negative sequence machine impedances,

$$[Z_1] = [Z_2]$$

Therefore, it is unnecessary to store these values in separate arrays, simplifying the program and reducing the computer storage requirements significantly. The error introduced by this approximation is usually not important. The methods previously discussed neglect the prefault, or load, component of current; that is, the usual assumption is that currents throughout the system were zero prior to the fault. This is almost never strictly true; however, the error produced is small since the fault currents are generally much larger than the load currents.

Also, the load currents and fault currents are out of phase with each other, making their sum more nearly equal to the larger components than would have been the case if the currents were in phase. In addition, selection of precise values for prefault currents is somewhat speculative, since there is no way of predicting what the loaded state of the system is when a fault occurs. When it is important to consider load currents, a power flow study is made to calculate currents throughout the system, and these values are superimposed on (added to) results from the fault study.

A term which has wide industrial use and acceptance is the *fault level* or fault MVA at a bus. It relates to the amount of current that can be expected to flow out of a bus into a three-phase fault. As such, it is an alternate way of providing positive sequence impedance information. Define

$$\begin{aligned}\text{Fault level in MVA at bus } i &= V_{i_{\text{pu}_{\text{nominal}}}} I_{i_{\text{pu}_{\text{fault}}}} s_{3\phi_{\text{base}}} \\ &= (1)\frac{1}{Z_{ii}^1} s_{3\phi_{\text{base}}} = \frac{s_{3\phi_{\text{base}}}}{Z_{ii}^1}\end{aligned}$$

Fault study results may be further refined by approximating the effect of dc offset.

The basic reason for making fault studies is to provide data that can be used to size and set protective devices. The role of such protective devices is to detect and remove faults to prevent or minimize damage to the power system.

Defining Terms

DC offset: The natural response component of the transient fault current, usually approximated with a first-order exponential expression.

Fault: An unintentional and undesirable conducting path in an electrical power system.

Fault MVA: At a specific location in a system, the initial symmetrical fault current multiplied by the prefault nominal line-to-neutral voltage ($\times 3$ for a three-phase system).

Sequence (012) quantities: Symmetrical components computed from phase (*abc*) quantities. Can be voltages, currents, and/or impedances.

References

P.M. Anderson, *Analysis of Faulted Power Systems,* Ames: Iowa State Press, 1973.
M.E. El-Hawary, *Electric Power Systems: Design and Analysis,* Reston, Va.: Reston Publishing, 1983.
M.E. El-Hawary, *Electric Power Systems,* New York: IEEE Press, 1995.
O.I. Elgerd, *Electric Energy Systems Theory: An Introduction,* 2nd ed., New York: McGraw-Hill, 1982.
General Electric, *Short-Circuit Current Calculations for Industrial and Commercial Power Systems,* Publication GET-3550.
C.A. Gross, *Power System Analysis,* 2nd ed., New York: Wiley, 1986.
S.H. Horowitz, *Power System Relaying,* 2nd ed, New York: Wiley, 1995.
I. Lazar, *Electrical Systems Analysis and Design for Industrial Plants,* New York: McGraw-Hill, 1980.
C.R. Mason, *The Art and Science of Protective Relaying,* New York: Wiley, 1956.
J.R. Neuenswander, *Modern Power Systems,* Scranton, Pa.: International Textbook, 1971.
G. Stagg and A.H. El-Abiad, *Computer Methods in Power System Analysis,* New York: McGraw-Hill, 1968.
Westinghouse Electric Corporation, *Applied Protective Relaying,* Relay-Instrument Division, Newark, N.J., 1976.
A.J. Wood, *Power Generation, Operation, and Control,* New York: Wiley, 1996.

Further Information

For a comprehensive coverage of general fault analysis, see Paul M. Anderson, *Analysis of Faulted Power Systems,* New York, IEEE Press, 1995. Also see Chapters 9 and 10 of *Power System Analysis* by C.A. Gross, New York: Wiley, 1986.

3.4 Protection

Arun G. Phadke

Fundamental Principles of Protection

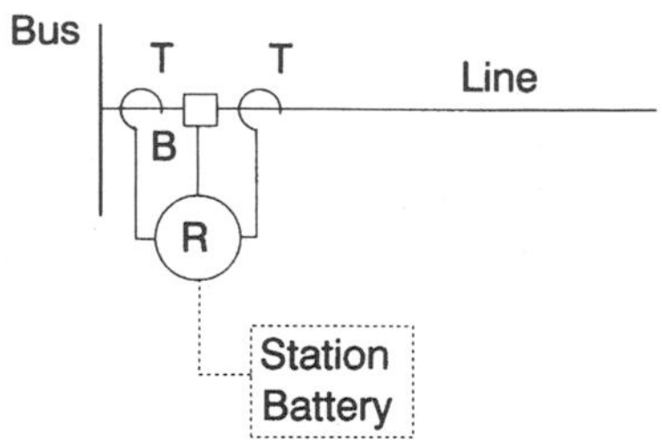

FIGURE 3.36 Elements of a protection system.

Protective equipment—**relays**—is designed to respond to system abnormalities (faults) such as short circuits. When faults occur, the relays must signal the appropriate circuit breakers to trip and isolate the faulted equipment. The protection systems not only protect the faulty equipment from more serious damage, they also protect the power system from the consequences of having faults remain on the system for too long. In modern high-voltage systems, the potential for damage to the power system—rather than to the individual equipment—is often far more serious, and power system security considerations dictate the design of the protective system. The protective system consists of four major subsystems as shown in Figure 3.36. The **transducers** (*T*) are current and voltage transformers, which transform high voltages and currents to a more manageable level. In the United States, the most common standard for current transformers is a secondary current of 5 A (or less) for steady-state conditions. In Europe, and in some other foreign countries, a 1-A standard is also common. The voltage transformer standard is 69.3 V line-to-neutral or 120 V line-to-line on the transformer secondary side. Standardization of the secondary current and voltage ratings of the transducers has permitted independent development of the transducers and relays. The power handling capability of the transducers is expressed in terms of the volt-ampere burden, which they can supply without significant waveform distortion. In general, the transient response of the transducers is much more critical in relaying applications.

The second element of the protection system is the relay (*R*). This is the device that, using the current, voltage, and other inputs, can determine if a fault exists on the system, for which action on the part of the relay is needed. We will discuss relays in greater detail in the following. The third element of the protection chain is the circuit breaker (*B*), which does the actual job of interrupting the flow of current to the fault. Modern high-voltage circuit breakers are capable of interrupting currents of up to 100,000 A, against system voltages of up to 800,000 V, in about 15 to 30 ms. Lower-voltage circuit breakers are generally slower in operating speed. The last element of the protection chain is the station battery, which powers the relays and circuit breakers. The battery voltage has also been standardized at 125 V, although some other voltage levels may prevail in generating stations and in older substations.

The relays and circuit breakers must remove the faulted equipment from the system as quickly as possible. Also, if there are many alternative ways of deenergizing the faulty equipment, the protection system must choose a strategy that will remove from service the minimum amount of equipment. These ideas are embodied in the concepts of zones of protection, relay speed, and reliability of protection.

Zones of Protection

To make sure that a protection system removes the minimum amount of equipment from the power system during its operation, the power system is divided into zones of protection. Each zone has its associated protection system. A fault inside the zone causes the associated protection system to operate. A fault in any other zone must not cause an operation. A zone of protection usually covers one piece of equipment, such as a transmission line. The zone boundary is defined by the location of transducers (usually current transformers) and also by circuit breakers that will operate to isolate the zone. A set of zones of protection is shown in Figure 3.37. Note that all zones are shown to overlap with their neighbors. This is to ensure that no point on the system is left unprotected. Occasionally, a circuit breaker may not exist at a zone boundary. In such cases, the tripping must be done at some other remote circuit breakers. For example, consider protection zone *A* in

Figure 3.37. A fault in that zone must be isolated by tripping circuit breakers *X* and *Y*. While the breaker *X* is near the transformer and can be tripped locally, *Y* is remote from the station, and some form of communication channel must be used to transfer the trip command to *Y*. Although most zones of protection have a precise extent, there are some zones that have a loosely defined reach. These are known as *open* zones and are most often encountered in transmission line protection.

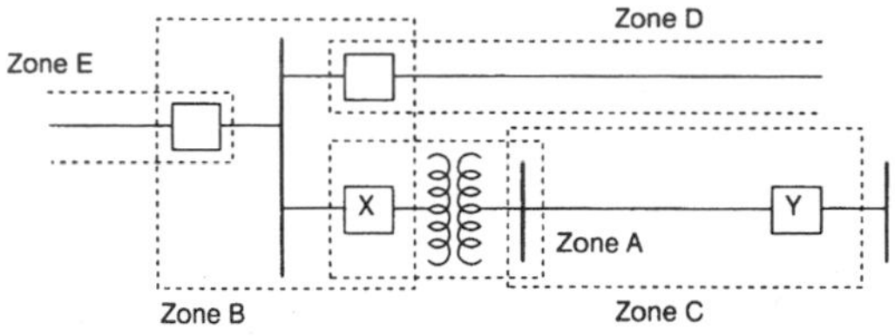

FIGURE 3.37 Zones of protection for a power system. Zones overlap; most zones are bounded by breakers.

Speed of Protection

The faster the operation of a protection function, the quicker is the prospect of removing a fault from the system. Thus, all protection systems are made as fast as possible. However, there are considerations that dictate against making the protection faster than a minimum limit. Also, occasionally, it may be necessary to slow down a protection system in order to satisfy some specific system need. In general, the fastest protection available operates in about 5 to 10 ms after the inception of a fault [Thorp et al., 1979]. If the protection is made faster than this, it is likely to become "trigger happy" and operate falsely when it should not. When a protection system is intended as a backup system for some other protection, it is necessary to deliberately slow it down so that the primary protection may operate in its own time before the backup system will operate. This calls for a deliberate slowing of the backup protection. Depending upon the type of backup system being considered, the protection may sometimes be slowed down to operate in up to several seconds.

Reliability of Protection

In the field of relaying, reliability implies certain very specific concepts [Mason, 1956]. A reliable protection system has two attributes: *dependability* and *security*. A dependable relay is one that always operates for conditions for which it is designed to operate. A secure relay is one that will not operate for conditions for which it is not intended to operate. In modern power systems, the failure to operate when a fault occurs—lack of dependability—has very serious consequences for the power system. Therefore, most protective systems are made secure by duplicating relaying equipment, duplicating relaying functions, and providing several levels of backup protection. Thus modern systems tend to be very dependable, i.e., every fault is cleared, perhaps by more than one relay. As a consequence, security is somewhat degraded: modern protection systems will, occasionally, act and trip equipment falsely. Such occurrences are rare, but not uncommon. As power systems become leaner, i.e., they have insufficient margins of reserve generation and transmission, lack of security can be quite damaging. This has led to recent reevaluation of the proper balance between security and dependability of the protection systems.

Overcurrent Protection

The simplest fault detector is a sensor that measures the increase in current caused by the fault. The fuse is the simplest overcurrent protection; in fact, it is the complete protection chain—sensor, relay, and circuit breaker—in one package. Fuses are used in lower-voltage (distribution) circuits. They are difficult to set in high-voltage circuits, where load and fault currents may be of the same order of magnitude. Furthermore, they must be replaced when blown, which implies a long duration outage. They may also lead to system unbalances. However, when applicable, they are simple and inexpensive.

Inverse-Time Characteristic

Overcurrent relays sense the magnitude of the current in the circuit, and when it exceeds a preset value (known as the *pickup setting* of the relay), the relay closes its output contact, energizing the trip coil of the appropriate circuit breakers. The pickup setting must be set above the largest load current that the circuit may carry and must be smaller than the smallest fault current for which the relay must operate. A margin factor of 2 to 3 between the maximum load on the one hand and the minimum fault current on the other and the pickup setting of the relay is considered to be desirable. The overcurrent relays usually have an *inverse-time* characteristic as shown in Figure 3.38. When the current exceeds the pickup setting, the relay operating time decreases in inverse proportion to the current magnitude. Besides this built-in feature in the relay mechanism, the relay also has a *time-dial* setting, which shifts the inverse-time curve vertically, allowing for more flexibility in setting the relays. The time dial has 11 discrete settings, usually labeled 1/2, 1, 2, ..., 10, the lowest setting providing the fastest operation. The inverse-time characteristic offers an ideal relay for providing primary and backup protection in one package.

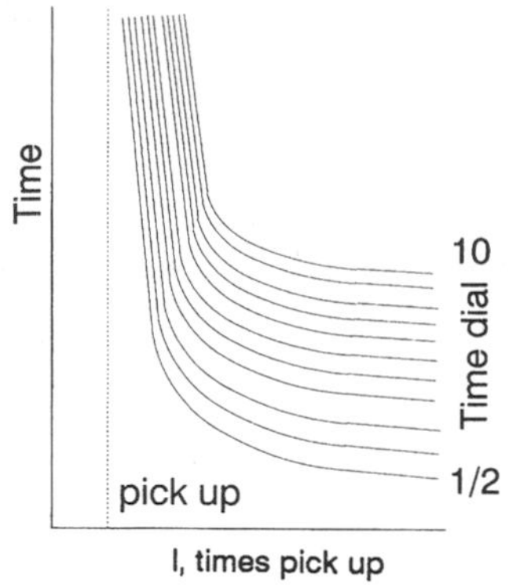

FIGURE 3.38 Inverse-time relay characteristic.

Coordination Principles

Consider the radial transmission system shown in Figure 3.39. The transformer supplies power to the feeder, which has four loads at buses *A*, *B*, *C*, and *D*. For a fault at F_1, the relay R_{cd} must operate to open the circuit breaker B_{cd}. The relay R_{bc} is responsible for a zone of protection, which includes the entire zone of R_{cd}. This constitutes a remote backup for the protection at bus *C*. The backup relay (R_{bc}) must be slower than the primary relay (R_{cd}), its associated circuit breaker, with a safety margin. This delay in operating of the backup relay is known as the *coordination delay* and is usually about 0.3 s. In a similar fashion, R_{ab} backs up R_{bc}. The magnitude of the fault current varies as shown in Figure 3.39(b), as the location of the fault is moved along the length of the feeder. We may plot the inverse time characteristic of the relay with the fault location as the abscissa, recalling that a smaller current magnitude gives rise to a longer operating time for the relay. The coordinating time delay between the primary and backup relays is also shown. It can be seen that, as we move from the far end of the feeder toward the source, the fault clearing time becomes progressively longer. The coordination is achieved by selecting relays with a time dial setting that will provide the proper separation in operating times.

The effect of cumulative coordination-time delays is slowest clearing of faults with the largest fault currents. This is not entirely satisfactory from the system point of view, and wherever possible, the inverse-time relays are supplemented by *instantaneous* overcurrent relays. These relays, as the name implies, have no intentional time delays and operate in less than one cycle. However, they cannot coordinate with the downstream relays and therefore must not operate ("see") for faults into the protection zone of the downstream relay.

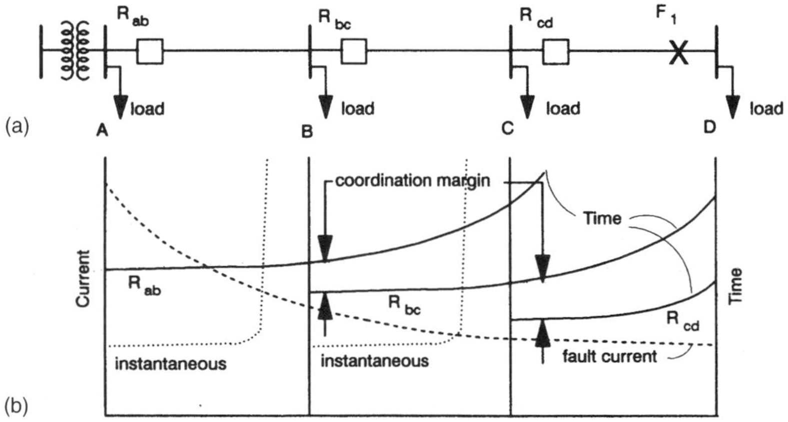

FIGURE 3.39 Coordination of inverse-time overcurrent and instantaneous relays for a radial system.

This criterion is not always possible to meet. However, whenever it can be met, instantaneous relays are used and provide a preferable compromise between fast fault clearing and coordinated backup protection.

Directional Overcurrent Relays

When power systems become meshed, as for most subtransmission and high-voltage transmission networks, inverse time overcurrent relays do not provide adequate protection under all conditions. The problem arises because the fault current can now be supplied from either end of the transmission line, and discrimination between faults inside and outside the zone of protection is not always possible. Consider the loop system shown in Figure 3.40. Notice that in this system there must be a circuit breaker at each end of the line, as a fault on the line cannot be interrupted by opening one end alone. Zone A is the zone of protection for the line A–D. A fault at F_1 must be detected by the relays R_{ad} and R_{da}. The current through the circuit breaker B_{da} for the fault F_1 must be the determining quantity for the operation of the relay R_{da}. However, the impedances of the lines may be such that the current through the breaker B_{da} for the fault F_2 may be higher than the current for the fault F_1. Thus, if current magnitude alone is the criterion, the relay R_{da} would operate for fault F_2, as well as for the fault F_1. Of course, operation of R_{da} for F_2 is inappropriate, as it is outside its zone of protection, zone A. This problem is solved by making the overcurrent relays directional. By this is meant that the relays will respond as overcurrent relays only if the fault is in the forward direction from the relays, i.e., in the direction in which their zone of protection extends. The directionality is provided by making the relay sensitive to the phase angle between the fault current and a reference quantity, such as the line voltage at the relay location. Other reference sources are also possible, including currents in the neutral of a transformer bank at the substation.

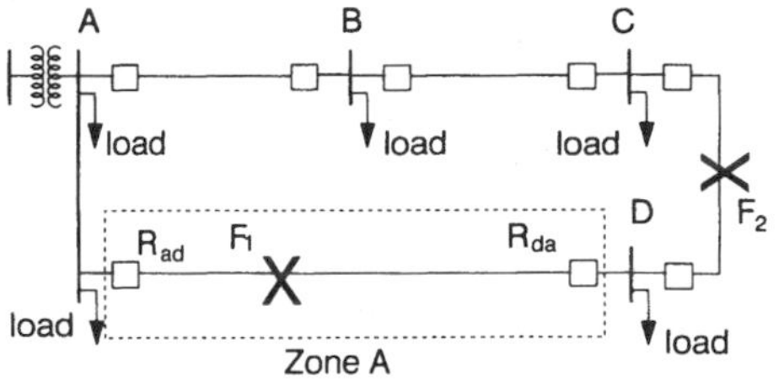

FIGURE 3.40 Protection of a loop (network) system with directional overcurrent relays.

Distance Protection

As the power networks become more complex, protection with directional overcurrent relays becomes even more difficult, if not impossible. Recall that the pickup setting of the relays must be set above the maximum load which the line is expected to carry. However, a network system has so many probable configurations due to various circuit breaker operations that the maximum load becomes difficult to define. For the same reason, the minimum fault current—the other defining parameter for the pickup setting—also becomes uncertain. Under these circumstances, the setting of the pickup of the overcurrent relays, and their reach, which will satisfy all the constraints, becomes impossible. Distance relays solve this problem.

Distance relays respond to a ratio of the voltage and current at the relay location. The ratio has the dimensions of an impedance, and the impedance between the relay location and fault point is proportional to the distance of the fault. As the zone boundary is related to the distance between the sending end and the receiving end of the transmission line, the distance to the fault forms an ideal relaying parameter. The distance is also a unique parameter in that it is independent of the current magnitude. It is thus free from most of the difficulties associated with the directional overcurrent relays mentioned above.

In a three-phase power system, 10 types of faults are possible: three single phase-to-ground faults, three phase-to-phase faults, three double phase-to-ground faults, and one three-phase fault. It turns out that relays responsive to the ratio of delta voltages and delta currents measure the correct distance to all multiphase faults. The delta quantities are defined as the difference between any two phase quantities; for example, $E_a - E_b$ is the delta voltage between a and b phases. Thus for a multiphase fault between phases x and y,

$$\frac{E_x - E_y}{I_x - I_y} = Z_1$$

where x and y can be a, b, or c and Z_1 is the positive sequence impedance between the relay location and the fault. For ground distance relays, the faulted phase voltage, and a compensated faulted phase current must be used

$$\frac{E_x}{I_x + mI_0} = Z_1$$

where m is a constant depending upon the line impedances and I_0 is the zero sequence current in the transmission line. A full complement of relays consists of three phase distance relays and three ground distance relays. As explained before, the phase relays are energized by the delta quantities, while the ground distance relays are energized by each of the phase voltages and the corresponding compensated phase currents. In many instances, ground distance protection is not preferred, and the time overcurrent relays may be used for ground fault protection.

Step-Distance Protection

The principle of distance measurement for faults is explained above. A relaying system utilizing that principle must take into account several features of the measurement principle and develop a complete protection scheme. Consider the system shown in Figure 3.41. The distance relay R_{ab} must protect line AB, with its zone of protection as indicated by the dashed line. However, the distance calculation made by the relay is not precise enough for it to be able to distinguish between a fault just inside the zone and a fault just outside the zone, near bus B. This problem is solved by providing a two-zone scheme, such that if a fault is detected to be in zone 1, the relay trips instantaneously, and if the fault is detected to be inside zone 2, the relay trips with a time delay of about 0.3 s. Thus for faults near the zone boundary, the fault is cleared with this time delay, while for near faults, the clearing is instantaneous. This arrangement is referred to as a *step-distance* protection scheme, consisting of an underreaching zone (zone 1), and an overreaching zone (zone 2). The relays of the neighboring line (BC) can also be backed up by a third zone of the relay, which reaches beyond the zone of protection of relay R_{bc}. Zone 3 operation is delayed further to allow the zone 1 or zone 2 of R_{bc} to operate and clear the fault on line BC.

The distance relays may be further subdivided into categories depending upon the shape of their protection characteristics. The most commonly used relays have a directional distance, or a mho characteristic. The two characteristics are shown in Figure 3.42. The directional impedance relay consists of two functions, a directional detection function and a distance measurement function. The mho characteristic is inherently

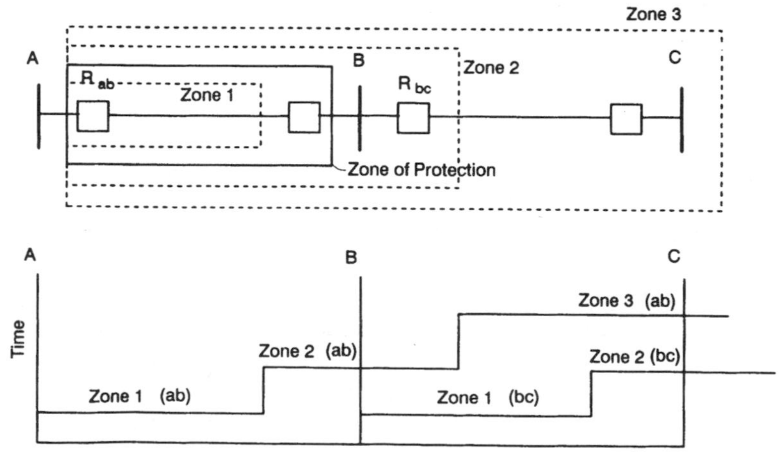

FIGURE 3.41 Zones of protection in a step-distance protection scheme. Zone 3 provides backup for the downstream line relays.

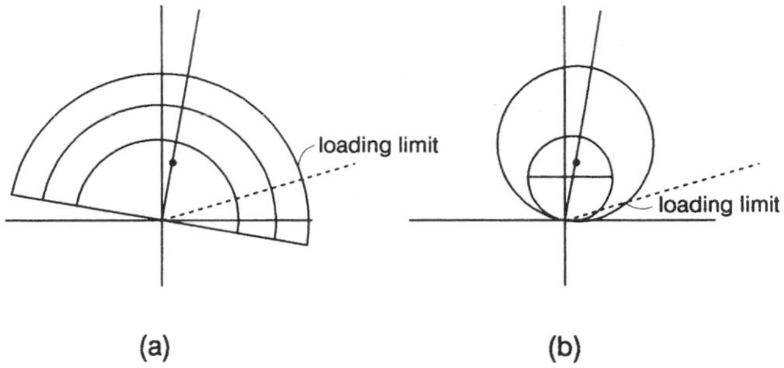

FIGURE 3.42 (a) Directional impedance characteristic. (b) Mho characteristic. Loadability limits as shown.

directional, as the mho circle, by relay design, passes through the origin of the *RX* plane. Figure 3.42 also shows the multiple zones of the step distance protection.

Loadability of Distance Relays

The load carried by a transmission line translates into an apparent impedance as seen by the relay, given by

$$Z_{\text{app}} = \frac{|E|^2}{P - jQ}$$

where $P - jQ$ is the load complex power and E is the voltage at the bus where a distance relay is connected. This impedance maps into the *RX* plane, as do all other apparent impedances, and hence the question arises whether this apparent load impedance could be mistaken for a fault by the distance relay. Clearly, this depends upon the shape of the distance relay characteristic employed. The loadability of a distance relay refers to the maximum load power (minimum apparent impedance) that the line can carry before a protective zone of a distance relay is penetrated by the apparent impedance. A typical load line is shown in Figure 3.42. It is clear from this figure that the mho characteristic has a higher loadability than the directional impedance relay. In fact, other relay characteristics can be designed so that the loadability of a relay increased even further.

Other Uses of Distance Relays

Although the primary use of distance relays is in protecting transmission lines, some other protection tasks can also be served by distance relays. For example, loss-of-field protection of generators is often based upon distance relays. Out-of-step relays and relays for protecting reactors may also be distance relays. Distance relays are also used in pilot protection schemes described next, and as backup relays for power apparatus.

Pilot Protection

Pilot protection of transmission lines uses communication channels (pilot channels) between the line terminals as an integral element of the protection system. In general, pilot schemes may be subdivided into categories according to the medium of communication used. For example, the pilot channels may be wire pilots, leased telephone circuits, dedicated telephone circuits, microwave channels, power line carriers, or fiber optic channels. Pilot protection schemes may also be categorized according to their function, such as a tripping pilot or a blocking pilot. In the former, the communication medium is used to send a tripping signal to a remote line terminal, while in the latter, the pilot channel is used to send a signal that prevents tripping at the remote terminal for faults outside the zone of protection of the relays. The power line carrier system is the most common system used in the United States. It uses a communication channel with a carrier signal frequency ranging between 30 and 300 kHz, the most common bands being around 100 kHz. The modulated carrier signal is coupled into one or more phases of the power line through coupling capacitors. In almost all cases, the capacitors of the capacitive-coupled voltage transformers are used for this function (see Figure 3.43).

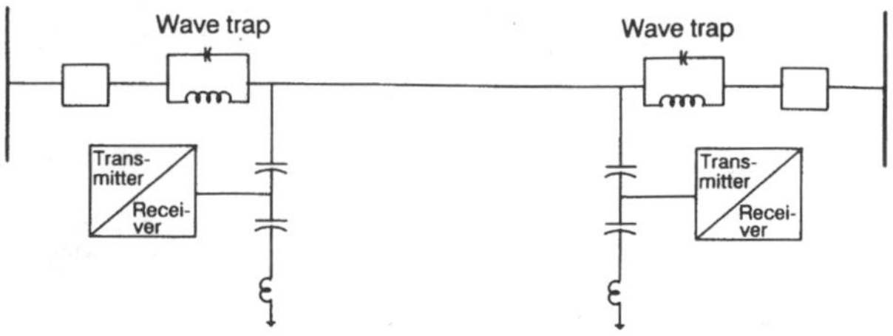

FIGURE 3.43 Carrier system for pilot protection of lines. Transmitter and receiver are connected to relays.

The carrier signal is received at both the sending and the receiving ends of the transmission line by tuned receivers. The carrier signal is blocked from flowing into the rest of the power system by blocking filters, which are parallel resonant circuits, known as *wave traps.*

Coverage of 100% of Transmission Line

The step-distance scheme utilizes the zone 1 and zone 2 combination to protect 100% of the transmission line. The middle portion of the transmission line, which lies in zone 1 of relays at the two ends of the line, is protected at high speed from both ends. However, for faults in the remaining portion of the line, the near end clears the fault at high speed, i.e., in zone 1 time, while the remote end clears the fault in zone 2 time. In effect, such faults remain on the system for zone 2 time, which may be of the order 0.3 to 0.5 s. This becomes undesirable in modern power systems where the margin of stability may be quite limited. In any case, it is good protection practice to protect the entire line with high-speed clearing of all internal faults from both ends of the transmission line. Pilot protection accomplishes this task.

Directional Comparison Blocking Scheme

Consider the fault at F_2 shown in Figure 3.44. As discussed above, this fault will be cleared in zone 1 time by the step-distance relay at bus *B*, while the relay at bus *A* will clear the fault in zone 2 time. Since the relays at bus *B* can determine, with a high degree of certainty, that a fault such as F_2 is indeed inside the zone of protection of the relays, one could communicate this knowledge to terminal *A*, which can then cause the local circuit breaker to trip for the fault F_2. If the entire relaying and communication task can be accomplished quickly, 100% of the line can be protected at high speed. One of the most commonly used methods of achieving this function is to use overreaching zones of protection at both terminals, *and* if a fault is detected to be inside this zone, and if the remote terminal confirms that the fault is inside the zone of protection, then the

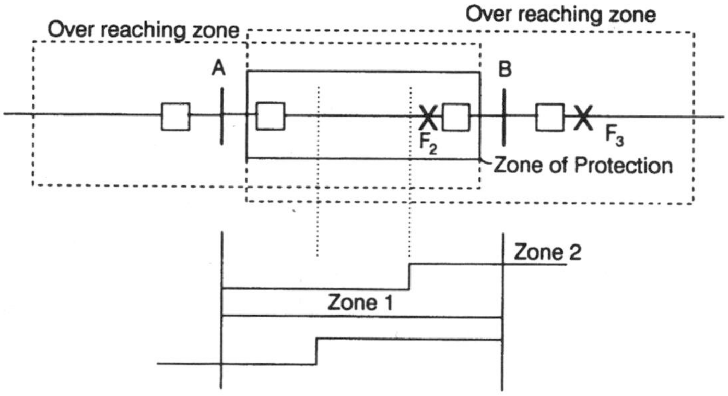

FIGURE 3.44 Pilot protection with overreaching zones of protection. This is most commonly used in a directional comparison blocking scheme.

local relay may be allowed to trip. In actual practice, the complement of this information is used to block the trip at the remote end. Thus, the remote end, terminal *B* in this case, detects faults that are outside the zone of protection and, for those faults, sends a signal which asks the relay at terminal *A* to block the tripping command. Thus, for a fault such as F_3, the relay at *A* will trip, unless the communication is received from terminal *B* that this particular fault is outside the zone of protection—as indeed fault F_3 happens to be. This mode, known as a blocking carrier, is preferred, since a loss of the carrier signal created by an internal fault, or due to causes that are unrelated to the fault, will not prevent the trip at the remote end. This is a highly dependable protection system, and precisely because of that it is somewhat less secure. Nevertheless, as discussed previously, most power systems require that a fault be removed as quickly as possible, even if in doing so for a few faults an unwarranted trip may result.

Other Pilot Protection Schemes

Several other types of pilot protection schemes are available. The choice of a specific scheme depends upon many factors. Some of these factors are importance of the line to the power system, the available communication medium, dependability of the communication medium, loading level of the transmission line, susceptibility of the system to transient stability oscillations, presence of series or shunt compensating devices, multiterminal lines, etc. A more complete discussion of all these issues will be found in the references [Westinghouse, 1982; Blackburn, 1987; Horowitz and Phadke, 1992].

Computer Relaying

Relaying with computers began to be discussed in technical literature in the mid-1960s. Initially, this was an academic exercise, as neither the computer speeds nor the computer costs could justify the use of computers for relaying. However, with the advent of high-performance microprocessors, computer relaying has become a very practical and attractive field of research and development. All major manufacturers of electric power equipment have computer relays to meet all the needs of power system engineers. Computer relaying is also being taught in several universities and has provided a very fertile field of research for graduate students. Computer relaying has also uncovered new ways of measuring power system parameters and may influence future development of power system monitoring and control functions.

Incentives for Computer Relaying

The acceptance of computer relays has been due to economic factors which have made microcomputers relatively inexpensive and computationally powerful. In addition to this economic advantage, the computer relays are also far more versatile. Through their self-diagnostic capability, they provide an assurance of availability. Thus, even if they should suffer the same (or even greater) number of failures in the field as traditional relays, their failures could be communicated to control centers and a maintenance crew called to repair the failures immediately. This type of diagnostic capability was lacking in traditional protection systems and often led to failures of relays, which went undetected for extended periods. Such hidden failures have been identified as one of the main sources of power system blackouts.

The computing power available with computer relays has also given rise to newer and better protection functions in several instances. Improved protection of transformers, multiterminal lines, fault location, and reclosing are a few of the protection functions where computer relaying is likely to have a significant impact. Very significant developments in the computer relaying field are likely to occur in the coming years.

Architecture for a Computer Relay

There are many ways of implementing computer-based relays. Figure 3.45 is a fairly typical block diagram of a computer relay architecture. The input signals consisting of voltage and currents and contact status are filtered to remove undesired frequency components and potentially damaging surges. These signals are

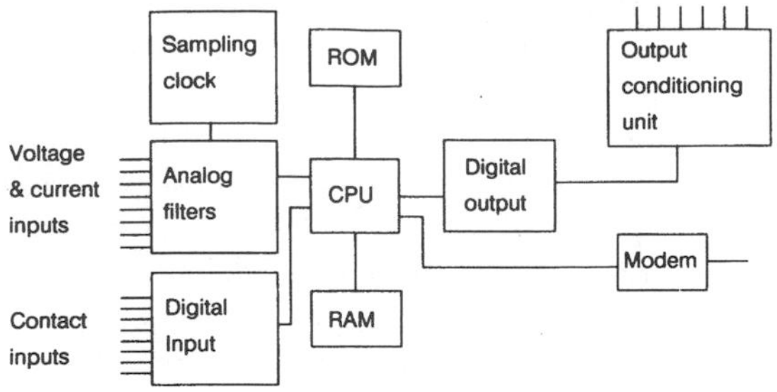

FIGURE 3.45 Block diagram of a computer relay architecture.

sampled by the CPU under the control of a sampling clock. Typical sampling frequency used in a modern digital relay varies between 4 and 32 times the nominal power system frequency. The sampled data is processed by the CPU with a digital filtering algorithm, which estimates the appropriate relaying quantity. A typical relaying quantity may be the rms value of a current, the voltage or current phasor, or the apparent impedance. The estimated parameters are then compared with prestored relay characteristics, and the appropriate control action is initiated. The decision of the relay is communicated to the substation equipment, such as the circuit breaker, through the output ports. These outputs must also be filtered to block any surges from entering the relay through the output lines. In most cases, the relay can also communicate with the outside world through a modem. The data created by a fault is usually saved by the relaying computer and can be used for fault analysis or for sequence-of-event analysis following a power system disturbance. The user may interface with the relay through a keyboard, a control panel, or a communication port. In any case, provision must be made to enter relay settings in the relay and to save these settings in case the station power supply fails. Although the block diagram in Figure 3.45 shows different individual subsystems, the actual hardware composition of the subsystems is dependent on the computer manufacturer. Thus, we may find several microprocessors in a given implementation, each controlling one or more subsystems. Also, the hardware technology is in a state of flux, and in a few years, we may see an entirely different realization of the computer relays.

Experience and Future Trends

Field experience with the computer relays has been excellent so far. The manufacturers of traditional relays have adopted this technology in a big way. As more experience is gained with the special requirements of computer relays, it is likely that other—nontraditional—relay manufacturers will enter the field.

It seems clear that in computer relaying, power system engineers have obtained a tool with exciting new possibilities. Computers, with the communication networks now being developed, can lead to improved monitoring, protection, and control of power systems. An entirely new field, adaptive relaying, has been introduced recently [Phadke and Horowitz, 1990]. The idea is that protection systems should adapt to changing conditions of the power networks. In doing so, protection systems become more sensitive and reliable. Another development, which can be traced to computer relaying, is that of synchronized phasor measurements in power systems [Phadke and Thorp, 1991]. The development of the Global Positioning System (GPS) satellites has made possible the synchronization of sampling clocks used by relays and other measuring devices across the power system. This technology is expected to have a major impact on static and dynamic state estimation and on control of the electric power networks.

Defining Terms

Computer relays: Relays that use digital computers as their logic elements.

Distance protection: Relaying principle based upon estimating fault location (distance) and providing a response based upon the distance to the fault.

Electromechanical relays: Relays that use electromechanical logic elements.
Pilot: A communication medium used by relays to help reach a reliable diagnosis of certain faults.
Relays: Devices that detect faults on power equipment and systems and take appropriate control actions to deenergize the faulty equipment.
Reliability: For relays, reliability implies *dependability*, i.e., certainty of operating when it is supposed to, and *security*, certainty of not operating when it is not supposed to.
Solid state relays: Relays that use solid state analog components in their logic elements.
Transducers: Current and voltage transformers that reduce high-magnitude signals to standardized low-magnitude signals which relays can use.

References

J.L. Blackburn, "Protective relaying," Marcel Dekker, 1987.

S.H. Horowitz and A.G. Phadke, *Power System Relaying*, Research Studies Press, New York: Wiley & Sons, 1992.

C.R. Mason, *The Art and Science of Protective Relaying*, New York: Wiley & Sons, 1956.

A.G. Phadke and S.H. Horowitz, "Adaptive relaying," *IEEE Computer Applications in Power*, vol. 3, no. 3, pp. 47–51, July 1990.

A.G. Phadke and J.S. Thorp, "Improved control and protection of power systems through synchronized phasor measurements," in *Analysis and Control System Techniques for Electric Power Systems*, part 3, C.T. Leondes, Ed., San Diego: Academic Press, pp. 335–376, 1991.

J.S. Thorp, A.G. Phadke, S.H. Horowitz, and J.E. Beehler, "Limits to impedance relaying," *IEEE Trans. PAS*, vol. 98, no. 1, pp. 246–260, January/February 1979.

Westinghouse Electric Corporation, "Applied Protective Relaying," 1982.

Further Information

In addition to the references provided, papers sponsored by the Power System Relaying Committee of the IEEE and published in the *IEEE Transactions on Power Delivery* contain a wealth of information about protective relaying practices and systems. Publications of CIGRÉ also contain papers on relaying, through their Study Committee 34 on protection. Relays and relaying systems usually follow standards, issued by IEEE in this country, and by such international bodies as the IEC in Europe. The field of computer relaying has been covered in *Computer Relaying for Power Systems*, by A.G. Phadke and J.S. Thorp (New York: Wiley, 1988).

3.5 Transient Operation of Power Systems

R.B. Gungor

Stable operations of power transmission systems have been a great concern of utilities since the beginning of early power distribution networks. The transient operation and the stability under transient operation are studied for existing systems, as well as the systems designed for future operations.

Power systems must be stable while operating normally at steady state for slow system changes under switching operations, as well as under emergency conditions, such as lightning strikes, loss of some generation, or loss of some transmission lines due to **faults**.

The tendency of a power system (or a part of it) to develop torques to maintain its stable operation is known as **stability**. The determination of the stability of a system then is based on the static and dynamic characteristics of its synchronous generators. Although large induction machines may contribute energy to the system during the *subtransient* period that lasts one or two cycles at the start of the **disturbance**, in general, induction machine loads are treated as static loads for **transient stability** calculations. This is one of the simplification considerations, among others.

The per-phase model of an ideal synchronous generator with nonlinearities and the stator resistance neglected is shown in Figure 3.46, where E_g is the generated (excitation) voltage and X_s is the steady-state direct axis *synchronous reactance*. In the calculation of transient and subtransient currents, X_s is replaced by *transient reactance* $X_s^{\prime}$ and *subtransient reactance* $X_s^{\prime\prime}$, respectively.

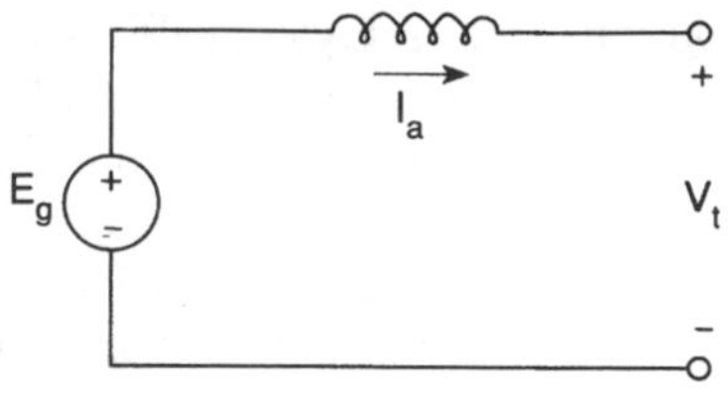

FIGURE 3.46 Per-phase model of an ideal synchronous generator. (*Source*: R.B. Gungor, *Power Systems*, San Diego: Harcourt Brace Jovanovich, 1988, chap. 11. With permission.)

Per-phase electrical power output of the generator for this model is given by Equation (3.32).

$$P_e = \frac{E_g V_t}{X_s} \sin\delta = P_{max} \sin\delta \tag{3.32}$$

where δ is the power angle, the angle between the generated voltage and the terminal voltage.

The simple power-angle relation of Equation (3.32) can be used for real power flow between any two voltages separated by a reactance. For the synchronous machine, the total real power is three times the value calculated by Equation (3.32), when voltages in volts and the reactance in ohms are used. On the other hand, Equation (3.32) gives *per-unit* power when per-unit voltages and reactance are used.

Figure 3.47 shows a sketch of the power-angle relation of Equation (3.32). Here the power P_1 is carried by the machine under δ_1, and P_2 under δ_2. For gradual changes in the output power up to P_{max} for $\delta = 90°$, the machine will be stable. So we can define the **steady-state stability** limit as

$$\delta \leqslant 90° \quad \frac{\partial P}{\partial \delta} > 0 \tag{3.33}$$

A sudden change in the load of the generator, e.g., from P_1 to P_2, will cause the rotor to slow down so that the power angle δ is increased to supply the additional power to the load. However, the deceleration of the rotor cannot stop instantaneously. Hence, although at δ_2 the developed power is sufficient to supply the load, the

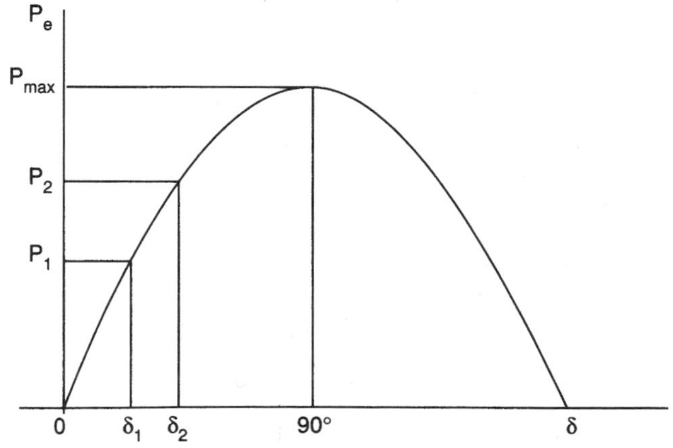

FIGURE 3.47 Power-angle characteristics of ideal synchronous generator. (*Source:* R.B. Gungor, *Power Systems*, San Diego: Harcourt Brace Jovanovich, 1988, chap. 11. With permission.)

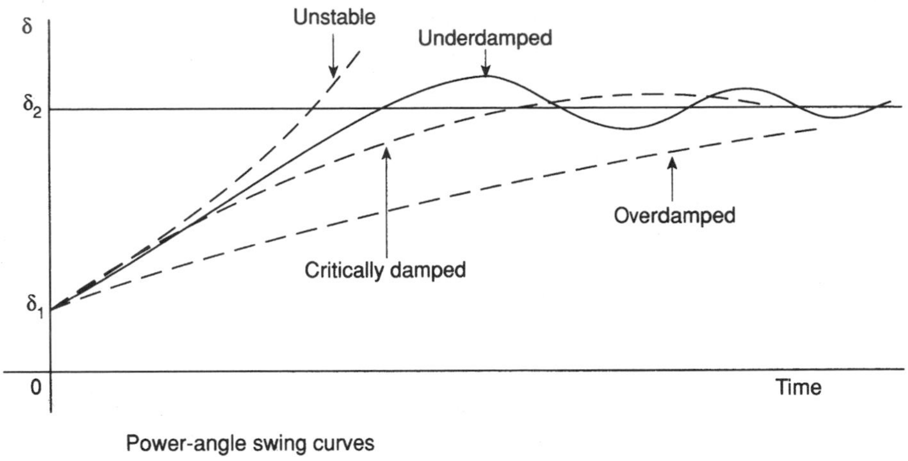

FIGURE 3.48 Typical power angle–time relations. (*Source:* R.B. Gungor, *Power Systems,* San Diego: Harcourt Brace Jovanovich, 1988, chap. 11. With permission.)

rotor will overshoot δ_2 until a large enough opposite torque is built up to stop deceleration. Now the excess energy will start accelerating the rotor to decrease δ. Depending on the inertia and damping, these oscillations will die out or the machine will become unstable and lose its synchronism to drop out of the system. This is the basic transient operation of a synchronous generator. Note that during this operation it may be possible for δ to become larger than 90° and the machine still stay stable. Thus $\delta = 90°$ is not the transient stability limit.

Figure 3.48 shows typical power-angle versus time relations.

In the discussions to follow, the damping (stabilizing) effects of (1) the excitation systems; (2) the speed governors; and (3) the damper windings (copper squirrel-cage embedded into the poles of the synchronous generators) are omitted.

Stable Operation of Power Systems

Figure 3.49 shows an *N*-bus power system with *G* generators.

To study the stability of multimachine transmission systems, the resistances of the transmission lines and transformers are neglected and the reactive networks are reduced down to the generator internal

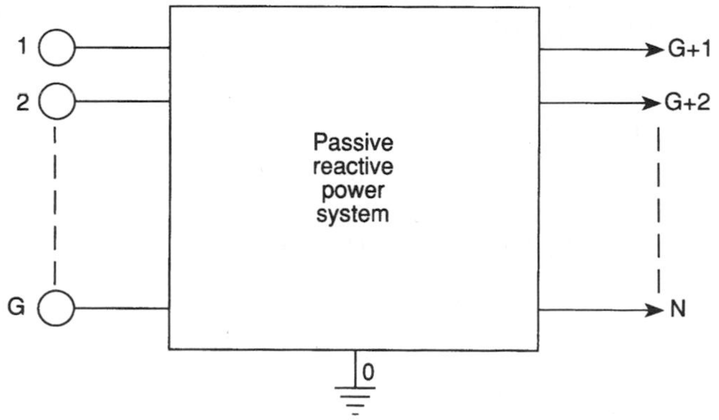

FIGURE 3.49 A multimachine reactive power system. (*Source:* R.B. Gungor, *Power Systems,* San Diego: Harcourt Brace Jovanovich, 1988, chap. 11. With permission.)

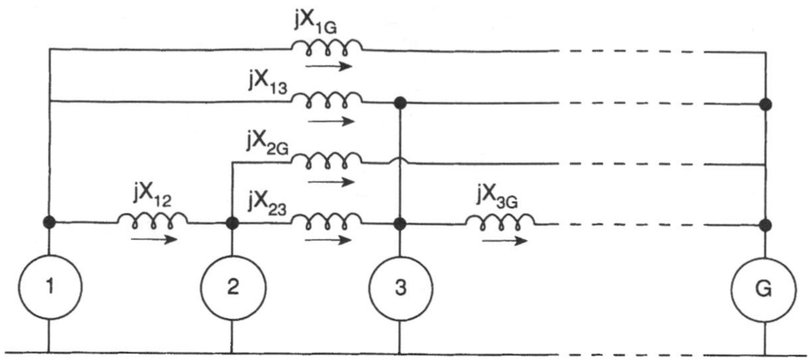

FIGURE 3.50 Multiport reduced reactive network. (*Source:* R.B. Gungor, *Power Systems,* San Diego: Harcourt Brace Jovanovich, 1988, chap. 11. With permission.)

voltages by dropping the loads and eliminating the load buses. One such reduced network is sketched in Figure 3.50.

The power flow through the reactances of a reduced network are

$$P_{ij} = \frac{E_i E_j}{X_{ij}} \sin \delta_{ij} \quad i, j = 1, 2, \ldots, G \tag{3.34}$$

The generator powers are

$$P_i = \sum_{k=1}^{G} P_{ik} \tag{3.35}$$

The system will stay stable for

$$\frac{\partial P_i}{\partial \delta_{ij}} > 0 \quad i = 1, 2, \ldots, G \tag{3.36}$$

Equation (3.36) is observed for two machines at a time by considering all but two (say k and n) of the powers in Equation (3.35) as constants. Since the variations of all powers but k and n are zero, we have

$$\mathrm{d}P_i = \frac{\partial P_i}{\partial \delta_{i1}} \mathrm{d}\delta_{i1} + \frac{\partial P_i}{\partial \delta_{i2}} \mathrm{d}\delta_{i2} + \cdots + \frac{\partial P_i}{\partial \delta_{iG}} \mathrm{d}\delta_{iG} = 0 \tag{3.37}$$

These G-2 equations are simultaneously solved for G-2 $\mathrm{d}\delta_{ij}$s, then these are substituted in $\mathrm{d}P_k$ and $\mathrm{d}P_n$ equations to calculate the partial derivatives of P_k and P_n with respect to δ_{kn} to see if Equations (3.36) for $i = k$ and $i = n$ are satisfied. Then the process is repeated for the remaining pairs.

Although the procedure outlined seems complicated, it is not too difficult to produce a computer algorithm for a given system.

To study the transient stability, dynamic operations of synchronous machines must be considered. An ideal generator connected to an infinite bus (an ideal source) through a reactance is sketched in Figure 3.51.

The so-called *swing equation* relating the accelerating (or decelerating) power (difference between shaft power and electrical power as a function of δ) to the second derivative of the power angle is given in Equation (3.38).

$$\begin{aligned} P_a &= P_s - P_e \\ M \frac{\mathrm{d}^2 \delta}{\mathrm{d}t^2} &= P_s - \frac{E_d E_i}{X} \sin \delta \end{aligned} \tag{3.38}$$

where $M = HS/180f$ (MJ/electrical degree); H is the inertia constant (MJ/MVA); S is the machine rating (MVA); f is the frequency (Hz); P_s is the shaft power (MW).

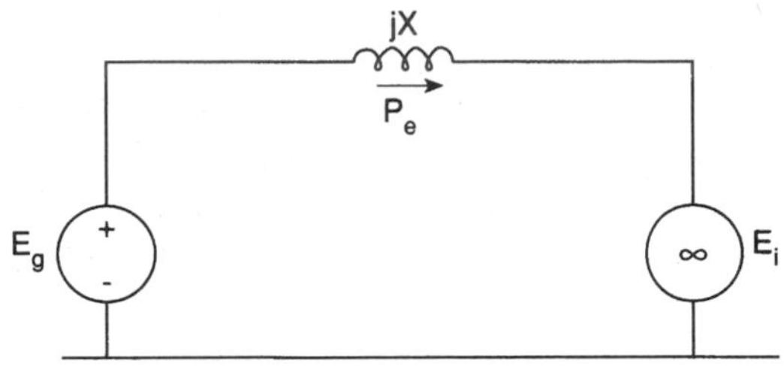

FIGURE 3.51 An ideal generator connected to an infinite bus. (*Source:* R.B. Gungor, *Power Systems,* San Diego: Harcourt Brace Jovanovich, 1988, chap. 11. With permission.)

For a system of G machines, a set of G swing equations as given in Equation (3.39) must be solved simultaneously.

$$M_i \frac{d^2\delta_i}{dt^2} = P_{s_i} - P_{\max_i} \sin \delta_i \quad i = 1, 2, \ldots, G \tag{3.39}$$

The swing equation of the single-machine system of Figure 3.51 can be solved either graphically or analytically. For graphical integration, which is called *equal-area criterion,* we represent the machine by its subtransient reactance, assuming that electrical power can be calculated by Equation (3.32), and during the transients the shaft power P_s remains constant. Then, using the power-angle curve(s), we sketch the locus of operating point on the curve(s) and equate the areas for stability. Figure 3.52 shows an example for which the shaft power of the machine is suddenly increased from the initial value of P_o to P_s.

The excess energy (area A_1) will start to accelerate the rotor to increase δ from δ_o to δ_m for which the area (A_2) above P_s equals the area below. These areas are

$$\begin{aligned} A_1 &= P_s(\delta_s - \delta_o) - \int_{\delta_o}^{\delta_s} P_{\max} \sin \delta d\delta \\ A_2 &= \int_{\delta_s}^{\delta_m} P_{\max} \sin \delta d\delta - P_s(\delta_m - \delta_s) \end{aligned} \tag{3.40}$$

Substituting, the values of P_o, P_s, δ_o, and δ_s, δ_m can be calculated.

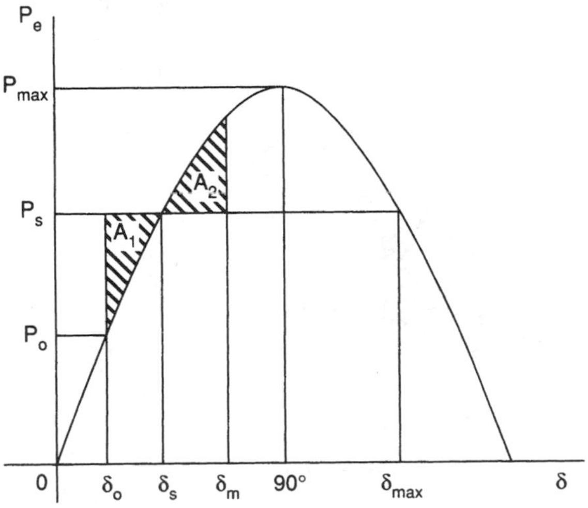

FIGURE 3.52 A sudden loading of a synchronous generator. (*Source:* R.B. Gungor, *Power Systems,* San Diego: Harcourt Brace Jovanovich, 1988, chap. 11. With permission.)

Figure 3.53 illustrates another example, where a three-phase fault reduces the power transfer to infinite bus to zero. δ_{cc} is the critical clearing angle beyond which the machine will not stay stable.

The third example, shown in Figure 3.54, indicates that the power transfers before, during, and after the fault are different. Here the system is stable as long as $\delta_m \leq \delta_{max}$.

For the analytical solution of the swing equation a *numerical integration* technique is used (Euler's method, modified Euler's method, Runge-Kutta method, etc.). The latter is most commonly used for computer algorithms.

The solution methods developed are based on various assumptions. As before, machines are represented by subtransient reactances, electrical powers can be calculated by Equation (3.32), and the shaft power

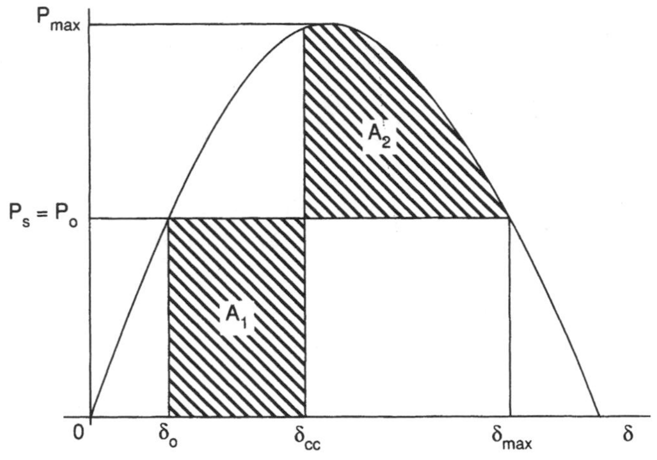

FIGURE 3.53 Critical clearing angle for stability. (*Source:* R.B. Gungor, *Power Systems,* San Diego: Harcourt Brace Jovanovich, 1988, chap. 11. With permission.)

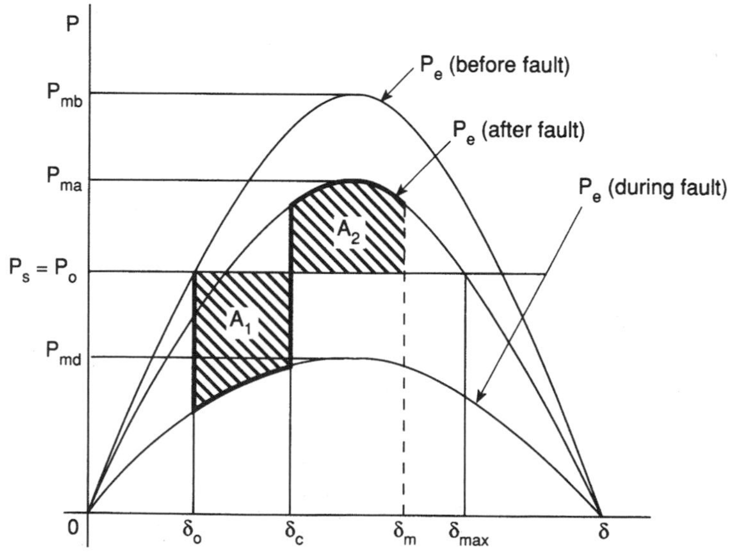

FIGURE 3.54 Power-angle relation for power transfer during fault. (*Source:* R.B. Gungor, *Power Systems,* San Diego: Harcourt Brace Jovanovich, 1988, chap. 11. With permission.)

does not change during transients. In addition, the velocity increments are assumed to start at the beginning of time increments, and acceleration increments start at the middle of time increments; finally, an average acceleration can be used where acceleration is discontinuous (e.g., where circuit breakers open or close).

Figure 3.55 shows a sketch of angle, velocity, and acceleration changes related to time as outlined above. Under these assumptions the next value of the angle δ can be obtained from the previous value as

$$\delta_{k+1} = \delta_k + \Delta_{k+1}\delta = \delta_k + \Delta_k\delta + \frac{(\Delta t)^2}{M} P_{\text{ak}} \tag{3.41}$$

where the accelerating power is

$$P_{\text{ak}} = P_{\text{s}} - P_{ek}$$

and

$$P_{ek} = P_{\text{max}k} \sin \delta_k$$

For hand calculations a table, as shown in Table 3.3, can be set up for fast processing.

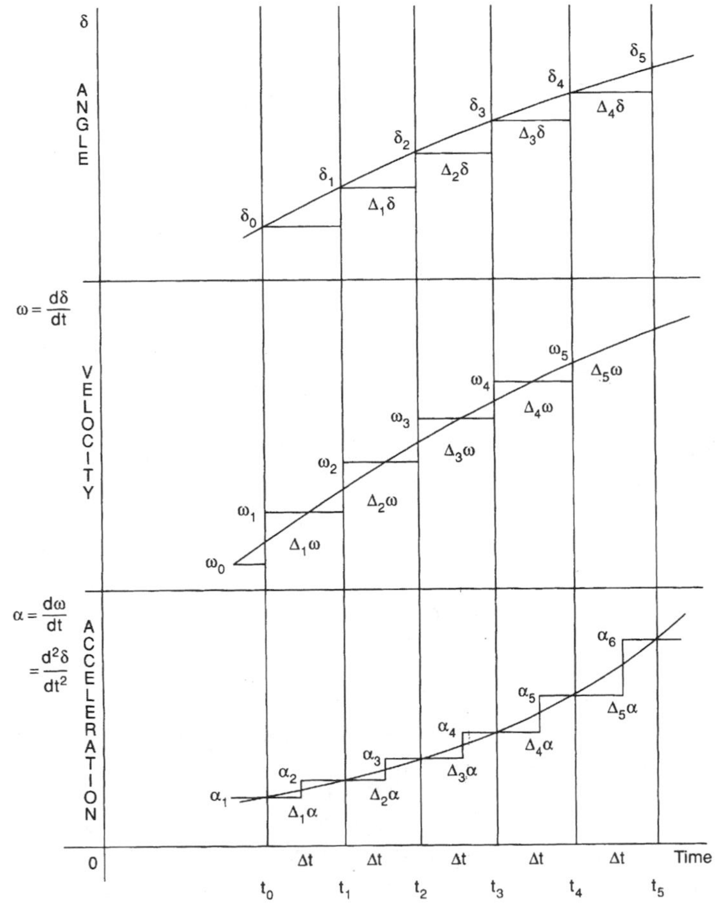

FIGURE 3.55 Incremental angle, velocity, and acceleration changes versus time. (*Source:* R.B. Gungor, *Power Systems*, San Diego: Harcourt Brace Jovanovich, 1988, chap. 11. With permission.)

Table 3.3 Numerical Calculations of Swing Equations

n	t	P_{max}	P_e	P_{ak}	$\frac{(\Delta t)^2 P_a}{M}$	$\Delta_{k+1}\delta$	δ_k
0	0_-						
0	0_+						
0	0_{av}						
1	Δt						
2	$2\Delta t$						
3	$3\Delta t$						
4	$4\Delta t$						
5	$5\Delta t$						
6	$6\Delta t$						

Computer algorithms are developed by using the before-fault, during-fault, and after-fault Z_{BUS} matrix of the reactive network reduced to generator internal voltages with generators represented by their subtransient reactances. Each generator's swing curve is obtained by numerical integration of its power angle for a specified condition, then a set of swing curves is tabulated or graphed for observation of the transient stability.

Defining Terms

Critical clearing angle: Power angle corresponding to the critical clearing time.

Critical clearing time: The maximum time at which a fault must be cleared for the system to stay transiently stable.

Disturbance (fault): A sudden change or a sequence of changes in the components or the formation of a power system.

Large disturbance: A disturbance for which the equations for dynamic operation cannot be linearized for analysis.

Power angle: The electrical angle between the generated and terminal voltages of a synchronous generator.

Small disturbance: A disturbance for which the equations for dynamic operation can be linearized for analysis.

Stability: The tendency of a power system (or a part of it) to develop torques to maintain its stable operation for a disturbance.

Steady-state stability: A power system is steady-state stable if it reaches another steady-state operating point after a small disturbance.

Transient operation: A power system operating under abnormal conditions because of a disturbance.

Transient stability: A power system is transiently stable if it reaches a steady-state operating point after a large disturbance.

References

J. Arrillaga, C.P. Arnold, and B.J. Harker, *Computer Modeling of Electrical Power Systems*, New York: Wiley, 1983.

A.R. Bergen, *Power System Analysis*, Englewood Cliffs, N.J.: Prentice-Hall, 1986.

H.E. Brown, *Solution of Large Networks by Matrix Methods,* New York: Wiley, 1985.

A.A. Fouad and V. Vittal, *Power System Transient Stability Analysis*, Englewood Cliffs, N.J.: Prentice-Hall, 1992.

J.D. Glover and M. Sarma, *Power System Analysis and Design*, Boston: PWS Publishers, 1987.

C.A. Gross, *Power System Analysis*, 2nd ed., New York: Wiley, 1986.

R.B. Gungor, *Power Systems*, San Diego: Harcourt Brace Jovanovich, 1988.

G.T. Heydt, *Computer Analysis Methods for Power Systems*, New York: Macmillan, 1986.

W.D. Stevenson, *Elements of Power System Analysis*, 4th ed., New York: McGraw-Hill, 1982.
Y. Wallach, *Calculations & Programs for Power System Networks*, Englewood Cliffs, N.J.: Prentice-Hall, 1986.

Further Information

In addition to the references listed above, further and more recent information can be found in IEEE publications, such as *IEEE Transactions on Power Systems, IEEE Transactions on Power Delivery, IEEE Transactions on Energy Conversion*, and *IEEE Transactions on Automatic Control.*

Power Engineering Review and *Computer Applications in Power* of the IEEE are good sources for paper summaries.

Finally, *IEEE Transactions on Power Apparatus and Systems* dating back to the 1950s can be consulted.

3.6 Planning

J. Duncan Glover

Introduction

An electric utility transmission system performs three basic functions: delivers outputs from generators to the system; supplies power to the distribution system; and provides for power interchange with other utilities. The electric utility industry has developed planning principles and criteria to ensure that the transmission system reliably performs these basic functions.

The North American Electric Reliability Council (NERC) has provided definitions of the terms **reliability**, **adequacy**, and **security** (see Defining Terms at the end of this section).

System reliability may be viewed from two perspectives: short-term reliability and long-term reliability. The system operator is primarily concerned with real-time security aspects in the short term, that is, supplying steady, uninterrupted service under existing operating conditions and as they occur over the next few minutes, hours, days, or months. The transmission planning engineer, however, is concerned not only with security aspects in the short term but also adequacy and security aspects in the long term, as many as 25 or more years into the future.

The actual construction of a major transmission facility requires three to five years or more, depending largely on the siting and certification process. As such, the planning process requires up to ten years prior to operation of these facilities to ensure that they are available when required. The long lead times, environmental impacts, and high costs required for new transmission facilities require careful and near-optimal planning. Future changes in system operating conditions, such as changes in spatial load and generation patterns, create uncertainties that challenge the transmission planning engineer to select the best technical solution among several alternatives with due consideration of nontechnical factors. Transmission planning strives to maintain an optimal balance between system reliability, environmental impacts, and cost under future uncertainties.

Before transmission planning is started, long-term load forecasting and generation planning are completed. In long-term load forecasting, peak and off-peak loads in each area of the system under study are projected, year by year, from the present up to 25 years into the future. Such forecasts are based on present and past load trends, population growth patterns, and economic indicators. In generation planning, generation resources are selected with sufficient generation reserve margins to meet projected customer loads with adequate quality and reliability in an economic manner. New generating units both at new plant sites and at existing plants are selected, and construction schedules are established to ensure that new generation goes online in time to meet projected loads.

The results of long-term load forecasting and generation planning are used by transmission planning engineers to design the future transmission system so that it performs its basic functions. The following are selected during the transmission planning process:

- Routes for new lines
- Number of circuits for each route or right-of-way
- EHV vs. HVDC lines

- Overhead vs. underground line construction
- Types of towers for overhead lines
- Voltage levels
- Line ratings
- Shunt reactive and series capacitive line compensation
- Number and locations of substations
- Bus and circuit-breaker configurations at substations
- Circuit-breaker ratings
- Number, location, and ratings of bulk power system transformers
- Number, location, and ratings of voltage-regulating transformers and phase-shifting transformers
- Number, location, and ratings of shunt capacitor banks and synchronous condensers for voltage control
- Number, location, and ratings of flexible ac transmission system (FACTS) modules, including static VAR compensators (SVCs) and thyristor-controlled series compensators for control of voltage and power flows
- Basic insulation levels (BILs)
- Surge arrester locations and ratings
- Protective relaying schemes
- Communications facilities
- Upgrades of existing circuits
- Reinforcements of system interconnections.

Planning Tools

As electric utilities have grown in size and the number of interconnections has increased, making the above selections during the planning process has become increasingly complex. The increasing cost of additions and modifications has made it imperative that planning engineers consider a wide range of design options and perform detailed studies on the effects on the system of each option based on a number of assumptions: normal and emergency operating conditions, peak and off-peak loadings, and present and future years of operation. A large volume of network data must be collected and accurately handled. To assist the planning engineer, the following digital computer programs are used [Glover and Sarma, 2002]:

1. *Power-flow programs.* Power-flow (also called load-flow) programs compute voltage magnitudes, phase angles, and transmission line power flows for a power system network under steady-state operating conditions. Other output results, including transformer tap settings, equipment losses, and reactive power outputs of generators and other devices, are also computed. To do this, the locations, sizes, and operating characteristics of all loads and generation resources of the system are specified as inputs. Other inputs include the network configuration as well as ratings and other characteristics of transmission lines, transformers, and other equipment. Today's computers have sufficient storage and speed to compute in less than one minute power-flow solutions for networks with more than 100,000 buses and 100,000 transmission lines. High-speed printers then print out the complete solution in tabular form for analysis by the planning engineer. Also available are interactive power-flow programs, whereby power-flow results are displayed on computer screens in the form of interactive and animated single-line diagrams. The engineer uses these to modify the network from a keyboard or with a mouse and can readily visualize the results. Spreadsheet analyses are also used. The computer's large storage and high-speed capabilities allow the engineer to run the many different cases necessary for planning.
2. *Transient stability programs.* Transient stability programs are used to study power systems under disturbance conditions to predict whether synchronous generators remain in synchronism and system stability is maintained. System disturbances can be caused by the sudden loss of a generator or a transmission line, by sudden load increases or decreases, and by short circuits and switching operations. The stability program combines power-flow equations and generator dynamic equations to compute the angular swings of machines during disturbances. The program also computes critical clearing times for

network faults and allows the planning engineer to investigate the effects of various network modifications, machine parameters, disturbance types, and control schemes.

3. *Short-circuits programs.* Short-circuits programs compute three-phase and line-to-ground fault currents in power system networks in order to evaluate circuit breakers and relays that detect faults and control circuit breakers. Minimum and maximum short-circuit currents are computed for each circuit breaker and relay location under various system operating conditions, such as lines or generating units out of service, in order to specify circuit breaker ratings and protective relay schemes.
4. *Transients programs.* Transients programs compute the magnitudes and shapes of transient overvoltages and currents that result from switching operations and lightning strikes. Planning engineers use the results of transients programs to specify BILs for transmission lines, transformers, and other equipment and to select surge arresters that protect equipment against transient overvoltages.

Research efforts aimed at developing computerized, automated transmission planning tools are ongoing. Examples and references are given in Overbye and Weber (2001). Other programs for transmission planning include production cost, investment cost, relay coordination, power-system database management, transformer thermal analysis, and transmission line design programs. Some of the vendors that offer software packages for transmission planning are given as follows:

- ABB Network Control Ltd., Switzerland (www.abb.com)
- CYME International, St. Bruno, Quebec, Canada (www.cyme.com)
- Electrocon International, Inc., Ann Arbor, MI. (www.electrocon.com)
- GE Energy, Schenectady, NY (www.gepower.com)
- PowerWorld Corporation, Champaign, IL (www.powerworld.com)
- Shaw Power Technologies, Inc., Schenectady, NY (www.shawgrp.com)
- Operation Technology, Inc., Irvine, CA (www.etap.com)

Basic Planning Principles

The electric utility industry has established basic planning principles intended to provide a balance among all power system components so as not to place too much dependence on any one component or group of components. Transmission planning criteria are developed from these principles along with actual system operating history and reasonable contingencies. These planning principles are given as follows:

1. Maintain a balance among power system components based on size of load, size of generating units and power plants, the amount of power transfer on any transmission line or group of lines, and the strength of interconnections with other utilities. In particular:

   a. Avoid excessive generating capacity at one unit, at one plant, or in one area.
   b. Avoid excessive power transfer through any single transformer, through any transmission line, circuit, tower, or right-of-way, or though any substation.
   c. Provide interconnection capacity to neighboring utilities, which is commensurate with the size of generating units, power plants, and system load.

2. Provide transmission capability with ample margin above that required for normal power transfer from generators to loads in order to maintain a high degree of flexibility in operation and to meet a wide range of contingencies.
3. Provide for power system operation such that all equipment loadings remain within design capabilities.
4. Utilize switching arrangements, associated relay schemes, and controls that permit:

   a. Effective operation and maintenance of equipment without excessive risk of uncontrolled power interruptions.
   b. Prompt removal and isolation of faulted components.
   c. Prompt restoration in the event of loss of any part of the system.

Equipment Ratings

Transmission system loading criteria used by planning engineers are based on equipment ratings. Both normal and various emergency ratings are specified. Emergency ratings are typically based on the time required for either emergency operator actions or equipment repair times. For example, up to two hours may be required following a major event such as loss of a large generating unit or a critical transmission facility in order to bring other generating resources online and to perform appropriate line-switching operations. The time to repair a failed transmission line typically varies from two to ten days, depending on the type of line (overhead, underground cable in conduit, or pipe-type cable). The time required to replace a failed bulk-power-system transformer is typically 30 days. As such, ratings of each transmission line or transformer may include normal, two-hour emergency, two- to ten-day emergency, and in some cases 30-day emergency ratings.

The rating of an overhead transmission line is based on the maximum temperature of the conductors. Conductor temperature affects the conductor sag between towers and the loss of conductor tensile strength due to annealing. If the temperature is too high, proscribed conductor-to-ground clearances [ANSI-C2, 2002] may not be met, or the elastic limit of the conductor may be exceeded such that it cannot shrink to its original length when cooled. Conductor temperature depends on the current magnitude and its time duration, as well as on ambient temperature, wind velocity, solar radiation, and conductor surface conditions. Standard assumptions on ambient temperature, wind velocity, etc. are selected, often conservatively, to calculate overhead transmission line ratings [ANSI/IEEE Std. 738–93, 1993]. It is common practice to have summer and winter normal line ratings, based on seasonal ambient temperature differences. Also, in locations with higher prevailing winds, such as coastal areas, larger normal line ratings may be selected. Emergency line ratings typically vary from 110 to 120% of normal ratings. Real-time monitoring of actual conductor temperatures along a transmission line has also been used for online dynamic transmission line ratings [Henke and Sciacca, 1989].

Normal ratings of bulk power system transformers are determined by manufacturers' nameplate ratings. Nameplate ratings are based on the following ANSI/IEEE standard conditions:

1. Continuous loading at nameplate output
2. 30°C average ambient temperature (never exceeding 40°C)
3. 110°C average hot-spot conductor temperature (never exceeding 120°C) for 65°C average-winding-rise transformers [ANSI/IEEE C57.91, 1995].

For 55°C average-winding-rise transformers, the hot-spot temperature limit is 95°C average (never exceeding 105°C). The actual output that a bulk power system transformer can deliver at any time with normal life expectancy may be more or less than the nameplate rating, depending on the ambient temperature and actual temperature rise of the windings. Emergency transformer ratings typically vary from 130 to 150% of nameplate ratings.

Planning Criteria

Transmission system planning criteria have been developed from the above planning principles and equipment ratings as well as from actual system operating data, probable operating modes, and equipment failure rates. These criteria are used to plan and build the transmission network with adequate margins to ensure a reliable supply of power to customers under reasonable equipment-outage contingencies. The transmission system should perform its basic functions under a wide range of operating conditions. Transmission planning criteria include equipment loading criteria, transmission voltage criteria, stability criteria, and regional planning criteria.

Equipment Loading Criteria

Typical equipment loading criteria are given in Table 3.4. With no equipment outages, transmission equipment loadings should not exceed normal ratings for all realistic combinations of generation and interchange. Operation of all generating units, including base-loaded and peaking units during peak load periods as well as operation of various combinations of generation and interchange during off-peak periods,

Table 3.4 Typical Transmission Equipment Loading Criteria

Equipment Out of Service	Rating Not to Be Exceeded	Comment
None	Normal	
Any generator	Normal	
Any line or transformer	Two-hour emergency	Before switching
Any line and any transformer*	Two- to ten-day emergency	After switching required for both outages. Line repair time.
Any line and any generator	Two- to ten-day emergency	After switching required for both outages. Line repair time.
Any transformer and any generator	30-day emergency	After switching required for both outages. Install spare transformer.

*Some utilities do not include double-contingency outages in transmission system loading criteria.

should be considered. Also, normal ratings should not be exceeded with all transmission lines and transformers in service and with any generating unit out of service.

With any single-contingency outage, emergency ratings should not be exceeded. One loading criterion is not to exceed two-hour emergency ratings when any transmission line or transformer is out of service. This gives time to perform switching operations and change generation levels, including use of peaking units, to return to normal loadings.

With some of the likely double-contingency outages, the transmission system should supply all system loads without exceeding emergency ratings. One criterion is not to exceed two- to ten-day emergency ratings when any line and any transformer are out of service or when any line and any generator are out of service. This gives time to repair the line. With the outage of any transformer and any generator, 30-day emergency ratings should not be exceeded, which gives time to install a spare transformer.

The loading criteria in Table 3.4 do not include all types of double-contingency outages. For example, the outage of a double-circuit transmission line or two transmission lines in the same right-of-way is not included. Also, the loss of two transformers in the same load area is not included. Under these double-contingency outages, it may be necessary to shed load at some locations during heavy load periods. Although experience has shown that these outages are relatively unlikely, their consequences should be evaluated in specific situations. Factors to be evaluated include the size of load served, the degree of risk, and the cost of reinforcement.

Specific loading criteria may also be required for equipment serving critical loads and critical load areas. One criterion is to maintain service to critical loads under a double-contingency outage with the prior outage of any generator.

Transmission Voltage Criteria

Transmission voltages should be maintained within suitable ranges for both normal and reasonable emergency conditions. Abnormal transmission voltages can cause damage or malfunction of transmission equipment such as circuit breakers or transformers and adversely affect many customers. Low transmission voltages tend to cause low distribution voltages, which in turn cause increased distribution losses as well as higher motor currents at customer loads and at power plant auxiliaries. Transmission voltage planning criteria are intended to be conservative.

Maximum planned transmission voltage is typically 105% of rated nominal voltage for both normal and reasonable emergency conditions. Typical minimum planned transmission voltages are given in Table 3.5. System conditions in Table 3.5 correspond to equipment out of service in Table 3.4. Single-contingency outages correspond to the loss of any line, any transformer, or any generator. Double-contingency outages correspond to the loss of any transmission line and transformer, any transmission line and generator, any transformer and generator, or any two generators.

Table 3.5 Typical Minimum Transmission Voltage Criteria

	Planned Minimum Transmission Voltage at Substations, % of Normal		
System Condition	Generator Station	EHV Station	HV Station
Normal	102	98	95–97.5
Single-contingency outage	100	96	92.5–95
Double-contingency outage*	98	94	92.5

*Some utilities do not include double-contingency outages in planned minimum voltage criteria.

Typical planned minimum voltage criteria shown in Table 3.5 for EHV (345 kV and higher) substations and for generator substations are selected to maintain adequate voltage levels at interconnections, at power plant auxiliary buses, and on the lower-voltage transmission systems. Typical planned minimum voltage criteria for lower HV (such as, 138 kV, 230 kV) transmission substations vary from 95 to 97.5% of nominal voltage under normal system conditions to as low as 92.5% of nominal under double-contingency outages.

Equipment used to control transmission voltages includes voltage regulators at generating units (excitation control), tap-changing transformers, regulating transformers, synchronous condensers, shunt reactors, shunt capacitor banks, and FACTS modules. When upgrades are selected during the planning process to meet planned transmission voltage criteria, some of this equipment should be assumed out of service.

Stability Criteria

System stability is the ability of all synchronous generators in operation to stay in synchronism with each other while moving from one operating condition to another. Steady-state stability refers to small changes in operating conditions, such as normal load changes. Transient stability refers to larger, abrupt changes, such as the loss of the largest generator or a short circuit followed by circuit breakers opening, where synchronism or loss of synchronism occurs within a few seconds. Dynamic stability refers to longer time periods, from minutes up to half an hour following a large, abrupt change, where steam generators (boilers), automatic generation control, and system operator actions affect stability.

In the planning process, steady-state stability is evaluated via power-flow programs by the system's ability to meet equipment loading criteria and transmission voltage criteria under steady-state conditions. Transient stability is evaluated via stability programs by simulating system transient response for various types of disturbances, including short circuits and other abrupt network changes. The planning engineer designs the system to remain stable for the following typical disturbances:

1. With all transmission lines in service, a permanent three-phase fault (short circuit) occurs on any transmission line, on both transmission lines on any double-circuit tower, or at any bus; the fault is successfully cleared by primary relaying.
2. With any one transmission line out of service, a permanent three-phase fault occurs on any other transmission line; the fault is successfully cleared by primary relaying.
3. With all transmission lines in service, a permanent three-phase fault occurs on any transmission line; backup relaying clears the fault after a time delay, due to a circuit-breaker failure.

Regional Planning Criteria

The North American Electric Reliability Council (NERC) is a nonprofit company formed in 1968, as a result of the 1965 northeast blackout, to promote the reliability of bulk electric systems that serve North America.

NERC works with all segments of the electric industry, to "keep the lights on" by developing and encouraging compliance with rules for the reliable operation of the electric grid. Its mission is to ensure reliability of the North American power grid (www.nerc.com).

As shown in Figure 3.56, NERC comprises ten regional reliability councils that account for virtually all the electricity supplied in the United States, Canada, and a portion of Mexico [www.nerc.com]. The interconnected bulk electric systems in this overall region comprise many individual systems, each with its own electrical characteristics, customers, as well as geography, weather, economy, regulations, and political climates [NERC, 1997]. All electric systems within an integrated network are electrically connected, and whatever one system does can affect the reliability of the other systems. In order to maintain the reliability of the interconnected bulk electric systems, the regional councils and their members strive to comply with the NERC Planning Standards for Transmission Systems, which are excerpted from [NERC, 1997] as follows.

S1. The interconnected transmission systems shall be planned, designed, and constructed such that with all transmission facilities in service and with normal (precontingency) operating procedures in effect, the network can deliver generator unit output to meet projected customer demands and provide contracted firm (nonrecallable reserved) transmission services, at all demand levels, under the conditions defined in Category A of Table 3.6.

S2. The interconnected transmission systems shall be planned, designed, and constructed such that the network can be operated to supply projected customer demands and contracted firm (nonrecalled reserved) transmission services, at all demand levels, under the conditions of the contingencies as defined in Category B of Table 3.6.

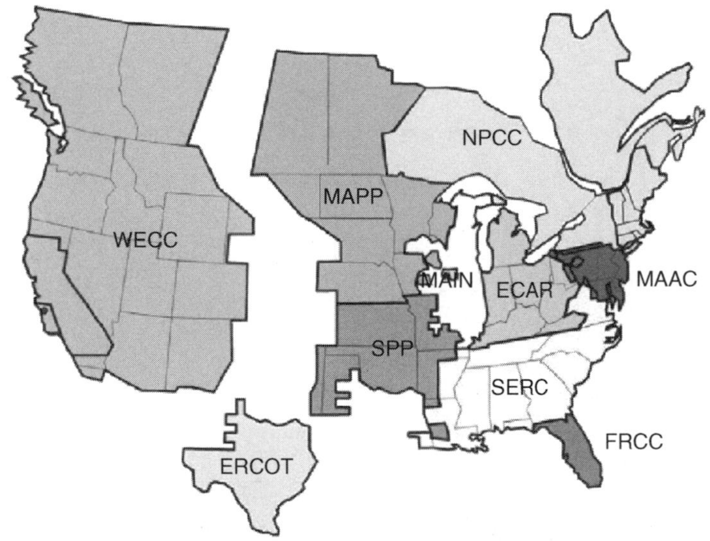

ECAR - East Central Area Reliability Coordination Agreement
FRCC - Florida Reliability Coordinating Council
MAAC - Mid-Atlantic Area Council
MAIN - Mid-America Interconnected Network, Inc.
MAAP - Mid-Continent Area Power Pool
NPCC - Northeast Power Coordinating Council
SERC - Southeastern Electric Reliability Council
SPP - Southwest Power Pool, Inc.
WECC - Western Electricity Coordinating Council
ERCOT - Electric Reliability Council of Texas, Inc.

FIGURE 3.56 Ten regional reliability councils established by NERC. (*Source*: www.nerc.com. With permission.)

Table 3.6 Transmission Systems Standards — Normal and Contingency Conditions [NERC, 1997, with permission]

Category	Contingencies: Initiating Events and Contingency Components	Components Out of Service	System Limits or Impacts: Thermal Limits	Voltage Limits	System Stable	Loss of Demand or Curtailed Firm Transfers	Cascading[c] Outages
A – No Contingencies	All facilities in service	None	Normal Applicable	Normal Applicable	Yes	No	No
B – Event resulting in the loss of a single component	Single line ground (SLG) or three-phase (3∅Fault), with normal clearing:	Single	Rating[a] (A/R)	Rating[a] (A/R)	Yes	No[b]	No
	1. Generator	Single	A/R	A/R	Yes	No[b]	No
	2. Transmission circuit	Single	A/R	A/R	Yes	No[b]	No
	3. Transformer	Single	A/R	A/R	Yes	No[b]	No
	Loss of a component without a fault						
	Single pole block, with normal clearing:						
	4. Single pole (dc) line	Single	A/R	A/R	Yes	No[b]	No
C – Events resulting in the loss of two or more (multiple) components	SLG fault, with normal clearing:						
	1. Bus section	Multiple	A/R	A/R	Yes	Planned[d]	No
	2. Breaker (failure or internal fault)	Multiple	A/R	A/R	Yes	Planned[d]	No
	SLG or 3∅Fault, with normal clearing, manual system adjustments, followed by another SLG or 3∅Fault, with normal clearing:						

	3. Category B (B1, B2, B3, or B4) contingency, manual system adjustments, followed by another Category B (B1, B2, B3, or B4) contingency	Multiple	A/R	A/R	Yes	Planned[d]	No
	Bipolar block, with normal clearing:						
	4. Bipolar (dc) line fault non (3∅, with normal clearing:	Multiple	A/R	A/R	Yes	Planned[d]	No
	5. Double circuit towerline SLG fault, with delayed clearing:	Multiple	A/R	A/R	Yes	Planned[d]	No
	6. Generator	Multiple	A/R	A/R	Yes	Planned[d]	No
	7. Transmission circuit	Multiple	A/R	A/R	Yes	Planned[d]	No
	8. Transformer						
	9. Bus section						
D[e] – Extreme event resulting in two or more (multiple) components removed or cascading out of service	3∅Fault, with delayed clearing (stuck breaker or protection system failure):	Evaluate for risks and consequences					
	1. Generator 2. Transmission circuit	• May involve substantial loss of customer demand and generation in a widespread area or areas					
	3. Transformer 4. Bus section						
	3∅Fault, with normal clearing:						
	5. Breaker (failure or internal fault)	• Portions or all of the interconnected systems may or may not achieve a new, stable operating point					
	Other:	Evaluation of these events may require joint studies with neighboring systems					

(continued)

Table 3.6 Continued

	Contingencies		System Limits or Impacts				
Category	Initiating Events and Contingency Components	Components Out of Service	Thermal Limits	Voltage Limits	System Stable	Loss of Demand or Curtailed Firm Transfers	Cascading[c] Outages
	6. Loss of towering with three or more circuits						
	7. All transmission lines on a common right-of-way	Document measures or procedures to mitigate the extent and effects of such events					
	8. Loss of a substation (one voltage level plus transformers)						
	9. Loss of a switching station (one voltage level plus transformers)	Mitigation or elimination of the risks and consequences of these events shall be at the discretion of the entities responsible for the reliability of the interconnected transmission systems					
	10. Loss of all generating units at a station						
	11. Loss of a large load or major load center						
	12. Failure of a fully redundant special protection system (or remedial action scheme) to operate when required						
	13. Operation, partial operation, or misoperation of a fully redundant special protection system (or remedial action scheme) for an event or condition for which it was not intended to operate						

	14. Impact of severe power swings or oscillations from disturbances in another regional council		

[a]Applicable rating (A/R) refers to the applicable normal and emergency facility thermal rating or system voltage limit as determined and consistently applied by the system or facility owner.

[b]Planned or controlled interruption of generators or electric supply to radial customers or some local network customers, connected to or supplied by the faulted component or by the affected area, may occur in certain areas without impacting the overall security of the interconnected transmission systems. To prepare for the next contingency, system adjustments are permitted, including curtailments of contracted firm (nonrecallable reserved) electric power transfers.

[c]Cascading is the uncontrolled successive loss of system elements triggered by an incident at any location. Cascading results in widespread service interruption which cannot be restrained from sequentially spreading beyond an area predetermined by appropriate studies.

[d]Depending on system design and expected system impacts, the controlled interruption of electric supply to customers (load shedding), the planned removal from service of certain generators, or the curtailment of contracted firm (nonrecallable reserved) electric power transfers may be necessary to maintain the overall security of the interconnected transmission systems.

[e]A number of extreme contingencies that are listed under Category D and judged to be critical by the transmission planning entities will be selected for evaluation. It is not expected that all possible facility outages under each listed contingency of Category D will be evaluated.

The transmission systems shall also be capable of accommodating planned bulk electric equipment maintenance outages and continuing to operate within thermal, voltage, and stability limits under the conditions of the contingencies as defined in Category B of Table 3.6.

S3. The interconnected transmission systems shall be planned, designed, and constructed such that the network can be operated to supply projected customer demands and contracted firm (nonrecalled reserved) transmission services, at all demand levels, under the conditions of the contingencies as defined in Category C of Table 3.6. The controlled interruption of customer demand, the planned removal of generators, or the curtailment of firm (nonrecalled reserved) power transfers may be necessary to meet this standard.

The transmission systems also shall be capable of accommodating bulk electric equipment maintenance outages and continuing to operate within thermal, voltage, and stability limits under the conditions of the contingencies as defined in Category C of Table 3.6.

S4. The interconnected transmission systems shall be evaluated for the risks and consequences of a number of each of the extreme contingencies that are listed under Category D of Table 3.6.

Following the blackout that occurred on August 14, 2003, in the northeastern United States and Canada, the U.S.–Canada Power System Outage Task Force recommended that NERC reliability standards be made mandatory and enforceable [U.S.—Canada Power System Outage Task Force, 2004, and www.nerc.com]. Mandatory compliance with NERC Planning Standards is required to maintain the reliability of interconnected bulk electric systems.

Value-Based Transmission Planning

Some utilities have used a value-of-service concept in transmission planning [EPRI, 1986]. This concept establishes a method of assigning a dollar value to various levels of reliability in order to balance reliability and cost. For each particular outage, the amount and dollar value of unserved energy are determined. Dollar value of unserved energy is based on rate surveys of various types of customers. If the cost of the transmission project required to eliminate the outage exceeds the value of service, then that project is given a lower priority. As such, reliability is quantified and benefit-to-cost ratios are used to compare and prioritize planning options.

Defining Terms

The NERC defines reliability and the related terms adequacy and security as follows [NERC, 1988]:

Adequacy: The ability of the bulk-power electric system to supply the aggregate electric power and energy requirements of the consumers at all times, taking into account scheduled and unscheduled outages of system components.

Reliability: In a bulk-power electric system, reliability is the degree to which the performance of the elements of that system results in power being delivered to consumers within accepted standards and in the amount desired. The degree of reliability may be measured by the frequency, duration, and magnitude of adverse effects on consumer service.

Security: The ability of the bulk-power electric system to withstand sudden disturbances such as electric short circuits or unanticipated loss of system components.

References

ANSI C2–2002, *National Electrical Safety Code*, 2002 ed., New York: IEEE, 2002.

ANSI/IEEE C57.91–1995, *IEEE Guide for Loading Mineral-Oil Immersed Power Transformers*, New York: IEEE, 1995.

ANSI/IEEE Std. 738–1993, *IEEE Standard for Calculating the Current-Temperature of Bare Overhead Conductors*, New York: IEEE, 1993.

H. Back et al., "PLATINE — A new computerized system to help in planning the power transmission networks," *IEEE Trans. Power Systems*, vol. 4, no. 1, pp. 242–247, 1989.

Electric Power Research Institute (EPRI), Value of Service Reliability to Consumers, Report EA-4494, Palo Alto, CA: EPRI, March 1986.

J.D. Glover and M.S. Sarma, *Power System Analysis and Design with Personal Computer Applications*, 3rd ed., Pacific Grove, CA: Brooks/Cole, 2002.

R.K. Henke and S.C. Sciacca, "Dynamic thermal rating of critical lines — A study of real-time interface requirements," *IEEE Computer Applications in Power*, pp. 46–51, July 1989.

NERC, *NERC Planning Standards*, Princeton, NJ: North American Electric Reliability Council, www.nerc.com, September 1997.

NERC, *Reliability Concepts*, Princeton, NJ: North American Electric Reliability Council, www.nerc.com, February 1985.

NERC, *Overview of Planning Reliability Criteria*, Princeton, NJ: North American Electric Reliability Council, www.nerc.com, April 1988.

NERC, *Electricity Transfers and Reliability*, Princeton, NJ: North American Electric Reliability Council, www.nerc.com, October 1989.

NERC, *A Survey of the Voltage Collapse Phenomenon*, Princeton, NJ: North American Electric Reliability Council, www.nerc.com, 1991.

T.J. Overbye and J.D. Weber, "Visualizing the Grid," *IEEE Spectrum*, February 2001.

H.A. Smolleck et al., "Translation of Large Data-Bases for Microcomputer-Based Application Software: Methodology and a Case Study," *IEEE Comput. Appl. Power*, pp. 40–45, July 1989.

U.S.–Canada Power System Outage Task Force, *Final Report on the August 14, 2003, Blackout in the United States and Canada — Causes and Recommendations*, www.nerc.com, April 2004.

Further Information

The NERC was formed in 1968, in the aftermath of the November 9, 1965, northeast blackout, to promote the reliability of bulk electric power systems of North America. Transmission planning criteria presented here are partially based on NERC criteria as well as on specific criteria used by transmission planning departments from three electric utility companies: American Electric Power Service Corporation, Commonwealth Edison Company, and Pacific Gas & Electric Company. NERC's publications, developed by utility experts, have become standards for the industry. In most cases, these publications are available at no charge from NERC, Princeton, NJ (www.nerc.com).

4

Power Quality

Jos Arrillaga
University of Canterbury

Ideally, power should be supplied without interruptions at constant frequency, constant voltage and with perfectly sinusoidal and, in the case of three-phase, symmetrical waveforms. Supply reliability constitutes a recognized independent topic, and is not usually discussed under power quality. The specific object of power quality is the "*pureness*" of the supply including voltage variations and waveform **distortion**.

Power system **disturbances** and the continually changing demand of consumers give rise to voltage variations. Deviation from the sinusoidal voltage supply can be due to transient phenomena or to the presence of non-linear components.

The power network is not only the main source of energy supply but also the conducting vehicle for possible interferences between consumers. This is a subject that comes under the general heading of electromagnetic compatibility (EMC).

EMC refers to the ability of electrical and electronic components, equipment, and systems to operate satisfactorily without causing interference to other equipment or systems, or without being affected by other operating systems in that electromagnetic environment.

EMC is often perceived as interference by electromagnetic radiation between the various elements of a system. The scope of EMC, however, is more general and it also includes conductive propagation and coupling by capacitance, inductance (self and mutual) encompassing the whole frequency spectrum.

A power quality problem is any occurrence manifested in voltage, current, or frequency deviation that results in failure or misoperation of equipment. The newness of the term reflects the newness of the concern. Decades ago, power quality was not a worry because it had no effect on most loads connected to electric distribution systems.

Therefore, power quality can also be defined as the ability of the electrical power system to transmit and deliver electrical energy to the consumers within the limits specified by EMC standards.

4.1 Power Quality Disturbances

Following standard criteria [IEC, 1993], the main deviations from a perfect supply are

- periodic waveform distortion (harmonics, interharmonics)
- voltage fluctuations, flicker
- short voltage interruptions, dips (sags), and increases (swells)
- three-phase unbalance
- transient overvoltages

The main causes, effects and possible control of these disturbances are considered in the following sections.

Periodic Waveform Distortion

Harmonics are sinusoidal voltages or currents having frequencies that are whole multiples of the frequency at which the supply system is designed to operate (e.g., 50 Hz or 60 Hz). An illustration of fifth harmonic distortion is shown in Figure 4.1. When the frequencies of these voltages and currents are not an integer of the fundamental they are termed interharmonics.

Both harmonic and interharmonic distortion is generally caused by equipment with non-linear voltage/current characteristics.

In general, distorting equipment produces harmonic currents that, in turn, cause harmonic voltage drops across the impedances of the network. Harmonic currents of the same frequency from different sources add vectorially.

The main detrimental effects of harmonics are [Arrillaga et al., 1985]

- maloperation of control devices, main signalling systems, and protective relays
- extra losses in capacitors, transformers, and rotating machines
- additional noise from motors and other apparatus
- telephone interference
- The presence of power factor correction capacitors and cable capacitance can cause shunt and series resonances in the network producing voltage amplification even at a remote point from the distorting load.

As well as the above, interharmonics can perturb **ripple control signals** and at sub-harmonic levels can cause flicker.

To keep the harmonic voltage content within the recommended levels, the main solutions in current use are

- the use of high pulse rectification (e.g., smelters and **HVdc** converters)
- passive filters, either tuned to individual frequencies or of the band-pass type
- active filters and conditioners

The harmonic sources can be grouped in three categories according to their origin, size, and predictability, i.e., small and predictable (domestic and residential), large and random (arc furnaces), and large and predictable (static converters).

Small Sources

The residential and commercial power system contains large numbers of single-phase converter-fed power supplies with capacitor output smoothing, such as TVs and PCs, as shown in Figure 4.2. Although their individual rating is insignificant, there is little diversity in their operation and their combined effect produces

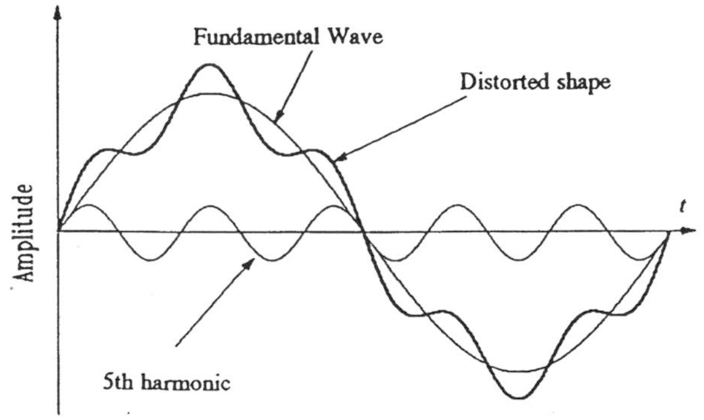

FIGURE 4.1 Example of a distorted sine wave.

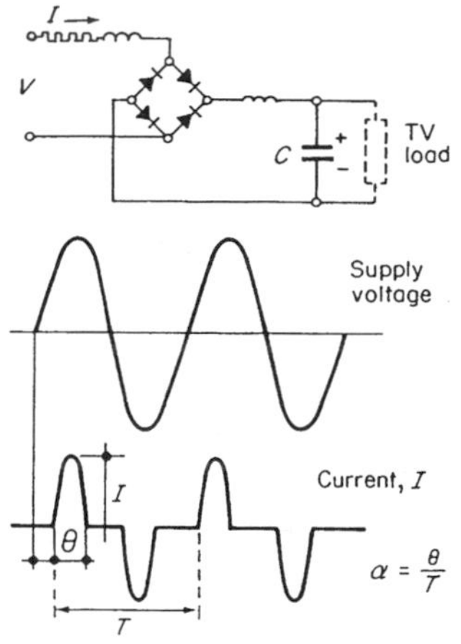

FIGURE 4.2 Single-phase bridge supply for a TV set.

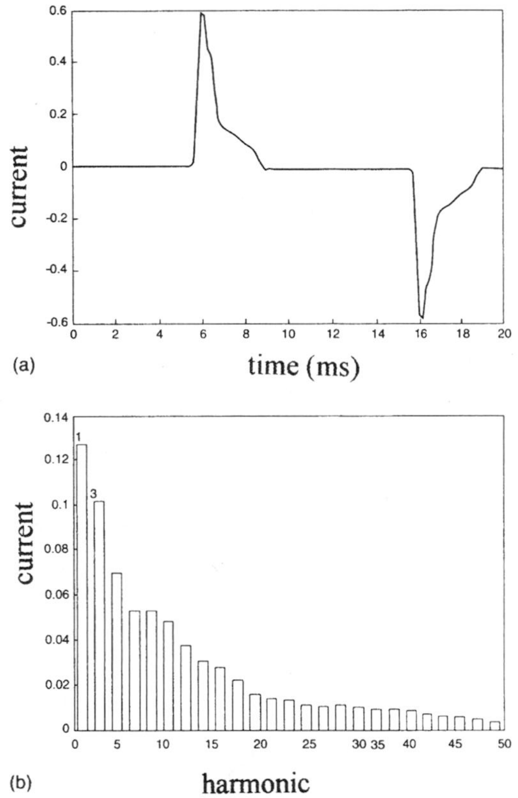

FIGURE 4.3 Current waveform (a) and harmonic spectrum (b) of a high efficiency lamp.

considerable odd-harmonic distortion. The gas discharge lamps add to that effect as they produce the same harmonic components.

Figure 4.3 illustrates the current waveform and harmonic spectrum of a typical high efficiency lamp. The total harmonic distortion (THD) of such lamps can be between 50 and 150%.

Large and Random Sources

The most common and damaging load of this type is the arc furnace. Arc furnaces produce random variations of harmonic and interharmonic content which is uneconomical to eliminate by conventional filters.

Figure 4.4 shows a snap-shot of the frequency spectra produced by an arc furnace during the melting and refining processes, respectively. These are greatly in excess of the recommended levels.

These loads also produce voltage fluctuations and flicker. Connection to the highest possible voltage level and the use of series reactances are among the measures currently taken to reduce their impact on power quality.

Static Converters

Large power converters, such as those found in smelters and HVdc transmission, are the main producers of harmonic current and considerable thought is given to their local elimination in their design.

The standard configuration for industrial and HVdc applications is the twelve-pulse converter, shown in Figure 4.5. The "*characteristic*" harmonic currents for the configuration are of orders $12\ K \pm 1$ and their amplitudes are inversely proportional to the harmonic order, as shown by the spectrum of Figure 4.6(b) which correspond to the time waveform of Figure 4.6(a). These are, of course, maximum levels for ideal system conditions, i.e., with an infinite (zero impedance) ac system and a perfectly flat direct current (i.e., infinite smoothing reactance).

When the ac system is weak and the operation not perfectly symmetrical, **uncharacteristic harmonics** appear [Arrillaga, 1983].

While the characteristic harmonics of the large power converter are reduced by filters, it is not economical to reduce in that way the uncharacteristic harmonics and, therefore, even small injection of these harmonic currents can, via parallel resonant conditions, produce very large voltage distortion levels.

An example of uncharacteristic converter behavior is the presence of fundamental frequency on the dc side of the converter, often induced from ac transmission lines in the proximity of the dc line, which produces second harmonic and direct current on the ac side.

Even harmonics, particularly the second, are very disruptive to power electronic devices and are, therefore, heavily penalized in the regulations.

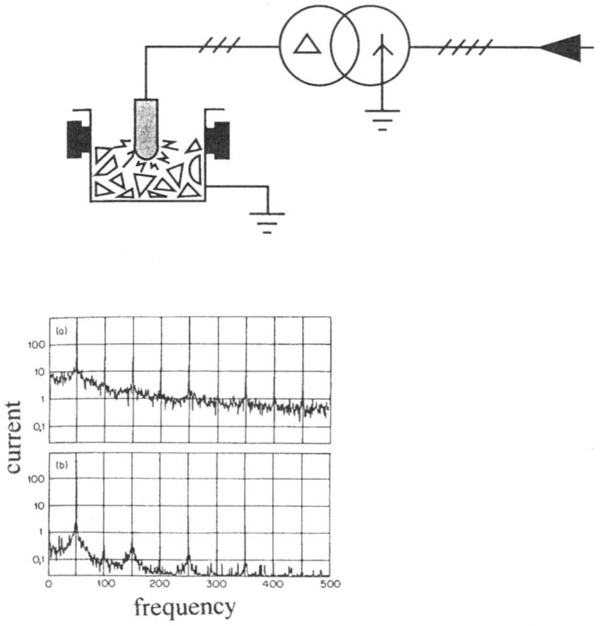

FIGURE 4.4 Typical frequency spectra of arc furnace operation. (a) During fusion; (b) during refining.

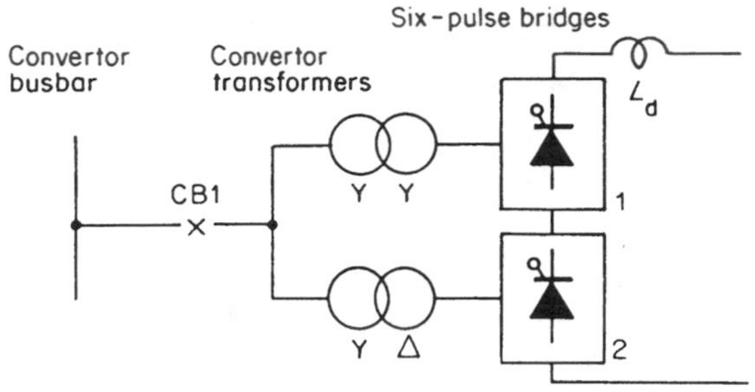

FIGURE 4.5 Twelve-pulse converter.

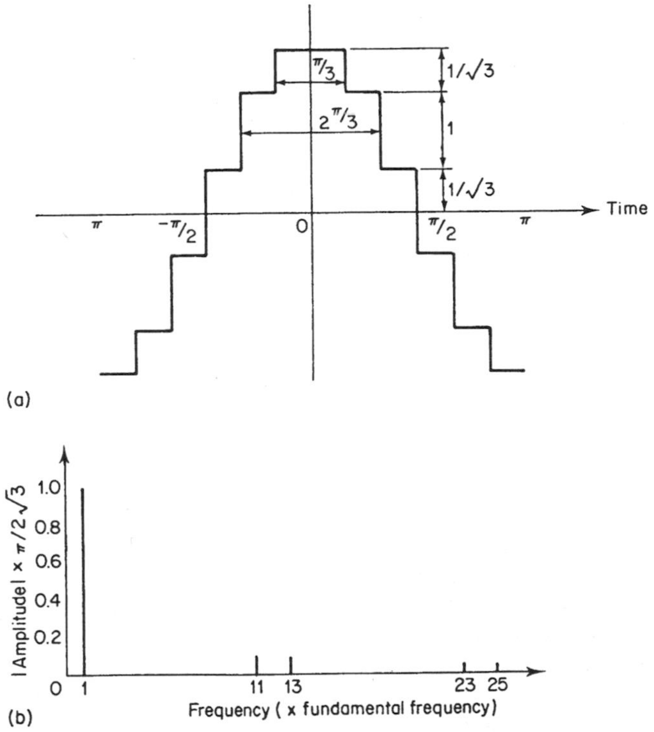

FIGURE 4.6 Twelve-pulse converter current. (a) Waveform; (b) spectrum.

The flow of dc current in the ac system is even more distorting, the most immediate effect being asymmetrical saturation of the converters or other transformers with a considerable increase in even harmonics which, under certain conditions, can lead to **harmonic instabilities** [Chen et al., 1996].

Another common example is the appearance of triplen harmonics. Asymmetrical voltages, when using a common firing angle control for all the valves, result in current pulse width differences between the three phases which produce triplen harmonics. To prevent this effect, modern large power converters use the equidistant firing concept instead [Ainsworth, 1968]. However, this controller cannot eliminate second harmonic amplitude modulation of the dc current which, via the converter modulation process, returns third harmonic current of positive sequence. This current can flow through the converter transformer regardless of its connection and penetrate far into the ac system. Again, the presence of triplen harmonics is discouraged by stricter limits in the regulations.

Voltage Fluctuations and Flicker

This group includes two broad categories, i.e.,

- step voltage changes, regular or irregular in time, such as those produced by welding machines, rolling mills, mine winders, etc. [Figures 4.7(a) and (b)].
- cyclic or random voltage changes produced by corresponding variations in the load impedance, the most typical case being the arc furnace load (Figure 4.7(c)).

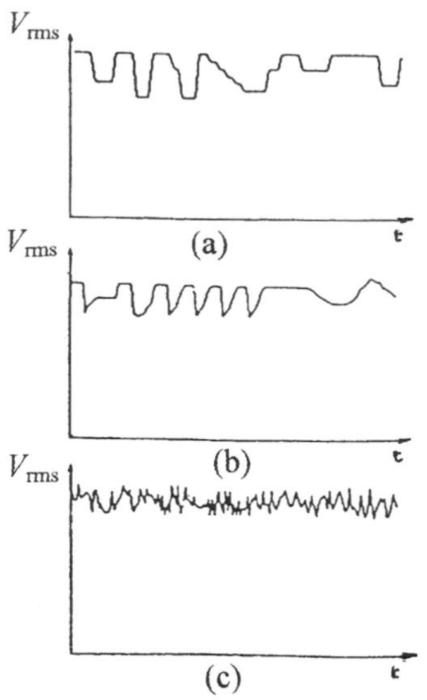

FIGURE 4.7 Voltage fluctuations.

Generally, since voltage fluctuations have an amplitude not exceeding ±10%, most equipment is not affected by this type of disturbance. Their main disadvantage is flicker, or fluctuation of luminosity of an incandescent lamp. The important point is that it is impossible, in practice, to change the characteristics of the filament. The physiological discomfort associated with this phenomenon depends on the amplitude of the fluctuations, the rate of repetition for voltage changes, and the duration of the disturbance. There is, however, a perceptibility threshold below which flicker is not visible.

Flicker is mainly associated with the arc furnaces because they draw different amounts of current each power cycle. The upshot is a modulation of the system voltage magnitude in the vicinity of the furnace. The modulation frequency is in the band 0 to 30 Hz, which is in the range that can cause noticeable flicker of light bulbs.

The flicker effect is usually evaluated by means of a flickermeter (IEC Publication 868). Moreover, the amplitude of modulation basically depends on the ratio between the impedance of the disturbing installation and that of the supply network.

Brief Interruptions, Sags, and Swells

Voltage Dips (SAGS)

A voltage dip is a sudden reduction (between 10 and 90%) of the voltage, at a point in the electrical system, such as that shown in Figure 4.8, and lasting for 0.5 cycle to several seconds.

Dips with durations of less than half a cycle are regarded as transients.

A voltage dip may be caused by switching operations associated with temporary disconnection of supply, the flow of heavy current associated with the start of large motor loads or the flow of fault currents. These events may emanate from customers' systems or from the public supply network.

The main cause of momentary voltage dips is probably the lightning strike. In the majority of cases, the voltage drops to about 80% of its nominal value. In terms of duration, dips tend to cluster around three values: 4 cycles (the typical clearing time for faults), 30 cycles (the instantaneous reclosing time for breakers), and 120 cycles (the delayed reclosing time of breakers). The effect of a voltage dip on equipment depends on both its magnitude and its duration; in about 42% of the cases observed to date they are severe enough to exceed the tolerance standard adopted by computer manufacturers.

Possible effects are:

- extinction of discharge lamps
- incorrect operation of control devices
- speed variation or stopping of motors
- tripping of contactors
- computer system crash or measuring errors in instruments equipped with electronic devices
- commutation failure in HVdc converters [Arrillaga, 1983]

Brief Interruptions

Brief interruptions can be considered as voltage sags with 100% amplitude (see Figure 4.9). The cause may be a blown fuse or breaker opening and the effect an expensive shutdown. For instance, a five-cycle interruption at a glass factory has been estimated as $200,000, and a major computer center reports that a 2-second outage can cost approximately $600,000. The main protection of the customer against such events is the installation of uninterruptible power supplies or power quality conditioners (discussed later).

Brief Voltage Increases (SWELLS)

Voltage swells, shown in Figure 4.10, are brief increases in rms voltage that sometimes accompany voltage sags. They appear on the unfaulted phases of a three-phase circuit that has developed a single-phase short circuit. They also occur following load rejection.

Swells can upset electric controls and electric motor drives, particularly common adjustable-speed drives, which can trip because of their built-in protective circuitry. Swells may also stress delicate computer components and shorten their life.

Possible solutions to limit this problem are, as in the case of sags, the use of uninterruptible power supplies and conditioners.

Unbalances

Unbalance describes a situation, as shown in Figure 4.11, in which the voltages of a three-phase voltage source are not identical in magnitude, or the phase differences between them are not 120 electrical

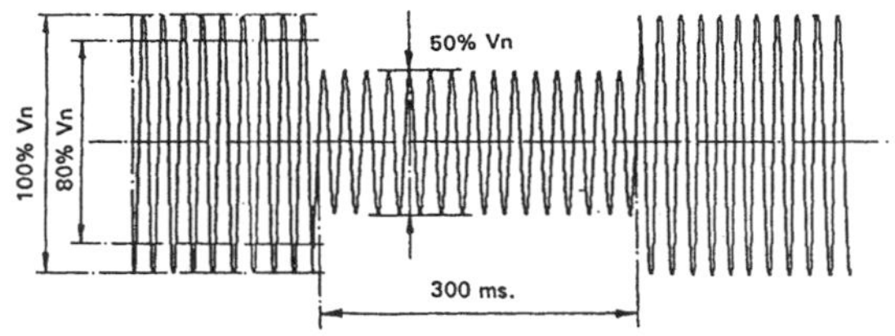

FIGURE 4.8 Voltage sag.

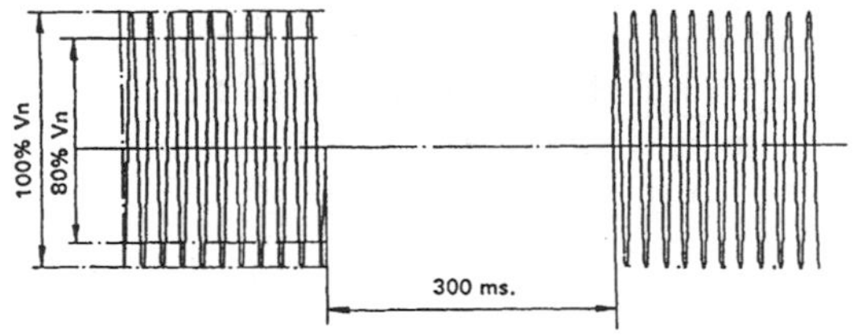

FIGURE 4.9 Voltage interruption.

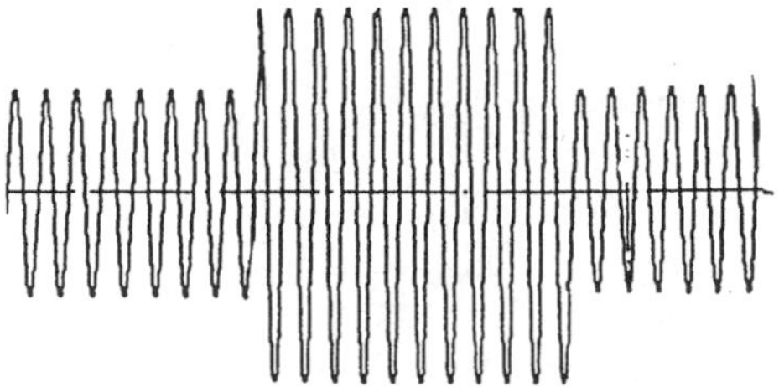

FIGURE 4.10 Voltage swell.

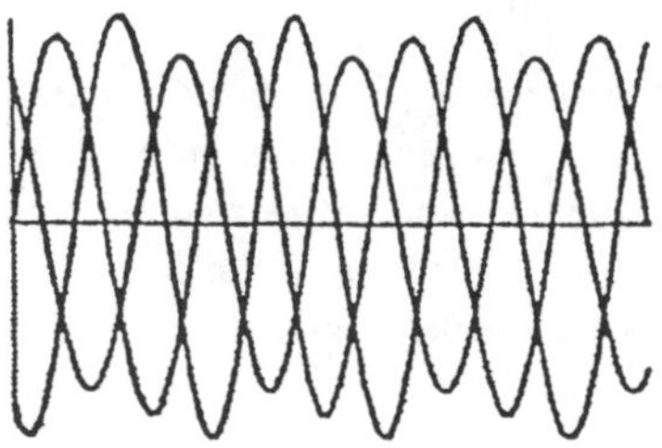

FIGURE 4.11 Voltage unbalance.

degrees, or both. It affects motors and other devices that depend on a well-balanced three-phase voltage source.

The degree of unbalances is usually defined by the proportion of negative and zero **sequence components**.

The main causes of unbalance are single-phase loads (such as electric railways) and untransposed overhead transmission lines.

A machine operating on an unbalanced supply will draw a current with a degree of unbalance several times that of the supply voltage. As a result, the three-phase currents may differ considerably and temperature rise in the machine will take place.

Motors and generators, particularly the large and more expensive ones, may be fitted with protection to detect extreme unbalance. If the supply unbalance is sufficient, the "*single-phasing*" protection may respond to the unbalanced currents and trip the machine.

Polyphase converters, in which the individual input phase voltages contribute in turn to the dc output, are also affected by an unbalanced supply, which causes an undesirable ripple component on the dc side, and non-characteristic harmonics on the ac side.

Transients

Voltage disturbances shorter than sags or swells are classified as transients and are caused by sudden changes in the power system [Greenwood, 1971]. They can be impulsive, generally caused by lightning and load switching, and oscillatory, usually due to capacitor-bank switching.

Capacitor switching can cause resonant oscillations leading to an overvoltage some three to four times the nominal rating, causing tripping or even damaging protective devices and equipment. Electronically based controls for industrial motors are particularly susceptible to these transients.

According to their duration, transient overvoltages can be divided into:

- switching surge (duration in the range of *ms*)
- impulse, spike (duration in the range of μs)

Surges are high-energy pulses arising from power system switching disturbances, either directly or as a result of resonating circuits associated with switching devices. They also occur during step load changes.

Impulses in microseconds, as shown in Figure 4.12, result from direct or indirect lightning strokes, arcing, insulation breakdown, etc.

Protection against surges and impulses is normally achieved by surge-diverters and arc-gaps at high voltages and avalanche diodes at low voltages.

Faster transients in nanoseconds due to electrostatic discharges, an important category of EMC, are not normally discussed under Power Quality.

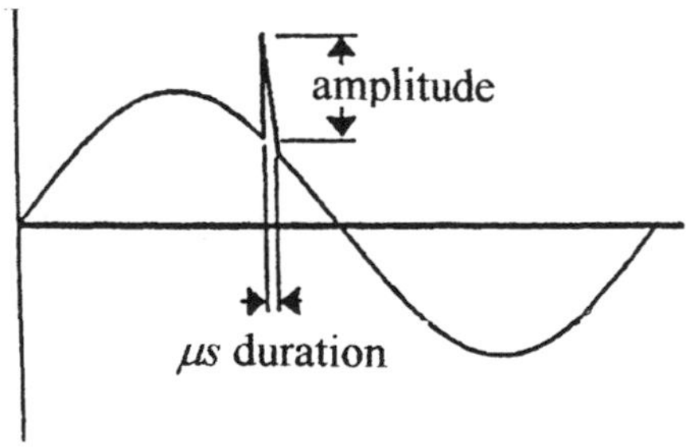

FIGURE 4.12 Impulse.

4.2 Power Quality Monitoring

Figure 4.13 illustrates the various components of a power quality detection system, i.e., voltage and current transducers, information transmission, instrumentation, and displays.

The most relevant information on power quality monitoring requirements can be found in the document IEC 1000–4.7. This document provides specific recommendations on monitoring accuracy in relation to the operating condition of the power system.

With reference to monitoring of individual frequencies, the maximum recommended relative errors for the magnitude and phase are 5% and 5°, respectively, under normal operating conditions and with constant voltage or current levels. However, such precision must be maintained for voltage variations of up to 20% (of nominal value) and 100% (peak value). For current measurements, the precision levels apply for overcurrents of up to 20% and peaks of 3 times rms value (on steady state) and 10 times the nominal current for a 1-sec duration.

Errors in the frequency response of current transformers occur due to capacitive effects, which are not significant in the harmonic region (say, up to the 50th harmonic), and also due to magnetizing currents. The latter can be minimized by reducing the current transformer load and improving the power factor; the ideal load being a short-circuited secondary with a *clamp* to monitor the current. Alternative transducers are being proposed for high frequency measurements using optical, magneto-optical, and Hall effect principles.

The iron-core voltage transformers respond well to harmonic frequencies for voltages up to 11 kV. Due to insulation capacitance, these transformers are not recommended for much higher voltages. The conventional capacitive voltage transformers (CVTs) are totally inadequate due to low frequency resonances between the capacitive divider and the output magnetic transformer; special portable capacitive dividers, without the output transformers, are normally used for such measurements. Again, alternative transducer principles, as for the current transformer, are being proposed for future schemes.

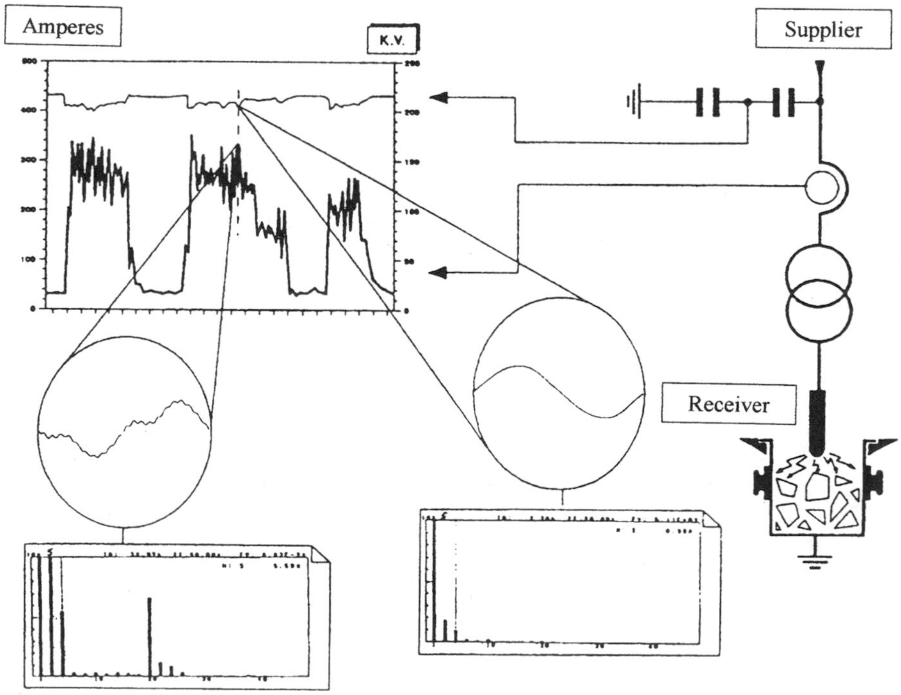

FIGURE 4.13 Power quality monitoring components.

The signal transmission from the transducers to the control room passes through very noisy electromagnetic environments and the tendency is to use fiber optic cables, designed to carry either analog or digital samples of information in the time domain.

The time domain information is converted by signal or harmonic analyzers into the frequency domain; the instrumentation is also programmed to derive any required power quality indexes, such as THD (total harmonic distortion), EDV (equivalent distortion voltage), EDI (equivalent distortion current), etc.

The signal processing is performed by either analog or digital instrumentation, though the latter is gradually displacing the former. Most digital instruments in existence use the **FFT** (fast Fourier transform). The processing of information can be continuous or discontinuous depending on the characteristic of the signals under measurement with reference to waveform distortion. Document IEC 1000–4.7 lists the following types:

- quasi stationary harmonics
- fluctuating harmonics
- intermittent harmonics
- interharmonics

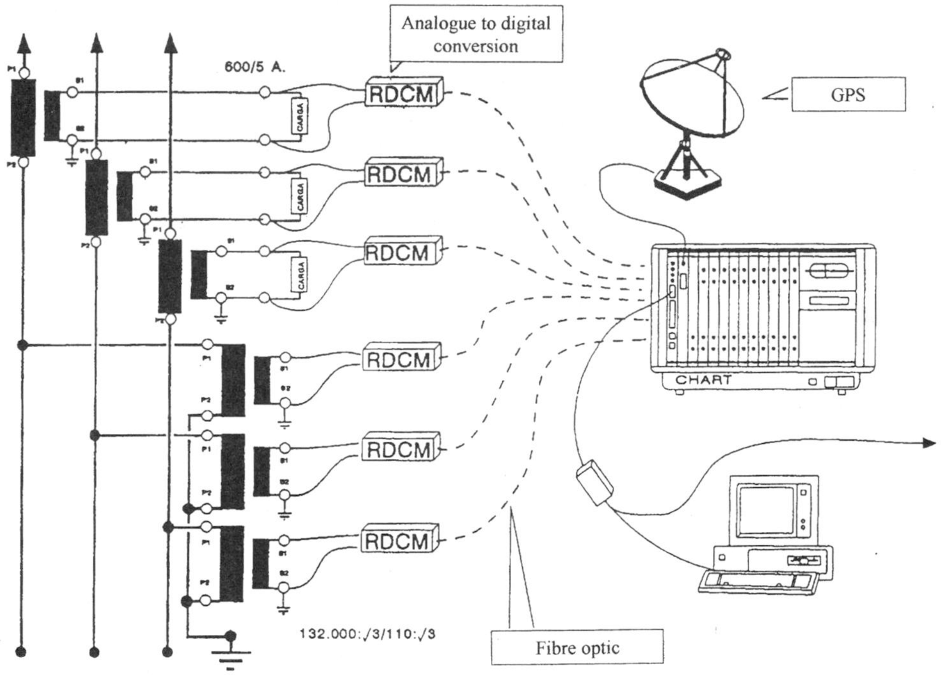

FIGURE 4.14 Simultaneous measurement of voltages and currents in a three-phase line.

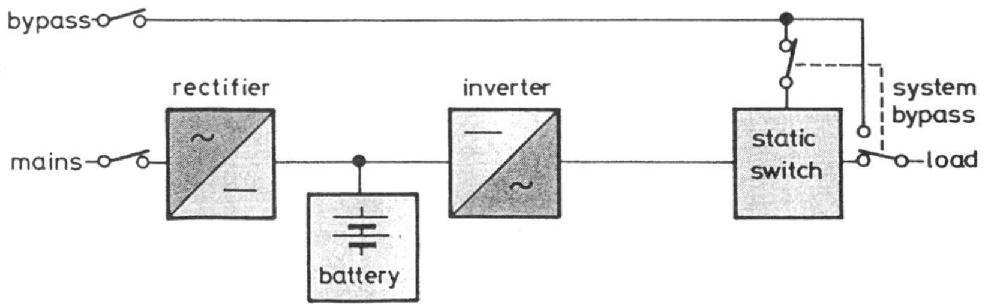

FIGURE 4.15 Uninterruptible power supply.

Only in the case of quasi stationary waveforms can the use of discontinuous monitoring be justified; examples of this type are the well-defined loads such as TV and PC sets.

In the remaining categories, it is necessary to perform real time continuous monitoring; examples of loads producing non-stationary distortion are arc furnaces and rolling mills.

Most of the instruments commercially available are not designed specifically for power system application, i.e., they are not multi-phase and cannot process continuous information. At the time of writing, the only system capable of multi-channel three-phase real time continuous monitoring is CHART [Miller and Dewe, 1992] which, although originally designed for harmonic monitoring, is capable of deriving continuous information of other power quality indexes such as flicker. It is based on the Intel Multi-bus II architecture and the RMX 386 operating system. An illustration of the system, shown in Figure 4.14, includes remote data conversion modules, digital fiber optic transmission, **GPS** synchronization, central parallel processing, and ethernet-connected PCs for distant control and display.

4.3 Power Quality Conditioning

A common device in current use to ensure supply continuity for critical loads is the UPS or uninterruptible power supply. For brief interruptions, the UPS are of the static type, using batteries as the energy source and involving a rectifier/inverter system. A block diagram of a typical UPS is shown in Figure 4.15 [Heydt, 1991].

In the next few years power quality enhancements, in terms of reduced interruptions and voltage variations, can be expected by the application of power electronic controllers to utility distribution systems and/or at the supply end of many industrial and commercial customers.

Among the solutions already available are the solid state circuit breaker, the static condensers (or statcon), and the dynamic voltage restorer [Hingorani, 1995].

In a solid state circuit breaker, thyristors connected back-to-back form an ac switch module, several of which are, in turn, connected in series to acquire the required voltage rating. The breaker will interrupt the circuit at the first zero of the ac current. This means a delay of a few milliseconds, which should be acceptable for most applications.

Figure 4.16 shows a simplified illustration of a statcon which is made up of GTOs (gate turn off) or similar devices such as insulated-gate bipolar transistors (IGBTs) or MOS-controlled thyristors (MCTs). The converter is driven by a dc storage device such as a dc capacitor, battery, or superconducting magnetic storage, and an ac transformer. The dynamic voltage restorer, shown schematically in Figure 4.17, turns a distorted waveform, including voltage dips, into the required waveform. The device injects the right amount of voltage by way of a series-connected transformer into the distribution feeder between the power supply side and load side.

The dynamic voltage restorer is similar to the statcon, with a transformer, converter, and storage, except that the transformer is connected in series with the busbar feeding the sensitive load. Compensation occurs in both directions, making up for the voltage dips and reducing the overvoltage. The response is very fast, occurring within a few milliseconds.

The capacity of the dc storage capacitor, in both the statcon and the dynamic voltage restorer, determines the duration of the correction provided for individual voltage dips. It can be a few cycles or seconds long. To enhance the load support capability, a storage battery with a booster electronic circuit can be connected in parallel with the capacitor.

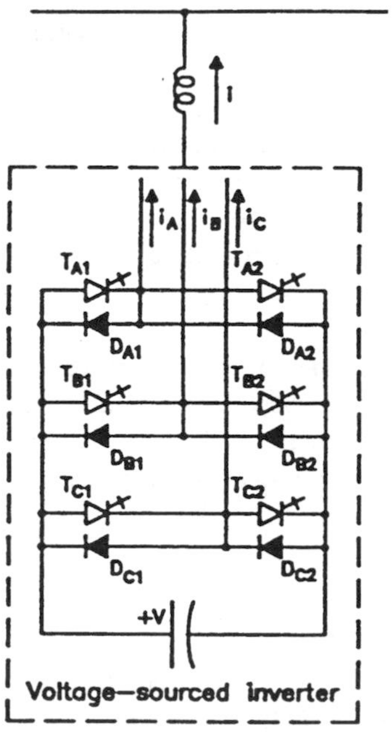

FIGURE 4.16 Static condenser.

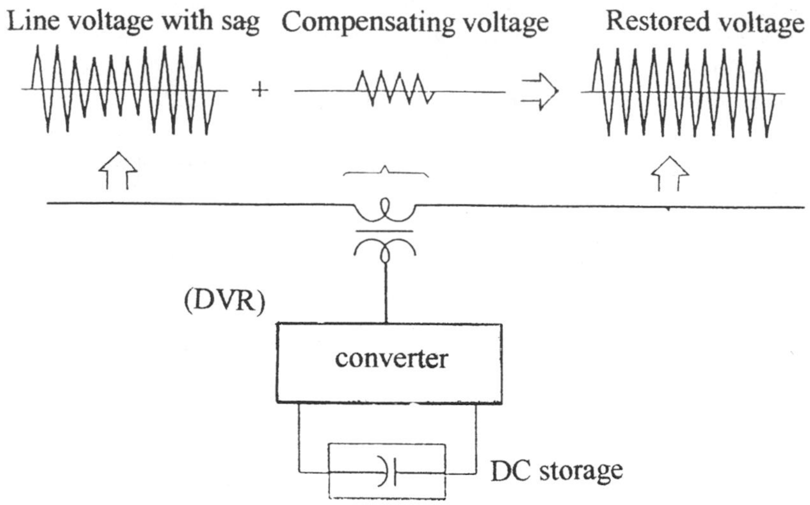

FIGURE 4.17 Dynamic voltage restorer.

Superconducting magnetic energy storage can be very effective to provide power for short periods. When the storage is not supporting the load, the converter will automatically charge the storage from the utility system, to be ready for the next event.

Defining Terms

Distortion: Any deviation from a perfectly sinusoidal wave.
Disturbance: Any sudden change in the intended power, voltage, or current supply.
FFT (fast Fourier transform): Efficient computation of the discrete Fourier transform.
GPS (global positioning satellite): Used for time stamping and synchronization of multi-measurements at different geographical locations.
Harmonic instability: Extreme distortion of the voltage waveform at a particular frequency that causes inverter maloperation.
HVdc: High voltage direct current transmission.
Ripple control signal: A burst of pulses at a fixed non-harmonic frequency injected into the power system for the purpose of load management control.
Sequence components: Three symmetrical sets of voltages or currents equivalent to an asymmetrical three-phase unbalanced set.
THD (total harmonic distortions): The ratio of rms value of the harmonic content to the rms value of the generated frequency (in %).
Uncharacteristic harmonics: Static converter harmonics of orders different from $Pk \pm 1$ where P is the pulse number.

References

J. Ainsworth, "The phase-locked oscillator. A new control system for controlled static convertors", *Trans. IEEE*, PAS-87, pp. 859–865, 1968.
J. Arrillaga, *High Voltage Direct Current Transmission*, London: IEE-Peter Peregrinus, 1983.
J. Arrillaga, D.A. Bradley, and P.S. Bolger, *Power System Harmonics*, London: John Wiley & Sons, 1985.

S. Chen, A.R. Wood, and J. Arrillaga, "HVdc converter transformer core saturation instability: a frequency domain analysis", *IEE Proc.—Gener. Transm. Distrib,* 143(1), 75–81, 1996.
A. Greenwood, *Electrical Transients in Power Systems*, New York: Wiley Interscience, 1971.
J. Heydt, *Electric Power Quality*, Stars in a Circle Publications, 1991.
N.G. Hingorani, "Introducing custom power", *IEEE Spectrum*, 41–48, June 1995.
International Electrotechical Commission Group, IEC TL 77, 1993.
A.J. Miller and M.B. Dewe, "Multichannel continuous harmonic analysis in real time", *Trans. IEEE Power Delivery*, Vol. 7, no. 4, pp. 1913–1919, 1992.

Further Information

Electric Power Quality by J. Heydt and *Power System Harmonics* by J. Arrillaga et al. are the only texts discussing the topic, though the latter is currently out of print. Two international conferences take place biennially specifically related to Power Quality; these are PQA (Power Quality: end use applications and perspectives) and the IEEE sponsored ICHQP (International Conference on Harmonics and Quality of Power).

Important information can also be found in the regular updates of the IEC and CENELEC standards, CIGRE, CIRED, UIC, and UNIPEDE documents and national guides such as the IEEE 519–1992.

Finally, the IEE and IEEE Journals on Power Transmission and Delivery, respectively, publish regularly important contributions in this area.

5

Power System Analysis

Andrew Hanson
ABB, Inc.

Leo Grigsby
Auburn University

5.1 Introduction

The equivalent circuit parameters of many power system components are described in Chapters 3, 6, and 8. The interconnection of the different elements allows development of an overall power system model. The system model provides the basis for computational simulation of the system performance under a wide variety of projected operating conditions. Additionally, "post mortem" studies, performed after system disturbances or equipment failures, often provide valuable insight into contributing system conditions. The different types of power system analyses are discussed below; the type of analysis performed depends on the conditions to be assessed.

5.2 Types of Power System Analysis

Power Flow Analysis

Power systems typically operate under slowly changing conditions that can be analyzed using steady-state analysis. Further, transmission systems operate under balanced or near-balanced conditions, allowing per-phase analysis to be used with a high degree of confidence in the solution. Power flow analysis provides the starting point for most other analyses. For example, the small-signal and transient stability effects of a given disturbance are dramatically affected by the "predisturbance" operating conditions of the power system. (A disturbance resulting in instability under heavily loaded system conditions may not have any adverse effects under lightly loaded conditions.) Additionally, fault analysis and transient analysis can also be impacted by the "predisturbance" operating point of a power system (although, they are usually affected much less than transient stability and small-signal stability analyses).

Fault Analysis

Fault analysis refers to power-system analysis under severely unbalanced conditions. (Such conditions include downed or open conductors.) Fault analysis assesses the system behavior under the high current and severely unbalanced conditions typical during faults. The results of fault analyses are used to size and apply system protective devices (breakers, relays, etc.). Fault analysis is discussed in more detail in Section 3.3.

Transient Stability Analysis

Transient stability analysis, unlike the analyses previously discussed, assesses the system's performance over a period of time. The system model for transient stability analysis typically includes not only the transmission network parameters, but also the dynamics data for the generators. Transient stability analysis is most often used to determine if individual generating units will maintain synchronism with the power system following a disturbance (typically a fault).

Extended Stability Analysis

Extended stability analysis deals with system stability beyond the generating units' "first swing." In addition to the generator data required for transient stability analysis, extended stability analysis requires excitation system, speed governor, and prime-mover dynamics data. Often, extended stability analysis will also include dynamics data for control devices such as tap changing transformers, switched capacitors, and relays. The addition of these elements to the system model complicates the analysis, but provides comprehensive simulation of nearly all major system components and controls. Extended stability analyses complement small-signal stability analyses by verifying the existence of persistent oscillations and establishing the magnitudes of power and voltage oscillations.

Small-Signal Stability Analysis

Small-signal stability assesses the stability of the power system when subjected to "small" perturbations. Small-signal stability uses a linearized model of the power system that includes generator, prime mover, and control device dynamics data. The system of nonlinear equations describing the system is linearized about a specific operating point, and eigenvalues and eigenvectors of the linearized system found. The imaginary part of each eigenvalue indicates the frequency of the oscillations associated with the eigenvalue. The real part indicates damping of the oscillation. Usually, small-signal stability analysis attempts to find disturbances and system conditions that can lead to sustained oscillations (indicated by small damping factors) in the power system. Small signal stability analysis does not provide oscillation magnitude information since the eigenvalues only indicate oscillation frequency and damping. Additionally, the controllability matrices (based on the linearized system) and the eigenvectors can be used to identify candidate generating units for application of new or improved controls (i.e., power system stabilizers and new or improved excitation systems).

Transient Analysis

Transient analysis involves the analysis of the system (or at least several components of the system) when subjected to "fast" transients (i.e., lightning and switching transients). Transient analysis requires detailed component information that is often not readily available. Typically, only system components in the immediate vicinity of the area of interest are modeled in transient analyses. Specialized software packages (most notably EMTP) are used to perform transient analyses.

Operational Analyses

Several additional analyses used in the day-to-day operation of power systems are based on the results of the analyses described above. *Economic dispatch* analyses determine the most economic real power output for each generating unit based on cost of generation for each unit and the system losses. *Security or contingency* analyses assess the system's ability to withstand the sudden loss of one or more major elements without overloading the remaining system. *State estimation* determines the "best" estimate of the real-time system state based on a redundant set of system measurements.

5.3 The Power Flow Problem

Power flow analysis is fundamental to the study of power systems; in fact, power flow forms the core of power system analysis. A power flow study is valuable for many reasons. For example, power flow analyses play a key role in the planning of additions or expansions to transmission and generation facilities. A power flow solution is often the starting point for many other types of power system analyses. In addition, power flow analysis and many of its extensions are an essential ingredient of the studies performed in power system operations. In this latter case, it is at the heart of contingency analysis and the implementation of real-time monitoring systems.

The power flow problem (popularly known as the load flow problem) can be stated as follows:

> For a given power network, with known complex power loads and some set of specifications or restrictions on power generations and voltages, solve for any unknown bus voltages and unspecified generation and finally for the complex power flow in the network components.

Additionally, the losses in individual components and the total network as a whole are usually calculated. Furthermore, the system is often checked for component overloads and voltages outside allowable tolerances.

Balanced operation is assumed for most power flow studies and will be assumed in this chapter. Consequently, the positive sequence network is used for the analysis. In the solution of the power flow problem, the network element values are almost always taken to be in per unit. Likewise, the calculations within the power flow analysis are typically in per unit. However, the solution is usually expressed in a mixed format. Solution voltages are usually expressed in per unit; powers are most often given with kVA or MVA.

The "given network" may be in the form of a system map and accompanying data tables for the network components. More often, however, the network structure is given in the form of a one-line diagram (such as shown in Figure 5.1).

Regardless of the form of the given network and how the network data is given, the steps to be followed in a power flow study can be summarized as follows:

1. Determine element values for passive network components.
2. Determine locations and values of all complex power loads.
3. Determine generation specifications and constraints.
4. Develop a mathematical model describing power flow in the network.
5. Solve for the voltage profile of the network.
6. Solve for the power flows and losses in the network.
7. Check for constraint violations.

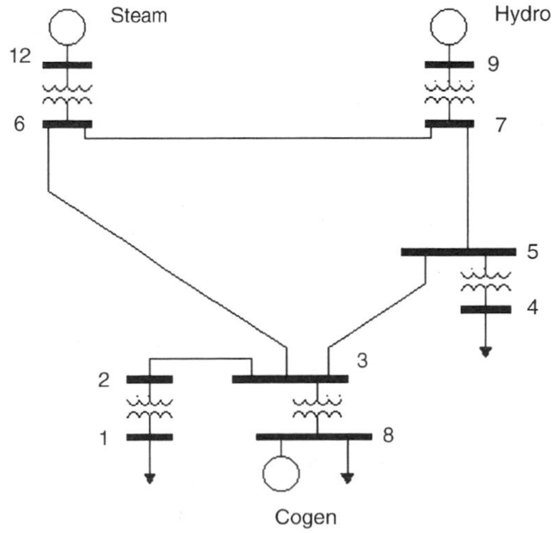

FIGURE 5.1 The one-line diagram of a power system.

5.4 Formulation of the Bus Admittance Matrix

The first step in developing the mathematical model describing the power flow in the network is the formulation of the bus admittance matrix. The bus admittance matrix is an $n \times n$ matrix (where n is the number of buses in the system) constructed from the admittances of the equivalent circuit elements of the segments making up the power system. Most system segments are represented by a combination of shunt elements (connected between a bus and the reference node) and series elements (connected between two system buses). Formulation of the bus admittance matrix follows two simple rules:

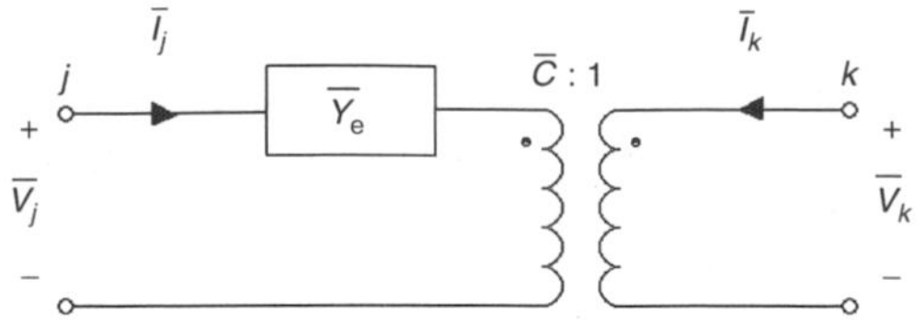

FIGURE 5.2 Off-nominal turns ratio transformer.

1. The admittance of elements connected between node k and reference is added to the (k, k) entry of the admittance matrix.
2. The admittance of elements connected between nodes j and k is added to the (j, j) and (k, k) entries of the admittance matrix. The *negative* of the admittance is added to the (j, k) and (k, j) entries of the admittance matrix.

Off-nominal transformers (transformers with transformation ratios different from the system voltage bases at the terminals) present some special difficulties. Figure 5.2 shows a representation of an off-nominal turns ratio transformer.

The admittance matrix base mathematical model of an isolated off-nominal transformer is:

$$\begin{bmatrix} \bar{I}_j \\ \bar{I}_k \end{bmatrix} = \begin{bmatrix} \bar{Y}_e & -\bar{c}\bar{Y}_e \\ -\bar{c}^*\bar{Y}_e & |\bar{c}|^2\bar{Y}_e \end{bmatrix} \begin{bmatrix} \bar{V}_j \\ \bar{V}_k \end{bmatrix} \tag{5.1}$$

where

$\bar{Y}_e$ is the equivalent series admittance (referred to node j)
c is the complex (off-nominal turns ratio)
$\bar{I}_j$ is the current injected at node j
$\bar{V}_j$ is the voltage at node j (with respect to reference)

Off-nominal transformers are added to the bus admittance matrix by adding the corresponding entry of the isolated off-nominal transformer admittance matrix to the system bus admittance matrix.

5.5 Example Formulation of the Power Flow Equations

Considerable insight into the power flow problem and its properties and characteristics can be obtained by consideration of a simple example before proceeding to a general formulation of the problem. This simple case will also serve to establish some notation.

A conceptual representation of a one-line diagram for a four-bus power system is shown in Figure 5.3. For generality, we have shown a generator and a load connected to each bus. The following notation applies:

$$\bar{S}_{G1} = \text{Complex power flow into bus 1 from the generator}$$

$$\bar{S}_{D1} = \text{Complex power flow into the load from bus 1}$$

Comparable quantities for the complex power generations and loads are obvious for each of the three other buses.

The positive sequence network for the power system represented by the one-line diagram of Figure 5.3 is shown in Figure 5.4.

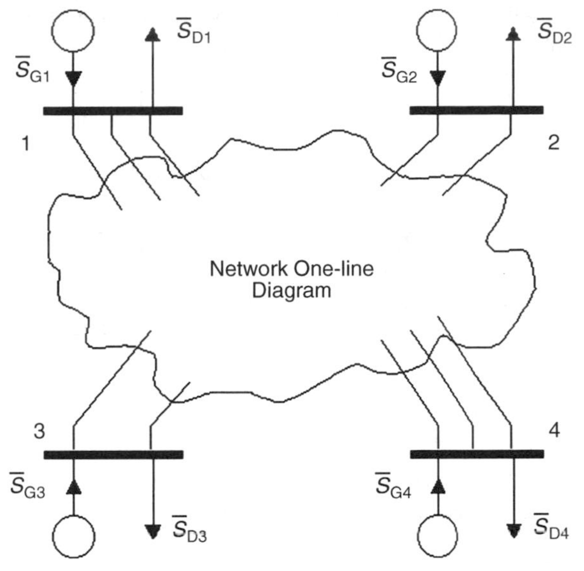

FIGURE 5.3 Conceptual one-line diagram of a four-bus power system.

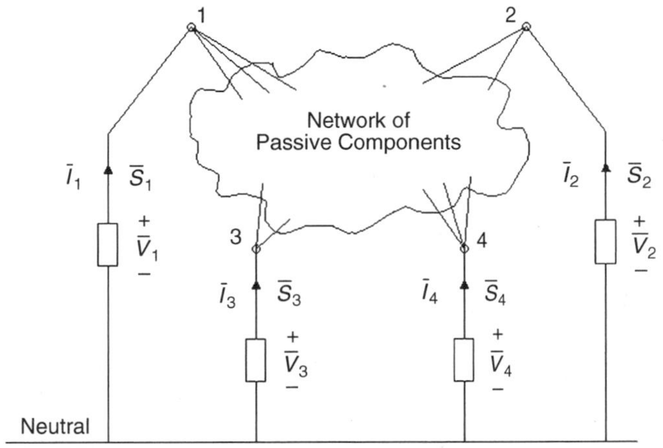

FIGURE 5.4 Positive sequence network for the system of Figure 5.3.

The boxes symbolize the combination of generation and load. Network texts refer to this network as a five-node network. (The balanced nature of the system allows analysis using only the positive sequence network, reducing each three-phase bus to a single node. The reference or ground represents the fifth node.) However, in power systems literature it is usually referred to as a four-bus network or power system.

For the network of Figure 5.4, we define the following additional notation:

$$\bar{S}_1 = \bar{S}_{G1} - \bar{S}_{D1} = \text{Net complex power injected at bus 1}$$

$$\bar{I}_1 = \text{Net positive sequence phasor current injected at bus 1}$$

$$\bar{V}_1 = \text{Positive sequence phasor voltage at bus 1}$$

The standard node voltage equations for the network can be written in terms of the quantities at bus 1 (defined above) and comparable quantities at the other buses.

$$\bar{I}_1 = \bar{Y}_{11}\bar{V}_1 + \bar{Y}_{12}\bar{V}_2 + \bar{Y}_{13}\bar{V}_3 + \bar{Y}_{14}\bar{V}_4 \tag{5.2}$$

$$\bar{I}_2 = \bar{Y}_{21}\bar{V}_1 + \bar{Y}_{22}\bar{V}_2 + \bar{Y}_{23}\bar{V}_3 + \bar{Y}_{24}\bar{V}_4 \tag{5.3}$$

$$\bar{I}_3 = \bar{Y}_{31}\bar{V}_1 + \bar{Y}_{32}\bar{V}_2 + \bar{Y}_{33}\bar{V}_3 + \bar{Y}_{34}\bar{V}_4 \tag{5.4}$$

$$\bar{I}_4 = \bar{Y}_{41}\bar{V}_1 + \bar{Y}_{42}\bar{V}_2 + \bar{Y}_{43}\bar{V}_3 + \bar{Y}_{44}\bar{V}_4 \tag{5.5}$$

The admittances in Equation (5.2) to Equation (5.5), $\bar{Y}_{ij}$, are the *ij*th entries of the bus admittance matrix for the power system. The unknown voltages could be found using linear algebra if the four currents $\bar{I}_1 \ldots \bar{I}_4$ were known. However, these currents are not known. Rather, something is known about the complex power and voltage at each bus. The complex power injected into bus k of the power system is defined by the relationship between complex power, voltage, and current given by Equation (5.6).

$$\bar{S}_k = \bar{V}_k \bar{I}_k^* \tag{5.6}$$

Therefore,

$$\bar{I}_k = \frac{\bar{S}_k^*}{\bar{V}_k^*} = \frac{\bar{S}_{Gk}^* - \bar{S}_{Dk}^*}{\bar{V}_k^*} \tag{5.7}$$

By substituting this result into the nodal equations and rearranging, the basic power flow equations for the four-bus system are given as Equation (5.8) to Equation (5.11)

$$\bar{S}_{G1}^* - \bar{S}_{D1}^* = \bar{V}_1^*[\bar{Y}_{11}\bar{V}_1 + \bar{Y}_{12}\bar{V}_2 + \bar{Y}_{13}\bar{V}_3 + \bar{Y}_{14}\bar{V}_4] \tag{5.8}$$

$$\bar{S}_{G2}^* - \bar{S}_{D2}^* = \bar{V}_2^*[\bar{Y}_{21}\bar{V}_1 + \bar{Y}_{22}\bar{V}_2 + \bar{Y}_{23}\bar{V}_3 + \bar{Y}_{24}\bar{V}_4] \tag{5.9}$$

$$\bar{S}_{G3}^* - \bar{S}_{D3}^* = \bar{V}_3^*[\bar{Y}_{31}\bar{V}_1 + \bar{Y}_{32}\bar{V}_2 + \bar{Y}_{33}\bar{V}_3 + \bar{Y}_{34}\bar{V}_4] \tag{5.10}$$

$$\bar{S}_{G4}^* - \bar{S}_{D4}^* = \bar{V}_4^*[\bar{Y}_{41}\bar{V}_1 + \bar{Y}_{42}\bar{V}_2 + \bar{Y}_{43}\bar{V}_3 + \bar{Y}_{44}\bar{V}_4] \tag{5.11}$$

Examination of Equation (5.8) to Equation (5.11) reveals that except for the trivial case where the generation equals the load at every bus, the complex power outputs of the generators cannot be arbitrarily selected. In fact, the complex power output of at least one of the generators must be calculated last since it must take up the unknown "slack" due to the, as yet, uncalculated network losses. Further, losses cannot be calculated until the voltages are known. These observations are a result of the principle of conservation of complex power. (That is, the sum of the injected complex powers at the four system buses is equal to the system complex power losses.)

Further examination of Equation (5.8) to Equation (5.11) indicates that it is not possible to solve these equations for the absolute phase angles of the phasor voltages. This simply means that the problem can only be solved to some arbitrary phase angle reference.

In order to alleviate the dilemma outlined above, suppose $\bar{S}_{G4}$ is arbitrarily allowed to float or swing (in order to take up the necessary slack caused by the losses) and that $\bar{S}_{G1}$, $\bar{S}_{G2}$, and $\bar{S}_{G3}$ are specified (other cases will be considered shortly). Now, with the loads known, Equation (5.8) to Equation (5.11) are seen as four simultaneous nonlinear equations with complex coefficients in five unknowns $\bar{V}_1$, $\bar{V}_2$, $\bar{V}_3$, $\bar{V}_4$, and $\bar{S}_{G4}$.

The problem of too many unknowns (which would result in an infinite number of solutions) is solved by specifying another variable. Designating bus 4 as the slack bus and specifying the voltage $\bar{V}_4$ reduces the problem to four equations in four unknowns. The slack bus is chosen as the phase reference for all phasor calculations, its magnitude is constrained, and the complex power generation at this bus is free to take up the slack necessary in order to account for the system real and reactive power losses.

The specification of the voltage $\bar{V}_4$ decouples Equation (5.11) from Equation (5.8) to Equation (5.10), allowing calculation of the slack bus complex power after solving the remaining equations. (This property carries over to larger systems with any number of buses.) The example problem is reduced to solving only three equations simultaneously for the unknowns $\bar{V}_1$, $\bar{V}_2$, and $\bar{V}_3$. Similarly, for the case of n buses it is necessary to solve $n - 1$ simultaneous, complex coefficient, nonlinear equations.

The formulation can be further complicated by the presence of generators at other buses since, in all realistic cases, the voltage magnitude at a generator bus is controlled by the generating unit exciter. Specifying the voltage magnitude at a generator bus requires a variable specified in the above analysis to become an unknown (in order to bring the number of unknowns back into correspondence with the number of equations). Normally, the reactive power injected by the generator becomes a variable, leaving the real power and voltage magnitude as the specified quantities at the generator bus. Specifying the voltage magnitude at a bus and treating the bus reactive power injection as a variable effectively results in retention of the same number of complex unknowns. For example, if the voltage magnitude of bus 1 of the earlier four-bus system is specified and the reactive power injection at bus 1 becomes a variable, Equation (5.8) to Equation (5.10) again effectively have three complex unknowns. (The phasor voltages $\bar{V}_2$ and $\bar{V}_3$ at buses 2 and 3 are two complex unknowns, and the angle δ_1 of the voltage at bus 1 plus the reactive power generation Q_{G1} at bus 1 result in the equivalent of a third complex unknown.)

Systems of nonlinear equations, such as Equation (5.8) to Equation (5.10), cannot (except in rare cases) be solved by closed-form techniques. Direct simulation was used extensively for many years; however, essentially all power flow analyses today are performed using iterative techniques on digital computers.

5.6 Bus Classifications

There are four quantities of interest associated with each bus:

1. Real power, P
2. Reactive power, Q
3. Voltage magnitude, V
4. Voltage angle, δ

At every bus of the system, two of these four quantities will be specified and the remaining two will be unknowns. Each of the system buses may be classified in accordance with which of the two quantities are specified. The following classifications are typical:

Slack Bus — The slack bus for the system is *a single bus for which the voltage magnitude and angle are specified.* The real and reactive power are unknowns. The bus selected as the slack bus must have a source of both real and reactive power, since the injected power at this bus must "swing" to take up the "slack" in the solution. The best choice for the slack bus (since, in most power systems, many buses have real and reactive power sources) requires experience with the particular system under study. The behavior of the solution is often influenced by the bus chosen. (In the earlier discussion, the last bus was selected as the slack bus for convenience.)

Load Bus (P-Q Bus) — A load bus is defined as *any bus of the system for which the real and reactive power are specified.* Load buses may contain generators with specified real and reactive power outputs; however, it is often convenient to designate any bus with specified injected complex power as a load bus.

Voltage Controlled Bus (P-V Bus) — *Any bus for which the voltage magnitude and the injected real power are specified is classified as a voltage-controlled (or P-V) bus.* The injected reactive power is a variable (with specified upper and lower bounds) in the power flow analysis. Typically, all generator buses are treated as voltage-controlled buses.

5.7 Generalized Power Flow Development

The more general (n bus) case is developed by extending the results of the simple four-bus example. Consider the case of an n-bus system and the corresponding $n+1$ node positive sequence network. Assume that the buses are numbered such that the slack bus is numbered last. Direct extension of the earlier equations (writing the node voltage equations and making the same substitutions as in the four-bus case) yields the basic power flow equations (PFE) in the general form.

$$\bar{S}_k^* = P_k - jQ_k = \bar{V}_k^* \sum_{i=1}^{n} \bar{Y}_{ki}\bar{V}_i \qquad \text{for } k = 1, 2, 3, \ldots, n-1 \tag{5.12}$$

and

$$P_n - jQ_n = \bar{V}_n^* \sum_{i=1}^{n} \bar{Y}_{ni}\bar{V}_i \tag{5.13}$$

Equation (5.13) is the equation for the slack bus. Equation (5.12) represents $n-1$ simultaneous equations in $n-1$ complex unknowns if all buses (other than the slack bus) are classified as load buses. Thus, given a set of specified loads, the problem is to solve Equation (5.12) for the $n-1$ complex phasor voltages at the remaining buses. Once the bus voltages are known, Equation (5.13) can be used to calculate the slack bus power.

Bus j is normally treated as a P-V bus if it has a directly connected generator. The unknowns at bus j are then the reactive generation, Q_{Gj}, and δ_j because the voltage magnitude, V_j, and the real power generation, P_{Gj}, have been specified.

The next step in the analysis is to solve Equation (5.12) for the bus voltages using some iterative method. Once the bus voltages have been found, the complex power flows and complex power losses in all of the network components are calculated.

5.8 Solution Methods

The solution of the simultaneous nonlinear power flow equations requires the use of iterative techniques for even the simplest power systems. Although there are many methods for solving nonlinear equations, only two methods are discussed here.

The Newton–Raphson Method

The Newton–Raphson algorithm has been applied in the solution of nonlinear equations in many fields. The algorithm will be developed using a general set of two equations (for simplicity). The results are easily extended to an arbitrary number of equations.

A set of two nonlinear equations is shown in Equation (5.14) and Equation (5.15).

$$f_1(x_1, x_2) = k_1 \tag{5.14}$$

$$f_2(x_1, x_2) = k_2 \tag{5.15}$$

Now, if $x_1^{(0)}$ and $x_2^{(0)}$ are inexact solution estimates, and $\Delta x_1^{(0)}$ and $\Delta x_2^{(0)}$ are the corrections to the estimates to achieve an exact solution, Equation (5.14) and Equation (5.15) can be rewritten as

$$f_1(x_1 + \Delta x_1^{(0)}, x_2 + \Delta x_2^{(0)}) = k_1 \tag{5.16}$$

$$f_2(x_1 + \Delta x_1^{(0)}, x_2 + \Delta x_2^{(0)}) = k_2 \tag{5.17}$$

Expanding Equation (5.16) and Equation (5.17) in a Taylor series about the estimate yields

$$f_1(x_1^{(0)}, x_2^{(0)}) + \left.\frac{\partial f_1}{\partial x_1}\right|^{(0)} \Delta x_1^{(0)} + \left.\frac{\partial f_1}{\partial x_2}\right|^{(0)} \Delta x_2^{(0)} + \text{h.o.t.} = k_1 \tag{5.18}$$

$$f_2(x_1^{(0)}, x_2^{(0)}) + \left.\frac{\partial f_2}{\partial x_1}\right|^{(0)} \Delta x_1^{(0)} + \left.\frac{\partial f_2}{\partial x_2}\right|^{(0)} \Delta x_2^{(0)} + \text{h.o.t.} = k_2 \tag{5.19}$$

where the subscript, (0), on the partial derivatives indicates evaluation of the partial derivatives at the initial estimate and h.o.t. indicates the higher-order terms.

Neglecting the higher-order terms (an acceptable approximation if $\Delta x_1{}^{(0)}$ and $\Delta x_2{}^{(0)}$ are small), Equation (5.18) and Equation (5.19) can be rearranged and written in matrix form.

$$\begin{bmatrix} \left.\frac{\partial f_1}{\partial x_1}\right|^{(0)} & \left.\frac{\partial f_1}{\partial x_2}\right|^{(0)} \\ \left.\frac{\partial f_2}{\partial x_1}\right|^{(0)} & \left.\frac{\partial f_2}{\partial x_2}\right|^{(0)} \end{bmatrix} \begin{bmatrix} \Delta x_1^{(0)} \\ \Delta x_2^{(0)} \end{bmatrix} \approx \begin{bmatrix} k_1 - f_1(x_1^{(0)}, x_2^{(0)}) \\ k_2 - f_2(x_1^{(0)}, x_2^{(0)}) \end{bmatrix} \tag{5.20}$$

The matrix of partial derivatives in Equation (5.20) is known as the Jacobian matrix and is evaluated at the initial estimate. Multiplying each side of Equation (5.20) by the inverse of the Jacobian yields an approximation of the required correction to the estimated solution. Since the higher-order terms were neglected, addition of the correction terms to the original estimate will not yield an exact solution, but will provide an improved estimate. The procedure may be repeated, obtaining sucessively better estimates, until the estimated solution reaches a desired tolerance. Summarizing, correction terms for the lth iterate are given in Equation (5.21), and the solution estimate is updated according to Equation (5.22).

$$\begin{bmatrix} \Delta x_1^{(\ell)} \\ \Delta x_2^{(\ell)} \end{bmatrix} = \begin{bmatrix} \left.\frac{\partial f_1}{\partial x_1}\right|^{(\ell)} & \left.\frac{\partial f_1}{\partial x_2}\right|^{(\ell)} \\ \left.\frac{\partial f_2}{\partial x_1}\right|^{(\ell)} & \left.\frac{\partial f_2}{\partial x_2}\right|^{(\ell)} \end{bmatrix}^{-1} \begin{bmatrix} k_1 - f_1(x_1^{(\ell)}, x_2^{(\ell)}) \\ k_2 - f_2(x_1^{(\ell)}, x_2^{(\ell)}) \end{bmatrix} \tag{5.21}$$

$$x^{(l+1)} = x^{(l)} + \Delta x^{(l)} \tag{5.22}$$

The solution of the original set of nonlinear equations has been converted to a repeated solution of a system of linear equations. This solution requires evaluation of the Jacobian matrix (at the current solution estimate) in each iteration.

The power flow equations can be placed into the Newton–Raphson framework by separating the power flow equations into their real and imaginary parts, and taking the voltage magnitudes and phase angles as the unknowns. Writing Equation (5.21) specifically for the power flow problem:

$$\begin{bmatrix} \Delta \underline{\delta}^{(\ell)} \\ \Delta \underline{V}^{(\ell)} \end{bmatrix} = \begin{bmatrix} \left.\frac{\partial P}{\partial \delta}\right|^{(\ell)} & \left.\frac{\partial P}{\partial V}\right|^{(\ell)} \\ \left.\frac{\partial Q}{\partial \delta}\right|^{(\ell)} & \left.\frac{\partial Q}{\partial V}\right|^{(\ell)} \end{bmatrix}^{-1} \begin{bmatrix} P(\text{sched}) - P^{(\ell)} \\ Q(\text{sched}) - Q^{(\ell)} \end{bmatrix} \tag{5.23}$$

The underscored variables in Equation (5.23) indicate vectors (extending the two-equation Newton–Raphson development to the general power-flow case). The (sched) notation indicates the scheduled real and reactive powers injected into the system. $P^{(l)}$ and $Q^{(l)}$ represent the calculated real and reactive power injections based on the system model and the lth voltage phase angle and voltage magnitude estimates. The bus voltage phase angle and bus voltage magnitude estimates are updated, the Jacobian reevaluated, and the mismatch between the scheduled and calculated real and reactive powers evaluated in each iteration of the Newton–Raphson algorithm. Iterations are performed until the estimated solution reaches an acceptable tolerance or a maximum number of allowable iterations is exceeded. Once a solution (within an acceptble tolerance) is reached, *P-V* bus reactive power injections and the slack bus complex power injection may be evaluated.

Fast-Decoupled Power Flow Solution

The fast-decoupled power flow algorithm simplifies the procedure presented for the Newton–Raphson algorithm by exploiting the strong coupling between real power/ bus voltage phase angles and reactive power/ bus voltage magnitudes commonly seen in power systems. The Jacobian matrix is simplified by approximating the partial derivatives of the real power equations with respect to the bus voltage magnitudes as zero. Similarly, the partial derivatives of the reactive power equations with respect to the bus voltage phase angles are approximated as zero. Further, the remaining partial derivatives are often approximated using only the imaginary portion of the bus admittance matrix. These approximations yield the following correction equations

$$\Delta\delta^{(l)} = [B'][P(\text{sched}) - P^{(l)}] \tag{5.24}$$

$$\Delta V^{(l)} = [B''][Q(\text{sched}) - Q^{(l)}] \tag{5.25}$$

where B' is an approximation of the matrix of partial derviatives of the real power flow equations with respect to the bus voltage phase angles, and B'' is an approximation of the matrix of partial derivatives of the reactive power flow equations with respect to the bus voltage magnitudes. B' and B'' are typically held constant during the iterative process, eliminating the necessity of updating the Jacobian matrix (required in the Newton–Raphson solution) in each iteration.

The fast-decoupled algorithm has good convergence properties despite the many approximations used during its development. The fast-decoupled power flow algorithm has found widespread use since it is less computationally intensive (requires fewer computational operations) than the Newton–Raphson method.

5.9 Component Power Flows

The positive sequence network for components of interest (connected between buses *i* and *j*) will be of the form shown in Figure 5.5.

An admittance description is usually available from earlier construction of the nodal admittance matrix. Thus,

$$\begin{bmatrix} \bar I_i \\ \bar I_j \end{bmatrix} = \begin{bmatrix} \bar Y_a & \bar Y_b \\ \bar Y_c & \bar Y_d \end{bmatrix} \begin{bmatrix} \bar V_i \\ \bar V_j \end{bmatrix} \tag{5.26}$$

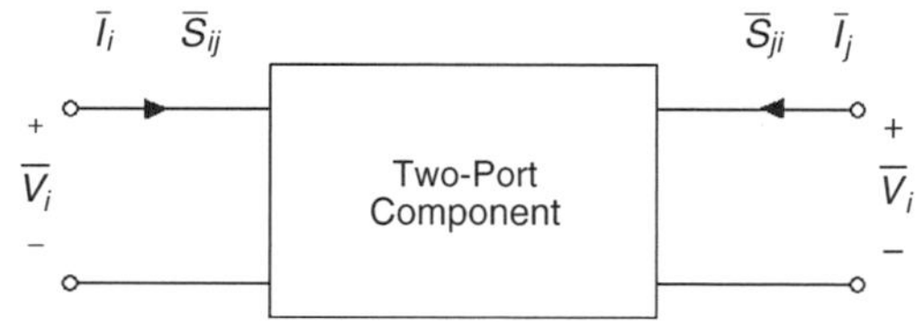

FIGURE 5.5 Typical power-system component.

Therefore the complex power flows and the component loss are

$$\bar S_{ij} = \bar V_i \bar I_i^* = \bar V_i\left[\bar Y_a \bar V_i + \bar Y_b \bar V_j\right]^* \tag{5.27}$$

$$\bar S_{ji} = \bar V_j \bar I_j^* = \bar V_i\left[\bar Y_c \bar V_i + \bar Y_d \bar V_j\right]^* \tag{5.28}$$

$$\bar S_{\text{loss}} = \bar S_{ij} + \bar S_{ji} \tag{5.29}$$

The calculated component flows combined with the bus voltage magnitudes and phase angles provide extensive information about the power systems operating point. The bus voltage magnitudes may be checked to ensure operation within a prescribed range. The segment power flows can be examined to ensure that no equipment ratings are exceeded. Additionally, the power-flow solution may be used as the starting point for other analyses.

An elementary discussion of the power-flow problem and its solution is presented in this chapter. The power-flow problem can be complicated by the addition of further constraints such as generator real and reactive power limits. However, discussion of such complications is beyond the scope of this chapter. The references provide detailed development of power flow formulation and solution under additional constraints. The references also provide some background in the other types of power-system analysis discussed at the begining of the chapter.

References

Bergen, A.R. and Vittal. V., *Power Systems Analysis*, 2nd ed., Prentice-Hall, Inc., Englewood Cliffs, NJ, 2000.
Elgerd, O.I., *Electric Energy Systems Theory — An Introduction*, 2nd ed., McGraw-Hill, New York, 1982.
Glover, J.D. and Sarma, M., *Power System Analysis and Design*, 3rd ed., PWS Publishing, Boston, 2002.
Gross, C.A., *Power System Analysis*, 2nd ed., John Wiley & Sons, New York, 1986.
Stevenson, W.D., *Elements of Power System Analysis*, 4th ed., McGraw-Hill, New York, 1982.

Further Information

The references provide clear introductions to the analysis of power systems. An excellent review of many issues involving the use of computers for power system analysis is provided in July 1974, *Proceedings of the IEEE* (special issue on computers in the power industry). The quarterly journal *IEEE Transactions on Power Systems* provides excellent documentation of more recent research in power system analysis.

6

Power Transformers

Charles A. Gross
Auburn University

6.1 Transformer Construction

The Transformer Core

The core of the power transformer is usually made of laminated cold-rolled magnetic steel that is grain oriented such that the rolling direction is the same as that of the flux lines. This type of core construction tends to reduce the eddy current and hysteresis losses. The eddy current loss P_e is proportional to the square of the product of the maximum flux density $B_M(T)$, the frequency f (Hz), and thickness $t(m)$ of the individual steel lamination.

$$P_e = K_e(B_M t f)^2 \quad \text{(W)} \tag{6.1}$$

K_e is dependent upon the core dimensions, the specific resistance of a lamination sheet, and the mass of the core. Also,

$$P_h = K_h f\, B_M^n \quad \text{(W)} \tag{6.2}$$

In Equation (6.2), P_h is the hysteresis power loss, n is the Steinmetz constant ($1.5 < n < 2.5$) and K_h is a constant dependent upon the nature of core material and varies from $3\times10^{-3}m$ to $20\times10^{-3}m$, where m = core mass in kilograms.

The core loss therefore is

$$P_e = P_e + P_h \tag{6.3}$$

FIGURE 6.1 230kVY:17.1kVΔ 1153-MVA 3ϕ power transformer. (Photo courtesy of General Electric Company.)

Core and Shell Types

Transformers are constructed in either a shell or a core structure. The shell-type transformer is one where the windings are completely surrounded by transformer steel in the plane of the coil. Core-type transformers are those that are not shell type. A power transformer is shown in Figure 6.1.

Multiwinding transformers, as well as polyphase transformers, can be made in either shell- or core-type designs.

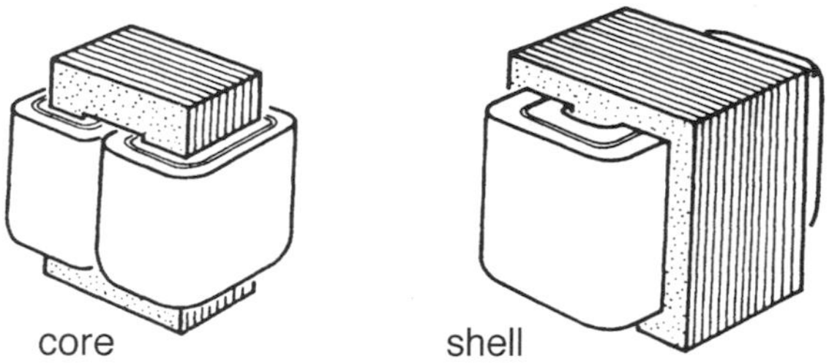

Transformer Windings

The windings of the power transformer may be either copper or aluminum. These conductors are usually made of conductors having a circular cross section; however, larger cross-sectional area conductors may require a rectangular cross section for efficient use of winding space.

The life of a transformer insulation system depends, to a large extent, upon its temperature. The total temperature is the sum of the ambient and the temperature rise. The temperature rise in a transformer is intrinsic to that transformer at a fixed load. The ambient temperature is controlled by the environment the transformer is subjected to. The better the cooling system that is provided for the transformer, the higher the "kVA" rating for the same ambient. For example, the kVA rating for a transformer can be increased with forced air (fan) cooling. Forced oil and water cooling systems are also used. Also, the duration of operating time at high temperature directly affects insulation life.

Other factors that affect transformer insulation life are vibration or mechanical stress, repetitive expansion and contraction, exposure to moisture and other contaminants, and electrical and mechanical stress due to overvoltage and short-circuit currents.

Paper insulation is laid between adjacent winding layers. The thickness of this insulation is dependent on the expected electric field stress. In large transformers oil ducts are provided using paper insulation to allow a path for cooling oil to flow between coil elements.

The short-circuit current in a transformer creates enormous forces on the turns of the windings. The short-circuit currents in a large transformer are typically 8 to 10 times larger than rated and in a small transformer are 20 to 25 times rated. The forces on the windings due to the short-circuit current vary as the square of the current, so whereas the forces at rated current may be only a few newtons, under short-circuit conditions these forces can be tens of thousands of Newtons. These mechanical and thermal stresses on the windings must be taken into consideration during the design of the transformer. The current-carrying components must be clamped firmly to limit movement. The solid insulation material should be precompressed and formed to avoid its collapse due to the thermal expansion of the windings.

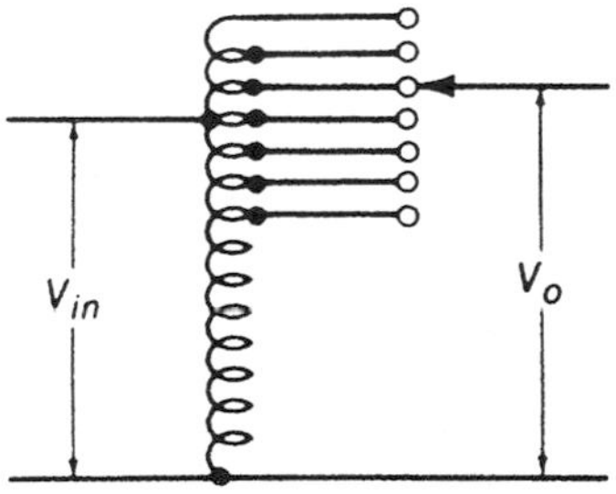

Taps

Power transformer windings typically have taps, as shown. The effect on transformer models is to change the turns ratio.

6.2 Power Transformer Modeling

The electric power **transformer** is a major power system component which provides the capability of reliably and efficiently changing (transforming) ac voltage and current at high power levels. Because electrical power is proportional to the product of voltage and current, for a specified power level, low current levels can exist only at high voltage, and vice versa.

The Three-Winding Ideal Transformer Equivalent Circuit

Consider the three coils wrapped on a common core as shown in Figure 6.2(a). For an infinite core permeability (μ) and windings made of material of infinite conductivity (σ):

$$v_1 = N_1 \frac{d\phi}{dt} \quad v_2 = N_2 \frac{d\phi}{dt} \quad v_3 = N_3 \frac{d\phi}{dt} \tag{6.4}$$

where ϕ is the core flux. This produces:

$$\frac{v_1}{v_2} = \frac{N_1}{N_2} \quad \frac{v_2}{v_3} = \frac{N_2}{N_3} \quad \frac{v_3}{v_1} = \frac{N_3}{N_1} \tag{6.5}$$

For sinusoidal steady state performance:

$$\overline{V}_1 = \frac{N_1}{N_2}\overline{V}_2 \quad \overline{V}_2 = \frac{N_2}{N_3}\overline{V}_3 \quad \overline{V}_3 = \frac{N_3}{N_1}\overline{V}_1 \tag{6.6}$$

where $\bar{V}$, etc. are complex phasors.

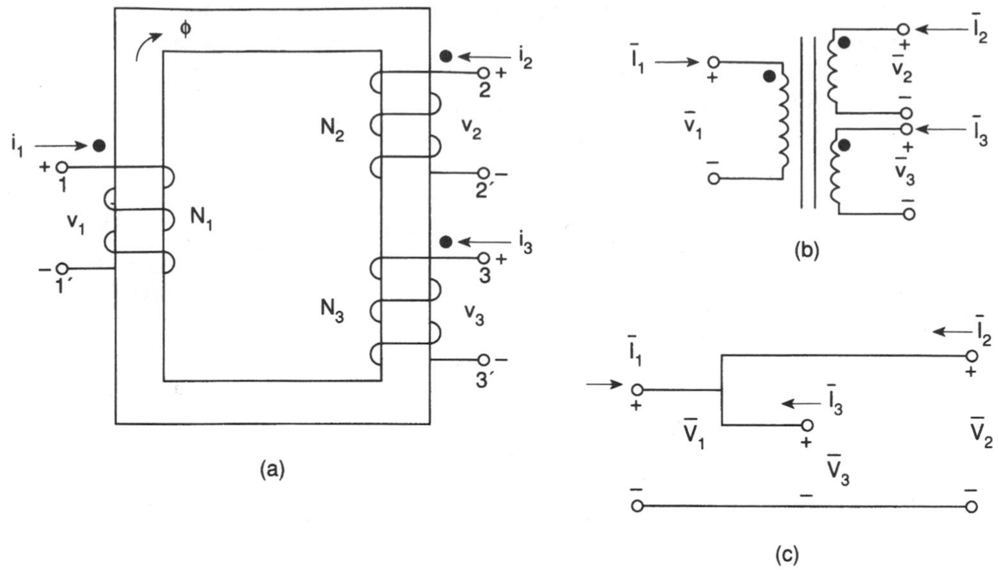

FIGURE 6.2 Ideal three-winding transformer. (a) Ideal three-winding transformer; (b) schematic symbol; (c) per-unit equivalent circuit.

The circuit symbol is shown in Figure 6.2(b). Ampere's law requires that

$$\oint \hat{H} \cdot \hat{d}l = i_{\text{enclosed}} = 0 \tag{6.7}$$

$$0 = N_1 i_1 + N_2 i_2 + N_3 i_3 \tag{6.8}$$

Transform Equation (6.8) into phasor notation:

$$N_1 \bar{I}_1 + N_2 \bar{I}_2 + N_3 \bar{I}_3 = 0 \tag{6.9}$$

Equation (6.6) and Equation (6.9) are basic to understanding transformer operation. Consider Equation (6.6). Also note that $-V_1$, $-V_2$, and $-V_3$ must be in phase, with dotted terminals defined positive. Now consider the total input complex power $-S$.

$$\bar{S} = \bar{V}_1 \bar{I}_1^* + \bar{V}_2 \bar{I}_2^* + \bar{V}_3 \bar{I}_3^* = 0 \tag{6.10}$$

Hence, ideal transformers can absorb neither real nor reactive power.

It is customary to scale system quantities (V, I, S, Z) into dimensionless quantities called per-unit values. The basic per-unit scaling equation is

$$\text{Per-unit value} = \frac{\text{actual value}}{\text{base value}}$$

The base value always carries the same units as the actual value, forcing the per-unit value to be dimensionless. Base values normally selected arbitrarily are V_{base} and S_{base}. It follows that:

$$I_{\text{base}} = \frac{S_{\text{base}}}{V_{\text{base}}}$$

$$Z_{\text{base}} = \frac{V_{\text{base}}}{I_{\text{base}}} = \frac{V_{\text{base}}^2}{S_{\text{base}}}$$

When per-unit scaling is applied to transformers V_{base} is usually taken as V_{rated} as in each winding. S_{base} is common to all windings; for the two-winding case S_{base} is S_{rated}, since S_{rated} is common to both windings.

Per-unit scaling simplifies transformer circuit models. Select two primary base values, $V_{1\text{base}}$ and $S_{1\text{base}}$. Base values for windings 2 and 3 are:

$$V_{2_{\text{base}}} = \frac{N_2}{N_1} V_{1_{\text{base}}} \qquad V_{3_{\text{base}}} = \frac{N_3}{N_1} V_{1_{\text{base}}} \tag{6.11}$$

and

$$S_{1_{\text{base}}} = S_{2_{\text{base}}} = S_{3_{\text{base}}} = S_{\text{base}} \tag{6.12}$$

By definition:

$$I_{1_{\text{base}}} = \frac{S_{\text{base}}}{V_{1_{\text{base}}}} \qquad I_{2_{\text{base}}} = \frac{S_{\text{base}}}{V_{2_{\text{base}}}} \qquad I_{3_{\text{base}}} = \frac{S_{\text{base}}}{V_{3_{\text{base}}}} \tag{6.13}$$

It follows that

$$I_{2_{\text{base}}} = \frac{N_1}{N_2} I_{1_{\text{base}}} \qquad I_{3_{\text{base}}} = \frac{N_1}{N_3} I_{1_{\text{base}}} \tag{6.14}$$

Thus, Equation (6.3) and Equation (6.6) scaled into per-unit become:

$$\overline{V}_{1_{\text{pu}}} = \overline{V}_{2_{\text{pu}}} = \overline{V}_{3_{\text{pu}}} \tag{6.15}$$

$$\overline{I}_{1_{\text{pu}}} + \overline{I}_{2_{\text{pu}}} + \overline{I}_{3_{\text{pu}}} = 0 \tag{6.16}$$

The basic per-unit equivalent circuit is shown in Figure 6.2(c). The extension to the *n*-winding case is clear.

A Practical Three-Winding Transformer Equivalent Circuit

The circuit of Figure 6.2(c) is reasonable for some power system applications, since the core and windings of actual transformers are constructed of materials of high μ and σ, respectively, though of course not infinite. However, for other studies, discrepancies between the performance of actual and ideal transformers are too great to be overlooked. The circuit of Figure 6.2(c) may be modified into that of Figure 6.3 to account for the most important discrepancies. Note:

R_1,R_2,R_3 Since the winding conductors cannot be made of material of infinite conductivity, the windings must have some resistance.

X_1,X_2,X_3 Since the core permeability is not infinite, not all of the flux created by a given winding current will be confined to the core. The part that escapes the core and seeks out parallel paths in surrounding structures and air is referred to as *leakage* flux.

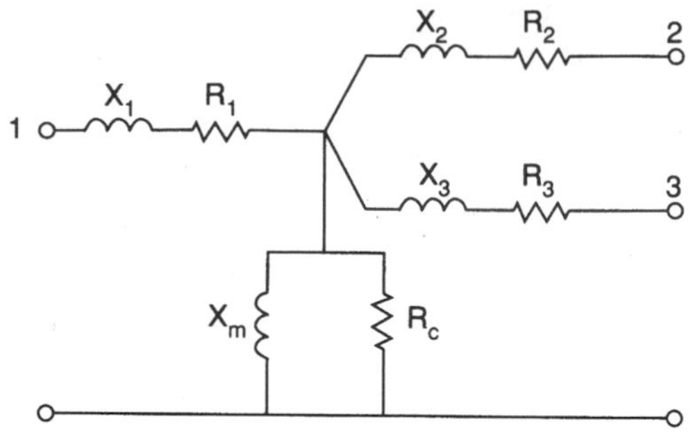

FIGURE 6.3 A practical equivalent circuit.

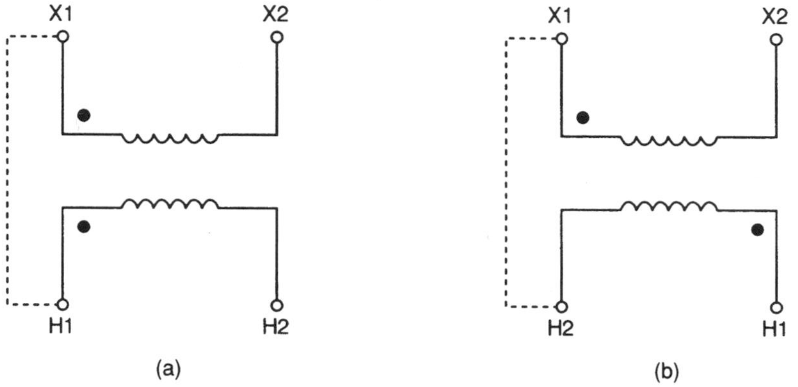

FIGURE 6.4 Transformer polarity terminology: (a) subtractive; (b) additive.

R_c,X_m Also, since the core permeability is not infinite, the magnetic field intensity inside the core is not zero. Therefore, some current flow is necessary to provide this small H. The path provided in the circuit for this "magnetizing" current is through X_m. The core has internal power losses, referred to as *core loss*, due to hystereses and eddy current phenomena. The effect is accounted for in the resistance R_c. Sometimes R_c and X_m are neglected.

The circuit of Figure 6.3 is a refinement on that of Figure 6.2(c). The values R_1, R_2, R_3, X_1, X_2, X_3 are all small (less than 0.05 per-unit) and R_c, X_m, large (greater than 10 per-unit). The circuit of Figure 6.3 requires that all values be in per-unit. Circuit data are available from the manufacturer or obtained from conventional tests. It must be noted that although the circuit of Figure 6.3 is commonly used, it is not rigorously correct because it does not properly account for the mutual couplings between windings.

The terms **primary** and **secondary** refer to source and load sides, respectively (i.e., energy flows from primary to secondary). However, in many applications energy can flow either way, in which case the distinction is meaningless. Also, the presence of a third winding (tertiary) confuses the issue. The terms *step up* and *step down* refer to what the transformer does to the voltage from source to load. ANSI standards require that for a two-winding transformer the high-voltage and low-voltage terminals be marked as H1-H2 and X1-X2, respectively, with H1 and X1 markings having the same significance as *dots* for **polarity** markings. [Refer to ANSI C57 for comprehensive information.] *Additive* and *subtractive transformer polarity* refer to the physical positioning of high-voltage, low-voltage *dotted* terminals as shown in Figure 6.4. If the dotted terminals are adjacent, then the transformer is said to be *subtractive*, because if these adjacent terminals (H1-X1) are connected together, the voltage between H2 and X2 is the *difference* between primary and secondary. Similarly, if adjacent terminals X1 and H2 are connected, the voltage (H1-X2) is the *sum* of primary and secondary values.

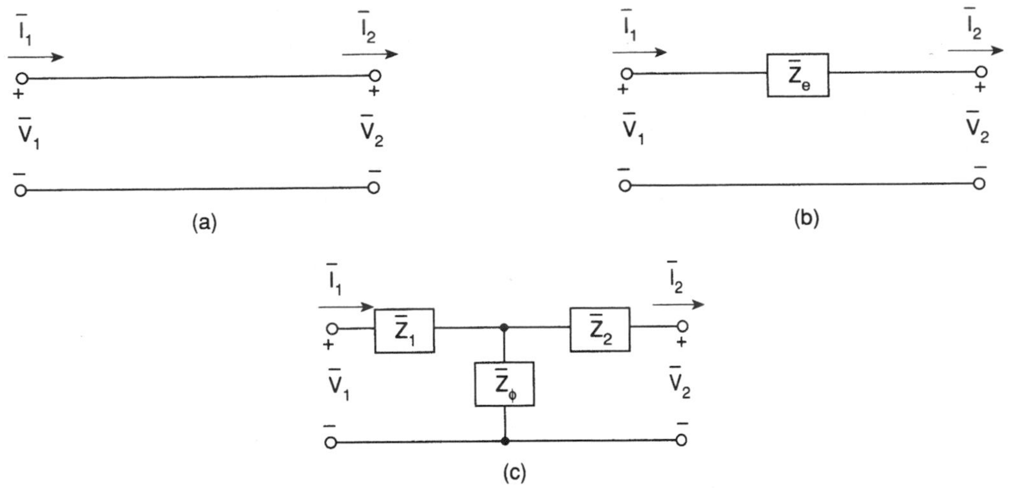

FIGURE 6.5 Two-winding transformer-equivalent circuits. All values in per-unit. (a) Ideal case; (b) no load current negligible; (c) precise model.

The Two-Winding Transformer

The device can be simplified to two windings. Common two-winding transformer circuit models are shown in Figure 6.5.

$$\bar{Z}_e = \bar{Z}_1 + \bar{Z}_2 \tag{6.17}$$

$$\bar{Z}_m = \frac{R_c(jX_m)}{R_c + jX_m} \tag{6.18}$$

Circuits (a) and (b) are appropriate when $-Z_m$ is large enough that magnetizing current and core loss is negligible.

6.3 Transformer Performance

There is a need to assess the quality of a particular transformer design. The most important measure for performance is the concept of efficiency, defined as follows:

$$\eta = \frac{P_{out}}{P_{in}} \tag{6.19}$$

where P_{out} is output power in watts (kW, MW) and P_{in} is input power in watts (kW, MW).

The situation is clearest for the two-winding case where the output is clearly defined (i.e., the secondary winding), as is the input (i.e., the primary). Unless otherwise specified, the output is understood to be rated power at rated voltage at a user-specified power factor. Note that

$$\Sigma L = P_{in} - P_{out} = \text{sum of losses}$$

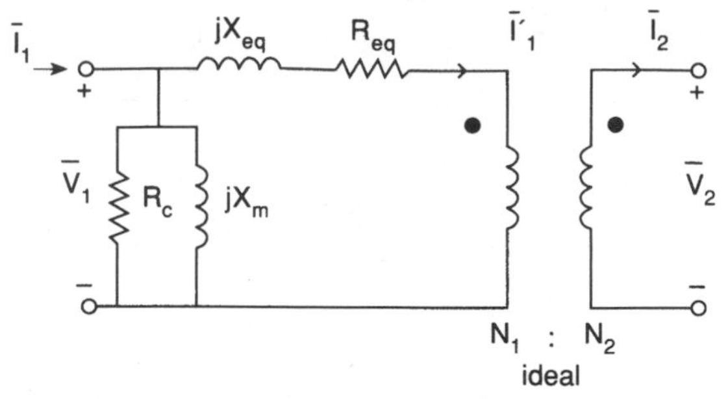

FIGURE 6.6 Transformer circuit model.

The transformer is frequently modeled with the circuit shown in Figure 6.6. Transformer losses are made up of the following components:

Electrical losses :

$$I_1'^2 R_{eq} = I_1^2 R_1 + I_2^2 R_2 \tag{6.20a}$$

$$\text{Primary winding loss} = I_1^2 R_1 \tag{6.20b}$$

$$\text{Secondary winding loss} = I_2^2 R_2 \tag{6.20c}$$

Magnetic (core) loss :

$$\begin{aligned} P_c &= P_e + P_h = V_1^2/R_c \\ \text{Core eddy current loss} &= P_e \\ \text{Core hysterisis loss} &= P_h \end{aligned} \tag{6.21}$$

Hence:

$$\Sigma L = I_1'^2 R_{eq} + V_1^2/R_c \tag{6.22}$$

A second concern is fluctuation of secondary voltage with load. A measure of this situation is called *voltage regulation*, which is defined as follows:

$$\text{Voltage Regulation (VR)} = \frac{V_{2NL} - V_{2FL}}{V_{2FL}} \tag{6.23}$$

where V_{2FL} = rated secondary voltage, with the transformer supplying rated load at a user-specified power factor, and V_{2NL} = secondary voltage with the load removed (set to zero), holding the primary voltage at the full load value.

A complete performance analysis of a 100 kVA 2400/240 V single-phase transformer is shown in Table 6.1.

6.4 Transformers in Three-Phase Connections

Transformers are frequently used in three-phase connections. For three identical three-winding transformers, nine windings must be accounted for. The three sets of windings may be individually connected in wye or delta in any combination. The symmetrical component transformation can be used to produce the sequence equivalent circuits shown in Figure 6.7 which are essentially the circuits of Figure 6.3 with R_c and X_m neglected.

The positive and negative sequence circuits are valid for both wye and delta connections. However, Y–Δ connections will produce a phase shift which is not accounted for in these circuits.

TABLE 6.1 Analysis of a Single-Phase 2400:240V 100-kVA Transformer

Voltage and Power Ratings

HV (Line-V)	LV (Line-V)	S (Total-kVA)
2400	240	100

Test Data

Short Circuit (HV) Values	Open Circuit (LV) Values
Voltage = 211.01	240.0 volts
Current = 41.67	22.120 amperes
Power = 1400.0	787.5 watts

Equivalent Circuit Values (in ohms)

Values referred to		HV Side	LV Side	Per-Unit
Series Resistance	=	0.8064	0.008064	0.01400
Series Reactance	=	4.9997	0.049997	0.08680
Shunt Magnetizing Reactance	=	1097.10	10.9714	19.05
Shunt Core Loss Resistance	=	7314.30	73.1429	126.98

Power Factor (—)	Efficiency (%)	Voltage Regulation (%)	Power Factor (—)	Efficiency (%)	Voltage Regulation (%)
0.0000 lead	0.00	−8.67	0.9000 lag	97.54	5.29
0.1000 lead	82.92	−8.47	0.8000 lag	97.21	6.50
0.2000 lead	90.65	−8.17	0.7000 lag	96.81	7.30
0.3000 lead	93.55	−7.78	0.6000 lag	96.28	7.86
0.4000 lead	95.06	−7.27	0.5000 lag	95.56	8.26
0.5000 lead	95.99	−6.65	0.4000 lag	94.50	8.54
0.6000 lead	96.62	−5.89	0.3000 lag	92.79	8.71
0.7000 lead	97.07	−4.96	0.2000 lag	89.56	8.79
0.8000 lead	97.41	−3.77	0.1000 lag	81.09	8.78
0.9000 lead	97.66	−2.16	0.0000 lag	0.00	8.69
1.0000 —	97.83	1.77			

Rated load performance at power factor = 0.866 lagging.

Secondary Quantities; LOW Voltage Side

	SI Units	Per-Unit
Voltage	240 volts	1.0000
Current	416.7 amperes	1.0000
Apparent power	100.0 kVA	1.0000
Real power	86.6 kW	0.8660
Reactive power	50.0 kvar	0.5000
Power factor	0.8660 lag	0.8660

Primary Quantities; HIGH Voltage Side

	SI Units	Per-Unit
Voltage	2539 volts	1.0577
Current	43.3 amperes	1.0386
Apparent power	109.9 kVA	1.0985
Real power	88.9 kW	0.8888
Reactive power	64.6 kvar	0.6456
Power factor	0.8091 lag	0.8091

Efficiency = 97.43%; voltage regulation = 5.77%.

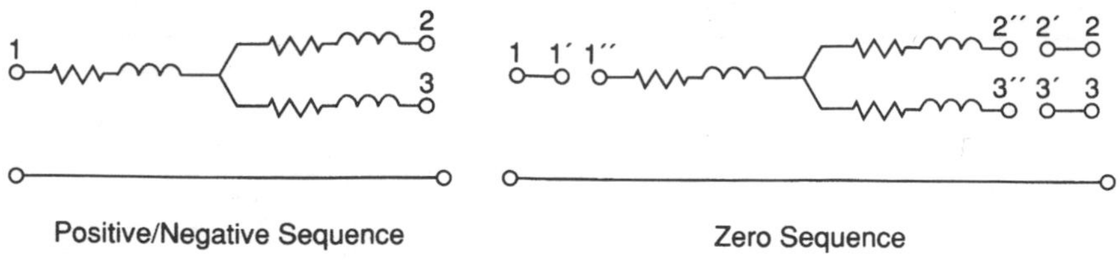

FIGURE 6.7 Sequence equivalent transformer circuits.

The zero sequence circuit requires special modification to account for wye, delta connections. Consider winding 1:

1. Solid grounded wye — short 1′ to 1″.
2. Ground wye through $-Z_n$ — connect 1′ to 1″ through $3\text{–}Z_n$.
3. Ungrounded wye — leave 1′ to 1″ open.
4. Delta — short 1″ to reference.

Winding sets 2 and 3 interconnections produce similar connection constraints at terminals 2′–2″ and 3′–3″, respectively.

Example

Three identical transformers are to be used in a three-phase system. They are connected at their terminals as follows:

Winding set 1 wye, grounded through $-Z_n$
Winding set 2 wye, solid ground
Winding set 3 delta

The zero sequence network is as shown.

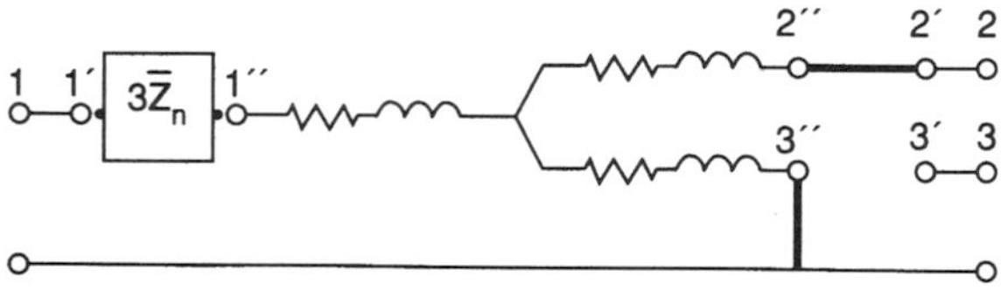

Phase Shift in Y–Δ Connections

The positive and negative sequence networks presented in Figure 6.7 are misleading in one important detail. For Y–Y or Δ–Δ connections, it is always possible to label the phases in such a way that there is no phase shift between corresponding primary and secondary quantities. However, for Y–Δ or Δ–Y connections, it is impossible to label the phases in such a way that no phase shift between corresponding quantities is introduced. ANSI standard C57.12.10.17.3.2 is as follows:

For either wye-delta or delta-wye connections, phases shall be labeled in such a way that positive sequence quantities on the high voltage side lead their corresponding positive sequence quantities on the low voltage side by 30°. The effect on negative sequence quantities may be the reverse, i.e., HV values lag *LV values by 30°.*

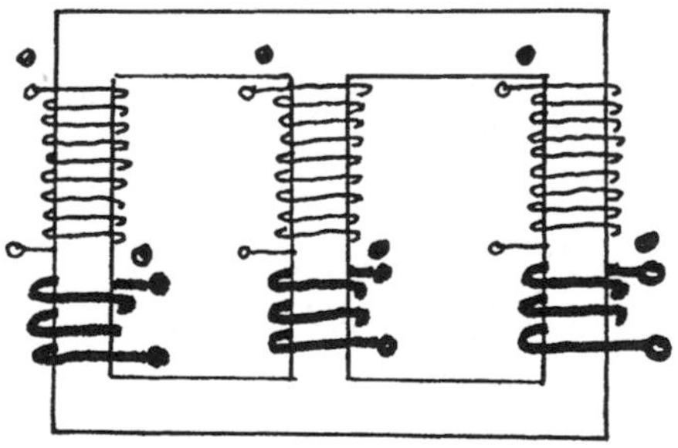

This 30° phase shift is *not* accounted for in the sequence networks of Figure 6.7. The effect only appears in the positive and negative sequence networks; the zero sequence network quantities are unaffected.

The Three-Phase Transformer

It is possible to construct a device (called a three-phase transformer) which allows the phase fluxes to share common magnetic return paths. Such designs allow considerable savings in core material, and corresponding economies in cost, size, and weight. Positive and negative sequence impedances are equal; however, the zero sequence impedance may be different. Otherwise the circuits of Figure 6.7 apply as discussed previously.

Determining per-Phase Equivalent Circuit Values for Power Transformers

One method of obtaining such data is through testing. Consider the problem of obtaining transformer equivalent circuit data from short-circuit tests. A numerical example will clarify per-unit scaling considerations.

The short-circuit test circuit arrangement is shown in Figure 6.8. The objective is to derive equivalent circuit data from the test data provided in Figure 6.8. Note that measurements are made in winding "*i*", with winding "*j*" shorted, and winding "*k*" left open. The short circuit impedance, looking into winding "*i*" with the transformer so terminated is designated as Z_{ij}. The indices *i*, *j*, and *k*, can be 1, 2, or 3.

The impedance calculations are done in per-unit; base values are provided in Figure 6.8(c). The transformer ratings of the transformer of Figure 6.2(a) would conventionally be provided as follows:

3ϕ 3W Transformer
15kVY/115kVY/4.157kVΔ
100/100/20 MVA

where 3ϕ means that the transformer is a three-phase piece of equipment (as opposed to an interconnection of three single-phase devices). 3W means three three-phase windings (actually nine windings). Usually the schematic is supplied also. The 15 kV rating is the *line* (phase-to-phase) value; three-phase apparatus is always rated in *line* values. "Y" means winding No. 1 is internally wye connected. 115kVY means that 115 kV is the

(a)

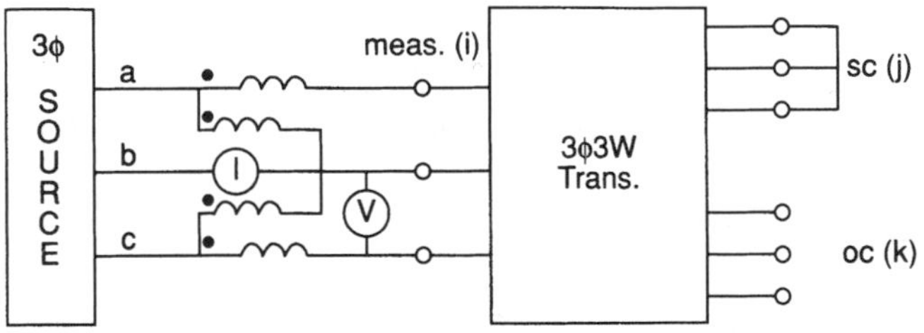

(b)

Line Voltage		3ph Power Ratings	S3ph Base = 100 MVA
Primary	15 kV	100 MVA	Zbase = 2.25
Secondary	115 kV	100 MVA	Zbase = 132.25
Tertiary	4.157 kV	20 MVA	Zbase = 0.1728

(c)

Voltage (line)	Current (line)	Power (3ph)	1	2	3
1200.0 V	3849.0 A	889.0 kW	meas	sc	oc
1840.0 V	100.0 A	100.0 kW	oc	meas	sc
33.0 V	2776.0 A	120.0 kW	sc	oc	meas

(d)

	R	*X*	*Z*		*R*	*X*	*Z*
Z12:	0.00889	0.07950	0.08000	Z1:	0.00686	0.20145	0.20157
Z23:	0.02520	0.07627	0.08033	Z2:	0.00203	–0.12195	0.12196
Z31:	0.03004	0.39967	0.40080	Z3:	0.02318	0.19822	0.19957

FIGURE 6.8 Transformer circuit data from short-circuit tests. (a) Setup for transformer short-circuit tests; (b) transformer data; (c) short-circuit test data; (d) short-circuit impedance values in per-unit.

line voltage rating, and winding No. 2 is wye connected. In 4.157kVΔ, again, "4.157kV" is the line voltage rating, and winding No. 3 is delta connected. 100/100/20 MVA are the *total* (3ϕ) power ratings for the primary, secondary, and tertiary winding, respectively; three-phase apparatus is always rated in three-phase terms.

The per-unit bases for $S_{3\phi\text{base}} = 100$ MVA are presented in Figure 6.8(b). Calculating the short-circuit impedances from the test data in Figure 6.8(c):

$$Z_{ij} = \frac{V_{i_{\text{line}}}/\sqrt{3}}{I_{i_{\text{line}}}}$$

$$R_{ij} = \frac{R_{3\phi}/3}{I_{1_{\text{line}}}^2}$$

$$X_{ij} = \sqrt{Z_{ij}^2 - R_{ij}^2}$$

Now calculate the transformer impedances from the short-circuit impedances:

$$\overline{Z}_1 = \frac{1}{2}(\overline{Z}_{12} - \overline{Z}_{23} + \overline{Z}_{31})$$

$$\overline{Z}_2 = \frac{1}{2}(\overline{Z}_{23} - \overline{Z}_{13} + \overline{Z}_{12})$$

$$\overline{Z}_3 = \frac{1}{2}(\overline{Z}_{31} - \overline{Z}_{12} + \overline{Z}_{23})$$

Results are shown in Figure 6.8(d). Observe that the Y–Δ winding connections had no impact on the calculations.

Another detail deserves mention. Although the real and reactive parts of the short-circuit impedances ($-Z_{12}, -Z_{23}, -Z_{31}$) will always be positive, this is not true for the transformer impedances ($-Z_1, -Z_2, -Z_3$). One or more of these can be, and frequently is, negative for actual short-circuit data. Negative values underscore that the circuit of Figure 6.7 is a *port equivalent* circuit, producing correct values at the winding terminals.

6.5 Autotransformers

Transformer windings, though magnetically coupled, are electrically isolated from each other. It is possible to enhance certain performance characteristics for transformers by electrically interconnecting primary and secondary windings. Such devices are called **autotransformers.** The benefits to be realized are lower cost, smaller size and weight, higher efficiency, and better voltage regulation. The basic connection is illustrated in Figure 6.9. The issues will be demonstrated with an example.

Consider the conventional connection, shown in Figure 6.9(a).

$$\overline{V}_2 = a\overline{V}_1$$

$$\overline{I}_2 = \frac{1}{a}\overline{I}_1$$

$$S_{\text{rating}} = V_1 I_1 = V_2 I_2 = S_{\text{load}}$$

Now for the autotransformer:

$$\overline{V}_2 = \overline{V}_1 + b\overline{V}_1 = (1+b)\overline{V}_1$$

$$\overline{I}_1 = \overline{I}_2 + b\overline{I}_2 = (1+b)\overline{I}_2$$

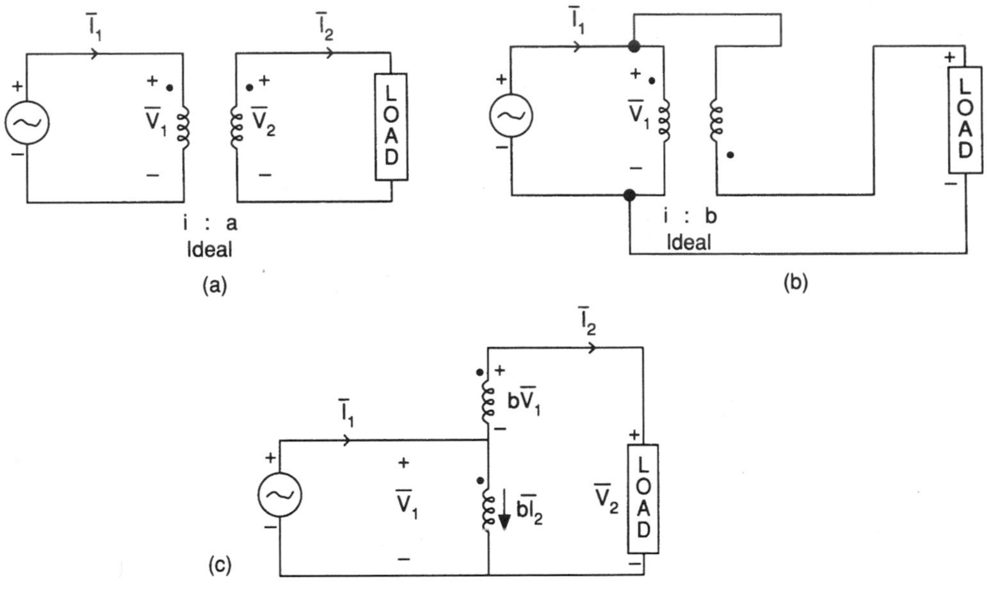

FIGURE 6.9 Autotransformer connection. (a) Conventional step-up connection; (b) autotransformer connection; (c) part (b) redrawn.

For the same effective ratio

$$1 + b = a$$

Therefore each winding rating is:

$$S_{\text{rated}} = S_{\text{load}}\left(\frac{b}{1+b}\right)$$

For example if $b = 1$ $(a = 2)$

$$S_{\text{rating}} = 1/2\, S_{\text{load}}$$

meaning that the transformer rating is only 50% of the load.

The principal advantage of the autotransformer is the increased power rating. Also, since the losses remain the same, expressed as a percentage of the new rating, they go down, and correspondingly, the efficiency goes up. The machine impedances in per unit drop for similar reasons. A disadvantage is the loss of electrical isolation between primary and secondary. Also, low impedance is not necessarily good, as we shall see when we study faults on power systems. Autotransformers are used in three-phase connections and in voltage control applications.

Defining Terms

Autotransformer: A transformer whose primary and secondary windings are electrically interconnected.
Polarity: Consideration of in-phase or out-of-phase relations of primary and secondary ac currents and voltages.
Primary: The source-side winding.
Secondary: The load-side winding.

Tap: An electrical terminal that permits access to a winding at a particular physical location.

Transformer: A device which converts ac voltage and current to different levels at essentially constant power and frequency.

References

ANSI Standard C57, New York: American National Standards Institute.

S.J. Chapman, *Electric Machinery Fundamentals,* 2nd ed, New York: McGraw-Hill, 1991.

V. Del Toro, *Basic Electric Machines*, Englewood Cliffs, NJ.: Prentice-Hall, 1990.

M.E. El-Hawary, *Electric Power Systems: Design and Analysis*, Reston, VA.: Reston Publishing, 1983.

O.I. Elgerd, *Electric Energy Systems Theory: An Introduction,* 2nd ed., New York: McGraw-Hill, 1982.

R. Feinburg, *Modern Power Transformer Practice,* New York: Wiley, 1979.

A.E. Fitzgerald, C. Kingsley, and S. Umans, *Electric Machinery*, 5th ed., New York: McGraw-Hill, 1990.

C.A. Gross, *Power Systems Analysis*, 2nd ed., New York: Wiley, 1986.

N.N. Hancock, *Matrix Analysis of Electrical Machinery*, 2nd ed., Oxford: Pergamon, 1974.

E. Lowden, *Practical Transformer Design Handbook,* 2nd ed, Blue Ridge Summit, PA.: TAB, 1989.

G. McPherson, *An Introduction to Electrical Machines and Transformers*, New York: Wiley, 1981.

A.J. Pansini, *Electrical Transformers,* Englewood Cliffs, NJ: Prentice-Hall, 1988.

G.R. Slemon, *Magnetoelectric Devices*, New York: Wiley, 1966.

R. Stein and W.T. Hunts, Jr., *Electric Power System Components: Transformers and Rotating Machines*, New York: Van Nostrand Reinhold, 1979.

Further Information

For a comprehensive coverage of general transformer theory, see Chapter 2 of *Electric Machines* by G.R. Slemon and A. Straughen (Addison-Wesley, 1980). For transformer standards, see ANSI Standard C57. For a detailed explanation of transformer per-unit scaling, see Chapter 5 of *Power Systems Analysis* by C.A. Gross (John Wiley, 1986). For design information see *Practical Transformer Design Handbook* by E. Lowden (TAB, 1989).

7

Energy Distribution

George G. Karady
Arizona State University

Tom Short
EPRI Solutions, Inc.

7.1 Introduction

Distribution is the last section of the electrical power system. Figure 7.1 shows the major components of the electric power system. The power plants convert the energy stored in the fuel (coal, oil, gas, nuclear) or hydro into electric energy. The energy is supplied through step-up transformers to the electric network. To reduce energy transportation losses, step-up transformers increase the voltage and reduce the current. The high-voltage network, consisting of transmission lines, connects the power plants and high-voltage **substations** in parallel. The typical voltage of the high-voltage transmission network is between 240 and 765 kV. The high-voltage substations are located near the load centers, for example, outside a large town. This network permits load sharing among power plants and assures a high level of reliability. The failure of a line or power plant will not interrupt the energy supply.

The subtransmission system connects the high-voltage substations to the distribution substations. These stations are directly in the load centers. For example, in urban areas, the distance between the distribution stations is around 5 to 10 miles. The typical voltage of the subtransmission system is between 138 and 69 kV. In high load density areas, the subtransmission system uses a network configuration that is similar to the high-voltage network. In medium and low load density areas, the loop or radial connection is used. Figure 7.1 shows a typical radial connection.

The distribution system has two parts, primary and secondary. The primary distribution system consists of overhead lines or underground cables, which are called **feeders**. The feeders run along the streets and supply the distribution transformers that step the voltage down to the secondary level (120–480 V). The secondary

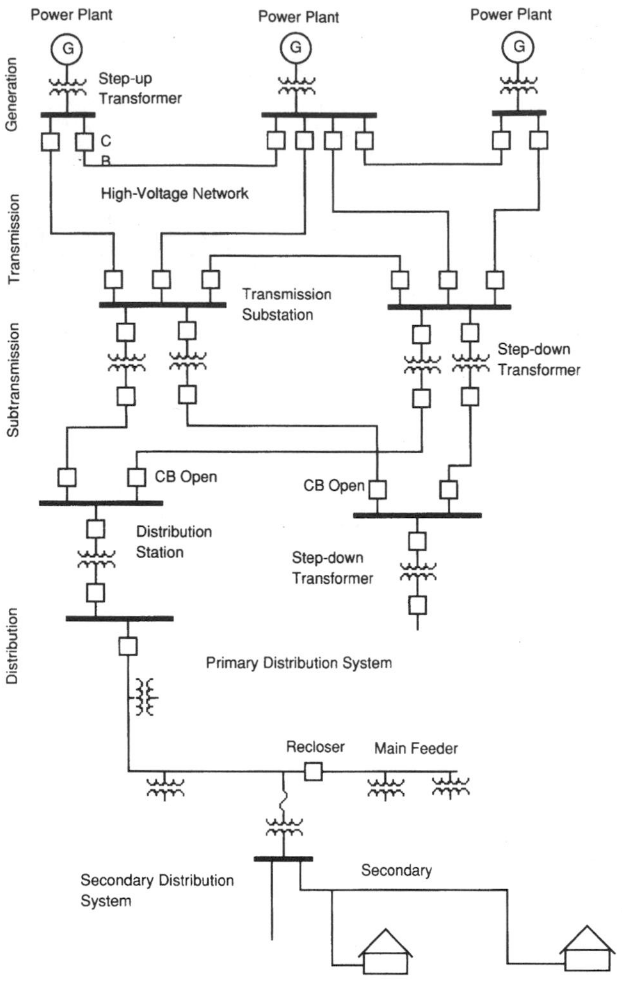

FIGURE 7.1 Electric energy system.

distribution system contains overhead lines or underground cables supplying the consumers directly (houses, light industry, shops, etc.) by single- or three-phase power. Separate, dedicated primary feeders supply industrial customers requiring several megawatts of power. The subtransmission system directly supplies large factories consuming over 50 MW.

7.2 Primary Distribution System

The most frequently used voltages and wiring in the primary distribution system are listed in Table 7.1.

Primary distribution, in low load density areas, is a radial system. This is economical but yields low reliability. In large cities, where the load density is very high, a primary cable network is used. The distribution substations are interconnected by the feeders (lines or cables). Circuit breakers (CBs) are installed at both ends of the feeder for short-circuit protection. The loads are connected directly to the feeders through fuses. The connection is similar to the one-line

TABLE 7.1 Typical Primary Feeder Voltages

Class, kV	Voltage, kV	Wiring
2.5	2.4	3-wire delta
5	4.16	4-wire Y
8.66	7.2	4-wire Y
15	12.47	3-wire delta/4-wire Y
25	22.9	4-wire Y
35	34.5	4-wire Y

diagram of the high-voltage network shown in Figure 7.1. The high cost of the network limits its application. A more economical and fairly reliable arrangement is the loop connection, when the main feeder is supplied from two independent distribution substations. These stations share the load. The problem with this connection is the circulating current that occurs when the two supply station voltages are different. The loop arrangement significantly improves system reliability.

The circulating current can be avoided by using the open-loop connection. This is a popular, frequently used circuit. Figure 7.2 shows a typical open-loop primary feeder. The distribution substation has four outgoing main feeders. Each feeder supplies a different load area and is protected by a reclosing CB.

The three-phase four-wire main feeders supply single-phase lateral feeders. A **recloser** and a sectionalizing switch divide the main feeder into two parts. The normally open tie-switch connects the feeder to the adjacent

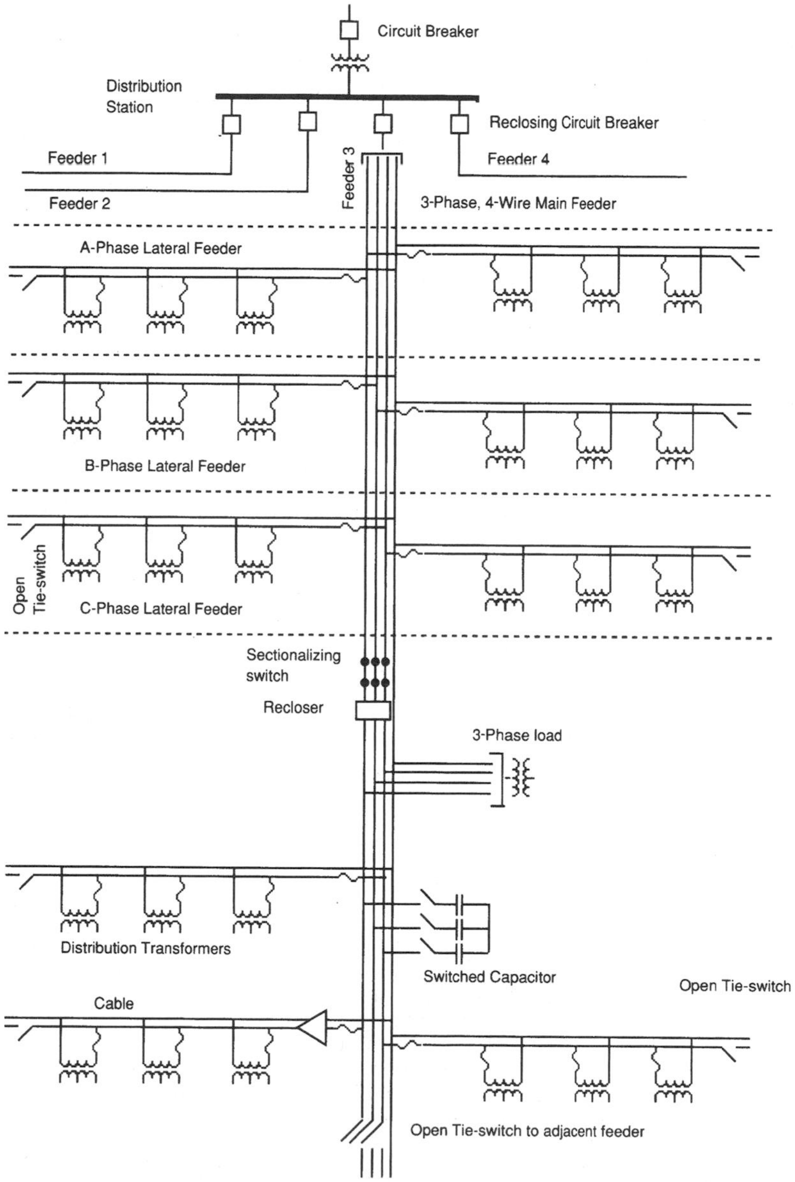

FIGURE 7.2 Radial primary distribution system.

distribution substation. The fault between the CB and recloser opens the reclosing CB. The CB recloses after a few cycles. If the fault is not cleared, the opening and reclosing process is repeated two times. If the fault has not been cleared before the third reclosing, the CB remains open. Then the sectionalizing switch opens and the tie-switch closes. This energizes the feeder between the recloser and the tie-switch from the neighboring feeder. Similarly, the fault between the recloser and tie-switch activates the recloser. The recloser opens and recloses three times. If the fault is not cleared, the recloser remains open and separates the faulty part of the feeder. This method is particularly effective in overhead lines where temporary faults are often caused by lightning and wind.

A three-phase switched **capacitor bank** is rated two-thirds of the total average reactive load and installed two-thirds of the distance out on the feeder from the source. The capacitor bank improves the power factor and reduces voltage drop at heavy loads. However, at light loads, the capacitor is switched off to avoid overvoltages.

Some utilities use voltage regulators at the primary feeders. The voltage regulator is an autotransformer. The secondary coil of the transformer has 32 taps, and a switch connects the selected tap to the line to regulate the voltage. The problem with the **tap changer** is that the lifetime of the switch is limited. This permits only a few operations per day.

The lateral single-phase feeders are supplied from different phases to assure equal phase loading. Fuse cutouts protect the lateral feeders. These fuses are coordinated with the fuses protecting the distribution transformers. The fault in the distribution transformer melts the transformer fuse first. The lateral feeder fault operates the cutout fuse before the recloser or CB opens permanently.

A three-phase line supplies the larger loads. These loads are protected by CBs or high-power fuses.

Most primary feeders in rural areas are overhead lines using pole-mounted distribution transformers. The capacitor banks and the reclosing and sectionalizing switches are also pole-mounted. Overhead lines reduce the installation costs but reduce aesthetics.

In urban areas, an underground cable system is used. The switchgear and transformers are placed in underground vaults or ground-level cabinets. The underground system is not affected by weather and is highly reliable. Unfortunately, the initial cost of an underground cable is significantly higher than an overhead line with the same capacity. The high cost limits the underground system to high-density urban areas and housing developments.

7.3 Secondary Distribution System

The secondary distribution system provides electric energy to the customers through the distribution transformers and secondary cables. Table 7.2 shows the typical voltages and wiring arrangements.

In residential areas, the most commonly used is the single-phase three-wire 120/240-V radial system, where the lighting loads are supplied by the 120 V and the larger household appliances (air conditioner, range, oven, and heating) are connected to the 240-V lines. Depending on the location, either underground cables or overhead lines are used for this system.

TABLE 7.2 Secondary Voltages and Connections

Class	Voltage	Connection	Application
1-phase	120/240	Three-wire	Residential
3-phase	208/120	Four-wire	Commercial/residential
3-phase	480/277	Four-wire	High-rise buildings
3-phase	380/220	Four-wire	General system, Europe
3-phase	120/240	Four-wire	Commercial
3-phase	240	Three-wire	Commercial/industrial
3-phase	480	Three-wire	Industrial
3-phase	240/480	Four-wire	Industrial

In urban areas, with high-density mixed commercial and residential loads, the three-phase 208/120-V four-wire network system is used. This network provides higher reliability but has significantly higher costs. Underground cables are used by most secondary networks.

High-rise buildings are supplied by a three-phase four-wire 480/277-V spot network. The fluorescent lighting is connected to 277 V, and the motor loads are supplied by a 480-V source. A separate local 120-V system supplies the outlets in the various rooms. This 120-V radial system is supplied by small transformers from the 480-V network.

7.4 Radial Distribution System

A typical overhead single-phase three-wire 120/240-V secondary system is shown in Figure 7.3. The three distribution transformers are mounted on separate primary feeder poles and supplied from different phases. Each transformer supplies 6 to 12 houses. The transformers are protected by fuses. The secondary feeders and the service drops are not protected individually. The secondary feeder uses insulated No. 1/0 or 4/0 aluminum conductors. The average secondary length is from 200 to 600 ft. The typical load is from 15 to 30 W/ft.

The underground distribution system is used in modern suburban areas. The transformers are pad-mounted or placed in an underground vault. A typical 50-kVA transformer serves 5 to 6 houses, with each house supplied by an individual cable.

The connection of a typical house is shown in Figure 7.4. The incoming secondary service drop supplies the kW and kWh meter. The modern, mostly electronic meters measure 15-min kW demand and the kWh energy consumption, and they record the maximum power demand and energy consumption. The electrical utility maintains the distribution system up to the secondary terminals of the meter. The homeowner is responsible for the service panel and house wiring. The typical service panel is equipped with a main switch and circuit

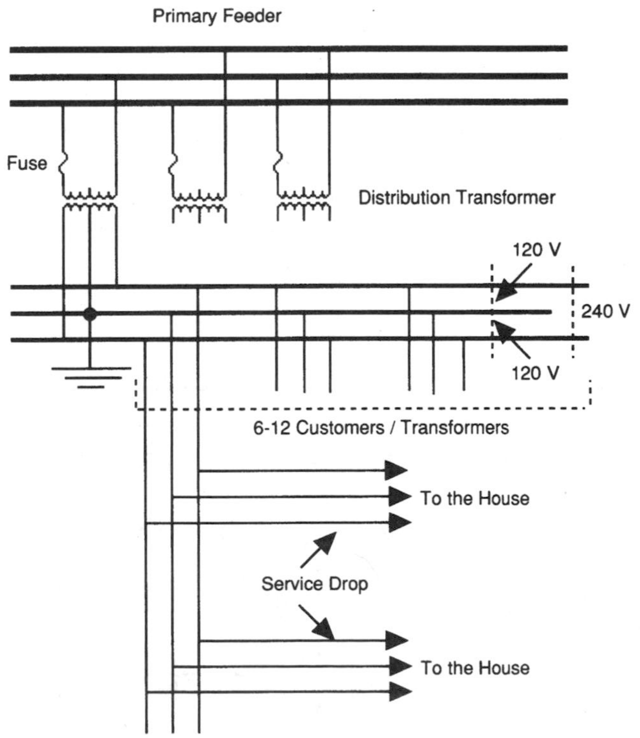

FIGURE 7.3 Typical 120/240-V radial secondary system.

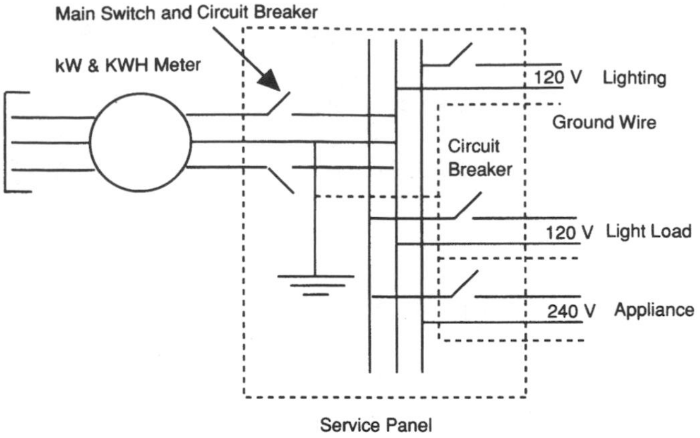

FIGURE 7.4 Residential electrical connection.

breaker. The main switch permits the deenergization of the house and protects against short circuits. The smaller loads are supplied by 120 V and the larger loads by 240 V. Each outgoing line is protected by a circuit breaker. The neutral has to be grounded at the service panel, just past the meter. The water pipe was used for grounding in older houses. In new houses a metal rod, driven in the earth, provides proper grounding. In addition, a separate bare wire is connected to the ground. The ground wire connects the metal parts of the appliances and service panel box together to protect against ground-fault-produced electric shocks.

7.5 Secondary Networks

The secondary network is used in urban areas with high load density. Figure 7.5 shows a segment of a typical secondary network.

The secondary feeders form a mesh or grid that is supplied by transformers at the node points. The multiple supply assures higher reliability and better load sharing. The loads are connected directly to the low-voltage grid, without any protection equipment. The network is protected by fuses and network protector circuit breakers installed at the secondary transformers. A short circuit blows the fuses and limits the current. The network protectors automatically open on reverse current and reclose when the voltage on the primary feeder is restored after a fault.

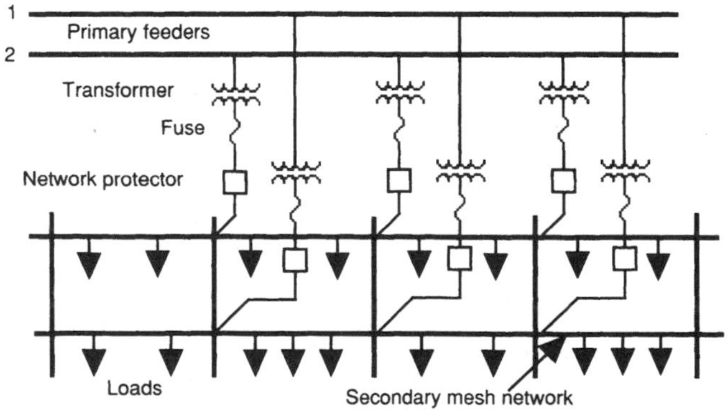

FIGURE 7.5 Typical segment of a secondary distribution network.

7.6 Load Characteristics

The distribution system load varies during the day. The maximum load occurs in the early evening or late afternoon, and the minimum load occurs at night. The design of the distribution system requires both values, because the voltage drop is at the maximum during the peak load, and overvoltages may occur during the minimum load. The power companies continuously study the statistical variation of the load and can predict the expected loads on the primary feeders with high accuracy. The feeder design considers the expected peak load or maximum demand and the future load growth.

The economic conductor cross-section calculation requires the determination of average losses. The average loss is calculated by the loss factor (LSF), which is determined by statistical analyses of load variation.

$$\mathrm{LSF} = \frac{\text{average loss}}{\text{loss at peak load}}$$

The average load is determined by the load factor (LF), which is the ratio of average load to peak load. The load factor for an area is determined by statistical analyses of the load variation in past years. The approximate relation between the loss factor and load factor is

$$\mathrm{LSF} = 0.3\mathrm{LF} + 0.7\mathrm{LF}^2$$

This equation is useful because the load factor is measured continuously by utilities, and more accurate values are available for the load factor than for the loss factor. Typical values are given in Table 7.3.

The connected load or demand can be estimated accurately in residential and industrial areas. The connected load or demand is the sum of continuous ratings of apparatus connected to the system. However, not all equipment is used simultaneously. The actual load in a system is significantly lower than the connected load. The demand factor (DF) is used to estimate the actual or maximum demand. DF is defined as

$$\mathrm{DF} = \frac{\text{maximum demand}}{\text{total connected demand}}$$

The demand factor depends on the number of customers and the type of load. Typical demand factor values are given in Table 7.4.

TABLE 7.3 Typical Annual Load Factor Values

Type of Load	Load Factor
Residential	0.48
Commercial	0.66
Industrial	0.72

TABLE 7.4 Typical Demand Factors for Multi-family Dwellings

Number of Dwellings	Demand Factor, %
3 to 5	45
18 to 20	38
39 to 42	28
62 & over	23

Adapted from Article 220–32, Table 202–32, *National Electrical Code 1987*, Quincy, Mass.: National Fire Protection Association, 1986. With permission.

7.7 Voltage Regulation

The voltage supplied to each customer should be within the ±5% limit, which, at 120 V, corresponds to 114 and 126 V. Figure 7.6 shows a typical voltage profile for a feeder at light and heavy load conditions. The figure shows that at heavy load, the voltage at the end of the line will be less than the allowable minimum voltage. However, at the light load condition the voltage supplied to each customer will be within the allowable limit. Calculation of the voltage profile, voltage drop, and feeder loss is one of the major tasks in

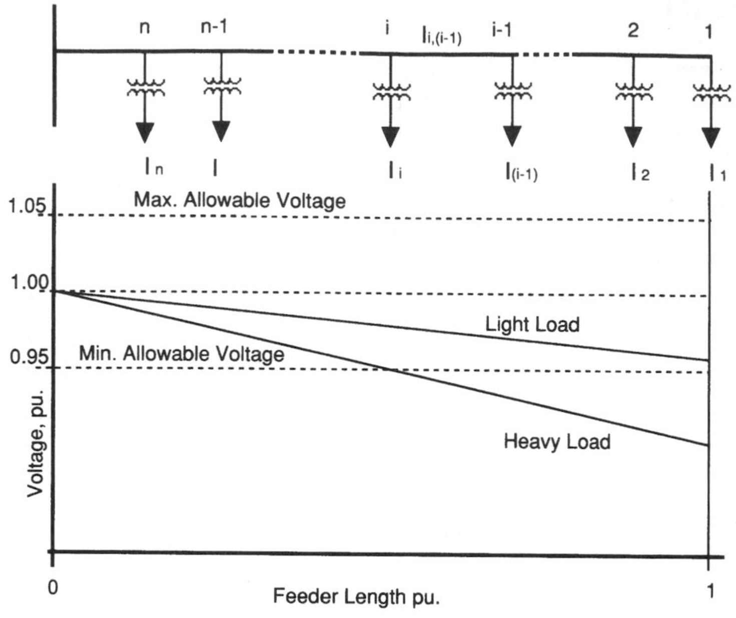

FIGURE 7.6 Feeder voltage profiles.

distribution system design. The concept of voltage drop and loss calculation is demonstrated using the feeder shown in Figure 7.6.

To calculate the voltage drop, the feeder is divided into sections. The sections are determined by the loads. Assuming a single-phase system, the load current is calculated by Equation (7.1):

$$|I_i| = \frac{P_i}{V\cos\varphi_i}, \quad I_i = |I_i|(\cos\varphi_i + \sin\varphi_i) \tag{7.1}$$

where P is the power of the load, V is the rated voltage, and ϕ is the power factor.

The section current is the sum of the load currents. Equation (7.2) gives the section current between load i and $i-1$:

$$I_{(i,i-1)} = \sum_{1}^{i=1} I_i \tag{7.2}$$

The electrical parameters of the overhead feeders are the resistance and reactance, which are given in Ω/mi.

The underground feeders have significant capacitance in addition to the reactance and resistance. The capacitance is given in μF/mi. The actual values for overhead lines can be calculated using the conductor diameter and phase-to-phase and phase-to-ground distance. The residential underground system generally uses single-conductor cables with polyethylene insulation. The older systems use rubber insulation with neoprene jacket. Circuit parameters should be obtained from manufacturers. The distribution feeders are short transmission lines. Even the primary feeders are only a few miles long. This permits the calculation of the section resistance and reactance by multiplying the Ω/mi values by the length of the section. The length of the section in a single-phase two-wire system is two times the actual length. In a balanced three-phase system, it is the simple length. In a single-phase three-wire system, the voltage drop on the neutral conductor must be calculated.

Equation (7.3) gives the voltage drop, with a good approximation, for section i, $(i-1)$. The total voltage drop is the sum of the sections' voltage drops.

$$e_{i,(i-1)} = |I_{i,(i-1)}|(R_{i,(i-1)}\cos\varphi_{i,(i-1)} + X_{i,(i-1)}\sin\varphi_{i,(i-1)}) \tag{7.3}$$

Equation (7.4) gives the losses on the line:

$$\text{Loss}_i = \sum_{1}^{i-1} (I_{i,(i-1)})^2 R_{i,(i-1)} \tag{7.4}$$

The presented calculation method describes the basic concept of feeder design; more details can be found in the literature.

7.8 Capacitors and Voltage Regulators

The voltage drop can be reduced by the application of a shunt capacitor. As shown in Figure 7.7, a properly selected and located shunt capacitor assures that the voltage supplied to each of the customers will be within the allowable limit at the heavy load condition. However, at light load, the same capacitor will increase the voltage above the allowable limit. Most capacitors in the distribution system use switches. The capacitor is switched off during the night when the load is light and switched on when the load is heavy. The most frequent use of capacitors is on the primary feeders. In an overhead system, three-phase capacitor banks with vacuum switches are installed on the poles. Residential underground systems require less shunt capacitance for voltage control due to the reduced reactance. Even so, shunt capacitors are used for power factor correction and loss reduction.

The optimum number, size, and location of capacitor banks on a feeder is determined by detailed computer analyses. The concept of optimization includes the minimization of the operation, installation, and investment costs. The most important factor that affects the selection is the distribution and power factor of loads.

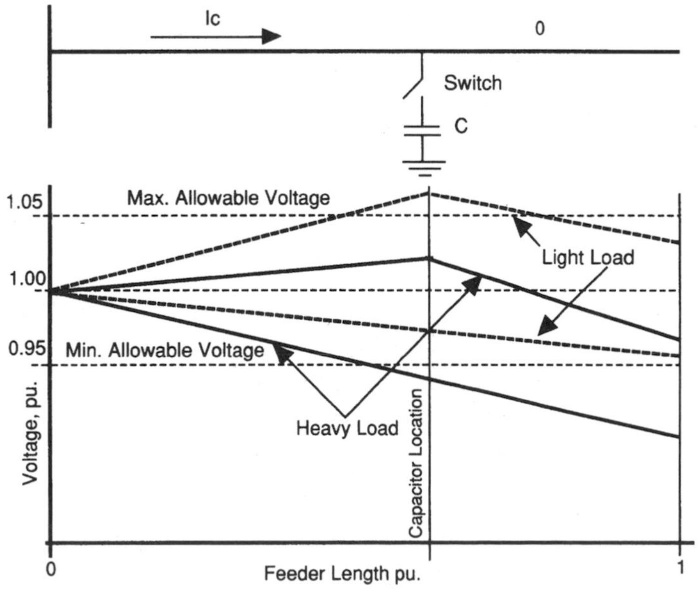

FIGURE 7.7 Capacitor effect on voltage profile.

In residential areas, the load is uniformly distributed. In this case the optimum location of the capacitor bank is around two-thirds of the length of the feeder.

The effect of a capacitor bank can be studied by adding the capacitor current to the load current. The capacitor current flows between the supply and the capacitor as shown in Figure 7.7. Its value can be calculated from Equation (7.5) for a single-phase system:

$$I_c = \mathrm{j}\omega CV, \quad \omega = 2\pi f \tag{7.5}$$

where C is the capacitance, f is the frequency (60 Hz), and V is the voltage to ground.

The capacitive current is added to the inductive load current, reducing the total current, the voltage drop, and losses. The voltage drop and loss can be calculated using Equations (7.2) to (7.5).

The voltage regulator is a tap-changing transformer, which is located, in most cases, at the supply end of the feeder. The tap changer increases the supply voltage, which in turn increases the voltage above the allowable minimum at the last load. The tap-changer transformer has two windings. The excitation winding is connected in parallel. The regulating winding is connected in series with the feeder. The latter has taps and a tap-changer switch. The switch changes the tap position according to the required voltage. The frequent tap changes reduce the lifetime of the tap-changer switch. This problem limits the number of tap-changer operations to one to three per day.

7.9 Distribution System Hardware

The major components of the distribution system include overhead distribution lines, distribution cables, transformers, capacitor banks, circuit breakers and reclosers, and fuse cutouts and disconnects.

Overhead distribution lines. A typical distribution line is a three-phase line built with wooden poles. The most frequently used are pine and fir poles. These poles must be treated against fungus, which cause decay of the structural strength of the wood. The decay starts at the ground line if the pole is not treated. For pole treatment, commonly used pesticides include creosote, pentachlorophenol, and chromated copper arsenate. The most frequently used treatment is creosote. Several methods have been developed to treat wooden poles. Typically, the wood is kiln dried for several hours and the treating material is injected into the wood using vacuum.

Figure 7.8 shows a typical 15-kV-class distribution pole with pole-mounted transformer. The figure shows a wooden cross arm with three small porcelain insulators. The conductors are attached to the insulator by wires. The distances between the conductors are normally more than 61 cm (24 in.) at 15 kV and 1 m (40 in.) at 33 kV. The typical span is 200 ft to 500 ft.

In a four-wire system the neutral conductor is placed in a fourth insulator on the cross arm or under the cross arm. Most frequently, the pole is directly buried in a hole without any foundation. Distribution poles are frequently anchored by steel cables. The anchoring balances the loads in the case of a change of direction or dead-end structure.

Cables. In cities and many suburban areas, underground cables are used. The cables are insulated conductors buried directly in the ground or placed in conduit or concrete-buried cable ducts. A cable duct carries 4 to 8 cables and permits the easy localization of cable fault and repair.

Figure 7.9 shows a typical distribution cable. At the center of a cable is the phase conductor, then a semiconducting conductor shield, the insulation, a semiconducting insulation shield, the neutral or shield, and finally a covering jacket. Most distribution cables are single conductor. Cable insulation is normally tree-retardant cross-linked polyethylene (XLPE) or ethylene-propylene rubber (EPR) compounds.

Transformers. Distribution systems have oil-cooled and -insulated transformers. The transformers are pole mounted on a distribution line or placed in metal boxes in the yard or underground vaults in a cable system. Figure 7.10 shows a pole-mounted distribution transformer. This transformer is protected against lightning by a surge arrester. The surge arrester is equipped with a plastic cup, which prevents squirrels and other animals

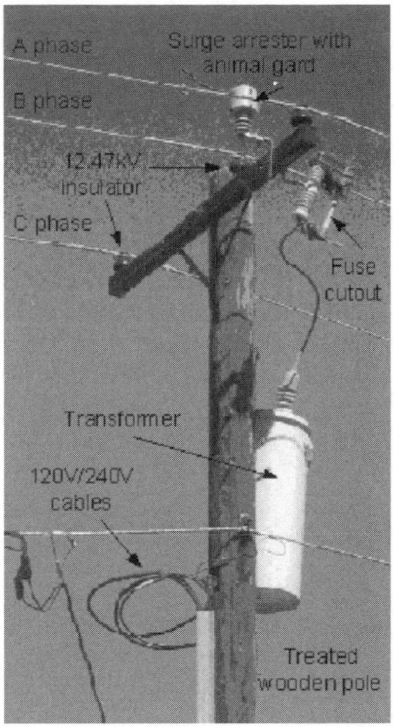

FIGURE 7.8 Overhead distribution line with a pole-mounted distribution transformer.

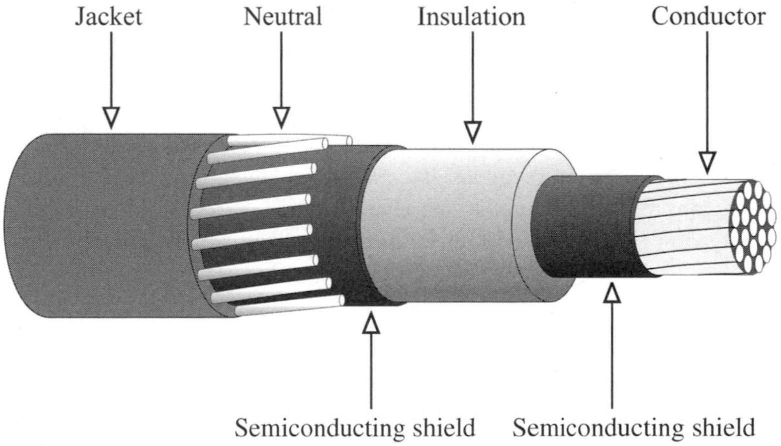

FIGURE 7.9 A concentric neutral cable, typically used for underground residential power delivery.

from electrocution. The surge arrester is grounded, together with the transformer tank and other metal parts. A fuse cutout is connected in series with the transformer for short-circuit protection.

Capacitor banks. The switched capacitor banks are also pole mounted. Figure 7.10 shows a 12.47-kV pole-mounted capacitor bank. The figure shows three capacitors connected in a wey. A vacuum switch and a fuse cutout are connected in series with each capacitor. The fuse cutout protects the capacitor against short circuit; the switch permits switching the capacitor on and off. The capacitor is switched off when the load is light at night and switched on in the morning when the load is increasing.

FIGURE 7.10 Three-phase switched capacitor bank.

Circuit breakers and reclosers. The distribution system uses circuit breakers for the protection of primary feeders. Pole-mounted reclosers are also used. The short-circuit current triggers the switch to open and interrupt the current. After a few cycles of delay, the switch closes and attempts to reenergize the line. If the fault is not cleared (a permanent fault), the recloser opens and remains open. The operator must dispatch a crew to repair the line.

The vacuum circuit breaker is equipped with a vacuum bottle, which has a fixed and a moving electrode. The current is interrupted by the fast separation of the electrodes in the vacuum. The current is interrupted at a zero crossing. The high dielectric strength of the vacuum prevents reignition.

Fuse cutouts and disconnects (tie switch). The fuse cutout shown in Figure 7.8 is used for short-circuit protection. The figure shows a pivoted fuse supported by a porcelain insulator. The fuse is a thin wire. The short-circuit current melts the wire, and the compression and force of the arc interrupt the current. Line workers can use cutouts as convenient sectionalizing switches operated by pulling the fuse out with a long insulated stick. The disconnect or tie-switch is a mechanical blade switch operated from the ground. Many disconnects cannot interrupt the load current; they can be operated only when the line current is zero.

7.10 Overhead versus Underground

Both overhead and underground designs have advantages (Table 7.5). The major advantage of overhead circuits is cost; an underground circuit typically costs anywhere from 1 to 2.5 times the equivalent overhead circuit. But the cost differences vary wildly, and it is often difficult to define "equivalent" systems in terms of performance. Under the right conditions, some estimates of cost report that cable installations can be less expensive than overhead lines. If the soil is easy to dig, if the soil has few rocks, and if the ground has no other obstacles like water pipes or telephone wires, then crews may be able to plow in cable faster and for less cost than an overhead circuit. In urban areas, underground is almost the only choice; just too many circuits are needed, and aboveground space is too expensive or just not available. But, urban duct-bank construction is expensive on a per-length basis (fortunately, circuits are short in urban applications). On many rural applications, the cost of underground circuits is difficult to justify, especially on long, lightly loaded circuits, given the small number of customers that these circuits feed.

Aesthetics is the main driver towards underground circuits. Especially in residential areas, parks, wildlife areas, and scenic areas, visual impact is important. Undergrounding removes a significant amount

TABLE 7.5. Overhead versus Underground: Advantages of Each

Overhead	Underground
• *Cost* — Overhead's number one advantage; significantly less cost, especially initial cost • *Longer life* — 30 to 50 years versus 20 to 40 for new underground works • *Reliability* — Shorter outage durations because of faster fault finding and faster repair • *Loading* — Overhead circuits can more readily withstand overloads	• *Aesthetics* — Underground's number one advantage; much less visual clutter • *Safety* — Less chance for public contact • *Reliability* — Significantly fewer short- and long-duration interruptions • *O&M* — Notably lower maintenance costs (no tree trimming) • *Longer reach* — Less voltage drop because reactance is lower

of visual clutter. Overhead circuits are ugly. It is possible to make overhead circuits less ugly—tidy construction practices, fiberglass poles instead of wood, keeping poles straight, tight conductor configurations, joint use of poles to reduce the number of poles, and so on. Even the best, though, are still ugly, and many older circuits look awful (weathered poles tipped at odd angles, crooked crossarms, rusted transformer tanks, etc.).

Underground circuits get rid of all that mess, with no visual impact in the air. Trees replace wires, and trees do not have to be trimmed. At ground level, instead of poles every 150 feet, many having one or more guy wires, urban construction has no obstacles, and user requirements documents (URD)-style construction just has pad-mounted transformers spaced much less frequently. Of course, for maximum benefit, all utilities must be underground. There is little improvement from undergrounding electric circuits if phone and cable television are still strung on poles (if the telephone wires are overhead, you might as well have the electric lines there too).

While underground circuits are certainly more appealing when finished, during installation construction is messier than overhead installation. Lawns, gardens, sidewalks, and driveways are dug up; construction lasts longer; and the installation "wounds" take time to heal. These factors do not matter much when installing circuits into land that is being developed, but it can be upsetting to customers in an existing, settled community.

Underground circuits are more reliable. Overhead circuits typically fault about 90 times per 100 miles per year; underground circuits fail less than 10 times/100 miles/year. Because overhead circuits have more faults, they cause more voltage sags, more momentary interruptions, and more long-duration interruptions. Even accounting for the fact that most overhead faults are temporary, overhead circuits have more permanent faults that lead to long-duration circuit interruptions. The one disadvantage of underground circuits is that when they do fail, finding the failure is harder, and fixing the damage or replacing the equipment takes longer. This can partially be avoided by using loops capable of serving customers from two directions, by using conduits for faster replacement, and by using better fault location techniques. Underground circuits are much less prone to the elements. A major hurricane may drain an overhead utility's resources, crews are completely tied up, customer outages become very long, and cleanup costs are a major cost to utilities. A predominantly underground utility is more immune to the elements, but underground circuits are not totally immune to the elements. In "heat storms," underground circuits are prone to rashes of failures. Underground circuits have less overload capability than overhead circuits, and failures increase with operating temperature.

Underground circuits are safer for the public than overhead circuits. Overhead circuits are more exposed to the public. Kites, ladders, downed wires, truck booms—despite the best public awareness campaigns, these still expose the public to electrocution from overhead lines. Do not misunderstand—underground circuits still have dangers, but they are much less than on overhead circuits. For the public, digging is the most likely source of contact. For utility crews, both overhead and underground circuits offer dangers that proper work practices must address to minimize risks.

We cannot assume that underground infrastructure will last as long as overhead circuits. Early URD systems failed at a much higher rate than expected. While most experts believe that modern underground equipment is more reliable, it is still prudent to believe that an overhead circuit will last 40 years, while an underground circuit will only last 30 years.

7.11 Faults

Faults kill—faults start fires—faults force interruptions—faults create voltage sags. Tree trimming, surge arresters, animal guards, cable replacements—these tools reduce faults. We cannot eliminate all faults, but appropriate standards and maintenance practices help in the battle. When faults occur, we have ways to reduce their impacts. A fault is normally a short circuit between conductors.

When a short-circuit fault occurs, the fault path explodes in an intense arc. Local customers endure an interruption, and for customers farther away, a voltage sag; faults cause most reliability and power quality problems. Faults kill and injure line operators: crew operating practices, equipment, and training must account for where fault arcs are likely to occur and must minimize crew exposure.

There are many causes of faults on distribution circuits. A large EPRI study was done to characterize distribution faults in the 1980s at 13 utilities monitoring 50 feeders [Burke and Lawrence, 1984; EPRI 1209–1, 1983]. The distribution of permanent fault causes found in the EPRI study is shown in Figure 7.11. Approximately 40% of faults in this study occurred during periods of "adverse" weather, which included rain, snow, and ice.

Distribution faults occur on one phase, on two phases, or on all three phases. Single-phase faults are the most common. Almost 80% of the faults measured involved only one phase either in contact with the neutral or with ground. Most faults are single phase because most of the overall length of distribution lines is single phase, so any fault on single-phase sections would only involve one phase. Also, on three-phase sections, many types of faults tend to occur from phase to ground. Equipment faults and animal faults tend to cause line-to-ground faults. Trees can also cause line-to-ground faults on three-phase structures, but line-to-line faults are more common. Lightning faults tend to be two or three phases to ground on three-phase structures.

Ninety faults per 100 miles per year (55 faults/100 km/year) is common for utilities with moderate lightning.

Limiting fault current has many benefits that improve the safety and reliability of distribution systems:

- *Failures*—Overhead line burndowns are less likely, cable thermal failures are less likely, violent equipment failures are less likely.
- *Equipment ratings*—We can use reclosers and circuit breakers with less interrupting capability and switches and elbows with less momentary and fault close ratings. Lower fault currents reduce the need for current-limiting fuses and for power fuses and allows the use of cutouts and under-oil fuses.
- *Shocks*—Step and touch potentials are less severe during faults.

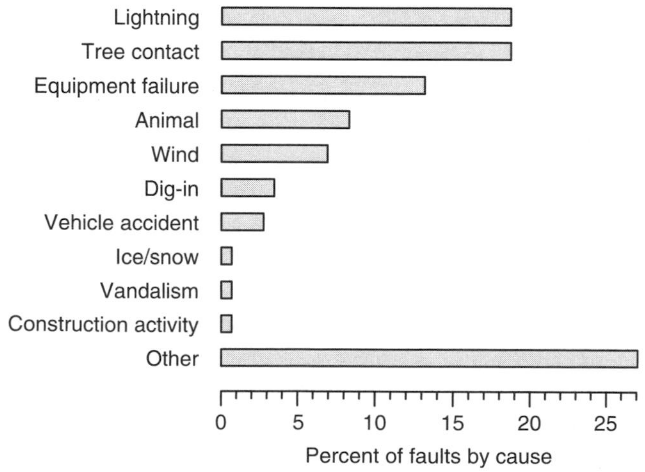

FIGURE 7.11 Fault causes measured in the EPRI fault study. (Data from Burke and Lawrence [1984], EPRI 1209–1 [1983].)

- *Conductor movement*—Conductors move less during faults (this provides more safety for workers in the vicinity of the line and makes conductor slapping faults less likely).
- *Coordination*—Fuse coordination is easier. Fuse saving is more likely to work.

At most distribution substations, three-phase fault currents are limited to less than 10 kA, with many sites achieving limits of 7 to 8 kA. The two main ways that utilities manage fault currents are:

- Transformer impedance—Specifying a higher-impedance substation transformer limits the fault current. Normal transformer impedances are around 8%, but utilities can specify impedances as high as 20% to reduce fault currents.
- Split substation bus—Most distribution substations have an open tie between substation buses, mainly to reduce fault currents (by a factor of two).

Line reactors and a neutral reactor on the substation transformer are also used to limit fault currents, especially in large urban stations where fault currents may exceed 40 kA.

When a conductor comes in physical contact with the ground but does not draw enough current to operate typical protective devices, you have a *high-impedance* fault. In the most common scenario, an overhead wire breaks and falls to the ground (a *downed wire*). If the phase wire misses the grounded neutral or another ground as it falls, the circuit path is completed by the high-impedance path provided by the contact surface and the earth.

The return path for a conductor laying on the ground can be a high impedance. The resistance varies depending on the surface of the ground. Table 7.6 shows typical current values measured for conductors on different surfaces (for 15-kV class circuits).

On distribution circuits, high-impedance faults are still an unsolved problem. It is not for lack of effort—considerable research has been done to find ways to detect high-impedance faults, and progress has been made (see [IEEE Tutorial Course 90EH0310–3-PWR, 1990] for a more in-depth summary). Research has identified many characteristics of high-impedance faults that have been tested as ways of detecting high-impedance faults.

TABLE 7.6. Typical High-Impedance Fault Current Magnitudes

Surface	Current, A
Dry asphalt	0
Concrete (no rebar)	0
Dry sand	0
Wet sand	15
Dry sod	20
Dry grass	25
Wet sod	40
Wet grass	50
Concrete (with rebar)	75

Source: IEEE Tutorial Course 90EH0310–3-PWR, "Detection of Downed Conductors on Utility Distribution Systems," 1990. With permission. © 1990 IEEE.

7.12 Short-Circuit Protection

Overcurrent protection or short-circuit protection is very important on any electrical power system, and the distribution system is no exception. Circuit breakers and reclosers, expulsion fuses and current-limiting fuses—these protective devices interrupt fault current, a vitally important task. Short-circuit protection is the selection of equipment, placement of equipment, selection of settings, and coordination of devices to efficiently isolate and clear faults with as little impact on customers as possible.

Of highest importance, good fault protection clears faults quickly to prevent:

- Fires and explosions
- Further damage to utility equipment such as transformers and cables

Secondary goals of protection include practices that help reduce the impact of faults on:

- *Reliability* (long-duration interruptions)—In order to reduce the impact on customers, reclosing of circuit breakers and reclosers automatically restore service to customers. Having more protective devices that are properly coordinated assures that the fewest customers possible are interrupted and makes fault finding easier.

- *Power quality* (voltage sags and momentary interruptions)—Faster tripping reduces the duration of voltage sags. Coordination practices and reclosing practices impact the number and severity of momentary interruptions.

Circuit interrupters should only operate for faults, not for inrush, cold-load pickup, or transients. Additionally, protective devices should coordinate to interrupt as few customers as possible.

The philosophies of distribution protection differ from transmission-system protection and industrial protection. In distribution systems, protection is not normally designed to have backup—if a protective device fails to operate, the fault may burn and burn (until an upstream device is manually opened). Of course, protection coverage should overlap, so that if a protective device fails due to an internal short circuit (which is different than fails to open), an upstream device operates for the internal fault in the downstream protector. Backup is not a mandatory design constraint (and is impractical to achieve in all cases).

Most often, we base distribution protection on standardized settings, standardized equipment, and standardized procedures. Standardization makes operating a distribution company easier if designs are consistent. The engineering effort to do a coordination study on every circuit reduces considerably.

Fusing

Expulsion fuses are the most common protective device on distribution circuits. Fuses are low-cost interrupters that are easily replaced (when in cutouts). Interruption is relatively fast and can occur in half of a cycle for large currents.

Distribution transformer fuses are primarily there to disconnect the transformer from the circuit if it fails. Engineers most commonly pick fuse sizes for distribution transformers from a fusing table developed by the utility, transformer manufacturer, or fuse manufacturer. These tables are developed based on criteria for applying a fuse such that the fuse should not have false operations from inrush and cold-load pickup. Current-limiting fuses are regularly used on transformers in high fault-current areas to provide protection against violent transformer failure.

When an electrical distribution system energizes, components draw a high, short-lived inrush; the largest component magnetizes the magnetic material in distribution transformers (in most cases, it is more accurate to say remagnetize, because the core likely is magnetized in a different polarity if the circuit is energized following a short-duration interruption). Cold-load pickup is the extra load following an extended interruption due to loss of the normal diversity between customers. Following an interruption, the water in water heaters cools down, refrigerators warm up. When the power is restored, all appliances that need to catch up energize at once.

Fuses are also used on lateral taps. Utilities use the two main philosophies to apply tap fuses: fusing based on load and standardized fusing schedules. The fuse should not operate for cold-load pickup or inrush to prevent nuisance operations and should coordinate with upstream and downstream devices.

Reclosing

Automatic reclosing is a universally accepted practice on most overhead distribution feeders. On overhead circuits, 50 to 80% of faults are temporary, so if a circuit breaker or recloser clears a fault, and it *recloses*, most of the time the fault is gone, and customers do not lose power for an extended period of time.

On underground circuits, since virtually all faults are permanent, we do not reclose. A circuit might be considered underground if something like 60 to 80% of the circuit is underground. Utility practices vary considerably relative to the exact percentage [IEEE Working Group on Distribution Protection, 1995]. A significant number of utilities treat a circuit as underground if as little as 20% is underground, while some others put the threshold over 80%.

The first reclose usually happens with a very short delay, either an immediate reclose, which means a 1/3 to 1/2-second dead time, or with a one to five-second delay. Subsequent reclose attempts follow longer delays. The nomenclature is usually stated as 0–15–30, meaning there are three reclose attempts: the first reclose indicated by the "0" is made after no intentional delay (this is an immediate reclose), the second attempt is made following a 15-second dead time, and the final attempt is made after a 30-second dead time. If the fault is still present, the circuit opens and locks open. We also find this specified using circuit-breaker terminology as O - 0 sec - CO - 15 sec - CO - 30 sec - CO where "C" means close and "O" means open. Other common cycles that utilities use are 0–30–60–90 and 5–45.

With reclosers and reclosing relays on circuit breakers, the reclosing sequence is reset after an interval that is normally adjustable. This interval is generally set somewhere in the range of 10 seconds to two minutes. Only a few utilities have reported excessive operations without lockout [IEEE Working Group on Distribution Protection, 1995].

A recloser is a specialty distribution protective device capable of interrupting fault current and automatically reclosing. Like a circuit breaker, interruption occurs at a natural current zero. The interrupting medium of a recloser is most commonly vacuum or oil. The insulating medium is generally oil, air, a solid dielectric, or SF_6.

Fuse Saving versus Fuse Blowing

Fuse saving is a protection scheme where a circuit breaker or recloser is used to operate before a lateral tap fuse. A fuse does not have reclosing capability; a circuit breaker (or recloser) does. Fuse saving is usually

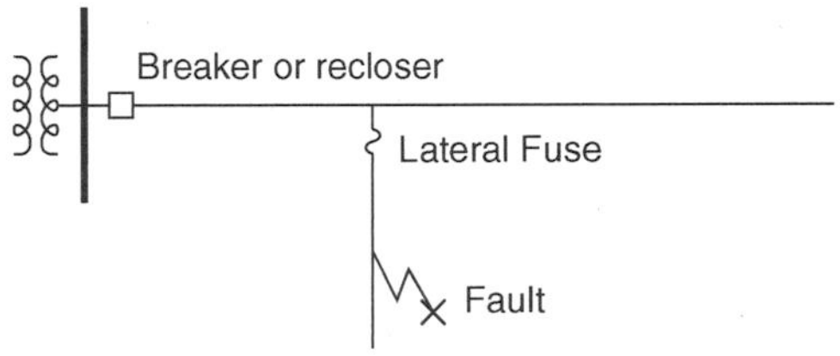

Fuse Saving

Temporary fault

The circuit breaker operates on the instantaneous relay trip (before the fuse operates).
The breaker recloses.
The fault is gone, so no other action is necessary.

Permanent fault

The circuit breaker operates on the instantaneous relay trip (before the fuse operates).
The breaker recloses.
The fault is still there.
The instantaneous relay is disabled, so the fuse operates.
Crews must be sent out to fix the fault and replace the fuse.

Fuse Blowing

Temporary fault

The fuse operates.
Crews must be sent out to replace the fuse.

Permanent fault

The fuse operates.
Crews must be sent out to fix the fault and replace the fuse.

FIGURE 7.12 Comparison of the sequence of events for fuse saving and fuse blowing for a fault on a lateral.

implemented with an instantaneous relay on a breaker (or the fast curve on a recloser). The instantaneous trip is disabled after the first fault, so after the breaker recloses, if the fault is still there, the system is time coordinated, so the fuse blows. Because most faults are temporary, fuse saving prevents a number of lateral fuse operations.

The main disadvantage of fuse saving is that all customers on the circuit see a momentary interruption for lateral faults. Because of this, many utilities are switching to a fuse-blowing scheme. The instantaneous relay trip is disabled, and the fuse is always allowed to blow. The fuse-blowing scheme is also called trip saving or breaker saving. Figure 7.12 shows a comparison of the sequence of events of each mode of operation. Fuse saving is primarily directed at reducing sustained interruptions, and fuse blowing is primarily aimed at reducing the number of momentary interruptions.

7.13 Reliability and Power Quality

Power outages disrupt more businesses than any other factor. End users expect good reliability, and expectations keep rising. Interruptions and voltage sags cause most disruptions. Reliability statistics, based on long-duration interruptions, are the primary benchmark used by utilities and regulators to identify service quality. Faults on the distribution system cause most long-duration interruptions; a fuse, breaker, recloser, or sectionalizer locks out the faulted section.

Utilities most commonly use two indices, SAIFI and SAIDI, to benchmark reliability. These characterize the frequency and duration of interruptions [IEEE Std. 1366–2000].

SAIFI, System average interruption frequency index

$$\text{SAIFI} = \frac{\text{Total number of customer interruptions}}{\text{Total number of customers served}}$$

Typically, a utility's customers average between one and two sustained interruptions per year. SAIFI is also the average failure rate, which is often labeled λ. Another useful measure is the mean time between failure (MTBF), which is the reciprocal of the failure rate: MTBF in years $= 1/\lambda$.

SAIDI, System average interruption duration frequency index

$$\text{SAIDI} = \frac{\text{Sum of all customer interruption durations}}{\text{Total number of customers served}}$$

SAIDI quantifies the average total duration of interruptions. SAIDI is cited in units of hours or minutes per year. Survey results for SAIFI and SAIDI are shown in Table 7.7.

TABLE 7.7. Reliability Indices Found by Industry Surveys

	SAIFI, number of interruptions per year*			SAIDI, hours of interruption per year*		
	25%	50%	75%	25%	50%	75%
[IEEE Std. 1366–2000]	0.90	1.10	1.45	0.89	1.50	2.30
[EEI, 1999] (excludes storms)	0.92	1.32	1.71	1.16	1.74	2.23
[EEI, 1999] (with storms)	1.11	1.33	2.15	1.36	3.00	4.38
[CEA, 2001] (with storms)	1.03	1.95	3.16	0.73	2.26	3.28
[PA Consulting, 2001] (with storms)	—	—	—	1.55	3.05	8.35
IP&L Large City Comparison [Indianapolis Power & Light, 2000]	0.72	0.95	1.15	1.02	1.64	2.41

*The three columns represent the lower quartile, the median, and the upper quartile.

Utility indices vary widely because of many differing factors, mainly: weather, physical environment (mainly the amount of tree coverage), load density, distribution voltage, age, percent underground, and the methods of recording interruptions. Within a utility, performance of circuits varies widely for many of the same reasons causing the spread in utility indices.

Much of the reliability data reported to regulators excludes major storm or major event interruptions. There are pros and cons to excluding storm interruptions. The argument for excluding storms is that storm interruptions significantly alter the duration indices to the extent that restoration performance dominates the index. Further, a utility's performance during storms does not necessarily represent the true performance of the distribution system. Including storms also adds considerable year-to-year variation in results. On the other hand, from the customer point of view an interruption is still an interruption. Also, the performance of a distribution system is reflected in the storm performance—for example if a utility does more tree trimming and puts more circuits underground, their circuits will have less interruptions when a storm hits.

We have many different methods of reducing long-duration interruptions, including:

- *Reduce faults* — tree trimming, tree wire, animal guards, arresters, circuit patrols
- *Find and repair faults faster* — faulted circuit indicators, outage management system, crew staffing, better cable fault finding
- *Limit the number of customers interrupted* — more fuses, reclosers, sectionalizers
- *Only interrupt customers for permanent faults* — reclosers instead of fuses, fuse-saving schemes

Besides long-duration interruptions, shorter-duration disruptions can also affect utility customers. Momentary interruptions and voltage sags are the most common. Different customers are affected differently. Most residential customers are affected by sustained and momentary interruptions. For commercial and industrial customers, sags and momentaries are the most common problems. Each circuit is different, and each customer responds differently to power quality disturbances. These three power quality problems are caused by faults on the utility power system, with most of them on the distribution system.

Momentary interruptions primarily result from reclosers or reclosing circuit breakers attempting to clear temporary faults, first opening and then reclosing after a short delay. The devices are usually on the distribution system; but at some locations, momentary interruptions also occur for faults on the subtransmission system. Terms for short-duration interruptions include short interruptions, momentary interruptions, instantaneous interruptions, and transient interruptions, all of which are used with more or less the same meaning. The dividing line for duration between sustained and momentary interruptions is most commonly thought of as five minutes. Table 7.8 shows the number of momentary interruptions based on surveys of the reliability index MAIFI. MAIFI is the same as SAIFI, but it is for short-duration rather than long-duration interruptions.

Voltage sags cause some of the most common and hard-to-solve power quality problems. Sags can be caused by faults some distance from a customer's location. The same voltage sag affects different customers and different equipment differently. Solutions include improving the ride-through capability of equipment, adding additional protective equipment (such as an uninterruptible power supply), or making improvements or changes in the power system.

A voltage sag is defined as an rms reduction in the ac voltage, at the power frequency, for durations from a half cycle to a few seconds [IEEE Std. 1159–1995]. Sags are also called *dips* (the preferred European term). Faults in the utility transmission or distribution system cause most sags. Utility system protective devices clear most faults, so the duration of the voltage sag is the clearing time of the protective device.

Voltage-sag problems are a contentious issue between customers and utilities. Customers complain that the problems are due to events on the power system (true), and that is the utilities' responsibility. The utility responds that the customer has overly-sensitive equipment, and the power system can never be designed to be disturbance free. Utilities, customers, and the manufacturers of equipment all share some of the responsibility for voltage sag problems. There are almost no industry standards or regulations to govern these disputes, and most are worked out in negotiations between a customer and the utility.

TABLE 7.8. Surveys of MAIFI

Survey	Median
1995 IEEE [IEEE Std. 1366–2000]	5.42
1998 EEI [EEI, 1999]	5.36
2000 CEA [CEA, 2001]	4.0

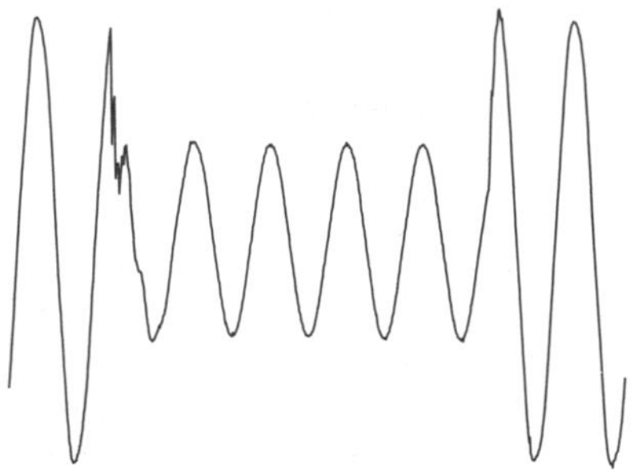

FIGURE 7.13 Example of voltage sag caused by a fault.

TABLE 7.9. Annual Number of Power Quality Events (Upper Quartile, Median, and Lower Quartile) for the EPRI DPQ Feeder Sites with a One-Minute Filter

	Duration, seconds						
Voltage	0	0.02	0.05	0.1	0.2	0.5	1
0.9	$_{32.8}57.5_{104.8}$	$_{30.8}49.0_{95.1}$	$_{24.4}35.3_{65.6}$	$_{13.6}22.7_{38.7}$	$_{7.6}13.2_{24.0}$	$_{3.3}7.3_{14.2}$	$_{1.4}3.2_{8.9}$
0.8	$_{16.4}31.6_{54.1}$	$_{14.8}26.0_{50.1}$	$_{12.1}20.9_{37.9}$	$_{8.1}15.0_{25.1}$	$_{4.9}9.6_{16.9}$	$_{2.4}5.3_{11.0}$	$_{0.9}2.7_{7.5}$
0.7	$_{10.1}20.5_{33.8}$	$_{8.6}18.8_{32.7}$	$_{8.1}15.3_{27.6}$	$_{5.8}11.3_{18.8}$	$_{4.0}7.8_{13.5}$	$_{1.8}4.5_{9.3}$	$_{0.9}2.5_{7.0}$
0.5	$_{4.7}9.7_{19.2}$	$_{4.5}9.0_{17.4}$	$_{4.2}7.7_{14.3}$	$_{3.5}5.9_{11.2}$	$_{2.3}5.0_{9.6}$	$_{1.4}3.3_{7.7}$	$_{0.8}2.2_{5.7}$
0.3	$_{2.1}4.8_{12.8}$	$_{1.8}4.5_{11.0}$	$_{1.6}4.2_{9.5}$	$_{1.4}3.6_{8.6}$	$_{1.1}3.5_{8.3}$	$_{0.8}2.8_{6.6}$	$_{0.5}1.6_{5.1}$
0.1	$_{0.9}3.2_{8.3}$	$_{0.9}2.9_{7.8}$	$_{0.8}2.8_{7.8}$	$_{0.8}2.7_{7.8}$	$_{0.7}2.7_{7.8}$	$_{0.5}2.2_{6.1}$	$_{0.3}1.6_{4.9}$

Figure 7.13 shows a voltage sag that caused the system voltage to fall to approximately 45% of nominal voltage for 4.5 cycles.

The voltage during the fault at the substation bus is given by the voltage-divider expression in Figure 7.14 based on the source impedance (Z_s), the feeder line impedance (Z_f), and the prefault voltage (V).

The voltage sags deeper for faults electrically closer to the bus (smaller Z_f). Also, as the available fault current decreases (larger Z_s) the sag becomes deeper. The source impedance includes the transformer impedance plus the subtransmission source impedance (often, subtransmission impedance is small enough to be ignored). The impedances used in the equation depend on the type of fault it is. For a three-phase fault (giving the most severe voltage sag), use the positive-sequence impedance ($Z_f = Z_{f1}$). For a line-to-ground fault (the least severe voltage sag), use the loop impedance, which is $Z_f = (2Z_{f1}+Z_{f0})/3$. A good approximation is one ohm for the substation transformer (which represents a 7- to 8-kA bus fault current) and one ohm per mile (0.6 Ω/km) of overhead line for ground faults.

Several power quality monitoring studies have characterized the frequency of voltage sags. EPRI's Distribution Power Quality (DPQ) project recorded power quality in distribution substations and on distribution feeders measured on the primary at voltages from 4.16 to 34.5 kV [EPRI TR-106294-V2, 1996; EPRI TR-106294-V3, 1996]. Two hundred seventy-seven sites resulted in 5691 monitor-months of data. In most cases three monitors were installed for each randomly selected feeder, one at the substation and two at randomly selected places along the feeder. Table 7.9 shows cumulative numbers of voltage sags measured at sites during the DPQ study.

7.14 Grounding

Grounding is one of the main defenses against hazardous electric shocks and hazardous overvoltages. Good equipment grounding helps reduce the chances that line workers and the public receive shocks from internal failures of the equipment. System grounding determines how loads are connected and how line-to-ground

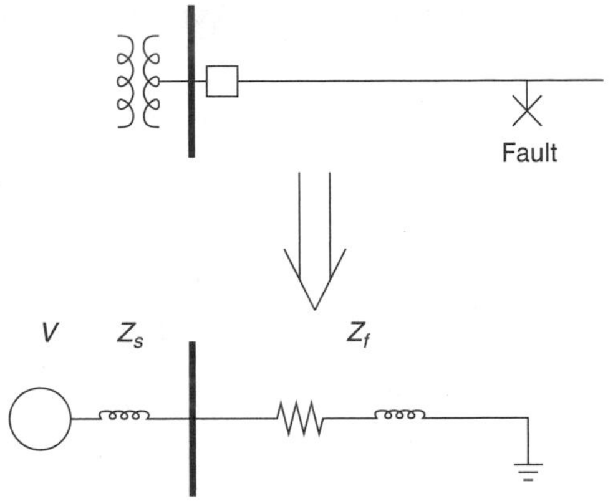

$$V_{bus} = \frac{Z_f}{Z_f + Z_s}$$

FIGURE 7.14 Voltage divider equation giving the voltage at the bus for a fault downstream. (This can be the substation bus or another location on the power system.)

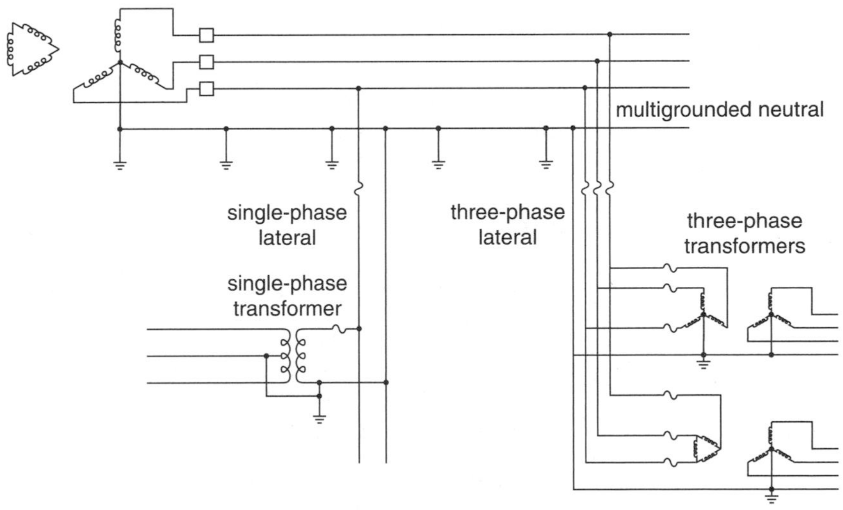

FIGURE 7.15 A four-wire multigrounded distribution system.

faults are cleared. Most North American distribution systems have effective grounding; they have a neutral that acts as a return conductor and as an equipment safety ground.

There are several grounding configurations for three-phase power distribution systems. On distribution systems in North America, the four-wire, multigrounded neutral system predominates. Figure 7.15 shows how loads normally connect to the four-wire system. One phase and the neutral supply single-phase loads. The neutral carries unbalanced current and provides a safety ground for equipment. Low cost for supplying single-phase loads is a major reason the four-wire system evolved in North America. More than half of most distribution systems consist of single-phase circuits, and most customers are single phase. On a multigrounded neutral system, the earth serves as a return conductor for part of the unbalanced current during normal and fault conditions.

Four-wire multigrounded systems have several advantages over three-wire systems. Four-wire systems provide low cost for serving single-phase loads.

- Single cables for underground single-phase load (and pot heads and elbows and pad-mounted transformer bushings)
- Single-phase overhead lines are less expensive
- Single-bushing transformers
- One arrester and one fuse for a single-phase transformer
- The neutral can be located lower on the pole
- Lower-rated arresters
- Lower insulation required; transformer insulation can be graded

Extensive use of single-phase lines provides significant cost savings. The primary neutral can be shared with the secondary neutral for further cost savings. And flexibility increases: upgrading a single-phase line to two- or three-phase service on overhead lines is a cost-effective way to upgrade a circuit following load growth.

Most other distribution configurations are three-wire systems, either grounded or unigrounded. A unigrounded system is grounded at one place, normally the substation. In North America, many older distribution system are three-wire systems, including 2400-, 4160-, and 4800-V systems. European systems are primarily three-wire unigrounded systems.

Defining Terms

Capacitor bank: Consists of capacitors connected in parallel. Each capacitor is placed in a metal can and equipped with bushings.

Feeder: Overhead lines or cables that are used to distribute the load to the customers. They interconnect the distribution substations with the loads.

Recloser: A circuit breaker that is designed to interrupt short-circuit current and reclose the circuit after interruption.

Substation: A junction point in the electric network. The incoming and outgoing lines are connected to a bus bar through circuit breakers.

Tap changer: A transformer. One of the windings is equipped with taps. The usual number of taps is 32. Each tap provides a 1% voltage regulation. A special circuit breaker is used to change the tap position.

References

Brown, R.E., *Electric Power Distribution Reliability*, Marcel Dekker, New York, 2002.

Burke, J.J., *Power Distribution Engineering: Fundamentals and Applications*, Marcel Dekker, New York, 1994.

Burke, J. J. and Lawrence, D. J., "Characteristics of fault currents on distribution systems," *IEEE Transactions on Power Apparatus and Systems*, vol. PAS-103, no. 1, pp. 1–6, January 1984.

CEA, *CEA 2000 Annual Service Continuity Report on Distribution System Performance in Electric Utilities*, Canadian Electrical Association, Ottawa, 2001.

Dugan, R.C., McGranaghan, M., Santosa, S., and Beaty, H.W., *Electrical Power Systems Quality*, 2nd ed., McGraw Hill, New York, 2003.

EEI, "EEI Reliability Survey," minutes of the 8th meeting of the Distribution Committee, Mar. 28–31, 1999.

EPRI 1209–1, *Distribution Fault Current Analysis*, Electric Power Research Institute, Palo Alto, California, 1983.

EPRI TR-106294-V2, *An Assessment of Distribution System Power Quality: Volume 2: Statistical Summary Report*, Electric Power Research Institute, Palo Alto, California, 1996.

EPRI TR-106294-V3, *An Assessment of Distribution System Power Quality: Volume 3: Library of Distribution System Power Quality Monitoring Case Studies*, Electric Power Research Institute, Palo Alto, California, 1996.

IEEE Std. 1159–1995, *IEEE Recommended Practice for Monitoring Electric Power Quality.*

IEEE Std. 1366–2000, *IEEE Guide for Electric Power Distribution Reliability Indices.*

IEEE Tutorial Course 90EH0310–3-PWR, "Detection of downed conductors on utility distribution systems," 1990.

IEEE Working Group on Distribution Protection, "Distribution line protection practices industry survey results," *IEEE Transactions on Power Delivery*, vol. 10, no. 1, pp. 176–86, January 1995.

Indianapolis Power & Light, "Comments of Indianapolis Power & Light Company to Proposed Discussion Topic, Session 7, Service Quality Issues," 2000.

Kersting, W.H., *Distribution System Modeling and Analysis*, CRC Press, Boca Raton, Florida, 2002.

PA Consulting, "Evaluating Utility Operations and Customer Service," FMEA-FMPA Annual Conference, Boca Raton, Florida, July 23–26, 2001.

Short, T.A., *Electric Power Distribution Handbook*, CRC Press, Boca Raton, Florida, 2004.

Willis, H.L., *Power Distribution Planning Reference Book*, Marcel Dekker, New York, 1997.

Further Information

Other recommended publications include J.M. Dukert, *A Short Energy History of the United States*, Edison Electric Institute, Washington, D.C., 1980. Also, the *IEEE Transactions on Power Delivery* publishes distribution papers sponsored by the Transmission and Distribution Committee. These papers deal with the latest developments in the distribution area. Everyday problems are presented in two magazines: *Transmission & Distribution* and *Electrical World.*

8

Electrical Machines

Ioan Serban
University Politehnica of Timisoara

Mehdi Ferdowsi
University of Missouri–Rolla

Elias G. Strangas
Michigan State University

8.1 Generators

Ioan Serban

Introduction

Electric generators are devices that convert energy from a mechanical form to an electrical form. This process, known as electromechanical energy conversion, involves magnetic fields that act as an intermediate *medium.* There are two types of generators: alternating current (ac) and direct current (dc). This section explains how these devices work and how they are modeled in analytical or numerical studies. The input to the machine can be derived from a number of energy sources. For example, in the generation of large-scale electric power, coal can produce steam that drives the shaft of the machine. Typically, for such a thermal process, only about one-third of the raw energy (i.e., from coal) is converted into mechanical energy. The final step of the energy conversion is quite efficient, with an efficiency close to 100%. The generator's operation is based on Faraday's law of electromagnetic induction. In brief, if a coil (or winding) is linked to a varying magnetic field, then an electromotive force (emf) or voltage is induced across the coil. Thus, generators have two essential parts, one creates a magnetic field and the other is where the emf is induced. The magnetic field is typically generated by electromagnets (thus, the field intensity can be adjusted for control purposes) whose windings are referred to as field windings or **field circuits**. The coils where the emf is induced are called *armature* windings or **armature circuits**. One of these two components is stationary (stator) and the other is a rotational part (rotor) driven by an external torque. Conceptually, it is immaterial which of the two components is to rotate because, in either case, the armature circuits always "see" a varying magnetic field. However, practical considerations lead to the common design that for ac generators the field windings are mounted on the rotor and the armature windings on the stator. In contrast, for dc generators, the field windings are on the stator and armature on the rotor.

Principles of Electric Generators

Electric generators are electromagnetic devices made of electric and magnetic circuits coupled together electrically and magnetically, where mechanical energy at the shaft is converted to electric energy at the terminals.

The extremely large power/unit span, from milliwatts to hundreds of MW and more, and the wide diversity of applications, from electric power plants to car alternators, should have led to numerous electric generator configurations and their control. And so they did. To bring some order to our exposure we need some classifications.

The Three Types of Electric Generators

Electric generators may be classified many ways but the following are deemed as fully representative:

By principle
By applications domain

The applications domain implies the power level. The classifications by principle unfolded here include commercial (widely used) types together with new configurations still in the laboratory (although advanced) stages.

By principle there are three main types of electric generators:

Synchronous (Figure 8.1)
Induction (Figure 8.2)
Parametric (with magnetic anisotropy and permanent magnets, Figure 8.3)

Parametric generators have in most configurations doubly salient magnetic circuit structures, so they may be called also doubly salient electric generators.

Synchronous generators (SGs) (Bödefeld and Sequenz, 1938; Concordia, 1951; Richter, 1963; Kostenko, and Pitrovski, 1974) have in general a stator magnetic circuit made of laminations provided with uniform slots that house a three-phase (sometimes single- or two-phase) winding and a rotor. It is the rotor design that leads to a cluster of SG configurations as seen in Figure 8.1.

They are all characterized by the rigid relationship between speed n, frequency f_1, and the number of poles $2p$

$$n = f/p \tag{8.1}$$

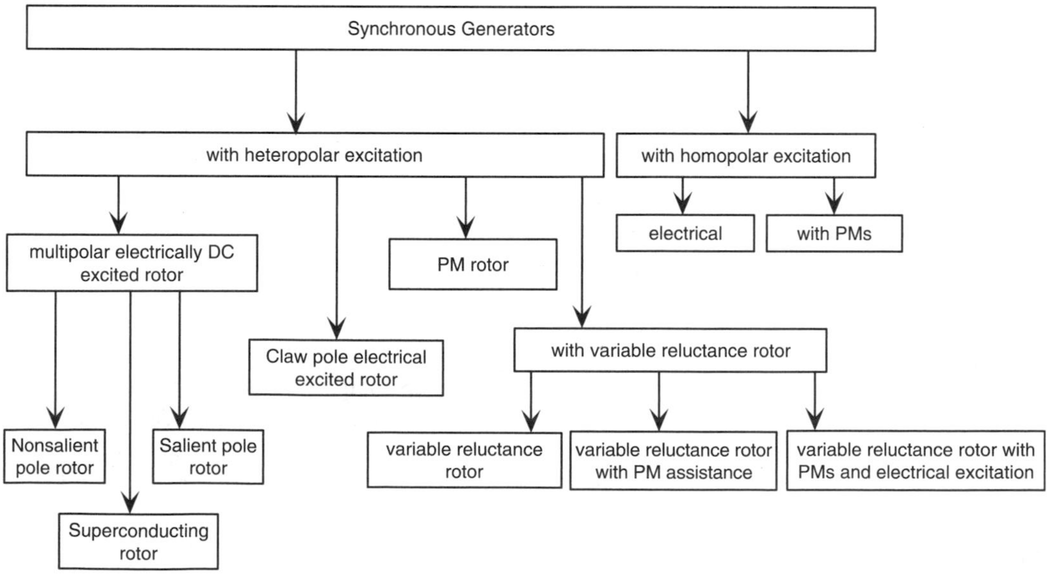

FIGURE 8.1 Synchronous generators.

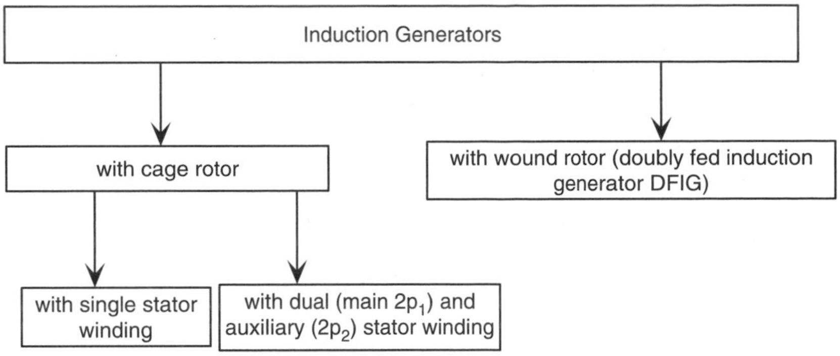

FIGURE 8.2 Induction generators.

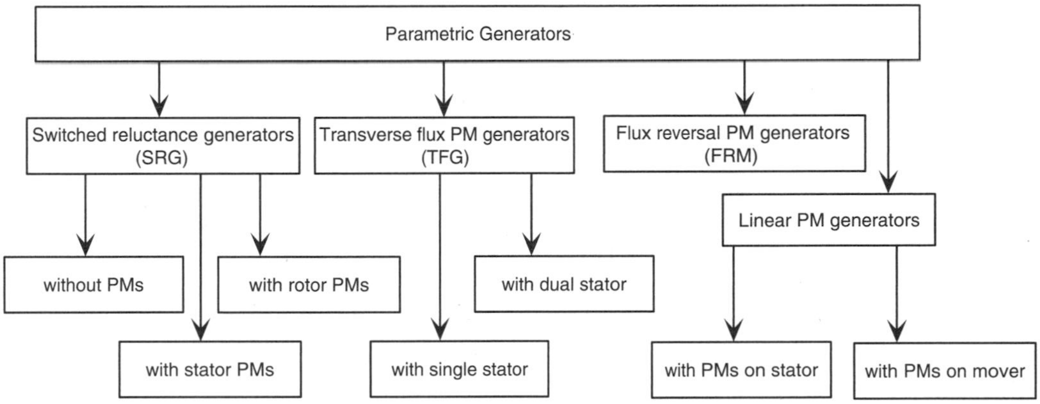

FIGURE 8.3 Parametric generators.

The dc-excited require power electronics excitation control, while those with PMs or variable reluctance rotors have to use full power electronics in the stator to operate at adjustable speed. Finally, even electronically excited SGs may be provided with full power electronics in the stator when they work alone or in power grids with high-voltage dc cable transmission lines.

Each of these configurations will be presented in terms of its principles later on in this chapter.

For powers in the MW/unit range and less, induction generators have been introduced also. They are (Figure 8.2):

With cage rotor and single stator winding
With cage rotor and dual (main and additional) stator winding with different number of poles
With wound rotor

PWM converters are connected to the stator for the single-stator winding and to the auxiliary stator winding for the case of dual-stator winding.

The principle of the induction generator with single-stator winding relies on the equation

$$f_1 = pn + f_2 \tag{8.2}$$

with
$f_1 > 0$
$f_2 <> 0$ slip (rotor) frequency

n = speed (rps)
p = pole pairs

f_2 may be either positive or negative in Equation (8.2), even zero, provided the PWM converter in the wound rotor is capable of supporting bidirectional power flow for speeds n above and below f_1/p.

Notice that for $f_2 = 0$ (dc rotor excitation) the SG operation mode is obtained with the doubly fed induction generator (IG).

The slip S definition is

$$S = \frac{f_2}{f_1} <> 0 \tag{8.3}$$

The slip is zero as $f_2 = 0$ (dc) for the SG mode.

For the dual-stator winding the frequency–speed relationship is applied twice

$$\begin{aligned} f_1 &= p_1 n + f_2; p_2 > p_1 \\ f_1' &= p_2 n + f_2' \end{aligned} \tag{8.4}$$

So the rotor bars experience, in principle, currents of two distinct (rather low) frequencies f_2 and f_2'. In general $p_2 > p_1$ to cover lower speeds.

The PWM converter feeds the auxiliary winding. Consequently its rating is notably lower than that of the full power of the main winding and is proportional to speed variation range.

As it may work in the pure synchronous mode also, the doubly fed IG may be used up to highest levels of powers for SGs (400-MW units have been in use for some years, since 1995, in Japan) and a 2 × 300 MW pump storage plant is being commissioned in 2004 in Germany.

On the contrary, the cage-rotor IG is more suitable for powers in the MW and lower power range.

Parametric generators rely on the variable reluctance principle but may also use permanent magnets to enhance the power/volume and reduce generator losses.

There are quite a few configurations that suit this category, such as the switched reluctance generator (SRG), the transverse flux PM generator (TFG), and the flux reversal generator (FRG). In general their principle relies on coenergy variation due to magnetic anisotropy (with or without PMs on rotor or on stator) in the absence of a traveling field with constant speed (f_1/p), which is so characteristic for synchronous and induction generators (machines).

Synchronous Generators

SGs (classifications in Figure 8.1) are characterized by a uniformly slotted stator laminated core that hosts a three-, two-, or single-phase ac winding and a dc current-excited, PM-excited, or variable saliency rotor (Bödefeld and Sequenz, 1938; Concordia, 1951; Richter, 1963; Kostenko and Pitrovski, 1974; Walker, 1981).

As only two traveling fields — of the stator and rotor — at relative standstill interact to produce a rippleless torque, the speed n is igidly tied to stator frequency f_1, because the rotor-produced magnetic field is typical in such SGs.

They are built with a nonsalient pole-distributed excitation rotor (Figure 8.4) for $2p = 2, 4$ (that is, high-speed or turbo-generators) or with a salient-pole concentrated excitation rotor (Figure 8.5) for $2p > 4$ (in general for low-speed or hydro-generators).

As power increases, the rotor peripheral speed increases also. In large turbogenerators it may reach more than 150 m/s (in a 200-MVA machine with $D_r = 1.2$ m rotor diameter at $n = 3600$ rpm, $2p = 2$, $U = \pi D_r n = \pi \times 1.2 \times 3600/60 > 216$ m/s). The dc excitation placement in slots, with dc coil end connections protected against centrifugal forces by rings of highly resilient resin materials, becomes thus necessary. Also the dc rotor current air-gap field distribution is closer to a sinusoid.

Consequently, the harmonics content of the stator motion induced voltage (emf or no-load voltage) is smaller, thus complying with the strict rules (standards) of commercial large power grids.

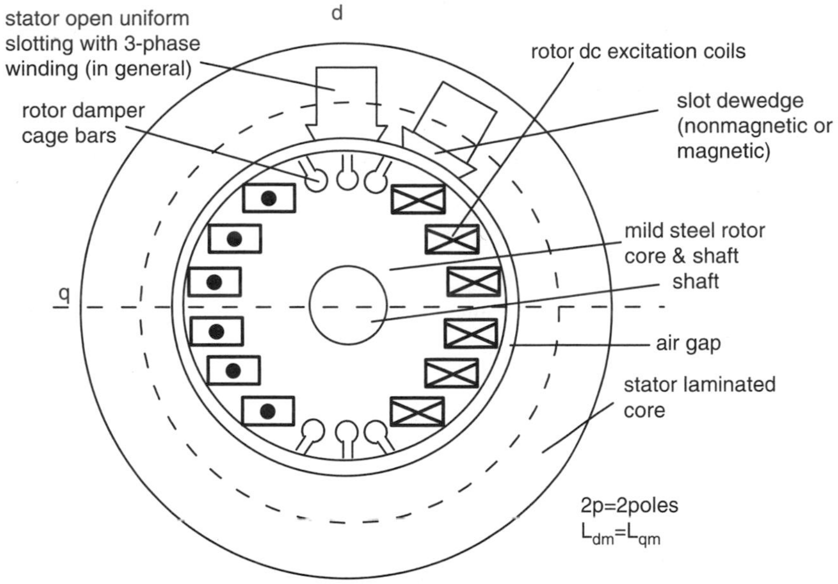

FIGURE 8.4 Synchronous generator with nonsalient pole heteropolar dc-distributed excitation.

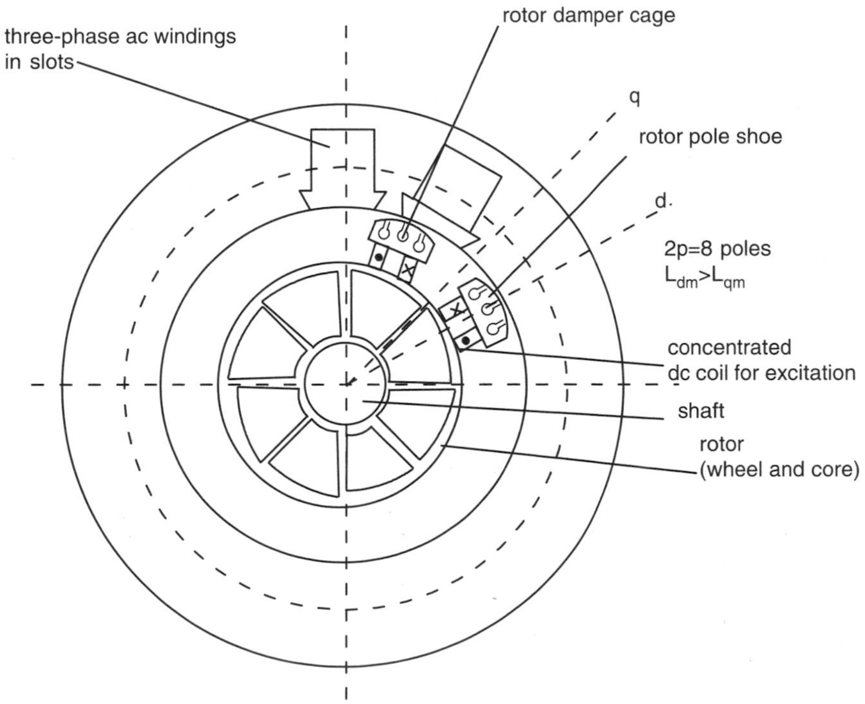

FIGURE 8.5 Synchronous generator with salient pole heteropolar dc-concentrated excitation.

The rotor body is made of solid iron for better mechanical rigidity and heat transmission.

The stator slots in large SGs are open (Figure 8.4 and Figure 8.5), and they are provided sometimes with magnetic wedges to further reduce the field space harmonics and thus reduce the emf harmonics content and rotor additional losses in the rotor damper cage. For steady state (sinusoidal symmetric stator currents of constant amplitude), the rotor damper cage currents are zero. However, should any load or mechanical transients occur, eddy currents show up in the damper cage to attenuate the rotor oscillations when the stator is connected to a constant frequency and voltage (high power) grid.

The rationale neglects the stator magneto-motive-force space harmonics due to the placement of windings in slots and due to slot openings. These space harmonics induce voltages and thus produce eddy currents in the rotor damper cage even during steady state.

Also, even during steady state, if the stator phase currents are not symmetric, their inverse components produce currents of $2f_1$ frequency in the damper cage.

Consequently, to limit the rotor temperature, the degree of current (load) unbalance permitted is limited by standards.

Nonsalient pole dc-excited rotor SGs have been manufactured for $2p = 2$, 4 poles high-speed turbogenerators that are driven by gas or steam turbines.

For lower speed SGs with a large number of poles ($2p > 4$), the rotors are made of salient rotor poles provided with concentrated dc excitation coils. The peripheral speeds are lower than for turbogenerators even for large power hydro generators (for 200-MW, 14-m rotor diameter at 75 rpm, and $2p = 80$, $f_1 = 50$ Hz, the peripheral speed $U = \pi \times D_r \times n = \pi \times 14 \times 75/60 > 50$ m/s). About 80 m/s is the limit, in general, for salient pole rotors. Still, the excitation coils have to be protected against centrifugal forces.

The rotor pole shoes may be made of laminations — to reduce additional rotor losses — but the rotor pole bodies and core are made of mild magnetic solid steel.

With a large number of poles, the stator windings are built with a smaller number of slots per pole: between 6 and 12, in many cases. The number of slots per pole and phase is thus between two and four. The smaller is q, the larger is the space harmonics presence in the emf. A fractionary q might be preferred, say 2.5, which also avoids the subharmonics and leads to a cleaner (more sinusoidal) emf, to comply with the current standards.

The rotor pole shoes are provided with slots that house copper bars short-circuited by copper rings to form a rather complete squirrel cage. A stronger damper cage has been thus obtained.

DC excitation power on the rotor is transmitted either by:

Copper slip-rings and brushes (Figure 8.6)
Brushless excitation systems (Figure 8.7)

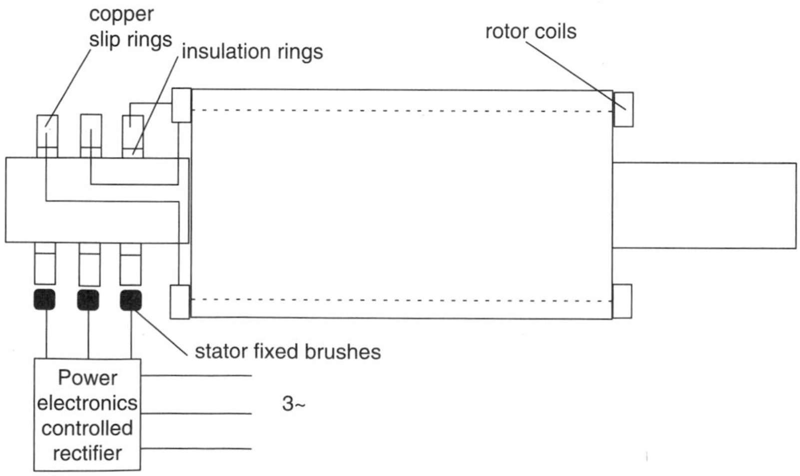

FIGURE 8.6 Slip-ring-brush power electronics rectifier dc excitation system.

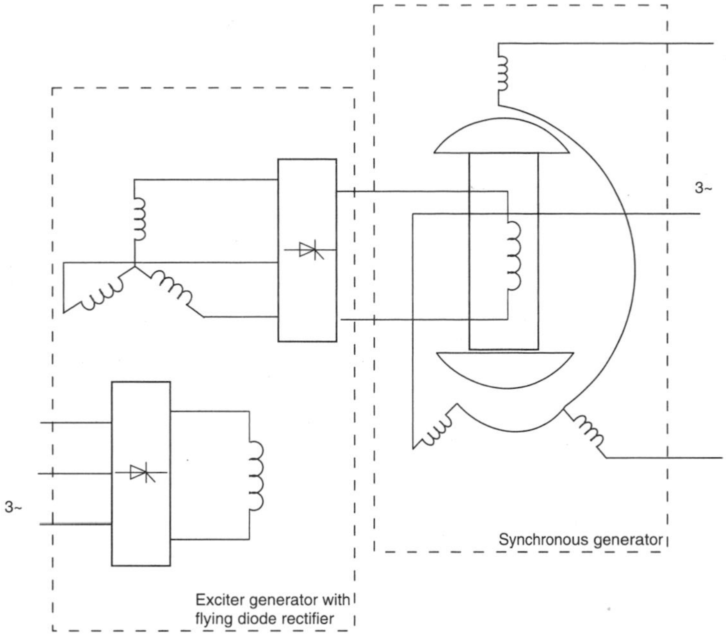

FIGURE 8.7 Brushless exciter with "flying diode" rectifier for synchronous generators.

The controlled rectifier, whose power is around 3% of generator rated power, and which has a sizeable voltage reserve to force the current in the rotor quickly, controls the dc excitation current according to needs of generator voltage and frequency stability.

Alternatively, an inverted SG (whose three-phase ac windings and diode rectifier are placed on the rotor and the dc excitation in the stator) may play the role of a brushless exciter (Figure 8.7). The field current of the exciter is controlled through a low-power half-controlled rectifier. Unfortunately the electrical time constant of the exciter generator slows down notably the response in the main SG excitation current control.

Claw-pole (Lundell) SGs are built now mostly as car alternators. The excitation winding power is reduced considerably for the multiple rotor construction ($2p = 10, 12, 14$), preferred to reduce external diameter and machine volume.

The claw-pole rotor solid cast-iron structure (Figure 8.8) is less costly to manufacture. Also the single ring-shape excitation coil produces a multipolar air-gap field with a three-dimensional field path to reduce copper volume and dc power losses in the rotor.

The stator holds a simplified three-phase/single-phase single-layer winding with three slots per pole, in general. Though slip rings and brushes are used, the power transmitted through them is small (in the order of 60 to 120 W for car and truck alternators) and thus small power electronics is used to control the output. The total costs of the claw-pole generator for automobiles, including the field current control and the diode full power rectifier is low. And so is the specific volume.

However the total efficiency, including the diode rectifier and excitation losses, is low. At 14-Vdc output it is below 50%. To blame are the diode losses (at 14 Vdc), the mechanical losses, and the eddy currents induced in the claw poles by the space and time harmonics of the stator currents mmf. Increasing the voltage to 42 Vdc would reduce the diode losses in relative terms, while the building of the claw poles from composite magnetic materials would reduce drastically the claw-pole eddy current losses. A notably high efficiency would result even if the excitation power might slightly increase due to the lower permeability (500 μ_0) of today's best composite magnetic materials. Also higher power levels might be obtained.

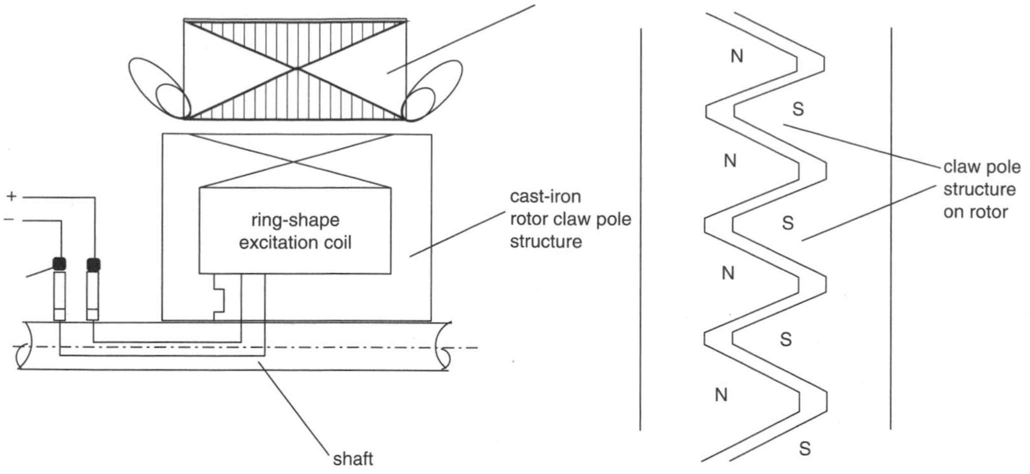

FIGURE 8.8 The claw-pole rotor synchronous generator.

Though the claw-pole SG could be built with the excitation on the stator, to avoid the brushes, the configuration is bulky and the arrival of high-energy permanent magnets for rotor dc excitation have put it apparently to a rest.

Mathematical Models

The steady-state voltage/current equation of the SG is (Bödefeld and Sequenz, 1938; Concordia, 1951; Richter, 1963; Kostenko and Pitrovski, 1974)

$$V_s = E_f - R_s I_s - jX_d I_d - jI_q X_q \tag{8.5}$$

where V_s, I_S are the stator phase voltage and current phasors, respectively; I_d, I_q are the stator current components whose magnetic field is aligned with rotor pole (d) and q axis, respectively; X_d and X_q are the d, q axis synchronous reactances, respectively; R_s is the stator phase resistance; and E_f is the emf produced by the rotor dc excitation current (or permanent magnet) in a stator phase. There is one equation for each phase, and all variables are sinusoidal in time, which explains why phasors are used. When the machine is overexcited (large excitation currents), $E_f > V_s$ and in some conditions the generator produces reactive power to its loads or to the power grid. The active power delivered by the generator comes from the rotor shaft and is increased by increasing the fuel input into the prime mover.

The electromechanical equation described by the so-called "swing equation" is

$$J\frac{d^2\theta}{dt^2} + D\frac{d\theta}{dt} = T_{mech} - T_{el} \tag{8.6}$$

where J and D are the rotor inertia and damping (mechanical and electrical), respectively, and θ denotes the angle between the emf E_f and the output voltage V_s of the generator, which is constant under steady state. Also, θ is related to the rotor position variation with respect to a certain rotating frame.

T_{mech} is the prime mover shaft torque and T_{el} is the generator electromagnetic torque that tends to slow down the rotor. Stable operation occurs when the two torques are equal to each other. Careful control of T_{mech} according to T_{el} produces stable operation. The generator equations for transients are derived in the so-called "orthogonal, dq" model whose axes are fixed to the rotor poles in order to eliminate the dependence of stator-to-rotor mutual inductances on rotor position, and they are

$$e_\mathrm{d} = \frac{\mathrm{d}}{\mathrm{dt}}\lambda_\mathrm{d} - \lambda_\mathrm{q}\frac{\mathrm{d}}{\mathrm{dt}}\theta - ri_\mathrm{d}$$
$$e_\mathrm{q} = \frac{\mathrm{d}}{\mathrm{dt}}\lambda_\mathrm{q} + \lambda_\mathrm{d}\frac{\mathrm{d}}{\mathrm{dt}}\theta - ri_\mathrm{q} \qquad (8.7)$$

$$\lambda_\mathrm{d} = G(s)i_\mathrm{F} - X_\mathrm{d}(s)i_\mathrm{d}$$
$$\lambda_\mathrm{q} = -X_\mathrm{q}(s)i_\mathrm{q} \qquad (8.8)$$

Under steady state e_d, e_q (stator voltage components), λ_d, λ_q (stator flux linkage components), and i_d, i_q are dc variables as the dq model orthogonal axes are tied to the rotor. The true phase voltages and currents are obtained from the above dq components by the so-called "Park transformation"; this way they become ac variables. $X_\mathrm{d}(s)$ and $X_\mathrm{q}(s)$ are the so-called "operational reactances" of the synchronous machine. In effect, during transients the machine behaves as a variable reactance in time. This is how the transient and subtransient reactances of the synchronous machine have been defined (Table 8.1). Consequently rather large transient currents occur in SGs. They warrant various protection measures clearly defined in international standards.

Permanent Magnet SGs

The rapid development of high-energy PMs with a rather linear demagnetization curve has led to widespread use of PM synchronous motors in variable-speed drives. As electric machines are reversible by principle, the generator regime is available also and, for directly driven wind generators in the hundreds of kW or MW range, such solutions are being proposed. Superhigh-speed gas-turbine-driven PM SGs in the 100-kW range at 60 to 80,000 rpm are also introduced. Finally PM SGs are being considered as starter-generators for the cars in the near future.

There are two main types of rotors for PM SGs:

With rotor-surface PMs (Figure 8.9) — nonsalient pole rotor (SPM)
With interior PMs (Figure 8.10) — salient pole rotor (IPM)

The configuration in Figure 8.9 shows a PM rotor made with parallelepipedic PM pieces such that each pole is patched with quite a few of them, circumferentially and axially.

TABLE 8.1 Typical Synchronous Generator Parameters[a]

Parameter	Symbol	Round Rotor	Salient-Pole Rotor with Damper Windings
Synchronous reactance			
d-axis	X_d	1.0–2.5	1.0–2.0
q-axis	X_q	1.0–2.5	0.6–1.2
Transient reactance			
d-axis	X'_d	0.2–0.35	0.2–0.45
q-axis	X'_q	0.5–1.0	0.25–0.8
Subtransient reactance			
d-axis	X''_d	0.1–0.25	0.15–0.25
q-axis	X''_q	0.1–0.25	0.2–0.8
Time constants			
Transient			
Stator winding open-circuited	T'_{do}	4.5–13	3.0–8.0
Stator winding short-circuited	T'_d	1.0–1.5	1.5–2.0
Subtransient			
Stator winding short-circuited	T''_d	0.03–0.1	0.03–0.1

[a]Reactances are per unit, i.e., normalized quantities. Time constants are in seconds.
Source: M.A. Laughton and M.G. Say, Eds., *Electrical Engineer's Reference Book*, Stoneham, Mass.: Butterworth, 1985.

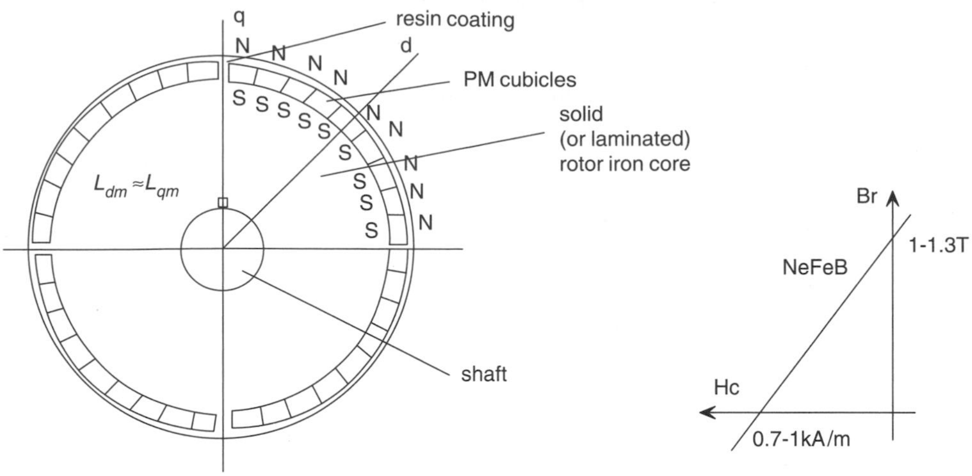

FIGURE 8.9 Surface PM rotor ($2p = 4$ poles).

The PMs are held tight to the solid (or laminated) rotor iron core by special adhesives, but also a highly resilient resin coating is added for mechanical rigidity.

The stator contains a laminated core with uniform slots (in general) that house a three-phase winding with distributed (standard) coils or concentrated (fractionary) coils.

The rotor is practically isotropic from the magnetic point of view. There is some minor difference between d- and q-axis magnetic permeances because the PM recoil permeability ($\mu_{rec} \approx [1.04 \text{ to } 1.07]\mu_0$ at 20°C) increases somewhat with temperature for NeFeB and SmCo high-energy PMs.

So the rotor may be considered as nonsalient magnetically (the magnetization inductances L_{dm} and L_{qm} are almost equal to each other).

To protect the PMs mechanically and produce reluctance torque, interior PM pole rotors have been introduced. Two such typical configurations are shown in Figure 8.10

Figure 8.10b shows a practical solution for two-pole interior PM (IPM) rotors. A practical $2p = 4, 6, \ldots$ IPM rotor as shown in Figure 8.10a has an inverse saliency: $L_{dm} < L_{qm}$, as usual in IPM machines. Finally, a high-saliency rotor ($L_{dm} > L_{qm}$) obtained with multiple flux barriers and PMs, acting along axis q (rather than axis d) is presented in Figure 8.10c. It is a typical IPM machine but with large magnetic saliency. In such a machine the reluctance torque may be larger than the PM interactive torque. The PMs' field saturates first the flux bridges and then overcompensates the stator-created field in axis q. This way the stator flux along axis q decreases with current in axis q. For flux weakening the I_d current component is reduced. Wide constant power (flux weakening) speed range of more than 5:1 has been obtained this way. Starter generators in cars are a typical application for this rotor.

As the PM role is limited, lower grade (lower B_r) PMs, at lower costs, may be used.

It is also possible to use the variable-reluctance rotor with high magnetic saliency (Figure 8.10c) without permanent magnets. With the reluctance generator, either power-grid or stand-alone mode operation is feasible. For stand-alone operation capacitor self-excitation is needed. The performance is moderate but the rotor cost is moderate. Standby power sources would be a good application for reluctance SGs with high saliency $L_{dm}/L_{qm} > 4$.

PM synchronous generators are characterized by high torque (power) density and high efficiency (excitation losses are zero). However, the cost of high energy PMs is still above 50 USD/kg. Also, to control the output, full power electronics is needed in the stator (Figure 8.11)

A bidirectional power flow PWM converter, with adequate filtering and control, may run the PM machine either as a motor (for starting the gas turbine) or as a generator, with controlled output at variable speed. The generator may work in the power-grid mode or in stand-alone mode. These flexibility features, together with

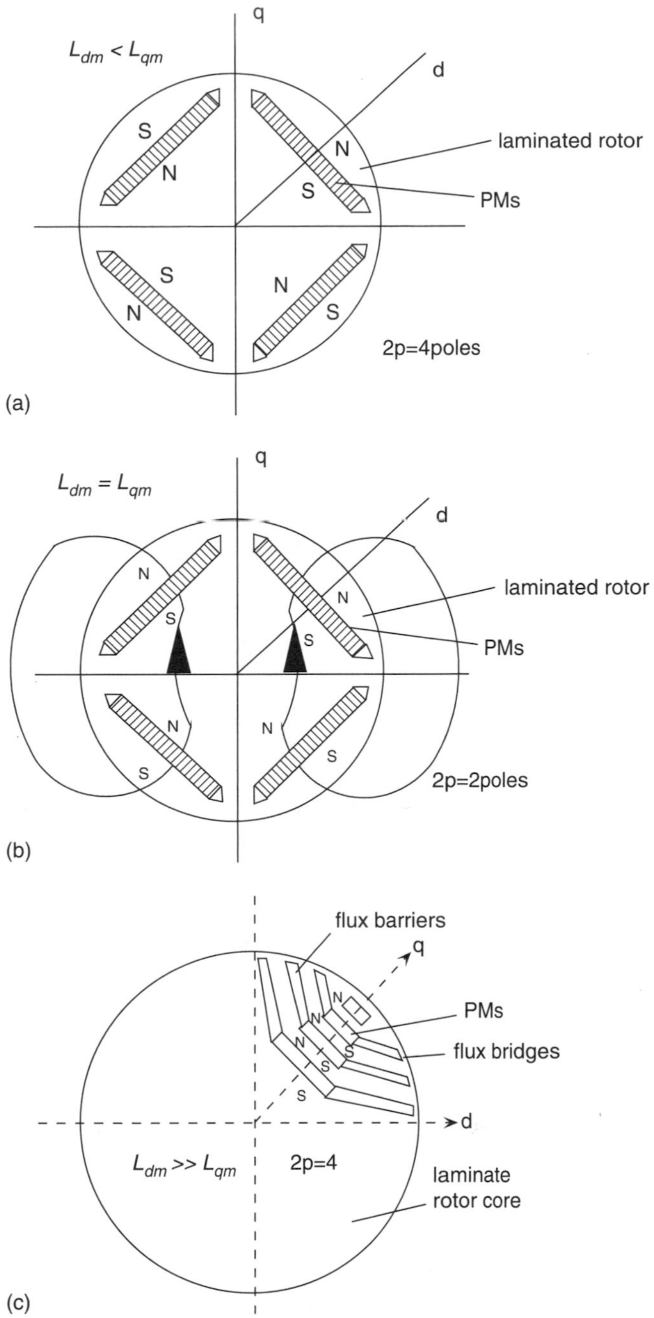

FIGURE 8.10 Interior PM rotors: (a) $2p = 4$ poles, (b) $2p = 2$, (c) with rotor flux barriers (IPM reluctance).

fast power-active and reactive-decoupled control at variable speed, may make such solutions a way of the future, at least in the tens and hundreds of kW range.

Many other PM SG configurations have been introduced, such as those with axial air gap. Among them we will mention one that is typical in the sense that it uses the IPM reluctance rotor (Figure 8.10c) but it adds an electrical excitation (Figure 8.12) (Boldea et al., 2000).

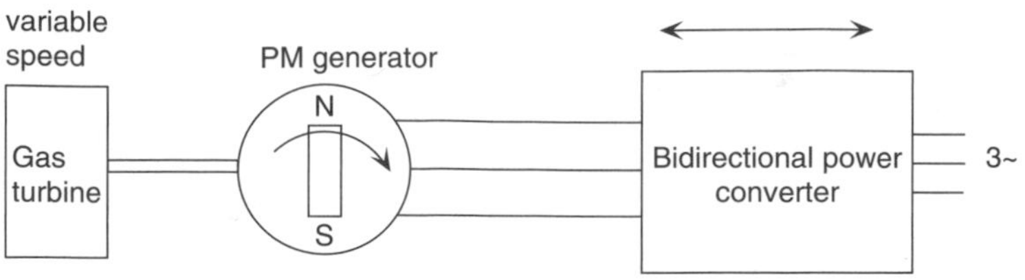

FIGURE 8.11 Bidirectional full-power electronics control.

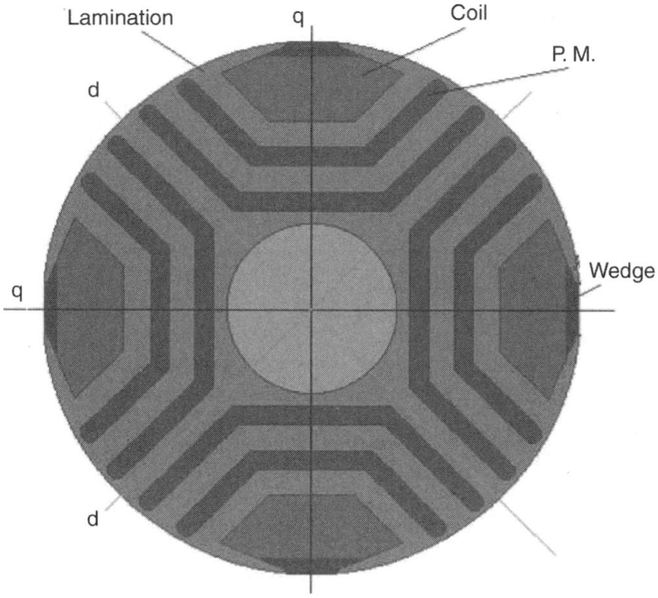

FIGURE 8.12 Biaxial excitation PM reluctance generator (BEGA).

Induction Generators

The cage-rotor induction machine is known to work as a generator provided:

The frequency f_1 is smaller than $n \times p$ (speed × pole pairs): $S < 0$ (Figure 8.13a).
There is a source to magnetize the machine.

An induction machine working as a motor, supplied to fixed frequency and voltage (f_1, V_1) power grid, becomes a generator if it is driven by a prime mover above the no-load ideal speed f_1/p

$$n > \frac{f_1}{p} \tag{8.9}$$

Alternatively, the induction machine with cage-rotor may self-excite on a capacitor at its terminals (Figure 8.13b)

For an induction generator connected to a strong (constant frequency and voltage) power grid, when the speed n increases (above f_1/p), the active power delivered to the power grid increases, but so does the reactive power drawn from the power grid.

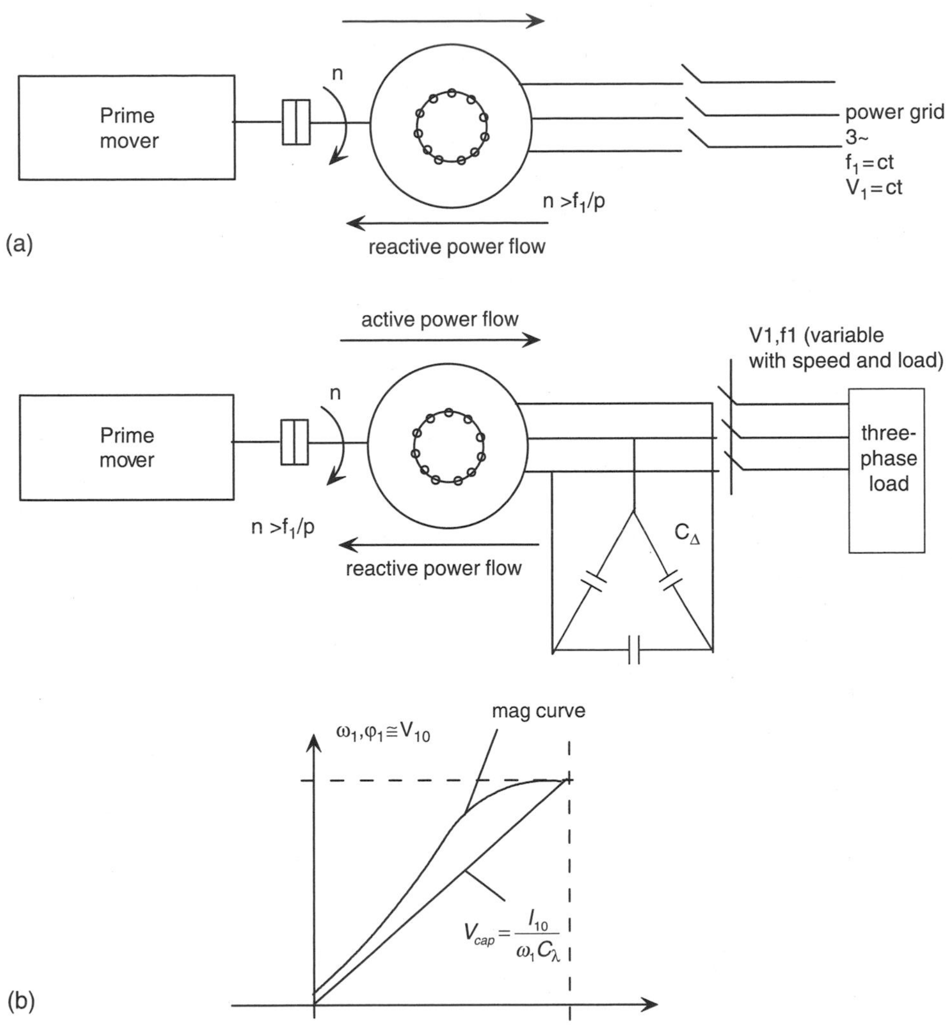

FIGURE 8.13 Cage-rotor induction generator: (a) at power grid: $V_1 = ct, f_1 = ct$, (b) stand alone (capacitor excited): V_1, f_1, variable with speed and load.

Many existing wind generators use such cage-rotor IGs connected to the power grid. The control is mechanical only. The blade pitch angle is adjusted according to wind speed and power delivery requirements. However such IGs tend to be rigid, as they are stable only until n reaches the value

$$n_{\max} = \frac{f_1}{p}(1 + |S_K|) \tag{8.10}$$

Where S_K is the critical slip, which decreases with power and is below 0.08 for IGs in the hundreds of kW. Also, additional parallel capacitors at power grid are required to compensate for the reactive power drained by the IG.

Alternatively, the reactive power may be provided by paralleled (plus series) capacitors (Figure 8. 13b). In this case we do have a self-excitation process, which requires some remnant flux in the rotor (from past operation) and the presence of magnetic saturation. The frequency f_1 of self-excitation voltage (on no load) depends on the capacitor value and on the magnetization curve of the induction machine $\Psi_1(I_{10})$

$$V_{10} \approx \psi_1(I_{10}) \cdot 2 \cdot \pi \cdot f_1 \approx V_{C_Y} = I_{10} \frac{1}{3 \cdot C_\Delta \cdot 2 \cdot \pi \cdot f_1} \tag{8.11}$$

The trouble is that on load, even if the speed is constant through prime-mover speed control, the output voltage and frequency vary with load. For constant speed, if frequency reduction under load of 1 Hz is acceptable, then voltage control suffices. A three-phase ac chopper (variac) supplying the capacitors would do it, but the harmonics introduced by it have to be filtered. In simple applications, a combination of parallel and series capacitors would provide rather constant (with 3 to 5% reduction) voltage up to rated load.

Now if variable speed is to be used then, for *constant voltage and frequency*, PWM converters are needed. These configurations are illustrated in Figure 8.14. A bidirectional power flow PWM converter (Figure 8.14a) provides for both generating and motoring functions at variable speed. The capacitor in the dc line of the converter may lead not only to active but also reactive power delivery to the grid without large transients decoupled, and fast active and reactive power control is implicit.

The stand-alone configuration in Figure 8.14b is less expensive, but it provides only unidirectional power flow. A typical V_1/f_1 converter for drives is used. It is possible to inverse the connections, that is, to connect the diode rectifier and capacitors to the grid and the converter to the machine. This way the system works as a variable-speed drive for pumping, etc., if a power grid is available. This commutation may be done automatically, though it would take one to two minutes. For variable speed — in a limited range — an excitation capacitor in two stages would provide the diode rectifier slightly variable output dc link voltage.

Provided the minimum and maximum converter voltage limits are met, the former would operate over the entire speed range. Now the converter is V_1/f_1 controlled for constant voltage and frequency.

A transformer (Y, Y_0) may be needed to accommodate unbalanced (or single-phase) loads.

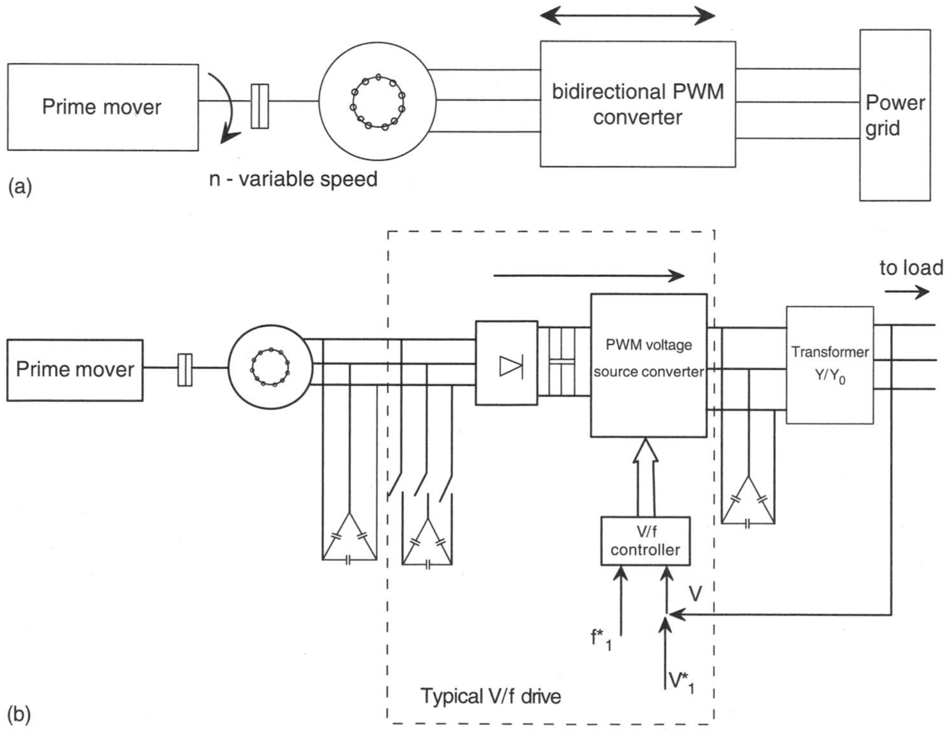

FIGURE 8.14 Cage-rotor IGs for variable speed: (a) at power grid, (b) stand alone.

The output voltage may be close-loop controlled through the PWM converter. On the other end, the bidirectional PWM converter configuration may be provided with a reconfigurable control system so as to work not only on the power grid, but to separate itself smoothly and operate as stand alone or wait on standby and then be reconnected smoothly to the power grid. Thus multifunctional power generation at variable speed is produced. As evident in Figure 8.14, full power electronics is required.

The Doubly Fed Induction Generator

It all started between 1907 and 1913, with the Scherbius and Kraemer cascade configurations, which are both slip-power recovery schemes of wound-rotor induction machines. A. Leonhard analyzed it pertinently in 1928, but adequate power electronics for it was then unavailable. A slip recovery scheme with thyristor power electronics is shown in Figure 8.15a. Unidirectional power flow, from IG rotor to the converter, is only feasible because of the diode rectifier. A step-up transformer is necessary for voltage adaptation, while the thyristor inverter produces constant voltage and frequency output. The principle of operation is based on the frequency theorem of traveling fields.

$$f_1 = np + f_2; f_2 <> 0, \text{ and variable } f_1 = ct \tag{8.12}$$

Negative frequency means that the sequence of rotor phases is different from the sequence of stator phases. Now, if f_2 is variable, n may also be variable, and as long as Equation (8.12) is fulfilled, f_1 is constant.

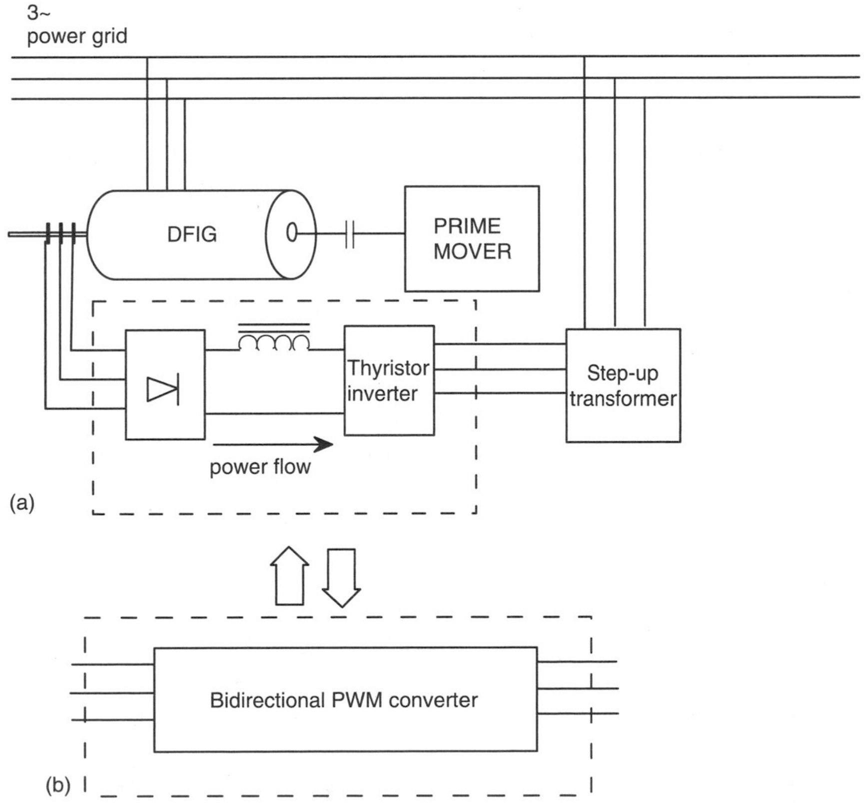

FIGURE 8.15 Doubly fed induction generator (DFIG): (a) with diode rectifier (slip recovery system), (b) with bidirectional PWM converter.

That is, constant frequency f_1 is provided in the stator for adjustable speed. The system may work at the power grid or even as a stand alone, although with reconfigurable control. When $f_2 > 0$, $n < f_1/p$, we do have subsynchronous operation. The case for $f_2 < 0$, $n > f_1/p$ corresponds to hypersynchronous operation. Synchronous operation takes place at $f_2 = 0$, which is not feasible with the diode rectifier current source inverter but it is with the bidirectional PWM converter.

The slip recovery system can work as a subsynchronous ($n < f_1/p$) motor or as a supersynchronous ($n > f_1/p$) generator. The DFIG with bidirectional PWM converter may work as a motor and generator both for subsynchronous and supersynchronous speed.

The power flow directions for such a system are shown in Figure 8.16

The converter rating is commensurable with speed range, that is, to maximum slip S_{max}

$$KVA_{rating} = K\frac{f_{2max}}{f_1} \times 100[\%] \tag{8.13}$$

$K = 1$ to 1.4 depending on the reactive power requirements from the converter.

Notice that, being placed in the rotor circuit through slip rings and brushes, the converter rating is around $|S_{max}|$ in percentage. The larger the speed range, the larger the rating and the costs of the converter. The fully bidirectional PWM converter as a back-to-back voltage source multilevel PWM converter system may provide fast and continuous decoupled active and reactive power control operation even at synchronism ($f_2 = 0$, dc rotor excitation). And it may perform self-starting as well. Self-starting is done by short-circuiting first the stator, previously disconnected from the power grid, and supplying the rotor through the PWM converter in the subsynchronous motoring mode. The rotor accelerates up to a prescribed speed corresponding to $f_2' > f_1(1-S_{max})$. Then the stator winding is opened and, with the rotor freewheeling the stator, no-load voltage and frequency are adjusted to coincide with those of the power grid, by adequate PWM converter control. Finally the stator winding is connected to the power grid without notable transients.

This kind of rotor starting requires $f_2' \approx (0.8$ to $1)f_1$, which means that the standard cycloconverter is out of the question. So it is only the back-to-back voltage PWM multilevel converter or the matrix converter that are suitable for full exploitation of motoring/generating at sub- and supersynchronous speeds, so typical in pump storage applications.

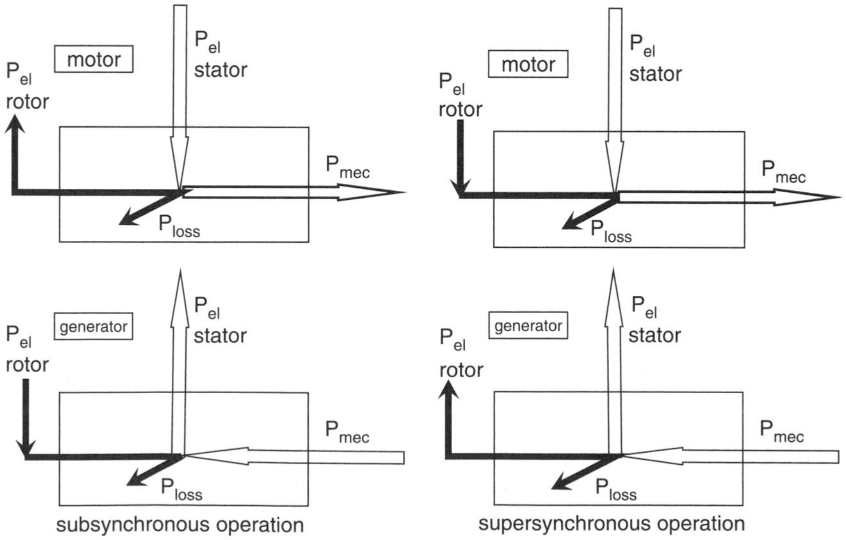

FIGURE 8.16 Operation modes of DFIG with bidirectional PWM converter (in the rotor) (subsynchronous operation, supersynchronous operation).

Parametric Generators

Parametric generators exploit the magnetic anisotropy of both stator and rotor. PMs may be added on the stator or on the rotor. Single magnetic saliency with PMs on the rotor is also used in some configurations. The parametric generators use nonoverlapping (concentrated) windings to reduce end-connection copper losses on the stator.

As the stator mmf does not produce a pure traveling field, there are core losses both in the stator and in the rotor. The simplicity and ruggedness of such generators made them adequate for some applications.

Among parametric generators, some of the most representative are detailed here:

- The SRGs (Miller, 1993):
  - without PMs
  - with PMs on stator or on rotor (Luo and Lipo, 1994; Radulescu et al., 2002; Blaabjerg et al., 1996)
- The TFGs (Weh et al., 1988; Henneberger and Viorel, 2001):
  - with-rotor PMs
  - with-stator PMs
- The FRGs (Rauch and Johnson, 1955; Bausch et al., 1998):
  - with PMs on the stator
  - with PMs on the rotor (and flux concentration)
- The linear motion PM alternators (LMAs) (Boldea and Nasar, 1997):
  - with coil mover and PMs on the stator
  - with PM mover; tubular or flat (with PM flux concentration)
  - with iron-mover and PMs on the stator

LMA

The microphone is the classical example of LMA with a moving coil. The loudspeaker illustrates its motoring operation mode.

Though there are very many potential LMA configurations (or actuators), they all use PMs and fall into three main categories (Boldea and Nasar 1997):

With moving coil (and stator PMs); Figure 8.17a
With moving PMs (and stator coil); Figure 8.17b
With moving iron (and stator PMs); Figure 8.17c

In essence the PM flux linkage in the coil changes sign when the mover travels the excursion length l_{stroke}, which serves as a kind of pole pitch. So they are all, in a way, single-phase flux-reversal machines. The average speed U_{av} is

$$U_{av} = 2 \cdot l_{stroke} \cdot f_1 \tag{8.14}$$

f_1 is frequency of mechanical oscillations

To secure high efficiency, beryllium-copper flexured springs are used to store the kinetic energy of the mover at excursion ends. They also serve as linear bearings. The frequency of these mechanical springs f_m should be equal to electrical frequency

$$f_e = f_m = \frac{1}{2\pi}\sqrt{\frac{K}{m}} \tag{8.15}$$

K is spring rigidity coefficient, m is moving mass.

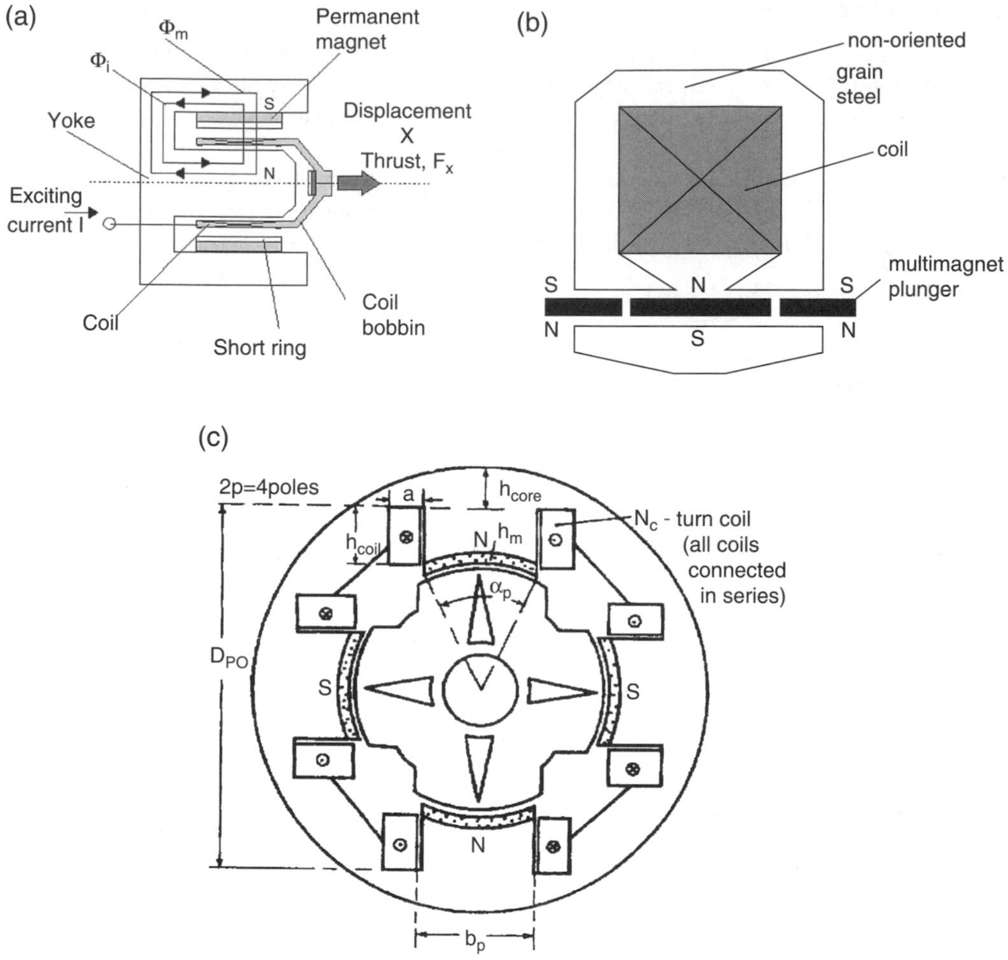

FIGURE 8.17 Commercial linear motion alternators: (a) with moving coil, (b) with moving PMs, (c) with moving iron.

Electric Generator Applications

The applications domains for electric generators embrace almost all industries — traditional and new — with powers from mW to hundreds of MW/unit and more (Boldea and Nasar, 1997; Kundur, 1994; Vu and Agee, 1993; Djukanovic et al., 1995; Müller et al., 2000; Kudo, 1994)

Table 8.2 summarizes our view of electric generator main applications and the competitive types that may suit them.

Summary

In this subchapter we have presented most representative — in use and newly proposed — types of electric generators by principle, configuration and application.

A few concluding remarks are in order:

- The power-per-unit range varies from a few mW to a few hundred MW (even 1500 MVA) per unit.
- Large power generators — above a few MW — are electrically excited on the rotor either dc, as in conventional SGs, or in three-phase ac, as in the doubly fed induction generator (DFIG).
- While conventional dc rotor-excited SG requires tightly controlled constant speed to produce constant frequency output, the DFIG can work with adjustable speed.

TABLE 68.2 Electric Generator Applications

Application	Large power systems (gas, coal, nuclear turbogens)	Distributed power systems (wind, hydrogens)	Standby diesel-driven EGs	Automotive starter-generators	Diesel locomotives
Suitable generator	Excited rotor synchronous generators, doubly fed induction generators (up to hundreds of MW/unit)	Excited rotor synchronous generators, cage-rotor induction generators, PM synchronous generators, Parametric generators (up to 10 MW power/unit)	PM synchronous generators, cage-rotor induction generators	IPM synchronous generators, induction generators, transverse flux PM generators	Excited-rotor synchronous generators
Application	Home electricity production	Spacecraft applications	Aircraft applications	Ship applications	
Suitable generator	PM synchronous generators, LMAs	Linear motion alternators (LMAs)	PM synchronous, cage-rotor induction generators (up to 500 kW/unit)	Excited synchronous generators (power in the order of a few MWs)	
Application	Small power telemetry-based vibration monitoring	Inertial batteries	Super-high-speed gas turbine generators	Super-high-speed gas turbine generators	Super-high-speed gas turbine generators
Suitable generator	LMAs 20–50 mW at 5 V	Axial-air gap PM synchronous generators up to hundreds of MJ/unit	PM synchronous generators up to 150 kW and 70,000 rpm (higher powers at lower speeds: 3 MW at 15,000 rpm).	PM synchronous generators up to 150 kW and 70,000 rpm (higher powers at lower speeds: 3 MW at 15,000 rpm).	PM synchronous generators up to 150 kW and 70,000 rpm (higher powers at lower speeds: 3 MW at 15,000 rpm).

- The rating of the rotor-connected PWM converter in DFIG is about equal to the adjustable speed range (slip), in general around or less than 20%. This implies reasonable costs for a more flexible generator with fast active and reactive power (or frequency and voltage) control.
- DFIG seems the way of the future in electric generation at adjustable speed for powers above a few MW, in general, per unit.
- PM SGs are emerging for kW, tenth of kW, even hundreds of kW or 1 to 3 MW/unit in special applications, such as automotive starter-alternators, superhigh-speed-gas-turbine generators, or in directly driven wind generators
- Linear motion alternators are emerging for power operation up to 15, even 50 kW, for home or special-series hybrid vehicles with linear gas combustion engines and electric propulsion.
- Parametric generators are being investigated for special applications: SRGs for aircraft jet engine starter-alternators and TFGs/motors for hybrid or electrical vehicle propulsion, or directly driven wind generators.
- Electric generators are driven by different prime movers having their own characteristics, performance, and mathematical models, which, in turn, influence the generator operation, since speed control is enacted upon the prime mover.

References

T. Bödefeld and H. Sequenz, *Elektrische Maschinen*, Vienna: Springer, 1938 (in German).

C. Concordia, "Synchronous machines," *Theory and Performance*, New York: Wiley, 1951.

R. Richter, "Electrical Machines," *Synchronous Machines*, vol. 2, Verlag Birkhäuser: Bassel, 1963 (in German).

M. Kostenko and L. Pitrovski, "Electrical Machines," *AC Machines*, vol. 2, Moscow: Mir Publishers, 1974.

J.H. Walker, *Large Synchronous Machines*, Oxford: Clarendon Press, 1981.

T.J.E. Miller, *Brushless PM and Reluctance Motor Drives*, Oxford: Clarendon Press, 1989.

S.A. Nasar, I. Boldea, and L. Unnewher, *Permanent Magnet, Reluctance and Selfsynchronous Motors*, Boca Raton, Florida: CRC Press, 1993.

D.C. Hanselman, *Brushless PM Motor Design*, New York: McGraw Hill, 1994.

D.R. Hendershot Jr. and T.J.E. Miller, *Design of Brushless PM*, Oxford: Magna Physics Publishing & Clarendon Press, 1994.

J. Gieras, F. Gieras, and M. Wuig, *PM Motor Technologies*, 2nd ed., New York: Marcel Dekker, 2002.

I. Boldea, S. Scridon, and L. Tutelea, "BEGA: biaxial excitation generator for automobiles," *Record of OPTIM-2000*, vol. 2, Romania: Poiana Brasov, 2000, pp. 345–352.

T. Miller, *Switched Reluctance Motors and Their Control*, Oxford, U.K.: Oxford University Press, 1993.

Y. Luo and T.A. Lipo, "A new doubly salient PM motor for adjustable speed drives," *EMPS*, vol. 22, no. 3, pp. 259–270, 1994.

M. Radulescu, C. Martis, and I.Husain, "Design and performance of small doubly salient rotor PM motor," *EPCS*, 30, 523–532, 2002 (former *EMPS*).

F. Blaabjerg, I. Christensen, P.O. Rasmussen, and L. Oestergaard, "New advanced control methods for doubly salient PM motor," *Record of IEEE-IAS-1996*, 786–793.

H. Weh, H. Hoffman, and J. Landrath, "New permanent excited synchronous machine with high efficiency at low speeds," *Proc. ICEM-1988*, Pisa, Italy, pp. 1107–1111.

G. Henneberger and I.A. Viorel, *Variable Reluctance Electric Machines*, Braunschweig: Shaker Verlag, 2001 (Chapter 6).

S.E. Rauch and L.J. Johnson, "Design principles of flux switch alternator," *IEEE Trans.*, 74, III, 1261–1268, 1955.

H. Bausch, A. Grief, K. Kanelis, and A. Milke, "Torque control of battery supplied switched reluctance drives for electric vehicle," *Record of ICEM*, Istambul, 1, 229–234, 1998.

I. Boldea and S.A. Nasar, *Linear Electric Actuators and Generators*, Cambridge: Cambridge University Press, 1997.

P. Kundur, *Power System Stability and Control*, New York: McGraw-Hill, 1994

H. Vu and J.C. Agee, "Comparison of power system stabilizers for damping local mode oscillations," *IEEE Trans.*, vol. EC-8, no. 3, pp. 533–538, 1993.

M. Djukanovic, M. Novicevic, D. Dobrojevic, R. Babic, D. Babic, and Y. Par, "Neural-net base coordinated stabilizing control for exciter and governor loops of low head hydroelectric power plants," *IEEE Trans*, vol. EC-10, no. 4, pp. 760–767, 1995.

S. Müller, M. Deicke, and R.W. De Doncker, "Adjustable speed generators for wind turbines based on double-fed induction machines and 4-quadrant IGBT converters linked to the rotor," *IEEE-IAS Annu. Meeting*, 2000.

K. Kudo, "Japanese experience with a converter fed variable speed pump-storage system," *Hydropower and Dams*, no. 3, pp. 67–71, 1994.

8.2 Motors

Mehdi Ferdowsi

Electric motors are the most commonly used prime movers in industry. They are usually classified based on the form of the electrical signal applied to them. The classification of different types of motors commonly used in industrial applications is shown in Figure 8.18.

Motor Applications

DC Motors

Permanent magnet (PM) field motors occupy the low end of the horsepower (hp) range and are commercially available up to about 10 hp. Below 1 hp they are used for servo applications, such as in machine tools, for robotics, and in high-performance computer peripherals.

Wound field motors are used above about 10 hp and represent the highest horsepower range of dc motor application. They are commercially available up to several hundred horsepower and are commonly used in traction, hoisting, and other applications, where a wide range of speed control is needed. The shunt wound dc motor is commonly found in industrial applications such as grinding and machine tools and in elevator and hoist applications. Compound wound motors have both a series and shunt field component to provide specific torque-speed characteristics. Propulsion motors for transit vehicles are usually compound wound dc motors.

AC Motors

Single-phase ac motors occupy the low end of the horsepower spectrum and are offered commercially up to about 5 hp. Single-phase synchronous motors are only used below about 1/10 of a horsepower. Typical applications are timing and motion control, where low torque is required at fixed speeds. Single-phase induction motors are used for operating household appliances and machinery from about 1/3 to 5 hp.

Polyphase ac motors are primarily three-phase and are by far the largest electric prime mover in all of industry. They are offered in ranges from 5 up to 50,000 hp and account for a large percentage of the total motor industry in the world. In number of units, the three-phase squirrel cage induction motor is the most common. It is commercially available from 1 hp up to several thousand horsepower and can be used on conventional ac power or in conjunction with adjustable speed ac drives. Fans, pumps, and material handling are the most common applications.

When the torque-speed characteristics of a conventional ac induction motor need to be modified, the wound rotor induction motor is used. These motors replace the squirrel cage rotor with a wound rotor and slip rings. External resistors are used to adjust the torque-speed characteristics for speed control in such applications as ac cranes, hoists, and elevators.

Three-phase synchronous motors can be purchased with PM fields up to about 5 hp and are used for applications such as processing lines and transporting film and sheet materials at precise speeds.

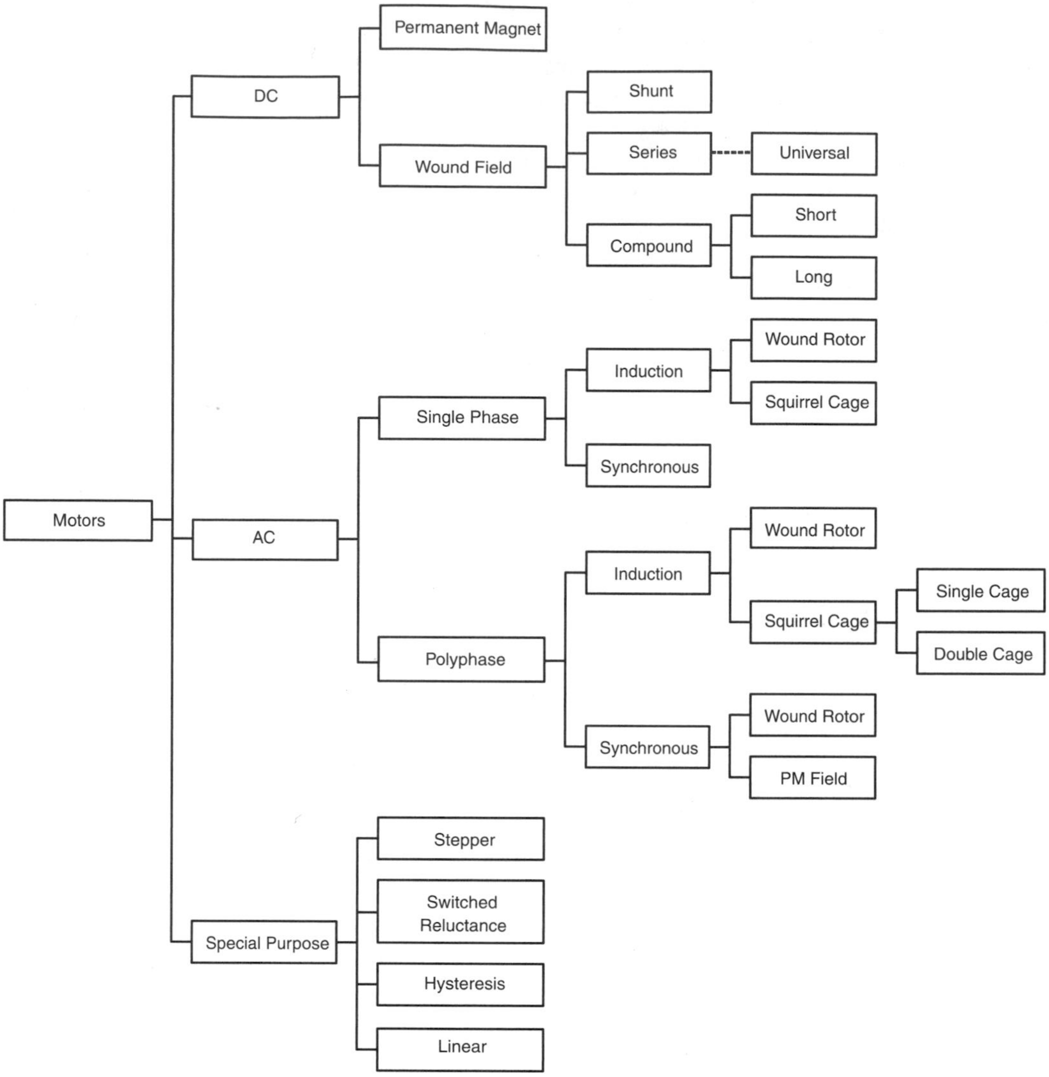

FIGURE 8.18 Classification of different types of motors for industrial applications.

In the horsepower range above about 10,000 hp, three-phase synchronous motors with wound fields are used rather than large squirrel cage induction motors. Starting current and other characteristics can be controlled by the external field exciter. Three-phase synchronous motors with wound fields are available up to about 50,000 hp.

Motor Analysis

DC Motors

The equivalent circuit of separately excited, shunt, and series dc motors (Guru and Hiziroglu, 2001; Sen, 1981) are depicted in Figure 8.19(a), Figure 8.19(b), and Figure 8.19(c), respectively. R_a and L_a represent the effective armature-winding resistance and self-inductance, respectively, whereas R_f and L_f represent the same parameters corresponding to the field winding. L_a and L_f only have effect on the transient response of the motor and are not considered at the steady-state analysis. E_a represents the induced electromagnetic force

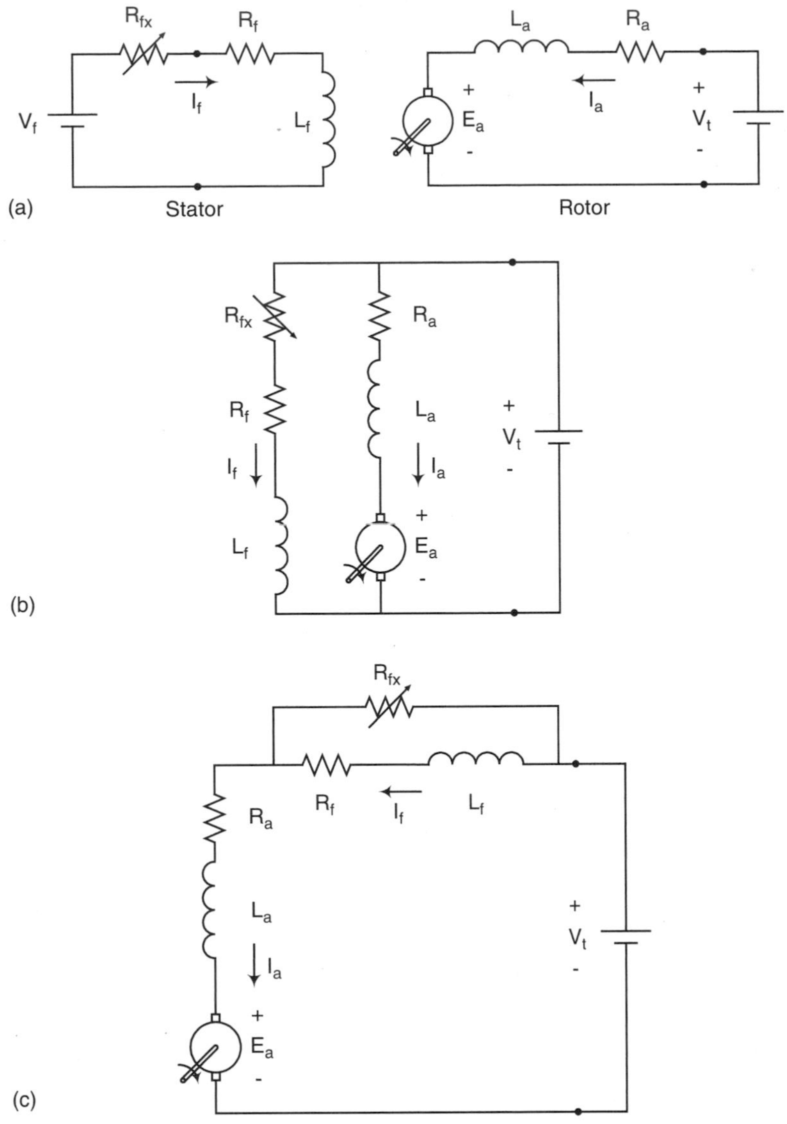

FIGURE 8.19 Equivalent circuit diagram of (a) separately excited, (b) shunt, and (c) series dc motors.

(emf) and can be described as

$$E_a = k_a k_f I_f \omega_m \tag{8.16}$$

where k_a and k_f are constants and depend on the structure of the motor, and ω_m is the mechanical speed in rad/s. Field current I_f can be controlled by either using a variable voltage source of V_f or using variable field resistance R_{fx} externally. The developed mechanical torque in an ideally linear motor is

$$T = k_a k_f I_f I_a \tag{8.17}$$

where I_a is the armature current.

Assuming there is no power loss in the magnetic circuit of the dc motor, we can write that

$$\omega_m T = E_a I_a \tag{8.18}$$

In order to find the torque-speed characteristic of an ideal separately excited dc motor, we can write that

$$\omega_{\mathrm{m}} = \frac{E_{\mathrm{a}} I_{\mathrm{a}}}{T} = \frac{(V_{\mathrm{t}} - R_{\mathrm{a}} I_{\mathrm{a}}) I_{\mathrm{a}}}{k_{\mathrm{a}} k_{\mathrm{f}} I_{\mathrm{f}} I_{\mathrm{a}}} = \frac{V_{\mathrm{t}}}{k_{\mathrm{a}} k_{\mathrm{f}} I_{\mathrm{f}}} - \frac{R_{\mathrm{a}} T}{(k_{\mathrm{a}} k_{\mathrm{f}} I_{\mathrm{f}} I_{\mathrm{a}})^2} \tag{8.19}$$

The efficiency of a separately excited dc motor can be described as

$$\eta = \frac{P_{\mathrm{out}}}{P_{\mathrm{in}}} = \frac{E_{\mathrm{a}} I_{\mathrm{a}}}{V_{\mathrm{t}} I_{\mathrm{a}} + V_{\mathrm{f}} I_{\mathrm{f}}} \tag{8.20}$$

The disadvantage of using a separately excited dc motor is the need for a secondary dc voltage source of V_{f}. This drawback has been resolved in shunt and series dc motors by putting the field winding in parallel or series with the armature winding, respectively. Using the same analysis, the torque-speed characteristics of ideal shunt and series dc motors can be described as

$$\omega_{\mathrm{m}} = \frac{R_{\mathrm{f}} + R_{\mathrm{fx}}}{k_{\mathrm{a}} k_{\mathrm{f}}} - \frac{R_{\mathrm{a}} (R_{\mathrm{f}} + R_{\mathrm{fx}})^2}{(k_{\mathrm{a}} k_{\mathrm{f}} V_{\mathrm{t}})^2} T \tag{8.21}$$

and

$$\omega_{\mathrm{m}} = \frac{V_{\mathrm{t}} I_{\mathrm{a}} - (R_{\mathrm{a}} + R_{\mathrm{f}} || R_{\mathrm{fx}}) I_{\mathrm{a}}^2}{T} \tag{8.22}$$

respectively. Figure 8.20 sketches the torque-speed characteristics of these three dc motors in ideal case, where armature reaction has been neglected.

Compound dc motors are basically a combination of shunt and series motor. Therefore, desired speed-torque characteristic can be obtained. Figure 8.21(a) and Figure 8.21(b) depict the equivalent circuit diagram of cumulative short- and long-shunt compound motors.

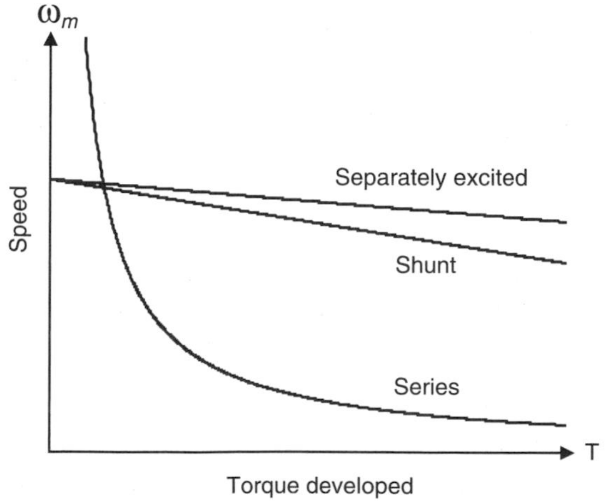

FIGURE 8.20 Torque-speed characteristics of ideal dc motors.

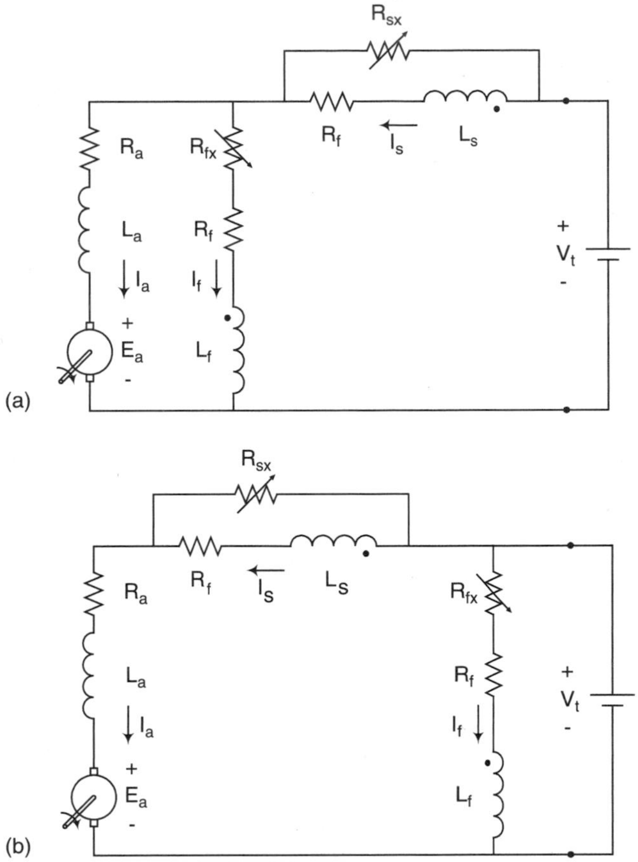

FIGURE 8.21 Equivalent circuit diagram of cumulative (a) short- and (b) long-shunt compound dc motors.

Synchronous Motors

Synchronous motors (Sen, 1997) run under constant speed and mainly have a three-phase stator structure that houses the armature windings. The rotor winding is fed by a dc voltage source and is in charge of magnetic field generation. The rotor has as many magnetic poles as the stator and is designed either cylindrical or salient pole. Only the cylindrical rotor (Boldea and Nasar, 1992; Slemon, 1992) is considered in the following discussion. Applying balanced three-phase sinusoidal voltages to the stator generates a rotating magnetic field, which rotates at the speed of

$$\omega_s = \frac{2}{P}\omega_e = \frac{120}{P}f_e \tag{8.23}$$

where ω_s is called the synchronous speed, P is the number of magnetic poles, ω_e is the electrical angular frequency, and f_e is the electrical frequency. Interaction of this rotating magnetic field with the field generated by the rotor develops mechanical torque.

Neglecting the armature reaction, the per-phase equivalent circuit diagram of cylindrical-rotor synchronous motor is depicted in Figure 8.22. R_f and L_f are the rotor winding equivalent resistance and inductance, respectively. R_a is the stator winding per phase, and X_s is the synchronous reactance, which is the summation of the stator winding leakage and magnetization reactances. E_a is the back emf and its root-mean square (rms) value is equal with

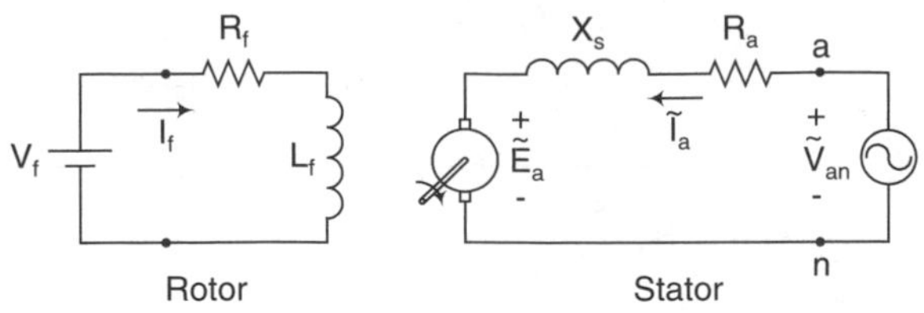

FIGURE 8.22 Per-phase equivalent circuit diagram of a cylindrical-rotor synchronous motor.

$$|\tilde{E}_a| = \frac{L_{af} I_f \omega_e}{\sqrt{2}} \tag{8.24}$$

where L_{af} is the mutual inductance of stator phase winding and rotor field winding. Using phasor representation of sinusoidal waveforms and from the equivalent circuit, for each phase of the synchronous machine we can write

$$\tilde{V}_{an} = (R_a + jX_s)\tilde{I}_a + \tilde{E}_a \tag{8.25}$$

Considering all three phases, total mechanical power developed by the motor is

$$P_{mech} = 3|\tilde{E}_a||\tilde{I}_a|\cos(\varphi) = T\omega_s \tag{8.26}$$

where φ is the phase angle of line current I_a with respect to E_a, T is the developed torque, and ω_s is the synchronous angular velocity. The input power is

$$P_{in} = 3|\tilde{V}_a||\tilde{I}_a|\cos(\theta) + V_f I_f \tag{8.27}$$

where θ is the phase angle of line current I_a with respect to line-to-neutral voltage V_a, V_f is the dc voltage applied to the rotor, and I_f is the rotor dc current. Total power loss in the motor can be described as

$$P_{loss} = P_{in} - P_{mech} = 3R_a|\tilde{I}_a|^2 + V_f I_f \tag{8.28}$$

Analysis is facilitated by use of the phasor diagram shown in Figure 8.23. Assuming that the stator impedance is in the form of

$$Z = R_a + jX_s = |Z| < Z_\varphi \tag{8.29}$$

then phase angle of the input voltage with the back emf voltage δ is

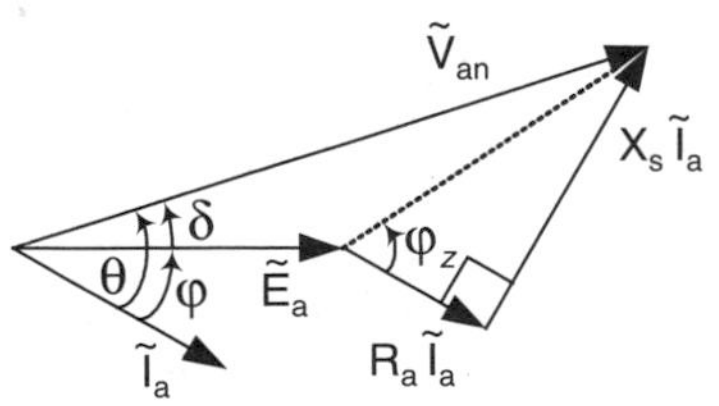

FIGURE 8.23 Phasor diagram for a cylindrical-rotor synchronous motor.

$$\delta = \tan^{-1}\left(\frac{|Z||\tilde{I}_a|\sin(\varphi_z - \varphi)}{|\tilde{E}_a| + |Z||\tilde{I}_a|\cos(\varphi_z - \varphi)}\right) \tag{8.30}$$

The input power factor is

$$\cos(\theta) = \frac{P_{in}}{3|\tilde{V}_a||\tilde{I}_a|} \tag{8.31}$$

and $\theta = \varphi + \delta$ as shown in Figure 8.23.
Total input complex power can be expressed as

$$S = 3\tilde{V}_{an}\tilde{I}_a^* = P_{in} + jQ_{in} \tag{8.32}$$

where * denotes the complex conjugate and Q_{in} is the total input reactive power. Using Equation (8.25) while neglecting R_a leads to

$$\frac{P_{in}^2}{9} + \left(\frac{Q_{in}}{3} + \frac{|\tilde{V}_{an}|^2}{X_s}\right)^2 = \left(\frac{|\tilde{V}_{an}||\tilde{E}_a|}{X_s}\right)^2 \tag{8.33}$$

Bearing Equation (8.24) in mind, while keeping the input voltage and total developed mechanical power, the amount of reactive power can be controlled based on Equation (8.33) using field current I_f. Keeping the input power and input voltage constant, adjusting the input reactive power by the field current, and considering Equation (8.33) lead to the idea of controlling the input current by means of the field current. Figure 8.24 depicts the relationship between the input current I_a and field current I_f for different power ratings. This graph is called a *V* curve due to its characteristic shape.

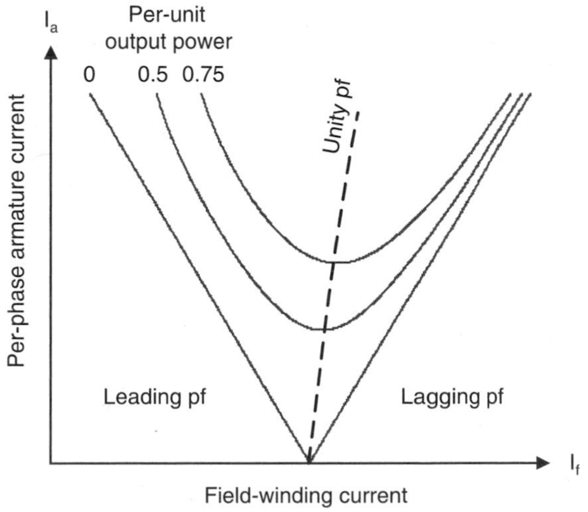

FIGURE 8.24 Synchronous motor *V* curves.

Three-Phase Induction Motors

The structure of the stator of a three-phase induction motor (Say and Taylor, 1986; Engelman and Middendorf, 1995; Novotny and Lipo, 1996) is quite similar to that of a three-phase synchronous motor. Yet, the structure of the rotor is quite different and so is the torque development mechanism. The rotor of an induction motor is not connected to a power source, and therefore the voltage signals appearing on the rotor side are a result of electromagnetic induction from the stator side, and that is where the term *induction* comes from. An induction motor is basically a transformer with a rotating secondary.

Similar to a synchronous machine, applying balanced three-phase sinusoidal voltages to the stator generates a rotating magnetic field, which rotates at the synchronous speed of ω_s. The generated magnetic field induces voltage signals to the rotor. Based on Faraday's law, the polarity of the induced voltages and the direction of the electric current that they generate in the rotor are in a way that opposes the revolving magnetic field of the stator. Therefore, the developed torque is in the direction that makes the rotor follow the magnetic field of the stator. It is obvious that the mechanical speed of the rotor will not be exactly equal with the synchronous speed, since in that case the rotor will see a stationary magnetic field and there will be no induced voltage on the rotor to generate any current and finally generate any torque. The rotor mechanical speed is usually expressed as a ratio of the synchronous speed using

$$s = 1 - \frac{\omega_m}{\omega_s} \tag{8.34}$$

where s is called fractional slip or simply slip, and ω_m is the rotor mechanical speed.

The steady-state analysis of three-phase induction motors is conducted based on the per-phase equivalent circuit of Figure 8.25. All voltages and currents are in sinusoidal steady state and represented as phasors.

In Figure 8.25, R_1 is the per-phase stator resistance and models the stator copper loss, X_1 is the stator winding leakage reactance, R_c is the per-phase equivalent core-loss resistance, X_m represents the magnetization reactance, R_2 is the rotor resistance, X_2 is the rotor leakage reactance, and finally $R_2(1\text{-}s)/s$ models the mechanical load.

In power, torque, and efficiency calculations, it is more convenient to use the Thevenin-equivalent circuit of the stator part, as depicted in Figure 8.26.

Based on the Thevenin's theorem, we can write that

$$R_{eq} + jX_{eq} = (R_1 + jX_1)||R_c||jX_m \tag{8.35}$$

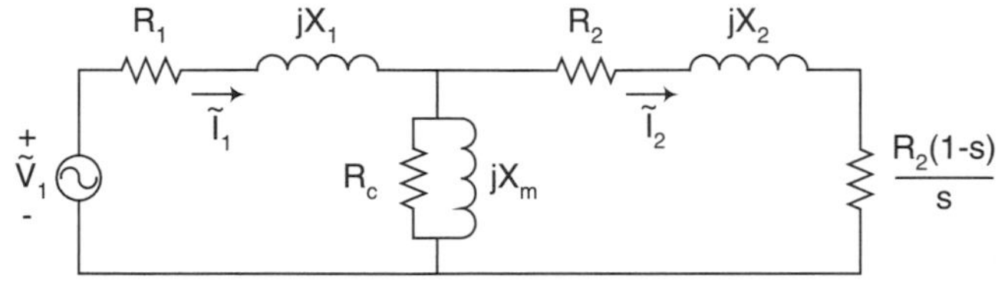

FIGURE 8.25 Per-phase equivalent circuit diagram of an induction motor.

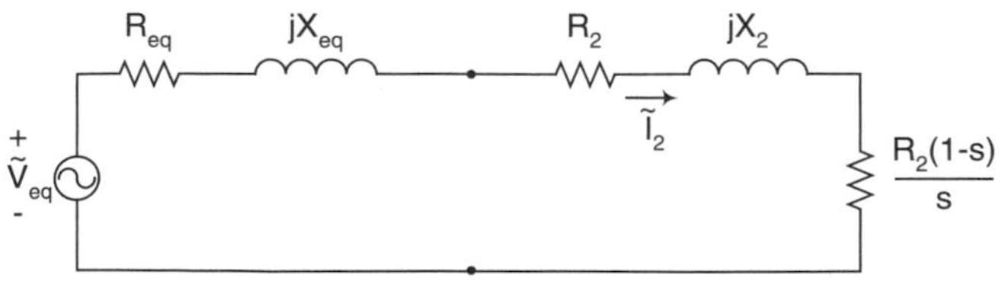

FIGURE 8.26 Thevenin-equivalent circuit of the stator of the induction motor.

$$\tilde{V}_{\text{eq}} = \tilde{V}_1 \left[\frac{R_c || jX_m}{R_1 + jX_1 + (R_c || jX_m)} \right] \tag{8.36}$$

Considering the Thevenin's equivalent circuit, the phasor form of the rotor current can be expressed as

$$\tilde{I}_2 = \frac{\tilde{V}_{\text{eq}}}{\left(R_{\text{eq}} + \frac{R_2}{s}\right) + j\left(X_{\text{eq}} + X_2\right)} \tag{8.37}$$

The developed mechanical power in the three-phase induction motor as a function of slip becomes

$$P_{\text{mech}}(s) = 3\frac{R_2}{s}(1-s)|\tilde{I}_2|^2 = \frac{3\frac{R_2}{s}(1-s)|\tilde{V}_{\text{eq}}|^2}{\left(R_{\text{eq}} + \frac{R_2}{s}\right)^2 + \left(X_{\text{eq}} + X_2\right)^2} \tag{8.38}$$

Taking the derivate of developed power with respect to slip leads to finding the speed at which the developed power is at its maximum level.

$$s_{P_{\max}} = \frac{R_2}{R_2 + \sqrt{\left(R_{\text{eq}} + R_2\right)^2 + \left(X_{\text{eq}} + X_2\right)^2}} \tag{8.39}$$

And also the maximum developed power by the three-phase induction motor is

$$P_{\max} = \frac{\frac{3}{2}|\tilde{V}_{\text{eq}}|^2}{R_2 + R_{\text{eq}} + \sqrt{\left(R_{\text{eq}} + R_2\right)^2 + \left(X_{\text{eq}} + X_2\right)^2}} \tag{8.40}$$

The developed torque by the three-phase induction motor as a function of slip is

$$T(s) = \frac{3\frac{R_2}{s}|\tilde{V}_{\text{eq}}|^2}{\left[\left(R_{\text{eq}} + \frac{R_2}{s}\right)^2 + \left(X_{\text{eq}} + X_2\right)^2\right]\omega_s} \tag{8.41}$$

Figure 8.27 draws the mechanical torque as a function of slip. The slip corresponding with the maximum torque is called the breakdown slip and is described as

$$s_{T_{\max}} = \frac{R_2}{\sqrt{R_{\text{eq}}^2 + \left(X_{\text{eq}} + X_2\right)^2}} \tag{8.42}$$

And the maximum value of the developed torque, which is called breakdown torque, is

$$T_{\max} = \frac{\frac{3}{2}|\tilde{V}_{\text{eq}}|^2}{\left[R_{\text{eq}} + \sqrt{R_{\text{eq}}^2 + \left(X_{\text{eq}} + X_2\right)^2}\right]\omega_s} \tag{8.43}$$

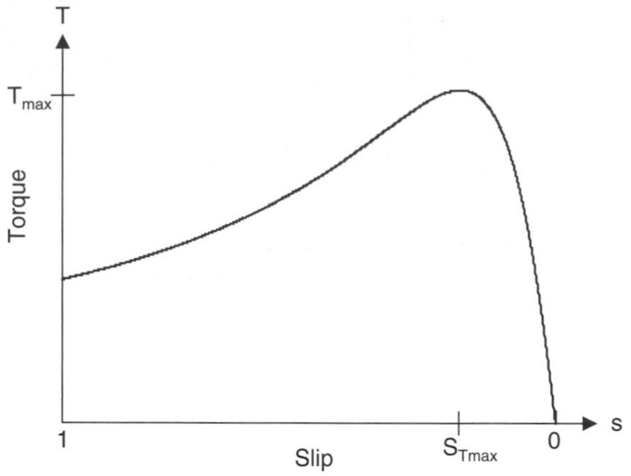

FIGURE 8.27 Torque-slip characteristic of a typical induction motor.

It is important to note that the breakdown torque, unlike the breakdown slip, is not a function of rotor resistance R_2. Using Equation (8.41) at the start-up when the slip is 1, we can find the start-up torque as

$$T_{\text{start-up}} = \frac{3R_2\left|\tilde{V}_{\text{eq}}\right|^2}{\left[\left(R_{\text{eq}} + \frac{R_2}{s}\right)^2 + \left(X_{\text{eq}} + X_2\right)^2\right]\omega_s} \tag{8.44}$$

As Equation (8.44) suggests, larger values of R_2 result in a higher start-up torque while causing more power dissipation in the rotor and reducing the efficiency. Figure 8.28 depicts the torque-speed characteristic of a three-phase induction motor for different values of the rotor resistance. The efficiency of the induction motor can be expressed as

$$\eta = \frac{P_{\text{out}}}{P_{\text{in}}} = \frac{3\frac{R_2}{s}(1-s)|\tilde{I}_2|^2}{3\,\text{Re}\left[\tilde{V}_{\text{eq}}\tilde{I}^*_{\text{eq}}\right]} \tag{8.45}$$

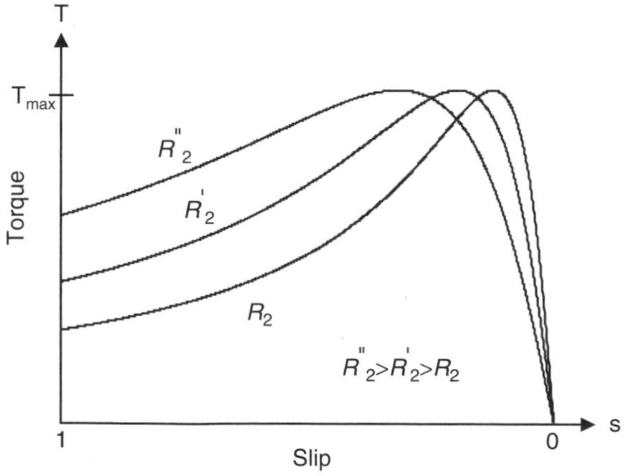

FIGURE 8.28 Effect of rotor resistance on the torque-slip characteristic of an induction motor.

AC and DC Motor Terms

General Terms

ω_m: Shaft angular velocity in radians/second
P_{out}: Mechanical output power
P_{in}: Electrical input power
η: Efficiency
T: Shaft torque

DC Motor Terms

I_a: Armature current
I_f: Field current
E_a: Back emf generated by the armature
V_t: Motor terminal voltage
V_f Field voltage
R_a: Armature resistance
L_a: Armature inductance
R_f: Field resistance
R_{fx}: Field variable resistance
L_f: Field reactance
K_a: Armature constant
K_f: Field constant

AC Synchronous Motor Terms

ω_s: Synchronous speed
V_{an}: Line-to-neutral input voltage
I_a: Line current
E_a: Back emf generated by the rotor
R_a: Per-phase stator circuit resistance
X_s: Per-phase synchronous reactance
Z: Per-phase stator impedance $Z = R_a + jX_s$
δ: Angle between V_{an} and E_a
φ: Angle between E_a and I_a
φ_z: Stator circuit reactance angle $\varphi_z = \tan^{-1} X_s/R_a$
θ: Power factor angle $\theta = \varphi + \delta$

AC Induction Motor Terms

R_1: Per-phase stator winding resistance
X_1: Per-phase stator winding leakage reactance
R_2: Per-phase rotor resistance
X_2: Per-phase rotor leakage reactance
R_c: Per-phase equivalent core loss resistance
X_m: Per-phase magnetizing reactance
P: Number of poles in the stator winding
ω_e: Electrical frequency of input voltage applied to the stator in radians/second
s: Slip $s = (\omega_s - \omega_m)/\omega_s$
T_{max}: Maximum developed torque
s_{Tmax}: Slip at the maximum torque

Defining Terms

DC motor: A dc motor consists of a stationary active part, usually called the field structure, and a moving active part, usually called the armature. Both the field and armature carry dc currents.

Induction motor: An ac motor in which a primary winding on the stator is connected to the power source and polyphase secondary winding on the rotor carries induced current.

Permanent magnet dc motor: A dc motor in which the magnetic field flux is supplied by permanent magnets instead of a wound field.

Rotor: The rotating member of a motor including the shaft. It is commonly called the armature on most dc motors.

Separately excited dc motor: A dc motor in which the field current is derived from a circuit that is independent of the armature circuitry.

Squirrel cage induction motor: An induction motor in which the secondary circuit (on the rotor) consists of bars, short-circuited by end rings. This forms a squirrel cage conductor structure, which is disposed in slots in the rotor core.

Stator: The portion of a motor that includes and supports the stationary active parts. The stator includes the stationary portions of the magnetic circuit and the associated windings and leads.

Synchronous motor: An ac motor in which the average speed of normal operation is exactly proportional to the frequency to which it is connected. A synchronous motor generally has rotating field poles that are excited by dc.

Wound rotor induction motor: An induction motor in which the secondary circuit consists of a polyphase winding or coils connected through a suitable circuit. When provided with slip rings, the term slip-ring induction motor is used.

References

B.S. Guru and H.R. Hiziroglu, *Electric Machinery and Transformers*, 3rd ed., New York: Oxford University Press, 2001.

P.C. Sen, *Thyristor DC Drives*, New York: Wiley, 1981.

P.C. Sen, *Principles of Electric Machines and Power Electronics*, 2nd ed., New York: Wiley, 1997.

G.R. Slemon, *Electric Machines and Drives*, Reading, MA: Addison-Wesley, 1992.

I. Boldea and S.A. Nasar, *Vector Control of AC Drives*, Boca Raton, FL: CRC Press, 1992.

M.G. Say and E.O. Taylor, *Direct Current Machines*, 2nd ed., London: Pitman Publishing, 1986.

R.H. Engelmann and W.H. Middendorf, *Handbook of Electric Motors*, New York: Marcel Dekker, 1995.

D.W. Novotny and T.A. Lipo, *Vector Control and Dynamics of AC Drives*, Oxford: Clarendon Press, 1996.

Further Information

The theory of ac motor drive operation is covered in the collection of papers edited by Bimal K. Bose, Adjustable Speed AC Drive Systems (IEEE, 1981). A good general text is *Electric Machinery*, by Fitzgerald, Kingsley, and Umans. The analysis of synchronous machines is covered in the book *Alternating Current Machines*, by M.G. Say (Wiley, 1984). *Three-Phase Electrical Machines — Computer Simulation* by J.R. Smith (Wiley, 1993) covers computer modeling and simulation techniques.

8.3 Small Electric Motors

Elias G. Strangas

Introduction

Small electrical machines carry a substantial load both in residential and industrial environments, where they are used mainly to control processes. To adapt to power limitations, cost requirements, and widely varying operating requirements, small motors are available in a variety of designs. Some small motors require electronics to start and operate, while others can start and run by directly connecting to the supply line.

Ac motors that can start directly from the line primarily are the induction type. Universal motors also are used extensively for small, ac-powered, handheld tools. They either can run directly from the line or have their speed adjusted through electronics.

Stepping motors of varying designs require electronics to operate. They are used primarily to position a tool or component and seldom are used to provide steady rotating motion.

Besides these motors, permanent-magnet ac motors rapidly are replacing dc and induction motors for accurate speed and position control, but they also decrease size and increase efficiency. They require power and control electronics to start and run.

Single-Phase Induction Motors

To produce rotation, a multi-phase stator winding often is used in an ac motor, supplied from a symmetric and balanced current system. The magnetomotive force of these windings interacts with the magnetic field of the rotor (induced or applied) to produce torque. In three-phase induction motors, the rotor field is created by currents that are induced by the relative speed of the rotor and the synchronously rotating stator field.

In an induction motor supplied by a single-phase stator current, it is not as clear how a rotating magnetomotive force can be created and torque produced. Two different concepts can be used to generate torque.

The first, conceptually simpler design concept involves the generation of a second current flowing in a second winding of the stator. This auxiliary winding is spatially displaced on the stator. This brings the motor design close to the multi-phase principle. The auxiliary winding current must be out of phase with the current in the main winding. This is accomplished through the use of increased resistance in it or through a capacitor in series with it. A motor can operate in this fashion over its entire speed range.

Once the motor is rotating, the second design concept allows the auxiliary phase to be disconnected. The current in the main winding produces only a pulsating flux, which can be analyzed as the sum of two rotating fields of equal amplitude but opposite direction. These fields, as seen from the moving rotor, rotate at different speeds, inducing in it currents of different frequency and amplitude. If the rotor speed is ω_r, the applied frequency to the stator is f_s, and the number of pole pairs in the motor is p; the frequencies of the currents induced in the rotor are $f_s - p\omega$ and $f_s + p\omega$. These unequal currents produce unequal torques in the two directions, leading to a nonzero net torque.

The various designs of single-phase induction motors result from the variety of ways the two phases are generated and by whether the auxiliary phase remains energized after starting.

Shaded-Pole Motors

These motors are simple, reliable, and inefficient. The stator winding is not distributed on the rotor surface, but rather it is concentrated on salient poles. The auxiliary winding, which must produce flux out of phase with the main winding, is nothing but a hard-wired, shorted turn around a portion of the main pole, as shown in Figure 8.29.

Due to the shorted turn, the flux from the shaded part of the pole lags behind the flux from the main pole. The motor always rotates from the main to the shaded pole, and it is not possible to change directions.

Shaded-pole motors are inefficient and have high starting and running currents and low starting torque.

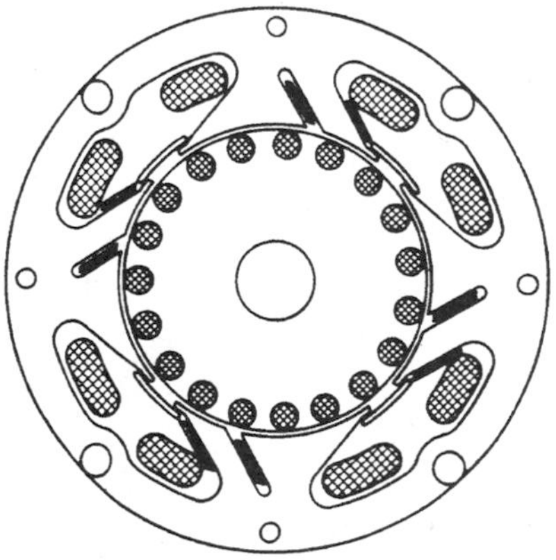

FIGURE 8.29 A shaded-pole motor with tapered poles and magnetic wedges. (*Source:* C.G. Veinott and J.E. Martin, *Fractional and Subfractional Horsepower Electric Motors*, New York: McGraw-Hill, 1986. With permission.)

They are used when reliability and cost are important. Their small size makes the overall effect of their disadvantages, such as small fans, unimportant. Their sizes range from 0.002 to 0.1 hp.

Resistance Split-Phase Motors

These motors have an auxiliary winding that has higher resistance than the main winding and is displaced spatially on the stator by about 90°. Both windings are distributed on the stator surface and are connected to the line voltage. But the time constants between them make the current in the auxiliary winding lead that of the main pole. This arrangement results in a nonzero but relatively low starting torque and high starting current.

The use of the auxiliary winding is limited only to starting; the motor runs more efficiently without it, as a single-phase motor described earlier. A switch, activated by centrifugal speed or by stator temperature, disconnects the auxiliary winding shortly after starting. Figure 8.30 shows a schematic of the connections for this type of motor.

These motors represent an improvement in efficiency and starting torque over shaded-pole motors, at the expense of increased cost and lower reliability. They are built to larger sizes, but their application is limited by high starting current.

Capacitor Motors

Another way to generate a phase angle of current in the auxiliary winding is to include a capacitor in series with it. The capacitor can be disconnected after starting, in a capacitor-start motor. Their operation is similar to that of the resistance split-phase motor, but they have better starting characteristics and can be as large as 5 hp. Figure 8.31 shows schematically the wiring diagram of the capacitor-start motor.

To optimize both starting and running, different capacitor values are used. One capacitor value is calculated to minimize starting current and maximize starting torque, while the other is designed to maximize operating point efficiency. A centrifugal switch handles the changeover. Such motors are built for as much as 10 hp, and their cost is relatively high because of the switch and the two capacitors. Figure 8.32 shows a schematic of the wiring diagram of the capacitor-start and -run motors.

A permanent split-capacitor motor uses the same capacitor throughout the motor's speed range. Its value requires a compromise between the values of the two-capacitor motors. The result is a motor design optimized for a particular application, such as a compressor or a fan. Figure 8.33 shows a schematic of the wiring diagram for the permanent split-capacitor motor.

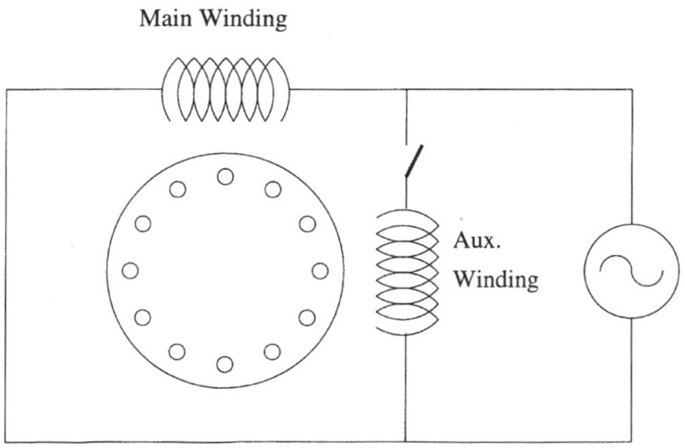

FIGURE 8.30 Connections of a resistive split-phase motor.

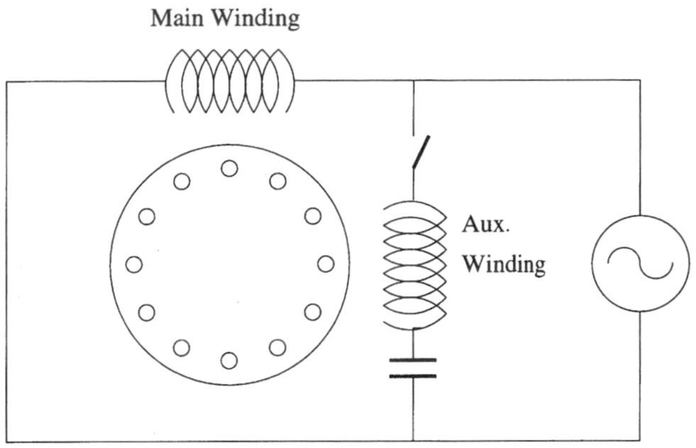

FIGURE 8.31 Connections of a capacitor-start motor.

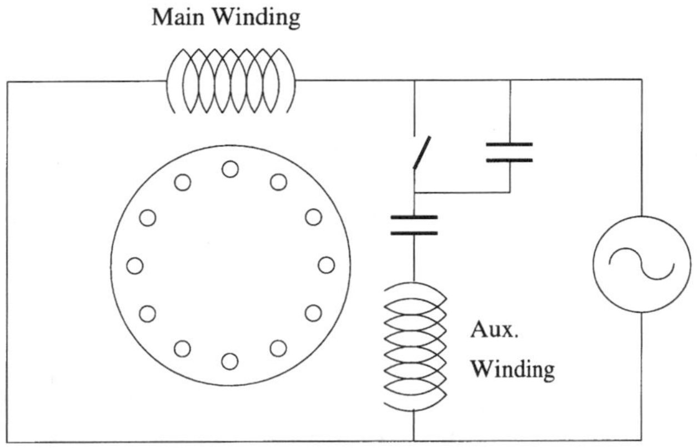

FIGURE 8.32 Connections of a capacitor-start, capacitor-run motor.

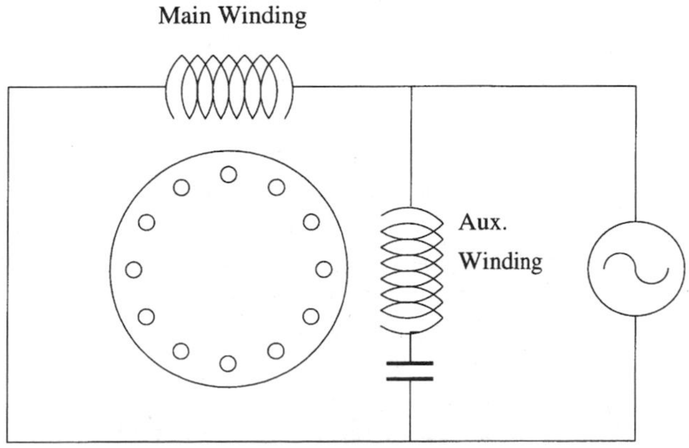

FIGURE 8.33 Connections of a permanent split-capacitor motor.

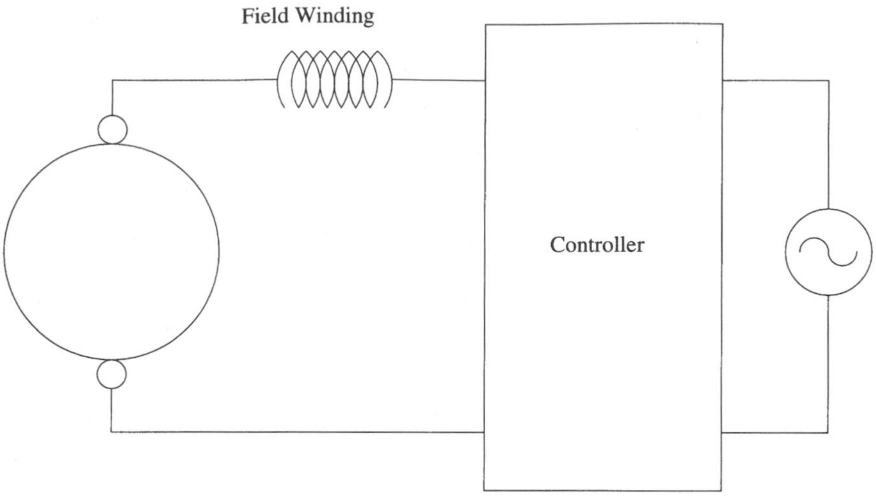

FIGURE 8.34 Connections of a universal motor.

Universal Motors

These motors can be supplied from either dc or ac power. Their design is essentially similar to a dc motor with series windings. When operated as an ac motor, supplied for example by a 60-Hz source, the current in the armature and the field windings reverses 120 times per second. As torque is roughly proportional to armature and field currents, connecting these windings in series guarantees that the current reverses in both at the same time. This retains the unidirectional torque. Figure 8.34 shows a schematic diagram of the connections for universal motors.

These motors can run at speeds as high as 20,000 rpm, a compact speed for a given horsepower. Their most popular applications include portable drills, food mixers, and fans.

Universal motors supplied from ac lend themselves easily to variable-speed applications. A potentiometer, placed across the line voltage, controls the firing of a TRIAC, helping vary the effective value of the voltage at the motor.

Permanent-Magnet AC Motors

When compared to induction motors, permanent-magnet motors offer higher steady-state torque for the same size and better efficiency. They have a polyphase winding in the stator that can be concentrated or distributed. The density distribution turns of these windings can be rectangular or can approximate a sinusoid. The rotor has a steel core, with permanent magnets mounted on its surface or inset. These magnets can be made from a variety of materials, including rare earth, ceramic, or ferrites.

Figure 8.35 and Figure 8.36 show schematics of permanent magnetic ac machines with the d-axis aligned with the direction of rotor magnetization and the q-axis perpendicular to it. The magnets can be mounted on the surface of the rotor as shown in Figure 8.35, or be inset or "buried" in it as shown in Figure 8.36. Figure 8.37 shows a schematic of a distributed winding in the stator, while Figure 8.38 shows a concentrated one. Motors with concentrated windings normally have lower winding losses, as the end turns are shorter. Also, the magnetic field they produce in the air gap is not sinusoidally distributed and can cause higher torque pulsations and eddy currents in the rotor magnets.

The rotors of permanent-magnet ac motors can be designed with a variety of objectives. Rare-earth magnetic materials such as Nd-Fe-B have a high energy density and can lead to compact designs. Ferrites and ceramic magnets have lower costs and lower energy density. On the other hand, rare-earth magnets feature high conductivity, which can cause high eddy-current losses. Since most magnetic materials have permeability

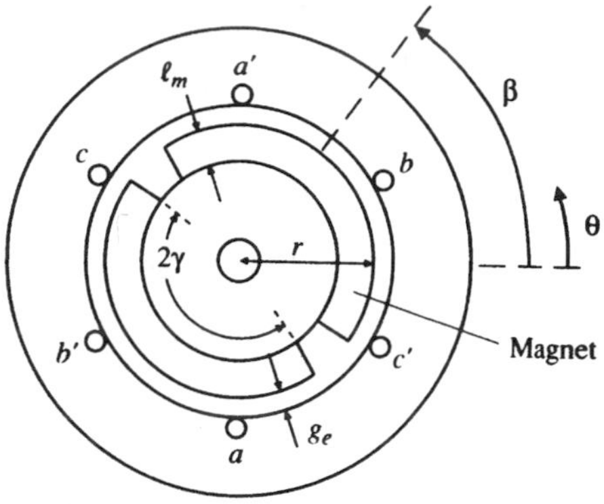

FIGURE 8.35 Surface-mounted magnets on a permanent-magnet ac motor. (*Source*: G.R. Slemon, *Electrical Machines and Drives*, Reading, MA: Addison-Wesley, 1992. With permission.)

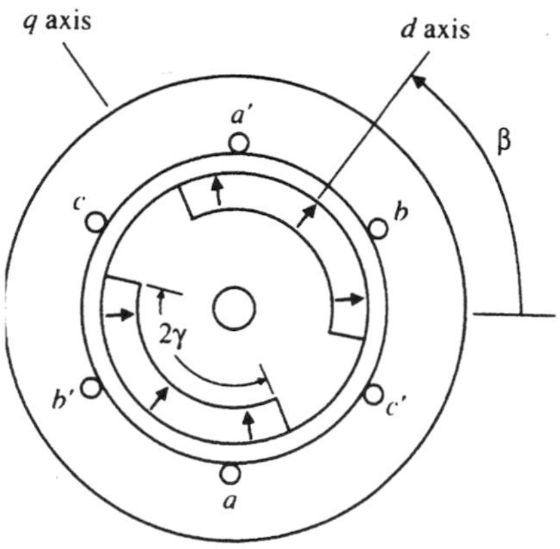

FIGURE 8.36 Inset (interior) magnets on a permanent-magnet ac motor. (*Source*: G.R. Slemon, *Electrical Machines and Drives*, Reading, MA: Addison-Wesley, 1992. With permission.)

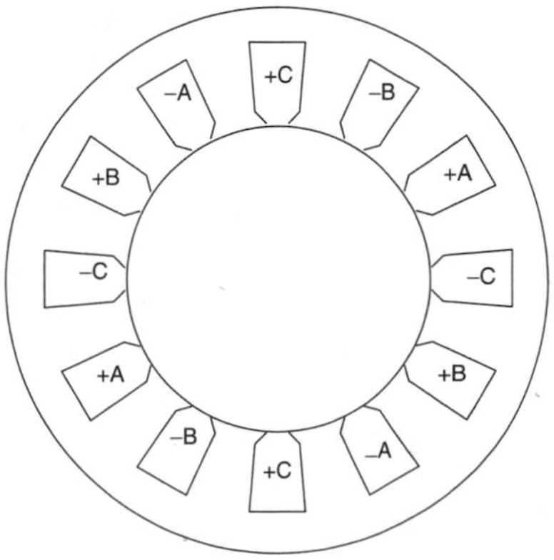

FIGURE 8.37 Permanent-magnet ac stator with distributed windings.

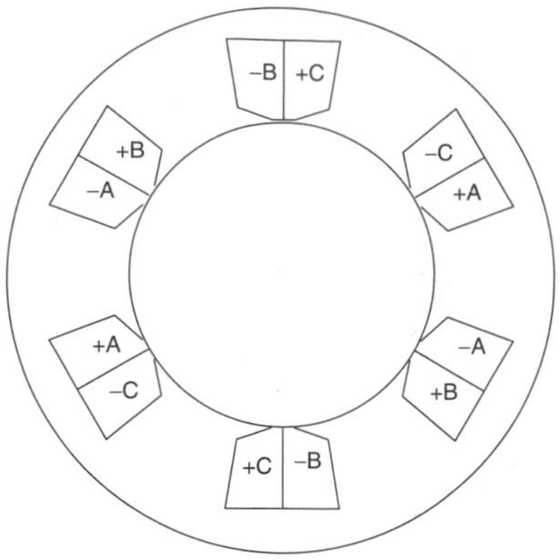

FIGURE 8.38 Permanent-magnet ac stator with concentrated windings.

similar to that of air, two basic topologies result: surface-mounted PMAC motors, as shown in Figure 8.39, and interior-magnet PMAC motors, the basic topology shown in Figure 8.40.

For surface-magnet motors, currents are applied to the stator windings, producing a current vector, $\mathbf{i}_q$, perpendicular to the rotor flux, $\mathbf{B}_m$. The torque then is

$$T = ki_q\lambda_m \tag{8.46}$$

where λ is the stator flux linkage per phase, due to $\mathbf{B}_m$. As the rotor speed, ω, is increased, the voltage, E, induced in the stator windings due to the rotor flux,

$$E = k\omega\lambda_m \tag{8.47}$$

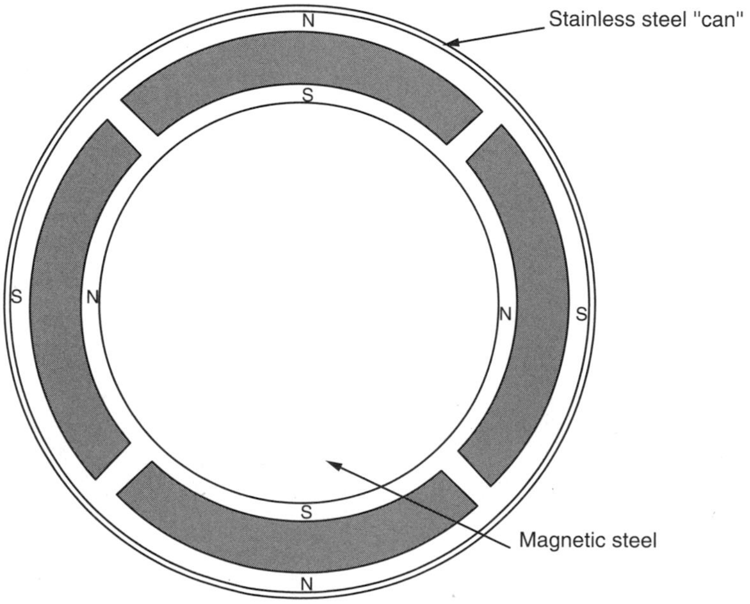

FIGURE 8.39 Permanent-magnet ac rotor with surface-mounted magnets.

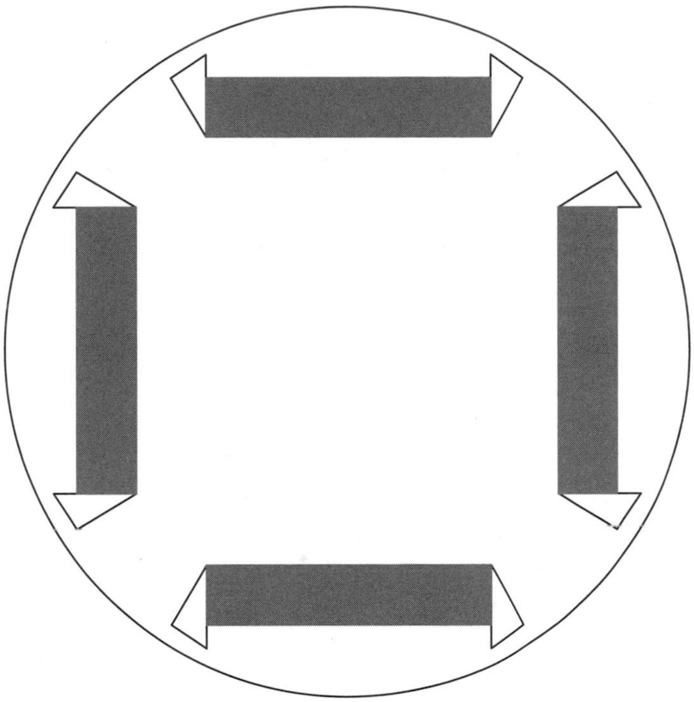

FIGURE 8.40 Permanent-magnet ac rotor with inset magnets.

increases as well. For a given load torque, a base speed ω_B allows the power supply to reach its upper voltage limit and cannot provide the stator voltage required. This prevents the motor from reaching a higher speed. To alleviate this problem, the current in the stator is advanced ahead of the rotor flux, beyond 90°. This essentially demagnetizes the motor and allows operation at higher speed. This region of operation, called field weakening, leads to lower efficiency and decreasing maximum available torque.

In the motor type with magnets buried in the rotor, magnetic permeability and reluctance are not rotationally constant. Although a number of designs can be based on this concept, the developed electromagnetic torque in all of them has two components: one similar to that of the motor with a cylindrical rotor, as described previously and resulting from the interaction of stator currents and rotor magnets, and another resulting from the variation of magnetic reluctance

$$T = k(i_q\lambda_m + (L_d - L_q)i_d i_q) \tag{8.48}$$

Figure 8.41 shows that the stator current space–vector is perpendicular to the rotor flux (indicated by I_F) below a certain speed w_b (base speed) determined by the maximum stator voltage that can be provided by the inverter. To attain speeds above w_b, the stator current has to be advanced beyond 90°, weakening the magnetic field and creating higher winding losses. The stator windings are supplied by a dc source through electronic switches that constitute an inverter. A controller determines which switches are to be closed at any instant to provide the appropriate currents to the stator. This controller uses as inputs a speed or torque command, a measurement of the currents, and a measurement or an estimate of the rotor position.

When stator windings are rectangular and are energized based only on the rotor position, the resulting set of PM motor, inverter, and controller is called a brushless dc motor. The developed torque is proportional to the air-gap flux, B_m, and the stator current, I_s

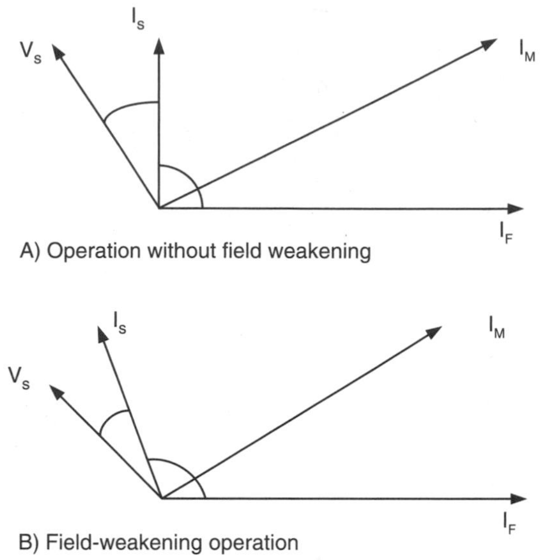

FIGURE 8.41 Permanent-magnet ac space vectors for minimal losses and for field weakening.

$$T = kB_{\mathrm{m}}I_{\mathrm{s}} \tag{8.49}$$

Due to the rotor speed, ω_s, a voltage, E, (back emf) is induced to the stator windings.

$$E = k\omega B_{\mathrm{m}} \tag{8.50}$$

Stepping Motors

Figure 8.42 and Figure 8.43 show the case of a three-phase brushless dc motor with each phase belt covering 60° and the stator magnets covering close to 180°. Only two phases should be energized, as determined by the rotor position. These motors convert a series of power pulses to a corresponding series of equal angular movements. These pulses can be delivered at a variable rate, allowing the accurate positioning of the rotor without feedback. They can develop torque up to 15 Nm and can handle 1500 to 2500 pulses per second. They have zero steady-state error in positioning and high torque density. An important characteristic of stepping motors is that when one phase is activated, they do not develop a rotating torque. Instead, they form a holding torque, helping them accurately retain their position, even under load.

Stepping motors are derived either from a variable-reluctance motor or from a permanent magnet synchronous motor.

One design of stepping motors, based on the doubly salient switched-reluctance motor, uses a large number of teeth in the rotor (typically 45) to create saliency, as shown in Figure 8.44. In this design, when the rotor teeth are aligned in Phase 1, they are misaligned in Phases 2 and 3. A pulse of current in Phase 2 will cause a rotation that leads to alignment. Instead, if a pulse is applied to Phase 3, the rotor will move the same distance in the opposite rotation.

The angle corresponding to a pulse is small, typically 3 to 5°. This results from alternatively exciting one stator phase at a time.

A permanent-magnet stepping motor uses permanent magnets in the rotor. Figure 8.45 shows the steps in the motion of a four-phase PM stepping motor.

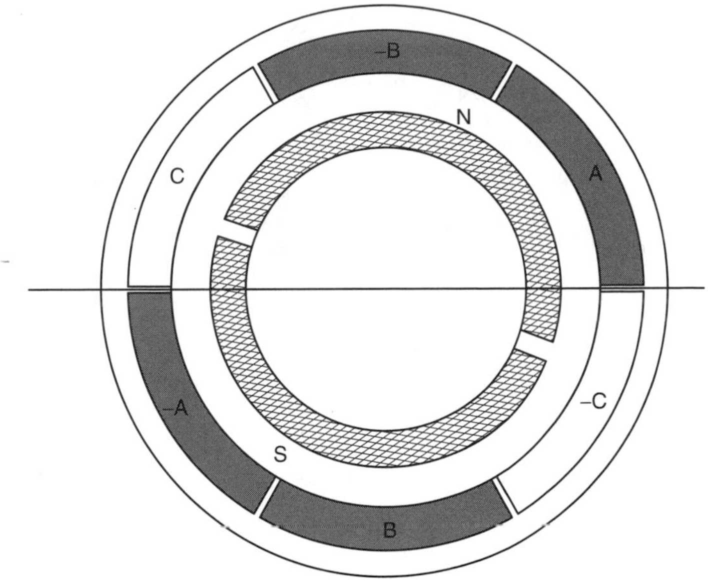

FIGURE 8.42 Brushless dc motor with energized windings A and B.

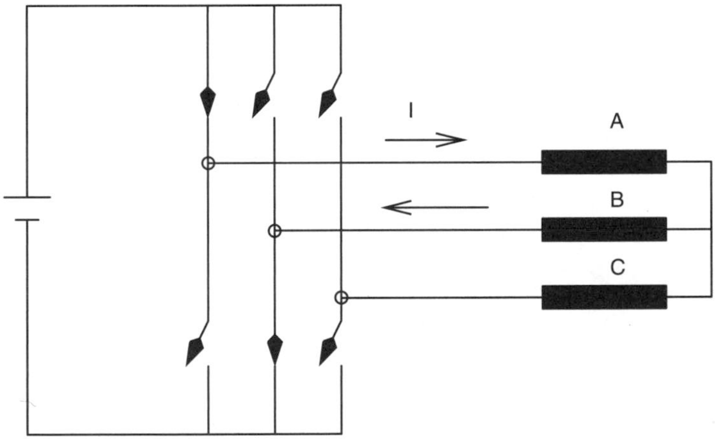

FIGURE 8.43 Inverter for the brushless dc with energized windings A and B.

Hybrid stepping motors come in many designs. One, shown in Figure 8.46, consists of two rotors mounted on the same shaft, displaced by a half-tooth. The permanent magnet is placed axially between the rotors, and the magnetic flux flows radially at the air gaps. This closes through the stator circuit. Torque is created by the interaction of two magnetic fields coming from the magnets and the stator currents. This design allows a finer step-angle control and higher torque, as well as smoother torque during a step.

A fundamental of the stepping-motor operation is the utilization of power electronic switches and of a circuit providing the timing and duration of the pulses. A specific stepping motor can operate at maximum frequency while starting or running without load. As the frequency of the pulses to a running motor is increased, the motor eventually loses synchronism. The relation between the frictional load torque and maximum pulse frequency is called the pull-out characteristic.

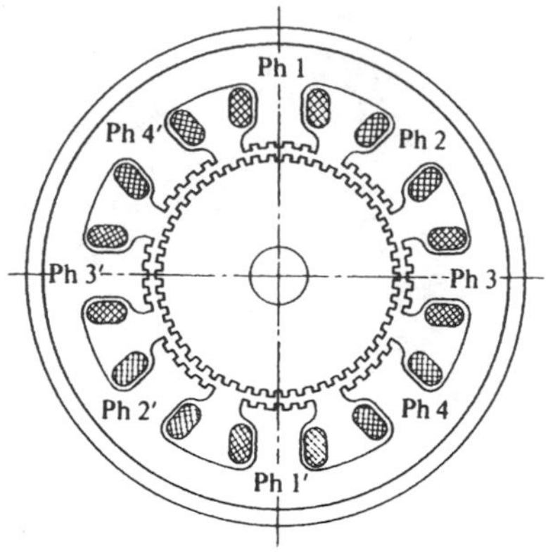

FIGURE 8.44 Cross-sectional view of a four-phase variable-reluctance motor. Number of rotor teeth 50, step number 200, step angle 1.8. (*Source*: Miller, T.J.E., *Brushless Permanent-Magnet and Reluctance Motor Drives*, Oxford: Oxford University Press, 1989. With permission.)

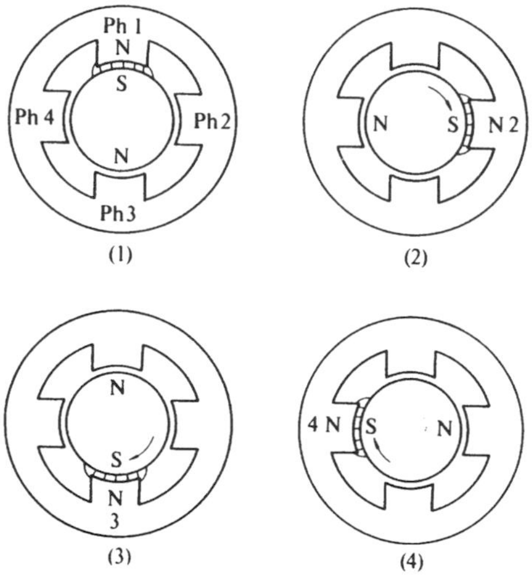

FIGURE 8.45 Steps in the operation of a permanent-magnet stepping motor. (*Source*: T. Kenjo, *Stepping Motors and Their Microprocessor Controls*, Oxford: Oxford University Press, 1989. With permission.)

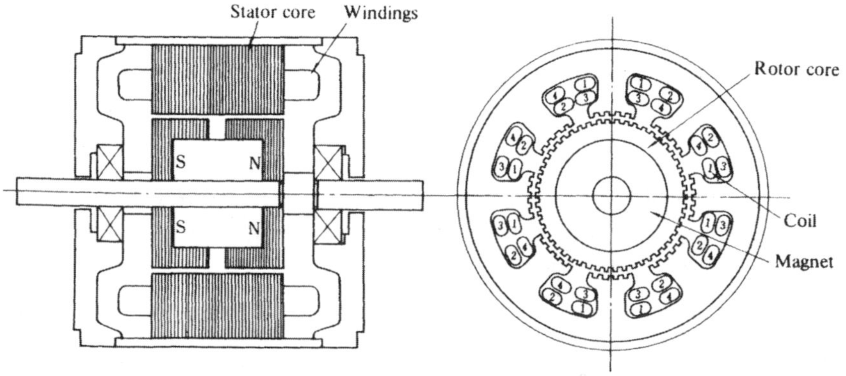

FIGURE 8.46 Construction of a hybrid stepping motor. (*Source*: Miller, T.J.E., *Brushless Permanent-Magnet and Reluctance Motor Drives*, Oxford: Oxford University Press, 1989. With permission.)

References

T. Kenjo, *Stepping Motors and Their Microprocessor Controls*, Oxford: Oxford University Press, 1984.

T.J.E. Miller, *Brushless Permanent-Magnet and Reluctance Motor Drives*, Oxford: Oxford University Press, 1989.

R. Miller and M.R. Miller, *Fractional Horsepower Electric Motors*, New York: Bobs Merrill Co., 1984.

S.A. Nasar, I. Boldea, and L.E. Unnewehr, *Permanent Magnet, Reluctance and Self-Synchronous Motors*, Boca Raton, FL: CRC Press, 1993.

G.R. Slemon, *Electrical Machines and Drives*, Reading, MA: Addison-Wesley, 1992.

C.G. Veinott and J.E. Martin, *Fractional and Subfractional Horsepower Electric Motors*, New York: McGraw-Hill, 1986.

Further Information

There is an abundance of books and literature on small electrical motors. *IEEE Transactions on Industry Applications*, *Power Electronics*, *Power Delivery*, and *Industrial Electronics* all have articles on the subject. In addition, *IEEE* and other publications and conference records can provide the reader with specific and useful information.

Electrical Machines and Drives (Slemon, 1992) is one of the many excellent textbooks on the subject. *Stepping Motors and their Microprocessor Controls* (Kenjo, 1984) offers a thorough discussion of stepping motors, while *Fractional and Subfractional Horsepower Electric Motors* (Veinott and Martin, 1986) covers small ac and dc motors. *Brushless Permanent-Magnet and Reluctance-Motor Drives* (Miller, 1989) and *Permanent Magnet, Reluctance and Self-Synchronous Motors* (Nasar et al., 1993) reflect the increased interest in reluctance and brushless dc motors and provide information on their theory of operation, design, and control. *The Handbook of Electrical Motors* (Engelman and Middendorf, 1995) and *Fractional Horsepower Electric Motors* (Miller and Miller, 1984) offer practical information about the application of small motors.

9

Energy Management

K. Neil Stanton
Stanton Associates

Jay C. Giri
AREVA T&D Corporation

Anjan Bose
Washington State University

9.1 Introduction

Energy management is the process of monitoring, coordinating, and controlling the generation, transmission, and distribution of electrical energy. The physical plant to be managed includes generating plants that produce energy fed through transformers to the high-voltage transmission network (grid), interconnecting generating plants and load centers. Transmission lines terminate at substations that perform switching, voltage transformation, measurement, and control. Substations at load centers transform to subtransmission and distribution levels. These lower-voltage circuits typically operate radially, i.e., no normally closed paths between substations through subtransmission or distribution circuits. (Underground cable networks in large cities are an exception.)

Since transmission systems provide negligible energy storage, supply and demand must be balanced by either generation or load. Production is controlled by turbine governors at generating plants, and automatic generation control is performed by control center computers remote from generating plants. Load management, sometimes called demand-side management, extends remote supervision and control to subtransmission and distribution circuits, including control of residential, commercial, and industrial loads.

Events such as lightning strikes, short circuits, equipment failure, or accidents may cause a system fault. Protective relays actuate rapid, local control through operation of circuit breakers before operators can respond. The goal is to maximize safety, minimize damage, and continue to supply load with the least inconvenience to customers. Data acquisition provides operators and computer control systems with status and measurement information needed to supervise overall operations. **Security** control analyzes the consequences of faults to establish operating conditions that are both robust and economical.

Energy management is performed at control centers, typically called system control centers, by computer systems called *energy management systems* (EMS). Data acquisition and remote control is performed by computer systems called *supervisory control and data acquisition* (SCADA) systems. These latter systems may

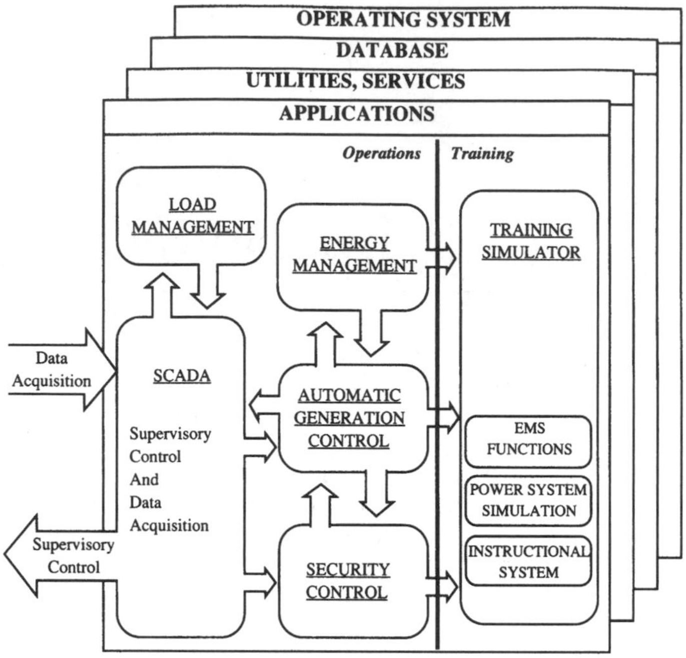

FIGURE 9.1 Layers of a modern EMS.

be installed at a variety of sites including system control centers. An EMS typically includes a SCADA "front-end" through which it communicates with generating plants, substations, and other remote devices.

Figure 9.1 illustrates the **applications** layer of modern EMS as well as the underlying layers on which it is built: the operating system, a database manager, and a utilities/services layer.

9.2 Power System Data Acquisition and Control

A SCADA system consists of a master station that communicates with **remote terminal units** (RTUs) for the purpose of allowing operators to observe and control physical plants. Generating plants and transmission substations certainly justify RTUs, and their installation is becoming more common in distribution substations as costs decrease. RTUs transmit device status and measurements to, and receive control commands and setpoint data from, the master station. Communication is generally via dedicated circuits operating in the range of 600 to 4800 bits/s, with the RTU responding to periodic requests initiated from the master station (polling) every 2 to 10 s, depending on the criticality of the data.

The traditional functions of SCADA systems are summarized:

- Data acquisition: Provides telemetered measurements and status information to operator.
- Supervisory control: Allows operator to remotely control devices, e.g., open and close circuit breakers. A "select before operate" procedure is used for greater safety.
- Tagging: Identifies a device as subject to specific operating restrictions and prevents unauthorized operation.
- Alarms: Informs operator of unplanned events and undesirable operating conditions. Alarms are sorted by criticality, area of responsibility, and chronology. Acknowledgment may be required.
- Logging: Logs all operator entry, all alarms, and selected information.
- Load shed: Provides both automatic and operator-initiated tripping of load in response to system emergencies.
- Trending: Plots measurements on selected time scales.

Since the master station is critical to power system operations, its functions are generally distributed among several computer systems, depending on specific design. A dual computer system configured in primary and standby modes is most common. SCADA functions are listed below without stating which computer has specific responsibility.

- Manage communication circuit configuration
- Downline load RTU files
- Maintain scan tables and perform polling
- Check and correct message errors
- Convert to engineering units
- Detect status and measurement changes
- Monitor abnormal and out-of-limit conditions
- Log and time-tag sequence of events
- Detect and annunciate alarms
- Respond to operator requests to:
  - Display information
  - Enter data
  - Execute control action
  - Acknowledge alarms
- Transmit control action to RTUs
- Inhibit unauthorized actions
- Maintain historical files
- Log events and prepare reports
- Perform load shedding

9.3 Automatic Generation Control

Automatic generation control (AGC) consists of two major and several minor functions that operate on-line in real time to adjust the generation against load at minimum cost. The major functions are load frequency control and economic **dispatch,** each of which is described below. The minor functions are reserve monitoring, which assures enough reserve on the system, **interchange** scheduling, which initiates and completes scheduled interchanges, and other similar monitoring and recording functions.

Load Frequency Control

Load frequency control (LFC) has to achieve three primary objectives, which are stated below in priority order:

1. To maintain frequency at the scheduled value
2. To maintain net power interchanges with neighboring control areas at the scheduled values
3. To maintain power allocation among units at economically desired values

The first and second objectives are met by monitoring an error signal, called *area control error* (ACE), which is a combination of net interchange error and frequency error and represents the power imbalance between generation and load at any instant. This ACE must be filtered or smoothed such that excessive and random changes in ACE are not translated into control action. Since these excessive changes are different for different systems, the filter parameters have to be tuned specifically for each control area. The filtered ACE is then used to obtain the proportional plus integral control signal. This control signal is modified by limiters, deadbands, and gain constants that are tuned to the particular system. This control signal is then divided among the generating units under control by using participation factors to obtain *unit control errors* (UCE).

These participation factors may be proportional to the inverse of the second derivative of the cost of unit generation so that the units would be loaded according to their costs, thus meeting the third objective. However, cost may not be the only consideration because the different units may have different response rates, and it may be necessary to move the faster generators more to obtain an acceptable response. The UCEs are then sent to the various units under control and the generating units monitored to see that the corrections take place. This control action is repeated every 2 to 6 s.

In spite of the integral control, errors in frequency and net interchange do tend to accumulate over time. These time errors and accumulated interchange errors have to be corrected by adjusting the controller settings according to procedures agreed upon by the whole interconnection. These accumulated errors as well as ACE serve as performance measures for LFC.

The main philosophy in the design of LFC is that each system should follow its own load very closely during normal operation, while during emergencies each system should contribute according to its relative size in the interconnection without regard to the locality of the emergency. Thus, the most important factor in obtaining good control of a system is its inherent capability of following its own load. This is guaranteed if the system has adequate regulation margin as well as adequate response capability. Systems that have mainly thermal generation often have difficulty in keeping up with the load because of the slow response of the units.

The design of the controller itself is an important factor, and proper tuning of the controller parameters is needed to obtain "good" control without "excessive" movement of units. Tuning is system-specific, and although system simulations are often used as aids, most of the parameter adjustments are made in the field using heuristic procedures.

Economic Dispatch

Since all the generating units that are on-line have different costs of generation, it is necessary to find the generation levels of each of these units that would meet the load at the minimum cost. This has to take into account the fact that the cost of generation in one generator is not proportional to its generation level but is a nonlinear function of it. In addition, since the system is geographically spread out, the transmission losses are dependent on the generation pattern and must be considered in obtaining the optimum pattern.

Certain other factors have to be considered when obtaining the optimum generation pattern. One is that the generation pattern provide adequate reserve margins. This is often done by constraining the generation level to a lower boundary than the generating capability. A more difficult set of constraints to consider are the transmission limits. Under certain real-time conditions it is possible that the most economic pattern may not be feasible because of unacceptable line flows or voltage conditions. The present-day economic dispatch (ED) algorithm cannot handle these security constraints. However, alternative methods based on optimal power flows have been suggested but have not yet been used for real-time dispatch.

The minimum cost dispatch occurs when the incremental cost of all the generators is equal. The cost functions of the generators are nonlinear and discontinuous. For the equal marginal cost algorithm to work, it is necessary for them to be convex. These incremental cost curves are often represented as monotonically increasing piecewise-linear functions. A binary search for the optimal marginal cost is conducted by summing all the generation at a certain marginal cost and comparing it with the total power demand. If the demand is higher, a higher marginal cost is needed, and vice versa. This algorithm produces the ideal setpoints for all the generators for that particular demand, and this calculation is done every few minutes as the demand changes.

The losses in the power system are a function of the generation pattern, and they are taken into account by multiplying the generator incremental costs by the appropriate penalty factors. The penalty factor for each generator is a reflection of the sensitivity of that generator to system losses, and these sensitivities can be obtained from the transmission loss factors (Section 9.6).

This ED algorithm generally applies to only thermal generation units that have cost characteristics of the type discussed here. The hydro units have to be dispatched with different considerations. Although there is no cost for the water, the amount of water available is limited over a period, and the displacement of fossil fuel by this water determines its worth. Thus, if the water usage limitation over a period is known, say from a previously computed hydro optimization, the water worth can be used to dispatch the hydro units.

LFC and the ED functions both operate automatically in real time but with vastly different time periods. Both adjust generation levels, but LFC does it every few seconds to follow the load variation, while ED does it every few minutes to assure minimal cost. Conflicting control action is avoided by coordinating the control errors. If the unit control errors from LFC and ED are in the same direction, there is no conflict. Otherwise, a logic is set to either follow load (permissive control) or follow economics (mandatory control).

Reserve Monitoring

Maintaining enough reserve capacity is required in case generation is lost. Explicit formulas are followed to determine the spinning (already synchronized) and ready (10 min) reserves required. The availability can be assured by the operator manually, or, as mentioned previously, the ED can also reduce the upper dispatchable limits of the generators to keep such generation available.

Interchange Transaction Scheduling

The contractual exchange of power between utilities has to be taken into account by the LFC and ED functions. This is done by calculating the net interchange (sum of all the buy and sale agreements) and adding this to the generation needed in both the LFC and ED. Since most interchanges begin and end on the hour, the net interchange is ramped from one level to the new over a 10- or 20-min period straddling the hour. The programs achieve this automatically from the list of scheduled transactions.

9.4 Distribution System Management

SCADA, with its relatively expensive RTUs installed at distribution substations, can provide status and measurements for distribution feeders at the substation. Distribution automation equipment is now available to measure and control at locations dispersed along distribution circuits. This equipment can monitor sectionalizing devices (switches, interruptors, fuses), operate switches for circuit reconfiguration, control voltage, read customers' meters, implement time-dependent pricing (on-peak, off-peak rates), and switch customer equipment to manage load. This equipment requires significantly increased functionality at distribution control centers.

Distribution control center functionality varies widely from company to company, and the following list is evolving rapidly.

- Data acquisition: Acquires data and gives the operator control over specific devices in the field. Includes data processing, quality checking, and storage.
- Feeder switch control: Provides remote control of feeder switches.
- Tagging and alarms: Provides features similar to SCADA.
- Diagrams and maps: Retrieves and displays distribution maps and drawings. Supports device selection from these displays. Overlays telemetered and operator-entered data on displays.
- Preparation of switching orders: Provides templates and information to facilitate preparation of instructions necessary to disconnect, isolate, reconnect, and reenergize equipment.
- Switching instructions: Guides operator through execution of previously prepared switching orders.
- Trouble analysis: Correlates data sources to assess scope of trouble reports and possible dispatch of work crews.
- Fault location: Analyzes available information to determine scope and location of fault.
- Service restoration: Determines the combination of remote control actions which will maximize restoration of service. Assists operator to dispatch work crews.
- Circuit continuity analysis: Analyzes circuit topology and device status to show electrically connected circuit segments (either energized or deenergized).
- Power factor and voltage control: Combines substation and feeder data with predetermined operating parameters to control distribution circuit power factor and voltage levels.
- Electrical circuit analysis: Performs circuit analysis, single-phase or three-phase, balanced or unbalanced.

- Load management: Controls customer loads directly through appliance switching (e.g., water heaters) and indirectly through voltage control.
- Meter reading: Reads customers' meters for billing, peak demand studies, time of use tariffs. Provides remote connect/disconnect.

9.5 Energy Management

Generation control and ED minimize the current cost of energy production and transmission within the range of available controls. Energy management is a supervisory layer responsible for economically scheduling production and transmission on a global basis and over time intervals consistent with cost optimization. For example, water stored in reservoirs of hydro plants is a resource that may be more valuable in the future and should, therefore, not be used now even though the cost of hydro energy is currently lower than thermal generation. The global consideration arises from the ability to buy and sell energy through the interconnected power system; it may be more economical to buy than to produce from plants under direct control. Energy accounting processes transaction information and energy measurements recorded during actual operation as the basis of payment for energy sales and purchases.

Energy management includes the following functions:

- System load forecast: Forecasts system energy demand each hour for a specified forecast period of 1 to 7 days.
- Unit commitment: Determines start-up and shut-down times for most economical operation of thermal generating units for each hour of a specified period of 1 to 7 days.
- Fuel scheduling: Determines the most economical choice of fuel consistent with plant requirements, fuel purchase contracts, and stockpiled fuel.
- Hydro-thermal scheduling: Determines the optimum schedule of thermal and hydro energy production for each hour of a study period up to 7 days while ensuring that hydro and thermal constraints are not violated.
- Transaction evaluation: Determines the optimal incremental and production costs for exchange (purchase and sale) of additional blocks of energy with neighboring companies.
- Transmission loss minimization: Recommends controller actions to be taken in order to minimize overall power system network losses.
- Security-constrained dispatch: Determines optimal outputs of generating units to minimize production cost while ensuring that a network security constraint is not violated.
- Production cost calculation: Calculates actual and economical production costs for each generating unit on an hourly basis.

9.6 Security Control

Power systems are designed to survive all probable contingencies. A contingency is defined as an event that causes one or more important components such as transmission lines, generators, and transformers to be unexpectedly removed from service. Survival means the system stabilizes and continues to operate at acceptable voltage and frequency levels without loss of load. Operations must deal with a vast number of possible conditions experienced by the system, many of which are not anticipated in planning. Instead of dealing with the impossible task of analyzing all possible system states, security control starts with a specific state: the current state if executing the real-time network sequence; a postulated state if executing a study sequence. Sequence means sequential execution of programs that perform the following steps:

1. Determine the state of the system based on either current or postulated conditions.
2. Process a list of contingencies to determine the consequences of each contingency on the system in its specified state.
3. Determine preventive or corrective action for those contingencies that represent unacceptable risk.

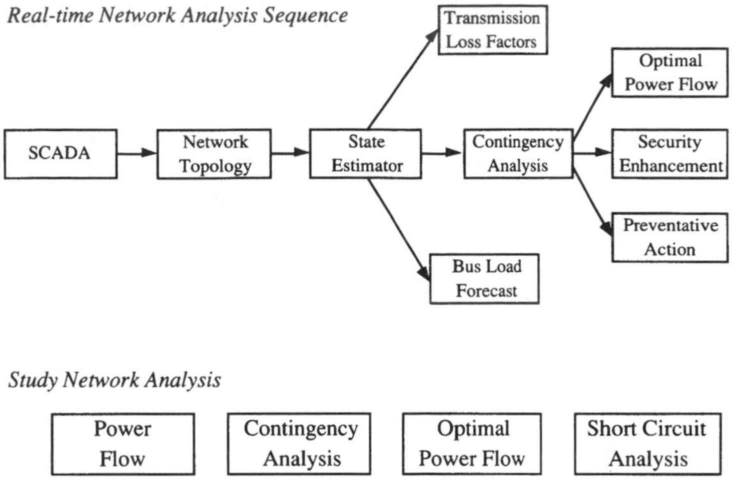

FIGURE 9.2 Real-time and study network analysis sequences.

Real-time and study network analysis sequences are diagramed in Figure 9.2.

Security control requires topological processing to build network models and uses large-scale ac network analysis to determine system conditions. The required applications are grouped as a network subsystem that typically includes the following functions:

- Topology processor: Processes real-time status measurements to determine an electrical connectivity (bus) model of the power system network.
- State estimator: Uses real-time status and analog measurements to determine the "best" estimate of the state of the power system. It uses a redundant set of measurements; calculates voltages, phase angles, and power flows for all components in the system; and reports overload conditions.
- Power flow: Determines the steady-state conditions of the power system network for a specified generation and load pattern. Calculates voltages, phase angles, and flows across the entire system.
- Contingency analysis: Assesses the impact of a set of contingencies on the state of the power system and identifies potentially harmful contingencies that cause operating limit violations.
- Optimal power flow: Recommends controller actions to optimize a specified objective function (such as system operating cost or losses) subject to a set of power system operating constraints.
- Security enhancement: Recommends corrective control actions to be taken to alleviate an existing or potential overload in the system while ensuring minimal operational cost.
- Preventive action: Recommends control actions to be taken in a "preventive" mode before a contingency occurs to preclude an overload situation if the contingency were to occur.
- Bus load forecasting: Uses real-time measurements to adaptively forecast loads for the electrical connectivity (bus) model of the power system network.
- Transmission loss factors: Determines incremental loss sensitivities for generating units; calculates the impact on losses if the output of a unit were to be increased by 1 MW.
- Short-circuit analysis: Determines fault currents for single-phase and three-phase faults for fault locations across the entire power system network.

9.7 Operator Training Simulator

Training simulators were originally created as generic systems for introducing operators to the electrical and dynamic behavior of power systems. Today, they model actual power systems with reasonable fidelity and are integrated with EMS to provide a realistic environment for operators and dispatchers to practice normal, everyday operating tasks and procedures as well as experience emergency operating situations. The various training activities can be safely and conveniently practiced with the simulator responding in a manner similar to the actual power system.

An operator training simulator (OTS) can be used in an investigatory manner to recreate past actual operational scenarios and to formulate system restoration procedures. Scenarios can be created, saved, and reused. The OTS can be used to evaluate the functionality and performance of new real-time EMS functions and also for tuning AGC in an off-line, secure environment.

The OTS has three main subsystems (Figure 9.3).

Energy Control System

The energy control system (ECS) emulates normal EMS functions and is the only part of the OTS with which the trainee interacts. It consists of the supervisory control and data acquisition (SCADA) system, generation control system, and all other EMS functions.

Power System Dynamic Simulation

This subsystem simulates the dynamic behavior of the power system. System frequency is simulated using the "long-term dynamics" system model, where frequency of all units is assumed to be the same. The prime-mover dynamics are represented by models of the units, turbines, governors, boilers, and boiler auxiliaries. The network flows and states (bus voltages and angles, topology, transformer taps, etc.) are calculated at periodic intervals. Relays are modeled, and they emulate the behavior of the actual devices in the field.

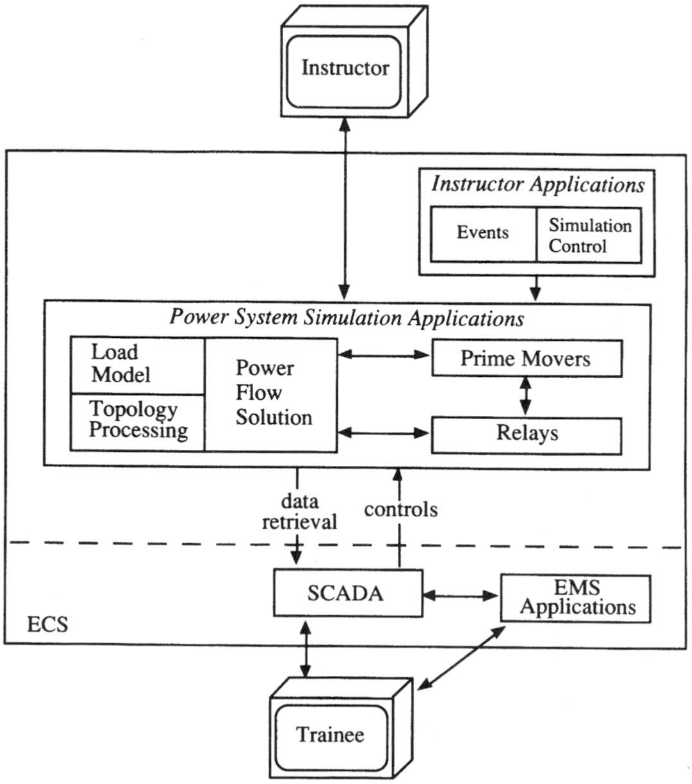

FIGURE 9.3 OTS block diagram.

Instructional System

This subsystem includes the capabilities to start, stop, restart, and control the simulation. It also includes making savecases, retrieving savecases, reinitializing to a new time, and initializing to a specific real-time situation.

It is also used to define event schedules. Events are associated with both the power system simulation and the ECS functions. Events may be deterministic (occur at a predefined time), conditional (based on a predefined set of power system conditions being met), or probabilistic (occur at random).

Defining Terms

Application: A software function within the energy management system that allows the operator to perform a specific set of tasks to meet a specific set of objectives.

Dispatch: The allocation of generation requirement to the various generating units that are available.

Distribution system: That part of the power system network that is connected to, and responsible for, the final delivery of power to the customer; typically the part of the network that operates at 33 kV and below, to 120 V.

Interchange or transaction: A negotiated purchase or sale of power between two companies.

Remote terminal unit (RTU): Hardware that gathers systemwide real-time data from various locations within substations and generating plants for telemetry to the energy management system.

Security: The ability of the power system to sustain and survive planned and unplanned events without violating operational constraints.

References

Application of Optimization Methods for Economy/Security Functions in Power System Operations, IEEE tutorial course, IEEE Publication 90EH0328–5-PWR, 1990.

Distribution Automation, IEEE Power Engineering Society, IEEE Publication EH0280–8-PBM, 1988.

C.J. Erickson, *Handbook of Electrical Heating,* IEEE Press, 1995.

Energy Control Center Design, IEEE tutorial course, IEEE Publication 77 TU0010–9 PWR, 1977.

Fundamentals of Load Management, IEEE Power Engineering Society, IEEE Publication EH0289–9-PBM, 1988.

Fundamentals of Supervisory Controls, IEEE tutorial course, IEEE Publication 91 EH0337–6 PWR, 1991.

M. Kleinpeter, *Energy Planning and Policy,* New York: Wiley, 1995.

"Special issue on computers in power system operations," *Proc. IEEE,* vol. 75, no. 12, 1987.

W.C. Turner, *Energy Management Handbook,* Lilburn, GA: Fairmont Press, 1997.

Further Information

Current innovations and applications of new technologies and algorithms are presented in the following publications:

- *IEEE Power Engineering Review* (monthly)
- *IEEE Transactions on Power Systems* (bimonthly)
- *Proceedings of the Power Industry Computer Application Conference* (biannual)
- *Power Systems Computational Conference* (biannual)
- *North American Electric Reliability Council* (NEXC) Web site: www.nerc.com

10

Power System Analysis Software

C.P. Arnold
University of Canterbury

N.R. Watson
University of Canterbury

10.1 Introduction

Power system software can be grouped in many different ways, e.g., functionality, computer platform, etc., but here it is grouped by end user. There are four major groups of end users for the software:

- Major utilities
- Small utilities and industry consumers of electricity
- Consultants
- Universities

Large comprehensive program packages are required by utilities. They are complex, with many different functions and must have very easy input/output (IO). They serve the needs of a single electrical system and may be tailor-made for the customer. They can be integrated with the electrical system using supervisory control and data acquisition (SCADA). It is not within the scope of this chapter to discuss the merits of these programs. Suffice to say that the component programs used in these packages usually have the same generic/development roots as the programs used by the other three end-user groups.

The programs used by the other three groups have usually been initially created in the universities. They start life as research programs and later are used for teaching and consultancy programs. Where the consultant is also an academic, the programs may well retain their crude research style IO. However, if they are to be used by others who are not so familiar with the algorithms, then usually they are modified to make them more user-friendly. Once this is achieved, the programs become commercial and are used by consultants, industry, and utilities. These are the types of programs that are now so commonly seen in the engineering journals quite often bundled together in a generic package.

10.2 Early Analysis Programs

Two of the earliest programs to be developed for power system analysis were the fault and load-flow (power flow) programs. Both were originally produced in the late 1950s. Many programs in use today are either based on these two types of program or have one or the other embedded in them.

Load Flow (Power Flow)

The need to know the flow patterns and voltage profiles in a network was the driving force behind the development of load-flow programs.

Although the network is linear, load-flow analysis is iterative because of nodal (busbar) constraints. At most busbars the active and reactive powers being delivered to customers are known but the voltage level is not. As far as the load-flow analysis is concerned, these busbars are referred to as PQ buses. The generators are scheduled to deliver a specific active power to the system, and usually the voltage magnitude of the generator terminals is fixed by automatic voltage regulation. These busbars are known as PV buses.

As losses in the system cannot be determined before the load-flow solution, one generator busbar only has its voltage magnitude specified. In order to give the required two specifications per node, this bus also has its voltage angle defined to some arbitrary value, usually zero. This busbar is known as the slack bus. The slack bus is a mathematical requirement for the program and has no exact equivalent in reality. However, in operating practice, the total load plus the losses are not known. When a system is not in power balance, i.e., when the input power does not equal the load power plus losses, the imbalance modifies the rotational energy stored in the system. The system frequency thus rises if the input power is too large and falls if the input power is too little. Usually a generating station and probably one machine are given the task of keeping the frequency constant by varying the input power. This control of the power entering a node can be seen to be similar to the slack bus.

The algorithms first adopted had the advantages of simple programming and minimum storage, but were slow to converge, requiring many iterations. The introduction of ordered elimination, which gives implicit inversion of the network matrix, and sparsity programming techniques, which reduces storage requirements, allowed much better algorithms to be used. The Newton–Raphson method gave convergence to the solution in only a few iterations. Using Newtonian methods of specifying the problem, a Jacobian matrix containing the partial derivatives of the system at each node can be constructed. The solution by this method has quadratic convergence. This method was followed quite quickly by the fast decoupled Newton–Raphson method. This exploited the fact that under normal operating conditions, and provided that the network is predominately reactive, the voltage angles are not affected by reactive power flow, and voltage magnitudes are not affected by real power flow. The fast decoupled method requires more iterations to converge but each iteration uses less computational effort than the Newton–Raphson method. A further advantage of this method is the robustness of the algorithm. However the emergence of FACTS (flexible ac transmission systems) devices has led to renewed interest in the full Newton–Raphson, as the decoupling of voltage magnitudes from real power flow and voltage angles from reactive power flow is no longer valid with these devices.

Further refinements can be added to a load-flow program to make it give more realistic results. Transformer onload tap changers, voltage limits, active and reactive power limits, plus control of the voltage magnitudes at buses other than the local bus help to bring the results close to reality. Application of these limits can slow down convergence.

The problem of obtaining an accurate load-flow solution with a guaranteed and fast convergence has resulted in more technical papers than any other analysis topic. This is understandable when it is realized that the load-flow solution is required during the running of many other types of power system analyses. While improvements have been made, there has been no major breakthrough in performance. It is doubtful if such an achievement is possible, as the time required to prepare the data and process the results represents a significant part of the overall time of the analysis.

Fault Analysis

A fault analysis program derives from the need to adequately rate switchgear and other busbar equipment for the maximum possible fault current that could flow through them.

Initially only three-phase faults were considered, and it was assumed that all busbars were operating at unity per unit voltage prior to the fault occurring. The load current flowing prior to the fault was also neglected.

By using the results of a load flow prior to performing the fault analysis, the load currents can be added to the fault currents, allowing a more accurate determination of the total currents flowing in the system.

Unbalanced faults can be included by using symmetrical components. The negative sequence network is similar to the positive sequence network, but the zero sequence network can be quite different primarily because of ground impedance and transformer winding configurations.

Transient Stability

After a disturbance, due usually to a network fault, the synchronous machine's electrical loading changes and the machines speed up (under very light loading conditions they can slow down). Each machine will react differently depending on its proximity to the fault, its initial loading, and its time constants. This means that the angular positions of the rotors relative to each other change. If any angle exceeds a certain threshold (usually between 140 and 160°) the machine will no longer be able to maintain synchronism. This almost always results in its removal from service.

Early work on transient stability had concentrated on the reaction of one synchronous machine coupled to a very large system through a transmission line. The large system can be assumed to be infinite with respect to the single machine and hence can be modeled as a pure voltage source. The synchronous machine is modeled by the three-phase windings of the stator plus windings on the rotor representing the field winding and the eddy current paths. These are resolved into two axes, one in line with the direct axis of the rotor and the other in line with the quadrature axis situated 90° (electrical) from the direct axis. The field winding is on the direct axis. Equations can be developed to determine the voltage in any winding depending on the current flows in all the other windings. A full set of differential equations can be produced, which allows the response of the machine to various electrical disturbances to be found. The variables must include rotor angle and rotor speed, which can be evaluated from a knowledge of the power from the turbine into, and power to the system out of, the machine. The great disadvantage with this type of analysis is that the rotor position is constantly changing as it rotates. As most of the equations involve trigonometrical functions relating to stator and rotor windings, the matrices must be constantly reevaluated. In the most severe cases of network faults the results, once the dc transients decay, are balanced. Further, on removal of the fault the network is considered to be balanced. There is thus much computational effort involved in obtaining detailed information for each of the three phases, which is of little value to the power system engineer. By contrast, this type of analysis is very important to machine designers. However, programs have been written for multimachine systems using this method.

Several power-system catastrophes in the U.S. and Europe in the 1960s gave a major boost to developing transient stability programs. What was required was a simpler and more efficient method of representing the machines in large power systems.

Initially, transient stability programs all ran in the time domain. A set of differential equations is developed to describe the dynamic behavior of the synchronous machines. These are linked together by algebraic equations for the network and any other part of the system that has a very fast response, i.e., an insignificant time constant, relative to the synchronous machines. All the machine equations are written in the direct and quadrature axes of the rotor so that they are constant regardless of the rotor position. The network is written in the real and imaginary axes similar to that used by the load-flow and faults programs. The transposition between these axes only requires knowledge of the rotor angle relative to the synchronously rotating frame of reference of the network.

Later work involved looking at the response of the system, not to major disturbances but to the build-up of oscillations due to small disturbances and poorly set control systems. As the time involved for these disturbances to occur can be large, time domain solutions are not suitable and frequency domain models of the system were produced. Lyapunov functions have also been used, but good models have been difficult to produce. However, they are now of sufficiently good quality to compete with time domain models where quick estimates of stability are needed, such as in the day-to-day operation of a system.

Fast Transients

While the transient stability program assumed that a fast transient response was equivalent to an instantaneous response and only concentrated on the slower response of the synchronous machines, the requirement to model the fast transient response of traveling waves on transmission lines brought about the development of programs that treated variables with large time constants as if they were constants and modeled the variables with very small time constants by differential equations.

The program is based on the equations governing voltage and current wave propagation along a lossless line. Attenuation is then included using suitable lumped resistances. A major feature of the method is that inductance and capacitance can both be represented by resistance in parallel with a current source. This allows a purely resistive network to be formed.

Whereas, with most of the other programs, source code was treated as intellectual property, the development of the fast transient program was done by many different researchers who pooled their ideas and programs. An electromagnetic transient program was developed quickly, and it probably became the first power systems analysis tool to be used for many different purposes throughout the world. From this base, numerous commercial packages have been developed.

In parallel with the development of electromagnetic transient programs, several state variable programs were produced to examine the fast transient behavior of parts of the electrical system, such as ac transmission lines and HVdc transmission systems. As these programs were specifically designed for the purpose they were intended, it gave them certain advantages over the general purpose electromagnetic transient program.

Reliability

Of constant concern to the operators of power systems is the reliability of equipment. This has become more important as systems are run harder. In the past, reliability was ensured by building in reserve equipment that was either connected in parallel with other similar devices or could be easily connected in the event of a failure. Not only that, knowledge of the capabilities of materials has increased so that equipment can be built with a more certain level of reliability.

However, reliability of a system is governed by the reliability of all the parts and their configuration. Much work has been done on the determination of the reliability of power systems, but work is still being done to fully model power system components and integrate them into system reliability models.

The information that is obtained from reliability analysis is very much governed by the nature of the system. The accepted breakdown of a power system containing generation, transmission, and distribution is into three hierarchical levels. The first level is for the generation facilities alone, the second level contains generation and transmission, while the third level contains generation, transmission, and distribution facilities. Much of the early work was focused on the generation facilities. The reasons for this were that, first, more information was available about the generation; second, the size of the problem was smaller; and, third, the emphasis of power systems was placed in generation. With the onset of deregulation, distribution and customer requirements are now considered paramount.

At the generation and transmission levels, the loss of load expectation and frequency and duration evaluation are prime reliability indicators. A power system component may well have several derated states along with the fully operational and nonoperational states. Recursive techniques are available to construct the system models, and they can include multistate components.

The usual method for evaluating reliability indices at the distribution level, such as the average interruption duration per customer per year, is an analytical approach based on a failure modes assessment and the use of equations for series and parallel networks.

Economic Dispatch and Unit Commitment

Many programs are devoted to power system operational problems. The minimization of the cost of production and delivery of energy are of great importance. Two types of programs that deal with this problem are economic dispatch and unit commitment.

Economic dispatch uses optimization techniques to determine the level of power each generator (unit) must supply to the system in order to meet the demand. Each unit must have its generating costs, which will be nonlinear functions of energy, defined along with the units' operational maximum and minimum power limits. The transmission losses of the system must also be taken into account to ensure an overall minimum cost.

Unit commitment calculates the necessary generating units that should be connected (committed) at any time in order to supply the demand and losses plus allowing sufficient reserve capability to withstand a load increase or accidental loss of a generating unit. Several operating restrictions must be taken into account when determining which machines to commit or decommit. These include maximum and minimum running times for a unit and the time needed to commit a unit. Fuel availability constraints must also be considered. For example, there may be limited fuel reserves such as coal stocks or water in the dam. Other fuel constraints may be minimum water flows below the dam or agreements to purchase minimum amounts of fuel. Determining unit commitment for a specific time cannot be evaluated without consideration of the past operational configuration or the future operating demands.

10.3 The Second Generation of Programs

It is not the intention to suggest that only the above programs were being produced initially. However, most of the other programs remained as either research tools or one-off analysis programs. The advent of the PC gave a universal platform on which most users and programs could come together. This process was further assisted when windowing reduced the need for such a high level of computer literacy on the part of users. For example, electromagnetic transient program's generality, which made it so successful, is also a handicap and it requires good programming skill to utilize it fully. This has led to several commercial programs that are loosely based on the methods of analysis first used in the electromagnetic transient program. They have the advantage of a much improved user interface.

Not all software is run on PCs. Apart from the Macintosh, which has a similar capability to a PC but which is less popular with engineers, more powerful workstations are available usually based on the Unix operating system. Mini computers and mainframe computers are also still in general use in universities and industry, even though it had been thought that they would be totally superseded.

Hardware and software for power system operation and control required at utility control centers is usually sold as a total package. These systems, although excellent, can only be alluded to here as the information is proprietary. The justification for a particular configuration requires input from many diverse groups within the utility.

Graphics

Two areas of improvement that stand out in this second wave of generally available programs are both associated with the graphical capabilities of computers. A good diagram can be more easily understood than many pages of text or tables.

The ability to produce graphical output of the results of an analysis has made the use of computers in all engineering fields, not just power system analysis, much easier. Tabulated results are never easy to interpret. They are also often given to a greater degree of accuracy than the input data warrants. A graph of the results, where appropriate, can make the results very easy to interpret, and if there is also an ability to generate a graph ahowing any variable with any other, or if three dimensions can be utilized, then new and possibly significant information can be quickly assimilated.

New packages became available for business and engineering that were based on either the spreadsheet or database principle. These also had the ability to produce graphical output. It was no longer essential to know a programming language to do even quite complex engineering analysis. The programming was usually inefficient, and obtaining results was more laborious, e.g., each iteration had to be started by hand. But, as engineers had to use these packages for other work, they became very convenient tools.

A word of caution here, be careful that the results are graphed in an appropriate manner. Most spreadsheet packages have very limited x-axis (horizontal) manipulation. Provided the x-axis data comes in regular steps, the results are acceptable. However, we have seen instances where very distorted graphs have been presented because of this problem.

Apart from the graphical interpretation of results, there are now several good packages that allow the analyst to enter the data graphically. It is a great advantage to be able to develop a one-line, or three-phase, diagram of a network directly with the computer. All the relevant system components can be included. Parameter data still require entry in a more orthodox manner, but by merely clicking on a component, a data form for that component can be made available. The chances of omitting a component are greatly reduced with this type of data entry. Further, the same system diagram can be used to show the results of some analyses.

An extension of the network diagram input is to make the diagram relate to the actual topography. In these cases, the actual routes of transmission lines are shown and can be superimposed on computer generated geographical maps. The lines in these cases have their lengths automatically established and, if the line characteristics are known, the line parameters can be calculated.

These topographical diagrams are an invaluable aid for power reticulation problems, for example, the minimum route length of reticulation given all the points of supply and the route constraints. Other optimization algorithms include determination of line sizes and switching operations.

The analysis techniques can be either linear or nonlinear. If successful, the nonlinear algorithm is more accurate, but these techniques suffer from larger data storage requirements, greater computational time, and possible divergence. There are various possible optimization techniques that can and have been applied to this problem. There is no definitive answer and each type of problem may require a different choice.

The capability chart represents a method of graphically displaying power system performance. These charts are drawn on the complex power plane and define the real and reactive power that may be supplied from a point in the system during steady-state operation. The power available is depicted as a region on the plane, and the boundaries of the region represent the critical operating limits of the system. The best-known example of a capability chart is the operating chart of a synchronous machine. The power available from the generator is restricted by limiting values of the rotor current, stator current, turbine power (if a generator), and synchronous stability limits. Capability charts have been produced for transmission lines and HVdc converters.

Where the capability chart is extended to cover more than one power system component, the two-dimensional capability chart associated with a single busbar can be regarded as being a single slice of an overall $2n$-dimensional capability chart for the n busbars that make up a general power system. If the system is small, a contour plotting approach can be used to gradually trace out the locus on the complex power plane. A load-flow algorithm is used to iteratively solve the operating equations at each point on the contour, without having to resort to an explicit closed-form solution.

The good contour behavior near the operating region has allowed a faster method to be adopted. A seed load-flow solution, corresponding to the nominal system state, is obtained to begin drawing the chart. A region-growing process is then used to locate the region in which all constrained variables are less than 10% beyond their limits. This process is similar to a technique used in computer vision systems to recognize shapes of objects. The region grows by investigating the six nearest lattice vertices to any unconstrained vertex. Linear interpolation along the edges between vertices is then used to estimate the points of intersection between the contour and the lattice. This method has a second advantage in that it can detect holes and islands in the chart. However, it should be noted that these regions are purely speculative and have not been found in practice.

Protection

The need to analyze protection schemes has resulted in the development of protection coordination programs. Protection schemes can be divided into two major groupings: unit and nonunit schemes.

The first group contains schemes that protect a specific area of the system, i.e., a transformer, transmission line, generator, or busbar. The most obvious example of unit protection schemes is based on Kirchhoff's current law: the sum of the currents entering an area of the system must be zero. Any deviation from this must indicate an abnormal current path. In these schemes, the effects of any disturbance or operating condition outside the area of interest are totally ignored, and the protection must be designed to be stable above the maximum possible fault current that could flow through the protected area. Schemes can be made to extend across all sides of a transformer to account for the different currents at different voltage levels. Any analysis of these schemes are thus of more concern to the protection equipment manufacturers.

The nonunit schemes, while also intended to protect specific areas, have no fixed boundaries. As well as protecting their own designated areas, the protective zones can overlap into other areas. While this can be very beneficial for backup purposes, there can be a tendency for too great an area to be isolated if a fault is detected by different nonunit schemes. The simplest of these schemes measures current and incorporates an inverse time characteristic into the protection operation to allow protection nearer to the fault to operate first. While this is relatively straightforward for radial schemes, in networks, where the current paths can be quite different depending on operating and maintenance strategies, protection can be difficult to set, and optimum settings are probably impossible to achieve. It is in these areas where protection software has become useful to manufacturers, consultants, and utilities.

The very nature of protection schemes has changed from electromechanical devices, through electronic equivalents of the old devices, to highly sophisticated system analyzers. They are computers in their own right and thus can be developed almost entirely by computer analysis techniques.

Other Uses for Load-Flow Analysis

It has already been demonstrated that load-flow analysis is necessary in determining the economic operation of the power system, and it can also be used in the production of capability charts. Many other types of analyses require load flow to be embedded in the program.

As a follow-on from the basic load-flow analysis, where significant unbalanced load or unbalanced transmission causes problems, a three-phase load flow may be required to study their effects. These programs require each phase to be represented separately and mutual coupling between phases to be taken into account. Transformer winding connections must be correctly represented, and the mutual coupling between transmission lines on the same tower or on the same right-of-way must also be included.

Motor starting can be evaluated using a transient stability program but in many cases this level of analysis is unnecessary. The voltage dip associated with motor start-up can be determined very precisely by a conventional load-flow program with a motor-starting module.

Optimal power system operation requires the best use of resources subject to a number of constraints over any specified time period. The problem consists of minimizing a scalar objective function (normally a cost criterion) through the optimal control of a vector of control parameters. This is subject to the equality constraints of the load-flow equations, inequality constraints on the control parameters, and inequality constraints of dependent variables and dependent functions. The programs to do this analysis are usually referred to as optimal power flow (OPF) programs.

Often optimal operation conflicts with the security requirements of the system. Load-flow studies are used to assess security (security assessment). This can be viewed as two separate functions. First, there is a need to detect any operating limit violations through continuous monitoring of the branch flows and nodal voltages. Second, there is a need to determine the effects of branch outages (contingency analysis). To reduce this to a manageable level, the list of contingencies is reduced by judicial elimination of most of the cases that are not expected to cause violations. From this, the possible overloading of equipment can be forecast. The program should be designed to accommodate the condition where generation cannot meet the load because of network islanding.

The conflicting requirements of system optimization and security require that they be considered together. The more recent versions of OPF interface with contingency analysis, and the computational requirements are enormous.

Extensions to Transient Stability Analysis

Transient stability programs have been extended to include many other system components, including FACTS (flexible ac transmission systems) and dc converters.

FACTS may be either shunt or branch devices. Shunt devices usually attempt to control busbar voltage by varying their shunt susceptance. The device is therefore relatively simple to implement in a time domain program. Series devices may be associated with transformers. Stability improvement is achieved by injecting a quadrature component of voltage derived from the other two phases rather than by a tap changer, which injects a direct component of voltage. Fast-acting power electronics can inject direct voltage, quadrature voltage, or a combination of both to help maintain voltage levels and improve stability margins.

Dc converters for HVdc links and rectifier loads have received much attention. The converter controls are very fast acting and therefore a quasi-steady-state (QSS) model can be considered accurate. That is, the model of the converter terminals contains no dynamic equations and in effect the link behaves as if it was in steady state for every time solution of the ac system. While this may be so some time after a fault has been removed, during and just after a fault the converters may well suffer from commutation failure or fire through. These events cannot be predicted or modeled with a QSS model. In this case, an appropriate method of analysis is to combine a state variable model of the converter, which can model the firing of the individual valves, with a conventional multimachine transient stability program containing a QSS model. During the period of maximum disturbance, the two models can operate together. Information about the overall system response is passed to the state variable model at regular intervals. Similarly the results from the detailed converter model are passed to the multimachine model, overriding its own QSS model. As the disturbance reduces, the results from the two different converter models converge, and it is then only necessary to run the computationally inexpensive QSS model within the multimachine transient stability program.

Voltage Collapse

Steady-state analysis of the problem of voltage instability and voltage collapse is often based on load-flow analysis programs. However, time solutions can provide further insight into the problem.

A transient stability program can be extended to include induction machines, which are associated with many of the voltage collapse problems. In these studies, it is the stability of the motors that is examined rather than the stability of the synchronous machines. The asynchronous nature of the induction machine means that rotor angle is not a concern, but instead the capability of the machines to recover after a fault has depressed the voltage and allowed the machines to slow down. The reaccelerating machines draw more reactive current, which can hold the terminal voltage down below that necessary to allow recovery. Similarly starting a machine will depress the voltage, which affects other induction machines, further lowering the voltage.

However, voltage collapse can also be due to longer-term problems. Transient stability programs then need to take into account controls that are usually ignored. These include automatic transformer tap adjustment and generator excitation limiters, which control the long-term reactive power output to keep the field currents within their rated values.

The equipment that can contribute to voltage collapse must also be more carefully modeled. Simple impedance models for loads ($P = P_o V_2$; $Q = Q_o V_2$) are no longer adequate. An improvement can be obtained by replacing the (mathematical) power 2 in the equations by more suitable values. Along with the induction machine models, the load characteristics can be further refined by including saturation effects.

SCADA

SCADA has been an integral part of system control for many years. A control center now has much real-time information available so that human and computer decisions about system operation can be made with a high degree of confidence.

In order to achieve high-quality input data, algorithms have been developed to estimate the state of a system based on the available online data (state estimation). These methods are based on weighted least squares techniques to find the best state vector to fit the scatter of data. This becomes a major problem when conflicting information is received. However, as more data becomes available, the reliability of the estimate can be improved.

Power Quality

One form of poor power quality, which has received a large amount of attention, is the high level of harmonics that can exist, and there are numerous harmonic analysis programs now available.

Recently, the harmonic levels of both currents and voltages have increased considerably due to the increasing use of nonlinear loads, such as arc furnaces, HVdc converters, FACTS equipment, dc motor drives, and ac motor speed control. Moreover, commercial sector loads now contain often-unacceptable levels of harmonics due to widespread use of rectifier-fed power supplies with capacitor output smoothing (e.g., computer power supplies and fluorescent lighting). The need to conserve energy has resulted in energy efficient designs that exacerbate the generation of harmonics. Although each source only contributes a very small level of harmonics, due to their small power ratings, widespread use of small nonlinear devices may create harmonic problems that are more difficult to remedy than one large harmonic source.

Harmonic analysis algorithms vary greatly in their formulas and features. However, almost all use the frequency domain. The most common technique is the direct method (also known as current injection method). Spectral analysis of the current waveform of the nonlinear components is performed and entered into the program. The network data is used to assemble a system admittance matrix for each frequency of interest. This set of linear equations is solved for each frequency to determine the node voltages and, hence, current flow throughout the system. This method assumes the nonlinear component is an ideal harmonic current source. A more advanced technique is to model the relationship between the harmonic currents injected by a component and its terminal voltage waveform. This then requires an iterative algorithm that does require excursion into the time domain for modeling this interaction. When the fundamental (load flow) is also included, thus simulating the interaction between fundamental and harmonic frequencies, it is termed a harmonic power flow. The most advanced technique, which is still only a research tool, is the harmonic domain. In this iterative technique, one Jacobian is built up representing all harmonic frequencies. This allows coupling between harmonics, which occurs, for example, in salient synchronous machines, to be modeled.

There are many other features that need to be considered, such as whether the algorithm uses symmetrical components or phase coordinates, or whether it is single- or three-phase. Data entry for single-phase typically requires the electrical parameters, whereas three-phase analysis normally requires the physical geometry of the overhead transmission lines, with cables and conductor details, so that a transmission line parameter program or cable parameter program can calculate the line or cable electrical parameters.

The communication link between the monitoring point and the control center can now be very sophisticated and can utilize satellites. This technology has led to the development of systems to analyze the power quality of a system. Harmonic measurement and analysis has now reached a high level of maturity. Many different pieces of information can be monitored and the results over time stored in a database. Algorithms based on the fast-Fourier transform can then be used to convert this data from the time domain to the frequency domain. Computing techniques coupled with fast and often parallel computing allows this information to be displayed in real time. By utilizing the time-stamping capability of the global positioning system (GPS), information gathered at remote sites can be linked together. Using the GPS time stamp, samples taken exactly simultaneously can be fed to a harmonic state estimator, which can even determine the position and magnitude of harmonics entering the system as well as the harmonic voltages and currents at points not monitored (provided enough initial monitoring points exist).

One of the most important features of harmonic analysis software is the ability to display the results graphically. The refined capabilities of present three-dimensional graphics packages have simplified the analysis considerably.

Finite Element Analysis

Finite element analysis is not normally used by power system engineers, although it is a common tool of high-voltage- and electrical-machine engineers. It is necessary, for example, where accurate machine representation is required. For example, in a unit-connected HVdc terminal the generators are closely coupled to the rectifier bridges. The ac system at the rectifier end is isolated from all but its generator. There is no need for costly filters to reduce harmonics. Models of the synchronous machine suitable for a transient stability study can be obtained from actual machine tests. For fast transient analysis, a three-phase generator model can be used, but it will not account for harmonics. A finite element model of the generator provides the means of allowing real-time effects such as harmonics and saturation to be directly included. Any geometric irregularities in the generator can be accounted for, and the studies can be done at the design stage rather than having to rely on measurements or extrapolation from manufactured machines to obtain circuit parameters. There is no reliance on estimated machine parameters. The disadvantages are the cost and time to run a simulation, and it is not suitable at present to integrate with existing transient stability programs as it requires a high degree of expertise. As the finite element model is in this case used in a time simulation, part of the air gap is left unmeshed in the model. At each time step the rotor is placed in the desired position, and the missing elements in the air-gap region are formed using the nodes on each side of the gap.

Grounding

The safe grounding of power system equipment is very important, especially as the short-circuit capability of power systems continues to grow. Programs have been developed to evaluate and design grounding systems in areas containing major power equipment, such as substations, and to evaluate the effects of fault current on remote, separately grounded equipment.

The connection to ground may consist of a ground mat of buried conductors, electrodes (earth rods), or both. The shape and dimensions of the electrodes, their locations, and the layout of a ground mat, plus the resistivity of the ground at different levels must be specified in order to evaluate the ground resistance. A grid of buried conductors and electrodes is usually considered to be all at the same potential. Where grid sections are joined by buried or aerial links, these links can have resistance allowing the grid sections to have different potentials. It is usual to consider a buried link as capable of radiating current into the soil.

Various methods of representing the fault current are available. The current can be fixed or it can be determined from the short-circuit MVA and the busbar voltage. A more complex fault path may need to be constructed for faults remote from the site being analyzed.

From the analysis, the surface potential over the affected area can be evaluated and, from that, step and touch potentials calculated. Three-dimensional graphics of the surface potentials are very useful in highlighting problem areas.

Market Software

As many countries have embraced electricity deregulation (a misnomer of grand proportions) and implemented an electricity market, new programs are being developed for this environment. Programs for data exchange, solving for the dispatch, setting prices, and assisting in the buying and selling of electricity have been developed. Although most electricity markets are based around some form of locational pricing, they all differ in their detail. The purest form is "nodal pricing" in which prices are set for each distinct electrical location ("bus" or "node") in the system. Many markets now also "cooptimize" generation and reserve dispatch, setting prices for both.

Traditionally "time of use" pricing has been implemented to reflect that more-expensive generation is used, and higher losses occurred, at times of high system loading. Market models implement this idea to allow "spot" prices to be set dynamically by the interaction of buyers and sellers. Nodal (spot) pricing extends this by incorporating different prices for electricity at different geographical locations in the network. Inherent in nodal pricing is marginal costing by which prices are set to match the cost to deliver the next increment in electrical power. Many believe marginal pricing provides the correct economic signals to the users of the transmission system. This marginal costing also incorporates generation, transmission, and often reserve constraints. Experimental models have used a full ac OPF to price both active and reactive power but, although some markets do employ ac models in a limited role, only active power is traded on the spot price. The traditional OPF is termed a *primal* problem in which the aim is to determine the optimal variable values given the costs of each decision and the amount of each resource available. The related *dual* problem determines the optimal prices given the same data. Linearization of the OPF formulation (or dc load-flow in some cases) is performed to allow a linear programming technique to be applied. While in the primal the objective is to minimize cost, the dual's objective is to maximize the system welfare (cumulative benefit of transactions). Hence the dual linear program can be stated as:

Maximize Welfare function
Subject to:
Linearized power-flow equations
Nodal prices defined by demand
Floor and ceiling prices set by generator costs
Voltage constraints
Generator ramp rates

In practice, solving the primal problem necessarily produces a solution to the dual problem, and dispatch and price variables are optimized simultaneously, and hence consistently.

In a nodal pricing regime, prices can be set before the event (*ex ante*) or after the event (*ex post*). In practice, a combination of *ex ante* and *ex post* pricing is often used. *Ex ante* prices are released so that generators and customers can respond to the expected price of electricity. If at a given time period the price is expected to be high, a generator ensures its availability and a customer may restrict its use of electricity. However, these are indicative prices, and the final prices are those calculated *ex post*, which, with the benefit of hindsight, are more accurate. It should be understood, though, that while these prices notionally apply to all electricity traded in the market, some form of contracting (either "physical" or "financial") is always used, so that average electricity prices are really dominated (and stabilized) by contract prices, with spot prices effectively applying only to deviations from contracted positions.

Other Programs

There are too many other programs available to be discussed. For example, neither automatic generator control nor load forecasting have been included. However, an example of a small program that can stand alone or fit into other programs is given here.

In order to obtain the electrical parameters of overhead transmission lines and underground cables, utility programs have been developed. Transmission line parameter programs use the physical geometry of the conductors, the conductor type, and ground resistivity to calculate the electrical parameters of the line. Cable parameter programs may use the physical dimensions of the cable, its construction, and its position in the ground. The results of these programs are usually fed directly to network analysis programs such as load flow or faults. The errors introduced during transfer are thus minimized. This is particularly true for three-phase analyses due to the volume of data involved.

Program Suites

As more users become involved with a program, its quirks become less acceptable and it must become easy to use, i.e., user-friendly, with a good graphical user interface (GUI) being an important part.

With the availability of many different types of programs, it became important to be able to transfer the results of one program to the input of another. If the user has access to the source code, this can often be done relatively quickly by generating an output file in a suitable format for the input of the second program. There has, therefore, been a great deal of attention devoted to creating common formats for data transfer as well as producing programs with easy data-entry formats and good result-processing capabilities. Forming suites of programs, either as a monolithic program or many individual programs linked in some way, overcomes this problem.

Another issue is the maintenance of the power system data needed as input to these programs. Previously, each software vendor had its own proprietary data format; however, when several different products are used, then the job of keeping each up to date becomes problematic. Therefore there is a trend to form program suites incorporating many of the above-mentioned analysis programs and to work from the same database. They benefit from the fact that much of the data requirements are common to each analysis. However, there is also data specific to each analysis, and this does complicate data entry in these types of programs. There is a trend towards interfacing programs to SQL (structured query language) databases to give more flexibility and power in data entry.

Many good "front end" programs are now available that allow the user to quickly write an analysis program and utilize the built-in IO features of the package. There are also several good general mathematical packages available. Much research work can now be done using tools such as these. By using these standard packages the researcher is freed from the chore of developing algorithms and IO routines. Not only that, extra software is being developed that can turn these general packages into specialist packages. It may well be that before long all software will be made to run on sophisticated developments of these types of packages and the stand-alone program will fall into oblivion.

No one approach suits all users; for example, the requirements of power companies, consultants, and researchers differ. From a power company perspective, the maintenance of an accurate database of the system it owns is paramount, hence only having one database to maintain for all types of analysis is important. Consultants and researchers perform a lot of one-off system studies; hence the ease of data entry and user-interface is very important. Moreover, some specialized studies are performed infrequently; hence the user interface is all-important.

10.4 Conclusions

There are many more programs available than can be discussed here. Those that have been discussed are not necessarily more significant than those omitted. There are programs to help you with almost every power system problem you have, and new software is constantly becoming available to solve the latest problems.

Make sure that the programs you use are designed to do the job you require. Some programs make assumptions that give satisfactory results in most cases but may not be adequate for your particular case. No matter how sophisticated and friendly the program may appear, the algorithm and processing of data are the most important parts. As programs become more complex and integrated, new errors (regressions) can be introduced. Wherever possible, check the answers and always make sure they feel right.

Further Information

There are several publications that can keep engineers up to date with the latest developments in power system analysis. The *IEEE Spectrum* (U.S.) and the *IEE Review* (U.K.) are the two most well-respected, general-interest, English language journals that report on the latest developments in electrical engineering. The *Power Engineering Journal* produced by the IEE regularly runs tutorial papers, many of which are of direct concern to power systems analysts. However, for magazine-style coverage of the developments in power system analysis, the *Power & Energy Magazine* is, in the authors' opinion, the most useful.

Finally, a few textbooks that provide a much greater insight into the programs discussed in the chapter have been included below.

J. Arrillaga and C.P. Arnold, *Computer Analysis of Power Systems*, London: John Wiley & Sons, 1990.

J. Arrillaga and N.R. Watson, *Computer Modeling of Electrical Power Systems*, 2nd ed., New York: John Wiley & Sons, 2001.

R. Billinton and R.N. Allan, *Reliability Evaluation of Power Systems*, New York: Plenum Press, 1984.

A.S. Debs, *Modern Power Systems Control and Operation*, New York: Kluwer Academic Publishers, 1988.

C.A. Gross, *Power System Analysis*, New York: John Wiley & Sons, 1986.

B.R. Gungor, *Power Systems*, New York: Harcourt Brace Jovanovich, 1988.

G.T. Heydt, *Computer Analysis Methods for Power Systems*, New York: Macmillan Publisher, 1986.

IEEE Brown Book, *Power Systems Analysis*, New York: IEEE, 1990.

G.L. Kusic, *Computer-Aided Power System Analysis*, Englewood Cliffs, N.J.: Prentice-Hall, 1986.

N.S. Rau, *Optimization Principles: Practical Applications to the Operation and Markets of the Electric Power Industry*, New York: IEEE Press, 2003.

B.M. Weedy and B.J. Cory, *Electric Power Systems*, 4th ed., New York: John Wiley & Sons, 1998.

A.J. Wood and B.F. Wollenberg, *Power Generation, Operation, and Control*, 2nd ed., New York: John Wiley & Sons, 1996

II

Systems

11

Control Systems

William L. Brogan
University of Nevada, Las Vegas

Gordon K.F. Lee
San Diego State University

Andrew P. Sage
George Mason University

Hitay Özbay
Bilkent University

Charles L. Phillips
Auburn University

Royce D. Harbor
University of West Florida

Raymond G. Jacquot
University of Wyoming

John E. McInroy
University of Wyoming

Derek P. Atherton
University of Sussex

11.1 Models

William L. Brogan

A naive trial-and-error approach to the design of a control system might consist of constructing a controller, installing it into the system to be controlled, performing tests, and then modifying the controller until satisfactory performance is achieved. This approach could be dangerous and uneconomical, if not impossible. A more rational approach to control system design uses mathematical models. A *model* is a mathematical description of system behavior, as influenced by input variables or initial conditions. The model is a stand-in for the actual system during the control system design stage. It is used to predict performance; to carry out stability, sensitivity, and trade-off studies; and answer various "what-if" questions in a safe and efficient manner. Of course, the validation of the model, and all conclusions derived from it, must ultimately be based upon test results with the physical hardware.

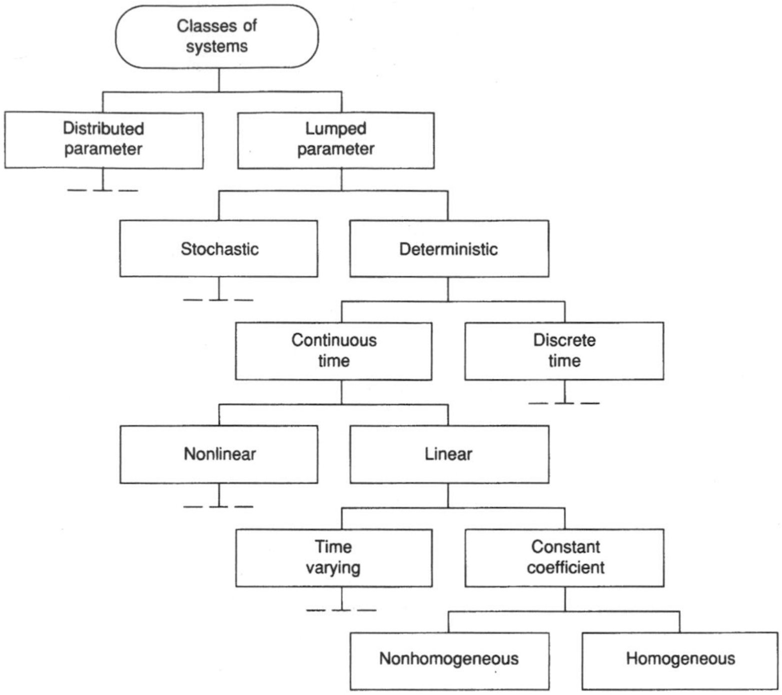

FIGURE 11.1 Major classes of system equations. (*Source*: W.L. Brogan, *Modern Control Theory*, 3rd ed., Englewood Cliffs, N.J.: Prentice-Hall, 1991, p. 13. With permission.)

The final form of the mathematical model depends upon the type of physical system, the method used to develop the model, and mathematical manipulations applied to it. These issues are discussed next.

Classes of Systems to Be Modeled

Most control problems are multidisciplinary. The system may consist of electrical, mechanical, thermal, optical, fluidic, or other physical components, as well as economic, biological, or ecological systems. Analogies exist between these various disciplines, based upon the similarity of the equations that describe the phenomena. The discussion of models in this section will be given in mathematical terms and therefore will apply to several disciplines.

Figure 11.1 shows the classes of systems that might be encountered in control systems modeling. Several branches of this tree diagram are terminated with a dashed line indicating that additional branches have been omitted, similar to those at the same level on other paths.

Distributed parameter systems have variables that are functions of both space and time (such as the voltage along a transmission line or the deflection of a point on an elastic structure). They are described by partial differential equations. These are often approximately modeled as a set of *lumped parameter* systems (described by ordinary differential or difference equations) by using modal expansions, finite element methods, or other approximations [Brogan, 1968]. The lumped parameter continuous-time and discrete-time families are stressed here.

Two Major Approaches to Modeling

In principle, models of a given physical system can be developed by two distinct approaches. Figure 11.2 shows the steps involved in *analytical modeling.* The real-world system is represented by an interconnection of idealized elements. Table 11.1 shows model elements from several disciplines and their elemental equations.

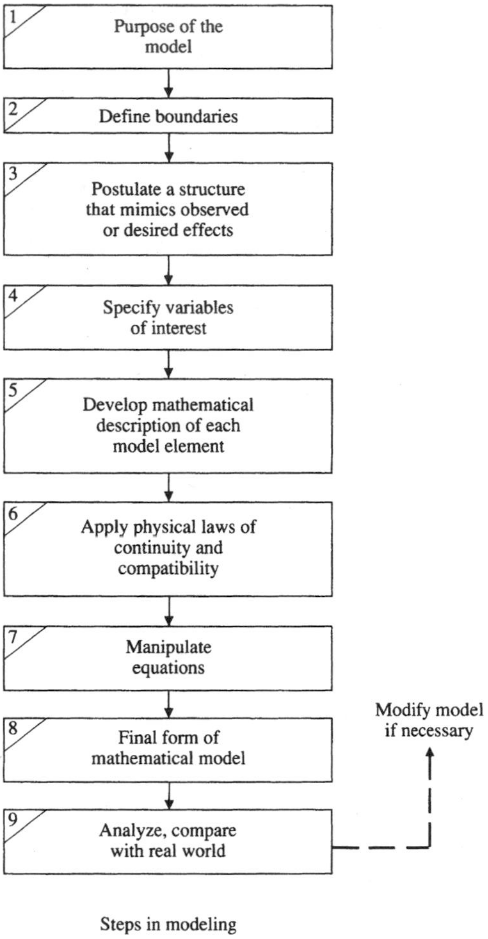

FIGURE 11.2 Modeling considerations. (*Source*: W.L. Brogan, *Modern Control Theory,* 3rd ed., Englewood Cliffs, N.J.: Prentice-Hall, 1991, p. 5. With permission.)

An electrical circuit diagram is a typical result of this physical modeling step (box 3 of Figure 11.2). Application of the appropriate physical laws (Kirchhoff, Newton, etc.) to the idealized physical model (consisting of point masses, ideal springs, lumped resistors, etc.) leads to a set of mathematical equations. For a circuit, these will be mesh or node equations in terms of elemental currents and voltages. Box 6 of Figure 11.2 suggests a generalization to other disciplines, in terms of continuity and compatibility laws, using through variables (generalization of current that flows through an element) and across variables (generalization of voltage, which has a differential value across an element).

Experimental or *empirical* modeling typically assumes an *a priori* form for the model equations and then uses available measurements to estimate the coefficient values that cause the assumed form to best fit the data. The assumed form could be based upon physical knowledge or it could be just a credible assumption. Time-series models include autoregressive (AR) models, moving average (MA) models, and the combination, called ARMA models. All are difference equations relating the input variables to the output variables at the discrete measurement times, of the form

$$\begin{aligned} y(k+1) = {} & a_0 y(k) + a_1 y(k-1) + a_2 y(k-2) + \ldots + a_n y(k-n) \\ & + b_0 u(k+1) + b_1 u(k) + \ldots + b_p u(k+1-p) + v(k) \end{aligned} \tag{11.1}$$

TABLE 11.1 Summary of Describing Differential Equations for Ideal Elements

Type of Element	Physical Element	Describing Equation	Energy E or Power P	Symbol
	Electrical inductance	$v_{21} = L\frac{di}{dt}$	$E = \frac{1}{2}Li^2$	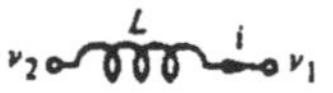
Inductive storage	Translational spring	$v_{21} = \frac{1}{K}\frac{dF}{dt}$	$E = \frac{1}{2}\frac{F^2}{K}$	
	Rotational spring	$\omega_{21} = \frac{1}{K}\frac{dT}{dt}$	$E = \frac{1}{2}\frac{T^2}{K}$	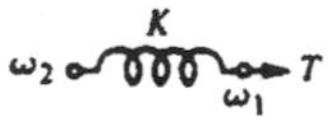
	Fluid inertia	$P_{21} = I\frac{dQ}{dt}$	$E = \frac{1}{2}IQ^2$	
	Electrical capacitance	$i = C\frac{dv_{21}}{dt}$	$E = \frac{1}{2}Cv_{21}^2$	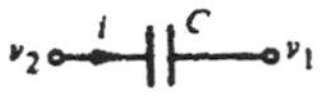
	Translational mass	$F = M\frac{dv_2}{dt}$	$E = \frac{1}{2}Mv_2^2$	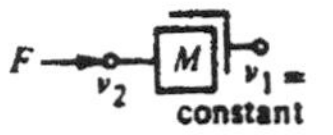
Capacitive storage	Rotational mass	$T = J\frac{d\omega_2}{dt}$	$E = \frac{1}{2}J\omega_2^2$	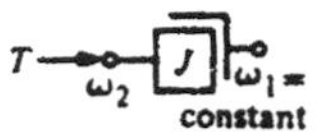
	Fluid capacitance	$Q = C_f\frac{dP_{21}}{dt}$	$E = \frac{1}{2}C_fP_{21}^2$	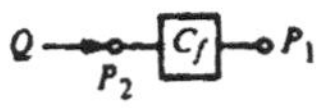
	Thermal capacitance	$q = C_t\frac{d\tau_2}{dt}$	$E = C_t\tau_2$	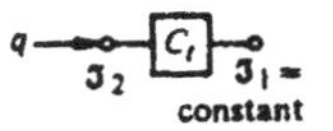
	Electrical resistance	$i = \frac{1}{R}v_{21}$	$\mathcal{P} = \frac{1}{R}v_{21}^2$	
	Translational damper	$F = fv_{21}$	$\mathcal{P} = fv_{21}^2$	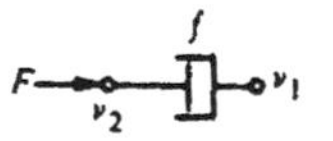
Energy dissipators	Rotational damper	$T = f\omega_{21}$	$\mathcal{P} = f\omega_{21}^2$	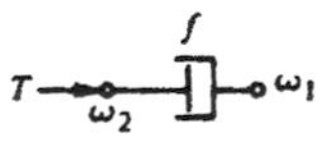
	Fluid resistance	$Q = \frac{1}{R_f}P_{21}$	$\mathcal{P} = \frac{1}{R_f}P_{21}^2$	
	Thermal resistance	$q = \frac{1}{R_t}\tau_{21}$	$\mathcal{P} = \frac{1}{R_t}\tau_{21}$	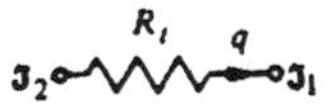

Source: R.C. Dorf, *Modern Control Systems,* 5th ed., Reading, Mass.: Addison-Wesley, 1989, p. 33. With permission.

where $v(k)$ is a random noise term. The z-transform transfer function relating u to y is

$$\frac{y(z)}{u(z)} = \frac{b_0 + b_1z^{-1} + \ldots + b_pz^{-p}}{1 - (a_0z^{-1} + \ldots + a_{n-1}z^{-n})} = H(z) \tag{11.2}$$

In the MA model all $a_i = 0$. This is alternatively called an all-zero model or a finite impulse response (FIR) model. In the AR model all b_j terms are zero except b_0. This is called an all-pole model or an infinite impulse response (IIR) model. The ARMA model has both poles and zeros and also is an IIR model.

Adaptive and learning control systems have an experimental modeling aspect. The data fitting is carried out on-line, in real time, as part of the system operation. The modeling described above is normally done off-line [Astrom and Wittenmark, 1989].

Forms of the Model

Regardless of whether a model is developed from knowledge of the physics of the process or from empirical data fitting, it can be further manipulated into several different but equivalent forms. This manipulation is box 7 in Figure 11.2. The class that is most widely used in control studies is the deterministic lumped-parameter continuous-time constant-coefficient system. A simple example has one input u and one output y. This might be a circuit composed of one ideal source and an interconnection of ideal resistors, capacitors, and inductors. The equations for this system might consist of a set of mesh or node equations. These could be reduced to a single nth-order linear ordinary differential equation by eliminating extraneous variables.

$$\frac{d^n y}{dt^n} + a_{n-1}\frac{d^{n-1}y}{dt^{n-1}} + \cdots + a_1\frac{dy}{dt} + a_0 y = b_0 u + b_1\frac{du}{dt} + \cdots + b_m\frac{d^m u}{dt^m} \tag{11.3}$$

This nth-order equation can be replaced by an input-output transfer function

$$\frac{Y(s)}{U(s)} = H(s) = \frac{b_m s^m + b_{m-1}s^{m-1} + \cdots + b_1 s + b_0}{s^n + a_{n-1}s^{n-1} + \cdots + a_1 s + a_0} \tag{11.4}$$

The inverse Laplace transform $\mathcal{L}^{-1}\{H(s)\} = h(t)$ is the system impulse response function. Alternatively, by selecting a set of n internal **state variables,** Equation (11.3) can be written as a coupled set of first-order differential equations plus an algebraic equation relating the states to the original output y. These equations are called state equations, and one possible choice for this example is, assuming $m = n$,

$$\dot{\mathbf{x}}(t) = \begin{bmatrix} -a_{n-1} & 1 & 0 & 0 & \cdots & 0 \\ -a_{n-2} & 0 & 1 & 1 & \cdots & 0 \\ \vdots & \vdots & \vdots & \vdots & \vdots & \vdots \\ -a_1 & 0 & 0 & 0 & \cdots & 1 \\ -a_0 & 0 & 0 & 0 & \cdots & 0 \end{bmatrix}\mathbf{x}(t) + \begin{bmatrix} b_{n-1} - a_{n-1}b_n \\ b_{n-2} - a_{n-2}b_n \\ \vdots \\ b_1 - a_1 b_n \\ b_0 - a_0 b_n \end{bmatrix} u(t)$$

and

$$y(t) = [1 \quad 0 \quad 0 \quad \ldots \quad 0\,]\mathbf{x}(t) + b_n u(t) \tag{11.5}$$

In matrix notation these are written more succinctly as

$$\dot{\mathbf{x}} = \mathbf{Ax} + \mathbf{Bu} \quad \text{and} \quad y = \mathbf{Cx} + \mathbf{Du} \tag{11.6}$$

Any one of these six possible model forms, or others, might constitute the result of box 8 in Figure 11.2. Discrete-time system models have similar choices of form, including an nth-order difference equation as given in Equation (11.1) or a z-transform input–output transfer function as given in Equation (11.2). A set of n first-order difference equations (state equations) analogous to Equation (11.5) or Equation (11.6) also can be written.

Extensions to systems with r inputs and m outputs lead to a set of m coupled equations similar to Equation (11.3), one for each output y_i. These higher-order equations can be reduced to n first-order state differential equations and m algebraic output equations as in Equation (11.5) or Equation (11.6). The **A** matrix is again of dimension $n \times n$, but **B** is now $n \times r$, **C** is $m \times n$, and **D** is $m \times r$. In all previous discussions, the number of state variables, n, is the order of the model. In transfer function form, an $m \times r$ matrix $H(s)$ of transfer functions will describe the input-output behavior

$$Y(s) = H(s)U(s) \tag{11.7}$$

Other transfer function forms are also applicable, including the left and right forms of the matrix fraction description (MFD) of the transfer functions [Kailath, 1980]

$$H(s) = \mathbf{P}(s)^{-1}\mathbf{N}(s) \quad \text{or} \quad H(s) = \mathbf{N}(s)\mathbf{P}(s)^{-1} \tag{11.8}$$

Both **P** and **N** are matrices whose elements are polynomials in s. Very similar model forms apply to continuous-time and discrete-time systems, with the major difference being whether Laplace transform or z-transform transfer functions are involved.

When time-variable systems are encountered, the option of using high-order differential or difference equations versus sets of first-order state equations is still open. The system coefficients $a_i(t)$, $b_j(t)$ and the matrices $\mathbf{A}(t)$, $\mathbf{B}(t)$, $\mathbf{C}(t)$, and $\mathbf{D}(t)$ will now be time-varying. Transfer-function approaches lose most of their utility in time-varying cases and are seldom used. With nonlinear systems, all the options relating to the order and number of differential or difference equation still apply.

The form of the nonlinear state equations is

$$\begin{aligned} \dot{x} &= f(\mathbf{x}, \mathbf{u}, \mathbf{t}) \\ y &= h(\mathbf{x}, \mathbf{u}, \mathbf{t}) \end{aligned} \tag{11.9}$$

where the nonlinear vector-valued functions $f(\mathbf{x}, \mathbf{u}, \mathbf{t})$ and $h(\mathbf{x}, \mathbf{u}, \mathbf{t})$ replace the right-hand sides of Equation (11.6). The transfer function forms are of no value in nonlinear cases.

Stochastic systems [Maybeck, 1979] are modeled in similar forms, except the coefficients of the model and the inputs are described in probabilistic terms.

Nonuniqueness

There is not a unique correct model of a given system for several reasons. The selection of idealized elements to represent the system requires judgment based upon the intended purpose. For example, a satellite might be modeled as a point mass in a study of its gross motion through space. A detailed flexible structure model might be required if the goal is to control vibration of a crucial on-board sensor. In empirical modeling, the assumed starting form, Equation (11.1), can vary.

There is a trade-off between the complexity of the model form and the fidelity with which it will match the data set. For example, a pth-degree polynomial can exactly fit to $p + 1$ data points, but a straight line might be a better model of the underlying physics. Deviations from the line might be caused by extraneous measurement noise. Issues such as these are addressed in Astrom [1980].

The preceding paragraph addresses nonuniqueness in determining an input-output system description. In addition, state models developed from input-output descriptions are not unique. Suppose the transfer function of a single-input, single-output linear system is known exactly. The state variable model of this system is not unique for at least two reasons. An arbitrarily high-order state variable model can be found that will have this same transfer function. There is, however, a unique minimal or irreducible order $n_{\min}$ from among all state models that have the specified transfer function. A state model of this order will have the desirable properties of **controllability** and **observability.** It is interesting to point out that the minimal order may be less than the actual order of the physical system.

The second aspect of the nonuniqueness issue relates not to order, i.e., the *number* of state variables, but to *choice* of internal variables (state variables). Mathematical and physical methods of selecting state variables are available [Brogan, 1991]. An infinite number of choices exist, and each leads to a different set $\{A, B, C, D\}$, called a realization. Some state variable model forms are more convenient for revealing key system properties such as stability, controllability, observability, **stabilizability,** and **detectability.** Common forms include the controllable canonical form, the observable canonical form, the Jordan canonical form, and the Kalman canonical form.

The reverse process is unique in that every valid realization leads to the same model transfer function

$$H(s) = \mathbf{C}\{sI - \mathbf{A}\}^{-1}\mathbf{B} + \mathbf{D} \tag{11.10}$$

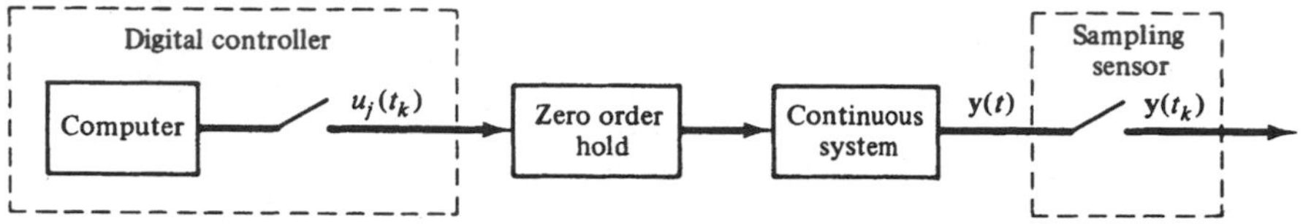

FIGURE 11.3 Model of a continuous system with digital controller and sensor. (*Source*: W.L. Brogan, *Modern Control Theory*, 3rd ed., Englewood Cliffs, N.J.: Prentice-Hall, 1991, p. 319. With permission.)

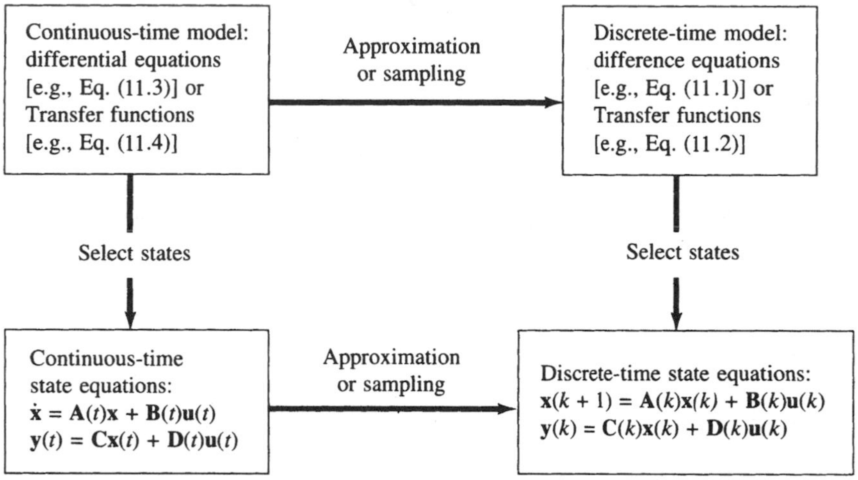

FIGURE 11.4 State variable modeling paradigm. (*Source*: W.L. Brogan, *Modern Control Theory*, 3rd ed., Englewood Cliffs, N.J.: Prentice-Hall, 1991, p. 309. With permission.)

Approximation of Continuous Systems by Discrete Models

Modern control systems often are implemented digitally, and many modern sensors provide digital output, as shown in Figure 11.3. In designing or analyzing such systems, discrete-time approximate models of continuous-time systems are frequently needed. There are several general ways of proceeding, as shown in Figure 11.4. Many choices exist for each path on the figure. Alternative choices of states or of approximation methods, such as forward or backward differences, lead to an infinite number of valid models.

Defining Terms

Controllability: A property that in the linear system case depends upon the **A,B** matrix pair, which ensures the existence of some control input that will drive any arbitrary initial state to zero in finite time.

Detectability: A system is detectable if all its unstable modes are observable.

Observability: A property that in the linear system case depends upon the **A,C** matrix pair, which ensures the ability to determine the initial values of all states by observing the system outputs for some finite time interval.

Stabilizable: A system is stabilizable if all its unstable modes are controllable.

State variables: A set of variables that completely summarize the system's status in the following sense. If all states x_i are known at time t_0, then the values of all states and outputs can be determined uniquely for any time $t_1 > t_0$, provided the inputs are known from t_0 onward. State variables are components in the state vector. State space is a vector space containing the state vectors.

References

K.J. Astrom, Maximum likelihood and prediction error methods, *Automatica*, vol. 16, pp. 551–574, 1980.
K.J. Astrom and B. Wittenmark, *Adaptive Control*, Reading, MA: Addison-Wesley, 1989.
W.L. Brogan, "Optimal control theory applied to systems described by partial differential equations," in *Advances in Control Systems*, vol. 6, C. T. Leondes (ed.), New York: Academic Press, 1968, chap. 4.
W.L. Brogan, *Modern Control Theory*, 3rd ed., Englewood Cliffs, NJ: Prentice-Hall, 1991.
R.C. Dorf, *Modern Control Systems*, 5th ed., Reading, MA: Addisson-Wesley, 1989.
R.C. Dorf, *Modern Control Systems*, 10th ed., Reading, MA: Addison-Wesley, 2004.
T. Kailath, *Linear Systems*, Englewood Cliffs, NJ: Prentice-Hall, 1980.
W.S. Levine, *Control Handbook*, Boca Raton, FL, CRC Press, 2001.
P.S. Maybeck, *Stochastic Models, Estimation and Control*, vol. 1, New York: Academic Press, 1979.

Further Information

The monthly *IEEE Control Systems Magazine* frequently contains application articles involving models of interesting physical systems.

The monthly *IEEE Transactions on Automatic Control* is concerned with theoretical aspects of systems. Models as discussed here are often the starting point for these investigations.

11.2 Dynamic Response

Gordon K.F. Lee (revised by Dariusz Uciński)

Components of the Dynamic System Response

An nth-order linear time-invariant dynamic system can be represented by a linear constant coefficient differential equation of the form

$$\begin{aligned}\frac{\mathrm{d}^n y(t)}{\mathrm{d}t^n} + a_{n-1}\frac{\mathrm{d}^{n-1} y(t)}{\mathrm{d}t^{n-1}} + \cdots + a_1\frac{\mathrm{d}y(t)}{\mathrm{d}t} + a_0 y(t) \\ = b_m\frac{\mathrm{d}^m u(t)}{\mathrm{d}t^m} + \cdots + b_1\frac{\mathrm{d}u(t)}{\mathrm{d}t} + b_0 u(t)\end{aligned} \tag{11.11}$$

where $y(t)$ is the response, and $u(t)$ is the input or forcing function. This description is accompanied by the initial conditions

$$y(0) = \zeta_0, \quad \frac{\mathrm{d}y(0)}{\mathrm{d}t} = \zeta_1, \quad \ldots, \quad \frac{\mathrm{d}^{n-1}y(0)}{\mathrm{d}t^{n-1}} = \zeta_{n-1} \tag{11.12}$$

in which the ζ_i's are given *a priori*.

Due to the linearity of the above initial value problem, its solution can be expressed as

$$y(t) = y_S(t) + y_I(t) \tag{11.13}$$

where: $y_S(t)$ is the so-called *zero-input response*, i.e., the component that results from solving Equation (11.11) and Equation (11.12) with the input assumed equal to zero, $u(t) \equiv 0$, and $y_I(t)$ is the *zero-state response*, i.e., the component that satisfies Equation (11.11) and Equation (11.12) where the initial conditions are all set to zero, $\zeta_0 = \zeta_1 = \cdots = \zeta_{n-1} = 0$. Thus $y_S(t)$ is the unforced part of the response, which is due to the initial conditions (or states) only, and $y_I(t)$ constitutes the forced part of the response produced by the input $u(t)$ only. When determining both the responses, the Laplace transformation turns out to be an extremely useful tool.

Zero-Input Response: $y_S(t)$

Here $u(t) = 0$, and thus Equation (11.11) becomes

$$\frac{d^n y(t)}{dt^n} + a_{n-1}\frac{d^{n-1}y(t)}{dt^{n-1}} + \cdots + a_1\frac{dy(t)}{dt} + a_0 y(t) = 0 \qquad (11.14)$$

Laplace transforming the above equation, we get

$$(s^n + a_{n-1}s^{n-1} + \cdots + a_1 s + a_0)Y(s) = \beta_{n-1}s^{n-1} + \cdots + \beta_1 s + \beta_0 \qquad (11.15)$$

where

$$\beta_i = \zeta_{n-1-i} + \sum_{j=0}^{n-2-i} a_{i+j+1}\zeta_j, \quad i = 0, \ldots, n-1$$

This calculation is easy to check if we recall that

$$L\left\{\frac{d^i y(t)}{dt^i}\right\} = s^i Y(s) - s^{i-1}y(0) - s^{i-1}\frac{dy(0)}{dt} - \cdots - \frac{d^{i-1}y(0)}{dt^{i-1}} \qquad (11.16)$$

Consequently,

$$Y(s) = \frac{IC(s)}{D(s)} \qquad (11.17)$$

where

$$IC(s) = \beta_{n-1}s^{n-1} + \cdots + \beta_1 s + \beta_0$$

$$D(s) = s^n + a_{n-1}s^{n-1} + \cdots + a_1 s + a_0$$

The distinct roots of $D(s) = 0$ can be categorized as either multiple or simple (nonmultiple), which yield the factorization

$$D(s) = \prod_{i=1}^{q}(s - \lambda_i)^{k_i}\prod_{i=1}^{r}(s - \lambda_{q+i}) \qquad (11.18)$$

where $\lambda_i, i = 1, \ldots, q$ are multiple roots (each has multiplicity $k_i > 1$) and $\lambda_{q+i}, i = 1, \ldots, r$ are simple roots (each has multiplicity 1). Note that $r + \sum_{i=1}^{q} k_i = n$. Converting the strictly proper rational function on the right-hand side of Equation (11.17) to partial fraction form, we obtain

$$Y(s) = \sum_{i=1}^{q}\sum_{j=1}^{k_i}\frac{c_{i,j}}{(s - \lambda_i)^j} + \sum_{i=1}^{r}\frac{d_i}{s - \lambda_{q+i}} \qquad (11.19)$$

where the $c_{i,j}$s and d_is are constant coefficients called residues. The inverse Laplace transform of Equation (11.19) then gives the sought zero-input response

$$y_S(t) = \sum_{i=1}^{q}\sum_{j=1}^{k_i} \frac{c_{i,j}}{(j-1)!} t^{j-1} e^{\lambda_i t} + \sum_{i=1}^{r} d_i e^{\lambda_{q+i} t} \tag{11.20}$$

In principle, the values of the residues could be computed using the method outlined in the next section on the zero-state response. However, it is easier to determine them making direct use of the initial conditions of Equation (11.12), which, additionally, does not involve the calculation of the coefficients $\beta_i, i = 0, \ldots, n-1$. In order to illustrate this technique, for simplicity we assume that all the roots of $D(s) = 0$ are simple. Substituting Equation (11.20) into Equation (11.12), we get the system of linear equations

$$\begin{bmatrix} 1 & 1 & \cdots & 1 \\ \lambda_1 & \lambda_2 & \cdots & \lambda_n \\ \vdots & \vdots & \vdots & \vdots \\ \lambda_1^{n-1} & \lambda_2^{n-1} & \cdots & \lambda_n^{n-1} \end{bmatrix} \begin{bmatrix} d_1 \\ d_2 \\ \vdots \\ d_n \end{bmatrix} = \begin{bmatrix} \zeta_0 \\ \zeta_1 \\ \vdots \\ \zeta_{n-1} \end{bmatrix} \tag{11.21}$$

whose solution gives the missing values of the d_is. Note that the square matrix of coefficients in Equation (11.21) constitutes a Vandermonde matrix, which can be exploited to make the numerical solution of Equation (11.21) more efficient than for the general linear problem.

Zero-State Response: $y_I(t)$

Assume the initial conditions are zero. Taking the Laplace transform of Equation (11.11) then yields

$$(s^n + a_{n-1}s^{n-1} + \cdots + a_1 s + a_0)Y(s) = (b_m s^m + \cdots + b_1 s + b_0)U(s) \tag{11.22}$$

Defining the Laplace transfer function (or simply the transfer function)

$$H(s) = \frac{b_m s^m + \cdots + b_1 s + b_0}{s^n + a_{n-1}s^{n-1} + \cdots + a_1 s + a_0} \tag{11.23}$$

i.e., the gain between the Laplace transform of the input and the Laplace transform of the output, we see that

$$Y(s) = H(s)U(s) \tag{11.24}$$

The transfer function of a system represents the relationship describing the dynamics of the system under consideration. If $b_m \neq 0$ and $m \leqslant n$, then $H(s)$ is said to be *proper*, otherwise it is called *improper*. If $b_m \neq 0$ and $m < n$, then $H(s)$ is *strictly proper*. Most physical systems have strictly proper transfer functions, and so we also assume that our system is strictly proper, i.e., $m < n$. We also make the standing assumption that there are no common roots between the numerator and the denominator of $H(s)$, i.e., that $H(s)$ is *coprime* (or irreducible), which is the case in most properly modeled physical systems.

The partial fraction expansion of $H(s)$ leads to

$$H(s) = \sum_{i=1}^{q}\sum_{j=1}^{k_i} \frac{f_{i,j}}{(s-\lambda_i)^j} + \sum_{i=1}^{r} \frac{g_i}{s-\lambda_{q+i}} \tag{11.25}$$

where the first term corresponds to the multiple roots of $D(s) = 0$, and the second term corresponds to the single roots of the same equation (cf. Equation (11.18)). The constant residues are evaluated as follows

$$f_{i,j} = \frac{1}{(k_i - j)!} \frac{d^{(k_i - j)}}{ds^{(k_i - j)}} \left\{ (s - \lambda_i)^{k_i} H(s) \right\} \Bigg|_{s=\lambda_i}$$

and

$$g_i = [(s - \lambda_{q+i}) H(s)]_{s=\lambda_{q+i}}$$

The inverse Laplace transform of Equation (11.25) gives

$$h(t) = \sum_{i=1}^{q} \sum_{j=1}^{k_i} \frac{f_{i,j}}{(j-1)!} t^{j-1} e^{\lambda_i t} + \sum_{i=1}^{r} g_i e^{\lambda_{q+i} t} \tag{11.26}$$

which is called the *impulse response* of Equation (11.11). This terminology comes from the fact that $h(t)$ equals the zero-state response to the input $u(t) = \delta(t)$, where $\delta(t)$ is the Dirac delta (impulse) function, which is clear from Equation (11.24) and the transformation $L\{\delta(t)\} = 1$.

The zero-state response to any input can be computed as the inverse Laplace transform of $Y(s) = H(s)U(s)$, which yields the impulse response convolved with the input

$$y_I(t) = L^{-1}\{H(s)U(s)\} = \int_0^t h(t - \tau) u(\tau) d\tau \tag{11.27}$$

Measures of the Dynamic System Response

A control system is an interconnection of components forming a system configuration that will provide a desired system response. Consider a typical configuration represented schematically in Figure 11.5. The process to be controlled is modeled by a proper transfer function $G(s)$ describing the cause-and-effect relationship between the input $u(t)$ and the output $y(t)$. The process is preceded by a controller that takes the difference between a given reference signal $r(t)$ and the output signal $y(t)$, also called the *feedback signal*, thus forming the error signal $e(t) = r(t) - y(t)$. The error signal is then used by the controller to produce the input to the process, $u(t)$, and the relationship between $e(t)$ and $u(t)$ is usually defined in terms of a proper transfer function $H(s)$. A control system that uses a measurement of the process output and feedback of this signal to compare it with the desired output (reference or command) is called the *closed-loop control system*. A proper selection of $H(s)$ makes the process track the desired reference input $r(t)$ with small error $e(t)$. Feedback implements a very natural idea of using the process output to correct the process input and constitutes a foundation of control system design.

Several measures can be employed to investigate the dynamic response performance of a control system. These include:

1. Speed of response—how quickly does the system reach its final value.
2. Accuracy—how close is the final response to the desired response.
3. Relative stability—how stable is the system or how close is the system to instability.
4. Sensitivity—what happens to the system response if the system parameters change.

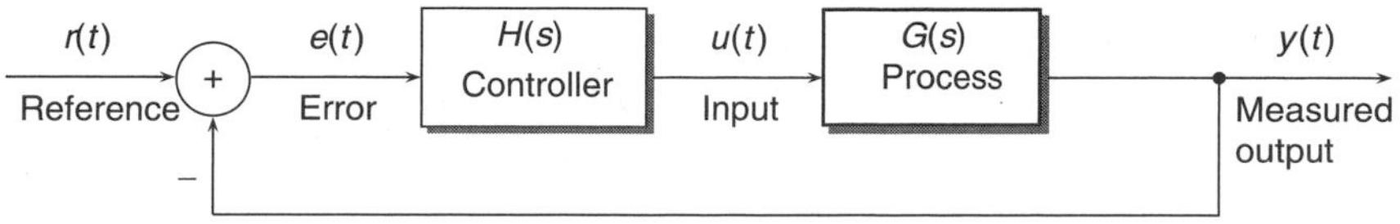

FIGURE 11.5 A closed-loop feedback control system.

Objectives 3 and 4 can be analyzed by frequency domain methods (Section 11.3). Time-domain measures classically analyze the dynamic response by partitioning the total response into its steady-state (Objective 2) and transient (Objective 1) components. The *steady-state response* is that part of the response that remains as time approaches infinity; the *transient response* is the part of the response that vanishes as time approaches infinity.

Measures of the Steady-State Response

In the steady state, the accuracy of the time response is an indication of how well the dynamic response follows a desired time trajectory. Ideally, the output, $y(t)$, equals the reference signal, $r(t)$, and the error, $e(t)$, is zero. This ideal situation is rarely met and, therefore, we have to determine the steady-state error for any system. Usually, a test reference signal is selected to measure accuracy. In the configuration of Figure 11.5, the objective is to force $y(t)$ to track a reference signal $r(t)$ as closely as possible. The *steady-state error* is a measure of the accuracy of the output $y(t)$ in tracking the reference input $r(t)$. It constitutes the error after the transient response has decayed. Other configurations with different performance measures would result in other definitions of the steady-state error between two signals.

From Figure 11.5, the error $e(t)$ is

$$e(t) = r(t) - y(t) \tag{11.28}$$

and the steady-state error is

$$e_{\text{SS}} = \lim_{t \to \infty} e(t) = \lim_{s \to 0} sE(s) \tag{11.29}$$

assuming that the limits exist, where $E(s)$ is the Laplace transform of $e(t)$. The last equality in Equation (11.21) results from the final-value theorem, being a fundamental property of the Laplace transformation. Here a simple pole of $E(s)$, i.e., a root of its denominator at the origin, is permitted, but poles on the imaginary axis and in the right half-plane and repeated poles at the origin are excluded.

The transfer function between $y(t)$ and $r(t)$ is found to be

$$T(s) = \frac{G(s)H(s)}{1 + G(s)H(s)} \tag{11.30}$$

with

$$E(s) = \frac{R(s)}{1 + G(s)H(s)} \tag{11.31}$$

where $R(s)$ is the Laplace transform of $r(t)$. Direct application of Equation (11.29) and Equation (11.31) for various test signals yields Table 11.2. This table can be extended to a higher-order polynomial reference $r(t) = At^m/m!$. Clearly, the steady-state error then depends on the structure of the product $P(s) = G(s)H(s)$. If $P(s)$ has no poles at the origin, then $P(0)$ is finite, which means that the steady-state step-response error is finite, and all response errors for polynomial inputs are infinite for $m > 0$. We define the *system type* as the order of the input polynomial that the closed-loop system can track with finite steady-state error. For $e(t)$ to go to zero with a reference signal $r(t) = At^m/m!$, the rational function $P(s)$ must have at least $m + 1$ poles at the origin (a type m system).

TABLE 11.2 Steady-State Error Constants

Test Signal	$r(t)$	$R(s)$	e_{ss}	Error Constant
Step	$r(t)=\begin{cases}A, t>0\\0, t<0\end{cases}$	$\frac{A}{s}$	$\frac{A}{1+K_p}$	$K_p=\lim_{s\to 0} G(s)H(s)$
Ramp	$r(t)=\begin{cases}At, t>0\\0, t<0\end{cases}$	$\frac{A}{s^2}$	$\frac{A}{K_v}$	$K_v=\lim_{s\to 0} sG(s)H(s)$
Parabolic	$r(t)=\begin{cases}At^2/2, t>0\\0, t<0\end{cases}$	$\frac{A}{s^3}$	$\frac{A}{K_a}$	$K_a=\lim_{s\to 0} s^2G(s)H(s)$

Measures of the Transient Response

In general, analysis of the transient response of a dynamic system to a reference input is difficult. Hence formulating a standard measure of performance becomes complicated. In many cases, the response is dominated by a pair of poles and thus acts like a second-order system.

Consider a reference unit step input to a dynamic system (Figure 11.6). Critical parameters that measure the transient response include:

1. M: maximum overshoot
2. %overshoot $= M/A \times 100\%$, where A is the final value of the time response
3. t_d: delay time, the time required to reach 50% of A
4. t_r: rise time, the time required to go from 10% of A to 90% of A
5. t_s: settling time, the time required for the response to reach and stay within 5% of A

To calculate these measures, consider a second-order system

$$T(s) = \frac{\omega_n^2}{s^2 + 2\xi\omega_n s + \omega_n^2} \tag{11.32}$$

where ξ is the damping coefficient and ω_n is the natural frequency of oscillation.

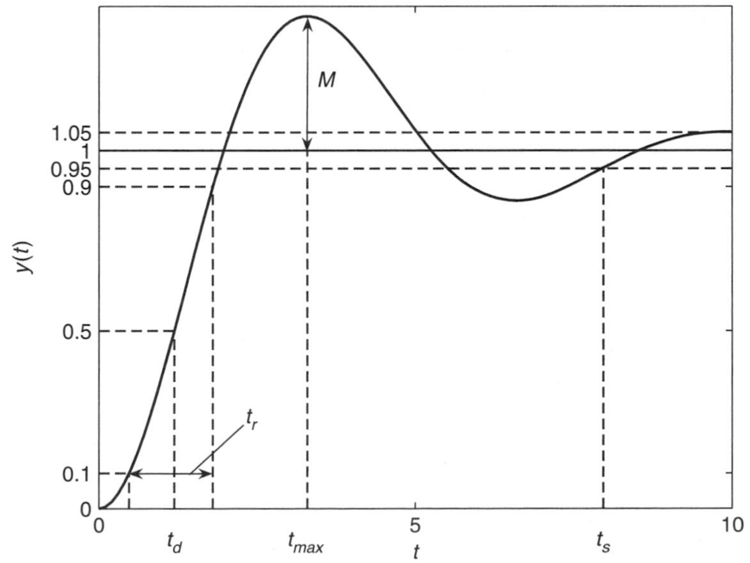

FIGURE 11.6 Step response of a second-order system ($e_{ss} = 1$).

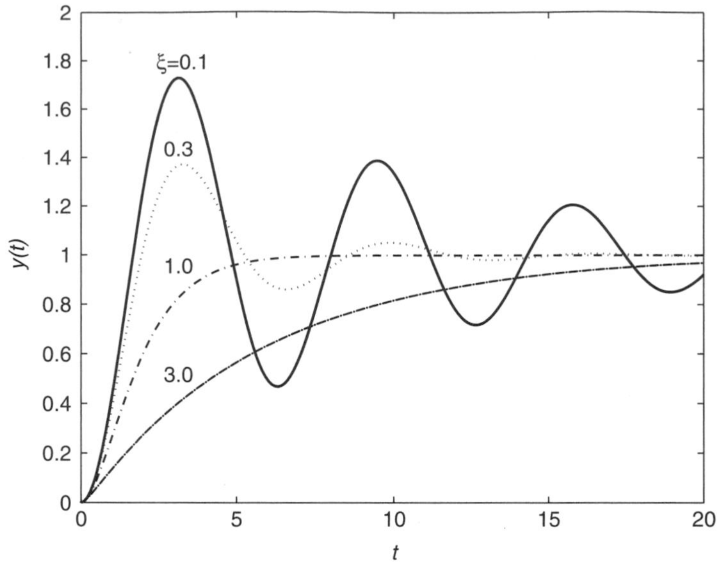

FIGURE 11.7 Effect of the damping coefficient on the dynamic response ($\omega_n = 1$).

For the range $0<\xi<1$, the system response is *underdamped*, resulting in a damped oscillatory output. For a unit step input we have $U(s) = 1/s$, so that the Laplace transform of the response is

$$Y(s) = \frac{1}{s}T(s) = \frac{\omega_{\mathrm{n}}^2}{s(s^2 + 2\xi\omega_{\mathrm{n}}s + \omega_{\mathrm{n}}^2)} \tag{11.33}$$

Partial fraction expansion and the inverse Laplace transformation yield

$$y(t) = 1 - \frac{e^{-\xi\omega_n t}}{\sqrt{1-\xi^2}}\sin(\omega_{\mathrm{n}}\sqrt{1-\xi^2}t + \arccos(\xi)) \quad (0<\xi<1) \tag{11.34}$$

The eigenvalues (poles) of the system (roots of the denominator of $T(s)$) provide some measure of the time constants of the system. For the system under study, the eigenvalues are at

$$\left(-\xi \pm j\sqrt{1-\xi^2}\right)\omega_{\mathrm{n}} \quad \text{where} \quad j = \sqrt{-1} \tag{11.35}$$

From the expression of $y(t)$, one sees that the term $\xi\omega_{\mathrm{n}}$ affects the rise time and exponential decay time. The effects of the damping coefficient on the transient response are seen in Figure 11.7. The effects of the natural frequency of oscillation ω_{n} on the transient response can be seen in Figure 11.8. As ω_{n} increases, the frequency of oscillation increases.

For the case when $0 < \xi < 1$, the underdamped case, one can analyze the critical transient response parameters.

To measure the peaks of Figure 11.6, one finds

$$y_{\mathrm{peak}}(t) = 1 + (-1)^{n-1}\exp\left(\frac{-n\pi\xi}{\sqrt{1-\xi^2}}\right), \quad n = 0, 1, \ldots \tag{11.36}$$

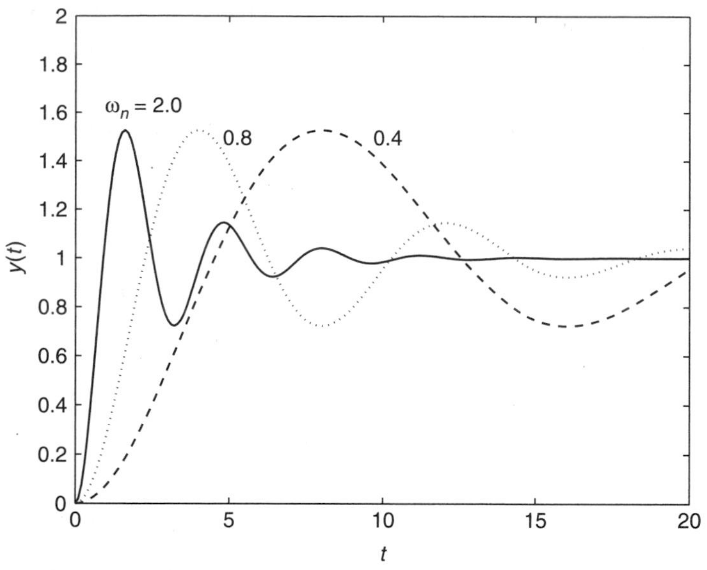

FIGURE 11.8 Effect of the natural frequency of oscillation on the dynamic response ($\xi = 0.2$).

occurring at

$$t = \frac{n\pi}{\omega_n\sqrt{1-\xi^2}} \qquad \begin{matrix} n: \text{odd (overshoot)} \\ n: \text{even (undershoot)} \end{matrix} \tag{11.37}$$

Hence

$$M = \exp\left(\frac{-\pi\xi}{\sqrt{1-\xi^2}}\right) \tag{11.38}$$

occurring at the peak time

$$t_{\max} = \frac{\pi}{\omega_n\sqrt{1-\xi^2}} \tag{11.39}$$

With these parameters, one finds

$$t_d \approx \frac{1+0.7\xi}{\omega_n}, \quad t_r \approx \frac{1+1.1\xi+1.4\xi^2}{\omega_n}, \quad t_s \approx \frac{3}{\xi\omega_n} \tag{11.40}$$

The swiftness of the response can be quantified by t_r and $t_{\max}$, while its closeness to the desired response is represented by M and t_s.

When $\xi = 1$, the system has a double pole at $s = -\omega_n$, resulting in a *critically damped* response. This is the point where the response just changes from oscillatory to exponential in form. For a unit step input, the response is then given by

$$y(t) = 1 - e^{-\omega_n t}(1 + \omega_n t) \quad (\xi = 1) \tag{11.41}$$

For the range $\xi > 1$, the system is *overdamped* due to two real system poles. For a unit step input, the response is given by

$$y(t) = 1 + \frac{1}{c_1 - c_2}(c_2 e^{c_1 \omega_n t} - c_1 e^{c_2 \omega_n t}) \quad (\xi > 1) \tag{11.42}$$

where

$$c_1 = -\xi + \sqrt{\xi^2 - 1}, \quad c_2 = -\xi - \sqrt{\xi^2 - 1} \tag{11.43}$$

Finally, when $\xi = 0$, the response is purely sinusoidal. For a unit step input, it is given by

$$y(t) = 1 - \cos(\omega_n t) \quad (\xi = 0) \tag{11.44}$$

Defining Terms

Impulse response: The response of a system when the input is the Dirac delta (impulse) function.
Steady-state error: The difference between the desired reference signal and the actual signal in steady state, i.e., when time approaches infinity.
Steady-state response: The part of the response that remains as time approaches infinity.
Transient response: The part of the response that vanishes as time approaches infinity.
Zero-input response: The part of the response due to the initial conditions only.
Zero-state response: The part of the response due to the input only.

Further Information

R.C. Dorf and R.H. Bishop, *Modern Control Systems*, 10th ed., Upper Saddle River, NJ: Pearson Education, 2005.
B. Friedland, *Advanced Control System Design*, Englewood Cliffs, NJ: Prentice Hall, 1996.
W. Levine, *The Control Handbook*, Boca Raton, FL: CRC Press, 1996.
N. Nise, *Control Systems Engineering*, Reading, MA: Addison-Wesley, 1995.
C. Phillips and R. Harbor, *Feedback Control Systems*, Englewood Cliffs, NJ: Prentice Hall, 1996.

11.3 Frequency Response Methods: Bode Diagram Approach

Andrew P. Sage

Our efforts in this section are concerned with analysis and design of linear control systems by frequency response methods. Design generally involves trial-and-error repetition of analysis until a set of design **specifications** has been met. Thus, analysis methods are most useful in the design process, which is one phase of the **systems engineering** life cycle. We will discuss one design method based on **Bode diagrams.** We will discuss the use of both simple **series equalizers** and composite equalizers as well as the use of minor-loop feedback in systems design.

Figure 11.9 presents a flowchart of the frequency response method design process and indicates the key role of analysis in linear systems control design. The flowchart of Figure 11.9 is applicable to control system design methods in general. There are several iterative loops, generally calling for trial-and-error efforts, that comprise the suggested design process. An experienced designer will often be able, primarily due to successful prior experience, to select a system structure and generic components such that the design specifications can be met

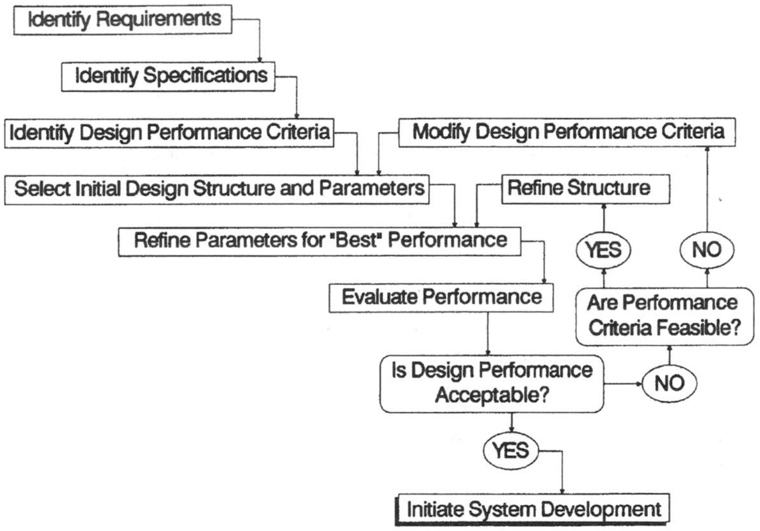

FIGURE 11.9 System design life cycle for frequency-response-based design.

with no or perhaps a very few iterations through the iterative loop involving adjustment of equalizer or compensation parameters to best meet specifications.

If the parameter optimization, or parameter refinement such as to lead to maximum phase margin, approach shows the specifications cannot be met, we are then assured that no **equalizer** of the specific form selected will meet specifications. The next design step, if needed, would consist of modification of the equalizer form or structure and repetition of the analysis process to determine equalizer parameter values to best meet specifications. If specifications still cannot be met, we will usually next modify generic fixed components used in the system. This iterative design and analysis process is again repeated. If no reasonable fixed components can be obtained to meet specifications, then structural changes in the proposed system are next contemplated. If no structure can be found that allows satisfaction of specifications, either the client must be requested to relax the frequency response specifications or the project may be rejected as infeasible using present technology. As we might suspect, economics will play a dominant role in this design process. Changes made due to iteration in the inner loops of Figure 11.9 normally involve little additional costs, whereas those made due to iterations in the outer loops will often involve major cost changes.

Frequency Response Analysis Using the Bode Diagram

The steady-state response of a stable linear constant-coefficient system has particular significance, as we know from an elementary study of electrical networks and circuits and of dynamics. We consider a stable linear system with input-output transfer function

$$H(s) = \frac{Z(s)}{U(s)}$$

We assume a sinusoidal input $u(t) = \cos \omega t$ so that we have for the Laplace transform of the system output

$$Z(s) = \frac{sH(s)}{s^2 + \omega^2}$$

We expand this ratio of polynomials using the partial-fraction approach and obtain

$$Z(s) = F(s) + \frac{a_1}{s + j\omega} + \frac{a_2}{s - j\omega}$$

In this expression, $F(s)$ contains all the poles of $H(s)$. All of these lie in the left half plane since the system, represented by $H(s)$, is assumed to be stable. The coefficients a_1 and a_2 are easily determined as

$$a_1 = \frac{H(-j\omega)}{2}$$

$$a_2 = \frac{H(j\omega)}{2}$$

We can represent the complex transfer function $H(j)$ in either of two forms,

$$H(j\omega) = B(\omega) + jC(\omega)$$

$$H(-j\omega) = B(\omega) - jC(\omega)$$

The inverse Laplace transform of the system transfer function will result in a transient term due to the inverse transform of $F(s)$, which will decay to zero as time progresses. A steady-state component will remain, and this is, from the inverse transform of the system equation, given by

$$z(t) = a_1 \mathrm{e}^{-j\omega t} + a_2^{j\omega t}$$

We combine several of these relations and obtain the result

$$z(t) = B(\omega)\left(\frac{\mathrm{e}^{j\omega t} + \mathrm{e}^{-j\omega t}}{2}\right) - C(\omega)\left(\frac{\mathrm{e}^{j\omega t} - \mathrm{e}^{-j\omega t}}{2j}\right)$$

This result becomes, using the Euler identity,[1]

$$\begin{aligned} z(t) &= B(\omega) \cos \omega t - C(\omega) \sin \omega t \\ &= [B^2(\omega) + C^2(\omega)]^{1/2} \cos(\omega + \beta) \\ &= |H(j\omega)| \cos(\omega t + \beta) \end{aligned}$$

where $\tan \beta(\omega) = C(\omega)/B(\omega)$.

As we see from this last result, there is a very direct relationship between the transfer function of a linear constant-coefficient system, the time response of a system to any known input, and the sinusoidal steady-state response of the system. We can always determine any of these if we are given any one of them. This is a very important result. This important conclusion justifies a design procedure for linear systems that is based only on sinusoidal steady-state response, as it is possible to determine transient responses, or responses to any given system input, from a knowledge of steady-state sinusoidal responses, at least in theory. In practice, this might be rather difficult computationally without some form of automated assistance.

[1]The Euler identity is $\mathrm{e}^{j\omega t} = \cos \omega t + j\sin \omega t$

Bode Diagram Design-Series Equalizers

In this subsection we consider three types of series equalization:

1. Gain adjustment, normally attenuation by a constant at all frequencies
2. Increasing the phase lead, or reducing the phase lag, at the **crossover frequency** by use of a phase **lead network**
3. Attenuation of the gain at middle and high frequencies, such that the crossover frequency will be decreased to a lower value where the phase lag is less, by use of a **lag network**

In the subsection that follows this, we will first consider use of a composite or **lag-lead network** near crossover to attenuate gain only to reduce the crossover frequency to a value where the phase lag is less. Then we will consider more complex composite equalizers and state some general guidelines for Bode diagram design. Here, we will use Bode diagram frequency domain design techniques to develop a design procedure for each of three elementary types of series equalization.

Gain Reduction

Many linear control systems can be made sufficiently stable merely by reduction of the open-loop system gain to a sufficiently low value. This approach ignores all performance specifications, however, except that of phase margin (PM) and is, therefore, usually not a satisfactory approach. It is a very simple one, however, and serves to illustrate the approach to be taken in more complex cases.

The following steps constitute an appropriate Bode diagram design procedure for compensation by gain adjustment:

1. Determine the required PM and the corresponding phase shift $\beta_c = -\pi + \text{PM}$.
2. Determine the frequency ω_c at which the phase shift is such as to yield the phase shift at crossover required to give the desired PM.
3. Adjust the gain such that the actual crossover frequency occurs at the value computed in step 2.

Phase-Lead Compensation

In compensation using a phase-lead network, we increase the phase lead at the crossover frequency such that we meet a performance specification concerning phase shift. A phase-lead-compensating network transfer function is

$$G_c(s) = \left(1 + \frac{s}{\omega_1}\right) \Big/ \left(1 + \frac{s}{\omega_2}\right) \qquad \omega_1 < \omega_2$$

Figure 11.10 illustrates the gain versus frequency and phase versus frequency curves for a simple lead network with the transfer function of the foregoing equation. The maximum phase lead obtainable from a phase-lead network depends upon the ratio ω_2/ω_1 that is used in designing the network. From the expression for the phase shift of the transfer function for this system, which is given by

$$\beta = \tan^{-1}\frac{\omega}{\omega_1} - \tan^{-1}\frac{\omega}{\omega_2}$$

we see that the maximum amount of phase lead occurs at the point where the first derivative with respect to frequency is zero, or

$$\left.\frac{d\beta}{d\omega}\right|_{\omega=\omega_m} = 0$$

or at the frequency where

$$\omega_m = (\omega_1\omega_2)^{0.5}$$

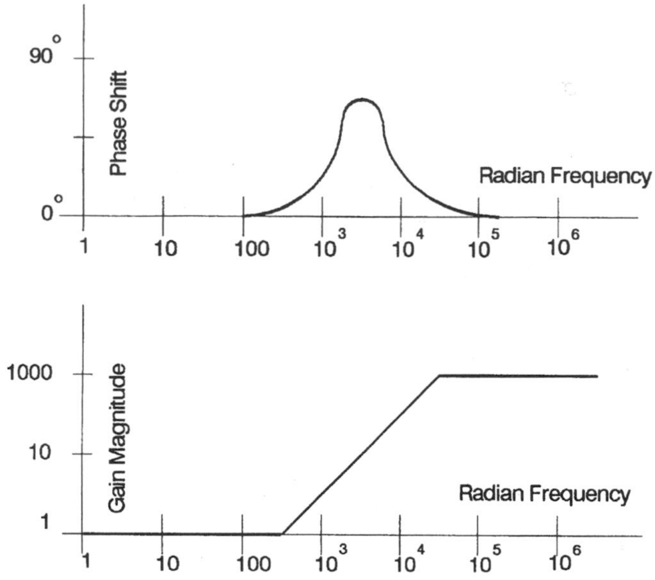

FIGURE 11.10 Phase shift and gain curves for a simple lead network.

This frequency is easily seen to be at the center of the two break frequencies for the lead network on a Bode log asymptotic gain plot. It is interesting to note that this is exactly the same frequency that we would obtain using an arctangent approximation[1] with the assumption that $\omega_1 < \omega < \omega_2$.

There are many ways of realizing a simple phase-lead network. All methods require the use of an active element, since the gain of the lead network at high frequencies is greater than 1. A simple electrical network realization is shown in Figure 11.11.

We now consider a simple design example. Suppose that we have an open-loop system with transfer function

$$G_f(s) = \frac{10^4}{s^2}$$

It turns out that this is often called a type-two system due to the presence of the double integration. This system will have a steady-state error of zero for a constant acceleration input. The crossover frequency, that is to say the frequency where the magnitude of the open-loop gain is 1, is 11 rad/s. The PM for the system without equalization is zero. We will design a simple lead network compensation for this zero-PM system. If uncompensated, the closed-loop transfer function will be such that the system is unstable and any disturbance at all will result in a sinusoidally oscillating output.

The asymptotic gain diagram for this example is easily obtained from the open-loop transfer function

$$G_f(s)G_c(s) = \frac{K(1 + s/\omega_1)}{(1 + s/\omega_2)s^2}$$

and we wish to select the break frequencies ω_1 and ω_2 such that the phase shift at crossover is maximum. Further, we want this maximum phase shift to be such that we obtain the specified PM. We use the procedure suggested in Figure 11.12.

[1]The arctangent approximation is $\tan^{-1}(\omega/\alpha) = \omega/\alpha$ for $\omega < \alpha$ and $\tan^{-1}(\omega/\alpha) = \pi/2 - \alpha/\omega$ for $\omega > \alpha$. This approximation is rather easily obtained through use of a Taylor series approximation.

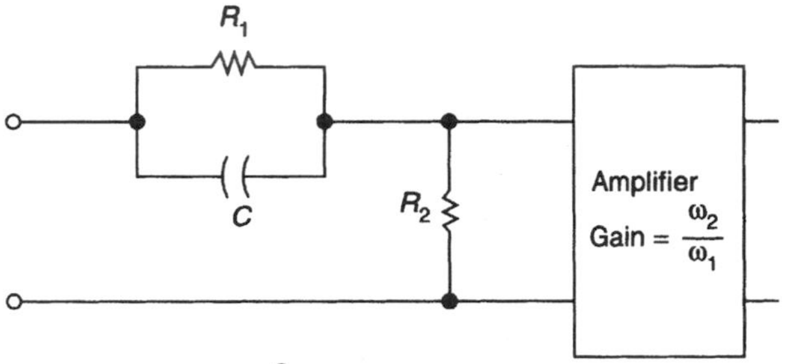

$$G_c = \frac{1 + \frac{s}{\omega_1}}{1 + \frac{s}{\omega_2}}$$

$$\omega_1 = \frac{1}{R_1 C}, \qquad \omega_2 = \left(1 + \frac{R_1}{R_2}\right)\omega_1$$

FIGURE 11.11 A simple electrical lead network.

Since the crossover frequency is such that $\omega_1 < \omega_c < \omega_2$, we have for the arctangent approximation to the phase shift in the vicinity of the crossover frequency

$$\begin{aligned}\beta(\omega) &= -\pi + \tan^{-1}\frac{\omega}{\omega_1} - \tan^{-1}\frac{\omega}{\omega_2} \\ &\approx \frac{-\pi}{2} - \frac{\omega_1}{\omega} - \frac{\omega}{\omega_2}\end{aligned}$$

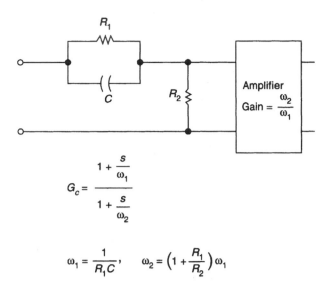

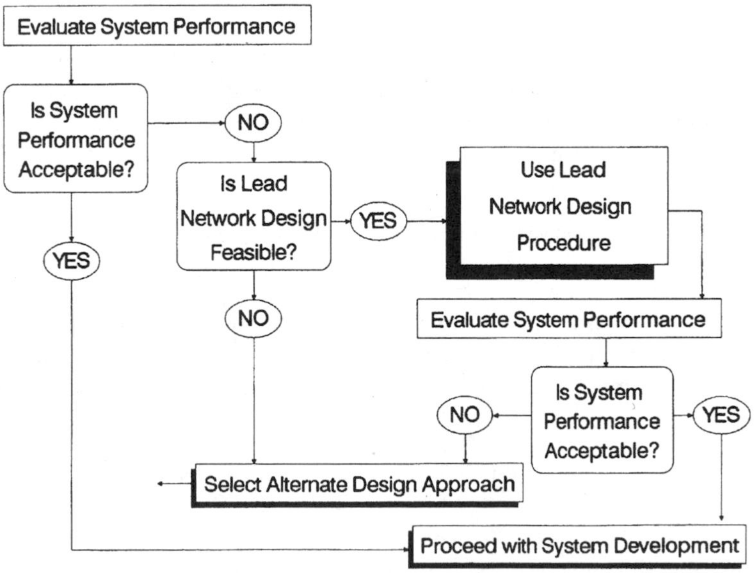

FIGURE 11.12 Life cycle of frequency domain design incorporating lead network compensation.

In order to maximize the phase shift at crossover, we set

$$\left.\frac{d\beta}{d\omega}\right|_{\omega=\omega_m} = 0$$

and obtain as a result

$$\omega_m = (\omega_1\omega_2)^{0.5}$$

We see that the crossover frequency obtained is halfway between the two break frequencies ω_1 and ω_2 on a logarithmic frequency coordinate. The phase shift at this optimum value of crossover frequency becomes

$$\beta_c = \beta(\omega_c) = \frac{-\pi}{2} - 2\left(\frac{\omega_1}{\omega_2}\right)^{0.5}$$

For a PM of $-3\pi/4$, for example, we have $-3\pi/4 = -\pi/2 - 2(\omega_1/\omega_2)^{0.5}$, and we obtain $\omega_1/\omega_2 = 0.1542$ as the ratio of frequencies. We see that we have need for a lead network with a gain of $\omega_1/\omega_2 = 6.485$. The gain at the crossover frequency is 1, and from the asymptotic gain approximation that is valid for $\omega_1 < \omega < \omega_2$, we have the expressions $|G(j\omega)| = K/\omega\omega_1$ and $|G(j\omega_c)| = 1 = K/\omega_c\omega_1$, which for a known K can be solved for ω_c and ω_1.

Now that we have illustrated the design computation with a very simple example, we are in a position to state some general results. In the direct approach to design for a specified PM we assume a single lead network equalizer such that the open-loop system to transfer function results. This approach to design results in the following steps that are applicable for Bode diagram design to achieve maximum PM within an experientially determined control system structure that comprises a fixed plant and a compensation network with adjustable parameters:

1. We find an equation for the gain at the crossover frequency in terms of the compensated open-loop system break frequency.
2. We find an equation of the phase shift at crossover.
3. We find the relationship between equalizer parameters and crossover frequency such that the phase shift at crossover is the maximum possible and a minimum of additional gain is needed.
4. We determine all parameter specifications to meet the PM specifications.
5. We check to see that all design specifications have been met. If they have not, we iterate the design process.

Figure 11.12 illustrates the steps involved in implementing this frequency domain design approach.

Phase-Lag Compensation

In the phase-lag-compensation frequency domain design approach, we reduce the gain at low frequencies such that crossover, the frequency where the gain magnitude is 1, occurs before the phase lag has had a chance to become intolerably large. A simple single-stage phase-lag-compensating network transfer function is

$$G_c(s) = \frac{1 + s/\omega_2}{1 + s/\omega_1} \qquad \omega_1 < \omega_2$$

Figure 11.13 illustrates the gain and phase versus frequency curves for a simple lag network with this transfer function. The maximum phase lag obtainable from a phase-lag network depends upon the ratio ω_2/ω_1 that is used in designing the network. From the expression for the phase shift of this transfer function,

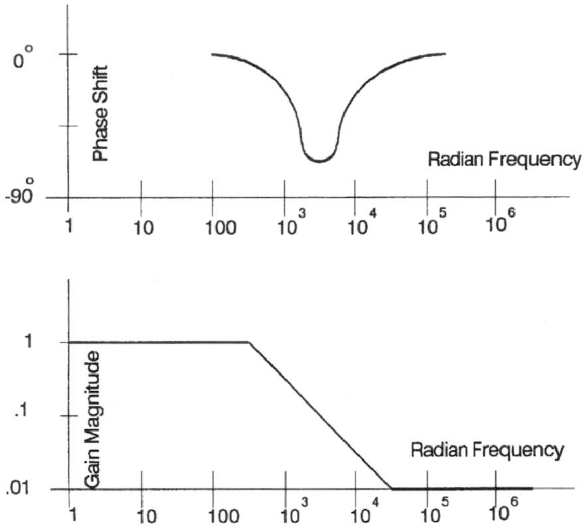

FIGURE 11.13 Phase shift and gain curves for a simple lag network.

$$\beta = \tan^{-1}\frac{\omega}{\omega_2} - \tan^{-1}\frac{\omega}{\omega_1}$$

we see that maximum phase lag occurs at that frequency ω_c where $d\beta/d\omega = 0$. We obtain for this value

$$\omega_m = (\omega_1\omega_2)^{0.5}$$

which is at the center of the two break frequencies for the lag network when the frequency response diagram is illustrated on a Bode diagram log-log asymptotic gain plot.

The maximum value of the phase lag obtained at $\omega = \omega_m$ is

$$\beta_m(\omega_m) = \frac{\pi}{2} - 2\tan^{-1}\left(\frac{\omega_2}{\omega_1}\right)^{0.5}$$
$$= \frac{\pi}{2} - 2\tan^{-1}\left(\frac{\omega_1}{\omega_2}\right)^{0.5}$$

which can be approximated in a more usable form, using the arctangent approximation, as

$$\beta_m(\omega_m) \approx \frac{\pi}{2}\sqrt{\frac{\omega_2}{\omega_1}}$$

The attenuation of the lag network at the frequency of minimum phase shift, or maximum phase lag, is obtained from the asymptotic approximation as

$$|G_c(\omega_m)| = \left(\frac{\omega_1}{\omega_2}\right)^{0.5}$$

Figure 11.13 presents a curve of attenuation magnitude obtainable at the frequency of maximum phase lag and the amount of the phase lag for various ratios ω_2/ω_1 for this simple lag network.

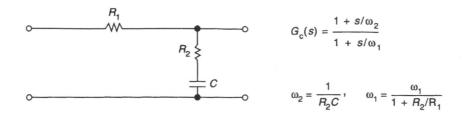

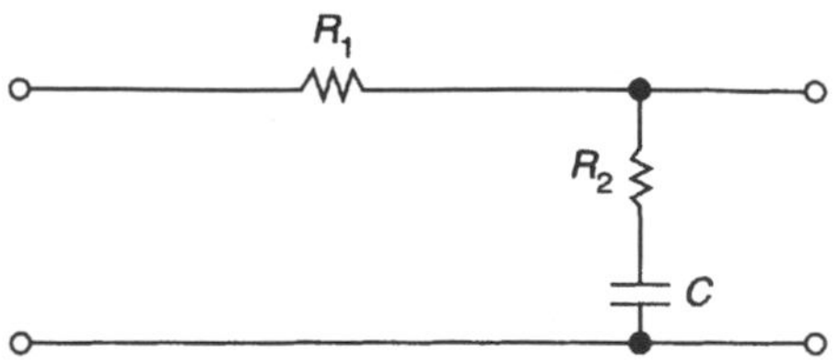

$$G_c(s) = \frac{1 + s/\omega_2}{1 + s/\omega_1}$$

$$\omega_2 = \frac{1}{R_2 C}, \quad \omega_1 = \frac{\omega_1}{1 + R_2/R_1}$$

FIGURE 11.14 A simple electrical lag network.

There are many ways to physically realize a lag network transfer function. Since the network only attenuates at some frequencies, as it never has a gain greater than 1 at any frequency, it can be realized with passive components only. Figure 11.14 presents an electrical realization of the simple lag network. Figure 11.15 presents a flowchart illustrating the design procedure envisioned here for lag network design. This is conceptually very similar to that for a lead network and makes use of the five-step parameter optimization procedure suggested earlier.

The object of lag network design is to reduce the gain at frequencies lower than the original crossover frequency in order to reduce the open-loop gain to unity before the phase shift becomes so excessive that the system PM is too small. A disadvantage of lag network compensation is that the attenuation introduced reduces the crossover frequency and makes the system slower in terms of its transient response. Of course, this would be advantageous if high-frequency noise is present and we wish to reduce its effect. The lag network is an entirely passive device and thus is more economical to instrument than the lead network.

In lead network compensation we actually insert phase lead in the vicinity of the crossover frequency to increase the PM. Thus we realize a specified PM without lowering the medium-frequency system gain. We see that the disadvantages of the lag network are the advantages of the lead network and the advantages of the lag network are the disadvantages of the lead network.

We can attempt to combine the lag network with the lead network into an all-passive structure called a lag-lead network. Generally we obtain better results than we can achieve using either a lead or a lag network.

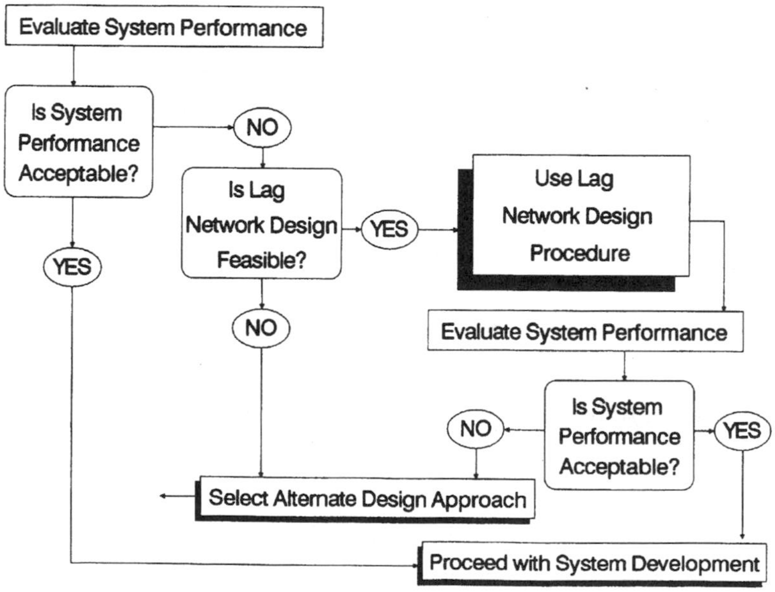

FIGURE 11.15 Life cycle of frequency domain design incorporating lag network compensation.

We will consider design using lag-lead networks in our next subsection as well as more complex composite equalization networks.

Composite Equalizers

In the previous subsection we examined the simplest forms of series equalization: gain adjustment, lead network compensation, and lag network compensation. In this subsection we will consider more complex design examples in which composite equalizers will be used for series compensation. The same design principles used earlier in this section will be used here as well.

Lag-Lead Network Design

The prime purpose of a lead network is to add phase lead near the crossover frequency to increase the PM. Accompanied with this phase lead is a gain increase that will increase the crossover frequency. This will sometimes cause difficulties if there is much phase lag in the uncompensated system at high frequencies. There may be situations where use of a phase-lead network to achieve a given PM is not possible due to too many high-frequency poles.

The basic idea behind lag network design is to reduce the gain at "middle" frequencies such as to reduce the crossover frequency to a lower value than for the uncompensated system. If the phase lag is less at this lower frequency, then the PM will be increased by use of the lag network. We have seen that it is not possible to use a lag network in situations in which there is not a frequency where an acceptable PM would exist if this frequency were the crossover frequency. Even if use of a lag network is possible, the significantly reduced crossover frequency resulting from its use may make the system so slow and sluggish in response to an input that system performance is unacceptable even though the relative stability of the system is acceptable.

Examination of these characteristics or attributes of lead network and lag network compensation suggests that it might be possible to combine the two approaches to achieve the desirable features of each approach. Thus we will attempt to provide attenuation below the crossover frequency to decrease the phase lag at crossover and phase lead closer to the crossover frequency in order to increase the phase lead of the uncompensated system at the crossover frequency.

The transfer function of the basic lag-lead network is

$$G_c(s) = \frac{(1 + s/\omega_2)(1 + s/\omega_3)}{(1 + s/\omega_1)(1 + s/\omega_4)}$$

where $\omega_4 > \omega_3 > \omega_2 > \omega_1$. Often it is desirable that $\omega_2\omega_3 = \omega_1\omega_4$ such that the high-frequency gain of the equalizer is unity. It is generally not desirable that $\omega_1\omega_4 > \omega_2\omega_3$ as this indicates a high-frequency gain greater than 1, and this will require an active network, or gain, and a passive equalizer. It is a fact that we should always be able to realize a linear minimum phase network using passive components only if the network has a rational transfer function with a gain magnitude that is no greater than 1 at any real frequency.

Figure 11.16 illustrates the gain magnitude and phase shift curves for a single-stage lag-lead network equalizer or compensator transfer function. Figure 11.17 illustrates an electrical network realization of a passive lag-lead network equalizer. Parameter matching can be used to determine the electrical network parameters that yield a specified transfer function. Because the relationships between the break frequencies and the equalizer component values are complex, it may be desirable, particularly in preliminary instrumentation of the control system, to use analog or digital computer programming techniques to construct the equalizer. Traditionally, there has been much analog computer simulation of control systems. The more contemporary approach suggests use of digital computer approaches that require numerical approximation of continuous-time physical systems.

Figure 11.18 presents a flowchart that we may use for lag-lead network design. We see that this flowchart has much in common with the charts and design procedures for lead network and lag network design and that each of these approaches first involves determining or obtaining a set of desired specifications for the control system. Next, the form of a trial compensating network and the number of break frequencies in the network

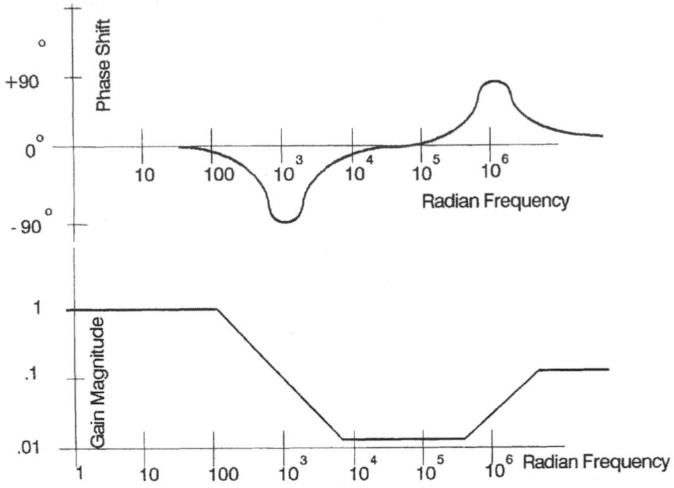

FIGURE 11.16 Phase shift and gain curves for a simple lag-lead network.

are selected. We must then obtain a number of equations, equal to the number of network break frequencies plus 1. One of these equations shows that the gain magnitude is 1 at the crossover frequency. The second equation will be an equation for the phase shift at crossover. It is generally desirable that there be at least two unspecified compensating network break frequencies such that we may use a third equation, the optimality of the phase shift at crossover equation, in which we set $d\beta/d\omega|_{\omega=\omega c}= 0$. If other equations are needed to represent the design situation, we obtain these from the design specifications themselves.

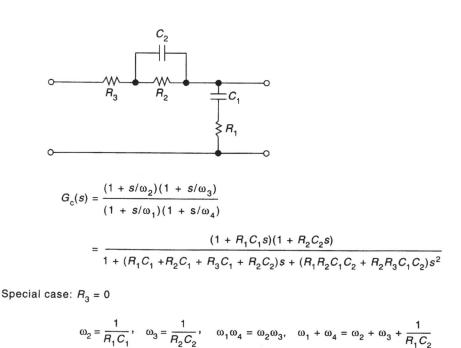

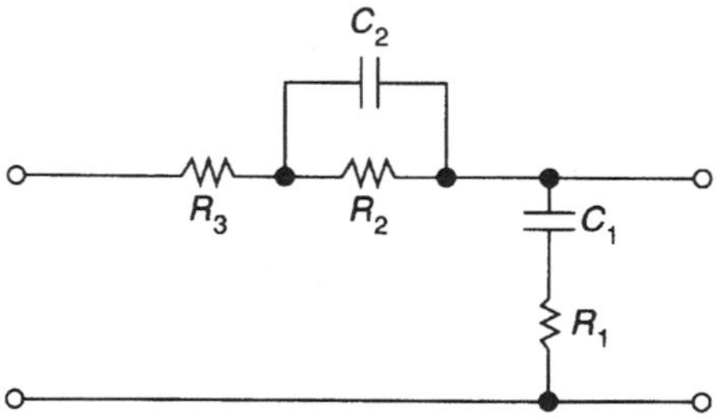

$$G_c(s) = \frac{(1 + s/\omega_2)(1 + s/\omega_3)}{(1 + s/\omega_1)(1 + s/\omega_4)}$$

$$= \frac{(1 + R_1C_1s)(1 + R_2C_2s)}{1 + (R_1C_1 + R_2C_1 + R_3C_1 + R_2C_2)s + (R_1R_2C_1C_2 + R_2R_3C_1C_2)s^2}$$

Special case: $R_3 = 0$

$$\omega_2 = \frac{1}{R_1C_1}, \quad \omega_3 = \frac{1}{R_2C_2}, \quad \omega_1\omega_4 = \omega_2\omega_3, \quad \omega_1 + \omega_4 = \omega_2 + \omega_3 + \frac{1}{R_1C_2}$$

FIGURE 11.17 Simple electrical lag-lead network.

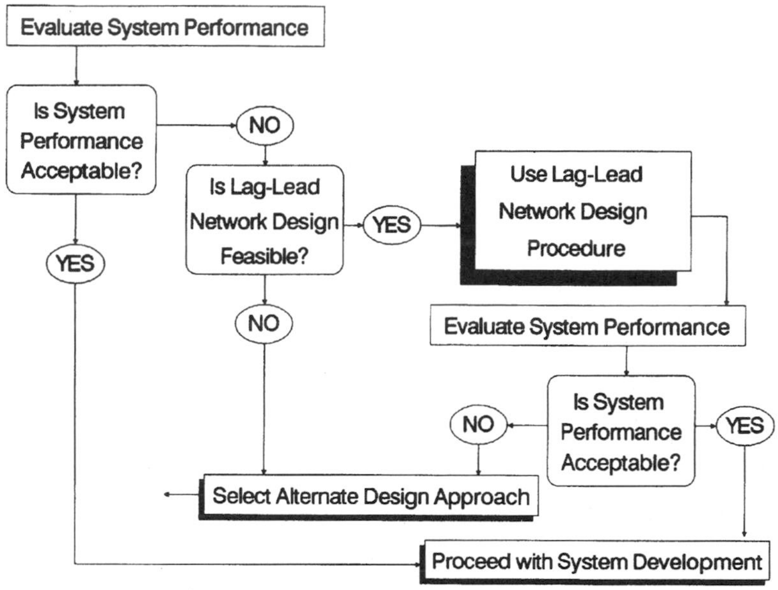

FIGURE 11.18 Life cycle of frequency domain design incorporating lag-lead network compensation.

General Bode Diagram Design

Figure 11.19 presents a flowchart of a general design procedure for Bode diagram design. As we will see in the next subsection, a minor modification of this flowchart can be used to accomplish design using minor-loop feedback or a combination of minor-loop and series equations. These detailed flowcharts for Bode diagram design are, of course, part of the overall design procedure of Figure 11.9.

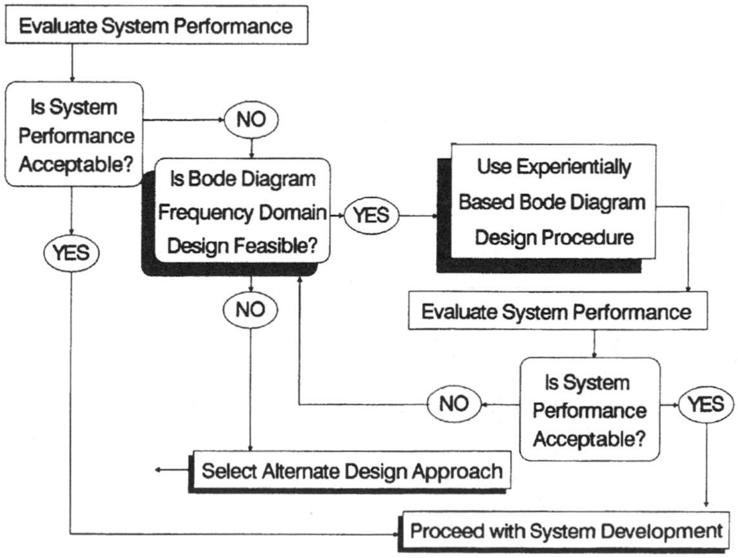

FIGURE 11.19 Life cycle of frequency domain design incorporating general Bode diagram compensation procedure.

Much experience leads to the conclusion that satisfactory linear systems control design using frequency response approaches is such that the crossover frequency occurs on a gain magnitude curve which has a –1 slope at the crossover frequency. In the vicinity of crossover we may approximate any minimum phase transfer function, with crossover on a –1 slope, by

$$G(s) = G_f(s)G_c(s) = \frac{\omega_c \omega_1^{n-1}(1 + s/\omega_1)^{n-1}}{s^n(1 + s/\omega_2)^{m-1}} \qquad \text{for } \omega_1 > \omega_c > \omega_2$$

Here ω_1 is the break frequency just prior to crossover and ω_2 is the break frequency just after crossover. It is easy to verify that we have $|G(j\omega_c)| = 1$ if $\omega_1 > \omega_c > \omega_2$. Figure 11.20 illustrates this rather general approximation to a compensated system Bode diagram in the vicinity of the crossover frequency. We will conclude this subsection by determining some general design requirements for a system with this transfer function and the associated Bode asymptotic gain magnitude diagram of Figure 11.20.

There are three unknown frequencies in the foregoing equation. Thus we need three requirements or equations to determine design parameters. We will use the same three requirements used thus far in all our efforts in this section, namely:

1. The gain at crossover is 1.
2. The PM is some specified value.
3. The PM at crossover is the maximum possible for a given ω_2/ω_1 ratio.

We see that the first requirement, that the gain is 1 at the crossover frequency, is satisfied by the foregoing equation if the crossover frequency occurs on the –1 slope portion of the gain curve as assumed in Figure 11.20. We use the arctangent approximation to obtain the phase shift in the vicinity of crossover as

$$\beta(\omega) = -\frac{n\pi}{2} + (n-1)\left(\frac{\pi}{2} - \frac{\omega_1}{\omega}\right) - (m-1)\frac{\omega}{\omega_2}$$

To satisfy requirement 3 we set

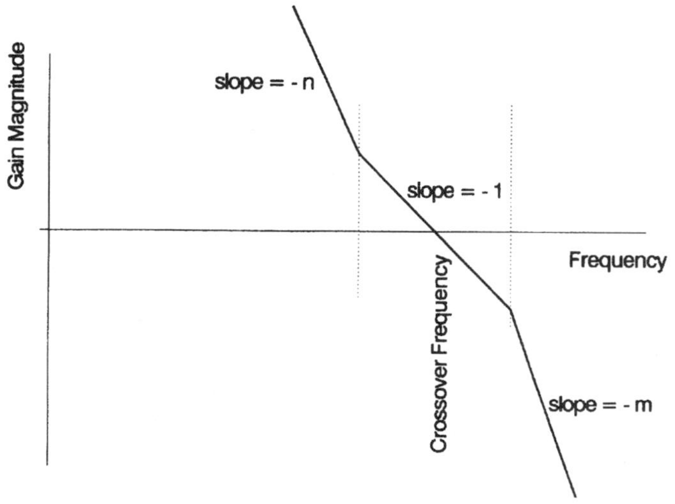

FIGURE 11.20 Illustration of generic gain magnitude in the vicinity of crossover.

$$\left.\frac{d\beta(\omega)}{d\omega}\right|_{\omega=\omega_c} = 0 = \frac{(n-1)\omega_1}{\omega_c^2} - \frac{m-1}{\omega_2}$$

and obtain

$$\omega_c^2 = \frac{n-1}{m-1}\omega_1\omega_2$$

as the optimum setting for the crossover frequency. Substitution of the "optimum" frequency given by the foregoing into the phase shift equation results in

$$\beta(\omega_c) = \frac{-\pi}{2} - 2\sqrt{(m-1)(n-1)}\sqrt{\frac{\omega_1}{\omega_2}}$$

We desire a specific PM here, and so the equalizer break frequency locations are specified. There is a single parameter here that is unspecified, and an additional equation must be found in any specific application. Alternatively, we could simply assume a nominal crossover frequency of unity or simply normalize frequencies ω_1 and ω_2 in terms of the crossover frequency by use of the normalized frequencies $\omega_1 = W_1\omega_c$ and $\omega_2 = W_2\omega_c$.

It is a relatively simple matter to show that for a specified PM expressed in radians, we obtain for the normalized break frequencies

$$W_1 = \frac{\omega_1}{\omega_c} = \frac{\text{PM}}{2(n-1)}$$

$$W_2 = \frac{\omega_2}{\omega_c} = \frac{2(m-1)}{\text{PM}}$$

It is relatively easy to implement this suggested Bode diagram design procedure, which is based upon considering only these break frequencies immediately above and below crossover and which approximate all others. Break frequencies far below crossover are approximated by integrations or differentiations, that is, poles or zeros at $s = 0$, and break frequencies far above the crossover frequency are ignored.

Minor-Loop Design

In our efforts thus far in this section we have assumed that compensating networks would be placed in series with the fixed plant and then a unity feedback ratio loop closed around these elements to yield the closed-loop system. In many applications it may be physically convenient, perhaps due to instrumentation considerations, to use one or more minor loops to obtain a desired compensation of a fixed plant transfer function.

For a single-input–single-output linear system there are no theoretical advantages whatever to any minor-loop compensation to series compensation, as the same closed-loop transfer function can be realized by all procedures. However, when there are multiple inputs or outputs, then there may be considerable advantages to minor-loop design as contrasted to series compensation design. Multiple inputs often occur when there is a single-signal input and one or more noise or disturbance inputs present, and a task of the system is to pass the signal inputs and reject the noise inputs. Also there may be saturation-type nonlinearities present, and we may be concerned not only with the primary system output but also with keeping the output at the saturation point within bounds such that the system remains linear. Thus there are reasons why minor-loop design may be preferable to series equalization.

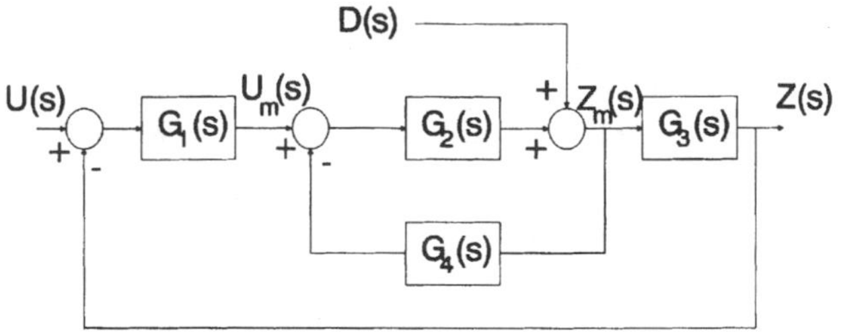

FIGURE 11.21 Feedback control system with a single minor loop and output disturbance.

We have discussed block diagrams elsewhere in this handbook. It is desirable here to review some concepts that will be of value for our discussion of minor-loop design. Figure 11.21 illustrates a relatively general linear control system with a single minor loop. This block diagram could represent many simple control systems. $G_1(s)$ could represent a discriminator and series compensation and $G_2(s)$ could represent an amplifier and that part of a motor transfer function excluding the final integration to convert velocity to position. $G_3(s)$ might then represent an integrator. $G_4(s)$ would then represent a minor-loop compensation transfer function, such as that of a tachometer.

The closed-loop transfer function for this system is given by

$$\frac{Z(s)}{U(s)} = H(s) = \frac{G_1(s)G_2(s)G_3(s)}{1 + G_2(s)G_4(s) + G_1(s)G_2(s)G_3(s)}$$

It is convenient to define several other transfer functions that are based on the block diagram in Figure 11.21. First there is the minor-loop gain

$$G_{\mathrm{m}}(s) = G_2(s)G_4(s)$$

which is just the loop gain of the minor loop only. The minor loop has the transfer function

$$\frac{Z_{\mathrm{m}}(s)}{U_{\mathrm{m}}(s)} = H_{\mathrm{m}}(s) = \frac{G_2(s)}{1 + G_2(s)G_4(s)} = \frac{G_2(s)}{1 + G_{\mathrm{m}}(s)}$$

There will usually be a range or ranges of frequency for which the minor-loop gain magnitude is much less than 1, and we then have

$$\frac{Z_{\mathrm{m}}(s)}{U_{\mathrm{m}}(s)} = H_{\mathrm{m}}(s) \approx G_2(s) \quad |G_{\mathrm{m}}(\omega)| \ll 1$$

There will also generally be ranges of frequency for which the minor-loop gain magnitude is much greater than 1, and we then have

$$\frac{Z_{\mathrm{m}}(s)}{U_{\mathrm{m}}(s)} = H_{\mathrm{m}}(s) \approx \frac{1}{G_4(s)} \quad |G_{\mathrm{m}}(\omega)| \gg 1$$

We may use these two relations to considerably simplify our approach to the minor-loop design problem. For illustrative purposes, we will use two major-loop gain functions. First we will consider the major-loop gain with the minor-loop-compensating network removed such that $G_4(s) = 0$. This represents the standard situation we have examined in the last subsection. This uncompensated major-loop transfer function is

$$G_{Mu}(s) = G_1(s)G_2(s)G_3(s)$$

With the minor-loop compensation inserted, the major-loop gain, the input-output transfer function with the unity ratio feedback open, is

$$G_{Mc}(s) = \frac{G_1(s)G_2(s)G_3(s)}{1 + G_m(s)}$$

We may express the complete closed-loop transfer function in the form

$$\frac{Z(s)}{U(s)} = H(s) = \frac{G_{Mc}(s)}{1 + G_{Mc}(s)}$$

A particularly useful relationship may be obtained by combining the last three equations into one equation of the form

$$G_{Mc}(s) = \frac{G_{Mu}(s)}{1 + G_m(s)}$$

We may give this latter equation a particularly simple interpretation. For frequencies where the minor-loop gain $G_m(s)$ is low, the minor-loop–closed major-loop transfer function $G_{Mc}(s)$ is approximately that of the minor-loop–open major-loop transfer function G_{Mu} in that

$$G_{Mc}(s) \approx G_{Mu}(s) \quad |G_m(\omega)| \ll 1$$

For frequencies where the minor-loop gain $G_m(s)$ is high, the minor-loop–closed major-loop transfer function is just

$$G_{Mc}(s) \approx \frac{G_{Mu}(s)}{G_m(s)} \quad |G_m(\omega)| \gg 1$$

This has an especially simple interpretation on the logarithmic frequency plots we use for Bode diagrams, for we may simply subtract the minor-loop gain $G_m(s)$ from the minor-loop–open major-loop gain $G_{Mu}(s)$ to obtain the compensated system gain as the transfer function $G_{Mc}(s)$.

The last several equations are the key relations for minor-loop design using this frequency response approach. These relations indicate that some forms of series compensation yield a given major-loop transfer function $G_{Mc}(s)$ that will not be appropriate for realization by minor-loop compensation. In particular, a lead network series compensation cannot be realized by means of equivalent minor-loop compensation. The gain of the fixed plant $G_{Mu}(s)$ is raised at high frequencies due to the use of a lead network compensation. Also, we see that $G_{Mc}(s)$ can only be lowered by use of a minor-loop gain $G_m(s)$.

A lag network used for series compensation will result in a reduction in the fixed plant gain $|G_{Mu}(\omega)|$ at all high frequencies. This can only be achieved if the minor-loop transfer gain $G_m(s)$ is constant for high frequencies. In some cases this may be achievable but often will not be. It is possible to realize the equivalent of lag network series equalization by means of a minor-loop equalization for systems where the low- and high-frequency behavior of $G_{Mu}(s)$, or $G_f(s)$, and $G_{Mc}(s)$ are the same and where the gain magnitude of the compensated system $|G_{Mc}(s)|$ is at no frequency any greater than is the gain magnitude of the fixed plant $|G_f(s)|$ or the minor-loop–open major-loop transfer function $|G_{Mu}(s)|$. Thus we see that lag-lead network series equalization is an ideal type of equalization to realize by means of equivalent minor-loop equalization. Figures 11.9 and 11.19 represent flowcharts of a suggested general design procedure for minor-loop compensator design as well as for the series equalization approaches we examined previously.

In our work thus far we have assumed that parameters were constant and known. Such is seldom the case, and we must naturally be concerned with the effects of parameter variations, disturbances, and nonlinearities upon system performance. Suppose, for example, that we design a system with a certain gain assumed as K_1. If the system operates open loop and the gain K_1 is in cascade or series with the other input-output components, then the overall transfer function changes by precisely the same factor as K_1 changes. If we have an amplifier with unity ratio feedback around a gain K_1, the situation is much different. The closed-loop gain would nominally be $K_1/(1 + K_1)$, and a change to $2K_1$ would give a closed-loop gain $2K_1/(1 + 2K_1)$. If K_1 is large, say 10_3, then the new gain is 0.99950025, which is a percentage change of less than 0.05% for a change in gain of 100%.

Another advantage of minor-loop feedback occurs when there are output disturbances such as those due to wind gusts on an antenna. We consider the system illustrated in Figure 11.21. The response due to $D(s)$ alone is

$$\frac{Z(s)}{D(s)} = \frac{1}{1 + G_2(s)G_4(s) + G_1(s)G_2(s)}$$

When we use the relation for the minor-loop gain

$$G_{\mathrm{m}}(s) = G_2(s)G_4(s)$$

and the major-loop gain

$$G_{\mathrm{Mc}}(s) = \frac{G_1(s)G_2(s)}{1 + G_2(s)G_4(s)}$$

we can rewrite the response due to $D(s)$ as

$$\frac{Z(s)}{D(s)} = \frac{1}{[1 + G_{\mathrm{m}}(s)]G_{\mathrm{Mc}}(s)}$$

Over the range of frequency where $|G_{\mathrm{Mc}}(j\omega)| >> 1$, such that the corrected loop gain is large, the attenuation of a load disturbance is proportional to the uncorrected loop gain. This is generally larger over a wider frequency range than the corrected loop gain magnitude $|G_{\mathrm{Mc}}(j\omega)|$, which is what the attenuation would be if series compensation were used.

Over the range of frequencies where the minor-loop gain is large but where the corrected loop gain is small, that is, where $|G_{\mathrm{m}}(j\omega)| > 1$ and $|G_{\mathrm{Mc}}(j\omega)| < 1$, we obtain for the approximate response due to the disturbance

$$\frac{Z(s)}{D(s)} \approx G_{\mathrm{m}}(s)$$

and the output disturbance is therefore seen to be attenuated by the minor-loop gain rather than unattenuated as would be the case if series compensation had been used. This is, of course, highly desirable.

At frequencies where both the minor-loop gain transfer and the major-loop gain are small, we have $Z(s)/D(s) = 1$, and over this range offrequencies neither minor-loop compensation nor series equalization is useful in reducing the effect of a load disturbance. Thus, we have shown here that there are quite a number of advantages to minor-loop compensation as compared with series equalization. Of course, there are limitations as well.

Summary

In this section, we have examined the subject of linear system compensation by means of the frequency response method of Bode diagrams. Our approach has been entirely in the frequency domain. We have discussed a variety of compensation networks, including:

1. Gain attenuation
2. Lead networks

3. Lag networks
4. Lag-lead networks and composite equalizers
5. Minor-loop feedback

Despite its age, the frequency domain design approach represents a most useful approach for the design of linear control systems. It has been tested and proven in a great many practical design situations.

Defining Terms

Bode diagram: A graph of the gain magnitude and frequency response of a linear circuit or system, generally plotted on log-log coordinates. A major advantage of Bode diagrams is that the gain magnitude plot will look like straight lines or be asymptotic to straight lines. H.W. Bode, a well-known Bell Telephone Laboratories researcher, published *Network Analysis and Feedback Amplifier Design* in 1945. The approach, first described there, has been refined by a number of other workers over the past half-century.

Crossover frequency: The frequency where the magnitude of the open-loop gain is 1.

Equalizer: A network inserted into a system that has a transfer function or frequency response designed to compensate for undesired amplitude, phase, and frequency characteristics of the initial system. Filter and equalizer are generally synonymous terms.

Lag network: In a simple phase-lag network, the phase angle associated with the input-output transfer function is always negative, or lagging. Figures 11.13 and 11.14 illustrate the essential characteristics of a lag network.

Lag-lead network: The phase shift versus frequency curve in a phase lag-lead network is negative, or lagging, for low frequencies and positive, or leading, for high frequencies. The phase angle associated with the input-output transfer function is always positive, or leading. Figures 11.16 and 11.17 illustrate the essential characteristics of a lag-lead network, or composite equalizer.

Lead network: In a simple phase-lead network, the phase angle associated with the input-output transfer function is always positive, or leading. Figures 11.10 and 11.11 illustrate the essential characteristics of a lead network.

Series equalizer: In a single-loop feedback system, a series equalizer is placed in the single loop, generally at a point along the forward path from input to output where the equalizer itself consumes only a small amount of energy. In Figure 11.21, $G_1(s)$ could represent a series equalizer. $G_1(s)$ could also be a series equalizer if $G_4(s) = 0$.

Specification: A statement of the design or development requirements to be satisfied by a system or product.

Systems engineering: An approach to the overall life cycle evolution of a product or system. Generally, the systems engineering process comprises a number of phases. There are three essential phases in any systems engineering life cycle: formulation of requirements and specifications, design and development of the system or product, and deployment of the system. Each of these three basic phases may be further expanded into a larger number. For example, deployment generally comprises operational test and evaluation, maintenance over an extended operational life of the system, and modification and retrofit (or replacement) to meet new and evolving user needs.

References and Bibliography

R.C. Dorf and R.H. Bishop, *Modern Control Systems*, 10th ed., Englewood Cliffs, NJ: Prentice Hall 2004.

N.S. Nise, *Control Systems Engineering*, 4th ed., Hoboken, NJ: Wiley, 2003.

K. Ortega, *Modern Control Engineering*, 4th ed., Englewood Cliffs, NJ: Prentice Hall, 2002.

R.T. Stefani, B. Shahian, C.J. Savant, and G.H. Hostetter, *Design of Feedback Control Systems*, New York: Oxford University Press, 2002.

L.C. Westphal, *Handbook of Control Systems Engineering*, 2nd ed., Dordrecht, The Netherlands: Kluwer, 2001.

S.K. Zak, *Systems and Control*, New York: Oxford University Press, 2003.

11.4 The Root Locus Method

Hitay Özbay

Introduction

The root locus technique is a graphical tool used in feedback control system analysis and design. It has been formally introduced to the engineering community by W. R. Evans [3,4], who received the Richard E. Bellman Control Heritage Award from the American Automatic Control Council in 1988 for this major contribution.

In order to discuss the root locus method, we must first review the basic definition of bounded input bounded output (BIBO) stability of the standard linear time invariant feedback system shown in Figure 11.22, where the plant, and the controller, are represented by their transfer functions $P(s)$ and $C(s)$, respectively[1] The plant, $P(s)$, includes the physical process to be controlled, as well as the actuator and the sensor dynamics.

The feedback system is said to be stable if none of the closed-loop transfer functions, from external inputs r and v to internal signals e and u, have any poles in the closed right half plane, $\overline{\mathbb{C}}_+ := \{s \in \mathbb{C} : \mathrm{Re}(s) \geqslant 0\}$. A necessary condition for feedback system stability is that the closed right half plane zeros of $P(s)$ (respectively $C(s)$) are distinct from the poles of $C(s)$ (respectively $P(s)$). When this condition holds, we say that there is no unstable pole–zero cancellation in taking the product $P(s)C(s) =: G(s)$, and then checking feedback system stability becomes equivalent to checking whether all the roots of

$$1 + G(s) = 0 \tag{11.45}$$

are in the open left half plane, $\mathbb{C}_- := \{s \in \mathbb{C} : \mathrm{Re}(s) < 0\}$. The roots of Equation (11.1) are the closed-loop system poles. We would like to understand how the closed-loop system pole locations vary as functions of a real parameter of $G(s)$. More precisely, assume that $G(s)$ contains a parameter K, so that we use the notation $G(s) = G_K(s)$ to emphasize the dependence on K. The *root locus* is the plot of the roots of Equation (11.45) on the complex plane, as the parameter K varies within a specified interval.

The most common example of the root locus problem deals with the uncertain (or adjustable) gain as the varying parameter: when $P(s)$ and $C(s)$ are fixed rational functions, except for a gain factor, $G(s)$ can be written as $G(s) = G_K(s) = KF(s)$, where K is the uncertain/adjustable gain, and

$$F(s) = \frac{N(s)}{D(s)} \quad \text{where} \quad \begin{matrix} N(s)=\prod_{j=1}^{m}(s-z_j) \\ D(s)=\prod_{i=1}^{n}(s-p_i), \end{matrix} \quad n \geqslant m \tag{11.46}$$

with $z_1,\ldots,z_m$, and $p_1,\ldots,p_n$ being the open-loop system zeros and poles. In this case, the closed-loop system

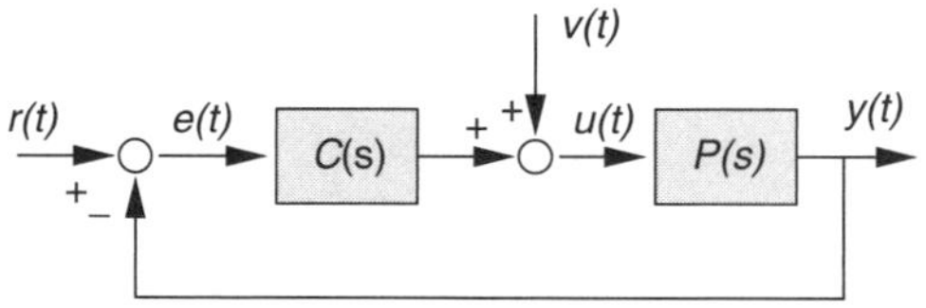

FIGURE 11.22 Standard unity feedback system.

[1]Here we consider the continuous time case; there is essentially no difference between the continuous time case and the discrete time case, as far as the root locus construction is concerned. In the discrete time case the desired closed-loop pole locations are defined relative to the unit circle, whereas in the continuous time case desired pole locations are defined relative to the imaginary axis.

poles are the roots of the characteristic equation

$$\chi(s) := D(s) + KN(s) = 0 \qquad (11.47)$$

The *usual root locus* is obtained by plotting the roots $r_1(K),\ldots,r_n(K)$ of the characteristic polynomial $\chi(s)$ on the complex plane, as K varies from 0 to $+\infty$. The same plot for the negative values of K gives the *complementary root locus*. With the help of the root locus plot the designer identifies the admissible values of the parameter K leading to a set of closed-loop system poles that are in the desired region of the complex plane. There are several factors to be considered in defining the "desired region" of the complex plane in which all the roots $r_1(K),\ldots,r_n(K)$ should lie. Those are discussed briefly in the next section. Section 11.3 contains the root locus construction procedure, and design examples are presented in section 11.4.

The root locus can also be drawn with respect to a system parameter other than the gain. For example, the characteristic equation for the system $G(s) = G_\lambda(s)$, defined by

$$G_\lambda(s) = P(s)C(s), \quad P(s) = \frac{(1-\lambda s)}{s(1+\lambda s)}, \quad C(s) = K_c\left(1 + \frac{1}{T_I s}\right)$$

can also be transformed into the form given in Equation (11.47). Here K_c and T_I are given fixed PI (Proportional plus Integral) controller parameters, and $\lambda > 0$ is an uncertain plant parameter. Note that the phase of the plant is

$$\angle P(j\omega) = -\frac{\pi}{2} - 2\tan^{-1}(\lambda\omega)$$

so the parameter λ can be seen as the uncertain phase lag factor (for example, a small uncertain time delay in the plant can be modeled in this manner, see [9]). It is easy to see that the characteristic equation is

$$s^2(\lambda s + 1) + K_c(1-\lambda s)\left(s + \frac{1}{T_I}\right) = 0$$

and by rearranging the terms multiplying λ this equation can be transformed to

$$1 + \frac{1}{\lambda}\frac{(s^2 + K_c s + K_c/T_I)}{s(s^2 - K_c s - K_c/T_I)} = 0$$

By defining $K = \lambda^{-1}$, $N(s) = (s^2 + K_c s + K_c/T_I)$, and $D(s) = s(s^2 + K_c s - K_c/T_I)$, we see that the characteristic equation can be put in the form of Equation (11.3). The root locus plot can now be obtained from the data $N(s)$ and $D(s)$ defined above; that shows how closed-loop system poles move as λ^{-1} varies from 0 to $+\infty$, for a given fixed set of controller parameters K_c and T_I. For the numerical example $K_c = 1$ and $T_I = 2.5$, the root locus is illustrated in Figure 11.23.

The root locus construction procedure will be given in section 11.3. Most of the computations involved in each step of this procedure can be performed by hand calculations. Hence, an approximate graph representing the root locus can be drawn easily. There are also several software packages to generate the root locus automatically from the problem data $z_1,\ldots,z_m$, and $p_1,\ldots,p_n$.

If a numerical computation program is available for calculating the roots of a polynomial, we can also obtain the root locus with respect to a parameter which enters into the characteristic equation nonlinearly. To illustrate this point let us consider the following example: $G(s) = G_{\omega_o}(s)$ where

$$G_{\omega_o}(s) = P(s)C(s), \quad P(s) = \frac{(s-0.1)}{(s^2 + 1.2\omega_o s + \omega_o^2)(s+0.1)}, \quad C(s) = \frac{(s-0.2)}{(s+2)}$$

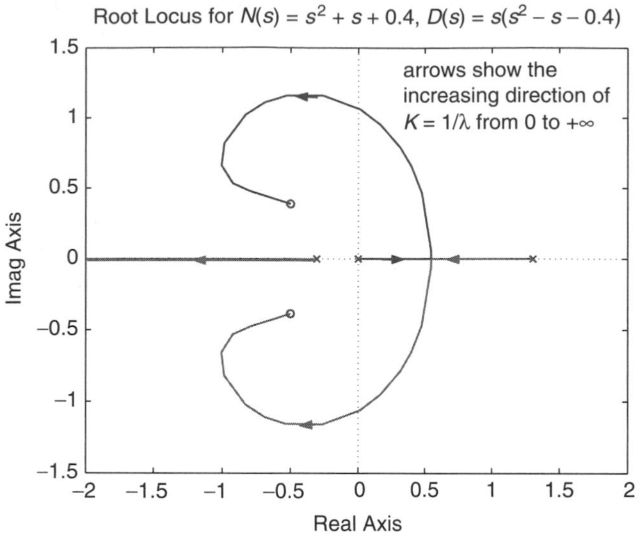

FIGURE 11.23 The root locus with respect to $K = 1/\lambda$.

Here $\omega_o \geq 0$ is the uncertain plant parameter. Note that the characteristic equation

$$1 + \frac{\omega_o(1.2s + \omega_o)(s + 0.1)(s + 2)}{s^2(s + 0.1)(s + 2) + (s - 0.2)(s - 0.1)} = 0 \tag{11.48}$$

cannot be expressed in the form of $D(s) + KN(s) = 0$ with a single parameter K. Nevertheless, for each ω_o we can numerically calculate the roots of Equation (11.48) and plot them on the complex plane as ω_o varies within a range of interest. Figure 11.24 illustrates all the four branches, $r_1(K), \ldots, r_4(K)$, of the root locus for

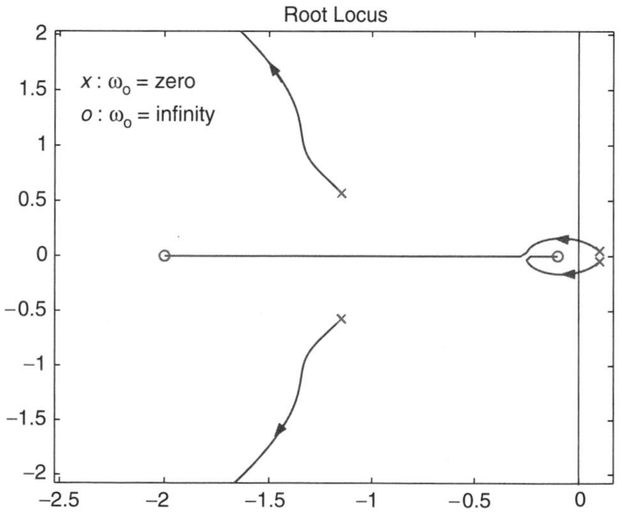

FIGURE 11.24 The root locus with respect to ω_o.

this system as ω_o increases from zero to infinity. The figure is obtained by computing the roots of Equation (11.48) for a set of values of ω_o by using MATLAB.

Desired Pole Locations

The performance of a feedback system depends heavily on the location of the closed-loop system poles $r_i(K) = 1,\ldots,n$. First of all, for stability we want $r_i(K) \in \mathbb{C}_-$ for all $i = 1,\ldots,n$. Clearly, having a pole "close" to the imaginary axis poses a danger, i.e., "small" perturbations in the plant might lead to an unstable feedback system. So the desired pole locations must be such that stability is preserved under such perturbations (or in the presence of uncertainties) in the plant. For second-order systems, we can define certain stability robustness measures in terms of the pole locations, which can be tied to the characteristics of the step response. For higher order systems, similar guidelines can be used by considering the dominant poles only.

In the standard feedback control system shown in Figure 11.22, assume that the closed-loop transfer function from $r(t)$ to $y(t)$ is in the form

$$T(s) = \frac{\omega_o^2}{s^2 + 2\zeta\omega_o s + \omega_o^2}, \qquad 0<\zeta<1, \quad \omega_o \in \mathbb{R}$$

and $r(t)$ is the unit step function. Then, the output is

$$y(t) = 1 - \frac{e^{-\zeta\omega_o t}}{\sqrt{1-\zeta^2}} \sin(\omega_d t + \theta), \quad t \geqslant 0$$

where $\omega_d := \omega_o\sqrt{1-\zeta^2}$ and $\theta := \cos^{-1}(\zeta)$. For some typical values of ζ, the step response $y(t)$ is as shown in Figure 11.25. The maximum *percent overshoot* is defined to be the quantity

$$\text{PO} := \frac{y_p - y_{ss}}{y_{ss}} \times 100\%$$

where y_p is the peak value. By simple calculations it can be seen that the peak value of $y(t)$ occurs at the time

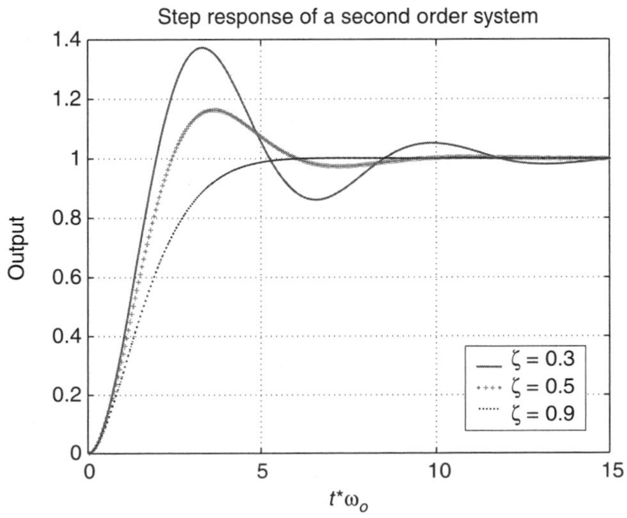

FIGURE 11.25 Step response of a second-order system.

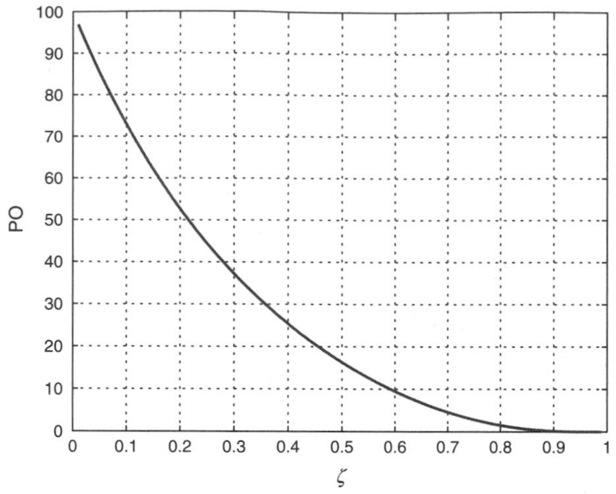

FIGURE 11.26 PO versus ζ.

instant $t_p = \pi/\omega_d$, and

$$\mathrm{PO} = \mathrm{e}^{-\pi\zeta/\sqrt{1-\zeta^2}} \times 100\%$$

Figure 11.26 shows PO versus ζ. The *settling time* is defined to be the smallest time instant t_s, after which the response $y(t)$ remains within 2% of its final value, i.e.,

$$t_s := \min\ \{t' : |y(t) - y_{ss}| \leqslant 0.02y_{ss} \forall t \geqslant t'\}$$

Sometimes 1% or 5% is used in the definition of settling time instead of 2%; conceptually, there is no difference. For the second-order system response, we have

$$t_s \approx \frac{4}{\zeta\omega_o}$$

So, in order to have a fast settling response, the product $\zeta\omega_o$ should be large.

The closed-loop system poles are

$$r_{1,2} = -\zeta\omega_o \pm j\omega_o\sqrt{1-\zeta^2}$$

Therefore, once the maximum allowable settling time and PO are specified, we can define the region of desired pole locations by determining the minimum allowable ζ and $\zeta\omega_o$. For example, let the desired PO and t_s be bounded by

$$\mathrm{PO} \leqslant 10\% \quad \text{and} \quad t_s \leqslant 8\text{s}$$

The PO requirement implies that $\zeta \geq 0.6$, equivalently $\theta \leq 53°$ (recall that $\cos(\theta) = \zeta$). The settling time requirement is satisfied if and only if $\mathrm{Re}(r_{1,2}) \leq -0.5$. Then, the region of desired closed-loop poles is the shaded area shown in Figure 11.27. The same figure also illustrates the region of desired closed-loop poles for similar design requirements in the discrete time case.

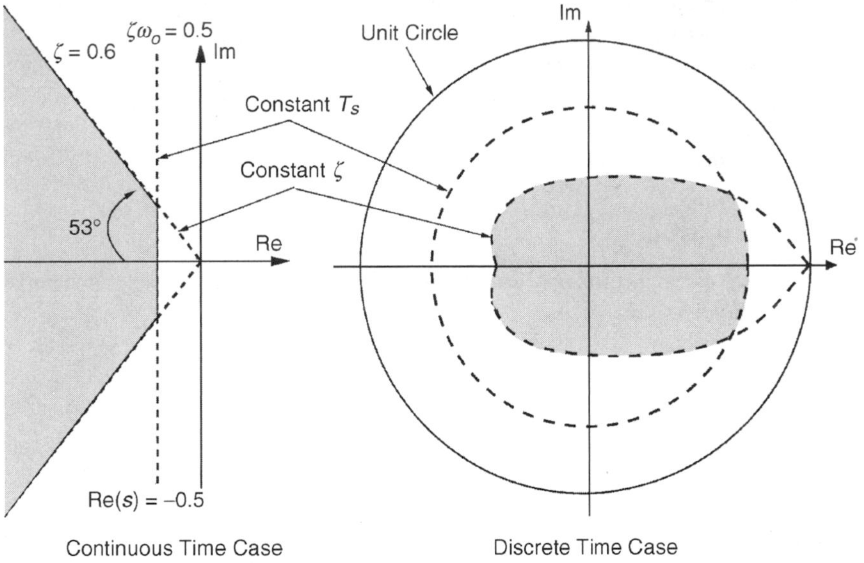

FIGURE 11.27 Region of the desired closed-loop poles.

If the order of the closed-loop transfer function $T(s)$ is higher than two, then, depending on the location of its poles and zeros, it may be possible to approximate the closed-loop step response by the response of a second-order system. For example, consider the third-order system

$$T(s) = \frac{\omega_o^2}{(s^2 + 2\zeta\omega_o s + \omega_o^2)(1 + s/r)} \quad \text{where } r \gg \zeta\omega_o$$

The transient response contains a term e^{-rt}. Compared with the envelope $e^{-\zeta\omega_o t}$ of the sinusoidal term, e^{-rt} decays very fast, and the overall response is similar to the response of a second-order system. Hence, the effect of the third pole $r_3 = -r$ is negligible.

Consider another example,

$$T(s) = \frac{\omega_o^2[1 + s/(r + \epsilon)]}{(s^2 + 2\zeta\omega_o s + \omega_o^2)(1 + s/r)} \quad \text{where } 0 < \epsilon \ll r$$

In this case, although r does not need to be much larger than $\zeta\omega_o$, the zero at $-(r + \epsilon)$ cancels the effect of the pole at $-r$. To see this, consider the partial fraction expansion of $Y(s) = T(s)R(s)$ with $R(s) = 1/s$:

$$Y(s) = \frac{A_0}{s} + \frac{A_1}{s - r_1} + \frac{A_2}{s - r_2} + \frac{A_3}{s + r}$$

where $A_0 = 1$ and

$$A_3 = \lim_{s \to -r} (s + r)Y(s) = \frac{\omega_o^2}{2\zeta\omega_o r - (\omega_o^2 + r^2)} \left(\frac{\epsilon}{r + \epsilon}\right)$$

Since $|A_3| \to 0$ as $\epsilon \to 0$ the term $A_3 e^{-rt}$ is negligible in $y(t)$.

In summary, if there is an approximate pole–zero cancellation in the left half plane, then this pole–zero pair can be taken out of the transfer function $T(s)$ to determine PO and t_s. Also, the poles closest to the imaginary axis dominate the transient response of $y(t)$. To generalize this observation, let $r_1,\ldots,r_n$ be the poles of $T(s)$, such that $\mathrm{Re}(r_k) \ll \mathrm{Re}(r_2) = \mathrm{Re}(r_1) < 0$, for all $k \geq 3$. Then, the pair of complex conjugate poles $r_{1,2}$ are called the *dominant poles*. We have seen that the desired transient response properties, e.g., PO and t_s, can be translated into requirements on the location of the dominant poles.

Root Locus Construction

As mentioned above, the root locus primarily deals with finding the roots of a characteristic polynomial that is an affine function of a single parameter, K,

$$\chi(s) = D(s) + KN(s) \tag{11.49}$$

where $D(s)$ and $N(s)$ are fixed monic polynomials (i.e., coefficient of the highest power is normalized to 1). If N and/or D are not monic, the highest coefficient(s) can be absorbed into K.

Root Locus Rules

Recall that the usual root locus shows the locations of the closed-loop system poles as K varies from 0 to $+\infty$. The roots of $D(s)$, $p_1,\ldots,p_n$, are the poles, and the roots of $N(s)$, $z_1,\ldots,z_m$, are the zeros, of the open-loop system, $G(s) = KF(s)$. Since $P(s)$ and $C(s)$ are proper, $G(s)$ is proper, and hence $n \geq m$. So the degree of the polynomial $\chi(s)$ is n and it has exactly n roots.

Let the closed-loop system poles, i.e., roots of $\chi(s)$, be denoted by $r_1(K),\ldots,r_n(K)$. Note that these are functions of K; whenever the dependence on K is clear, they are simply written as $r_1,\ldots,r_n$. The points in $\mathbb{C}$ that satisfy (Equation (11.49)) for some $K > 0$ are on the root locus. Clearly, a point $r \in \mathbb{C}$ is on the root locus if and only if

$$= -\frac{1}{F(r)} \tag{11.50}$$

The condition (Equation (11.50)) can be separated into two parts:

$$|K| = -\frac{1}{|F(r)|} \tag{11.51}$$

$$\angle K = 0^\circ = -(2\ell + 1) \times 180^\circ - \angle F(r), \quad \ell = 0, \pm 1, \pm 2, \ldots \,. \tag{11.52}$$

The phase rule (Equation (11.52)) determines the points in that are on the root locus. The magnitude rule (Equation (11.51)) determines the gain $K > 0$ for which the root locus is at a given point r. By using the definition of $F(s)$, (Equation (11.52)) can be rewritten as

$$(2\ell + 1) \times 180^\circ = \sum_{i=1}^{n} \angle(r - p_i) - \sum_{j=1}^{m} \angle(r - z_j) \tag{11.53}$$

Similarly, (Equation (11.51)) is equivalent to

$$K = \frac{\prod_{i=1}^{n} |r - p_i|}{\prod_{j=1}^{m} |r - z_j|} \tag{11.54}$$

Root Locus Construction

There are several software packages available for generating the root locus automatically for a given $F = N/D$. In particular, the related MATLAB commands are `rlocus` and `rlocfind`. In many cases, approximate root locus can be drawn by hand using the rules given below. These rules are determined from the basic definitions Equation (11.49), Equation (11.51), and Equation (11.52).

1. The root locus has n branches: $r_1(K),\ldots,r_n(K)$.
2. Each branch starts ($K \cong 0$) at a pole p_i and ends (as $K \to \infty$) at a zero z_j, or converges to an asymptote, $Me^{j\alpha\ell}$ where $M \to \infty$) and

$$\alpha_\ell = \frac{2\ell + 1}{n - m} \times 180^\circ, \quad \ell = 0, \ldots, (n - m - 1)$$

3. There are $(n - m)$ asymptotes with angles α_ℓ. The center of the asymptotes (i.e., their intersection point on the real axis) is

$$\sigma_a = \frac{\sum_{i=1}^{n} P_i - \sum_{j=1}^{m} z_j}{n - m}$$

4. A point $x \in \mathbb{R}$ is on the root locus if and only if the total number of poles p_i's and zeros z_j's to the right of x (i.e., total number of p_i's with $\text{Re}(p_i) > x$ plus total number of z_j's with $\text{Re}(z_j) > x$) is odd. Since $F(s)$ is a rational function with real coefficients, poles and zeros appear in complex conjugates, so when counting the number of poles and zeros to the right of a point $x \in \mathbb{R}$ we just need to consider the poles and zeros on the real axis.
5. The values of K for which the root locus crosses the imaginary axis can be determined from the Routh–Hurwitz stability test. Alternatively, we can set $s = j\omega$ in Equation (11.49) and solve for real ω and K satisfying

$$D(j\omega) + KN(j\omega) = 0$$

Note that there are two equations here, one for the real part and one for the imaginary part.
6. The break points (intersection of two branches on the real axis) are feasible solutions (satisfying rule 4) of

$$\frac{d}{ds}F(s) = 0 \tag{11.55}$$

7. Angles of departure ($K \cong 0$) from a complex pole, or arrival $K \to +\infty$ to a complex zero, can be determined from the phase rule. See example below.
Let us now follow the above rules step by step to construct the root locus for

$$F(s) = \frac{(s + 3)}{(s - 1)(s + 5)(s + 4 + j2)(s + 4 - j2)}$$

First, enumerate the poles and zeros as $p_1 = -4 + j2$, $p_2 = -4 - j2$, $p_3 = -5$, $p_4 = 1$, $z_1 = -3$. So, $n = 4$ and $m = 1$.

1. The root locus has four branches.
2. Three branches converge to the asymptotes whose angles are 60°, 180°, and –60°, and one branch converges to $z_1 = -3$.
3. The center of the asymptotes is $\sigma = (-12 + 3)/3 = -3$.
4. The intervals $(-\infty, -5]$ and $[-3, 1]$ are on the root locus.

5. The imaginary axis crossings are the feasible roots of

$$(\omega^4 - j12\omega^3 - 47\omega^2 + j40\omega - 100) + K(j\omega + 3) = 0 \tag{11.56}$$

for real ω and K. Real and imaginary parts of Equation (11.56) are

$$\omega^4 - 47\omega^2 - 100 + 3K = 0$$

$$j\omega(-12\omega^2 + 40 + K) = 0$$

They lead to two feasible pairs of solutions ($K = 100/3$, $\omega = 0$) and ($K = 215.83$, $\omega = \pm 4.62$).

6. Break points are the feasible solutions of

$$3s^4 + 36s^3 + 155s^2 + 282s + 220 = 0$$

Since the roots of this equation are $-4.55 \pm j1.11$ and $-1.45 \pm j1.11$, there is no solution on the real axis, hence no break points.

7. To determine the angle of departure from the complex pole $p_1 = -4 + j2$, let Δ represent a point on the root locus near the complex pole p_1, and define v_i, $i = 1,\ldots,5$, to be the vectors drawn from p_i, for $i = 1,\ldots,4$, and from z_1 for $i = 5$, as shown in Figure 11.28. Let $\theta_1,\ldots,\theta_5$ be the angles of $v_1,\ldots,v_5$. The phase rule implies

$$(\theta_1 + \theta_2 + \theta_3 + \theta_4) - \theta_5 = \pm 180° \tag{11.57}$$

As Δ approaches p_1, θ_1 becomes the angle of departure and the other θ_i's can be approximated by the angles of the vectors drawn from the other poles, and from the zero, to the pole p_1. Thus θ_1 can be solved from Equation (11.57), where $\theta_2 \approx 90°$, $\theta_3 \approx \tan^{-1}(2)$, $\theta_4 \approx 180° - \tan^{-1}\left(\frac{2}{5}\right)$, and $\theta_5 \approx 90° + \tan^{-1}\left(\frac{1}{2}\right)$. That yields $\theta_1 \approx -15°$.

The exact root locus for this example is shown in Figure 11.29. From the results of item 5 above, and the shape of the root locus, it is concluded that the feedback system is stable if

$$33.33 < K < 215.83$$

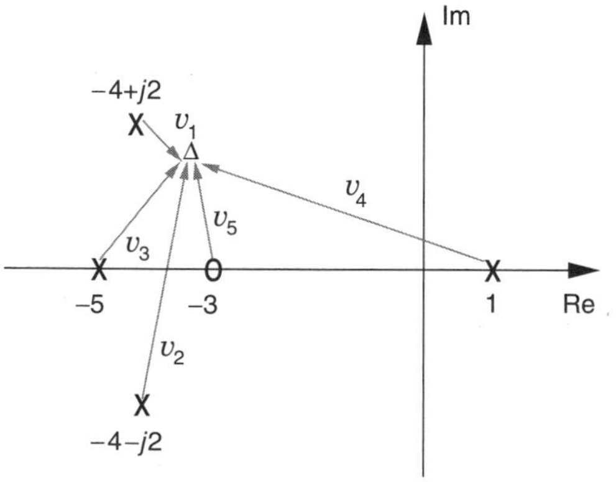

FIGURE 11.28 Angle of departure from $-4 + j2$.

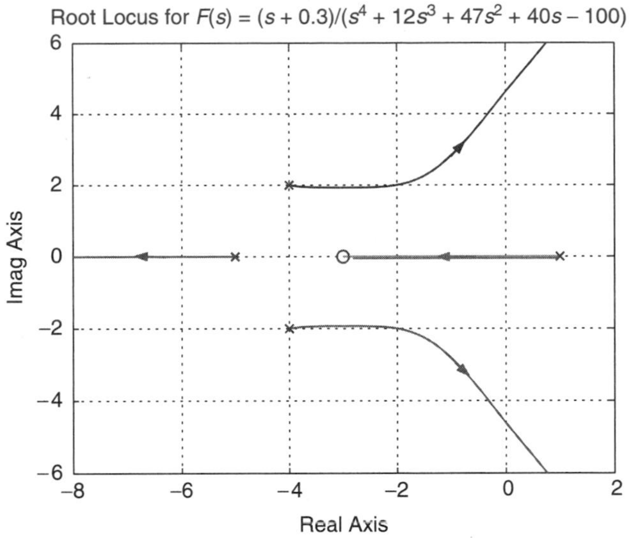

FIGURE 11.29 Root locus for $F(s) = \dfrac{(s+3)}{(s-1)(s+5)(s+4+j2)(s+4-j2)}$

i.e., by simply adjusting the gain of the controller, the system can be made stable. In some situations we need to use a dynamic controller to satisfy all the design requirements.

Design Examples

Example 11.1

Consider the standard feedback system with a plant

$$P(s) = \frac{1}{0.72}\frac{1}{(s+1)(s+2)}$$

and design a controller such that

- the feedback system is stable,
- PO $\leq$ 10%, $t_s \leq 4$ s, and steady state error is zero when $r(t)$ is unit step,
- steady state error is as small as possible when $r(t)$ is unit ramp.

It is clear that the second design goal cannot be achieved by a simple proportional controller. To satisfy this condition, the controller must have a pole at $s = 0$, i.e., it must have integral action. If we try an integral control of the form $C(s) = K_c/s$, with $K_c > 0$, then the root locus has three branches, the interval [–1, 0] is on the root locus; three asymptotes have angles {60°, 180°, –60°} with a center at $\sigma_a = -1$, and there is only one break point at $-1 + \frac{1}{\sqrt{3}}$, see Figure 11.30. From the location of the break point, center, and angles of the asymptotes, it can be deduced that two branches (one starting at $p_1 = -1$, and the other one starting at $p_3 = 0$) always remain to the right of p_1. On the other hand, the settling time condition implies that the real parts of the dominant closed-loop system poles must be less than or equal to –1. So, a simple integral control does not do the job. Now try a PI controller of the form

$$C(s) = K_c\left(\frac{s - z_c}{s}\right), \quad K_c > 0$$

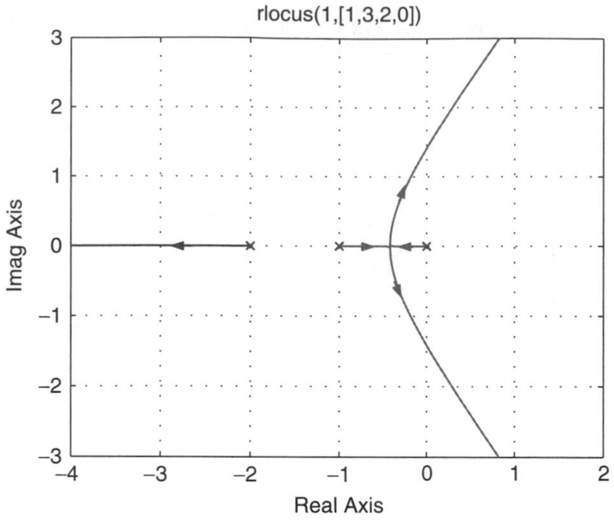

FIGURE 11.30 Root locus for Example 1.

In this case, we can select $z_c = -1$ to cancel the pole at $p_1 = -1$ and the system effectively becomes a second-order system. The root locus for $F(s) = 1/s(s+2)$ has two branches and two asymptotes, with center $\sigma_a = -1$ and angles $\{90°, -90°\}$; the break point is also at -1. The branches leave -2 and 0, and go toward each other, meet at -1, and tend to infinity along the line $\text{Re}(s) = -1$. Indeed, the closed-loop system poles are

$$r_{1,2} = -1 \pm \sqrt{1-K}, \quad \text{where } K = K_c/0.72$$

The steady state error, when $r(t)$ is unit ramp, is $2/K$. So K needs to be as large as possible to meet the third design condition. Clearly, $\text{Re}(r_{1,2}) = -1$ for all $K \geq 1$, which satisfies the settling time requirement. The percent overshoot is less than 10% if ζ of the roots $r_{1,2}$ is greater than 0.6. A simple algebra shows that $\zeta = 1/\sqrt{K}$, hence the design conditions are met if $K = 1/0.36$, i.e. $K_c = 2$. Thus a PI controller that solves the design problem is

$$C(s) = 2\left(\frac{s+1}{s}\right)$$

The controller cancels a stable pole (at $s = -1$) of the plant. If there is a slight uncertainty in this pole location, perfect cancellation will not occur and the system will be third-order with the third pole at $r_3 \cong -1$. Since the zero at $z_o = -1$ will approximately cancel the effect of this pole, the response of this system will be close to the response of a second-order system. However, we must be careful if the pole–zero cancellations are near the imaginary axis because in this case small perturbations in the pole location might lead to large variations in the feedback system response, as illustrated with the next example.

Example 11.2

A flexible structure with lightly damped poles has transfer function in the form

$$P(s) = \frac{\omega_1^2}{s^2(s^2 + 2\zeta\omega_1 s + \omega_1^2)}$$

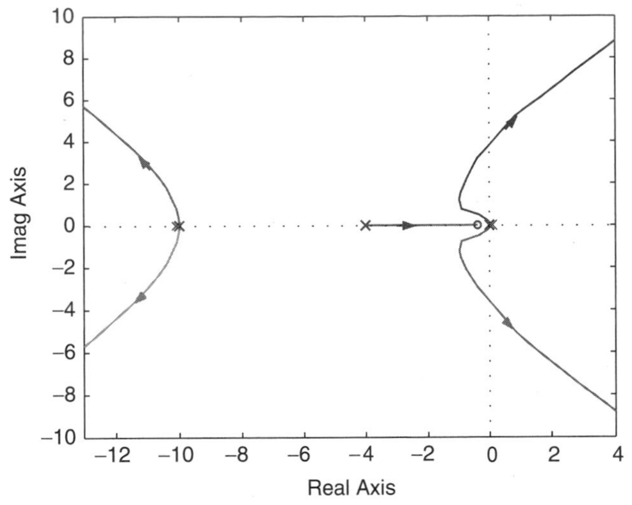

FIGURE 11.31 Root locus for Example 11.2(a).

By using the root locus, we can see that the controller

$$C(s) = K_c \frac{(s^2 + 2\zeta\omega_1 s + \omega_1^2)(s + 0.4)}{(s + r)^2(s + 4)}$$

stabilizes the feedback system for sufficiently large r and an appropriate choice of K_c. For example, let $\omega_1 = 2$, $\zeta = 0.1$, and $r = 10$. Then the root locus of $F(s) = P(s)C(s)/K$, where $K = K_c\omega_1^2$ is as shown in Figure 11.31. For $K = 600$, the closed-loop system poles are

$$\{-10.78 \pm j2.57, -0.94 \pm j1.61, -0.2 \pm j1.99, -0.56\}$$

Since the poles $-0.2 \pm j1.99$ are canceled by a pair of zeros at the same point in the closed-loop system transfer function $T = G(1 + G)^{-1}$, the dominant poles are at -0.56 and $-0.94 \pm j1.61$ (they have relatively large negative real parts and the damping ratio is about 0.5).

Now, suppose that this controller is fixed and the complex poles of the plant are slightly modified by taking $\zeta = 0.09$ and $\omega_1 = 2.2$. The root locus corresponding to this system is as shown in Figure 11.32. Since lightly damped complex poles are not perfectly canceled, there are two more branches near the imaginary axis. Moreover, for the same value of $K = 600$, the closed-loop system poles are

$$\{-10.78 \pm j2.57, -1.21 \pm j1.86, 0.05 \pm j1.93, -0.51\}$$

In this case, the feedback system is unstable.

Example 11.3

One of the most important examples of mechatronic systems is the DC motor. An approximate transfer function of a DC motor [8, pp. 141–143] is in the form

$$P_m(s) = \frac{K_m}{s(s + 1/\tau_m)}, \quad \tau_m > 0$$

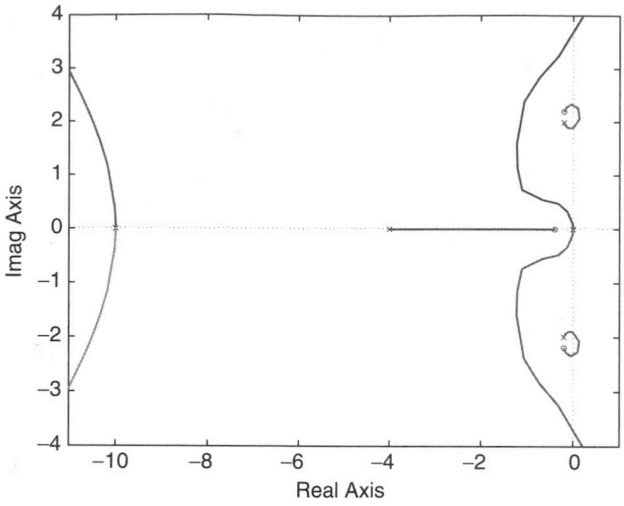

FIGURE 11.32 Root locus for Example 11.2(b).

Also note that if τ_m is large, then $P_m(s) \approx P_b(s)$, where $P_b(s) = K_b/s^2$, is the transfer function of a rigid beam. In this example, the general class of plants $P_m(s)$ will be considered. Assuming that $p_m = -1/\tau_m$ and K_m are given, a first-order controller

$$C(s) = K_c\left(\frac{s - z_c}{s - p_c}\right) \tag{11.58}$$

will be designed. The aim is to place the closed-loop system poles far from the Im-axis. Since the order of $F(s) = P_m(s)C(s)/K_mK_c$ is three, the root locus has three branches. Suppose the desired closed-loop poles are given as p_1, p_2, and p_3. Then, the pole placement problem amounts to finding $\{K_c, z_c, p_c\}$such that the characteristic equation is

$$\chi(s) = (s - p_1)(s - p_2)(s - p_3) = s^3 - (p_1 + p_2 + p_3)s^2 + (p_1p_2 + p_1p_3 + p_2p_3)s - p_1p_2p_3$$

But the actual characteristic equation, in terms of the unknown controller parameters, is

$$\chi(s) = s(s - p_m)(s - p_c) + k(s - z_c) = s^3 - (p_m + p_c)s^2 + (p_mp_c + K)s - Kz_c$$

where $K := K_mK_c$. Equating the coefficients of the desired $\chi(s)$ to the coefficients of the actual $\chi(s)$, three equations in three unknowns are obtained:

$$\begin{aligned} p_m + p_c &= p_1 + p_2 + p_3 \\ p_mp_c + K &= p_1p_2 + p_1p_3 + p_2p_3 \\ Kz_c &= p_1p_2p_3 \end{aligned}$$

From the first equation p_c is determined, then K is obtained from the second equation, and finally z_c is computed from the third equation.

For different numerical values of p_m, p_1, p_2, and p_3 the shape of the root locus is different. Below are some examples, with the corresponding root loci shown in Figure 11.33 to Figure 11.35.

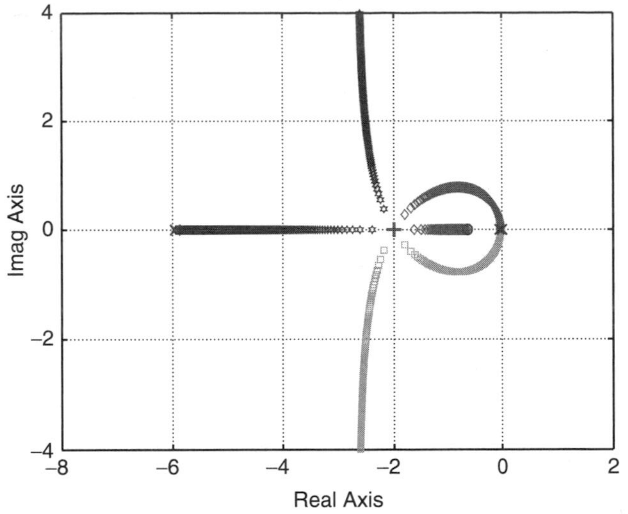

FIGURE 11.33 Root locus for Example 11.3(a).

(a) $p_m = -0.05,\ p_1 = p_2 = p_3 = -2 \Rightarrow$

$$K = 11.70, \quad p_c = -5.95, \quad z_c = -0.68$$

(b) $p_m = -0.5,\ p_1 = -1,\ p_2 = -2,\ p_3 = -3 \Rightarrow$

$$K = 8.25, \quad p_c = -5.50, \quad z_c = -0.73$$

(c) $p_m = -5,\ p_1 = -11,\ p_2 = -4 + j_1,\ p_3 = -4 - j_1 \Rightarrow$

$$K = 35, \quad p_c = -14, \quad z_c = -5.343$$

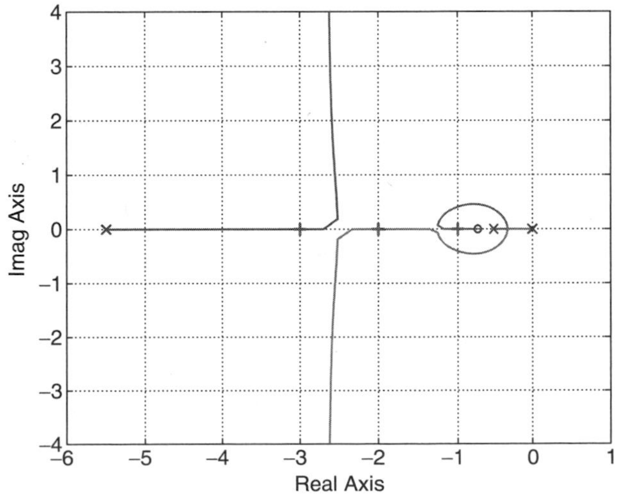

FIGURE 11.34 Root locus for Example 11.3(b).

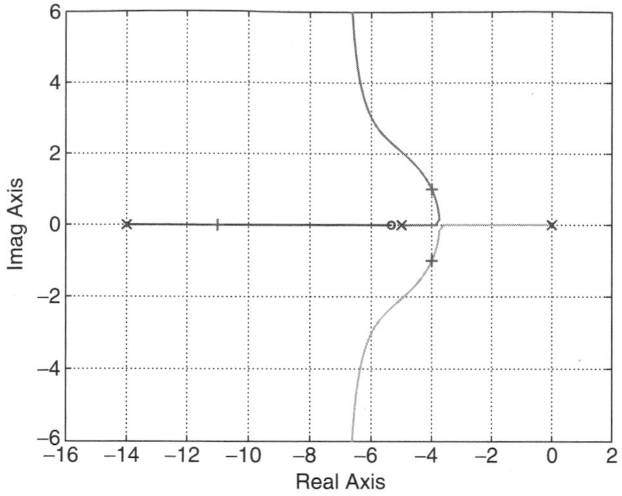

FIGURE 11.35 Root locus for Example 11.3(c).

Example 11.4

Consider the open-loop transfer function

$$P(s)C(s) = K_c \frac{(s^2 - 3s + 3)(s - z_c)}{s(s^2 + 3s + 3)(s - p_c)}$$

where K_c is the controller gain to be adjusted, and z_c and p_c are the controller zero and pole, respectively. Observe that the root locus has four branches except for the non-generic case $z_c = p_c$. Let the desired dominant closed-loop poles be $r_{1,2} = -0.4$. The steady state error for unit ramp reference input is

$$e_{ss} = \frac{p_c}{K_c z_c}$$

Accordingly, we want to make the ratio $K_c z_c / p_c$ as large as possible.

The characteristic equation is

$$\chi(s) = s(s^2 + 3s + 3)(s - p_c) + K_c(s^2 - 3s + 3)(s - z_c)$$

and it is desired to be in the form

$$\chi(s) = (s + 0.4)^2 (s - r_3)(s - r_4)$$

for some $r_{3,4}$ with $\text{Re}(r_{3,4}) < 0$, which implies that

$$\chi(s)|_{s=-0.4} = 0, \quad \frac{d}{ds}\chi(s)|_{s=-0.4} = 0 \tag{11.59}$$

Conditions (Equation (11.59)) give two equations:

$$0.784(0.4 + p_c) - 4.36K_c(0.4 + z_c) = 0$$

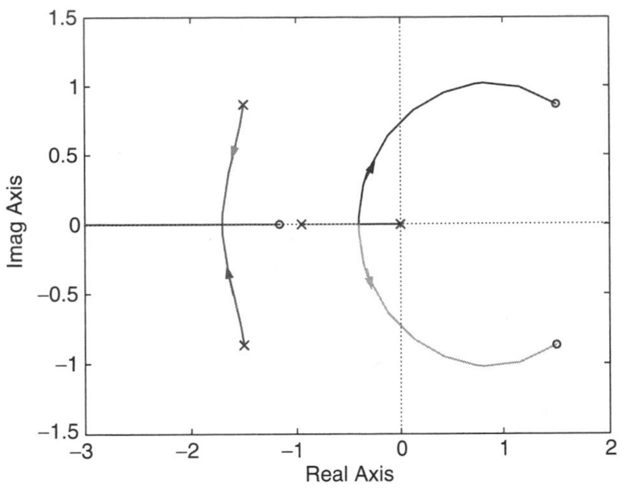

FIGURE 11.36 Root locus for Example 11.4.

$$4.36K_c - 0.784 - 1.08(0.4 + p_c) + 3.8K_c(0.4 + z_c) = 0$$

from which z_c and p_c can be solved in terms of K_c. Then, by simple substitutions, the ratio to be maximized, $K_c z_c / p_c$,can be reduced to

$$\frac{K_c z_c}{p_c} = \frac{3.4776K_c - 0.784}{24.2469K_c - 3.4776}$$

The maximizing value of K_c is 0.1297; it leads to $p_c = -0.9508$ and $z_c = -1.1637$. For this controller, the feedback system poles are

$$\{-1.64 + j0.37, -1.64 - j0.37, -0.40, -0.40\}$$

The root locus is shown in Figure 11.36.

11.4 Complementary Root Locus

In the previous section, the root locus parameter K was assumed to be positive and the phase and magnitude rules were established based on this assumption. There are some situations in which controller gain can be negative as well. Therefore, the complete picture is obtained by drawing the usual root locus (for $K > 0$) and the complementary root locus (for $K < 0$). The complementary root locus rules are

$$\ell \times 360^\circ = \sum_{i=1}^{n} \angle(r - p_i) - \sum_{j=1}^{n} \angle(r - z_j), \quad \ell = 0, \pm 1, \pm 2, \ldots \tag{11.60}$$

$$|K| = \frac{\Pi_{i=1}^{n} |r - p_i|}{\Pi_{j=1}^{m} |r - z_j|} \tag{11.61}$$

Since the phase rule (Equation (11.60)) is the 180° shifted version of (Equation (11.53)), the complementary root locus is obtained by simple modifications in the root locus construction rules. In particular, the number

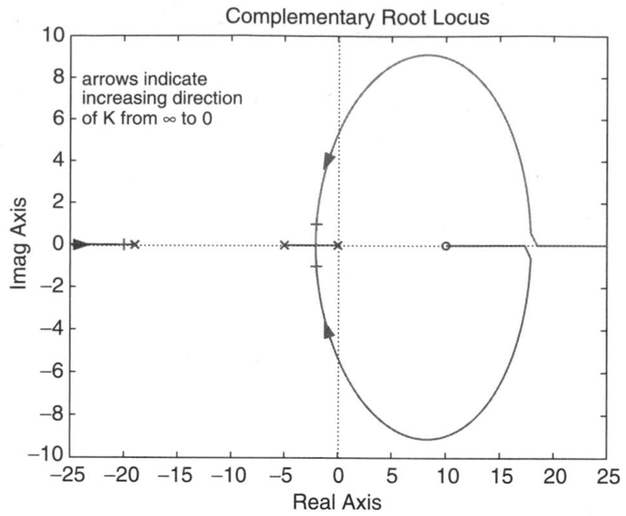

FIGURE 11.37 Complementary root locus for Example 11.3.

of asymptotes and their center are the same, but their angles α_ℓ's are given by

$$\alpha_\ell = \frac{2\ell}{(n-m)} \times 180^\circ, \quad \ell = 0, \ldots, (n-m-1)$$

Also, an interval on the real axis is on the complementary root locus if and only if it is not on the usual root locus.

Example 11.3 (revisited)

In the Example 11.3 given above, if the problem data is modified to $p_{\mathrm{m}} = -5$, $p_1 = -20$, and $p_{2,3} = -2 \pm j$, then the controller parameters become

$$K = -10, \quad p_{\mathrm{c}} = -19, \quad z_{\mathrm{c}} = 10$$

Note that the gain is negative. The roots of the characteristic equation as K varies between 0 and $-\infty$ form the complementary root locus; see Figure 11.37.

Example 11.4 (revisited)

In this example, if K increases from $-\infty$ to $+\infty$, the closed-loop system poles move along the complementary root locus, and then the usual root locus, as illustrated in Figure 11.38.

11.5 Root Locus for Systems with Time Delays

The standard feedback control system considered in this section is shown in Figure 11.39, where the controller C and plant P are in the form

$$C(s) = \frac{N_{\mathrm{c}}(s)}{D_{\mathrm{c}}(s)}$$

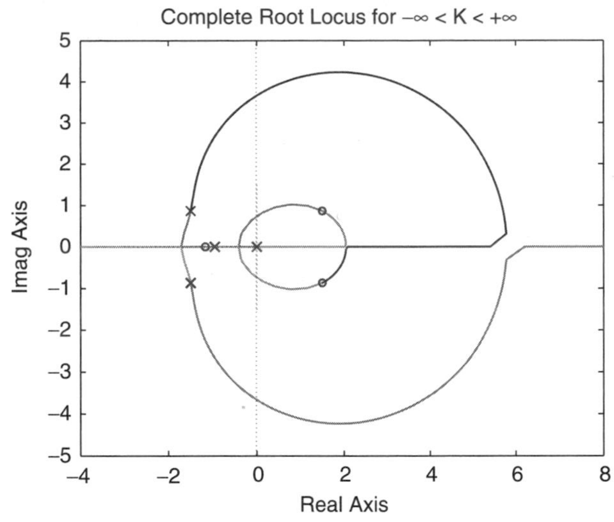

FIGURE 11.38 Complementary and usual root loci for Example 11.4.

and

$$P(s) = e^{-hs} P_0(s) \quad \text{where } P_0(s) = \frac{N_p(s)}{D_p(s)}$$

with (N_c, D_c) and (N_p, D_p) being coprime pairs of polynomials with real coefficients.[1] The term e^{-hs} is the transfer function of a pure delay element (in Figure 11.39 the plant input is delayed by h seconds). In general, time delays enter into the plant model when there is

- a sensor (or actuator) processing delay, and/or
- a software delay in the controller, and/or
- a transport delay in the process.

In this case the open-loop transfer function is

$$G(s) = G_h(s) = e^{-hs} G_0(s)$$

where $G_0(s) = P_0(s)C(s)$ corresponds to the no delay case, $h = 0$.

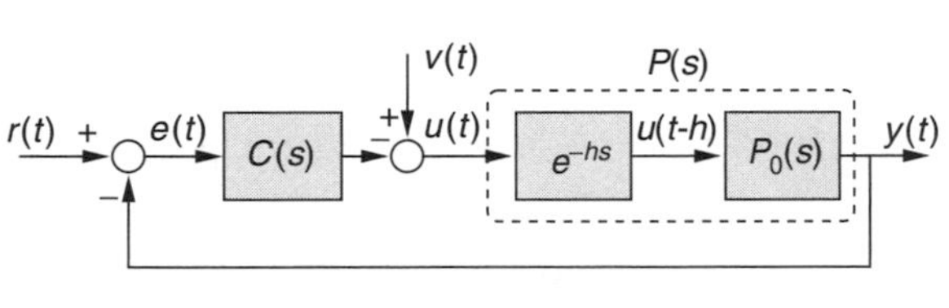

FIGURE 11.39 Feedback system a with time delay.

[1]A pair of polynomials is said to be coprime pair if they do not have common roots

Note that magnitude and phase of $G(j\omega)$ are determined from the identities

$$|G(j\omega)| = |G_0(j\omega)| \tag{11.62}$$

$$\angle G(j\omega) = -h\omega + \angle G_0(j\omega) \tag{11.63}$$

Stability of Delay Systems

Stability of the feedback system shown in Figure 11.39 is equivalent to having all the roots of

$$\chi(s) = D(s) + e^{-hs} N(s) \tag{11.64}$$

in the open left half plane, $\mathbb{C}_-$, where $D(s) = D_c(s)D_p(s)$ and $N(s) = N_c(s)N_p(s)$. We assume that there is no unstable pole–zero cancellation in taking the product $P_0(s)C(s)$, and that $\deg(D) > \deg(N)$ (here N and D need not be monic polynomials). Strictly speaking, $\chi(s)$ is not a polynomial because it is a transcendental function of s. The functions of the form (Equation (11.64)) belong to a special class of functions called *quasi-polynomials*. The closed-loop system poles are the roots of (Equation (11.64)).

Following are known facts (see [1,10]):

(i) If r_k is a root of Equation (20), then so is $\bar{r}_k$. (i.e., roots appear in complex conjugate pairs as usual)
(ii) There are infinitely many poles $r_k \in \mathbb{C}$, $k = 1, 2, \ldots$, satisfying $\chi(r_k) = 0$.
(iii) And r_k's can be enumerated in such a way that $R_e(r_k + 1) \leq R_e(r_k)$ moreover, $\mathrm{Re}(r_k) \rightarrow -\infty$ as $k \rightarrow \infty$.

Example 11.5

If $G_h(s) = e^{-hs}/s$, then the closed-loop system poles r_k, for $k = 1, 2, \ldots$, are the roots of

$$1 + \frac{e^{-h\sigma_k} e^{-jh\omega_k}}{\sigma_k + j\omega_k} e^{\pm j2k\pi} = 0 \tag{11.65}$$

where $r_k = \sigma_k + j\omega_k$ for some $\sigma_k, \omega_k \in \mathbb{R}$ Note that $e^{\pm j2k\pi} = 1$ for all $k = 1, 2, \ldots$. Equation (11.45) is equivalent to the following set of equations:

$$e^{-h\sigma_k} = |\sigma_k + j\omega_k| \tag{11.66}$$

$$\pm(2k-1)\pi = h\omega_k + \angle(\sigma_k + j\omega_k), \quad k = 1, 2, \ldots \tag{11.67}$$

It is quite interesting that for $h = 0$ there is only one root $r = -1$, but even for infinitesimally small $h > 0$ there are infinitely many roots. From the magnitude condition (Equation (11.66)), it can be shown that

$$\sigma_k \geqslant 0 \Rightarrow |\omega_k| \leqslant 1 \tag{11.68}$$

Also, for $\sigma_k \geq 0$ the phase $\angle(\sigma_k + j\omega_k)$ is between $-\pi/2$ and $+\pi/2$, therefore Equation (11.67) leads to

$$\sigma_k \geqslant 0 \Rightarrow h|\omega_k| \geqslant \frac{\pi}{2} \tag{11.69}$$

By combining Equation (11.68) and Equation (11.69), it can be proven that the feedback system has no roots in the closed right half plane when $h < \pi/2$. Furthermore, the system is unstable if $h \geq \pi/2$. In particular, for $h = \pi/2$ there are two roots on the imaginary axis, at $\pm j1$. It is also easy to show that, for any $h > 0$ as $k \rightarrow \infty$, the roots converge to

$$r_k \rightarrow \frac{1}{h}\left[-\ln\left(\frac{2k\pi}{h}\right) \pm j2k\pi\right]$$

As $h \rightarrow 0$, the magnitude of the roots converge to ∞.

As illustrated by the above example, property (iii) implies that for any given real number σ there are only finitely many r_k's in the region of the complex plane

$$\mathbb{C}_\sigma := \{s \in \mathbb{C} : \text{Re}(s) \geqslant \sigma\}$$

In particular, with $\sigma = 0$, this means that the quasi-polynomial $\chi(s)$ can have only finitely many roots in the right half plane. Since the effect of the closed-loop system poles that have very large negative real parts is negligible (as far as closed-loop systems' input–output behavior is concerned), only finitely many "dominant" roots r_k, for $k = 1,\ldots,m$, should be computed for all practical purposes.

Dominant Roots of a Quasi-Polynomial

Now we discuss the following problem: given $N(s)$, $D(s)$, and $h \geq 0$, find the dominant roots of the quasi-polynomial

$$\chi(s) = D(s) + \mathrm{e}^{-hs}N(s)$$

For each fixed $h > 0$, it can be shown that there exists $\sigma_{\max}$ such that $\chi(s)$ has no roots in the region, $\mathbb{C}_{\sigma_{\max}}$, see [11] for a simple algorithm to estimate $\sigma_{\max}$, based on Nyquist criterion. Given $h > 0$ and a region of the complex plane defined by $\sigma_{\min} \leq \text{Re}(s) \leq \sigma_{\max}$, the problem is to find the roots of $\chi(s)$ in this region.

Clearly, a point $r = \sigma + j\omega$ in $\mathbb{C}$ is a root of $\chi(s)$ if and only if

$$D(\sigma + j\omega) = -\mathrm{e}^{-h\sigma}\mathrm{e}^{-jh\omega}N(\sigma + j\omega)$$

Taking the magnitude square of both sides of the above equation, $\chi(r) = 0$ implies

$$A_\sigma(x) := D(\sigma + x)D(\sigma - x) - \mathrm{e}^{-2h\sigma}N(\sigma + x)N(\sigma - x) = 0$$

where $x = j\omega$. The term $D(\sigma + x)$ stands for the function $D(s)$ evaluated at $\sigma + x$. The other terms of $A_\sigma(x)$ are calculated similarly. For each fixed σ, the function $A_\sigma(x)$ is a polynomial in the variable x. By symmetry, if x is a zero of $A_\sigma(\cdot)$ then $(-x)$ is also a zero.

If $A_\sigma(x)$ has a root x_ℓ whose real part is zero, set $r_\ell = \sigma + x_\ell$. Next, evaluate the magnitude of $\chi(r_\ell)$; if it is zero, then n_ℓ is a root of $\chi(s)$. Conversely, if $A_\sigma(x)$ has no root on the imaginary axis, then $\chi(s)$ cannot have a root whose real part is the fixed value of σ from which $A_\sigma(\cdot)$ is constructed.

Algorithm

Given $N(s)$, $D(s)$, h, $\sigma_{\min}$, and $\sigma_{\max}$:

Step 1. Pick σ values $\sigma_1,\ldots,\sigma_M$ between $\sigma_{\min}$ and $\sigma_{\max}$ such that $\sigma_{\min} = \sigma_1$, $\sigma_i < \sigma_{i+1}$, and $\sigma_M = \sigma_{\max}$. For each σ_i perform the following.

Step 2. Construct the polynomial $A_i(x)$ according to

$$A_i(x) := D(\sigma_i + x)D(\sigma_i - x) - \mathrm{e}^{-2h\sigma_i}N(\sigma_i + x)N(\sigma_i - x)$$

Step 3. For each imaginary axis roots x_ℓ of A_i, perform the following test:
Check if $|\chi(\sigma_i + x_\ell)| = 0$; if yes, then $r = \sigma_i + x_\ell$ is a root of $\chi(s)$; if not, discard x_ℓ.

Step 4. If $i = M$, stop; else increase i by 1 and go to Step 2.

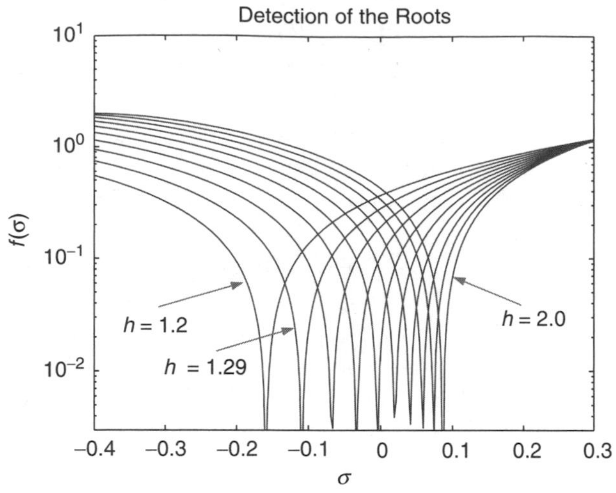

FIGURE 11.40 Detection of the dominant roots.

Example 11.6

We will find the dominant roots of

$$1 + \frac{e^{-hs}}{s} = 0 \tag{11.70}$$

for a set of critical values of h. Recall that Equation (11.70) has a pair of roots $\pm j1$ when $h = \pi/2 = 1.57$. Moreover, dominant roots of Equation (11.70) are in the right half plane if $h > 1.57$, and they are in the left half plane if $h < 1.57$. So, it is expected that for $h \in (1.2, 2.0)$ the dominant roots are near the imaginary axis. Take $\sigma_{\min} = -0.5$ and $\sigma_{\max} = 0.5$, with $M = 400$ linearly spaced σ_i's between them. In this case

$$A_i(x) = \sigma_i^2 - e^{-2h\sigma_i} - x^2$$

Whenever $e^{-2h\sigma_i} \geqslant \sigma_i^2$, $A_i(x)$ has two roots:

$$x_\ell = \pm j\sqrt{e^{-2h\sigma_i} - \sigma_i^2}, \quad \ell = 1, 2$$

For each fixed σ_i satisfying this condition, let $r_\ell = \sigma_i + x_\ell$ (note that x_ℓ is a function of σ_i, so r_ℓ is a function of σ_i) and evaluate

$$f(\sigma_i) := \left|1 + \frac{e^{-hr_\ell}}{r_\ell}\right|$$

If $f(\sigma_i) = 0$, then x_ℓ is a root of 11.70. For 10 different values of $h \in (1.2, 2.0)$, the function $f(\sigma)$ is plotted in Figure 11.40. This figure shows the feasible values of σ_i for which x_ℓ (defined from σ_i) is a root of Equation (11.70). The dominant roots of Equation (11.70), as h varies from 1.2 to 2.0, are shown in Figure 11.41. For $h < 1.57$, all the roots are in $\mathbb{C}_-$. For $h > 1.57$, the dominant roots are in $\mathbb{C}_{+}$,and for $h = 1.57$, they are at $\pm j1$.

Root Locus Using Padé Approximations

In this section we assume that $h > 0$ is fixed and we try to obtain the root locus, with respect to uncertain/adjustable gain K, corresponding to the dominant poles. The problem can be solved by numerically calculating

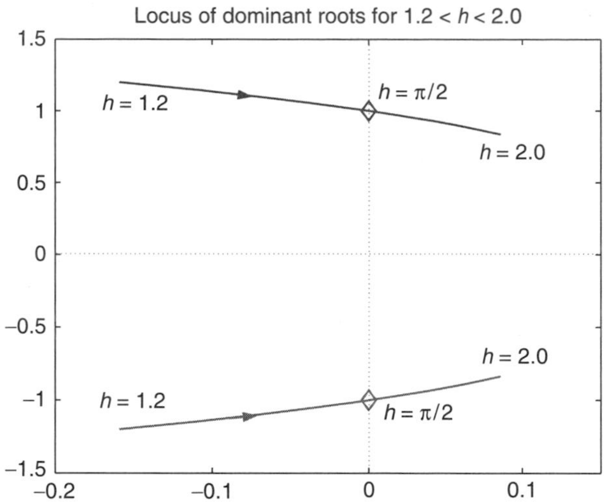

FIGURE 11.41 Dominant roots as h varies from 1.2 to 2.0.

the dominant roots of the quasi-polynomial

$$\chi(s) = D(s) + KN(s)e^{-hs} \tag{11.71}$$

for varying K, by using the methods presented in the previous section. In this section an alternative method is given that uses Padé approximation of the time delay term e^{-hs}. More precisely, the idea is to find polynomials $N_h(s)$ and $D_h(s)$ satisfying

$$e^{-hs} \approx \frac{N_h(s)}{D_h(s)} \tag{11.72}$$

so that the dominant roots

$$D(s)D_h(s) + KN(s)N_h(s) = 0 \tag{11.73}$$

closely match the dominant roots of $\chi(s)$, (11.71). How should we do the approximation (Equation (11.72)) for this match?

By using the stability robustness measures determined from the Nyquist stability criterion, we can show that for our purpose we may consider the following cost function in order to define a meaningful measure for the approximation error:

$$\Delta_h =: \sup_{\omega} \left|\frac{k_{\max}N(j\omega)}{D(j\omega)}\right| \left|e^{-jh\omega} - \frac{N_h(j\omega)}{D_h(j\omega)}\right|$$

where $K_{\max}$ is the maximum value of interest for the uncertain/adjustable parameter K.

The ℓ th order Padé approximation is defined as follows:

$$N_h(s) = \sum_{k=0}^{\ell} (-1)^k c_k h^k s^k$$

$$D_h(s) = \sum_{k=0}^{\ell} c_k h^k s^k$$

where coefficients c_k's are computed from

$$c_k = \frac{(2\ell - k)!\ell!}{2\ell!k!(\ell - k)!}, \quad k = 0, 1, \ldots, \ell$$

First-and second-order approximations are in the form

$$\frac{N_h(s)}{D_h(s)} = \begin{cases} \dfrac{1 - hs/2}{1 + hs/2}, & \ell = 1 \\ \dfrac{1 - hs/2 + (hs)^2/12}{1 + hs/2 + (hs)^2/12}, & \ell = 2 \end{cases}$$

Given the problem data $\{h, K_{\max}, N(s), D(s)\}$, how do we find the smallest degree, ℓ, of the Padé approximation, so that $\Delta_h \leq \delta$ (or $\Delta_h/K_{\max} \leq \delta'$) for a specified error δ, or a specified relative error δ'? The answer lies in the following result [7]: for a given degree of approximation ℓ we have

$$\left| e^{-jh\omega} - \frac{N_h(j\omega)}{D_h(j\omega)} \right| \leqslant \begin{cases} 2\left(\dfrac{eh\omega}{4\ell}\right)^{2\ell+1}, & \omega \leqslant \dfrac{4\ell}{eh} \\ 2, & \omega \geqslant \dfrac{4\ell}{eh} \end{cases}$$

In light of this result, we can solve the approximation order selection problem by using the following procedure:

1. Determine the frequency ω_x such that

$$\left| \frac{K_{\max} N(j\omega)}{D(j\omega)} \right| \leqslant \frac{\delta}{2}, \quad \text{for all } \omega \geqslant \omega_x$$

and initialize $\ell = 1$.

2. For each $\ell \geqslant 1$ define

$$\omega_\ell = \max\left\{\omega_X, \frac{4\ell}{eh}\right\}.$$

and plot the function

$$\Phi_\ell(\omega) := \begin{cases} 2\left|\dfrac{K_{\max} N(j\omega)}{D(j\omega)}\right| \left(\dfrac{eh\omega}{4\ell}\right)^{2\ell+1}, & \text{for } \omega \leqslant \dfrac{4\ell}{eh} \\ 2\left|\dfrac{K_{\max} N(j\omega)}{D(j\omega)}\right|, & \text{for } \omega_\ell \geqslant \omega \geqslant \dfrac{4\ell}{eh} \end{cases}$$

3. Check If

$$\max_{\omega \in [0, \omega_x]} \Phi_\ell(\omega) \leqslant \delta \tag{11.74}$$

If yes, stop, this value of ℓ satisfies the desired error bound: $\Delta_h \leq \delta$. Otherwise, increase ℓ by 1, and go to Step 2. Note that the left-hand side of the inequality (Equation (11.74)) is an upper bound of Δ_h.

Since we assumed $\deg(D) > \deg(N)$, the algorithm will pass Step 3 eventually for some finite $\ell \geqslant 1$. At each iteration, we have to draw the error function $\Phi_\ell(\omega)$ and check whether its peak value is less than δ. Typically, as δ decreases, ω_x increases, and that forces ℓ to increase. On the other hand, for very large values of ℓ, the relative magnitude c_0/c_ℓ of the coefficients becomes very large, in which case numerical difficulties arise in analysis and simulations. Also, as time delay h increases, ℓ should be increased to keep the level of the approximation error δ fixed. This is a fundamental difficulty associated with time delay systems.

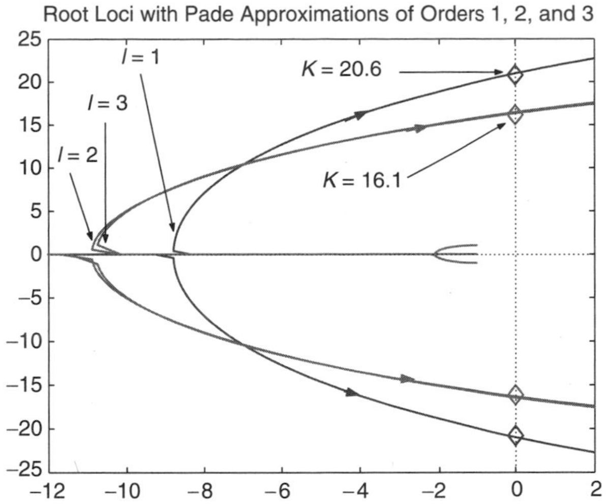

FIGURE 11.42 Dominant root for $\ell = 1$

Example 11.7

Let $N(s) = s+1$, $D(s) = s^2+2s+2$ and $h = 0.1$, and $K_{max} = 20$. Then, for $\delta' = 0.05$ applying the above procedure we calculate $\ell = 2$ as the smallest approximation degree satisfying $\Delta_h/K_{max} < \delta'$. Therefore, a second-order approximation of the time delay should be sufficient for predicting the dominant poles for $K \in [0, 20]$. Figure 11.42 shows the approximate root loci obtained from Padé approximations of degrees$\ell = 1, 2, 3$. There is a significant difference between the root loci for $\ell = 1$ and $\ell = 2$. In the region $\text{Re}(s) \geq -12$, the predicted dominant roots are approximately the same for $\ell = 2, 3$, for $K \in [0,20]$. So, we can safely say that using higher order approximations will not make any significant difference as far as predicting the behavior of the dominant poles for the given range of K.

Notes and References

This section in the chapter is an edited version of related parts of the author's book [9]. More detailed discussions of the root locus method can be found in all the classical control books, such as [2, 5, 6, 8]. As mentioned earlier, extension of this method to discrete time systems is rather trivial: the method to find the roots of a polynomial as a function of a varying real parameter is independent of the variable s (in the continuous time case) or z (in the discrete time case). The only difference between these two cases is the definition of the desired region of the complex plane: for the continuous time systems, this is defined relative to the imaginary axis, whereas for the discrete time systems the region is defined with respect to the unit circle, as illustrated in Figure 11.27.

References

1. Bellman, R.E., and Cooke, K.L., *Differential Difference Equations,* Academic Press, New York, 1963.
2. Dorf, R.C., and Bishop, R.H., *Modern Control Systems,* 9th ed., Prentice-Hall, Upper Saddle River, NJ, 2001.
3. Evans, W.R., "Graphical analysis of control systems," *Transac. Amer. Inst. Electrical Engineers,* vol. 67 (1948), pp. 547–551.
4. Evans, W.R., "Control system synthesis by root locus method," *Transac. Amer. Inst. Electrical Engineers,* vol. 69 (1950), pp. 66–69.
5. Franklin, G.F., Powell, J.D., and Emami-Naeini, A., *Feedback Control of Dynamic Systems,* 3rd ed., Addison Wesley, Reading, MA, 1994.

6. Kuo, B.C., *Automatic Control Systems*, 7th ed., Prentice-Hall, Upper Saddle River, NJ, 1995.
7. Lam, J., "Convergence of a class of Padé approximations for delay systems," *Int. J. Control*, vol. 52 (1990), pp. 989–1008.
8. Ogata, K., *Modern Control Engineering*, 3rd ed., Prentice-Hall, Upper Saddle River, NJ, 1997.
9. Özbay, H., *Introduction to Feedback Control Theory*, CRC Press LLC, Boca Raton, FL, 2000.
10. Stepan, G., *Retarded Dynamical Systems: Stability and Characteristic Functions*, Longman Scientific & Technical, New York, 1989.
11. Ulus, C., "Numerical computation of inner-outer factors for a class of retarded delay systems," *Int. J. Systems Sci.*, vol. 28 (1997), pp. 897–904.

11.5 Compensation

Charles L. Phillips and Royce D. Harbor

Compensation is the process of modifying a closed-loop control system (usually by adding a *compensator* or *controller*) in such a way that the compensated system satisfies a given set of design specifications. This section presents the fundamentals of compensator design; actual techniques are available in the references.

A single-loop control system is shown in Figure 11.43. This system has the transfer function from input $R(s)$ to output $C(s)$

$$T(s) = \frac{C(s)}{R(s)} = \frac{G_c(s)G_p(s)}{1 + G_c(s)G_p(s)H(s)} \tag{11.75}$$

and the characteristic equation is

$$1 + G_c(s)G_p(s)H(s) = 0 \tag{11.76}$$

where $G_c(s)$ is the *compensator* transfer function, $G_p(s)$ is the *plant* transfer function, and $H(s)$ is the *sensor* transfer function. The transfer function from the disturbance input $D(s)$ to the output is $G_d(s)/[1 + G_c(s)G_p(s)H(s)]$. The function $G_c(s)G_p(s)H(s)$ is called the *open-loop function.*

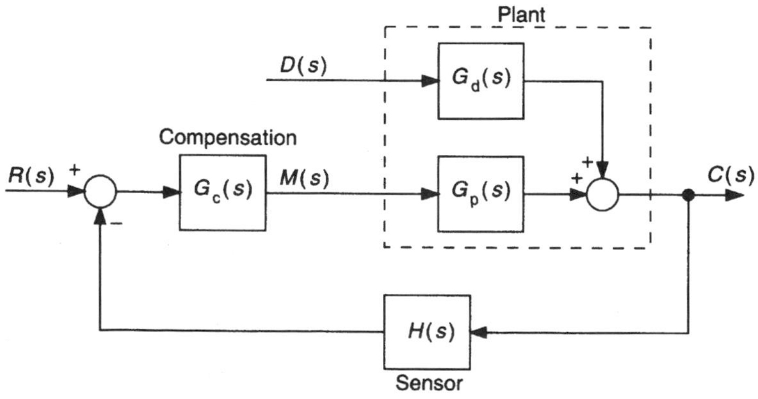

FIGURE 11.43 A closed-loop control system. (*Source*: C.L. Phillips and R.D. Harbor, *Feedback Control Systems*, 2nd ed., Englewood Cliffs, N.J.: Prentice-Hall, 1991, p. 161. With permission.)

Control System Specifications

The compensator transfer function $G_c(s)$ is designed to give the closed-loop system certain specified characteristics, which are realized through achieving one or more of the following:

1. Improving the transient response. Increasing the speed of response is generally accomplished by increasing the open-loop gain $G_c(j\omega)G_p(j\omega)H(j\omega)$ at higher frequencies such that the system bandwidth is increased. Reducing overshoot (ringing) in the response generally involves increasing the phase margin ϕ_m of the system, which tends to remove any resonances in the system. The phase margin ϕ_m occurs at the frequency ω_1 and is defined by the relationship

$$\left|G_c(j\omega_1)G_p(j\omega_1)H(j\omega_1)\right| = 1$$

   with the angle of $G_c(j\omega_1)G_p(j\omega_1)H(j\omega_1)$ equal to $(180° + \phi_m)$.
2. Reducing the steady-state errors. Steady-state errors are decreased by increasing the open-loop gain $G_c(j\omega)G_p(j\omega)H(j\omega)$ in the frequency range of the errors. Low-frequency errors are reduced by increasing the low-frequency open-loop gain and by increasing the type number of the system [the number of poles at the origin in the open-loop function $G_c(s)G_p(s)H(s)$].
3. Reducing the sensitivity to plant parameters. Increasing the open-loop gain $G_c(j\omega)G_p(j\omega)H(j\omega)$ tends to reduce the variations in the system characteristics due to variations in the parameters of the plant.
4. Rejecting disturbances. Increasing the open-loop gain $G_c(j\omega)G_p(j\omega)\mathrm{H}(j\omega)$ tends to reduce the effects of disturbances [D(s) in Figure 11.43] on the system output, provided that the increase in gain does not appear in the direct path from disturbance inputs to the system output.
5. Increasing the relative stability. Increasing the open-loop gain tends to reduce phase and gain margins, which generally increases the overshoot in the system response. Hence, a trade-off exists between the beneficial effects of increasing the open-loop gain and the resulting detrimental effects of reducing the stability margins.

Design

Design procedures for compensators are categorized as either *classical methods* or *modern methods.* Classical methods discussed are:

- Phase-lag frequency response
- Phase-lead frequency response
- Phase-lag root locus
- Phase-lead root locus

Modern methods discussed are:

- Pole placement
- State estimation
- Optimal

Frequency Response Design

Classical design procedures are normally based on the open-loop function of the uncompensated system, $G_p(s)H(s)$. Two compensators are used in classical design; the first is called a *phase-lag compensator,* and the second is called a *phase-lead compensator.*

The general characteristics of phase-lag-compensated systems are as follows:

1. The low-frequency behavior of a system is improved. This improvement appears as reduced errors at low frequencies, improved rejection of low-frequency disturbances, and reduced sensitivity to plant parameters in the low-frequency region.

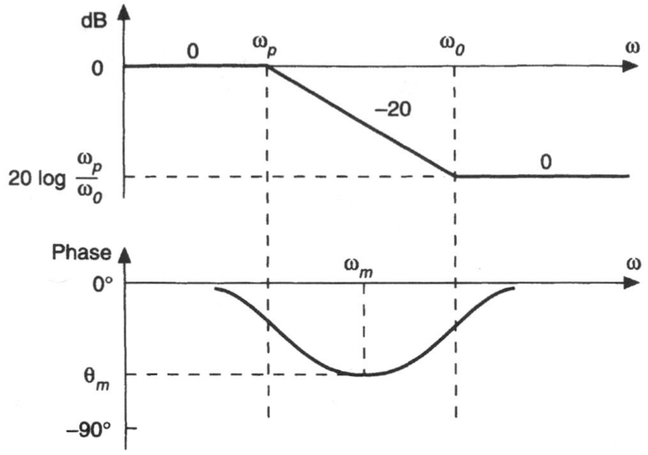

FIGURE 11.44 Bode diagram for a phase-lag compensator. (*Source*: C.L. Phillips and R.D. Harbor, *Feedback Control Systems*, 2nd ed., Englewood Cliffs, N.J.: Prentice-Hall, 1991, p. 358. With permission.)

2. The system bandwidth is reduced, resulting in a slower system time response and better rejection of high-frequency noise in the sensor output signal.

The general characteristics of phase-lead-compensated systems are as follows:

1. The high-frequency behavior of a system is improved. This improvement appears as faster responses to inputs, improved rejection of high-frequency disturbances, and reduced sensitivity to changes in the plant parameters.
2. The system bandwidth is increased, which can increase the response to high-frequency noise in the sensor output signal.

The transfer function of a first-order compensator can be expressed as

$$G_c(s) = \frac{K_c(s/\omega_0 + 1)}{s/\omega_p + 1} \tag{11.77}$$

where $-\omega_0$ is the compensator zero, $-\omega_p$ is its pole, and K_c is its dc gain. If $\omega_p < \omega_0$, the compensator is phase-lag. The Bode diagram of a phase-lag compensator is given in Figure 11.44 for $K_c = 1$.

It is seen from Figure 11.44 that the phase-lag compensator reduces the high-frequency gain of the open-loop function relative to the low-frequency gain. This effect allows a higher low-frequency gain, with the advantages listed above. The pole and zero of the compensator must be placed at very low frequencies relative to the compensated-system bandwidth so that the destabilizing effects of the negative phase of the compensator are negligible.

If $\omega_p > \omega_0$ the compensator is phase-lead. The Bode diagram of a phase-lead compensator is given in Figure 11.45 for $K_c = 1$.

It is seen from Figure 11.45 that the phase-lead compensator increases the high-frequency gain of the open-loop function relative to its low-frequency gain. Hence, the system has a larger bandwidth, with the advantages listed above. The pole and zero of the compensator are generally difficult to place, since the increased gain of the open-loop function tends to destabilize the system, while the phase lead of the compensator tends to stabilize the system. The pole-zero placement for the phase-lead compensator is much more critical than that of the phase-lag compensator.

A typical Nyquist diagram of an uncompensated system is given in Figure 11.46. The pole and the zero of a phase-lag compensator are placed in the frequency band labeled A. This placement negates the destabilizing effect of the negative phase of the compensator. The pole and zero of a phase-lead compensator are placed in

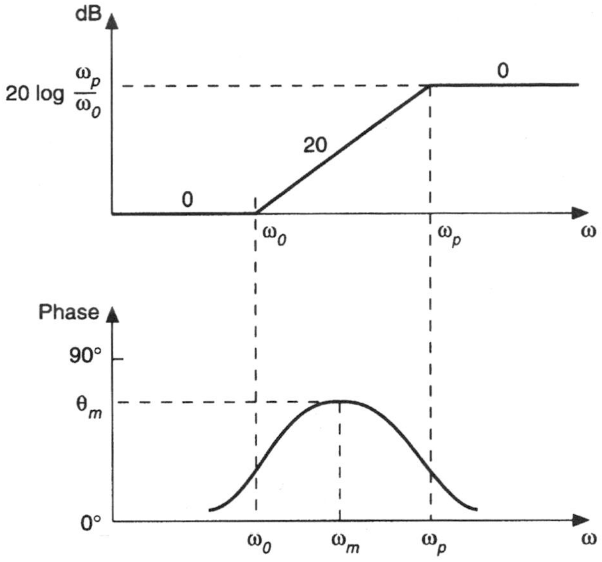

FIGURE 11.45 Bode diagram for a phase-lead compensator. (*Source*: C.L. Phillips and R.D. Harbor, *Feedback Control Systems,* 2nd ed., Englewood Cliffs, N.J.: Prentice-Hall, 1991, p. 363. With permission.)

the frequency band labeled B. This placement utilizes the stabilizing effect of the positive phase of the compensator.

PID Controllers

Proportional-plus-integral-plus-derivative (PID) compensators are probably the most utilized form for compensators. These compensators are essentially equivalent to a phase-lag compensator cascaded with a phase-lead compensator. The transfer function of this compensator is given by

$$G_c(s) = K_p + \frac{K_I}{s} + K_D s \tag{11.78}$$

A block diagram portrayal of the compensator is shown in Figure 11.47. The integrator in this compensator increases the system type by one, resulting in an improved low-frequency response. The Bode diagram of a PID compensator is given in Figure 11.48.

With $K_D = 0$, the compensator is phase-lag, with the pole in Equation (11.77) moved to $\omega_p = 0$. As a result the compensator is type one. The zero of the compensator is placed in the low-frequency range to correspond to the zero of the phase-lag compensator discussed above.

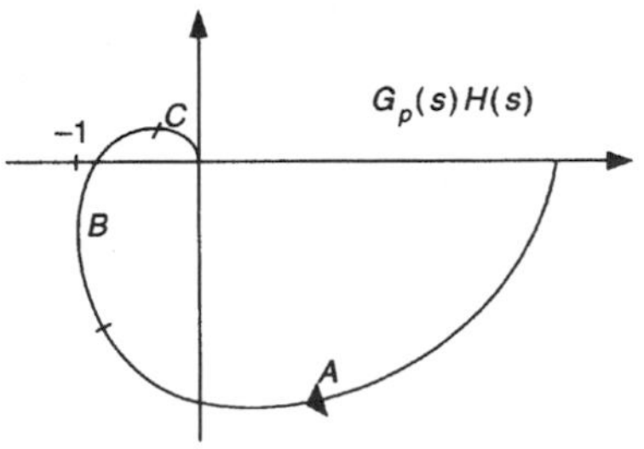

FIGURE 11.46 A typical Nyquist diagram for $G_p(s)H(s)$. (*Source*: C.L. Phillips and R.D. Harbor, *Feedback Control Systems,* 2nd ed., Englewood Cliffs, N.J.: Prentice-Hall, 1991, p. 364. With permission.)

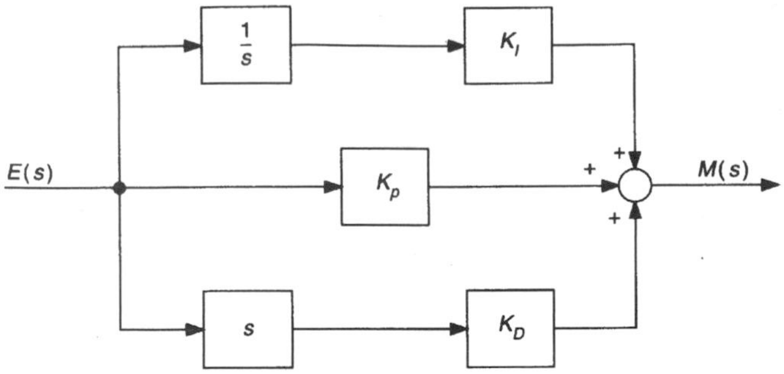

FIGURE 11.47 Block diagram of a PID compensator. (*Source*: C.L. Phillips and R.D. Harbor, *Feedback Control Systems*, 2nd ed., Englewood Cliffs, N.J.: Prentice-Hall, 1991, p. 378. With permission.)

With $K_I = 0$, the compensator is phase-lead, with a single zero and the pole moved to infinity. Hence, the gain continues to increase with increasing frequency. If high-frequency noise is a problem, it may be necessary to add one or more poles to the PD or PID compensators. These poles must be placed at high frequencies relative to the phase-margin frequency such that the phase margin (stability characteristics) of the system is not degraded. PD compensators realized using rate sensors minimize noise problems [Phillips and Harbor, 1991].

Root Locus Design

Root locus design procedures generally result in the placement of the two dominant poles of the closed-loop system transfer function. A system has two dominant poles if its behavior approximates that of a second-order system.

The differences in root locus designs and frequency response designs appear only in the interpretation of the control-system specifications. A root locus design that improves the low-frequency characteristics of the system will result in a phase-lag controller; a phase-lead compensator results if the design improves the

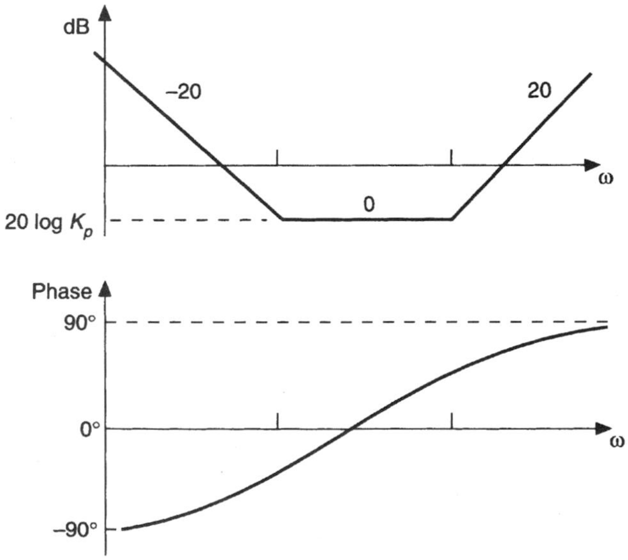

FIGURE 11.48 Bode diagram of a PID compensator. (*Source*: C.L. Phillips and R.D. Harbor, *Feedback Control Systems*, 2nd ed., Englewood Cliffs, N.J.: Prentice-Hall, 1991, p. 382. With permission.)

high-frequency response of the system. If a root locus design is performed, the frequency response characteristics of the system should be investigated. Also, if a frequency response design is performed, the poles of the closed-loop transfer function should be calculated.

Modern Control Design

The classical design procedures above are based on a transfer-function model of a system. Modern design procedures are based on a *state-variable model* of the plant. The plant transfer function is given by

$$\frac{Y(s)}{U(s)} = G_p(s) \tag{11.79}$$

where we use $u(t)$ for the plant input and $y(t)$ for the plant output. If the system model is nth order, the denominator of $G_p(s)$ is an nth-order polynomial.

The state-variable model, or state model, for a single-input–single-output plant is given by

$$\frac{d\mathbf{x}(t)}{dt} = \mathbf{Ax}(t) + \mathbf{B}u(t)$$

$$y(t) = \mathbf{Cx}(t) \tag{11.80}$$

$$y(t) = \mathbf{Cx}(t)$$

where $\mathbf{x}(t)$ is the $n \times 1$ state vector, $u(t)$ is the plant input, $y(t)$ is the plant output, $\mathbf{A}$ is the $n \times n$ *system matrix*, $\mathbf{B}$ is the $n \times 1$ *input matrix*, and $\mathbf{C}$ is the $1 \times n$ *output matrix*. The transfer function of Equation (11.79) is an input-output model; the state model of Equation (11.80) yields the same input-output model and in addition includes an internal model of the system. The state model of Equation (11.80) is readily adaptable to a multiple-input–multiple-output system (*a multivariable system*); for that case, $u(t)$ and $y(t)$ are vectors. We will consider only single-input–single-output systems.

The plant transfer function of Equation (11.79) is related to the state model of Equation (11.80) by

$$G_p(s) = \mathbf{C}(s\mathbf{I} - \mathbf{A})^{-1}\mathbf{B} \tag{11.81}$$

The state model is not unique; many combinations of the matrices **A, B,** and **C** can be found to satisfy Equation (11.81) for a given transfer function $G_p(s)$.

Classical compensator design procedures are based on the open-loop function $G_p(s)H(s)$ of Figure 11.43. It is common to present modern design procedures as being based on only the plant model of Equation (11.80). However, the models of the sensors that measure the signals for feedback must be included in the state model. This problem will become more evident as the modern procedures are presented.

Pole Placement

Probably the simplest modern design procedure is *pole placement.* Recall that root locus design was presented as placing the two dominant poles of the closed-loop transfer function at desired locations. The pole-placement procedure places *all* poles of the closed-loop transfer function, or equivalently, all roots of the closed-loop system characteristic equation, at desirable locations.

The system design specifications are used to generate the desired closed-loop system characteristic equation $\alpha_c(s)$, where

$$\alpha_c(s) = s^n + \alpha_{n-1}s^{n-1} + \ldots + \alpha_1 s + \alpha_0 = 0 \tag{11.82}$$

for an nth-order plant. This characteristic equation is realized by requiring the plant input to be a linear

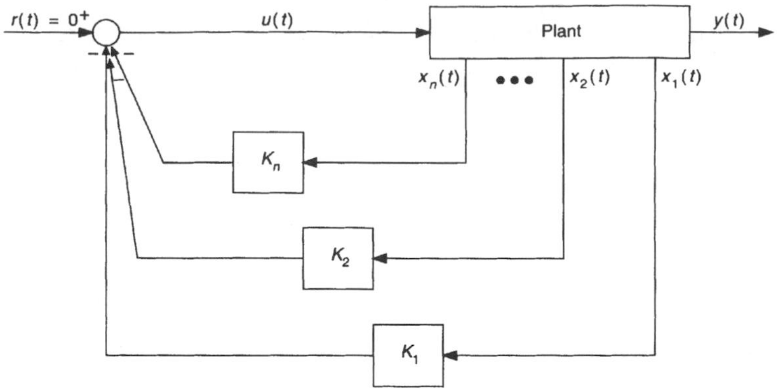

FIGURE 11.49 Implementation of pole-placement design. (*Source*: C.L. Phillips and R.D. Harbor, *Feedback Control Systems,* 2nd ed., Englewood Cliffs, N.J.: Prentice-Hall, 1991, p. 401. With permission.)

combination of the plant states, that is,

$$u(t) = -K_1x_1(t) - K_2x_2(t) - \ldots - K_nx_n(t) = -\mathbf{K}\mathbf{x}(t) \tag{11.83}$$

where $\mathbf{K}$ is the $1 \times n$ feedback-gain matrix. Hence *all* states must be measured and fed back. This operation is depicted in Figure 11.49.

The feedback-gain matrix $\mathbf{K}$ is determined from the desired characteristic equation for the closed-loop system of Equation (11.82):

$$\alpha_c(s) = |s\mathbf{I} - \mathbf{A} + \mathbf{B}\mathbf{K}| = 0 \tag{11.84}$$

The state feedback gain matrix $\mathbf{K}$ which yields the specified closed-loop characteristic equation $\alpha_c(s)$ is

$$\mathbf{K} = [\,0\;0\;\ldots\;\;0\;1\,][\,\mathbf{B}\;\mathbf{AB}\;\ldots\;\mathbf{A}^{n-1}\;\mathbf{B}\,]^{-1}\alpha_c(\mathbf{A}) \tag{11.85}$$

where $\alpha_c(\mathbf{A})$ is Equation (11.82) with the scalar s replaced with the matrix $\mathbf{A}$. A plant is said to be *controllable* if the inverse matrix in Equation (11.85) exists. Calculation of $\mathbf{K}$ completes the design process. A simple computer algorithm is available for solving Equation (11.85) for $\mathbf{K}$.

State Estimation

In general, modern design procedures require that the state vector $\mathbf{x}(t)$ be fed back, as in Equation (11.83). The measurement of all state variables is difficult to implement for high-order systems. The usual procedure is to estimate the states of the system from the measurement of the output $y(t)$, with the estimated states then fed back.

Let the estimated state vector be $\hat{\mathbf{x}}$. One procedure for estimating the system states is by an *observer,* which is a dynamic system realized by the equations

$$\frac{d\hat{\mathbf{x}}(t)}{dt} = (\mathbf{A} - \mathbf{G}\mathbf{C})\hat{\mathbf{x}}(t) + \mathbf{B}u(t) + \mathbf{G}y(t) \tag{11.86}$$

with the feedback equation of Equation (11.83) now realized by

$$u(t) = -\mathbf{K}\hat{\mathbf{x}}(t) \tag{11.87}$$

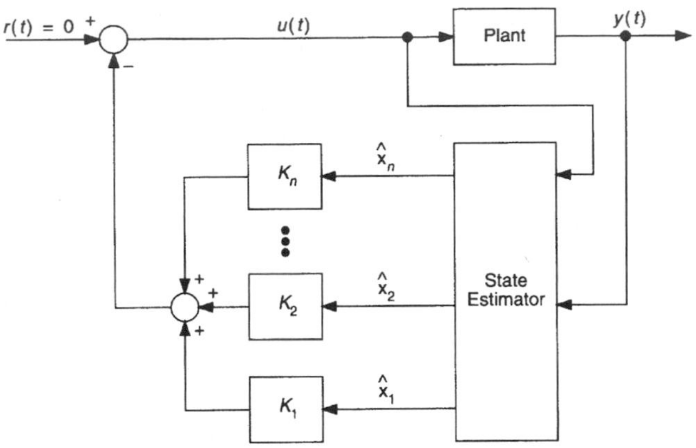

FIGURE 11.50 Implementation of observer-pole-placement design. (*Source*: C.L. Phillips and R.D. Harbor, *Feedback Control Systems,* 2nd ed., Englewood Cliffs, N.J.: Prentice-Hall, 1991, p. 417. With permission.)

The matrix **G** in Equation (11.86) is calculated by assuming an nth-order characteristic equation for the observer of the form

$$\alpha_e(s) = |s\mathbf{I} - \mathbf{A} + \mathbf{GC}| = 0 \tag{11.88}$$

The estimator gain matrix **G** which yields the specified estimator characteristic equation $\alpha_e(s)$ is

$$\mathbf{G} = \alpha_e(\mathbf{A})[\mathbf{C}\,\mathbf{CA}\,\ldots\,\mathbf{CA}^{n-1}]^{-T}[0\,0\,\ldots\,0\,1]^T \tag{11.89}$$

where $[\cdot]^T$ denotes the matrix transpose. A plant is said to be *observable* if the inverse matrix in Equation (11.89) exists. An implementation of the closed-loop system is shown in Figure 11.50. The observer is usually implemented on a digital computer. The plant and the observer in Figure 11.50 are both nth-order; hence, the closed-loop system is of order $2n$.

The observer-pole-placement system of Figure 11.50 is equivalent to the system of Figure 11.51, which is of the form of closed-loop systems designed by classical procedures. The transfer function of the controller-estimator (equivalent compensator) of Figure 11.51 is given by

$$\mathbf{G}_{ec}(s) = \mathbf{K}[s\mathbf{I} - \mathbf{A} + \mathbf{GC} + \mathbf{BK}]^{-1}\mathbf{G} \tag{11.90}$$

This compensator is nth-order for an nth-order plant; hence, the total system is of order $2n$. The characteristic equation for the compensated system is given by

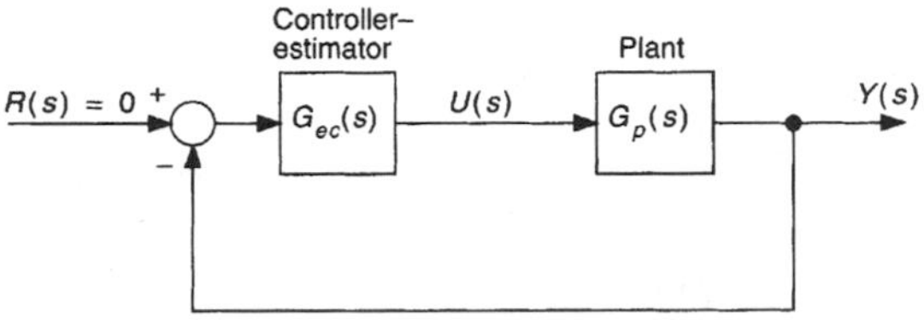

FIGURE 11.51 Equivalent system for pole-placement design. (*Source*: C.L. Phillips and R.D. Harbor, *Feedback Control Systems,* 2nd ed., Englewood Cliffs, N.J.: Prentice-Hall, 1991, p. 414. With permission.)

$$|s\mathbf{I} - \mathbf{A} + \mathbf{BK}||s\mathbf{I} - \mathbf{A} + \mathbf{GC}| = \alpha_c(s)\alpha_e(s) = 0 \tag{11.91}$$

The roots of this equation are the roots of the pole-placement design plus those of the observer design. For this reason, the roots of the characteristic equation for the observer are usually chosen to be faster than those of the pole-placement design.

Linear Quadratic Optimal Control

We define an optimal control system as one for which some mathematical function is minimized. The function to be minimized is called the *cost function*. For steady-state linear quadratic optimal control the cost function is given by

$$V_\infty = \int_t^\infty [\mathbf{x}^T(\tau)\mathbf{Q}\mathbf{x}(\tau) + Ru^2(\tau)]d\tau \tag{11.92}$$

where Q and R are chosen to satisfy the design criteria. In general, the choices are not straightforward. Minimization of Equation (11.92) requires that the plant input be given by

$$u(t) = -R^{-1}\mathbf{B}^T\mathbf{M}_\infty\mathbf{x}(t) \tag{11.93}$$

where the $n \times n$ matrix $\mathbf{M}_\infty$ is the solution to the *algebraic Riccati equation*

$$\mathbf{M}_\infty\mathbf{A} + \mathbf{A}^T\mathbf{M}_\infty - \mathbf{M}_\infty\mathbf{B}\mathbf{R}^{-1}\mathbf{B}^T\mathbf{M}_\infty + \mathbf{Q} = 0 \tag{11.94}$$

The existence of a solution for this equation is involved [Friedland, 1986] and is not presented here. Optimal control systems can be designed for cost functions other than that of Equation (11.92).

Other Modern Design Procedures

Other modern design procedures exist; for example, *self-tuning control systems* continually estimate certain plant parameters and adjust the compensator based on this estimation. These control systems are a type of *adaptive control systems* and usually require that the control algorithms be implemented using a digital computer. These control systems are beyond the scope of this book (see, for example, Astrom and Wittenmark, 1984).

Defining Term

Compensation: The process of physically altering a closed-loop system such that the system has specified characteristics. This alteration is achieved either by changing certain parameters in the system or by adding a physical system to the closed-loop system; in some cases both methods are used.

11.6 Digital Control Systems

Raymond G. Jacquot and John E. McInroy

The use of the **digital computer** to control physical processes has been a topic of discussion in the technical literature for over four decades, but the actual use of a digital computer for control of industrial processes was reserved only for massive and slowly varying processes such that the high cost and slow computing speed of available computers could be tolerated. The invention of the integrated-circuit microprocessor in the early 1970s radically changed all that; now microprocessors are used in control tasks in automobiles and household appliances, applications where high cost is not justifiable.

When the term *digital control* is used, it usually refers to the process of employing a digital computer to control some process that is characterized by continuous-in-time dynamics. The control can be of the open-loop variety where the control strategy output by the digital computer is dictated without regard to the status of the process variables. An alternative technique is to supply the digital computer with digital data about the process variables to be controlled, and thus the control strategy output by the computer depends on the process variables that are to be controlled. This latter strategy is a **feedback control** strategy wherein the computer, the process, and interface hardware form a closed loop of information flow.

Examples of dynamic systems that are controlled in such a closed-loop digital fashion are flight control of civilian and military aircraft, control of process variables in chemical processing plants, and position and force control in industrial robot manipulators. The simplest form of feedback control strategy provides an on-off control to the controlling variables based on measured values of the process variables. This strategy will be illustrated by a simple example in a following subsection.

In the past decade and a half many excellent textbooks on the subject of digital control systems have been written, and most of them are in their second edition. The texts in the References provide in-depth development of the theory by which such systems are analyzed and designed.

A Simple Example

Such a closed-loop or feedback control situation is illustrated in Figure 11.52, which illustrates the feedback control of the temperature in a simple environmental chamber that is to be kept at a constant temperature somewhat above room temperature.

Heat is provided by turning on a relay that supplies power to a heater coil. The on-off signal to the relay can be supplied by 1 bit of an output port of the microprocessor (typically the port would be 8 bits wide). A second bit of the port can be used to turn a fan on and off to supply cooling air to the chamber. An analog-to-digital (A/D) converter is employed to convert the amplified thermocouple signal to a digital word that is then supplied to the input port of the microprocessor. The program being executed by the microprocessor reads the temperature data supplied to the input port and compares the binary number representing the temperature to a binary version of the desired temperature and makes a decision whether or not to turn on the heater or the fan or to do nothing. The program being executed runs in a continuous loop, repeating the operations discussed above.

This simple on-off control strategy is often not the best when extremely precise control of the process variables is required. A more precise control may be obtained if the controlling variable levels can be adjusted to be somewhat larger if the deviation of the process variable from the desired value is larger.

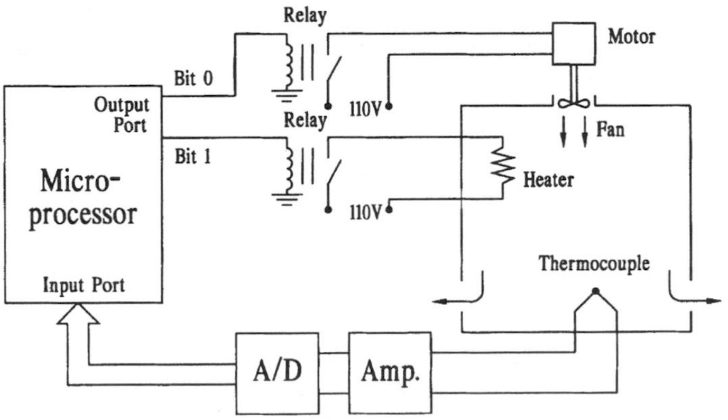

FIGURE 11.52 Microprocessor control of temperature in a simple environmental chamber.

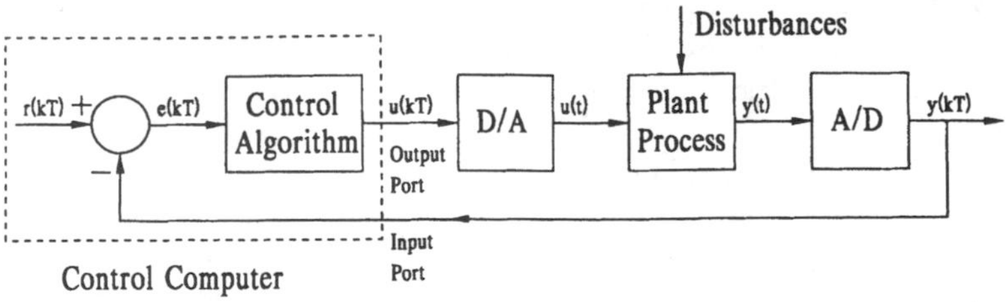

FIGURE 11.53 Closed-loop control of a single process variable.

Single-Loop Linear Control Laws

Consider the case where a single variable of the process is to be controlled, as illustrated in Figure 11.53. The output of the plant $y(t)$ is to be sampled every T seconds by an A/D converter, and this sequence of numbers will be denoted as $y(kT)$, $k = 0, 1, 2, \ldots$. The goal is to make the sequence $y(kT)$ follow some desired known sequence [the reference sequence $r(kT)$]. Consequently, the sequence $y(kT)$ is subtracted from $r(kT)$ to obtain the so-called error sequence $e(kT)$. The control computer then acts on the error sequence, using some control algorithms, to produce the control effort sequence $u(kT)$ that is supplied to the digital-to-analog (D/A) converter which then drives the actuating hardware with a signal proportional to $u(kT)$. The output of the D/A converter is then held constant on the current time interval, and the control computer waits for the next sample of the variable to be controlled, the arrival of which repeats the sequence. The most commonly employed control algorithm or control law is a linear difference equation of the form

$$\begin{aligned} u(kT) = {} & a_n e(kT) + a_{n-1} e((k-1)T) + \cdots + a_0 e((k-n)T) + b_{n-1} u((k-1)T) + \cdots \\ & + b_0 u((k-n)T) \end{aligned} \tag{11.95}$$

The question remains as to how to select the coefficients $a_0, \ldots, a_n$ and $b_0, \ldots, b_{n-1}$ in expression (Equation (11.95)) to give an acceptable degree of control of the plant.

Proportional Control

This is the simplest possible control algorithm for the digital processor wherein the most current control effort is proportional to the current error or using only the first term of relation (Equation (11.95))

$$u(kT) = a_n e(kT) \tag{11.96}$$

This algorithm has the advantage that it is simple to program, while, on the other hand, its disadvantage lies in the fact that it has poor disturbance rejection properties in that if a_n is made large enough for good disturbance rejection, the closed-loop system can be unstable (i.e., have transient responses that increase with time). Since the object is to regulate the system output in a known way, these unbounded responses preclude this regulation.

PID Control Algorithm

A common technique employed for decades in chemical process control loops is that of proportional-plus-integral-plus-derivative (PID) control wherein a continuous-time control law would be given by

$$u(t) = K_\mathrm{p} e(t) + K_i \int_0^t e(\tau)\mathrm{d}\tau + K_\mathrm{d} \frac{\mathrm{d}e}{\mathrm{d}t} \tag{11.97}$$

This would have to be implemented by an analog filter.

To implement the design in digital form the proportional term can be carried forward as in relation (Equation (11.96)); however, the integral can be replaced by trapezoidal integration using the error sequence, while the derivative can be replaced with the backward difference resulting in a computer control law of the form [Jacquot, 1995]

$$u(kT) = u((k-1)T) + \left(K_p + \frac{K_i T}{2} + \frac{K_d}{T}\right)e(kT) + \left(\frac{K_i T}{2} - K_p - \frac{2K_d}{T}\right)e((k-1)T) + \frac{K_d}{T}e((k-2)T) \tag{11.98}$$

where T is the duration of the sampling interval. The selection of the coefficients in this algorithm (K_i, K_d, and K_p) is best accomplished by the Ziegler-Nichols tuning process [Franklin et al., 1997].

The Closed-Loop System

When the plant process is linear or may be linearized about an operating point and the control law is linear as in expressions Equation (11.95), Equation (11.96), or Equation (11.98), then an appropriate representation of the complete closed-loop system is by the so-called z-transform. The z-transform plays the role for linear, constant-coefficient difference equations that the Laplace transform plays for linear, constant-coefficient differential equations. This z-domain representation allows the system designer to investigate system time response, frequency response, and stability in a single analytical framework.

If the plant can be represented by an s-domain transfer function $G(s)$, then the discrete-time (z-domain) transfer function of the plant, the analog-to-digital converter, and the driving digital-to-analog converter is

$$G(z) = \left(\frac{z-1}{z}\right)Z\left\{L^{-1}\left[\frac{G(s)}{s}\right]\right\} \tag{11.99}$$

where $Z(\cdot)$ is the z-transform and $L^{-1}(\cdot)$ is the inverse Laplace transform. The transfer function of the control law of Equation (11.95) is

$$D(z) = \frac{U(z)}{E(z)} = \frac{a_n z^n + a_{n-1}z^{n-1} + \cdots + a_0}{z^n - b_{n-1}z^{n-1} - \cdots - b_0} \tag{11.100}$$

For the closed-loop system of Figure 11.53 the closed-loop z-domain transfer function is

$$M(z) = \frac{Y(z)}{R(z)} = \frac{G(z)D(z)}{1 + G(z)D(z)} \tag{11.101}$$

where $G(z)$ and $D(z)$ are as specified above. The characteristic equation of the closed-loop system is

$$1 + G(z)D(z) = 0 \tag{11.102}$$

The dynamics and stability of the system can be assessed by the locations of the zeros of Equation (11.102) (the closed-loop poles) in the complex z plane. For stability the zeros of Equation (11.102) must be restricted to the unit circle of the complex z plane.

A Linear Control Example

Consider the temperature control of a chemical mixing tank shown in Figure 11.54. From a transient power balance, the differential equation relating the rate of heat added $q(t)$ to the deviation in temperature from the ambient $\theta(t)$ is given as

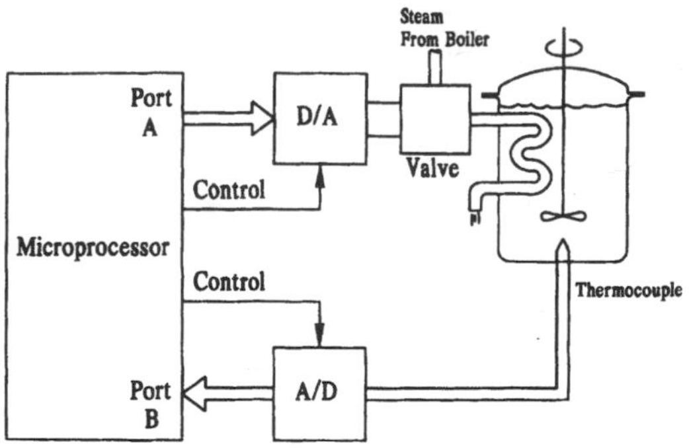

FIGURE 11.54 A computer-controlled thermal mixing tank.

$$\frac{\mathrm{d}\theta}{\mathrm{d}t} + \frac{1}{\tau}\theta = \frac{1}{mc}q(t) \tag{11.103}$$

where τ is the time constant of the process and mc is the heat capacity of the tank. The transfer function of the tank is

$$\frac{\Theta(s)}{Q(s)} = G(s) = \frac{1/mc}{s + 1/\tau} \tag{11.104}$$

The heater is driven by a D/A converter, and the temperature measurement is sampled with an A/D converter. The data converters are assumed to operate synchronously, so the discrete-time transfer function of the tank and the two data converters is from expression (Equation (11.99)):

$$G(z) = \frac{\Theta(z)}{Q(z)} = \frac{\tau}{mc}\,\frac{1 - \mathrm{e}^{-T/\tau}}{z - \mathrm{e}^{-T/\tau}} \tag{11.105}$$

If a proportional control law is chosen, the transfer function associated with the control law is the gain $a_n = K$ or

$$D(z) = K \tag{11.106}$$

The closed-loop characteristic equation is from Equation (11.102):

$$1 + \frac{K\tau}{mc}\,\frac{1 - \mathrm{e}^{-T/\tau}}{z - \mathrm{e}^{-T/\tau}} = 0 \tag{11.107}$$

If a common denominator is found, the resulting numerator is

$$z - \mathrm{e}^{-T/\tau} + \frac{K\tau}{mc}(1 - \mathrm{e}^{-T/\tau}) = 0 \tag{11.108}$$

The root of this equation is

$$z = \mathrm{e}^{-T/\tau} + \frac{K\tau}{mc}(\mathrm{e}^{-T/\tau} - 1) \tag{11.109}$$

If this root location is investigated as the gain parameter K is varied upward from zero, it is seen that the root starts at $z = \mathrm{e}^{-T/\tau}$ for $K = 0$ and moves to the left along the real axis as K increases. Initially it is seen that the

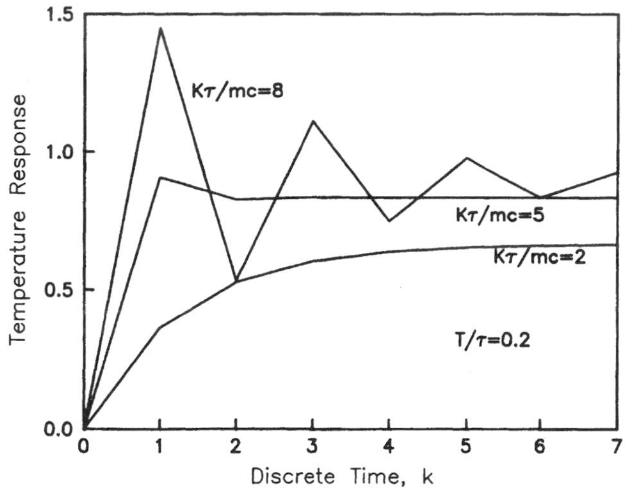

FIGURE 11.55 Step responses of proportionally controlled thermal mixing tank.

system becomes faster, but at some point the responses become damped and oscillatory, and as K is further increased the oscillatory tendency becomes less damped, and finally a value of K is reached where the oscillations are sustained at constant amplitude. A further increase in K will yield oscillations that increase with time. Typical unit step responses for $r(k) = 1$ and $T/\tau = 0.2$ are shown in Figure 11.55.

It is easy to observe this tendency toward oscillation as K increases, but a problem that is clear from Figure 11.55 is that in the steady state there is a persistent error between the response and the reference $[r(k) = 1]$. Increasing the gain K will make this error smaller at the expense of more oscillations. As a remedy for this steady-state error problem and control of the dynamics, a control law transfer function $D(z)$ will be sought that inserts integrator action into the loop while simultaneously canceling the pole of the plant. This dictates that the controller have a transfer function of the form

$$D(z) = \frac{U(z)}{E(z)} = \frac{K(z - e^{-T/\tau})}{z - 1} \tag{11.110}$$

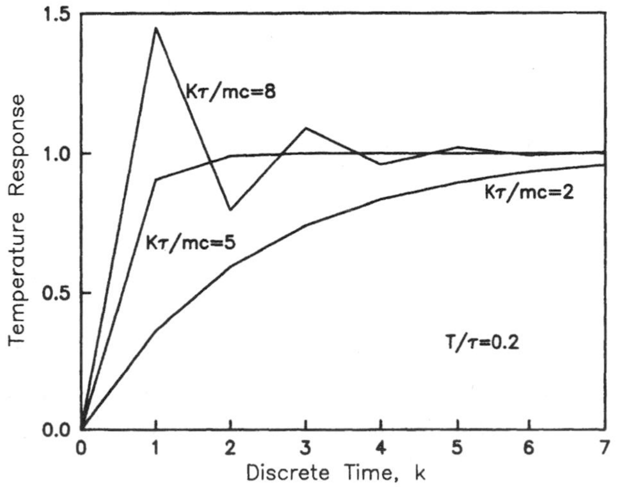

FIGURE 11.56 Step responses of the compensated thermal mixing tank.

Typical unit step responses are illustrated in Figure 11.56 for several values of the gain parameter. The control law that must be programmed in the digital processor is

$$u(kT) = u((k-1)T) + K[e(kT) - e^{-T/\tau}e((k-1)T)] \tag{11.111}$$

The additional effort to program this over that required to program the proportional control law of Equation (11.106) is easily justified since K and $e^{-T/\tau}$ are simply constants.

Defining Terms

Digital computer: A collection of digital devices including an arithmetic logic unit (ALU), read-only memory (ROM), random-access memory (RAM), and control and interface hardware.

Feedback control: The regulation of a response variable of a system in a desired manner using measurements of that variable in the generation of the strategy of manipulation of the controlling variables.

References

K.J. Astrom and B. Wittenmark, *Computer Controlled Systems: Theory and Design,* Englewood Cliffs, NJ: Prentice-Hall, 1984.

G.F. Franklin, J.D. Powell, and M.L. Workman, *Digital Control of Dynamic Systems,* 3rd ed., Reading, MA: Addison-Wesley, 1997.

R.G. Jacquot, *Modern Digital Control Systems,* 2nd ed., New York: Marcel Dekker, 1995.

W.S. Levine, *The Control Handbook,* Boca Raton, FL.: CRC Press, 2001.

C.L. Phillips and H.T. Nagle, *Digital Control System Analysis and Design,* 3rd ed., Englewood Cliffs, NJ: Prentice-Hall, 1995.

Further Information

The *IEEE Control Systems Magazine* is a useful information source on control systems in general and digital control in particular. Highly technical articles on the state of the art in digital control may be found in the *IEEE Transactions on Automatic Control,* the *IEEE Transactions on Control Systems Technology,* and the ASME *Journal of Dynamic Systems, Measurement and Control.*

11.7 Nonlinear Control Systems[1]

Derek P. Atherton

Introduction

So far in these articles on control systems, apart from a brief mention in the first on "models," only methods for linear systems have been discussed. In practice, however, nonlinearity always exists and no completely general analytical approach is available to assess its effect. The purpose of this article is to introduce the reader to the types of nonlinearity that occur in control systems, their possible effects on the system, and methods based on classical control concepts discussed in the earlier articles for their analysis and design. More recent theoretical developments are beyond the scope of this article but are

[1]The material in this section was previously published by CRC Press in *The Control Handbook*, William S. Levine, Ed., 1996.

commented on further in the final section on further information. Linear systems have the important property that they satisfy the superposition principle. This leads to many important advantages in methods for their analysis. For example, in a simple feedback loop with both setpoint and disturbance inputs, their effect on the output when they are applied simultaneously is the same as the sum of their individual effects when applied separately. This would not be the case if the system were nonlinear. Thus, mathematically a linear system may be defined as one which with input $x(t)$ and output $y(t)$ satisfies the property that the output for an input $ax_1(t) + bx_2(t)$ is $ay_1(t) + by_2(t)$, if $y_1(t)$ and $y_2(t)$ are the outputs in response to the inputs $x_1(t)$ and $x_2(t)$, respectively, and a and b are constants. A nonlinear system is defined as one that does not satisfy the superposition property. The simplest form of nonlinear system is the static nonlinearity where the output depends only on the current value of the input, say, $y(t) = ax(t) + bx^3(t)$ where the output is a linear plus cubed function of the input.

More commonly the relationship could involve both nonlinearity and dynamics so that it might be described by the nonlinear differential equation

$$\frac{\mathrm{d}^2 y}{\mathrm{d}t^2} + a\left[\frac{\mathrm{d}y}{\mathrm{d}t}\right]^3 + by(t) = x(t)$$

From an engineering viewpoint it may be desirable to think of this equation in terms of a block diagram consisting of linear dynamic elements and a static nonlinearity, which in this case is a cubic with input $\mathrm{d}y/\mathrm{d}t$ and output $a(\mathrm{d}y/\mathrm{d}t)^3$. A major point about nonlinear systems, however, is that their response is amplitude dependent so that if a particular form of response, or some measure of it, occurs for one input magnitude it may not result for some other input magnitude. This means that in a feedback control system with a nonlinear plant, if the controller designed does not produce a linear system, then to adequately describe the system behavior one needs to investigate the total allowable range of the system variables. For a linear system one can claim that a system has an optimum step response, assuming an optimum can be defined mathematically, for example, minimization of the integral squared error, using results obtained for a single input amplitude. On the other hand, for a nonlinear system the response to all input amplitudes must be investigated, and the optimum choice of parameters to minimize the criterion will be amplitude dependent. Perhaps the most interesting aspect of nonlinear systems is that they exhibit forms of behavior not possible in linear systems, and more will be said on this later.

Forms of Nonlinearity

Nonlinearity may be inherent in the dynamics of the plant to be controlled or in the components used to implement the control. For example, there will be a limit to the torque obtainable from an electric motor or the current that may be input to an electrical heater, and indeed good design will have circuits to ensure this to avoid destruction of a component. Sizing of components must take into account both the required performance and cost, so it is not unusual to find that a rotary position control system will develop maximum motor torque for a demanded step angle change of only a few degrees. Also, although component manufacture has improved greatly in the last decades, flow control valves possess a dead zone due to the effect of friction, and their characteristics on opening and closing are not identical due to the flow pressure. Improved design might produce more-linear components but at greater cost so that such strategies would not be justified economically. Alternatively one may have nonlinear elements intentionally introduced into a design in order to improve the system specifications, from either technical or economic viewpoint. A good example is the use of relay switching. Identifying the precise form of nonlinearity in a system component may not be easy, and like all modeling exercises the golden rule is to be aware of the approximations in a nonlinear model and the conditions for its validity.

Friction always occurs in mechanical systems and is very difficult to model, with many quite sophisticated models having been presented in the literature. The simplest is to assume the three components of *stiction*, an abbreviation for static friction, Coulomb friction, and viscous friction. As its name implies, stiction is assumed to exist only at zero differential speed between the two contact surfaces. Coulomb friction with a value less than stiction is assumed to be constant at all speeds, and viscous friction is a linear effect being directly proportional to speed. In practice there is often a term proportional to a higher power of speed, and this is also the situation for many rotating shaft loads, for example, a fan. Many mechanical loads are driven through gearing rather than directly. Although geared drives, like all areas of technology, have improved through the years, they always have some small backlash. This may be avoided by using antibacklash gears, but these are normally only employed for low torques. Backlash is a very complicated phenomenon involving impacts between surfaces and is often modeled in a very simplistic manner. For example, a simple approach often used consists of an input-output position characteristic of two parallel straight lines with possible horizontal movement between them. This makes two major assumptions. First is that the load shaft friction is high enough for contact to be maintained with the drive side of the backlash when the drive slows down to rest. Secondly when the drive reverses, the backlash is crossed and the new drive-side of the gear "picks up" the load instantaneously with no loss of energy in the impact, and both then move at the drive shaft speed. Clearly both these assumptions are never true in practice, so their limitations need to be understood and borne in mind when using such an approximation.

The most widely used intentional nonlinearity is the relay. The on-off type, which can be described mathematically by the signum function (that is, it switches on if its input exceeds a given value and off if it goes below the value), is widely used, often with some hysteresis between the switching levels. Use of this approach provides a control strategy where the controlled variable oscillates about the desired level. The switching mechanism varies significantly according to the application from electromechanical relays at low speed to fast electronic switches employing transistors or thyristors. A common usage of the relay is in temperature control of buildings, where typically the switching is provided by a thermostat having a pool of mercury on a metal expansion coil. As the temperature drops the coil contracts and this causes a change in angle of the mercury capsule so that eventually the mercury moves, closes a contact, and power is switched on. Electronic switching controllers are used in many modern electric motor drive systems, for example, to regulate phase currents in stepping motors and switched reluctance motors and to control currents in vector control drives for induction motors. Relays with a dead zone, that is, three-position relays giving positive, negative, and a zero output, are also used. The zero output allows for a steady-state position within the dead zone, but this affects the resulting steady-state control accuracy.

Theoretical analysis and design methods for nonlinear systems typically require a mathematical model for the nonlinearity, sometimes of a specific form, say a power-law or polynomial representation. The mathematical model used will be an approximation to the true situation. Typical models used are:

1. A single or simple mathematical function
2. A series approximation
3. A discontinuous set of functions, typically straight lines to produce a linear segmented characteristic

Stability and Behavior

A simple nonlinear feedback system is shown in Figure 11.57. The first question that usually has to be answered regarding any feedback control system is, is it stable? For a linear system, such as Figure 11.57 with the nonlinearity replaced by a linear gain K, necessary and sufficient conditions are known for the stability.

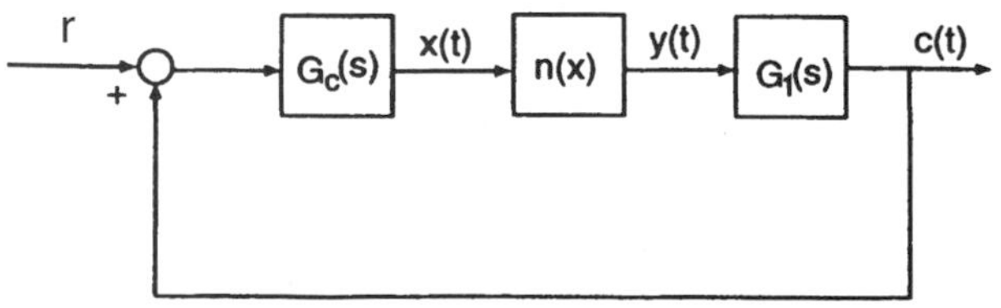

FIGURE 11.57 Block diagram of a nonlinear system.

The requirement is simply that all roots of the characteristic equation must lie in the left-hand side of the s-plane. The Hurwitz–Routh criterion provides a simple algebraic test for doing this, while for a characteristic equation with numerical values for its coefficients its, roots can now easily be determined using software such as MATLAB. A frequency response criterion for assessing stability was developed by Nyquist, and this can be extended for use with nonlinear systems. An early attempt at solving the nonlinear system stability problem was presented by Aizermann. He conjectured that the autonomous nonlinear system, that is, with $r = 0$, of Figure 11.57 would be stable for any nonlinearity lying within a sector defined by lines of slope k_1 and k_2 ($k_1 > k_2$), if the linear system were stable for a gain K replacing the nonlinearity, with $k_2 < K < k_1$. This is incorrect, as, for example, it may possess a limit cycle, which is a periodic motion, but its response cannot go unbounded.

The first work on the stability of nonlinear systems was by Lyapunov, and since that time there have been many attempts to determine necessary and sufficient conditions for the stability of the autonomous feedback system. Because frequency response methods play a major role in classical control, there has been much research on finding frequency domain conditions for stability. Several results have been found that give sufficient, but not necessary, conditions for stability. They use limited information about the nonlinearity, $n(x)$, typically its sector bounds, like the Aizermann conjecture, or the sector bounds of its slope. The nonlinearity $n(x)$ has sector bounds (k_1, k_2), that is, it is confined between the straight lines k_1x and k_2x if $k_1x^2 < xn(x) < k_2x^2$ for all x. Similarly it has slope bounds (k_1', k_2') if $k_1'x^2 < xn'(x) < k_2'x^2$, where $n'(x) = \mathrm{d}n(x)/\mathrm{d}x$.

Three results are briefly mentioned as they allow easy comparisons with those given by the describing function method mentioned later. The first, the Popov criterion, states that a sufficient condition for the autonomous system of Figure 11.57 to be stable if $G(s)$, equal to $G_c(s)G_1(s)$, is stable and $G(\infty) > -k^{-1}$ is that a real number $q > 0$ can be found such that for all ω

$$\mathrm{Re}[(1 + j\omega q)G(j\omega)] + k^{-1} > 0$$

where the nonlinearity $n(x)$ lies in the sector $(0, k)$. The theorem has the simple graphical interpretation shown in Figure 11.58, where for the system to be stable a line of slope q^{-1} can be drawn through the point $-k^{-1}$ so that the Popov locus $G^*(j\omega)$ lies to the right of the line. The Popov locus is given by

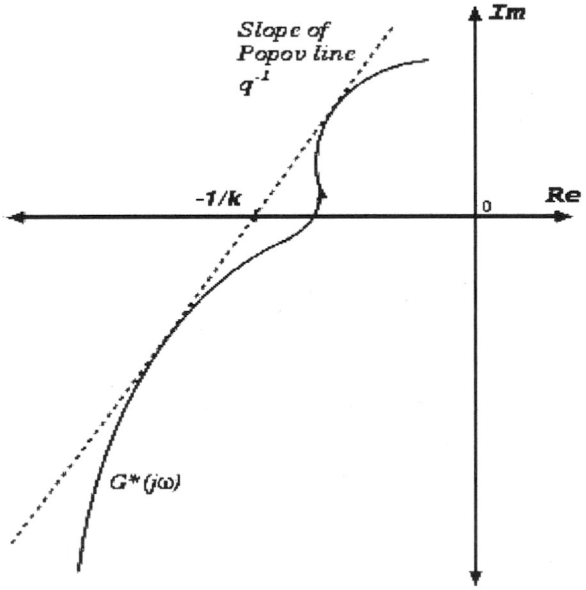

FIGURE 11.58 Graphical Interpretation of the Popov criterion.

$$G^*(j\omega) = \text{Re}[G(j\omega)] + j\omega\, \text{Im}[G(j\omega)]$$

The second, the circle criterion, can be obtained from the Popov criterion, but its validity, using different analytical approaches, has been extended to cover the situation of a bounded input, r, to the system of Figure 11.57. Satisfaction of the circle criterion guarantees that the autonomous system is absolutely stable and the system with bounded input has a bounded output. The criterion uses the Nyquist locus, $G(j\omega)$, and for stability of the system of Figure 11.57 with $n(x)$ in the sector (k_1, k_2) it is required that $G(j\omega)$ for all real ω has the following properties. If the circle C has its diameter from $-1/k_1$ to $-1/k_2$ on the negative real axis of the Nyquist diagram, then for (i) $k_1k_2 < 0$, $G(j\omega)$ should be entirely within C, (ii) if $k_1k_2 > 0$, $G(j\omega)$ should lie entirely outside and not encircle C, and (iii) if $k_1 = 0$ or $k_2 = 0$, $G(j\omega)$ lies entirely to the right of $-1/k_2$ or to the left of $-1/k_1$. The situation for stability in case (ii) is shown in Figure 11.59. Note that Aizermann's conjecture just "uses" the diameter of the circle.

The third result is for a monotonic nonlinearity with slope bounds (k_1', k_2') and $k_1'k_2' > 0$ for which an off-axis circle criterion exists. This states that the autonomous system of Figure 11.57 with a nonlinearity satisfying the aforementioned conditions will be absolutely stable if the Nyquist locus of a stable transfer function does not encircle any circle centered off the real axis and which intercepts it at $(-1/k_2', -1/k_2')$.

There are two viewpoints on the above absolute stability criteria. First are those of the engineer, who may argue that the criteria produce results that are too conservative if one wishes to apply them to a specific system with a well-defined nonlinearity, so that much more is known about it than its sector bounds. Second are those of the theoretician, who may argue that the results are very robust in the sense that they guarantee stability for a system with a poorly defined nonlinearity, that is, any nonlinearity satisfying certain sector properties.

Since Figure 11.57 may be regarded as the structure of many simple nonlinear feedback control loops, it is relevant to discuss further its possible forms of behavior. In doing so it will be assumed that any nonlinearity and any linear transfer function can exist in the appropriate blocks, but the form they must take so that the loop has one of the specific properties mentioned will not be discussed. As mentioned previously, the performance of the system, even for a specific type of input, will depend upon the amplitude of the input. If the autonomous system, that is, the system with no input, is released from several initial states, then the resulting behavior may be appreciably different for each state. In particular, instead of reaching a stationary equilibrium point, the system may move from some initial conditions into a limit cycle, a continuous

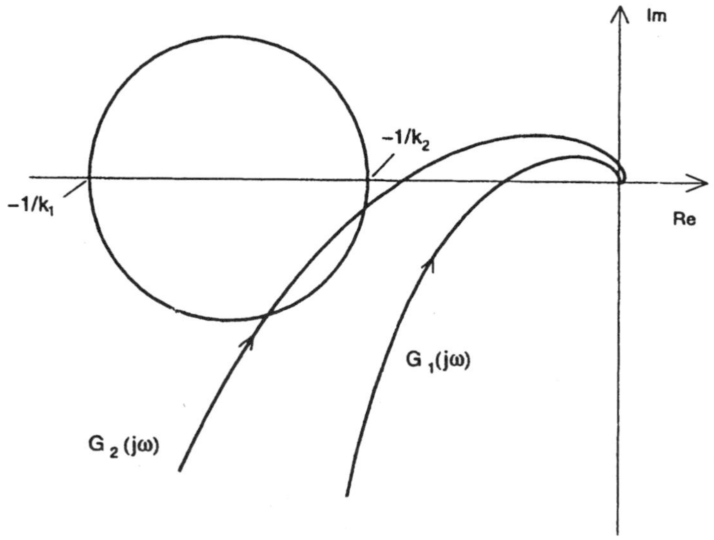

FIGURE 11.59 Circle criterion and stability.

oscillation that can be reached from several possible initial conditions. This behavior is distinct from an oscillation in an idealized linear system, which can be obtained by adjusting the gain K so that the characteristic equation has two purely imaginary roots, since the amplitude of this oscillation depends upon the initial energy input to the system. A limit cycle is a periodic motion, but its waveform may be significantly different from the sinusoid of an oscillation. The autonomous nonlinear system, as mentioned earlier, may also have a chaotic motion, a motion that is repeatable from given initial conditions, but that exhibits no easily describable mathematical form, is not periodic, and exhibits a spectrum of frequency components.

If a sinusoidal input is applied to the system, the output may be of the same frequency but will also contain harmonics or other components related to the input frequency. This output too, for certain frequencies and amplitudes of the input, may not be unique but have an amplitude dependent upon the past history of the input or the initial conditions of the system, the so-called jump phenomena in the frequency response. The sinusoidal input may also cause the system to oscillate at a related frequency so that the largest frequency component in the output is not the same as that of the input, which is known as a subharmonic oscillation. Also if, for example, the autonomous system had a limit cycle, then the addition of a sinusoidal input could cause the limit cycle frequency to change or cause synchronization of the limit cycle frequency with the input frequency or one of its harmonics.

In many instances the phenomena just mentioned are undesirable in a control system, so that one needs techniques to ensure that they do not exist. Control systems must be designed to meet specific performance objectives, and to do this one is required to design a control law that is implemented based on measurements or estimates of the system states, or by simple operations on the system, typically the error, signals. Many systems can be made to operate satisfactorily with the addition of a simple controller, $G_c(s)$, shown in Figure 11.57. Typical performance criteria, which the system may be required to meet, are that it be stable, have zero steady-state error and a good response to a step input, suitably reject disturbances, and be robust to parameter variations. Although one reason for using feedback control is to reduce sensitivity to parameter changes, specific design techniques can be used to ensure that the system is more robust to any parameter changes. If the process to be controlled is strongly nonlinear, then a nonlinear controller will have to be used if it is required to have essentially the same step response performance for different input step amplitudes. Some control systems, for example, simple temperature control systems, may work in a limit cycle mode, so that in these instances the designer is required to ensure that the frequency and amplitude variations of the controlled temperature are within the required specifications.

The Describing-Function Method

The describing-function method, abbreviated as DF, was developed in several countries in the 1940s [Atherton, 1982], to try and determine why limit cycles occurred in control systems. It was observed that, in many instances with structures such as Figure 11.57, the wave form of the oscillation at the input to the nonlinearity was almost sinusoidal. If, for example, the nonlinearity in Figure 11.57 is an ideal relay, that is has an on-off characteristic, so that an odd symmetrical input wave form will produce a square wave at its output, the output of $G(s)$ will be almost sinusoidal when $G(s)$ is a low pass filter that attenuates the higher harmonics in the square wave much more than the fundamental. It was, therefore, proposed that the nonlinearity should be represented by its gain to a sinusoid and that the conditions for sustaining a sinusoidal limit cycle be evaluated to assess the stability of the feedback loop. Because of the nonlinearity, this gain in response to a sinusoid is a function of the amplitude of the sinusoid and is known as the describing function. Because describing-function methods can be used other than for a single sinusoidal input, the technique is referred to as the single sinusoidal DF or sinusoidal DF.

If we assume in Figure 11.57 that $x(t) = a \cos \theta$, where $\theta = \omega t$ and $n(x)$ is a symmetrical odd nonlinearity, then the output $y(t)$ will be given by the Fourier series,

$$y(\theta) = \sum_{n=0}^{\infty} a_n \cos n\theta + b_n \sin n\theta, \tag{11.112}$$

where

$$a_0 = 0, \tag{11.113}$$

$$a_1 = (1/\pi)\int_0^{2\pi} y(\theta)\cos\theta \, d\theta, \tag{11.114}$$

and

$$b_1 = (1/\pi)\int_0^{2\pi} y(\theta)\sin\theta \, d\theta. \tag{11.115}$$

The fundamental output from the nonlinearity is $a_1 \cos \theta + b_1 \sin \theta$, so that the describing function, DF, defined as the fundamental output divided by the input amplitude, is complex and given by

$$N(a) = (a_1 - jb_1)/a \tag{11.116}$$

which may be written

$$N(a) = N_p(a) + jN_q(a) \tag{11.117}$$

where

$$N_p(a) = a_1/a \quad \text{and} \quad N_q(a) = -b_1/a. \tag{11.118}$$

Alternatively, in polar coordinates,

$$N(a) = M(a)e^{j\psi(a)} \tag{11.119}$$

where

$$M(a) = (a_1^2 + b_1^2)^{1/2}/a$$

and

$$\Psi(a) = -\tan^{-1}(b_1/a_1). \tag{11.120}$$

If $n(x)$ is single valued, then $b_1 = 0$ and

$$a_1 = (4/\pi)\int_0^{\pi/2} y(\theta)\cos\theta \, d\theta \tag{11.121}$$

giving

$$N(a) = a_1/a = (4/a\pi)\int_0^{\pi/2} y(\theta)\cos\theta \, d\theta \tag{11.122}$$

Although Equation (11.114) and Equation (11.115) are an obvious approach to evaluating the fundamental output of a nonlinearity, they are indirect, because one must first determine the output wave form $y(\theta)$ from the known nonlinear characteristic and sinusoidal input wave form. This is avoided if the substitution $\theta = \cos^{-1}(x/a)$ is made. After some simple manipulations,

$$a_1 = (4/a)\int_0^a x n_p(x)p(x)\mathrm{d}x \qquad (11.123)$$

and

$$b_1 = (4/a\pi)\int_0^a n_q(x)\mathrm{d}x \qquad (11.124)$$

The function $p(x)$ is the amplitude probability density function of the input sinusoidal signal given by

$$p(x) = (1/\pi)(a^2 - x^2)^{-1/2}. \qquad (11.125)$$

The nonlinear characteristics $n_p(x)$ and $n_q(x)$, called the inphase and quadrature nonlinearities, are defined by

$$n_p(x) = [n_1(x) + n_2(x)]/2 \qquad (11.126)$$

and

$$n_q(x) = [n_2(x) - n_1(x)]/2 \qquad (11.127)$$

where $n_1(x)$ and $n_2(x)$ are the portions of a double-valued characteristic traversed by the input for $\dot{x} > 0$ and $\dot{x} < 0$, respectively. When the nonlinear characteristic is single-valued, $n_1(x) = n_2(x)$, so $n_p(x) = n(x)$ and $n_q(x) = 0$. Integrating Equation (11.123) by parts yields

$$a_1 = (4/\pi)n(0^+) + (4/a\pi)\int_0^a n'(x)\left(a^2 - x^2\right)^{1/2}\mathrm{d}x \qquad (11.128)$$

where $n'(x) = \mathrm{d}n(x)/\mathrm{d}x$ and $n(0^+) = \lim_{\in\to\infty} n(\varepsilon)$, a useful alternative expression for evaluating a_1.

An additional advantage of using Equation (11.123) and Equation (11.124) is that they yield proofs of some properties of the DF for symmetrical odd nonlinearities. These include the following:

1. For a double-valued nonlinearity, the quadrature component $N_q(a)$ is proportional to the area of the nonlinearity loop, that is,

$$N_q(a) = -\left(1/a^2\pi\right)(\text{area of nonlinearity loop}) \qquad (11.129)$$

2. For two single-valued nonlinearities $n_\alpha(x)$ and $n_\beta(x)$, with $n_\alpha(x) < n_\beta(x)$ for all $0 < x < b$, $N_\alpha(a) < N_\beta(a)$ for input amplitudes less than b.
3. For a single-valued nonlinearity with $k_1 x < n(x) < k_2 x$ for all $0 < x < b$, $k_1 < N(a) < k_2$ for input amplitudes less than b. This is the sector property of the DF; a similar result can be obtained for a double-valued nonlinearity [Cook, 1973].

When the nonlinearity is single valued, from the properties of Fourier series, the DF, $N(a)$, may also be defined as:

1. The variable gain, K, having the same sinusoidal input as the nonlinearity, which minimizes the mean squared value of the error between the output from the nonlinearity and that from the variable gain, and
2. The covariance of the input sinusoid and the nonlinearity output divided by the variance of the input

Evaluation of the Describing Function

To illustrate the evaluation of the DF two simple examples are considered.

Saturation Nonlinearity

To calculate the DF, the input can alternatively be taken as $a \sin \theta$. For an ideal saturation characteristic, the nonlinearity output wave form $y(\theta)$ is as shown in Figure 11.60. Because of the symmetry of the nonlinearity, the fundamental of the output can be evaluated from the integral over a quarter period so that

$$N(a) = \frac{4}{a\pi}\int_0^{\pi/2} y(\theta)\sin\theta \, d\theta,$$

which, for $a > \delta$, gives

$$N(a) = \frac{4}{a\pi}\left[\int_0^{\alpha} ma\sin^2\theta \, d\theta + \int_{\alpha}^{\pi/2} m\delta \sin\theta \, d\theta\right]$$

where $a = \sin^{-1} \delta/a$. Evaluation of the integrals gives

$$N(a) = (4m/\pi)\left[\frac{\alpha}{2} - \frac{\sin 2\alpha}{4} + \delta \cos \alpha\right]$$

which, on substituting for δ, give the result

$$N(a) = (m/\pi)(2\alpha + \sin 2\alpha). \tag{11.130}$$

Because, for $a < \delta$, the characteristic is linear giving $N(a) = m$, the DF for ideal saturation is $mN_s(\delta/a)$ where

$$N_s(\delta/a) = \begin{cases} 1, & \text{for } a < \delta, \text{ and} \\ (1/\pi)[2\alpha + \sin 2\alpha], & \text{for } a > \delta, \end{cases} \tag{11.131}$$

where $a = \sin^{-1} \delta/a$.

Alternatively, one can evaluate $N(a)$ from Equation (11.128), yielding

$$N(a) = a_1/a = \left(4/a^2\pi \int_0^{\delta} m(a^2 - x^2)^{1/2}\right)dx.$$

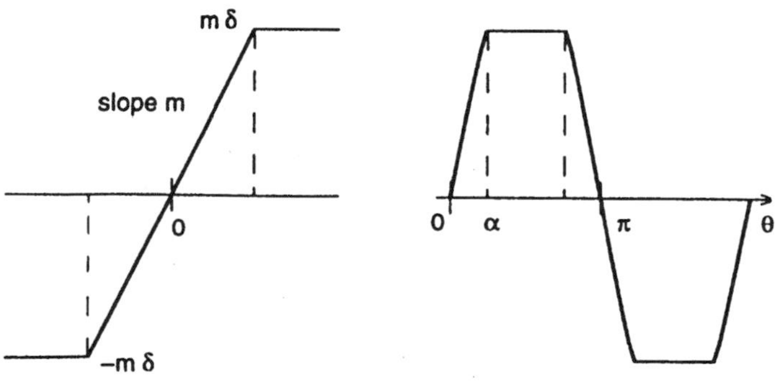

FIGURE 11.60 Saturation nonlinearity.

Using the substitution $x = a \sin \theta$,

$$N(a) = (4m/\pi)\int_0^{\alpha} \cos^2\theta \, d\theta = (m/\pi)(2\alpha + \sin 2\alpha)$$

as before.

Relay with Dead Zone and Hysteresis

The characteristic is shown in Figure 11.61 together with the corresponding input, assumed equal to $a \cos \theta$, and the corresponding output wave form. Using Equation (11.114) and Equation (11.115) over the interval $-\pi/2$ to $\pi/2$ and assuming that the input amplitude a is greater than $\delta+\Delta$,

$$\begin{aligned} a_1 &= \left(2/\pi \int_{-\alpha}^{\beta} h \cos\theta \, d\theta\right) \\ &= (2h/\pi)(\sin\beta + \sin\alpha), \end{aligned}$$

where $\alpha = \cos^{-1}[(\delta - \Delta)/a]$ and $\beta = \cos^{-1}[(\delta + \Delta)/a]$, and

$$\begin{aligned} b_1 &= (2/\pi)\int_{-\alpha}^{\beta} h \sin\theta \, d\theta \\ &= (-2h/\pi)\left(\frac{(\delta+\Delta)}{a} - \frac{(\delta-\Delta)}{a}\right) = 4h\Delta/a\pi. \end{aligned}$$

Thus

$$N(a) = \frac{2h}{a^2\pi}\left\{\left[a^2 - (\delta+\Delta)^2\right]^{1/2} + \left[a^2 - (\delta-\Delta)^2\right]^{1/2}\right\} - \frac{j4h\Delta}{a^2\pi} \qquad (11.132)$$

For the alternative approach, one must first obtain the in-phase and quadrature nonlinearities shown in Figure 11.62. Using Equation (11.123) and Equation (11.124),

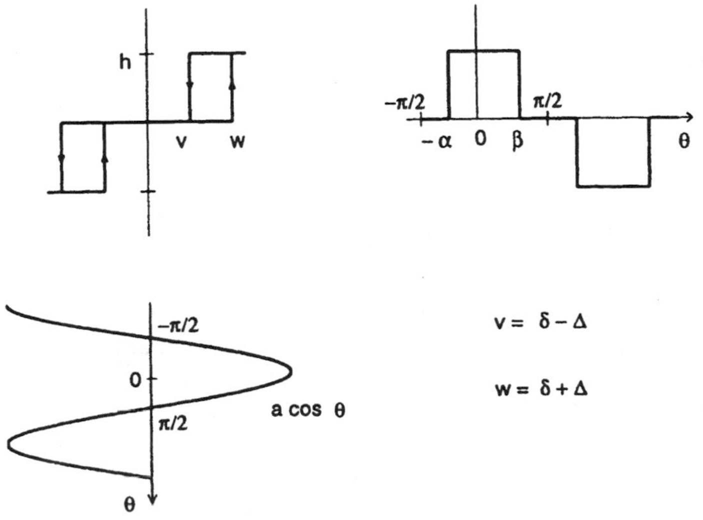

FIGURE 11.61 Relay with dead zone and hysteresis.

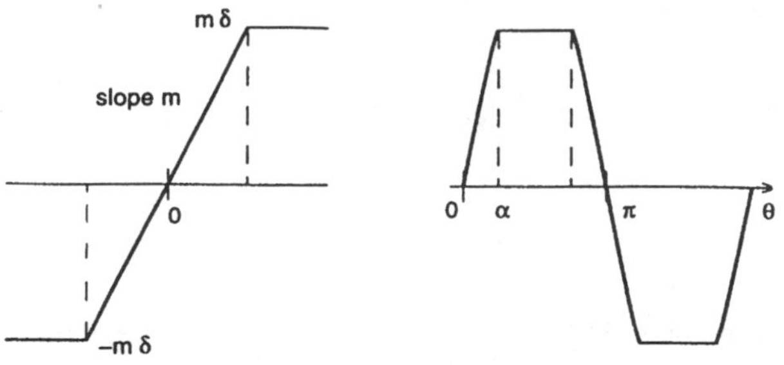

FIGURE 11.62 Function $n_p(x)$ and $n_q(x)$ for the relay of Figure 11.61.

$$a_1 = (4/a)\int_{\delta-\Delta}^{\delta+\Delta} x(h/2)p(x)\mathrm{d}x + \int_{\delta+\Delta}^{a} xhp(x)\mathrm{d}x,$$

$$= \frac{2h}{a\pi}\left\{\left[a^2-(\delta+\Delta)^2\right]^{1/2} + \left[a^2-(\delta-\Delta)^2\right]^{1/2}\right\},$$

and

$$b_1 = (4/a\pi)\int_{d-\Delta}^{\delta+\Delta}(h/2)\mathrm{d}x = 4h\Delta/a\pi = (\text{Area of nonlinearity loop})/a\pi$$

as before.

The DF of two nonlinearities in parallel equals the sum of their individual DFs, a result very useful for determining DFs, particularly of linear segmented characteristics with multiple break points. Several procedures [Atherton, 1982] are available for approximating the DF of a given nonlinearity either by numerical integration or by evaluating the DF of an approximating nonlinear characteristic defined, for example, by a quantized characteristic, linear segmented characteristic, or Fourier series. Table 11.3 gives a list of DFs for some commonly used approximations of nonlinear elements. Several of the results are in terms of the DF for an ideal saturation characteristic of unit slope, $N_s(\delta/a)$, defined in Equation (11.131).

Limit Cycles and Stability

To investigate the possibility of limit cycles in the autonomous closed-loop system of Figure 11.57, the input to the nonlinearity $n(x)$ is assumed to be a sinusoid so that it can be replaced by the amplitude-dependent DF gain $N(a)$. The open-loop gain to a sinusoid is thus $N(a)G(j\omega)$ and, therefore, a limit cycle exists if

$$N(a)G(j\omega) = -1 \tag{11.133}$$

where $G(j\omega) = G_c(j\omega)G_1(j\omega)$. As, in general, $G(j\omega)$ is a complex function of ω and $N(a)$ is a complex function of a, solving Equation (11.133) will yield both the frequency ω and amplitude a of a possible limit cycle.

A common procedure to examine solutions of Equation (11.132) is to use a Nyquist diagram, where the $G(j\omega)$ and $C(a) = -1/N(a)$ loci are plotted as in Figure 11.63, where they are shown intersecting for $a = a_0$ and $\omega = \omega_0$. The DF method indicates therefore that the system has a limit cycle with the input sinusoid to the nonlinearity, x, equal to $a_0 \sin(\omega_0 t + \phi)$, where ϕ depends on the initial conditions. When the $G(j\omega)$ and $C(a)$ loci do not intersect, the DF method predicts that no limit cycle will exist if the Nyquist stability criterion is satisfied for $G(j\omega)$ with respect to any point on the $C(a)$ locus. Obviously, if the nonlinearity has unit gain for small inputs, the point $(-1, j_0)$ will lie on $C(a)$ and may be used as the critical point, analogous to a linear system.

TABLE 11.3 DFs of Single-Valued Nonlinearities

General quantizer 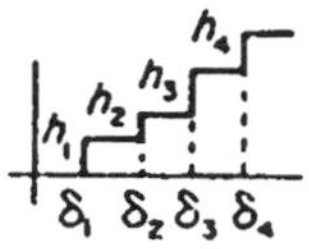	$a < \delta_1$ $\delta_{M+1} > a > \delta_M$	$N_P = 0$ $N_P = \left(4/a^2\pi\right)\sum_{m=1}^{M} h_m\left(a^2 - \delta_m^2\right)^{1/2}$
Uniform quantizer $h_1 = h_2 = \cdots h$ $\delta_m = (2m-1)\delta/2$	$a < d$ $(2M+1)\delta > a > (2M-1)\delta$ $n = (2m-1)/2$	$N_P = 0$ $N_P = \left(4h/a^2\pi\right)\sum_{m=1}^{M}\left(a^2 - n^2\delta^2\right)^{1/2}$
Relay with dead zone 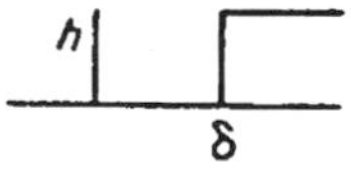	$a < \delta$ $a < \delta$	$N_P = 0$ $N_P = 4h(a^2 - \delta^2)^{1/2}/a^2\pi$
Ideal relay 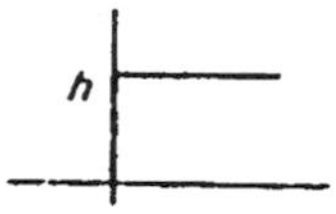		$N_P = 4h/a\pi$
Preload 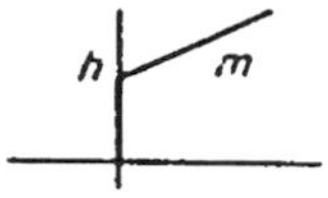		$N_P = (4h/a\pi) + m$
General piecewise linear 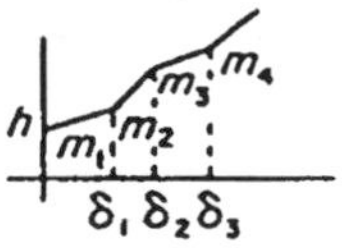	$a < \delta_1$ $\delta_{M+1} > a > \delta_M$	$N_P = (4h/a\pi) + m_1$ $N_P = (4h/a\pi) + m_{M+1} + \sum_{i=1}^{M}\left(m_j - m_{j+1}\right)N_s\left(\delta_j/a\right)$
Ideal saturation 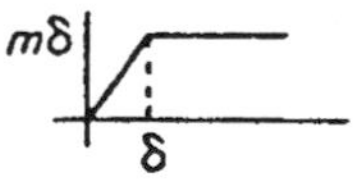		$N_P = mN_s(\delta/a)$
Dead zone 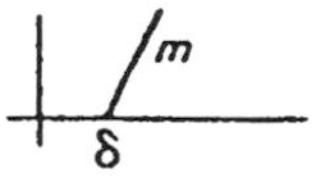		$N_P = m[1 - N_s(\delta/a)]$
Gain changing nonlinearity 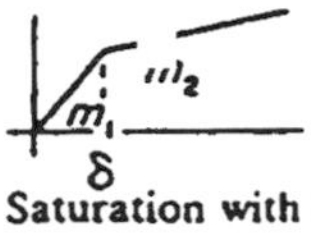		$N_p = (m_1 - m_2)N_s(\delta/a) + m_2$
Saturation with dead zone 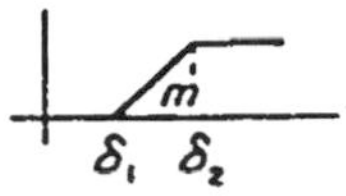		$N_p = m[N_s(\delta_2/a) - N_s(\delta_1/a)]$
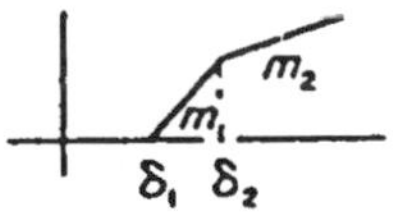		$N_p = -m_1N_s(\delta_1/a) + (m_1 - m_2)N_s(\delta_2/a) + m_2$

TABLE 11.3 (Continued)

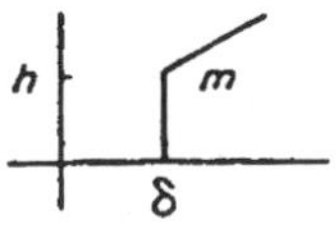	$a<\delta$ $a>\delta$	$N_p = 0$ $N_p = [4h(a^2 - \delta^2)^{1/2}/a_2\pi] + m - mN_s(\delta/a)$
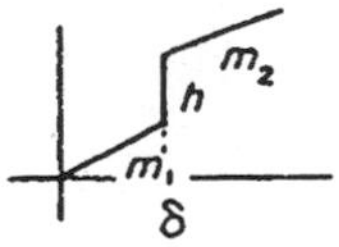	$a<\delta$ $a>\delta$	$N_p = m_1$ $N_p = (m_1 - m_2)N_s(\delta/a) + m_2 + 4h(a^2 - \delta^2)^{1/2}/a^2\pi$
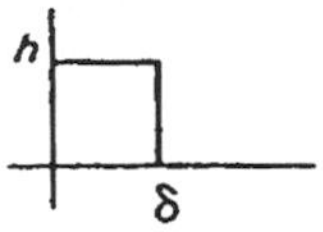	$a<\delta$ $a>\delta$	$N_p = 4h/a\pi$ $N_p = 4h/[a - (a^2 - \delta^2)^{1/2}]/a^2\pi$
Limited field of view		
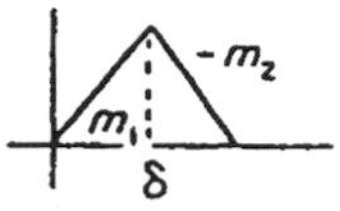		$N_p = (m_1 + m_2)N_s(\delta/a) - m_2N_s[(m_1 + m_2)\delta/m_2a]$
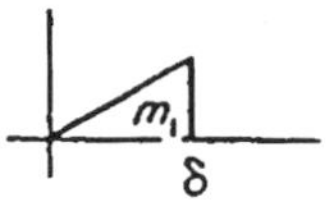	$a<\delta$ $a>\delta$	$N_p = m_1$ $N_p = m_1N_s(\delta/a) - 4m_1\delta(a^2 - \delta^2)^{1/2}/a^2\pi$
$y = x^m$	$m > -2\Gamma$ is the gamma function	$N_p = \dfrac{\Gamma(m+1)a^{m-1}}{2^{m-1}\Gamma[(3+m)/2]\Gamma[(1+m)/2]}$ $= \dfrac{2}{\sqrt{\pi}} \dfrac{\Gamma[(m+2)2]a^{m-1}}{\Gamma[(m+3)/2]}$

For a stable case, it is possible to use the gain and phase margin to judge the relative stability of the system. However, a gain and phase margin can be found for every amplitude a on the $C(a)$ locus, so it is usually appropriate to use the minimum values of the quantities [Atherton, 1982]. When the nonlinear block includes dynamics so that its response is both amplitude and frequency dependent, that is $N(a, \omega)$, then a limit cycle

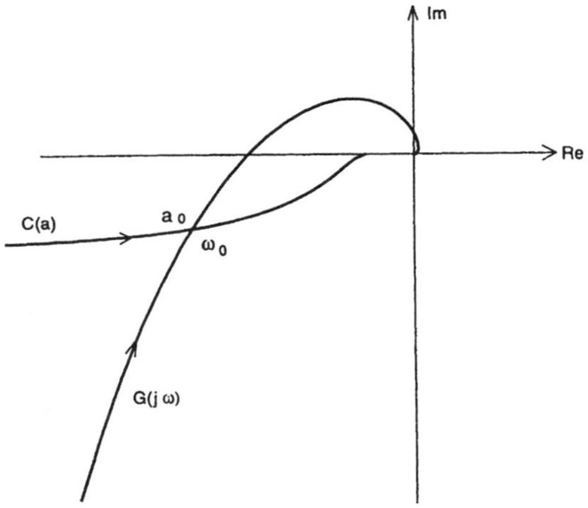

FIGURE 11.63 Nyquist plot showing solution for a limit cycle.

will exist if

$$G(j\omega) = -1/N(a,\omega) = C(a,\omega). \tag{11.134}$$

To check for possible solutions of this equation, a family of $C(a, \omega)$ loci, usually as functions of a for fixed values of ω, is drawn on the Nyquist diagram.

An additional point of interest is whether, when a solution to Equation (11.132) exists, the predicted limit cycle is stable. When there is only one intersection point, the stability of the limit cycle can be found using the Loeb criterion, which states that if the Nyquist stability criterion indicates instability (stability) for the point on $C(a)$ with $a < a_0$ and stability (instability) for the point on $C(a)$ with $a > a_0$, the limit cycle is stable (unstable).

When multiple solutions exist, the situation is more complicated and the criterion above is a necessary but not sufficient result for the stability of the limit cycle [Choudhury and Atherton, 1974].

Normally in these cases, the stability of the limit cycle can be ascertained by examining the roots of the characteristic equation

$$1 + N_{i\gamma}(a)G(s) = 0 \tag{11.135}$$

where $N_{i\gamma}(a)$ is known as the incremental describing function (IDF). $N_{i\gamma}(a)$ for a single-valued nonlinearity can be evaluated from

$$N_{i\gamma}(a) = \int_{-a}^{a} n'(x)p(x)\mathrm{d}x \tag{11.136}$$

where $n'(x)$ and $p(x)$ are as previously defined. $N_{i\gamma}(a)$ is related to $N(a)$ by the equation

$$N_{i\gamma}(a) = N(a) + (a/2)dN(a)/\mathrm{d}a. \tag{11.137}$$

Thus, for example, for an ideal relay, making $\delta = \Delta = 0$ in Equation (11.132) gives $N(a) = 4h/a\pi$, also found directly from Equation (11.128), and, substituting this value in Equation (11.137) yields $N_{i\gamma}(a) = 2h/a\pi$. Some examples of feedback system analysis using the DF follow.

Autotuning in Process Control

In 1943 Ziegler and Nichols [1943] suggested a technique for tuning the parameters of a PID controller. Their method was based on testing the plant in a closed loop with the PID controller in the proportional mode. The proportional gain was increased until the loop started to oscillate, and then the value of gain and the oscillation frequency were measured. Formulae were given for setting the controller parameters based on the gain named the critical gain, K_c, and the frequency called the critical frequency, ω_c.

Assuming that the plant has a linear transfer function $G_1(s)$, then K_c is its gain margin and ω_c the frequency at which its phase shift is 180°. Performing this test in practice may prove difficult. If the plant has a linear transfer function and the gain is adjusted too quickly, a large amplitude oscillation may start to build up. In 1984 Astrom and Hagglund [1984] suggested replacing the proportional control by a relay element to control the amplitude of the oscillation. Consider therefore the feedback loop of Figure 11.57 with $n(x)$ an ideal relay, $G_c(s) = 1$, and the plant with a transfer function $G_1(s) = 10/(s+1)^3$. The $C(a)$ locus, $-1/N(a) = -a\pi/4h$, and the Nyquist locus $G(j\omega)$ in Figure 11.64 intersect. The values of a and ω at the intersection can be calculated from

$$-a\pi/4h = \frac{10}{(1+j\omega)^3} \tag{11.138}$$

which can be written

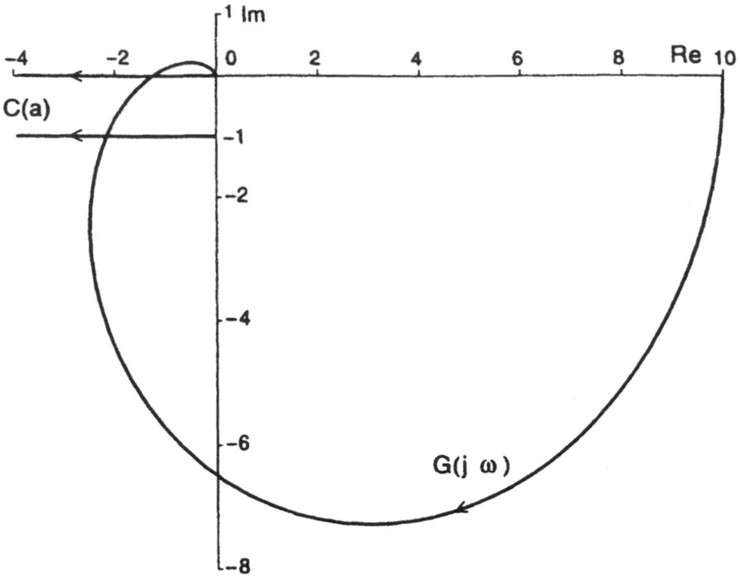

FIGURE 11.64 Nyquist plot $10/(s+1)^3$ and $C(a)$ loci for $\Delta = 0$ and $4h/\pi$.

$$\text{Arg}\left(\frac{10}{(1+j\omega)^3}\right) = 180°, \text{ and} \tag{11.139}$$

$$\frac{a\pi}{4h} = \frac{10}{(1+\omega^2)^{3/2}}. \tag{11.140}$$

The solution for ω_c from Equation (11.139) is $\tan^{-1}\omega_c = 60°$, giving $\omega_c = \sqrt{3}$. Because the DF solution is approximate, the actual measured frequency of oscillation will differ from this value by an amount that will be smaller the closer the oscillation is to a sinusoid. The exact frequency of oscillation in this case will be 1.708 rads/s in error by a relatively small amount. For a square wave input to the plant at this frequency, the plant output signal will be distorted by a small percentage. The distortion, d, is defined by

$$\text{d} = \left[\frac{\text{M.S. value of signal} - \text{M.S. value of fundamental harmonic}}{\text{M.S. value of fundamental harmonic}}\right]^{1/2} \tag{11.141}$$

Solving Equation (11.140) gives the amplitude of oscillation a as $5h/\pi$. The gain through the relay is $N(a)$ equal to the critical gain K_c. In the practical situation where a is measured, K_c equal to $4h/a\pi$, should be close to the known value of 0.8 for this transfer function.

If the relay has an hysteresis of Δ, then with $\delta = 0$ in Equation (11.132) gives

$$N(a) = \frac{4h(a^2-\Delta^2)^{1/2}}{a^2\pi} - j\frac{4h\Delta}{a^2\pi}$$

from which

$$C(a) = \frac{-1}{N(a)} = \frac{-\pi}{4h}\left[\left(a^2 - \Delta^2\right)^{1/2} + j\Delta\right].$$

Thus on the Nyquist plot, $C(a)$ is a line parallel to the real axis at a distance $\pi\Delta\delta/4h$ below it, as shown in Figure 11.64 for $\Delta = 1$ and $h = \pi/4$, giving $C(a) = -(a^2 - 1)^{1/2} - j$. If the same transfer function is used for the plant, then the limit cycle solution is given by

$$-\left(a^2 - 1\right)^{1/2} - j = \frac{10}{(1 + j\omega)^3} \tag{11.142}$$

where $\omega = 1.266$, which compares with an exact solution value of 1.254, and $a = 1.91$. For the oscillation with the ideal relay, Equation (11.135) with $N_{i\gamma}(a) = 2h/a\pi$ shows that the limit cycle is stable. This agrees with the perturbation approach, which also shows that the limit cycle is stable when the relay has hysteresis.

Feedback Loop with a Relay with Dead Zone

For this example the feedback loop of Figure 11.57 is considered with $n(x)$ a relay with dead zone and $G(s) = 2/s(s + 1)^2$. From Equation (11.132) with $\Delta = 0$, the DF for this relay, given by

$$N(a) = 4h\left(a^2 - \delta^2\right)^{1/2} / a^2\pi \text{ for } a > \delta. \tag{11.143}$$

is real because the nonlinearity is single valued. A graph of $N(a)$ against a is in Figure 11.65, and shows that $N(a)$ starts at zero, when $a = \delta$, increases to a maximum, with a value of $2h/\pi\delta$ at $a = \delta\sqrt{2}$, and then decreases toward zero for larger inputs. The $C(a)$ locus, shown in Figure 11.66, lies on the negative real axis starting at $-\infty$ and returning there after reaching a maximum value of $-\pi\delta/2h$. The given transfer function $G(j\omega)$ crosses the negative real axis, as shown in Figure 11.66, at a frequency of $\tan^{-1}\omega = 45°$, that is, $\omega = 1$ rad/s and, therefore, cuts the $C(a)$ locus twice. The two possible limit cycle amplitudes at this frequency can be found by solving

$$\frac{a^2\pi}{4h(a^2 - \delta^2)^{1/2}} = 1$$

which gives $a = 1.04$ and 3.86 for $\delta = 1$ and $h = \pi$. Using the perturbation method or the IDF criterion, the smaller amplitude limit cycle is unstable and the larger one is stable. If a condition similar to the lower amplitude limit cycle is excited in the system, an oscillation will build up and stabilize at the higher amplitude limit cycle.

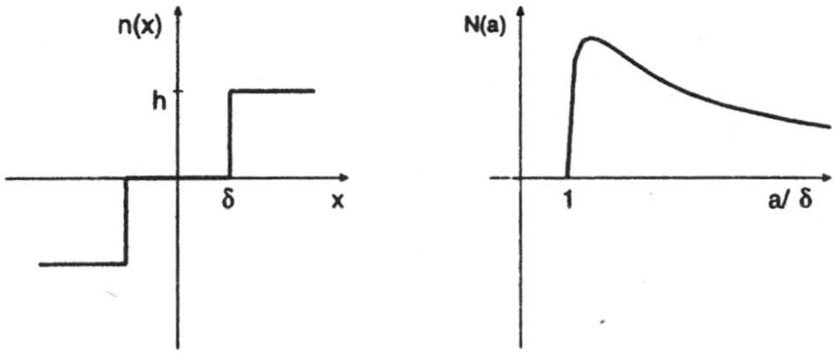

FIGURE 11.65 $N(a)$ for ideal relay with dead zone.

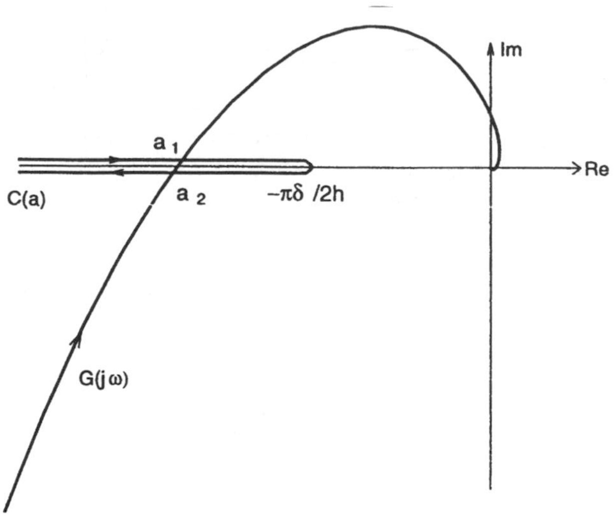

FIGURE 11.66 Two limit cycles: a_1, unstable; a_2, stable.

Other techniques show that the exact frequencies of the limit cycles for the smaller and larger amplitudes are 0.709 and 0.989, respectively. Although the transfer function is a good low-pass filter, the frequency of the smaller amplitude limit cycle is not predicted accurately because the output from the relay, a wave form with narrow pulses, is highly distorted.

If the transfer function of $G(s)$ is $K/s(s+1)^2$, then no limit cycle will exist in the feedback loop, and it will be stable if

$$\left.\frac{K}{\omega(1+\omega^2)}\right|_{\omega=1} < \frac{\pi d}{2h},$$

that is, $K < \pi\delta/h$. If $\delta = 1$ and $h = \pi$, $K < 1$, which may be compared with the exact result for stability of $K < 0.96$.

Stability and Accuracy

Because the DF method is an approximate procedure, it is desirable to judge its accuracy. Predicting that a system will be stable, when in practice it is not, may have unfortunate consequences. Many attempts have been made to solve this problem, but those obtained are difficult to apply or produce too conservative results [Mees and Bergen, 1975].

Since as mentioned previously the DF for a nonlinearity in a sector is the diameter of the circle of the circle criterion, then for a limit cycle in the system of Figure 11.57, errors in the DF method relate to its inability to predict a phase shift, which the fundamental harmonic may experience in passing through the nonlinearity, rather than an incorrect magnitude of the gain. When the input to a single-valued nonlinearity is a sinusoid together with some of its harmonics, the fundamental output is not necessarily in phase with the fundamental input, that is, the fundamental gain has a phase shift. The actual phase shift varies with the harmonic content of the input signal in a complex manner, because the phase shift depends on the amplitudes and phases of the individual input components.

From an engineering viewpoint one can judge the accuracy of DF results by estimating the distortion, *d*, in the input to the nonlinearity. This is straightforward when a limit-cycle solution is given by the DF method; the loop may be considered opened at the nonlinearity input, the sinusoidal signal corresponding to the DF solution can be applied to the nonlinearity, and the harmonic content of the signal fed back to the nonlinearity input can be calculated. Experience indicates that the percentage accuracy of the DF method in predicting the fundamental amplitude and frequency of the limit cycle is less than the percentage distortion in the fedback signal. As mentioned previously, the DF method may incorrectly predict stability. To investigate this problem, the procedure above can be used again, by taking, as the nonlinearity input, a sinusoid with amplitude and frequency corresponding to values of those parameters where the phase margin is small. If the calculated feedback distortion is high, say greater than 2% per degree of phase margin, the DF result should not be relied on.

The limit-cycle amplitude predicted by the DF is an approximation to the fundamental harmonic. The accuracy of this prediction cannot be assessed by using the peak value of the limit cycle to estimate an equivalent sinusoid. It is possible to estimate the limit cycle more accurately by balancing more harmonics, as mentioned earlier. Although this is difficult algebraically other than with loops whose nonlinearity is mathematically simply described, for example a cubic, software is available for this purpose [McNamara and Atherton, 1987]. The procedure involves solving sets of nonlinear algebraic equations, but good starting guesses can usually be obtained for the magnitudes and phases of the other harmonic components from the wave form that is fed back to the nonlinearity, assuming its input is the DF solution.

Compensator Design

Although the design specifications for a control system are often in terms of step-response behavior, frequency domain design methods rely on the premise that the correlation between the frequency and a step response yields a less oscillatory step response if the gain and phase margins are increased. Therefore the design of a suitable linear compensator for the system of Figure 11.57, using the DF method, is usually done by selecting for example a lead network to provide adequate gain and phase margins for all amplitudes. This approach may be used in example 2 of the previous section, where a phase lead network could be added to stabilize the system, say for a gain of 1.5, for which it is unstable without compensation. Other approaches are the use of additional feedback signals or modification of the nonlinearity $n(x)$ directly or indirectly [Atherton, 1982; Gelb and van der Velde, 1968].

When the plant is nonlinear, its frequency response also depends on the input sinusoidal amplitude represented as $G\,(j\omega, a)$. In recent years several approaches [Nanka-Bruce and Atherton, 1990; Taylor and Strobel, 1984] use the DF method to design a nonlinear compensator for the plant, with the objective of closed-loop performance approximately independent of the input amplitude.

Closed-Loop Frequency Response

When the closed-loop system of Figure 11.57 has a sinusoidal input $r(t) = R\sin(\omega t + \theta)$, it is possible to evaluate the closed-loop frequency response using the DF. If the feedback loop has no limit cycle when $r(t) = 0$ and, in addition, the sinusoidal input $r(t)$ does not induce a limit cycle, then, provided that $G_c(s)G_1(s)$ gives good filtering, $x(t)$, the nonlinearity input, almost equals the sinusoid $a\sin\omega t$. Balancing the components of frequency ω around the loop,

$$\begin{aligned} &g_c R\sin(\omega t + \theta\_\phi_c) - a g_1 g_c M(a) \\ &\sin[\omega t + \phi_1 + \phi_c + \Psi(a)] = a\sin\omega t \end{aligned} \tag{11.144}$$

where $G_c\,(j\omega) = g_c\,e^{j\phi_c}$ and $G_1(j\omega) = g_1\,e^{j\phi_1}$. In principle Equation (11.144), which can be written as two nonlinear algebraic equations, can be solved for the two unknowns a and θ, and the fundamental output signal can then be found from

$$c(t) = aM(a)g_1\,\sin[\omega t + \Psi(a) + \phi_1] \tag{11.145}$$

to obtain the closed-loop frequency for R and ω.

Various graphical procedures have been proposed for solving the two nonlinear algebraic equations resulting from Equation (11.144) [Levinson, 1953; Singh, 1965; West and Douce, 1954]. If the system is lightly damped, the nonlinear equations may have more than one solution, indicating that the frequency response of the system has a jump resonance. This phenomenon of a nonlinear system has been studied by many authors, both theoretically and practically [Lamba and Kavanagh, 1971; West and Douce, 1954].

The Phase-Plane Method

The phase-plane method was the first method used by control engineers for studying the effects of nonlinearity in feedback systems. The technique, which can only be used for systems with second-order models, was examined and further developed for control engineering purposes for several major reasons:

1. The phase-plane approach has been used for several studies of second-order nonlinear differential equations arising in fields such as planetary motion, nonlinear mechanics, and oscillations in vacuum tube circuits.
2. Many of the control systems of interest, such as servomechanisms, could be approximated by second-order nonlinear differential equations.
3. The phase plane was particularly appropriate for dealing with nonlinearities with linear segmented characteristics that were good approximations for the nonlinear phenomena encountered in control systems.

The next section considers the basic aspects of the phase-plane approach, but later concentration is focused on control engineering applications, where the nonlinear effects are approximated by linear segmented nonlinearities.

Background

Early analytical work [Andronov et al., 1966] on second-order models assumed the equations

$$\begin{aligned}\dot{x}_1 &= P(x_1, x_2)\\ \dot{x}_2 &= Q(x_1, x_2)\end{aligned} \tag{11.146}$$

for two first-order nonlinear differential equations. Equilibrium, or singular points, occur when

$$\dot{x}_1 = \dot{x}_2 = 0$$

and the slope of any solution curve, or trajectory, in the $x_1 - x_2$ state plane is

$$\frac{dx_2}{dx_1} = \frac{\dot{x}_2}{\dot{x}_1} = \frac{Q(x_1, x_2)}{P(x_1, x_2)} \tag{11.147}$$

A second-order nonlinear differential equation representing a control system can be written

$$\ddot{x} + f(x, \dot{x}) = 0 \tag{11.148}$$

If this is rearranged as two first-order equations, choosing the phase variables as the state variables, that is $x_1 = x$, $x_2 = \dot{x}$, then Equation (11.148) can be written as

$$\dot{x}_1 = \dot{x}_2 \quad \dot{x}_2 = -f(x_1, x_2) \tag{11.149}$$

which is a special case of Equation (11.147). A variety of procedures have been proposed for sketching state (phase) plane trajectories for Equation (11.147) and Equation (11.149). A complete plot showing trajectory motions throughout the entire state (phase) plane is known as a state (phase) portrait. Knowledge of these methods, despite the improvements in computation since they were originally proposed, can be particularly helpful for obtaining an appreciation of the system behavior. When simulation studies are undertaken, phase-plane graphs are easily obtained, and they are often more helpful for understanding the system behavior than displays of the variables x_1 and x_2 against time.

Many investigations using the phase-plane technique were concerned with the possibility of limit cycles in the nonlinear differential equations When a limit cycle exists, this results in a closed trajectory in the phase plane. Typical of such investigations was the work of Van der Pol, who considered the equation

$$\ddot{x} - \mu\left(1 - x^2\right)\dot{x} + x = 0 \tag{11.150}$$

where μ is a positive constant. The phase-plane form of this equation can be written as

$$\dot{x}_1 = x_2$$

$$\dot{x}_2 = -f(x_1, x_2) = \mu\left(1 - x_1^2\right)x_2 - x_1 \tag{11.151}$$

The slope of a trajectory in the phase plane is

$$\frac{dx_2}{dx_1} = \frac{\dot{x}_2}{\dot{x}_1} = \frac{\mu(1 - x_1^2)x_2 - x_1}{x_2} \tag{11.152}$$

and this is only singular (that is, at an equilibrium point) when the right-hand side of Equation (11.152) is 0/0, that is, $x_1 = x_2 = 0$.

The form of this singular point, which is obtained from linearization of the equation at the origin, depends upon μ being an unstable focus for $\mu < 2$ and an unstable node for $\mu > 2$. All phase-plane trajectories have a slope of r when they intersect the curve

$$rx_2 = \mu\left(1 - x_1^2\right)x_2 - x_1 \tag{11.153}$$

One way of sketching phase-plane behavior is to draw a set of curves given for various values of r by Equation (11.153) and marking the trajectory slope r on the curves. This procedure is known as the method of isoclines and has been used to obtain the limit cycles shown in Figure 11.67 for the Van der Pol equation with $\mu = 0.2$ and 4.

Piecewise Linear Characteristics

When the nonlinear elements occurring in a second-order model can be approximated by linear segmented characteristics, then the phase-plane approach is usually easy to use because the nonlinearities divide the phase plane into various regions within which the motion may be described by different linear second-order equations [Atherton, 1982]. The procedure is illustrated by the simple relay system in Figure 11.68.

The block diagram represents an "ideal" relay position control system with velocity feedback. The plant is a double integrator, ignoring viscous (linear) friction, hysteresis in the relay, or backlash in the gearing. If the system output is denoted by x_1 and its derivative by x_2, then the relay switches when $-x_1 - x_2 = \pm 1$; the equations of the dotted lines are marked switching lines on Figure 11.69.

Because the relay output provides constant values of ± 2 and 0 to the double integrator plant, if we denote the constant value by h, then the state equations for the motion are

$$\begin{aligned} \dot{x}_1 &= x_2 \\ \dot{x}_2 &= h \end{aligned} \tag{11.154}$$

which can be solved to give the phase-plane equation

$$x_2^2 - x_{20}^2 = 2h(x_1 - x_{10}) \tag{11.155}$$

which is a parabola for h finite and the straight line $x_2 = x_{20}$ for $h = 0$, where x_{20} and x_{10} are the initial values of x_2 and x_1. Similarly, more complex equations can be derived for other second-order transfer functions. Using Equation (11.155) with the appropriate values of h for the three regions in the phase plane, the step response for an input of 4.6 units can be obtained as shown in Figure 11.69.

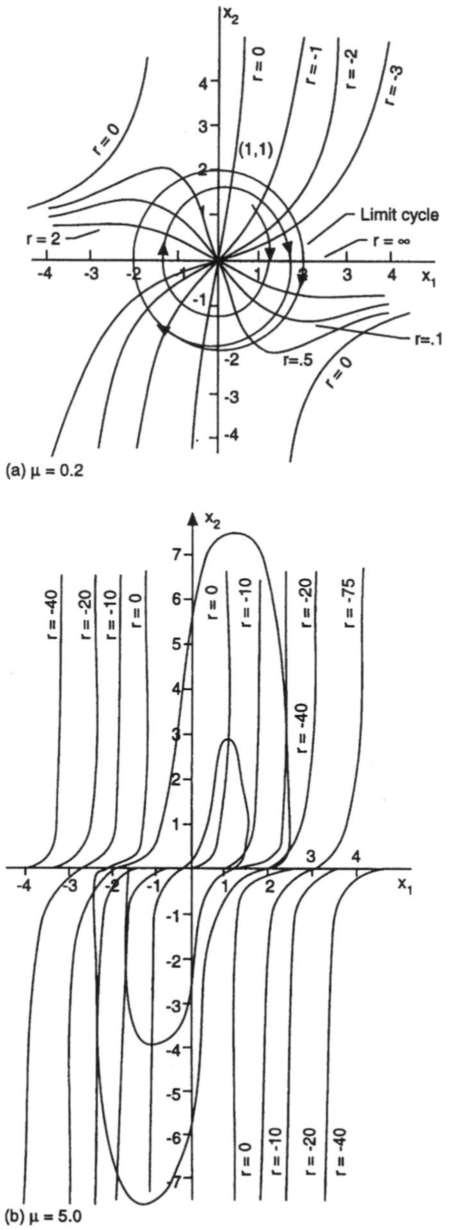

FIGURE 11.67 Phase portraits of the Van der Pol equation for different values of μ.

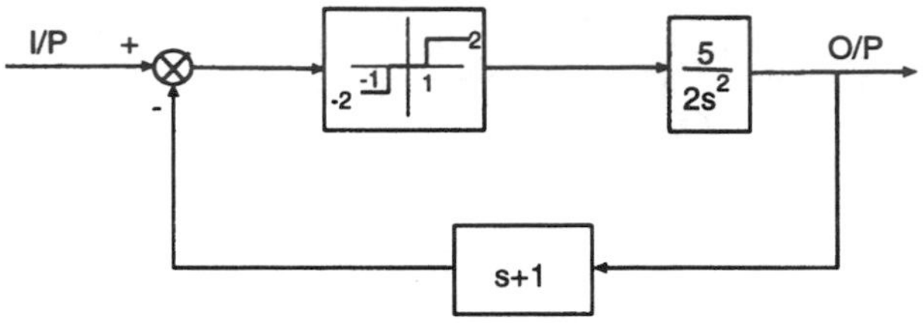

FIGURE 11.68 Relay system.

In the step response, when the trajectory meets the switching line $x_1 + x_2 = -1$ for the second time, trajectory motions at both sides of the line are directed towards it, resulting in a sliding motion down the switching line. Completing the phase portrait by drawing responses from other initial conditions shows that the autonomous system is stable and also that all responses will finally slide down a switching line to equilibrium at $x_1 = \pm 1$.

An advantage of the phase-plane method is that it can be used for systems with more than one nonlinearity and for those situations where parameters change as functions of the phase variables. For example, Figure 11.70 shows the block diagram of an approximate model of a servomechanism with nonlinear effects due to torque saturation and Coulomb friction.

The differential equation of motion in phase variable form is

$$\dot{x}_2 = f_s(-x_1) - (1/2)\ \mathrm{sgn}\ x_2 \tag{11.156}$$

where f_s denotes the saturation nonlinearity and sgn the signum function, which is +1 for $x_2 > 0$ and –1 for $x_2 < 0$. There are six linear differential equations describing the motion in different regions of the phase plane. For x_2 positive, Equation (11.156) can be written

$$\dot{x}_1 + f_s(x_1) + 1/2 = 0$$

so that for

(a) x_2+ve, $x_1 < -2$, we have $\dot{x}_1 = x_2, \dot{x}_2 = 3/2$, a parabola in the phase plane
(b) x_2+ve$|x_1|<2$, we have $\dot{x}_1 = x_2, \dot{x}_2 + x_1 + 1/2 = 0$

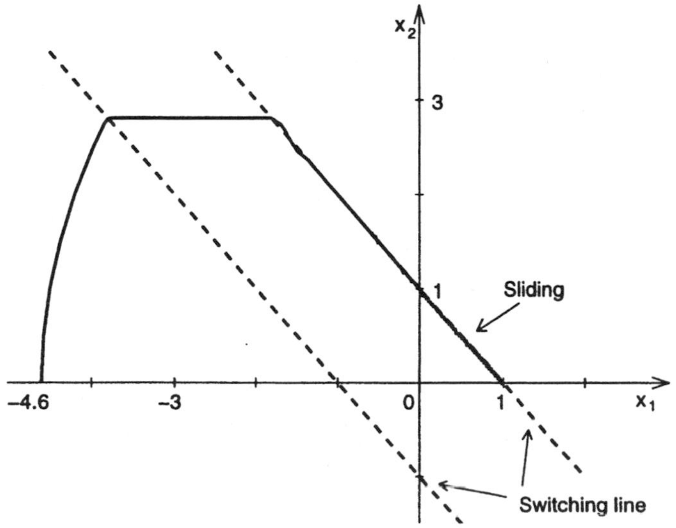

FIGURE 11.69 Phase plane for relay system.

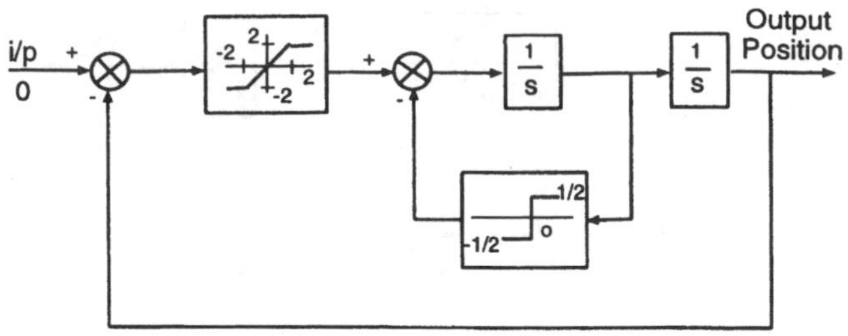

FIGURE 11.70 Block diagram of servomechanism.

(c) x_2+ve, $x_1 > 2$, we have $\dot{x}_1 = x_2, \dot{x}_2 = -5/2$, a parabola in the phase plane. Similarly for x_2 negative,
(d) x_2−ve, $x_1 - 2$, we have $\dot{x}_1 = x_2, \dot{x}_2 = -5/2$, a parabola in the phase plane
(e) x_2−ve, $|x_2| < 2$, we have $\dot{x}_1 = x_2, \dot{x}_2 + x_1 - 1/2 = 0$, a circle in the phase plane
(f) x_2−ve, $x_1 > 2$, we have $\dot{x}_1 = x_2, \dot{x}_2 = -3/2$, a parabola in the phase plane

Because all the phase-plane trajectories are described by simple mathematical expressions, it is straightforward to calculate specific phase-plane trajectories.

Discussion

The phase-plane approach is useful for understanding the effects of nonlinearity in second-order systems, particularly if it can be approximated by a linear segmented characteristic. Solutions for the trajectories with other nonlinear characteristics may not be possible analytically, so that approximate sketching techniques were used in early work on nonlinear control. These approaches are described in many books, for example, [Blaquiere, 1966; Cosgriff, 1958; Cunningham, 1958; Gibson, 1963; Graham and McRuer, 1961; Hayashi, 1964; Thaler and Pastel, 1962; West, 1960]. Although the trajectories are now easily obtained with modern simulation techniques, knowledge of the topological aspects of the phase plane are still useful for interpreting the responses in different regions of the phase plane and appreciating the system behavior.

References

A.A. Andronov, A.A. Vitt, and S.E. Khaikin, *Theory of Oscillators,* Reading, Mass.: Addison-Wesley, 1966. (First edition published in Russia in 1937.)

K.J. Astrom and T. Haggland, *Automatic Tuning of Single Regulators,* Budapest: Proc. IFAC Congress, Vol. 4, 267–272, 1984.

D.P. Atherton, *Nonlinear Control Engineering, Describing Function Analysis and Design,* London: Van Nostrand Reinhold, 1975.

D.P. Atherton, *Non Linear Control Engineering,* student ed., New York: Van Nostrand Reinhold, 1982.

A. Blaquiere, *Nonlinear Systems Analysis,* New York: Academic Press, 1966.

S.K. Choudhury and D.P. Atherton, "Limit cycles in high order nonlinear systems," *Proc. Inst. Electr. Eng.,* 121, 717–724, 1974.

P.A. Cook, "Describing function for a sector nonlinearity," *Proc. Inst. Electr. Eng.,* 120, 143–144, 1973.

R. Cosgriff, *Nonlinear Control Systems,* New York: McGraw-Hill, 1958.

W.J. Cunningham, *Introduction to Nonlinear Analysis,* New York: McGraw-Hill, 1958.

A. Gelb and W.E. van der Velde, *Multiple Input Describing Functions and Nonlinear Systems Design,* New York: McGraw-Hill, 1968.

J.E. Gibson, *Nonlinear Automatic Control,* New York: McGraw-Hill, 1963.

D. Graham and D. McRuer, *Analysis of Nonlinear Control Systems,* New York: John Wiley & Sons, 1961.

C. Hayashi, *Nonlinear Oscillations in Physical Systems,* New York, McGraw-Hill, 1964.

S.S. Lamba and R.J. Kavanagh, "The phenomenon of isolated jump resonance and its application," *Proc. Inst. Electr. Eng.,* 118, 1047–1050, 1971.

E. Levinson, "Some saturation phenomena in servomechanisms with emphasis on the tachometer stabilised system," *Trans. Am. Inst. Electr. Eng.,* Part 2, 72, 1–9, 1953.

O.P. McNamara and D.P. Atherton, "Limit cycle prediction in free structured nonlinear systems," *IFAC Congress,* Munich, 8, 23–28, July 1987.

A.I. Mees and A.R. Bergen, "Describing function revisited," *IEEE Trans. Autom. Control,* 20, 473–478, 1975.

O. Nanka-Bruce and D.P. Atherton, "Design of nonlinear controllers for nonlinear plants," *IFAC Congress,* Tallinn, 6, 75–80, 1990.

T.P. Singh, "Graphical method for finding the closed loop frequency response of nonlinear feedback control systems," *Proc. Inst. Electr. Eng.,* 112, 2167–2170, 1965.

J.H. Taylor and K.L. Strobel, "Applications of a nonlinear controller design approach based on the quasilinear system models," *Proc. ACC,* San Diego, 817–824, 1984.

G.J. Thaler and M.P. Pastel, *Analysis and Design of Nonlinear Feedback Control Systems,* New York: McGraw-Hill, 1962.

J.C. West, *Analytical Techniques of Nonlinear Control Systems,* London: E.U.P., 1960.

J.C. West and J.L. Douce, "The frequency response of a certain class of nonlinear feedback systems," *Br. J. Appl. Phys.,* 5, 201–210, 1954.

J.C. West, B.W. Jayawant, and D.P. Rea, "Transition characteristics of the jump phenomenon in nonlinear resonant circuits," *Proc. Inst. Electr. Eng.,* 114, 381–392, 1967.

J.G. Ziegler and N.B. Nichols, "Optimal setting for automatic controllers," *Trans. ASME,* 65, 433–444, 1943.

Further Information

Many control engineering textbooks contain material on nonlinear systems where the describing function is discussed. The coverage, however, is usually restricted to the basic sinusoidal DF for determining limit cycles in feedback systems. The basic DF method, which is one of quasilinearization, can be extended to cover other signals, such as random signals, and also to cover multiple input signals to nonlinearities and in feedback system analysis. The two books with the most comprehensive coverage of this are by Gelb and Van der Velde [1968] and Atherton [1975]. In recent years there have been many new approaches developed for the analysis and design of nonlinear systems, mainly based on state space methods. A good place for the interested reader to start would be Levine and Dorf's *The Control Handbook*, Boca Raton, FL: CRC Press, 1996.

12

Navigation Systems

Myron Kayton
Kayton Engineering Company

12.1 Introduction

"Navigation" is the determination of the position and velocity of a moving vehicle, on land, at sea, in the air, or in space. The three components of position and the three components of velocity make up a six-component state vector whose time variation fully describes the translational motion of the vehicle. With the advent of the Global Positioning System (GPS), surveyors use the same sensors as navigators but achieve higher accuracy as a result of longer periods of observation and more complex post-processing.

In the usual navigation system, the state vector is derived onboard, displayed to the crew, recorded on-board, or transmitted to the ground. Navigation information is usually sent to other on-board subsystems, for example to the waypoint steering, communication control, display, weapon-control, and electronic warfare (emission detection and jamming) computers. Some navigation systems, called *position-location systems*, measure a vehicle's state vector using sensors on the ground or in another vehicle (Section 12.6). The external sensors usually track passive radar returns or a transponder. Position-location systems usually supply information to a dispatch or control center.

Traditionally, *ship navigation* included the art of pilotage: entering and leaving port, making use of wind and tides, and knowing the coasts and sea conditions. However in modern usage, *navigation* is confined to the measurement of the state vector. The handling of the vehicle is called *conning* for ships, *flight control* for aircraft, and *attitude control* for spacecraft.

The term *guidance* has two meanings, both of which are different than *navigation*:

a. Steering toward a destination of known position from the vehicle's present position. The steering equations on a planet are derived from a plane triangle for nearby destinations or from a spherical triangle for distant destinations.
b. Steering toward a destination without calculating the state vector explicitly. A guided vehicle *homes* on radio, infrared, or visual emissions. Guidance toward a *moving* target is usually of interest to military tactical missiles in which a steering algorithm assures impact within the maneuver and fuel constraints

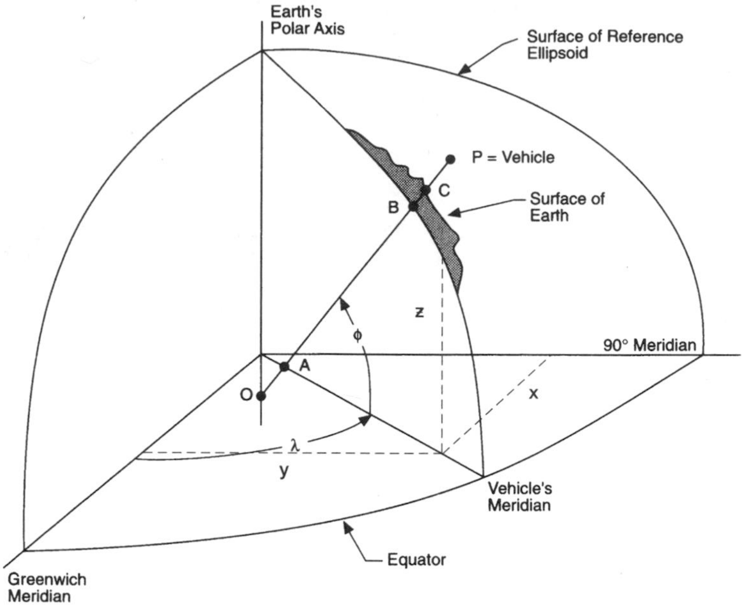

FIGURE 12.1 Latitude-longitude-altitude coordinate frame. φ = geodetic latitude; OP is normal to the ellipsoid at B; λ = geodetic longitude; h = BP = altitude above reference ellipsoid = altitude above mean sea level.

of the interceptor. Guidance toward a fixed target involves beam-riding, as in the Instrument Landing System, Section 12.5.

12.2 Coordinate Frames

Navigation is with respect to a coordinate frame of the designer's choice. Short-range robots navigate with respect to the local terrain or a building's walls. For navigation over hundreds of kilometers (e.g., automobiles and trucks), various map grids exist whose coordinates can be calculated from latitude-longitude. NATO land vehicles use a Universal Transverse Mercator grid. Long-range aircraft and ships navigate relative to an Earth-bound coordinate frame, the most common of which are (Figure 12.1):

a. Latitude-longitude-altitude. The most useful reference ellipsoid is described in [WGS-84, 1991].
b. Earth-centered rectangular (*xyz*).

Spacecraft in orbit around the Earth navigate with respect to an Earth-centered, inertially nonrotating coordinate frame whose *z*-axis coincides with the polar axis of the Earth and whose *x*-axis lies along the equator toward the first point of Aries. Interplanetary spacecraft navigate with respect to a Sun-centered, inertially nonrotating coordinate frame whose *z*-axis is perpendicular to the *ecliptic* and whose *x*-axis points to a convenient star [Battin, 1987].

12.3 Categories of Navigation

Navigation systems can be categorized as:

1. Absolute navigation systems that measure the state vector without regard to the path travelled by the vehicle in the past. These are of two kinds:

   - Radio systems (Section 12.5). They consist of a network of transmitters (sometimes transponders) on the ground or in satellites. A vehicle detects the transmissions and computes its position relative to the known positions of the stations in the navigation coordinate frame. The vehicle's velocity is measured from the Doppler shift of the transmissions or from a sequence of position measurements.

- Celestial systems (Section 12.7). They measure the elevation and azimuth of celestial bodies relative to the Earth's level and true north. Electronic star sensors are used in special-purpose high-altitude aircraft and in spacecraft. Manual celestial navigation was practiced at sea for millennia [Bowditch, 1995].

2. Dead-reckoning navigation systems that derive their state vector from a continuous series of measurements beginning at a known initial position. There are two kinds, those that measure vehicle heading and either speed or acceleration (Section 12.4) and those that measure emissions from continuous-wave radio stations whose signals create ambiguous "lanes" (Section 12.5). Dead-reckoning systems must be *updated* as errors accumulate and if electric power is lost.
3. Mapping navigation systems that observe and recognize images of the ground, profiles of altitude, sequences of turns, or external features (Section 12.8). They compare their observations with a stored data base.

12.4 Dead Reckoning

The simplest dead-reckoning systems measure vehicle heading and speed, resolve speed into the navigation coordinates, then integrate to obtain position, Figure 12.2. The oldest heading sensor is the magnetic compass, a magnetized needle or magnetometer, as shown in Figure 12.3. It measures the direction of the Earth's magnetic field to an accuracy of 2 degrees at a steady speed below 60-degrees magnetic latitude. The horizontal component of the magnetic field points toward magnetic north. The angle from true to *magnetic north* is called *magnetic variation* and is stored in the computers of modern vehicles as a function of position over the region of anticipated travel [Quinn, 1996]. Magnetic deviations caused by iron in the vehicle can exceed 30 degrees and must be compensated in the navigation computer or, in older ships, by placing compensating magnets near the compass [Bowditch, 1995].

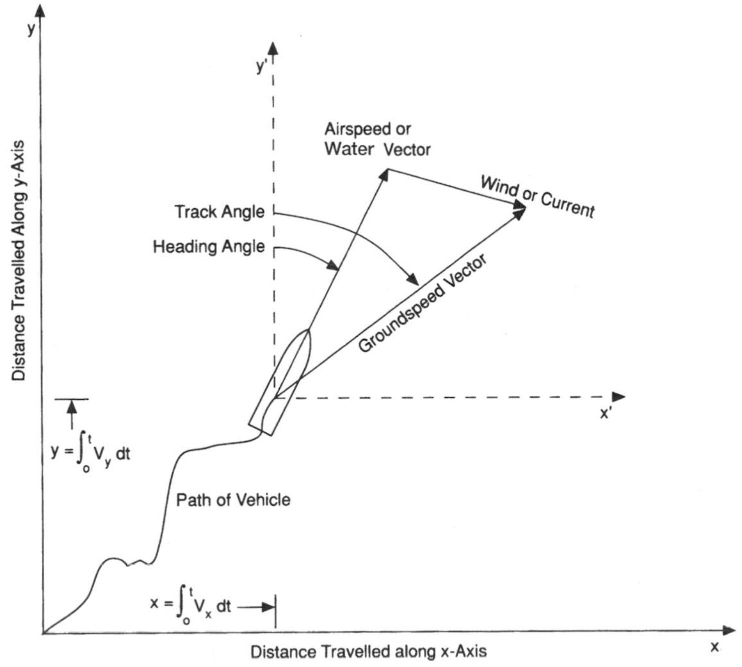

FIGURE 12.2 Geometry of dead reckoning.

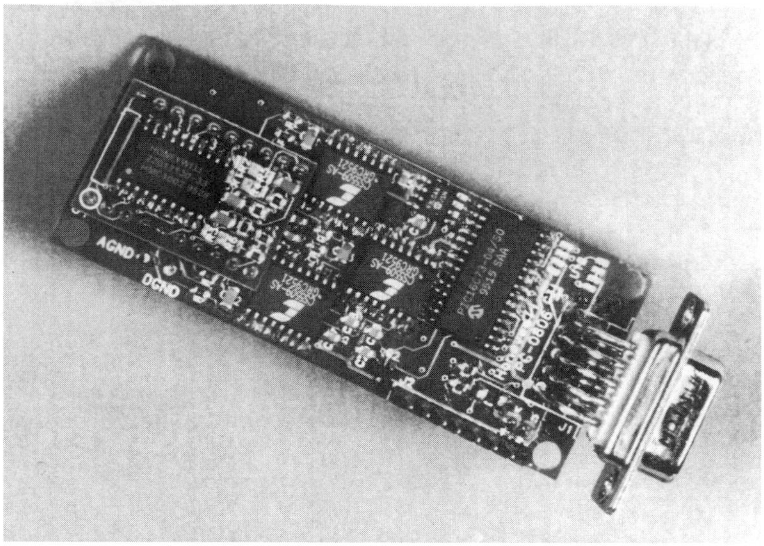

FIGURE 12.3 Circuit Board from three-axis digital magnetometer. A single-axis sensor chip and a two-axis sensor chip are mounted orthogonally at the end opposite the connector. The sensor chips are magneto-resistive bridges with analog outputs that are digitized on the board. (Photo courtesy of Honeywell, Copyright 2004.)

A more complex heading sensor is the *gyrocompass*, consisting of a spinning wheel whose axle is constrained to the horizontal plane (often by a pendulous weight). The ships' version points north, when properly compensated for vehicle motion, and exhibits errors less than a degree. The aircraft version (more properly called a *directional gyroscope*) holds any preset heading relative to Earth and drifts at 150 deg/hr or more. Inexpensive gyroscopes (some built on silicon chips as vibrating beams with on-chip signal conditioning) are often coupled to magnetic compasses to reduce maneuver-induced errors.

The simplest speed sensor is a wheel odometer on an automobile that generates electrical pulses. Ships use a dynamic-pressure probe or an electric-field sensor that measures the speed of the hull through the conductive water. Aircraft measure the dynamic pressure of the air stream from which they derive airspeed in an *air-data computer* in order to calculate ground speed. The velocity of the wind or sea-current must be vectorially added to that of the vehicle, as measured by a dynamic-pressure sensor (Figure 12.3). Hence, unpredicted wind or water current will introduce an error into the dead-reckoning computation. These sensors are insensitive to the component of airspeed or waterspeed normal to the sensor's axis (*leeway* in a ship, *drift* in an aircraft). A Doppler radar measures the frequency shift in radar returns from the ground or water below the aircraft, from which speed is inferred. A Doppler sonar measures a ship's speed relative to the water layer or ocean floor from which the beam reflects. Multibeam Doppler radars or sonars can measure all three components of the vehicle's velocity. Doppler radars are widely used on military helicopters; Doppler sonars in submarines.

The most accurate dead-reckoning system is an *inertial navigator* in which accelerometers measure the vehicle's acceleration while gyroscopes measure the orientation of the accelerometers. An onboard computer resolves the accelerations into navigation coordinates and integrates them to obtain velocity and position. The gyroscopes and accelerometers are mounted in either of two ways:

a. In servoed gimbals that angularly isolate them from rotations of the vehicle. The earliest inertial navigators used gimbals. In the 2000s, only super-precise star trackers and naval navigators do so.
b. Fastened directly to the vehicle (*strap-down*), whereupon the sensors are exposed to the angular rates and angular accelerations of the vehicle (Figure 12.4). In 2005, most inertial navigators are strap-down.

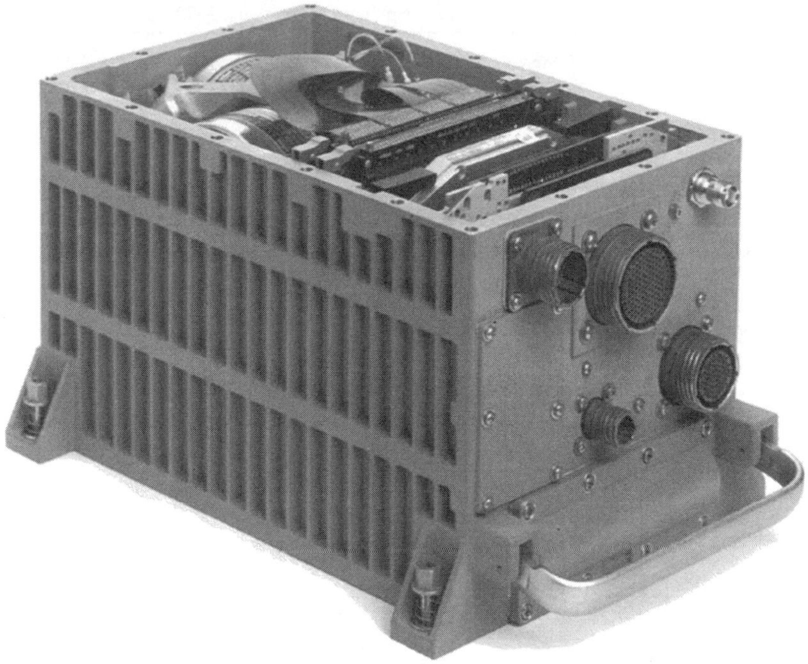

FIGURE 12.4 GPS-inertial navigator. The inertial instruments are mounted at the rear with two laser gyroscopes and electrical connectors visible. The input-output board is next from the rear; it includes MIL-STD-1553 and RS-422 buses. The computer board is next closest to the observer, and the power supply is in front. Between is the single-board shielded GPS receiver. Round connectors on the front are for signals and electric power. A battery is in the case behind the handle. Weight 10 kg, power consumption 40 watts. This navigation set is used in many military aircraft and, without the GPS receiver, in Airbus airliners. (Photo courtesy of Northrop-Grumman Corporation.)

Inertial-quality gyroscopes measure vehicle orientation within 0.1 degree for steering and pointing. Most accelerometers consist of a gram-sized proof-mass that is mounted on flexure pivots. The newest accelerometers, not yet of inertial grade, are etched into silicon chips. Older gyroscopes contained metal wheels rotating in ball bearings or gas bearings. The newest gyroscopes are evacuated cavities or optical fibers in which counter-rotating laser beams are compared in phase to measure the sensor's angular velocity relative to *inertial space* about an axis normal to the plane of the beams. Vibrating hemispheres and rotating, vibrating bars are the basis of some navigation-quality gyroscopes (drift rates less than 0.1 deg/hr).

Fault-tolerant configurations of cleverly oriented redundant gyroscopes and accelerometers (typically four to six of each) detect and correct sensor failures. Inertial navigators are used aboard naval ships, in long-range airliners, in business jets, in most military fixed-wing aircraft, in space boosters and entry vehicles, in manned spacecraft, in tanks, and on large mobile artillery pieces.

12.5 Radio Navigation

Scores of radio navigation aids have been invented, and many of them have been widely deployed, as summarized in Table 12.1. The most precise is the Global Positioning System (GPS), a network of 29 (in 2005) satellites and 16 ground stations for monitoring and control. A vehicle derives its three-dimensional position and velocity from one-way ranging signals at 1.575 GHz, received from four or more satellites (military users also receive 1.227 GHz) [U.S. Air Force, 2000]. GPS offers better than 100-meter ranging errors to civil users and 15-meter ranging errors to military users. Simple receivers were available

TABLE 12.1 Approximate Number of World-Wide Radio Navigation Aids in 2005

System	Frequency		Number of Stations	Number of Users			
	Hz	Band		Air	Marine	Space	Land
Omega	10–13 kHz	VLF	0	0	0	0	0
Loran-C/Chaika	100 kHz	LF	50	120,000	600,000	0	25,000
Decca	70–130 kHz	LF	0	0	0	0	0
Beacons*	200–1600 kHz	MF	4000	130,000	500,000	0	0
Instrument Landing System (ILS)*	108–112 MHz	VHF	2000	200,000	0	0	0
	329–335 MHz	UHF					
VOR*	108–118 MHz	VHF	2000	200,000	0	0	0
SARSAT/COSPAS	121.5 MHz	VHF	5 satellites	200,000	200,000	0	10^6
	243, 406 MHz	UHF					
Transit	150, 400 MHz	VHF	Network	0	0	0	0
PLRS	420–450 MHz	UHF	None	0	0	0	2,000
Russian Indentification Friend or Foe (IFF)	675 MHz	UHF	300	10,000	0	0	0
JTIDS	960–1213 MHz	L	Network	500	0	0	0
DME*	962–1213 MHz	L	1500	200,000	0	3	0
Tacan	962–1213 MHz	L	1000	15,000	0	3	0
Secondary Surveillance Radar (SSR)*and NATO Identification Friend or Foe (IFF)	1030, 1090 MHz	L	1000	250,000	0	0	0
GPS-GLONASS	1176, 1227, 1575 MHz	L	29 satellites	150,000	500,000	20	2×10^6
Satellite Control Network (SCN)	1760–1850 MHz	S	10	0	0	200	0
	2200–2300 MHz	S					
Spaceflight Tracking and Data Network (STDN)	2025–2150 MHz	S	3 satellites	0	0	50	0
	2200–2300 MHz		10 ground	0			
Radar Altimeter	4200 MHz	C	None	40,000	0	0	0
MLS*	5031–5091 MHz	C	25	50	0	0	0
Weather/map radar	10 GHz	X	None	40,000	0	0	0
Shuttle rendezvous radar	13.9 GHz	Ku	None	0	0	3	0
Airborne Doppler radar	13–16 GHz	Ku	None	20,000	0	0	0
SPN-41 carrier-landing monitor	15 GHz	Ku	25	1,600	0	0	0
SPN-42 carrier-landing radar	33 GHz	Ka	25	1,600	0	0	0

*Standardized by International Civil Aviation Organization or International Maritime Organization.

for less than $100 in 2005. GPS is used on highways, in low-rise cities, at sea, in aircraft, and in low-orbit spacecraft.

GPS provides continuous worldwide navigation for the first time in history. It is displacing precise dead reckoning on many vehicles and is reducing the cost of most navigation systems. Figure 12.5 is an artist's drawing of a GPS Block 2F spacecraft, scheduled to be launched before 2007. During the 1990s, Russia deployed a satellite navigation system, incompatible with GPS, called GLONASS, that they casually maintain. In 2005, the European Union was in the final stages of defining its own navigation satellite system, called Galileo, which will offer free and paid services [Hein, 2003]. The United States plans a major upgrade before 2015 to reduce vulnerability to jamming [Enge, 2004].

Differential GPS (DGPS) employs one or more ground stations at known locations that receive GPS signals and transmit measured errors on a radio link to nearby vehicles. DGPS improves accuracy (centimeters for fixed observers) and detects faults in GPS satellites immediately. Coast Guard authorities in several nations (including the United States, Canada, Norway, and Iceland, for example) operate marine DGPS stations along

FIGURE 12.5 Global Positioning Satellite, Block 2F (Courtesy of Rockwell.)

their coasts that transmit corrections on existing 300 kHz marine beacons to aid ships entering and leaving port. About 150 stations will cover the continental United States. Inland coverage will allow precise navigation on the Great Lakes, highways, and rail lines. Accuracy is about 10 meters, and failure detection occurs within 5 seconds.

In 2003, the United States created a nationwide aeronautical DGPS system consisting of about 25–50 stations and monitoring sites. This *Wide-Area Augmentation System* (WAAS) will eventually replace VORTAC on less-used airways. It transmits its corrections via geosynchronous communication satellites. In 2005, the United States was experimenting with a dense network of DGPS sites at airports called the *Local-Area Augmentation System* (LAAS). The intent is to replace several ILS and MLS landing aids with a small number of DGPS stations at each airport. The experiments show that in order to achieve Category II and III landing (*all-weather*), inertial aiding will be needed. WAAS and LAAS error detection is within 1 second.

In 2005, GPS had replaced Loran-C (see Table 12.1) as the most widely used marine radio aid. Loran's 100-kHz signals are usable within 1000 nmi of a chain, consisting of three or four stations. Chains cover the United States, parts of western Europe, Japan, Saudi Arabia, and a few other areas. Russia has a compatible system called *Chaika*. The vehicle-borne receiver measures the difference in time of arrival of pulses emitted by two ground stations, thus locating the vehicle on one branch of a hyperbola whose foci are at the stations. Two or more station pairs give a two-dimensional position fix whose typical accuracy is 0.25 nmi, limited by propagation uncertainties over the terrain between the transmitting station and the user. The measurement of a 100-microsecond time difference is made with a low-quality clock in the vehicle. Loran is also used by general aviation aircraft for en-route navigation and for nonprecision approaches to airports (in which the cloud bottoms are more than 400 feet above the runway). Loran might be retained as a diverse supplement to GPS or might be discontinued after 2010.

The most widely used aircraft radio aid is VORTAC, whose stations offer three services:

1. Analog bearing measurements at 108–118 MHz (called VOR). The vehicle compares the phases of a rotating cardioid pattern and an omnidirectional sinusoid emitted by the ground station.
2. Pulse distance measurements (called DME) at 1 GHz, by measuring the time delay for an aircraft to interrogate a VORTAC station and receive a reply,
3. TACAN bearing information, conveyed in the amplitude modulation of the DME replies from the VORTAC stations.

Throughout the Western world, civil aircraft use VOR/DME, whereas military aircraft use TACAN/DME for en-route navigation. In the 1990s, China and the successor states to the Soviet Union began to replace their direction-finding stations with ICAO-standard navigation aids (VOR, DME, and ILS) at their international airports and along the corridors that lead to them from the borders. Differential GPS sites will eventually replace most VORTACs; 50 DGPS sites could replace 1000 VORTACs, thus saving an immense sum of money for maintenance. VORTAC stations are likely to be retained on important routes through 2025.

Omega was a worldwide radio aid consisting of eight radio stations that emitted continuous sine waves at 10–13 KHz. Vehicles with precise clocks measured their range to a station by observing the absolute time of reception; other vehicles measured the range differences between two stations in the form of phase differences between the received sinusoids. Errors were about 2 nmi due to radio propagation irregularities. Omega was used by submarines, over-ocean general-aviation aircraft, and a few international air carriers. It was decommissioned in 1997.

Decca was a privately owned, short-range continuous-wave hyperbolic navigation aid at 100 kHz that was in widespread use by merchant ships from 1946 to 2000. At first, receivers were leased to ship operators and the revenues used to maintain the 42 chains, each of which consisted of a master station and three land-based slave transmitters. When the patents expired, low-cost receivers were marketed, and the revenue stream to operate the chains ceased. Some chains were abandoned and others taken over by governments for a few years. Errors were 100 meters within 200 nmi of the stations [Kayton first edition (1969) pp. 170–174]. Chains operated in all United Kingdom waters, the North Sea, Baltic, Persian Gulf, Japan, Nigeria, South Africa, and Australia, usually for the benefit of oil companies and their helicopters.

Landing guidance throughout the Western world and increasingly in China, India, and the former Soviet Union is with the Instrument Landing System (ILS) (Table 12.1). Transmitters adjacent to the runway create a horizontal guidance signal near 110 MHz and a vertical guidance signal near 330 MHz. Both signals are modulated such that the nulls intersect along a line in space that leads an aircraft from a distance of about 10 nmi to within 50 ft above the runway. ILS gives no information about where the aircraft is located along the beam except at two or three vertical *marker beacons*. Most ILS installations are certified to the International Civil Aviation Organization's *Category I*, where the pilot must abort the landing if the runway is not visible at an altitude of 200 feet while descending. About 100 ILSs are certified to *Category II*, which allows the aircraft to descend to 100 feet before aborting for lack of visibility. *Category III* allows an aircraft to land at still lower weather ceilings. About 50 ILSs are certified to Category III, mostly in Western Europe which has the worst flying weather in the developed world. Category II and III ILS detect their own failures and switch to a redundant channel within one second to protect aircraft that are flaring-out (within 50 feet above the runway) and can no longer execute a missed approach. Once above the runway, the aircraft's bottom-mounted radio altimeter measures altitude and either the electronics or the pilot guides the flare maneuver. Landing aids are described by Kayton and Fried [1997].

U.S. Navy aircraft use a microwave scanning system at 15.6 GHz to land on aircraft carriers; NASA's Space Shuttle uses the Navy system to land at its spaceports, but an inertially aided differential-GPS system may replace it. Another Microwave Landing System (MLS) at 5 GHz was supposed to replace the ILS in civil operations, especially for Categories II and III. However, experiments during the 1990s showed that differential GPS with a coarse inertial supplement could achieve an accuracy of one meter as a landing aid and could detect satellite errors within a second. Hence, it is likely that LAAS will replace or supplement ILS, which has been guaranteed to remain in service at least until the year 2010 (Federal Radionavigation Plan). NATO may use portable MLS or LAAS for flights into tactical airstrips.

All the spacefaring nations operate worldwide radio networks that track spacecraft, compute their state vectors, and predict future state vectors using complex models of gravity, atmospheric drag, and lunisolar perturbations. NASA operates three tracking and data relay satellites (TDRS) that track spacecraft in low Earth orbit with accuracies of 10 to 50 meters and 0.3 meter/sec. Specialized ground-based tracking stations monitor and reposition the world's many communication satellites [Berlin, 1988]. Other specialized stations track and communicate with deep-space probes. They achieve accuracies of 30 meters and a few centimeters per second, even at enormous interplanetary distances, due to long periods of observation and precise orbit equations [see Yuen, 1983].

12.6 Position-Reporting Systems

Position-location and position-reporting systems monitor the state vectors of many vehicles and usually display the data in a control room or dispatch center. Some vehicles derive their state vector from the ranging modulations; others merely report an independently derived position. Table 12.1 lists *secondary surveillance radars* that receive coded replies from aircraft so they can be identified by human air-traffic controllers and by collision-avoidance algorithms. The table also lists the United States's NASA and military spacecraft tracking networks (STDN and SCN). Tracking and reporting systems have long been in use at marine ports, for airplane traffic control, and for space vehicles. They are increasingly being installed in fire trucks, police cars, ambulances, and delivery-truck fleets that report to a control center. The aeronautical bureaucracy calls them *Automatic Dependent Surveillance* systems. The continuous broadcast of onboard-derived position (probably GPS-based) may be the basis of the worldwide air traffic control system of the early 21st century.

The location of a wire-line telephone can be easily identified. In 2005, there is interest in locating mobile phones that report emergencies and to sell position-related services. The coarsest way to locate a mobile phone is to identify the base station it is communicating with, usually within several kilometers. To achieve an accuracy of 100 meters, three approaches are used:

a. Include a GPS receiver in each cell phone, which reports its position when interrogated. This solution imposes costs on the phone user.
b. Send triangulation signals from base stations to cell phones and measure the time to reply. This solution requires that the mobile phone be in contact with several base stations. It requires complex software in the phone network to establish track files for each mobile phone. In urban areas, reflections off buildings worsen the ranging accuracy.
c. Let the mobile phone interrogate time differences between signals sent by three or more base stations, and measure time differences among the replies in the manner of Loran or Decca. This solution requires contact with several base stations and complex software in the cell phones.

Indoor mobile phones often cannot receive GPS signals. Hence, GPS repeaters or base stations inside the building locate the phone, perhaps not more accurately than the address of the building. Mobile-phone positioning can track children, disoriented elderly people, and criminals, for example. There are major privacy issues in maintaining tracks of individual mobile phones.

Several commercial communication satellites plan to offer digital-ranging services worldwide. The intermittent nature of commercial fixes would require that vehicles dead-reckon between fixes, perhaps using solid-state inertial instruments. Thus if taxpayers insist on collecting fees for navigation services, private comm-nav networks may replace the government-funded GPS and air-traffic communication network in the 21st century. Worldwide traffic control over oceans and undeveloped land areas would become possible.

Military communication-navigation systems measure the position of air, land, and naval vehicles on battlefields and report to headquarters; examples are the American Joint Tactical Information Distribution System (JTIDS) and the Position Location Reporting System (PLRS).

Approximately 40 SARSAT-COSPAS ground stations monitor signals from emergency location transmitters (on aircraft, ships, and land users), on 121.5, 243, and 406 MHz, the three international distress frequencies, relayed via low-orbit satellite-based transponders. Software at the listening stations calculates the position of the emergency location transmitters within 20 kilometers, based on the Doppler-shift history observed by the satellites, so that rescue vehicles can be dispatched. Some 406-MHz emergency location transmitters contain GPS sets; they transmit their position to geostationary satellites. SARSAT-COSPAS has saved more than 12,000 lives worldwide, from Arctic bush pilots to tropical fishermen, since 1982 [NASA/NOAA].

12.7 Celestial Navigation

Human navigators use sextants to measure the elevation angle of celestial bodies above the visible horizon. The peak elevation angle occurs at local noon or midnight:

$$\text{elev angle (degrees)} = 90 - \text{latitude} + \text{declination}$$

Thus at local noon or midnight, latitude can be calculated by simple arithmetic. Tables of declination (the angle of the sun or star above the Earth's equatorial plane) were part of the ancient navigator's proprietary lore. The declination of the Sun was first publicly tabulated in the 15th century in Spain. When time became measurable at sea, with a chronometer in the 19th century and by radio in the 20th century, off-meridian observations of the elevation of two or more celestial bodies were possible at any known time of night (cloud cover permitting). These fixes were hand-calculated using logarithms, then plotted on charts. In the 1930s, hand-held sextants were built that measured the elevation of celestial bodies from an aircraft using a bubble-level reference instead of the horizon. The accuracy of celestial fixes was 3 to 10 miles at sea and 5 to 30 miles in the air, limited by the uncertainty in the horizon and the inability to make precise angular measurements on a pitching, rolling vehicle. Kayton [1990] reviews the history of celestial navigation at sea and in the air.

The first automatic star trackers were built in the late 1950s. They measured the azimuth and elevation of stars relative to a gyroscopically stabilized platform. Approximate position measurements by dead reckoning allowed the telescope to point within a fraction of a degree of the desired star. Thus, a narrow field of view was possible, permitting the telescope and photodetector to track stars in the daytime. An onboard computer stored the right ascension and declination of 20 to 100 stars and computed the vehicle's position. Automatic star trackers are used in long-range military aircraft and on space shuttles, physically mounted on the stable element of a gimballed inertial navigator. Clever design of the optics and of stellar-inertial signal-processing filters achieves accuracies better than 500 feet [Kayton and Fried, 1997]. Future lower-cost systems may mount the star tracker directly to the vehicle.

Spacecraft use the line of sight to the sun and stars to measure orientation (for *attitude control*). Earth-pointing spacecraft usually carry horizon scanners that locate the center of the Earth's carbon-dioxide disc. All spacecraft navigate by radio tracking from Earth. When interplanetary spacecraft approach the target planet, the navigation computers (on Earth) transform from Sun-centered to planet-centered coordinates by observing star occultations and transmitting the images to Earth for human interpretation. During the Apollo translunar missions, crews experimentally measured the angle between celestial bodies and the Earth or Moon with a specially designed manual sextant coupled to a digital computer, which calculated the state vector. Other experiments have been made in which American and Soviet crews used manual sextants to observe the angle between celestial bodies and landmarks on Earth, from which state vectors were calculated. Autonomous land vehicles on other planets and certain military spacecraft may depend on celestial navigation.

12.8 Map-Matching Navigation

Mobile vehicles are increasingly being asked to navigate autonomously, whether on a planet, on a factory floor, underwater, or on a battlefield. The most complex systems observe their surroundings, usually by digitized video (often stereo), and create their own map of the navigated space. Tactile sensors detect collisions with solid objects. Guidance software then steers the vehicle toward its destination. NASA's 2004 Mars exploration vehicles allowed autonomous movement for a hundred meters before an Earth-bound crew reevaluated the visual images.

Delivery robots in buildings are furnished with a map and need only find their successive destinations while avoiding obstacles. They navigate by following stripes on the floor, by observing infrared beacons, or by observing the returns from on-board ultrasonic sonar or laser radar.

On aircraft, mapping radars and optical sensors present an image of the terrain to the crew. Since the 1960s, automatic map-matchers have been built that correlate the observed image to stored images, choosing the closest match to update the dead-reckoned state vector. In the 1980s and 1990s, aircraft and cruise missiles measured the vertical profile of the terrain below the vehicle and matched it to a stored profile. *Updating* with the matched profile, perhaps hourly, reduces the long-term drift of the inertial navigator. The profile of the terrain is measured by subtracting the readings of a baro-inertial altimeter (calibrated for altitude above sea level) and a radio altimeter (measuring terrain clearance). An onboard computer calculates the cross-correlation function between the measured profile and each of many stored profiles on possible parallel paths of the vehicle. The onboard inertial navigator usually contains a digital filter that corrects the drift of the azimuth gyroscope as a sequence of fixes is obtained. Hence the direction of flight through the stored map is known, saving the considerable computation time that would be needed to correlate for an unknown azimuth of the flight path. Marine map-matchers profile the seafloor with a sonar and compare the measured profile with stored bottom maps.

GPS is adequate for automotive navigation except in high-rise cities, in tunnels, and on roads with heavy foliage. To fill coverage gaps, map-matching software can take advantage of the fact that the vehicle remains on roads. On the highway, dead-reckoning or GPS errors can be rectified to the nearest road. In cities, turns can be correlated with the nearest street intersection of matching geometry. An accuracy of several meters is possible if all streets are included on the stored map (e.g., alleys, driveways, and parking garages).

12.9 Navigation Software

Navigation software is sometimes embedded in a central processor with other avionic-system software, sometimes confined to one or more navigation computers. The navigation software contains algorithms and data that process the measurements made by each sensor (e.g., GPS, inertial or air data). It contains calibration constants, initialization sequences, self-test algorithms, reasonability tests, and alternative algorithms for periods when sensors have failed or are not receiving information. In the simplest systems, the state vector is calculated independently from each sensor; more often, the navigation software contains multisensor algorithms that calculate the best estimate of position and velocity from several sensors. Prior to 1970, the best estimate was calculated from a least-squares algorithm with constant weighting functions or from a frequency-domain filter with constant coefficients. Now, a *Kalman filter* calculates the best estimate from mathematical models of the dynamics of each sensor.

Digital maps, often stored on compact disc, are carried on some aircraft and automobiles so position can be visually displayed to the crew. Military aircraft superimpose their navigated position on a stored map of terrain and cultural features to aid in the penetration of and escape from enemy territory. Algorithms for waypoint steering and for control of the vehicle's attitude are contained in the software of the *flight management* and *flight control* subsystems.

Specially equipped aircraft (sometimes ships) are often used for the routine calibration of radio navigation aids, speed and velocity sensors, heading sensors, and new algorithms.

12.10 Design Trade-Offs

The navigation-system designer conducts trade-offs for each vehicle to determine which navigation systems to use. Trade-offs consider the following attributes:

a. Cost, including the construction and maintenance of transmitter stations and the purchase of onboard electronics and software. Users are concerned only with the costs of onboard hardware and software.

FIGURE 12.6 Navigation displays in the U.S. Air Force C-5 transport showing flat-panel displays in front of each pilot; vertical situation display outboard and horizontal situation display inboard. Waypoints are entered on the horizontally mounted control-display unit just visible aft of the throttles. In the center of the instrument panel are status and engine displays and backup analog instruments. (Photo courtesy of Honeywell, Copyright 2004.)

b. Accuracy of position and velocity, which is specified as a circular error probable (CEP, in meters or nautical miles). The maximum allowable CEP is often based on the calculated risk of collision on a typical mission.
c. Autonomy, the extent to which the vehicle determines its own position and velocity without external aids. Autonomy is important to certain military vehicles and to civil vehicles operating in areas of inadequate radio-navigation coverage.
d. Time delay in calculating position and velocity, caused by computational and sensor delays. Ships and automobiles are less sensitive to delay than are aircraft.
e. Geographic coverage. Radio systems operating below 100 kHz can be received beyond line of sight on Earth; those operating above 100 MHz are confined to line of sight. On other planets, new navigation aids—perhaps navigation satellites or ground stations—will be installed,
f. Automation. The vehicle's operator (onboard crew or ground controller) receives a direct reading of position, velocity, and equipment status, usually without human intervention. The navigator's crew station disappeared in aircraft in the 1970s. Human navigators are becoming scarce, even on ships, because electronic equipment automatically selects stations, calculates waypoint-steering, and accommodates failures (Figure 12.6).

12.11 Animal Navigation

It is unlikely that lessons about navigation can be learned from migrating animals in the early 21st century. Fish, insects, and birds sometimes travel thousands of miles, seeming to navigate to an unknown accuracy while seeking food and mates. Deaths due to navigation errors seem to be high. For birds, terminal navigation is visual; for fish, terminal navigation is by odor. The midcourse sensors are presently unknown; there are data that support and that refute celestial cues, magnetic cues, odors, polarized light, etc. Young animals seem to follow experienced ones and memorize sensor data en route. Flocks pool sensor data to improve the accuracy of the individual animals. Five decades of tests have yet to determine the sensors and algorithms used in long-distance migrations.

Defining Terms

Circular error probable (CEP): Radius of a circle, centered at the destination, that contains 50% of the navigation measurements from a large sample.

Ecliptic: Plane of Earth's orbit around the Sun.

Inertial space: Any coordinate frame whose origin is on a freely falling (orbiting) body and whose axes are nonrotating relative to the fixed stars. It is definable within 10^{-7} deg/hr.

Lanes: Hyperbolic bands on the Earth's surface in which continuous-wave radio signals repeat in phase.

Nautical mile (nmi): 1852 meters, exactly. Approximately one minute of arc on the Earth's surface.

State vector: Six-component vector, three of whose elements are the position and three of whose elements are velocity.

Transponder: A radio that receives a signal and re-transmits it at a different frequency. Also, a radio that replies with a coded message when interrogated.

Update: The intermittent resetting of the dead-reckoned state vector based on absolute navigation measurements (see Section 12.3 and 12.8).

Bibliography

R.H. Battin, *An Introduction to the Mathematics and Methods of Astrodynamics*, Washington: AIAA Press, 1987, 796 pp.

P. Berlin, *The Geostationary Applications Satellite*, Cambridge: Cambridge University Press, 1988, 214 pp.

N. Bowditch, The American Practical Navigator, U.S. Government Printing Office, 1995, 873 pp. Reissued approximately every five years.

P. Enge, "Global Positioning System," *Scientific American*, 91–97, May 2004.

G.W. Hein et al., "Galileo frequency and signal design," *GPS World*, 30–45, 2003, January.

M. Kayton, *Navigation: Land, Sea, Air, and Space*, New York: IEEE Press, 1990. 461 pp.

M. Kayton and W.R. Fried, *Avionics Navigation Systems*, 2nd ed., New York: John Wiley, 1997, 650 pp.

R.A. Minzner, "The U.S. standard atmosphere 1976," NOAA report 76–1562, NASA SP-390. 1976, or latest edition, 227 pp.

P. Misra and P. Enge, *Global Positioning System*; Lincoln, MA: Ganga-Jamuna Press, 2001. 390 pp.

NASA, *Space Network Users Guide*, Greenbelt, Md.: Goddard Space Flight Center, or latest edition, 1988, 500 pp.

NASA/NOAA website, COSPAS-SARSAT Search and Rescue System, www.sarsat.noaa.gov

J. Quinn, "1995 Revision of Joint US/UK geomagnetic field models", *J. Geomagn. Geo-Electricity*, fall 1996.

U.S. Air Force, *NAVSTAR-GPS Space Segment/Navigation User Interfaces*, IRN-200c-004. Annapolis, Md: ARINC Research, 2000, 160 pp, or latest edition.

U.S. Defense Mapping Agency, World Geodetic System 1984 (WGS-84)—Its Definition and Relationship with Local Geodetic Systems, Washington, DC, 1991.

U.S. Government, *Federal Radionavigation Plan*, Departments of Defense and Transportation, issued biennially. 200 pp.

U.S. Government, *Federal Radionavigation Systems*, Departments of Defense and Transportation, issued biennially. 200 pp.

J. Yuen, *Deep Space Telecommunication System Engineering*, New York: Plenum Press, 1983, 603 pp.

Further Information

IEEE Transactions on Aerospace and Electronic Systems, bimonthly through 1991, now quarterly.

Proceedings of the IEEE Position Location and Navigation Symposium (PLANS), biennially.

NAVIGATION, journal of the U.S. Institute of Navigation, quarterly.

Journal of Navigation, royal Institute of Navigation (UK), quarterly.

AIAA Journal of Guidance, Control, and Dynamics, bimonthly.

Commercial aeronautical standards produced by International Civil Aviation Organization (ICAO, Montreal), Aeronautical Radio, Inc. (ARINC, Annapolis, Md.), Radio Technical Committee for Aeronautics (RTCA, Inc., Washington), and European Commission for Aviation Electronics (EUROCAE, Paris).

Websites: www.faa.gov, www.navcen.uscg.gov, www.gps.losangeles.af.mil, www.inmarsat.org, www.arinc.com, www.loran.org, www.imo.org.

13

Environmental Effects

Karen Blades
Lawrence Livermore National Laboratory

Braden Allenby
AT&T Environment, Health & Safety

Michele M. Blazek
AT&T Pleasanton

13.1 Introduction

Electronics technology is pervasive in the global economy and has been integral not only to communication and computing purposes but also to industrial controls and monitoring, e-commerce and education for individuals, schools, and businesses worldwide. As the industry has grown globally, it has faced challenges from innovation and increased competition as well as the global concerns for the environment and the need to improve the resource management. While perceived as a "clean" industry, the technological advances made by the industry create a significant demand on the Earth's resources. Over the past decade, the industry has made significant efforts to reduce environmental impacts through energy efficiency programs and pollution prevention initiatives, such as the phase out of the use of ozone-depleting substances from manufacturing and air-conditioning uses and the development of alternatives to solder technologies and materials.

Even with these successes, the industry still faces environmental challenges. These impacts can be scaled on global, regional, and local levels and can occur from the life cycle of the technology — from material extraction to manufacturing, use, and disposal. For example, the amount of water required in the production of semiconductors, the engines that motor most of today's electronic gadgets, is enormous — about 2000 gallons to process a single silicon wafer. The manufacture and use of electronics also require substantial energy resources. In the U.S., for example, the life-cycle power consumption has been estimated to be between 2 to 3% of the grid power. In addition, fabricating silicon chips requires the use of highly toxic materials, albeit in relatively low volumes, and other hazardous substances. Similarly, printed wiring boards (PWBs) present in most electronic products and produced in high volume use large amounts of solder and solvents or gases, which may contribute to multiple environmental impacts, including greenhouse effects and health impacts from the formation of ground-level ozone and exposure to hazardous chemical releases. The reuse and disposition of electronics have also been areas of concern due to material content of the equipment.

The compelling challenge for the industry is to deliver innovative products and services that people want, yet find creative solutions to minimize the environmental impact, enhance competitiveness, and address regulatory issues without impacting quality, productivity, or cost; in other words, to become an industry that is more "ecoefficient." Ecoefficiency is reached by the delivery of competitively priced goods and services that satisfy human needs and support a high quality of life, while progressively reducing ecological impacts and resource intensity to a level at least in line with the Earth's estimated carrying capacity.

Like sustainable development, a concept popularized by the Brundtland Report, *Our Common Future*, the notion of ecoefficiency requires a fundamental shift in the way environment is considered in industrial activity. Sustainable development — "development that meets the needs of the present without compromising the ability of future generations to meet their own needs" [World Commission on Environment and Development, 1987] — contemplates the integration of environmental, economic, and technological considerations to achieve continued human and economic development within the biological and physical constraints of the planet. Both ecoefficiency and sustainable development provide a useful direction, yet they prove difficult to operationalize and cannot guide technology development. Thus, the theoretical foundations for integrating technology and environment throughout the global economy are being provided by a multidisciplinary field known as "industrial ecology."

The ideas of industrial ecology, rooted in the engineering community, have helped to establish a framework within which the industry and its customers can move toward realizing sustainable development. In this context, the use of the electronics technology to replace the use of other, more polluting technologies has made this sector surprising enablers of sustainability. These technologies allow the provision of increasing quality of life using less material and energy, respectively, "dematerialization" and "decarbonization." This chapter will provide an introduction into industrial ecology and its implications for the electronics industry. Current activities, initiatives, and opportunities will also be explored, illustrating that the concomitant achievement of greater economic and environmental efficiency is indeed feasible in many cases.

13.2 Industrial Ecology

Industrial ecology is an emerging field that views manufacturing and other industrial activities, including forestry, agriculture, mining, and other extractive sectors, as an integral component of global natural systems. In doing so, it takes a systems view of design and manufacturing activities so as to reduce or, more desirably, eliminate the environmental impacts of materials, manufacturing processes, technologies, and products across their life cycles, including use and disposal. It incorporates, among other things, research involving energy supply and use, new materials, new technologies and technological systems, basic sciences, economics, law, management, and social sciences.

The study of industrial ecology will, in the long run, provide the means by which the human species can deliberately and rationally approach a desirable long-term global carrying capacity. Oversimplifying, it can be thought of as "the science of sustainability." The approach is "deliberate" and "rational," to differentiate it from other unplanned paths that might result, for example, in global pandemics, or economic and cultural collapse. The endpoint is "desirable," to differentiate it from other conceivable states such as a Malthusian subsistence world, which could involve much lower population levels, or oscillating population levels that depend on death rates to maintain a balance between resources and population levels. Figure 13.1 illustrates how industrial ecology provides a framework for operationalizing the vision of sustainable development.

As the term implies, industrial ecology is concerned with the evolution of technology and economic systems such that human economic activity mimics a mature biological system from the standpoint of being self-contained in its material and resource use. In such a system, little if any virgin material input is required, and little if any waste that must be disposed of outside of the economic system is generated. Energetically, the system can be open, just as biological systems are, although it is likely that overall energy consumption and intensity will be limited.

Although it is still a nascent field, a few fundamental principles are already apparent. Most importantly, the evolution of environmentally appropriate technology is seen as critical to reaching and maintaining a sustainable state. Unlike earlier approaches to environmental issues, which tended to regard technology as neutral at best, industrial ecology focuses on development of economically and environmentally efficient technology as key to any desirable, sustainable global state. It also addresses the use of the technology to

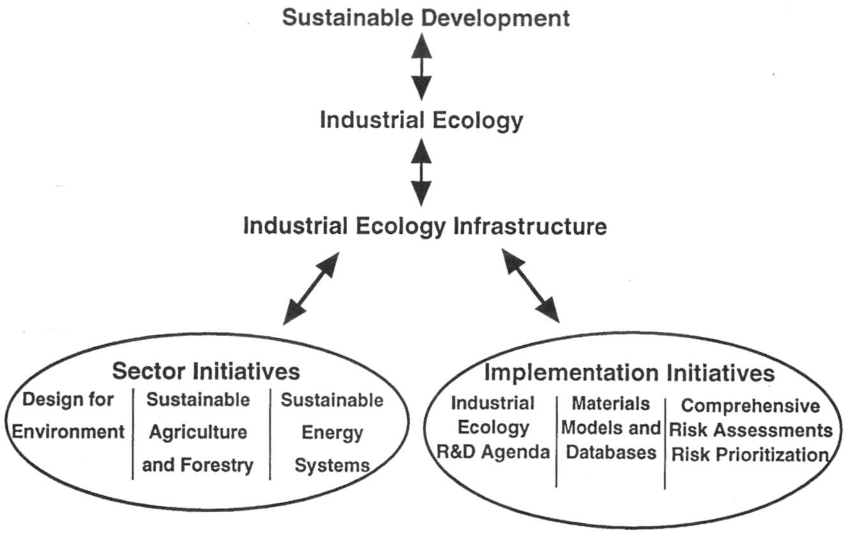

FIGURE 13.1 Industrial ecology framework.

enhance communication and media and extend economic and education institutions to areas that may have lacked traditional infrastructure to do so in the past.

Also, environmental considerations must be integrated into all aspects of economic behavior, especially product and process design, and the design of economic and social systems within which those products are used and disposed. Environmental concerns must be internalized into technological systems and economic factors. While it has been most common to deconstruct environmental issues to address one concern at a time, doing so may result in unintended trade-offs. It is not sufficient to design an energy efficient computer, for example; it is also necessary to look at the broader, systemwide consequences to ensure that the product, its components, or its constituent materials can be refurbished for reuse or recycled after the customer is through with it — all of this in a highly competitive and rapidly evolving market. This consideration implies a comprehensive and systems-based approach that requires sophisticated understanding of the relationship of the technology to the environment.

Industrial ecology requires an approach that is truly multidisciplinary. It is important to emphasize that industrial ecology is an objective field of study based on existing scientific and technological disciplines, not a form of industrial policy. It is profoundly a systems-oriented and comprehensive approach that can be difficult for most institutions to implement — the government, riddled with fiefdoms; academia, with rigid departmental lines; and private firms, with job slots defined by occupation. Nonetheless, it is all too frequent that industrial ecology is seen as an economic program by economists, a legal program by lawyers, a technical program by engineers, and a scientific program by scientists. It is in part each of these; more importantly, it is all of these.

Industrial ecology has an important implication, however, of special interest to electronics and telecommunications engineers, and thus is deserving of emphasis. The achievement of sustainability will, in part, require the substitution of intellectual and information capital for traditional physical capital, energy, and material inputs. Environmentally appropriate electronics, information management, telecommunications technologies and services, and the manufacturing base that supports them are therefore enabling technologies to achieve sustainable development. Recent examples include the dematerialization of the music media as seen by the ongoing replacement of CDs and other media with downloadable music files, and the decarbonization/dematerialization as indicated by the use of telecommuting and teleconferencing to reduce car and air miles. The possibility of innovative problem solving using technology to reduce impact and increase the quality of life offers unique opportunities for professional satisfaction. At the same time, it also places a unique responsibility on the community of electrical and electronics engineers. We cannot be simply satisfied with developing the newest gadgets, we should consider the consequences of the technology as well as its application in the economy. Even as the theory of industrial ecology is still emerging, the practical application of its foundations can be seen in everyday activities and everyday places, including schools and libraries, electronic billing and automatic banking, and multifunctional communications technologies.

13.3 Design for Environment

Design for environment (DFE) is the means by which the precepts of industrial ecology, as currently understood, have been widely implemented in the real world today. DFE requires that environmental objectives and constraints be driven into process and product design, and materials and technology choices.

The focus is on the design stage because, for many articles, that is where most, if not all, of their life-cycle environmental impacts are explicitly or implicitly established. Traditionally, electronics design has been based on a correct-by-verification approach, in which the environmental ramifications of a product (from manufacturing through disposition) are not considered until the product design is completed. DFE, by contrast, takes place early in a product's design phase as part of the concurrent engineering process to ensure that the environmental consequences of a product's life cycle are understood before manufacturing decisions are committed.

It is estimated that some 80 to 90% of the environmental impacts generated by product manufacture, use, and disposal are "locked-in" by the initial design. Materials choices, for example, ripple backwards towards environmental impacts associated with the extractive, smelting, and chemical industries. The design of a product and component selection control many environmental impacts associated with manufacturing, enabling, for example, substitution of no-clean or aqueous cleaning of printed wiring boards for previously used solvent that were associated with a number of effects including, but certainly not limited to, ozone layer depletion, smog, and elevated concentrations of hazardous substances in aquifers and airsheds. The design of products controls many aspects of environmental impacts during use — energy efficient design is one example. Product design also controls the ease with which a product may be refurbished, or disassembled for parts or materials reclamation, after consumer use. DFE tools and methodologies offer a means to address such concerns at the design stage. Some of the most successful strategies have been to design for modularity, upgradeability, and interchangeability, increased use of standardized components, and eliminating the need for new equipment through "tech refreshes" or enhanced services.

Obviously, DFE is not a panacea. It cannot, for example, compensate for failures of the current price structure to account for external factors, such as the real (i.e., social) cost of energy. It cannot compensate for deficiencies in sectors outside electronics, such as a poorly coordinated, polluting, or even nonexistent water and wastewater treatment, solid waste disposal, and material recycling infrastructure in some areas of the world. Moreover, it is important to realize that DFE recognizes environmental considerations as on par with other objectives and constraints, such as economic, technological, and market structure, not as superseding or dominating them. Nonetheless, if properly implemented, DFE programs represent a quantum leap forward in the way private firms integrate environmental concerns into their operations and technology.

It is useful to think of DFE within the firm as encompassing two different groups of activities as shown in Figure 13.2. In all cases, DFE activities require inclusion of life-cycle considerations in the analytical process. The first, which might be styled "generic DFE," involves the implementation of broad programs that make the company's operations more environmentally preferable across the board. This might include, for example, development and implementation of total cost of ownership and environmentally preferred procurement policies and practices that place emphasis on environmental supply-line management. The "standard components" lists maintained by many companies can be reviewed to ensure that they direct the use of environmentally appropriate components and products wherever possible. The use of contractual language and affidavits were important in the elimination of the use of CFCs from manufacturing supply lines.

Contract provisions can be reviewed to ensure that suppliers are being directed to use environmentally preferable technologies and materials where possible and to disclose the environmental management practices

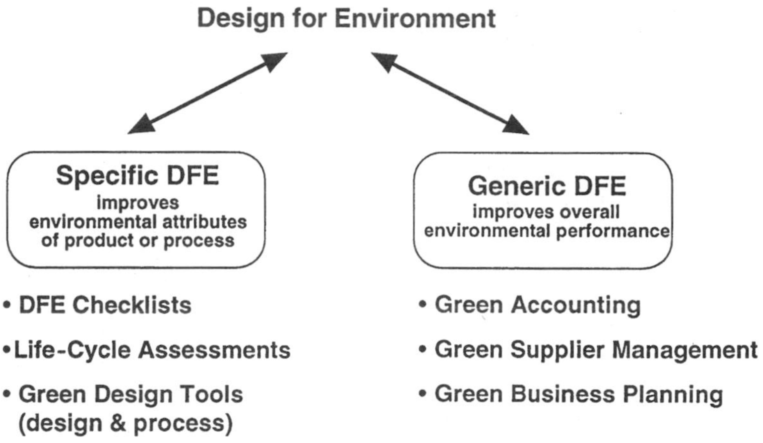

FIGURE 13.2 Examples of DFE activities within the firm. (Note: add the total cost of ownership; and environmentally preferred products and services procurement programs in generic DFE.)

and performance of their products and services. For example, are virgin materials being required where they are unnecessary? Do contracts, standards, and specifications clearly call for the use of recycled materials where they meet relevant performance requirements? Likewise, customer and internal standards and specifications can be reviewed with the same goal in mind. Does the supplier have programs in place to properly manage hazardous wastes? Can it verify that banned chemicals have not been used further up the supply chain?

The second group of DFE activities can be thought of as "specific DFE." Here, DFE is considered as a module of existing product realization processes, specifically the "design for X," or DFX, systems used by many electronics manufacturers. The method involves creation of software tools and checklists, similar to those used in other current design programs, such as Design for Manufacturability, Design for Testability, or Design for Safety modules. The tools, combined with standard procurement practices that can ensure relevant environmental considerations, are also included in the design process from the beginning. The challenge is to create modules that, in keeping with industrial ecology theory, are broad, comprehensive, and systems-based, yet can be defined well enough to be practically integrated into current design activities.

The successful application of DFE to the design of electronic systems requires the coordination of several design- and data-based activities, such as, environmental impact metrics, data and data management, design optimization including cost assessments, and others. Failure to address any of these aspects can limit the effectiveness and usefulness of DFE efforts. Data and methodological deficiencies abound, and the challenge is great, yet experience at world-class companies, such as AT&T, Digital, IBM, Motorola, Siemens Nixdorf, Volvo, and Xerox, indicate that it can be done. The examples of DFE implementation in the electronics industry are numerous and certainly differentiate the industry sector as innovators.

AT&T, for example, has had DFE programs since the early 1990s and has baselined the environmental attributes of a telephone at different life-cycle stages to determine where meaningful environmental improvements in design could be achieved [Seifert, 1995]; and with its supplier, Ericsson, it has extended the product life-cycle assessment (LCA) to a systems life-cycle-impact assessment of the telecommunication system for Stockholm and Sacramento, which included in its scope the broad operations of the network as well as business and customer use [Blazek et al., 1999]. From these studies, AT&T has implemented systemwide energy efficiency, life-cycle management and take-back of electronics, and environmentally preferable product procurement programs. In Sweden, the government and Volvo were among the first to develop systems-based environmental performance metrics through the Environment Priority Strategies for Environmental Design, or EPS, system that uses Environmental Load Units, or ELUs, to inform materials choices during the design process. In Germany, Siemens Nixdorf has developed an "Ecobalance" system to help it make design choices that reflect both environmental and economic requirements. Xerox is a world leader in designing their products for refurbishment using a product-life-extension approach. Dell Computers's innovation in this field was the idea of "tech-refresh" and long-term servicing contracts for the purpose of modular design. Hewlett Packard developed equipment take-back and recycling programs for its commercial customers and consumer products.

More broadly, the electronics industry has been actively pursuing the technology transfer and standardization of practices for DFE and LCA. The American Electronics Association (AEA) Design for Environment Task Force has created a series of white papers discussing various aspects of design for environment and its implementation. The Microelectronics and Computer Technology Corporation (MCC) has published a comprehensive study [Lipp et al., 1993] of the environmental impacts of a computer workstation, which is valuable not only for its technical findings, but for the substantial data and methodological gaps the study process identified. The Society of Environmental Toxicology and Chemistry (SETAC) and others, especially in Europe, are working on a number of comprehensive LCA methodologies designed to identify and prioritize environmental impacts of substances throughout their life cycle. In its Environmental Management System suite of standards (ISO 14000 series), the International Organization for Standardization (ISO) developed a suite of environmental LCA standards (ISO 14040, ISO 14041, ISO 14042, ISO 14043) to harmonize practice. This work has been further enhanced by the United Nations Environmental Program in conjunction with SETAC, which is standardizing life-cycle-impact indicators. The IEEE Environment, Health, and Safety Committee—was formed in July 1992 to support the integration of environmental, health, and safety considerations into electronics products and processes from design and manufacturing, to use, to recycling, refurbishing, or disposal—has held a series of annual symposia on electronics and the environment. The proceedings from these symposia are valuable resources to the practitioners of DFE. From these efforts as well as the many contributions of academia, the International Society of Industrial Ecology was formed to provide a platform to further develop the theory and practical application of industrial ecology.

13.4 Regulatory and Market Implications of Environmental Concerns for the Electronics Industry

The worldwide regulations and standards have been in development concurrent with the many voluntary industry voluntary initiatives. Global concerns and regulations associated with environmental issues are increasingly affecting the manufacturing and design of electronic products, their technology development, and marketing strategies. Among the first programs to illustrate this point was the German Blue Angel Ecolabeling scheme for personal computers (the Blue Angel is a quasigovernmental, multiattribute ecolabeling program). The Blue Angel requirements were numerous and span the complete life cycle of the computers. Examples of some the requirements included: modular design of the entire system, customer-replaceable subassemblies and modules, use of nonhalogenated flame retardants, and take-back by manufacturers at the end of the product life. Market requirements such as these, focused on products and integrating environmental and technology considerations, cannot possibly be met by continuing to treat environmental impact as an unavoidable result of industrial activity, i.e., as overhead. These requirements make environmental concerns truly strategic for the firm. From this and other country programs, the EU developed labeling programs for computers and other electronic equipment.

Perhaps the most familiar examples of "a new generation of environmental management" requirements that continue to have enormous effects on electronics design are "product take-back" and hazard lists or banned lists. The EU's law on Waste Electrical and Electronic Equipment (WEEE) mandates take-back as well as DFE initiatives. It generally requires that the firm take its products back once the consumer is through with them, recycle or refurbish the product, and assume responsibility for any remaining waste generated by the product. Japan and other countries are among others considering such take-back requirements. The EU law on the Restriction of Hazardous Substances (RoHS) codifies the use of DFE to phase out and eventually eliminate the use of listed hazardous substances from the manufacture and supply of a comprehensive list of electronics products.

The ISO standards for environmental management systems and labeling also present significant challenges to the industry. Like its quality predecessor, ISO 9000, companies have required that their suppliers conform to the environmental management system specification (ISO 14001). Though technically voluntary, in practice these standards in fact become requirements for firms wishing to engage in global commerce. The labeling standards have produced requirements for harmonization of programs for single and multiattribute claims as well as those product labels based on LCA results, also known as Type III labels. Several Japanese companies such as Sony have produced type III labels for many of its products. In addition, product declarations including material manifests have been required in some contracts. Motorola and Lucent have actively pursued material assays of their product lines.

13.5 Emerging Technology

New tools and technologies are emerging that will influence the environmental performance of electronic products and help the industry respond to the regulatory "push" and the market "pull" for environmentally responsible products. In the electronics industry, technology developments are important not only for the end-products, but for components, recycling, and materials technology as well. Below is a brief summary of technology developments and their associated environmental impacts as well as tools to address the many environmental concerns facing the industry.

The electronics industry has taken active steps toward environmental stewardship, evidenced by the formulation of the IEEE Environment, Safety, and Health Committee, the 1996 Electronics Industry Environmental Roadmap published by MCC, and chapters focused on environment in The National Technology Roadmap for Semiconductors. Moves such as this, taken together with the technical sophistication of control systems used in manufacturing processes, have allowed the electronics industry to maintain low emission levels relative to some other industries. Despite the industry's environmental actions, the projected growth in electronics over the next 10 to 20 years is dramatic, and continued technological innovation will be required to maintain historically low environmental impacts. Moreover, the rapid pace of technological change generates concomitantly high rates of product obsolescence and disposal, a factor that has led the European Union to implement the WEEE to focus on electronics products for environmental management.

Environmental considerations are not, of course, the only forces driving the technological evolution in the electronics industry. Major driving forces, as always, also include price, cost, performance, and market/regulatory requirements. Included in these market trends is the increased specialization of manufacturing, which has resulted in the industry trend to outsource several manufacturing steps to suppliers, sometimes in other areas around the world. This trend may have the consequence of reallocating resources, jobs, and environmental effects around the world.

However, to the extent that the trend is toward smaller devices, fewer processing steps, increased automation, and higher performance per device, such evolution will likely have a positive environmental impact at the unit production level, i.e., less materials, less chemicals, less waste related to each unit produced. Technological advances that have environmental implications at the upstream processing stage may well have significant benefits in the later stages of systems development and production. For example, material substitution in early production stages may decrease waste implications throughout the entire process. Since both semiconductors and printed wiring boards are produced in high volume and are present in virtually all electronic products ranging from electronic appliances to computers, automotive, aerospace, and military applications, we will briefly examine the impact of these two areas of the electronics industry.

Integrated Circuits

The complex process of manufacturing semiconductor integrated circuits (IC) often consists of over a hundred steps during which many copies of an individual IC are formed on a single wafer. Each of the major process steps used in IC manufacturing involves some combination of energy use, material consumption, and material waste. Water usage is high due to the many cleaning and rinsing process steps. Without process innovation, this trend will continue as wafer sizes increase, driving up the cost of water and wastewater fees, and increasing mandated water conservation. Intel is among the companies that have developed programs to reduce water use and increase its recycling.

Environmental issues that also require attention include the constituent materials for encapsulants, the metals used for connection and attachment, the energy consumed in high-temperature processes, and the chemicals and solvents used in the packaging process. Here, emerging packaging technologies will have the effect of reducing the quantity of materials used in the packaging process by shrinking IC package sizes. Increasing predominance of plastic packaging will reduce energy consumption associated with hermetic ceramic packaging.

Printed Wiring Boards

Printed wiring boards (PWB) represent the dominant interconnect technology on which chips will be attached and represent another key opportunity for making significant environmental advances. PWB manufacturing is a complicated process and uses large amounts of materials and energy (e.g., 1 MW of heat and 220 kW of energy are consumed during fabrication of prepeg for PWBs). On average, the waste streams constitute 92% — and the final product just 8% — of the total weight of the materials used in PWB production process. Approximately 80% of the waste produced is hazardous, and most of the waste is aqueous, including a range of hazardous chemicals. Although printed wiring boards had not been traditionally recycled, newer techniques have been developed to reduce the cost to recover and reuse certain components of the PWBs. After reclamation, the boards are incinerated and the residual ash buried in hazardous-waste landfills due to the lead content (from lead solder). To address possible long-term toxicity, the industry has been actively pursuing alternatives to lead solder and have developed a number of alloys that might exhibit less toxicity, albeit with tradeoffs of resource use.

13.6 Tools and Strategies for Environmental Design

The key to reducing the environmental impact of electronic products will be the application of DFE tools and methodologies. Development of Internet-based environmental impact metrics, materials selection data, cost, and product data management are examples of available or clearly foreseeable tools to assist firms in adopting DFE practices.

These tools will need to be based on LCA, the objective process used to evaluate the environmental impacts associated with a product and identify opportunities for improvement. LCA seeks to minimize the environmental impact of the manufacture, use, and eventual disposal of products without compromising essential product functions. Figure 13.3 shows the life stages that would be considered for electronic products (i.e., the life cycle considered has been bounded by product design activities). The ability of the electronics industry to operate in a more environmentally and economically efficient mode, use less chemicals and materials, and reduce energy consumption will require support tools that can be used to evaluate both product and process designs. To date, many firms are making immediate gains by incorporating basic tools like DFE checklists, design standards, and internal databases on chemicals and materials, while other firms are developing sophisticated software tools that give products environmental scores based on the product's compliance with a set of predetermined environmental attributes. These software tools rely heavily on environmental metrics (typically internal to the firm) to assess the environmental impact and then assign a score to the associated impact.

Other types of tools that will be necessary to implement DFE will include tools to characterize environmental risk, define and build flexible processes to reduce waste, and support dematerialization of processes and products. The following sections provide a brief review of design tools or strategies that can be employed.

Design Tools

Environmental design tools vary widely in the evaluation procedures offered in terms of the type of data used, method of analysis, and the results provided to the electronics designer. The tool strategies range in scope from

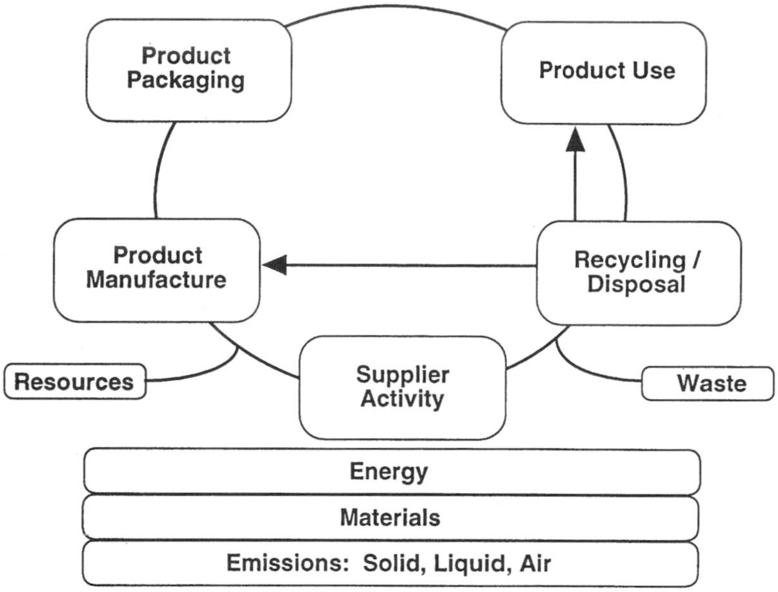

FIGURE 13.3 Design for environment: systems-based, life-cycle approach.

assessment of the entire product life cycle to the evaluation of a single aspect of its fabrication, use, or disposal. Today's DFE tools can be generally characterized as life-cycle analysis, recyclability analysis, manufacturing analysis, or process flow analysis tools.

The effectiveness of these design tools is based on both the tool's functionality as well as its corresponding support data. One of the biggest challenges designers face with regard to DFE is a lack of reliable data on materials, parts, and components needed to adequately convey the impact and trade-offs of their design decisions. To account for these data deficiencies, a number of environmental design tools attempt to use innovative, analytical methods to estimate environmental impacts; while necessary, this indicates they must be used with care and an understanding of their assumptions.

Although DFE provides a systems-based, life-cycle approach, its true value to the system designer is lost unless the impact of DFE decisions on other relevant economic and performance measures (i.e., cost, electrical performance, reliability, etc.) can be quickly and accurately assessed. Trade-off analysis tools that have DFE embedded can perform process-flow-based environmental analysis (energy/mass balance, waste stream analysis, etc.) concurrently with nonenvironmental cost and performance analysis so that system designers can accurately evaluate the impact of critical design decisions early.

Design Strategies

Design strategies such as lead minimization through component selection, and the reduction of waste resulting from rapid technological evolution through modular design, help to minimize the environmental impact of electronic products. Over the past decade, no-lead solder alloys have been slowly replacing the use of lead solder. These new solders have been combined with designs to minimize the lead content of electronic designs such as surface-mount technology, which requires less solder than through-hole technology. New interconnection technologies, such as microball grid array and direct chip attachment, also require less solder. The environmental benefits increase with the use of these advanced interconnection technologies.

The rapid advancement of the electronics industry has created a time when many products become obsolete in less than five years' time. Electronic products must be built to last, but only until it is time to take them apart for rebuilding or for reuse of material. This means employing modular design strategies to facilitate disassembly for recycle or upgrade of the product rather than replacement. Designers must extend their views to consider the full utilization of materials and the environmental impact of the material life cycle as well as the product life cycle.

Conclusion

The diverse product variety of the electronics industry offers numerous opportunities to curtail the environmental impact of the industry. These opportunities are multidimensional. Services made possible through telecommunications technology enable people to work from home, reducing emissions that would be generated by traveling to work. Smaller, faster computers and the Internet require less energy and material usage, thereby reducing the energy demand during processing and waste generated during fabrication. All these represent examples of how the electronics industry provides enablers of sustainability.

Global concerns and regulations associated with environmental issues are increasingly affecting the manufacturing and design of the electronics industry. Environmental management standards, electronic design, take-back programs such as the WEEE directive, and product-labeling requirements represent a sample of the initiatives driving the industry's move to more environmentally efficient practices. While the industry has initiated some activities to address environmental concerns, the future competitiveness of the industry will depend on improvements in environmental technology in manufacturing, accurate assessment of the environmental impact of products and processes, and design products that employ design for environment, reuse, and recyclability. Industrial ecology offers a framework for analyzing the environmental effects of the electronics industry that is complicated by the rapid pace of change.

Disclaimer

This document was prepared as an account of work sponsored by AT&T and by an agency of the United States government. AT&T, the United States government, or the University of California or any of their employees does not make any warranty, expressed or implied, or does not assume any legal liability or responsibility for the accuracy, completeness, or usefulness of any information, apparatus, product, or process disclosed, or does not represent that its use would not infringe privately owned rights. Reference herein to any specific commercial product, process, or service by trade name, trademark, manufacturer, or otherwise does not necessarily constitute or imply its endorsement, recommendation, or favoring by the United States government or the University of California. The views and opinions of authors expressed herein do not necessarily state or reflect those of the United States government or the University of California, and shall not be used for advertising or product endorsement purposes.

Defining Terms

Decarbonization: The reduction, over time, of carbon content per unit energy produced. Natural gas, for example, produces more energy per unit carbon than coal; equivalently, more CO_2 is produced from coal than from natural gas per unit energy produced.

Dematerialization: The decline, over time, in weight of materials used in industrial end products, or in the embedded energy of the products. Dematerialization is an extremely important concept for the environment because the use of less material translates into smaller quantities of waste generated in both production and consumption.

Design for environment (DFE): The systematic consideration of design performance with respect to environment over the full product and process life cycle from design through manufacturing, packaging, distribution, installation, use, and end of life. It is proactive to reduce environmental impact by addressing environmental concerns in the product or process design stage.

Ecolabel: Label or certificate awarded to a product that has met specific environmental performance requirements.

Industrial ecology: The objective, multidisciplinary study of industrial and economic systems and their linkages with fundamental natural systems.

ISO 14000: Series of international standards fashioned from the ISO 9000 standard which includes requirements for environmental management systems, environmental auditing and labeling guidelines, life-cycle-analysis guidelines, and environmental product standards.

Life-cycle assessment (LCA): The method for systematically assessing the material use, energy use, waste emissions, services, processes, and technologies associated with a product.

Product take-back: Program in which the manufacturer agrees to take back product at the end-of-life (typically at no cost to the consumer) and dispose of product in an environmentally responsible matter.

Acknowledgments

This work was supported by AT&T and the initial work performed under the auspices of the U.S. Department of Energy by Lawrence Livermore National Laboratory under contract W-7405-Eng-48.

References

R. Atkyns, M. Blazek, and J. Roitz, "Measurement of environmental impacts of telework adoption amidst change in complex organizations: AT&T survey methodology and results," *J. Resour. Conserv. Recycl.*, 3, 185–201, 2002.

M. Blazek and H. Lisa, Strategic Issues for Equipment End-of-Life, presented UC Irvine, March 2003.

M. Blazek, S. Rhodes, F. Kommonen, and E. Weidman, "Tale of two cities: environmental life cycle assessment for telecommunications systems: Stockholm, Sweden, and Sacramento, CA," *IEEE Int. Symp. Electron. Environ. Danvers, MA*, 11–13, 1999, May.

B.F. Dambach and B.A. Allenby, "Implementing design for environment at AT&T," *Total Quality Environmental Management*, 4, 51–62, 1995.

J. Mitchell-Jackson, J. Koomey, M. Blazek, and B. Nordman, "National and regional implications of internet data center growth," *J. Resour. Conserv. Recycl.*, 3, 175–185, 2002, (also LBNL-50534).

T.E. Graedel and B.R. Allenby, *Industrial Ecology*, Englewood Cliffs, N.J.: Prentice-Hall, 1995.

S. Lipp, G. Pitts, and F. Cassidy, Eds., "A life cycle environmental assessment of a computer workstation," Environmental Consciousness: A Strategic Competitiveness Issue for the Electronics and Computer Industry, Austin, Texas: Microelectronics and Computer Technology Corporation, 1993.

S. Pederson, C. Wilson, G. Pitts, and B. Stotesbery, Eds., *Electronics Industry Roadmap*, Austin, Texas: Microelectronics and Computer Technology Corporation, 1996.

L. Seifert, "AT&T technology and the environment," *AT&T Tech. J.*, 74, 4–7, 1995.

World Commission on Environment and Development, Our Common Future, Oxford: Oxford University Press, 1987.

Further Information

The IEEE Environment, Health, and Safety Committee annually sponsors and publishes the proceeding of the International Symposium on Electronics and the Environment. These proceedings are a valuable resource for practitioners of DFE.

The National Technology Roadmap for Semiconductors, published by the Semiconductor Industry Association contains information on the environmental impacts of semiconductor fabrication as well as initiatives begun to address these concerns.

The *AT&T Technical Journal* has a dedicated issue on industrial ecology and DFE entitled "AT&T Technology and the Environment," volume 74, no. 6, November/December 1995.

Other Suggested Reading

American Electronics Association, "The hows and whys of design for the environment," 1993.

Allenby, B.R. and Richards, D.J. (Eds.), *The Greening of Industrial Ecosystems*, Washington, D.C.: National Academy Press, 1994.

Eisenberger, P. (Ed.), *Basic Research Needs for Environmentally Responsive Technologies of the Future*, Princeton, N.J.: Princeton Materials Institute, 1996.

Graedel, T.E. and Allenby, B.R., *Industrial Ecology*, 2nd ed., Upper Saddle River, N.J.: Pearson Education, 2003.

Directive 2002/96/EC of the European Parliament and Council of 27 January 2003 on waste electrical and electronic equipment (WEEE), *Office Journal of the European Union*, 13.2.2003.

14

Robotics

Ty A. Lasky
University of California, Davis

Tien C. Hsia
University of California, Davis

Hodge E. Jenkins
Mercer University

Mark L. Nagurka
Marquette University

Thomas R. Kurfess
Clemson University

Nicholas G. Odrey
Lehigh University

14.1 Robot Configuration

Ty A. Lasky and Tien C. Hsia

Configuration is a fundamental classification for industrial robots. Configuration refers to the geometry of the robot manipulator, i.e., the manner in which the links of the manipulator are connected at each joint. The Robotic Industries Association (RIA, see http://www.roboticsonline.com/) defines a robot as "a manipulator designed to move material, parts, tools, or specialized devices, through variable programmed motions for the performance of a variety of tasks." With this definition, attention here is focused on industrial manipulator arms, typically mounted on a fixed pedestal or base. Mobile robots and hard automation (e.g., computer numerical control [CNC] machines) are excluded. The emphasis here is on serial-chain manipulator arms, which consist of a serial chain of linkages, where each link is connected to exactly two other links, with the exception of the first and last, which are connected to only one other link. Additionally, the focus is on the first three links, called the major linkages, with only a brief mention of the last three links, or wrist joints, also called the minor linkages.

Robot configuration is an important consideration in the selection of a manipulator. Configuration refers to the way the manipulator links are connected at each joint. Each link will be connected to the subsequent link by either a linear (sliding or prismatic) joint, which can be abbreviated with a P, or a revolute (or rotary) joint, abbreviated with an R. Using this notation, a robot with three revolute joints is abbreviated as RRR, while one with a rotary joint followed by two linear (prismatic) joints is denoted RPP. Each configuration type is well-suited to certain types of tasks and ill-suited to others. Some configurations are more versatile than others. In addition to the geometrical considerations, robot configuration affects the structural stiffness of the robot, which will be important in some applications. Also, configuration impacts the complexity of the forward and inverse kinematics, i.e., the mappings between the robot actuator (joint) space, and the Cartesian position and orientation of the robot end-effector, or tool.

There are six major robot configurations commonly used in industry. The simplest configuration is the Cartesian robot, which consists of three orthogonal, linear joints (PPP), so that the robot moves in the x, y, and z directions in the joint space. The cylindrical configuration consists of one revolute and two linear joints (RPP), so that the robot joints correspond to a cylindrical coordinate system. The spherical configuration consists of two revolute joints and one linear joint (RRP), so that the robot moves in a spherical, or polar, coordinate system. The articulated (a.k.a. arm-and-elbow) configuration consists of three revolute joints (RRR), giving the robot a human-like range of motion. The SCARA (selectively compliant assembly robot arm) configuration consists of two revolute joints and one linear joint (RRP), arranged in a different fashion than the spherical configuration. It may also be equipped with a revolute joint on the final sliding link. The gantry configuration is essentially a Cartesian configuration, with the robot mounted on an overhead track system. One can also mount other robot configurations on an overhead gantry system to give the robot an extended workspace and to free up valuable factory floor space. The percentage usage of the first five configuration types is listed in Table 14.1. This table does not include gantry robots, which are assumed to be included in the Cartesian category. Additionally, this information is from 1988, and clearly does not accurately represent current usage; however, it appears to be the most recent aggregate data available. For a current snapshot, Table 14.2 provides the number and percentage of industrial manipulator units installed in 2002. Note that the data is for selected (but representative) countries. In addition, [UN/IFR, 2003] aggregates Cartesian with gantry, and cylindrical with spherical configurations, and notes that many countries do not categorize by configuration.

TABLE 14.1 Robot Arm Geometry Usage (1988)

Configuration	Percentage
Cartesian	18%
Cylindrical	15%
Spherical	10%
Articulated	42%
SCARA	15%

Source: V.D. Hunt, *Robotics Sourcebook*, New York: Elsevier, 1988. With permission.

In general, robots with a rotary base have a speed advantage. However, they have more variation in resolution and dynamics over the workspace compared with Cartesian robots. This can lead to inferior performance if a fixed controller is used over the robot's entire workspace.

Cartesian Configuration

The Cartesian configuration consists of three orthogonal, linear axes, abbreviated as PPP, as shown in Figure 14.1. Thus, the joint space of the robot corresponds directly with the standard right-handed Cartesian xyz-coordinate system, yielding the simplest possible kinematic equations. The work envelope of the Cartesian robot is shown in Figure 14.2. The work envelope encloses all the points that can be reached by the mounting point for the end-effector or tool—the area reachable by an end-effector or tool is not considered part of the work envelope. All interaction with other machines, parts, or processes must take place within this volume. Herein, the workspace of a robot is equated with the work envelope.

TABLE 14.2 Installation of Industrial Robots in 2002 for Select Countries

Configuration	Units	Percentage
Cartesian or Gantry	11,899	17.4%
SCARA	4,038	5.9%
Articulated	34,467	50.3%
Cylindrical or Spherical	1,522	2.2%
Not Classified	16,623	24.2%
Total	68,549	100%

Source: U.N. and Intl. Fed. Robotics, *World Robotics*, 2003. With permission.

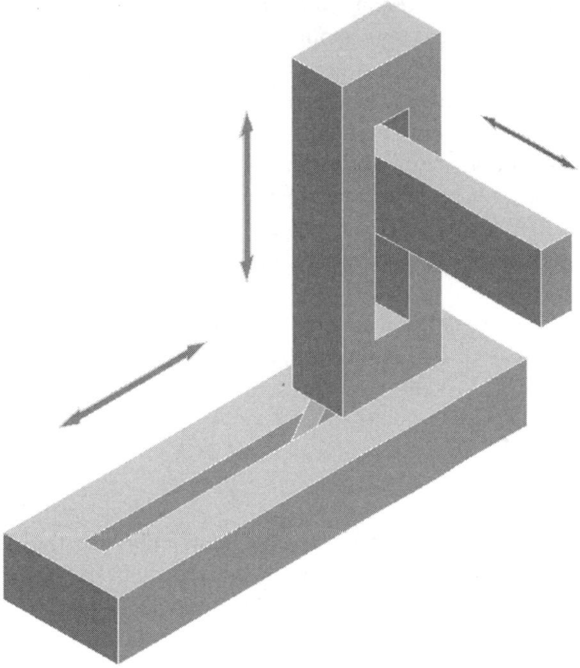

FIGURE 14.1 The Cartesian configuration.

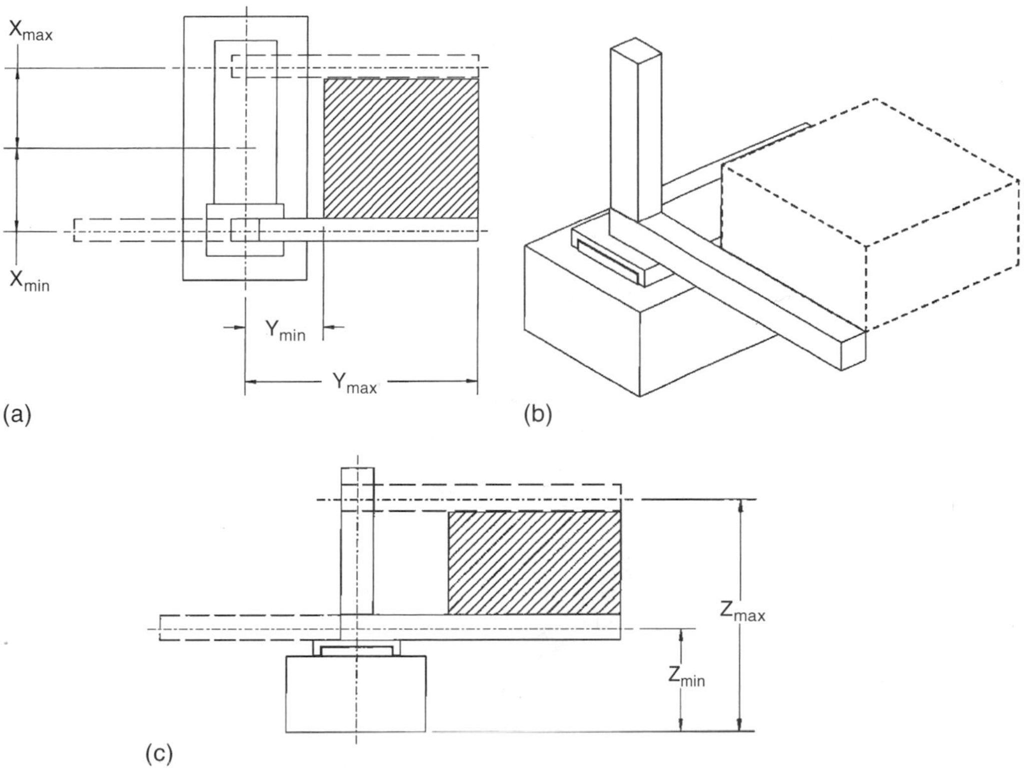

FIGURE 14.2 Cartesian robot work envelope.

There are several advantages to the Cartesian configuration. As noted above, the robot is kinematically simple, since motion on each Cartesian axis corresponds to motion of a single actuator, assuming the axes are aligned. This eases the programming of linear motions. In particular, it is easy to do a straight vertical motion, the most common assembly task motion. The Cartesian geometry also yields constant arm resolution throughout the workspace; i.e., for any configuration, the resolution for each axis corresponds directly to the resolution for that joint. The simple geometry of the Cartesian robot leads to correspondingly simple manipulator dynamics. The disadvantages of this configuration include inability to reach objects on the floor or points invisible from the base of the robot, and slow speed of operation in the horizontal plane compared to robots with a rotary base. Additionally, the Cartesian configuration requires a large operating volume for a relatively small workspace.

Cartesian robots are used for several applications. As noted above, they are well-suited for assembly operations, as they easily perform vertical straight-line insertions. Because of the ease of straight-line motions, they are also well-suited to machine loading and unloading. They are also used in clean-room tasks.

Cylindrical Configuration

The cylindrical configuration consists of one vertical revolute joint and two orthogonal linear joints (RPP), as shown in Figure 14.3. The resulting work envelope of the robot is a cylindrical toroid, as shown in Figure 14.4. This configuration corresponds with the cylindrical coordinate system.

As with the Cartesian robot, the cylindrical robot is well-suited for straight-line vertical and radial horizontal motions, so it is useful for assembly and machine loading operations. It is capable of higher speeds in the horizontal plane due to the rotary base. However, general horizontal straight-line motion is more complex and correspondingly more difficult to coordinate. Additionally, the horizontal end-point resolution of the cylindrical robot is not constant but depends on the extension of the horizontal linkage. A cylindrical robot cannot reach around obstacles. Finally, if a monomast construction is used on the horizontal linkage so that the linkage extends behind the robot when retracted, then there can be clearance problems.

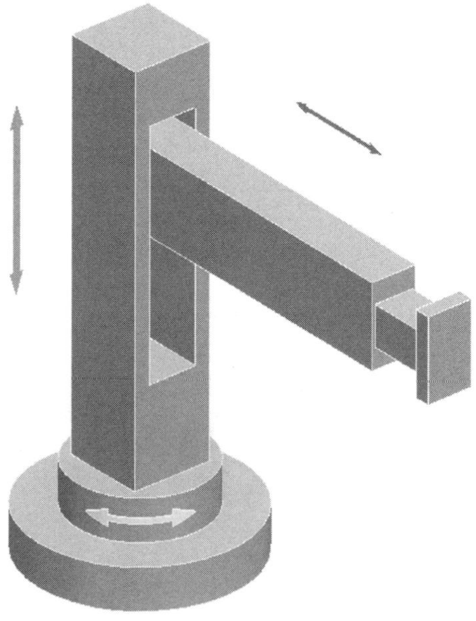

FIGURE 14.3 The cylindrical configuration.

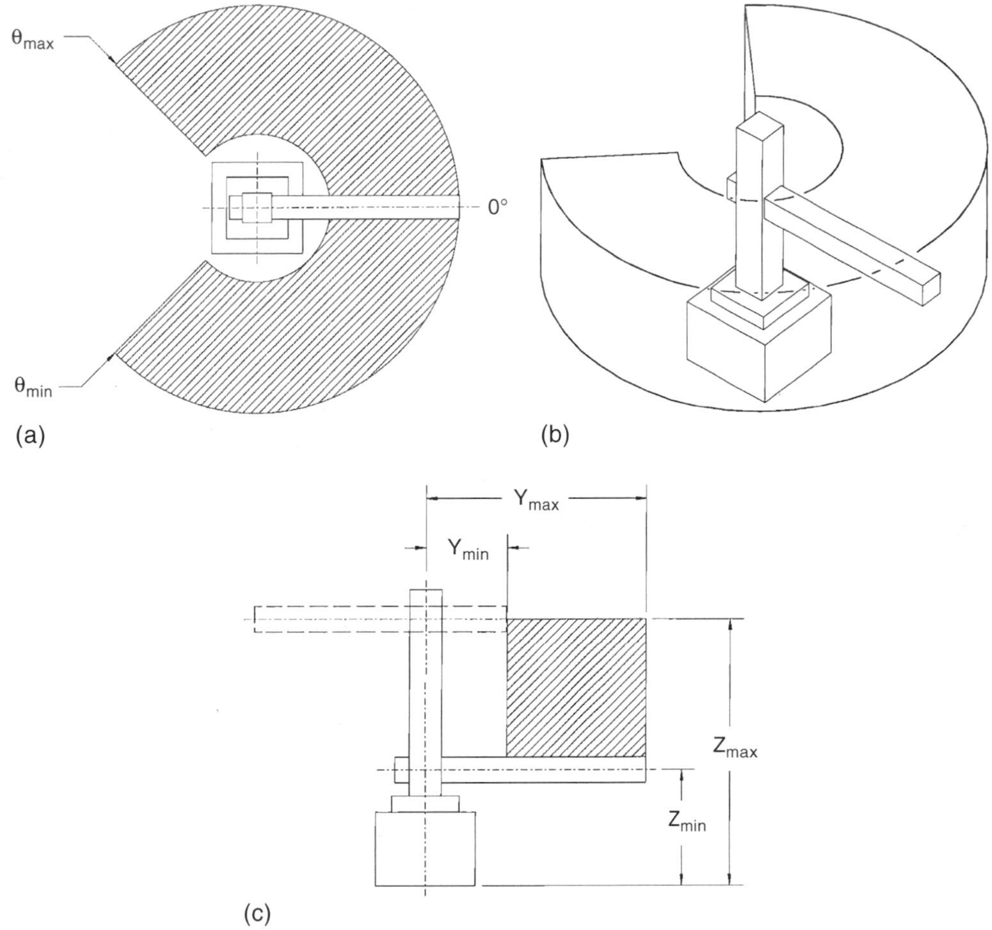

FIGURE 14.4 Cylindrical robot work envelope.

Spherical Configuration

The spherical (or polar) configuration consists of two revolute joints and one linear joint (RRP), as shown in Figure 14.5. This results in a set of joint coordinates that corresponds to the spherical coordinate system. A typical work envelope for a spherical robot is shown in Figure 14.6.

Spherical robots have the advantages of high speed, due to the rotary base, and large work volume, but are more kinematically complex than either Cartesian or cylindrical robots. Generally, they are used for heavy-duty tasks in, for example, automobile manufacturing. They do not have the dexterity to reach around obstacles in the workspace. Spherical robot resolution will vary throughout the workspace.

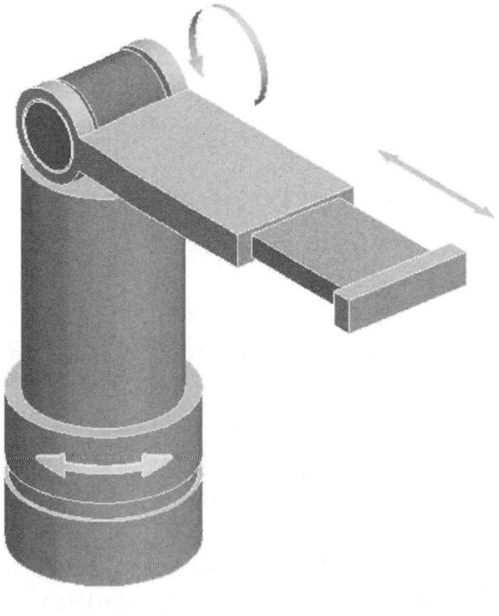

FIGURE 14.5 The spherical configuration.

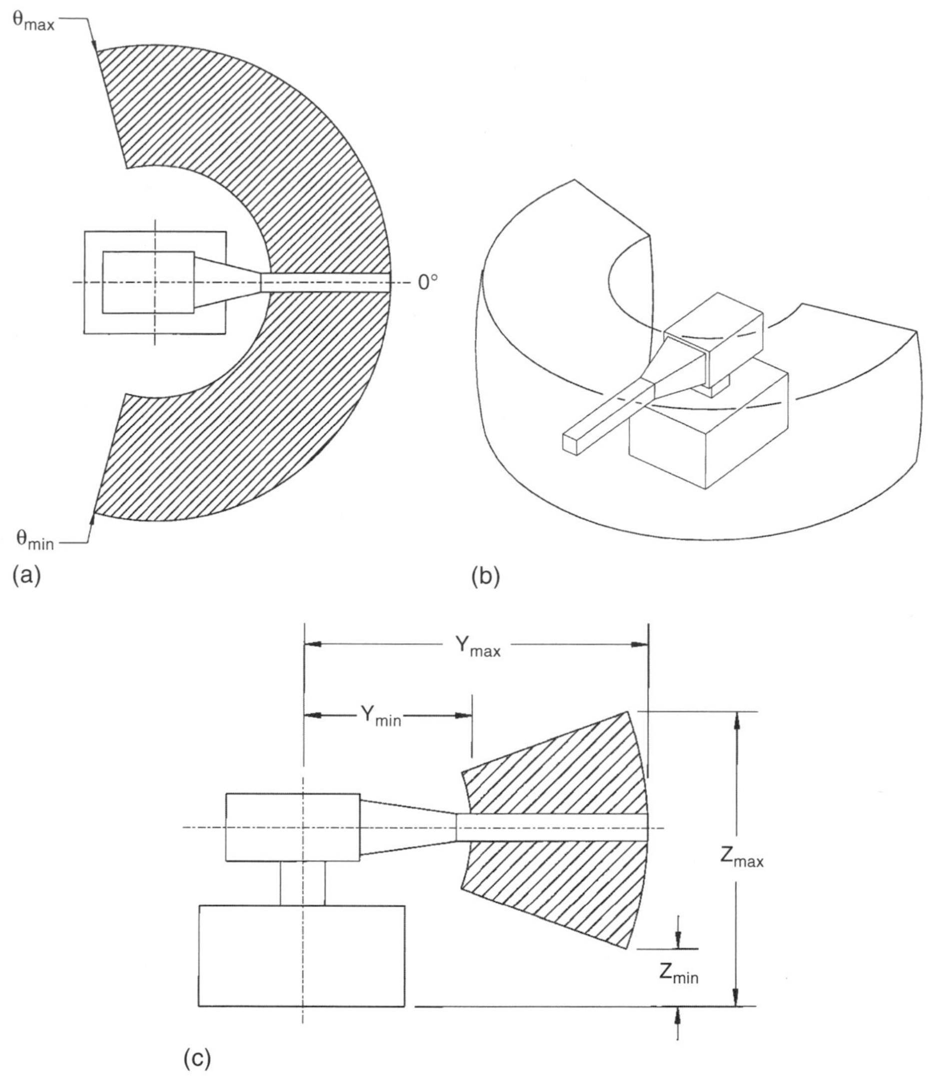

FIGURE 14.6 Spherical robot work envelope.

Articulated Configuration

The articulated (a.k.a. anthropomorphic, jointed, arm-and-elbow) configuration consists of three revolute joints (RRR), as shown in Figure 14.7. The resulting joint coordinates do not directly match any standard coordinate system. A slice of a typical work envelope for an articulated robot is shown in Figure 14.8.

The articulated robot is currently the most commonly used in research and industry. It has several advantages over other configurations. It comes closest to duplicating the motions of a human assembler's arm, so there is generally less need to redesign an existing workstation to utilize an articulated robot. It has a very large, dexterous work envelope; i.e., it can reach most points in its work envelope from a variety of orientations. Thus, it can more easily reach around or over obstacles in the workspace or into parts or machines. Because all the joints are revolute, high end-point speeds are possible. The articulated arm is good for tasks involving multiple insertions, complex motions, and varied tool orientations. The versatility of this configuration makes it applicable to a variety of tasks, so the user has fewer limitations on the use of the robot. However, the same features that give this robot its advantages lead to certain disadvantages. The geometry is

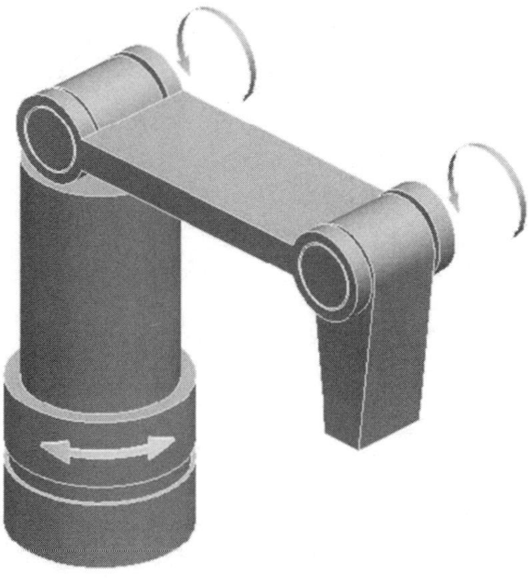

FIGURE 14.7 The articulated configuration.

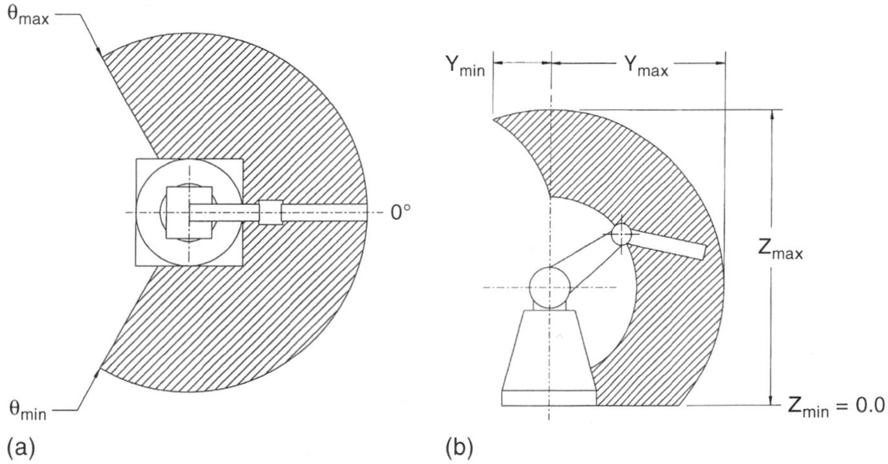

FIGURE 14.8 Articulated robot work envelope.

complex, and the resulting kinematic equations are intricate. Straight-line motion is more difficult to coordinate. Control is generally more difficult than for other geometries, with associated increase in cost. Here again, arm resolution is not fixed throughout the workspace. Additionally, the dynamics of an articulated arm vary widely throughout the workspace, so that performance will vary over the workspace for a fixed controller. In spite of these disadvantages, the articulated arm has been applied to a wide variety of research and industrial tasks, including spray painting, clean-room tasks, machine loading, and parts-finishing tasks. Recent research results and computational improvements have addressed some of the noted disadvantages.

SCARA Configuration

The SCARA (selectively compliant assembly robot arm) configuration consists of two revolute joints and a linear joint (RRP), as shown in Figure 14.9. This configuration is significantly different from the spherical

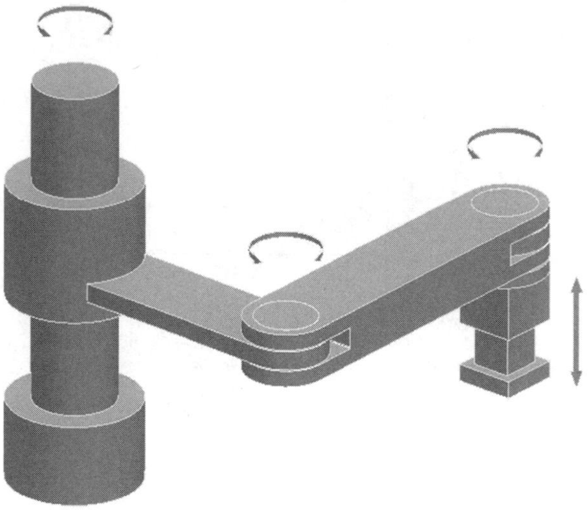

FIGURE 14.9 The SCARA configuration.

configuration, since the axes for all joints are always vertical. In addition to the first three degrees of freedom (DOF), the SCARA robot will often include an additional rotation about the last vertical link to aid in orientation of parts, as shown in Figure 14.9. The work envelope of the SCARA robot is illustrated in Figure 14.10. The SCARA configuration was developed by Professor Hiroshi Makino of Yamanashi University, Japan.

This configuration has many advantages and is quite popular in industry. The configuration was designed specifically for assembly tasks, so it has distinct advantages when applied in this area. Because of the vertical orientation of the joints, gravity does not affect the dynamics of the first two joints. In fact, for these joints, the actuators can be shut off and the arm will not fall, even without the application of brakes. As the name SCARA implies, this allows compliance in the horizontal directions to be selectively varied; therefore, the robot can comply to horizontal forces. Horizontal compliance is important for vertical assembly operations. Because of the vertical linear joint, straight-line vertical motions are simple. Also, SCARA robots typically have high positional repeatability. The revolute joints allow high-speed motion. On the negative side, the resolution of the arm is not constant throughout the workspace, and the kinematic equations are relatively complex. In addition, the vertical motion of the SCARA configuration is typically quite limited. While the SCARA robot can reach around objects, it cannot reach over them in the same manner as an articulated arm.

Gantry Configuration

The gantry configuration is geometrically equivalent to the Cartesian configuration, but is suspended from an overhead crane and typically can be moved over a large workspace. It consists of three linear joints (PPP), and is illustrated in Figure 14.11. In terms of work envelope, it will have a rectangular volume that sweeps out most of the inner area of the gantry system, with a height limited by the length of the vertical mast and the headroom above the gantry system. One consideration in the selection of a gantry robot is the type of vertical linkage employed in the z axis. A monomast design is more rigid, yielding tighter tolerances for repeatability and accuracy, but requires significant headroom above the gantry to have a large range of vertical motion. On the other hand, a telescoping linkage will require significantly less headroom but is less rigid, with corresponding reduction in repeatability and accuracy. Other robot configurations can be mounted on gantry systems, thus gaining many of the advantages of this geometry.

Gantry robots have many advantageous properties. They are geometrically simple, like the Cartesian robot, with corresponding kinematic and dynamic simplicity, and constant arm resolution throughout the

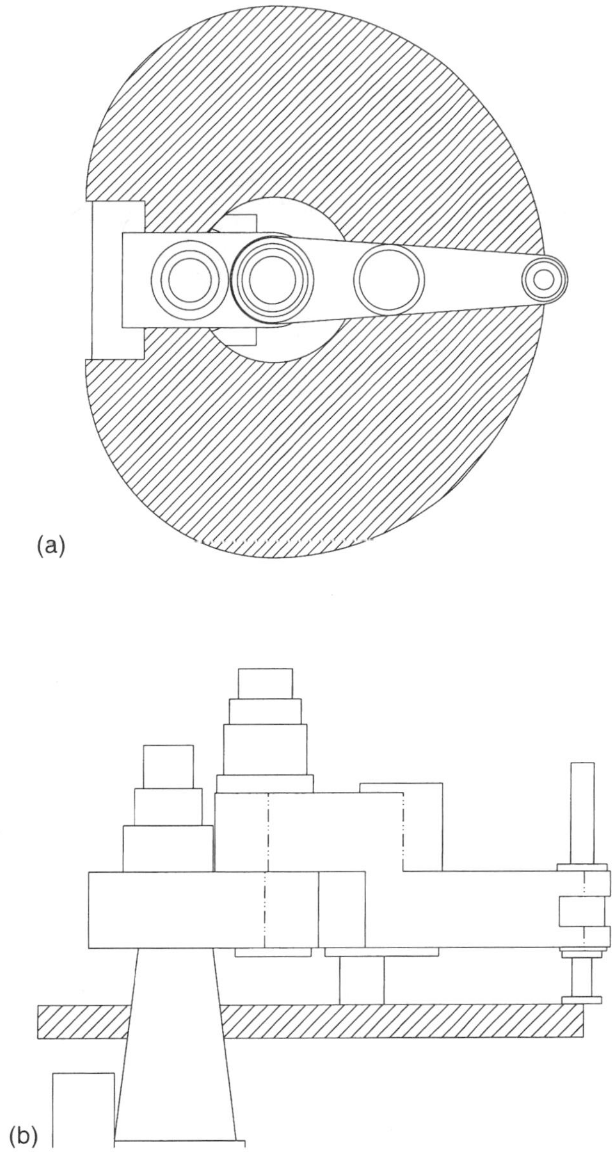

FIGURE 14.10 SCARA robot work envelope.

workspace. The gantry robot has better dynamics than the pedestal-mounted Cartesian robot, as its links are not cantilevered. One major advantage over revolute-base robots is that its dynamics vary much less over the workspace. This leads to less vibration and more even performance than typical pedestal-mounted robots in full extension. Gantry robots are much stiffer than other robot configurations, although they are still much less stiff than numerical control (NC) machines. The gantry robot can straddle a workstation, or several workstations for a large system, so that one gantry robot can perform the work of several pedestal-mounted robots. As with the Cartesian robot, the gantry robot's simple geometry is similar to that of an NC machine, so technicians should be more familiar with the system and require less training time. Also, there is no need for special path or trajectory computations. A gantry robot can be programmed directly from a computer-aided design (CAD) system with the appropriate interface (also true for the Cartesian configuration), and straight-line motions are particularly simple to program. Large gantry robots have a very high payload capacity and

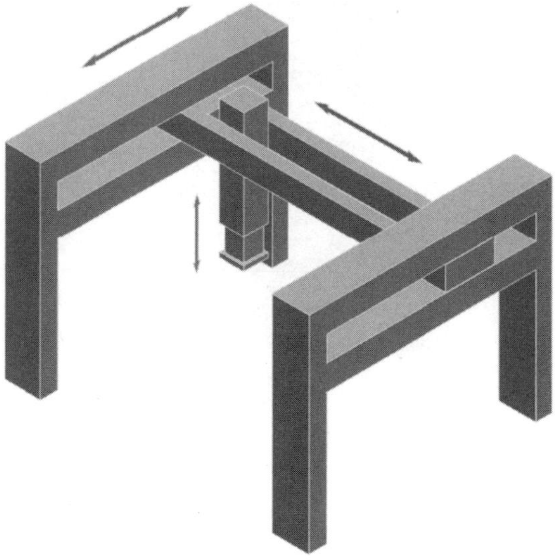

FIGURE 14.11 The gantry configuration.

can reach speeds of up to 40 in./s (1.0 m/s); however, due to the large scale, absolute accuracy will typically be lower. Very large-scale systems (500 ft × 40 ft × 25 ft, or 150 m × 12 m × 7.6 m) with capacity up to 15 tons (13,600 kg) have been reported [Nof, 1999]. Small table-top systems can achieve linear speeds of up to 40 in./s (1.0 m/s), with a payload capacity of 5.0 lb (2.3 kg), making them suitable for assembly operations. It can be more difficult to apply a gantry robot to an existing workstation, as workpieces must be brought into the gantry's work envelope, which may be harder to do than for a pedestal-mounted manipulator. The gantry configuration is generally the least space-efficient.

Gantry robots can be applied in many areas. They are used in the nuclear power industry to load and unload reactor fuel rods. Gantry robots are also applied to materials-handling tasks, such as parts transfer, machine loading, palletizing, materials transport, and some assembly applications. In addition, gantry robots are used for process applications such as welding, painting, drilling, routing, cutting, milling, inspection, and nondestructive testing.

Additional Information

The above six configurations are the main types currently used in industry. However, there are other configurations used in either research or specialized applications. Some of these configurations have found limited application in industry and may become more prevalent in the future.

All the above configurations are serial-chain manipulators. An alternative to this common approach is the parallel configuration, known as the Stewart platform [Waldron, 1990]. This manipulator consists of two platforms connected by six prismatic linkages. This arrangement yields the full six-DOF motion (three position, three orientation) that can be achieved with a six-axis serial configuration but has a comparably very high stiffness. It is used as a motion simulator for pilot training and virtual-reality applications. The negative aspects of this configuration are its relatively restricted motion capability and geometric complexity. Research in parallel platforms is quite active, and the configuration is commonly used in motion simulators—e.g., the national advanced driving simulator (NADS, http://www.nads-sc.uiowa.edu) at the University of Iowa. In fact, the NADS can be considered a hybrid configuration, consisting of a Stewart platform mounted on a large Cartesian manipulator (x-y motion, 64 ft × 64 ft).

The above configurations are restricted to a single manipulator arm. There are tasks that are either difficult or impossible to perform with a single arm. With this realization, there has been significant interest in the use of multiple arms to perform coordinated tasks. Possible applications include carrying loads that exceed the capacity of a single robot, and assembling objects without special fixturing. Multiple arms are particularly useful in zero-gravity environments. While there are significant advantages to the use of multiple robots, the complexity, in terms of kinematics, dynamics, and control, is quite high. However, the use of multiple robots is opening new areas of application for robots. One area of particular interest is robotic hands, which are effectively several serial manipulators attached to a common base and controlled in a coordinated fashion to provide human-like dexterity [Mason and Salisbury, 1985].

For any of the six standard robot configurations, the orientation capability of the major linkages is severely limited, so it is important to provide additional joints, known as the minor linkages, to provide the capability of varied orientations for a given position. Most robots include a 3-DOF revolute joint wrist connected to the last link of the major linkages. The three revolute axes are orthogonal and usually intersect in a common point, known as the wrist center point, so that the kinematic equations of the manipulator can be partitioned into locating the Cartesian position and orientation of the wrist center point and determining the orientation of a Cartesian frame fixed to the wrist axes. This provides a 6-DOF manipulator, so that the robot can, within its work envelope, reach arbitrary positions and orientations.

At the edge of the work envelope, a six-DOF robot can attain only one orientation. To increase the geometric dexterity of the manipulator, it is useful to consider robots with more than six DOF, i.e., redundant robots. In general, a redundant robot is defined as one with more DOF than the task-space dimensions—in this regard a 3-DOF robot is redundant for planar positioning tasks. Redundant robots are highly dexterous and can use the extra DOF in many ways: obstacle avoidance, joint torque minimization, kinematic singularity (points where the manipulator cannot move in certain directions) avoidance, bracing strategies where part of the arm is braced against a structure to raise the lowest structural resonant frequency of the arm, etc. While the redundant manipulator configuration has many desirable properties, the geometric complexity has previously limited their application in industry. More exotic redundant configurations have appeared in recent years. Snake-like robots with many DOF have been introduced for use in applications requiring highly redundant (hyper-redundant) configurations [e.g., Erkmen et al., 2002]. In some cases, these and similar systems use binary (i.e., two-position) actuators, rather than the more common continuously variable actuators. Reconfigurable or modular manipulators have also been developed, and such systems can be configured to provide redundancy as needed, and in general can be reconfigured based on evolving task requirements [Yim et al., 2002].

Mobile systems have seen increasing use in factory automation, military applications, security systems, transportation, and in specialized areas such as highway maintenance (see for example www.ahmct.ucdavis.edu for a variety of systems). Mobile platforms have been combined with manipulator arms to combine high mobility with the ability to dexterously manipulate the environment. Systems have been developed for many environments, including land-based, aerial, and underwater mobile robots [see *IEEE Robotics and Automation Magazine*, vol. 6, issue 2, June 1999; also Hebert, Thorpe, and Stentz, 1997]. Finally, legged mobile robots have been developed which can operate in rougher terrain—such systems can again be considered multiple serial manipulators attached to a common base and controlled in a coordinated fashion to provide human- or animal-like mobility [see *IEEE Robotics and Automation Magazine*, vol. 5, issue 2, June 1998; also Manko, 1992].

While the focus here has been on industrial robots that can be classified as macrorobots, there has also been considerable interest and development of microrobots in the past decade for applications in micromanipulation and microassembly. Typical fields of application are microelectronics, biotechnology, medicine, and nanotechnology. More recently, mobile microrobots have become increasingly important in such applications. In these applications, the robots are required to manipulate objects on a micrometer scale with microprecision. The basic challenge is how to scale down the traditional precision positioning systems. As of now, the most successful microrobot positioning configuration is of Cartesian design. More complex configurations are being investigated in wrist design to develop full 6-DOF microrobots.

Conclusions

Each of the six standard configurations has specific advantages and disadvantages. When choosing a manipulator for a task, the properties of the manipulator geometry are one of the most important considerations. If the manipulator will be used for a wide variety of tasks, one may need to trade off performance for any given task for the flexibility that will allow the manipulator to work for the various tasks. In such a case, a more flexible geometry should be considered. The future of robotics will be interesting. With the steady increase in computational capabilities, the more complex geometries, including redundant and multiple robots, are seeing increased applications in industry.

Defining Terms

Degrees of freedom: The number of degrees of freedom (DOF) of a manipulator is the number of independent position variables that must be specified in order to locate all parts of the manipulator. For a typical industrial manipulator, the number of joints equals the number of DOF.

Kinematics: The kinematics of the manipulator refers to the geometric properties of the manipulator. Forward kinematics is the computation of the Cartesian position and orientation of the robot end-effector given the set of joint coordinates. Inverse kinematics is the computation of the joint coordinates given the Cartesian position and orientation of the end-effector. The inverse kinematic computation may not be possible in closed form, may have no solution, or may have multiple solutions.

Redundant manipulator: A redundant manipulator contains more than six DOF. More generally, it is one with more DOF than the task-space dimensions.

Singularity: A singularity is a location in the workspace of the manipulator at which the robot loses one or more DOF in Cartesian space, i.e., there is some direction (or directions) in Cartesian space along which it is impossible to move the robot end-effector no matter which robot joints are moved.

References

J.J. Craig, *Introduction to Robotics: Mechanics and Control*, 2nd ed., Reading, MA: Addison-Wesley, 1989.

I. Erkmen, A.M. Erkmen, F. Matsuno, R. Chatterjee, and T. Kamegawa, "Snake robots to the rescue!" *IEEE Rob. Autom. Mag.*, vol. 9, no. 3, pp. 17–25, 2002.

M. Hebert, C. Thorpe, and A. Stentz, "Intelligent unmanned ground vehicles: autonomous navigation research at Carnegie Mellon," in *Kluwer International Series in Engineering and Computer Science: Robotics*, T. Kanade, Ed., Boston, MA: Kluwer Academic Publishers, 1997.

D. Manko, "A general model of legged locomotion on natural terrain," in *Kluwer International Series in Engineering and Computer Science: Robotics*, T. Kanade, Ed., Boston, MA: Kluwer Academic Publishers, 1992.

M.T. Mason and J.K. Salisbury, *Robot Hands and the Mechanics of Manipulation*, Cambridge, MA: MIT Press, 1985.

S.Y. Nof, Ed., *Handbook of Industrial Robotics*, 2nd ed., New York: John Wiley and Sons, 1999.

B. Siciliano and L. Villani, "Robot force control," in *Kluwer International Series in Engineering and Computer Science: Robotics*, T. Kanade, Ed., Boston, MA: Kluwer Academic Publishers, 1999.

United Nations and International Federation of Robotics, *World Robotics: Statistics, Market Analysis, Forecasts, Case Studies and Profitability of Robot Investment*, UN Economic Commission for Europe, October 2003.

K.J. Waldron, "Arm design," in *Concise International Encyclopedia of Robotics*, R.C. Dorf, Ed., New York: Wiley-Interscience, 1990.

Further Information

M. Yim, Y. Zhang, and D. Duff, "Modular robots", *IEEE Spectrum*, vol. 39, no. 2, pp. 30–34, 2002, Feb.

The journal *IEEE Transactions on Robotics and Automation* (TRA) is a valuable source for a wide variety of robotics research topics, occasionally including new robot configurations, while the *IEEE Robotics and Automation Magazine* provides a more applied forum. Additionally, IEEE's *Control Systems Magazine* occasionally publishes an issue devoted to robotic systems. The home page for the IEEE Robotics and Automation Society can be found at "http://www.ncsu.edu/IEEE-RAS/." Note that in mid-2004, the TRA split into two: the *IEEE Transactions on Automation Science and Engineering* (with a focus on more structured environments), and the *IEEE Transactions on Robotics* (with a focus on unstructured environments).

Another journal that often has robotics-related articles is the *ASME Journal of Dynamic Systems, Measurement and Control*. In addition, the *IEEE/ASME Transactions on Mechatronics* provides articles of interest to the robotics community, while the *IEEE/ASME Journal of Microelectromechanical Systems* covers MEMS.

Several conferences provide current robotics information, including the IEEE International Conference on Robotics and Automation, and the IEEE/RSJ (Robotics Society of Japan) International Conference on Intelligent Robots and Systems, both held annually.

Useful sources on the World Wide Web include the Robotics Resources page, located at http://www-robotics.cs.umass.edu/resources.html, and the Google Robotics Directory at http://directory.google.com/Top/Computers/Robotics/. The Internet newsgroup comp.robotics is another valuable source of information—its FAQ can be found at http://www.faqs.org/faqs/robotics-faq/. For any web resources listed here or in the article, addresses are subject to change, but the reader should be able to locate the resources through an appropriate Web search.

14.2 Robot Dynamics and Controls

Hodge E. Jenkins, Mark L. Nagurka, and Thomas R. Kurfess

This chapter provides the essential theoretical elements required for prescribing desired motion for robots: dynamical descriptors and control approaches. As the topics associated with the subject are vast, overviews of both dynamics and control are given in each section for clarity of content. To begin, the development of dynamic models (or equations of motion) for rigid-link robots is provided first as a foundation for control.

Robot Dynamics

Dynamics Overview

This chapter addresses the dynamics and controls of robots, modeled as lumped-parameter systems. The dynamic analysis of robots using the Newton–Euler method is presented in this section. This method is one of several approaches for deriving the governing equations of motion for a robotic system. It is assumed that the robots are characterized as open kinematic chains with actuators at the joints that can be controlled. The kinematic chain consists of discrete rigid members or links, with each link attached to neighboring links by means of revolute or prismatic joints, i.e., lower-pairs [Denavit and Hartenberg, 1955]. One end of the chain is fixed to a base (inertial) reference frame, and the other end of the chain, with an attached end-effector, tool, or gripper, can move freely in the robotic workspace and/or be used to apply forces and torques to objects being manipulated to accomplish a wide range of tasks.

Robot dynamics is predicated on an understanding of the associated kinematics, covered in a separate chapter in this handbook. A commonly adopted approach for robot kinematics is that of a 4×4 homogeneous transformation, sometimes referred to as the Denavit–Hartenberg (D–H) transformation [Denavit and Hartenberg, 1955]. The D–H transformations essentially determine the position of the origin and the rotation of one link coordinate frame with respect to another link coordinate frame. The D–H transformations can also be used in deriving the dynamic equations of robots.

Background

The field of robot dynamics has a rich history with many important developments. There is a wide literature base of reported work, with articles in professional journals and established textbooks [Shahinpoor, 1986] [Featherstone, 1987] [Spong and Vidyasagar, 1989] [Craig, 1989] [Yoshikawa, 1990] [McKerrow, 1991]

[Sciavicco and Siciliano, 2000] [Mason, 2001] [Niku, 2001] as well as research monographs [Vukobratovic, 2003]. A review of robot dynamic equations and computational issues is presented in [Featherstone and Orin, 2000]. A recent paper [Swain and Morris, 2003] formulates a unified dynamic approach for different robotic systems derived from first principles of mechanics.

The control of robot motions, forces, and torques requires a firm grasp of robot dynamics and plays an important role given the demand to move robots as fast as possible. Robot dynamic equations of motion can be highly nonlinear due to the configuration of links in the workspace and to the presence of Coriolis and centrifugal acceleration terms. In slow-moving and low-inertia applications, the nonlinear coupling terms are sometimes neglected, making the robot joints independent and simplifying the control problem.

Several different approaches are available for the derivation of the governing equations pertaining to robot-arm dynamics. These include the Newton–Euler (N–E) method, the Lagrange–Euler (L–E) method, Kane's method, bond graph modeling, as well as recursive formulations for both Newton–Euler and Lagrange–Euler methods. The N–E formulation is based on Newton's second law, and investigators have developed various forms of the N–E equations for open kinematic chains [Orin et al., 1979] [Luh et al., 1980] [Walker and Orin, 1982].

The N–E equations can be applied to a robot link-by-link and joint-by-joint either from the base to the end-effector, called forward recursion, or *vice versa*, called backward recursion. The forward-recursive N–E equations transfer kinematic information, such as the linear and angular velocities and the linear and angular accelerations, as well as the kinetic information of the forces and torques applied to the center of mass of each link, from the base reference frame to the end-effector frame. The backward-recursive equations transfer the essential information from the end-effector frame to the base frame. An advantage of the forward- and backward-recursive equations is that they can be applied to the robot links from one end of the arm to the other, providing an efficient means to determine the necessary forces and torques for real-time or near-real-time control.

Theoretical Foundations

Newton–Euler Equations. The Newton-Euler (N–E) equations relate forces and torques to the velocities and accelerations of the center of masses of the links. Consider an intermediate, isolated link n in a multibody model of a robot, with forces and torques acting on it. Fixed to link n is a coordinate system with its origin at the center of mass, denoted as centroidal frame C_n, that moves with respect to an inertial reference frame, R, as shown in Figure 14.12. In accordance with Newton's second law,

$$\mathbf{F}_n = \frac{d\mathbf{P}_n}{dt}, \quad \mathbf{T}_n = \frac{d\mathbf{H}_n}{dt} \tag{14.1}$$

where $\mathbf{F}_n$ is the net external force acting on link n, $\mathbf{T}_n$ is the net external torque about the center of mass of link n, and $\mathbf{P}_n$ and $\mathbf{H}_n$ are, respectively, the linear and angular momenta of link n,

$$\mathbf{P}_n = m_n {}^{R}\mathbf{v}_n, \quad \mathbf{H}_n = {}^{C}\mathbf{I}_n {}^{R}\boldsymbol{\omega}_n \tag{14.2}$$

In Equation (14.2) ${}^{R}\mathbf{v}_n$ is the linear velocity of the center of mass of link n as seen by an observer in R, m_n is the mass of link n, ${}^{R}\omega_n$ is the angular velocity of link n (or equivalently C_n) as seen by an observer in R, and ${}^{C}\mathbf{I}_n$ is the mass moment of inertia matrix about the center of mass of link n with respect to C_n.

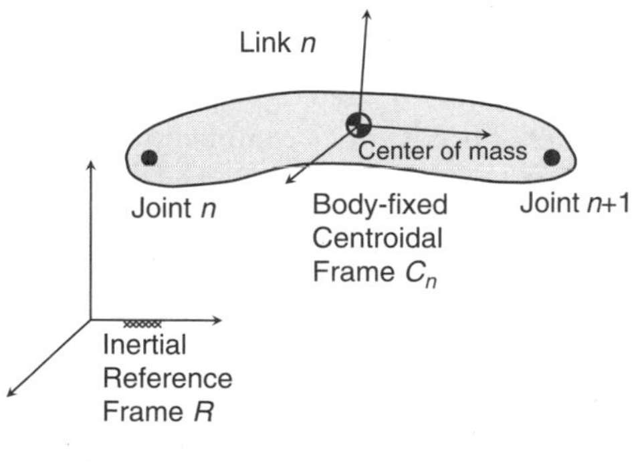

FIGURE 14.12 Coordinate systems associated with link n.

Two important equations result from substituting the momenta expressions 14.2 into 14.4 and taking the time derivatives (with respect to R):

$$ {}^{R}\mathbf{F}_n = m_n {}^{R}\mathbf{a}_n \tag{14.3}$$

$$ {}^{R}\mathbf{T}_n = {}^{C}\mathbf{I}_n {}^{R}\boldsymbol{\alpha}_n + {}^{R}\boldsymbol{\omega}_n \times \left({}^{C}\mathbf{I}_n {}^{R}\boldsymbol{\omega}_n\right) \tag{14.4}$$

where ${}^{R}\mathbf{a}_n$ is the linear acceleration of the center of mass of link n as seen by an observer in R and ${}^{R}\boldsymbol{\alpha}_n$ is the angular acceleration of link n (or equivalently C_n) as seen by an observer in R. In Equation (14.3) and Equation (14.4) superscript R has been appended to the external force and torque to indicate that these vectors are with respect to R.

Equation (14.3) is commonly known as Newton's equation of motion, or Newton's second law. It relates the linear acceleration of the link center of mass to the external force acting on the link. Equation (14.4) is the general form of Euler's equation of motion, and it relates the angular velocity and angular acceleration of the link to the external torque acting on the link. It contains two terms: an inertial torque term due to angular acceleration, and a gyroscopic torque term due to changes in the inertia as the orientation of the link changes. The dynamic equations of a link can be represented by these two equations: Equation (14.3) describing the translational motion of the center of mass and Equation (14.4) describing the rotational motion about the center of mass.

Force and Torque Balance on an Isolated Link. Using a "free-body" approach, as depicted in Figure 14.13, a single link in a kinematic chain can be isolated and the effect from its neighboring links can be accounted for by considering the forces and torques they apply. In general, three external forces act on link n: (i) link $n-1$ applies a force through joint n, (ii) link $n+1$ applies a force through joint $n+1$, and (iii) the force due to gravity acts through the center of mass. The dynamic equation of motion for link n is then

$$ {}^{R}\mathbf{F}_n = {}^{R}\mathbf{F}_{n-1,n} - {}^{R}\mathbf{F}_{n,n+1} + m_n {}^{R}\mathbf{g} = m_n {}^{R}\mathbf{a}_n \tag{14.5}$$

where ${}^{R}\mathbf{F}_{n-1,n}$ is the force applied to link n by link $n-1$, ${}^{R}\mathbf{F}_{n,n+1}$ is the force applied to link $n+1$ by link n, ${}^{R}\mathbf{g}$ is the acceleration due to gravity, and all forces are expressed with respect to the reference frame R. The negative sign before the second term in Equation (14.5) is used since the interest is in the force exerted on link n by link $n+1$.

For a torque balance, the forces from the adjacent links result in moments about the center of mass. External torques act at the joints, due to actuation and possibly friction. As the force of gravity acts through the center of mass, it does not contribute a torque effect. The torque equation of motion for link n is then

$$ \begin{aligned} {}^{R}\mathbf{T}_n &= {}^{R}\mathbf{T}_{n-1,n} - {}^{R}\mathbf{T}_{n,n+1} - {}^{R}\mathbf{d}_{n-1,n} \times {}^{R}\mathbf{F}_{n-1,n} + {}^{R}\mathbf{d}_{n,n} \times {}^{R}\mathbf{F}_{n,n+1} \\ &= {}^{C}\mathbf{I}_n {}^{R}\boldsymbol{\alpha}_n + {}^{R}\boldsymbol{\omega}_n \times \left({}^{C}\mathbf{I}_n {}^{R}\boldsymbol{\omega}_n\right) \end{aligned} \tag{14.6}$$

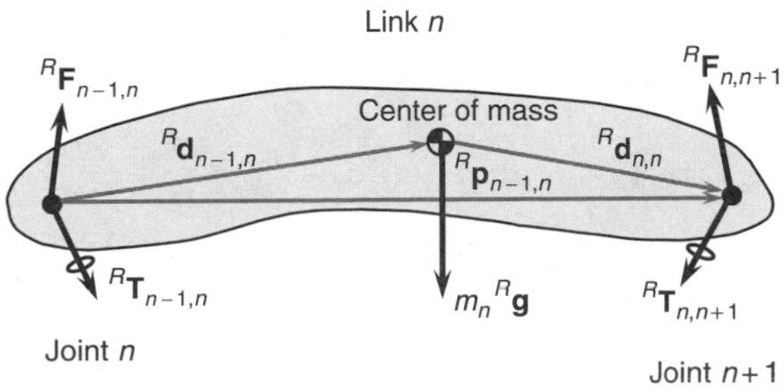

FIGURE 14.13 Forces and torques acting on link n.

where ${}^{R}\mathbf{T}_{n\text{-}1,n}$ is the torque applied to link n by link $n-1$ as seen by an observer in R and ${}^{R}\mathbf{d}_{n\text{-}1,n}$ is the position vector from the origin of the frame $n-1$ at joint n to the center of mass of link n as seen by an observer in R.

In dynamic equilibrium, the forces and torques on the link are balanced, giving a zero net force and torque. For example, a robot moving in free space may achieve torque balance when the actuator torques match the torques due to inertia and gravitation. If an external force or torque is applied to the tip of the robot, changes in the actuator torques may be required to regain torque balance. By rearranging Equation (14.5) and Equation (14.6) it is possible to write expressions for the joint force and torque corresponding to dynamic equilibrium.

$$ {}^{R}\mathbf{F}_{n-1,n} = m_n {}^{R}\mathbf{a}_n + {}^{R}\mathbf{F}_{n,n+1} - m_n {}^{R}\mathbf{g} \tag{14.7} $$

$$ {}^{R}\mathbf{T}_{n-1,n} = {}^{R}\mathbf{T}_{n,n+1} + {}^{R}\mathbf{d}_{n-1,n} \times {}^{R}\mathbf{F}_{n-1,n} - {}^{R}\mathbf{d}_{n,n} \times {}^{R}\mathbf{F}_{n,n+1} + {}^{C}\mathbf{I}_n {}^{R}\boldsymbol{\alpha}_n + {}^{R}\boldsymbol{\omega}_n \times \left({}^{C}\mathbf{I}_n {}^{R}\boldsymbol{\omega}_n\right) \tag{14.8} $$

Substituting Equation (14.7) into Equation (14.8) gives an equation for the torque at the joint with respect to R in terms of center-of-mass velocities and accelerations.

$$ \begin{aligned} {}^{R}\mathbf{T}_{n-1,n} = {}& {}^{R}\mathbf{T}_{n,n+1} + {}^{R}\mathbf{d}_{n-1,n} \times m_n {}^{R}\mathbf{a}_n + {}^{R}\mathbf{p}_{n-1,n} \times {}^{R}\mathbf{F}_{n,n+1} \\ & - {}^{R}\mathbf{d}_{n-1,n} \times m_n {}^{R}\mathbf{g} + {}^{C}\mathbf{I}_n {}^{R}\boldsymbol{\alpha}_n + {}^{R}\boldsymbol{\omega}_n \times \left({}^{C}\mathbf{I}_n {}^{R}\boldsymbol{\omega}_n\right) \end{aligned} \tag{14.9} $$

where ${}^{R}\mathbf{p}_{n\text{-}1,n}$ is the position vector from the origin of frame $n-1$ at joint n to the origin of frame n as seen by an observer in R. Equation (14.7) and Equation (14.9) represent one form of the Newton–Euler (N–E) equations. They describe the applied force and torque at joint n, supplied for example by an actuator, in terms of other forces and torques, both active and inertial, acting on the link.

Two-Link Robot Example. In this example, the dynamics of a two-link robot are derived using the N–E equations. From Newton's Equation (14.3) the net dynamic forces acting at the center of mass of each link are related to the mass center accelerations,

$$ {}^{0}\mathbf{F}_1 = m_1 {}^{0}\mathbf{a}_1, \quad {}^{0}\mathbf{F}_2 = m_2 {}^{0}\mathbf{a}_2 \tag{14.10} $$

and from Euler's Equation (14.4) the net dynamic torques acting around the center of mass of each link are related to the angular velocities and angular accelerations,

$$ {}^{0}\mathbf{T}_1 = \mathbf{I}_1 {}^{0}\boldsymbol{\alpha}_1 + {}^{0}\boldsymbol{\omega}_1 \times \left(\mathbf{I}_1 {}^{0}\boldsymbol{\omega}_1\right), \quad {}^{0}\mathbf{T}_2 = \mathbf{I}_2 {}^{0}\boldsymbol{\alpha}_2 + {}^{0}\boldsymbol{\omega}_2 \times \left(\mathbf{I}_2 {}^{0}\boldsymbol{\omega}_2\right) \tag{14.11} $$

where the mass moments of inertia are with respect to each link's centroidal frame. Superscript 0 is used to denote the inertial reference frame R.

Equation (14.9) can be used to find the torques at the joints, as follows:

$$ \begin{aligned} {}^{0}\mathbf{T}_{1,2} = {}& {}^{0}\mathbf{T}_{2,3} + {}^{0}\mathbf{d}_{1,2} \times m_2 {}^{0}\mathbf{a}_2 + {}^{0}\mathbf{p}_{1,2} \times {}^{0}\mathbf{F}_{2,3} \\ & - {}^{0}\mathbf{d}_{1,2} \times m_2 {}^{0}\mathbf{g} + \mathbf{I}_2 {}^{0}\boldsymbol{\alpha}_2 + {}^{0}\boldsymbol{\omega}_2 \times \left(\mathbf{I}_2 {}^{0}\boldsymbol{\omega}_2\right) \end{aligned} \tag{14.12} $$

$$ \begin{aligned} {}^{0}\mathbf{T}_{0,1} = {}& {}^{0}\mathbf{T}_{1,2} + {}^{0}\mathbf{d}_{0,1} \times m_1 {}^{0}\mathbf{a}_1 + {}^{0}\mathbf{p}_{0,1} \times {}^{0}\mathbf{F}_{1,2} \\ & - {}^{0}\mathbf{d}_{0,1} \times m_1 {}^{0}\mathbf{g} + \mathbf{I}_1 {}^{0}\boldsymbol{\alpha}_1 + {}^{0}\boldsymbol{\omega}_1 \times \left(\mathbf{I}_1 {}^{0}\boldsymbol{\omega}_1\right) \end{aligned} \tag{14.13} $$

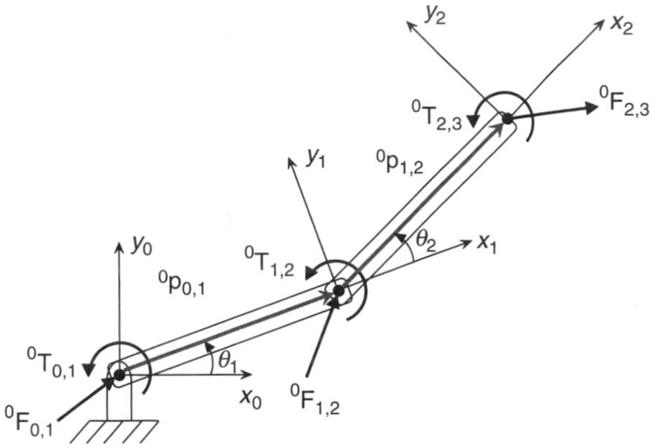

FIGURE 14.14 Two-link robot with two revolute joints.

An expression for $^0\mathbf{F}_{1,2}$ can be found from Equation (14.7) and substituted into Equation (14.13) to give

$$\begin{aligned}{}^0\mathbf{T}_{0,1} = {}^0\mathbf{T}_{1,2} + {}^0\mathbf{d}_{0,1} \times m_1{}^0\mathbf{a}_1 + {}^0\mathbf{p}_{0,1} \times m_2{}^0\mathbf{a}_2 + {}^0\mathbf{p}_{0,1} \times {}^0\mathbf{F}_{2,3} - {}^0\mathbf{p}_{0,1} \times m_2{}^0\mathbf{g} \\ - {}^0\mathbf{d}_{0,1} \times m_1{}^0\mathbf{g} + \mathbf{I}_1{}^0\boldsymbol{\alpha}_1 + {}^0\boldsymbol{\omega}_1 \times \left(\mathbf{I}_1{}^0\boldsymbol{\omega}_1\right)\end{aligned} \tag{14.14}$$

For a planar two-link robot with two revolute joints, as shown in Figure 14.14, moving in a horizontal plane, i.e., perpendicular to gravity, several simplifications can be made in Equation (14.11) to Equation (14.14): (i) the gyroscopic terms $^0\boldsymbol{\omega}_n \times ({}^C\mathbf{I}_n{}^0\boldsymbol{\omega}_n)$ for $n = 1, 2$ can be eliminated, since the mass moment of inertia matrix is a scalar and the cross product of a vector with itself is zero, (ii) the gravity terms $^0\mathbf{d}_{n-1,n} \times m_n{}^0\mathbf{g}$ disappear, since the robot moves in a horizontal plane, and (iii) the torques can be written as scalars and act perpendicular to the plane of motion. Equation (14.12) and Equation (14.14) can then be written in scalar form,

$${}^0\mathrm{T}_{1,2} = {}^0\mathrm{T}_{2,3} + \left({}^0\mathbf{d}_{1,2} \times m_2{}^0\mathbf{a}_2\right)\cdot\hat{\mathbf{u}}_z + \left({}^0\mathbf{p}_{1,2} \times {}^0\mathbf{F}_{2,3}\right)\cdot\hat{\mathbf{u}}_z + \mathrm{I}_2{}^0\boldsymbol{\alpha}_2 \tag{14.15}$$

$${}^0\mathrm{T}_{0,1} = {}^0\mathrm{T}_{1,2} + \left({}^0\mathbf{d}_{0,1} \times m_1{}^0\mathbf{a}_1\right)\cdot\hat{\mathbf{u}}_z + \left({}^0\mathbf{p}_{0,1} \times m_2{}^0\mathbf{a}_2\right)\cdot\hat{\mathbf{u}}_z + \left({}^0\mathbf{p}_{0,1} \times {}^0\mathbf{F}_{2,3}\right)\cdot\hat{\mathbf{u}}_z + \mathrm{I}_1{}^0\boldsymbol{\alpha}_1 \tag{14.16}$$

where the dot product is taken (with the terms in parentheses) with the unit vector $\hat{\mathbf{u}}_z$ perpendicular to the plane of motion.

A further simplification can be introduced by assuming the links have uniform cross sections and are made of homogeneous (constant density) material. The center of mass then coincides with the geometric center, and the distance to the center of mass is half the link length. The position vectors can be written as,

$${}^0\mathbf{p}_{0,1} = 2{}^0\mathbf{d}_{0,1}, \quad {}^0\mathbf{p}_{1,2} = 2{}^0\mathbf{d}_{1,2} \tag{14.17}$$

with the components,

$$\begin{bmatrix} {}^0\mathbf{p}_{0,1x} \\ {}^0\mathbf{p}_{0,1y} \end{bmatrix} = 2\begin{bmatrix} {}^0\mathbf{d}_{0,1x} \\ {}^0\mathbf{d}_{0,1y} \end{bmatrix} = \begin{bmatrix} l_1C_1 \\ l_1S_1 \end{bmatrix}, \quad \begin{bmatrix} {}^0\mathbf{p}_{1,2x} \\ {}^0\mathbf{p}_{1,2y} \end{bmatrix} = 2\begin{bmatrix} {}^0\mathbf{d}_{1,2x} \\ {}^0\mathbf{d}_{1,2y} \end{bmatrix} = \begin{bmatrix} l_2C_{12} \\ l_2S_{12} \end{bmatrix} \tag{14.18}$$

where the notation $S_1 = \sin\theta_1, C_1 = \cos\theta_1$, $S_{12} = \sin(\theta_1+\theta_2), C_{12} = \cos(\theta_1+\theta_2)$ is introduced, and l_1, l_2 are the lengths of links 1 and 2, respectively.

The accelerations of the center of mass of each link are needed in Equation (14.15) and Equation (14.16). These accelerations can be written as

$$ {}^0\mathbf{a}_1 = {}^0\boldsymbol{\alpha}_1 \times {}^0\mathbf{d}_{0,1} + {}^0\boldsymbol{\omega}_1 \times \left({}^0\boldsymbol{\omega}_1 \times {}^0\mathbf{d}_{0,1}\right) \tag{14.19} $$

$$ {}^0\mathbf{a}_2 = 2{}^0\mathbf{a}_1 + {}^0\boldsymbol{\alpha}_2 \times {}^0\mathbf{d}_{1,2} + {}^0\boldsymbol{\omega}_2 \times \left({}^0\boldsymbol{\omega}_2 \times {}^0\mathbf{d}_{1,2}\right) \tag{14.20} $$

Expanding Equation (14.19) and Equation (14.20) gives the acceleration components:

$$ \begin{bmatrix} {}^0\mathbf{a}_{1x} \\ {}^0\mathbf{a}_{1y} \end{bmatrix} = \begin{bmatrix} -\frac{1}{2}l_1 S_1 \ddot{\theta}_1 - \frac{1}{2}l_1 C_1 \dot{\theta}_1^2 \\ \frac{1}{2}l_1 C_1 \ddot{\theta}_1 - \frac{1}{2}l_1 S_1 \dot{\theta}_1^2 \end{bmatrix} \tag{14.21} $$

$$ \begin{bmatrix} {}^0\mathbf{a}_{2x} \\ {}^0\mathbf{a}_{2y} \end{bmatrix} = \begin{bmatrix} -\left(l_1 S_1 + \frac{1}{2}l_2 S_{12}\right)\ddot{\theta}_1 - \frac{1}{2}l_2 S_{12}\ddot{\theta}_2 - l_1 C_1 \dot{\theta}_1^2 - \frac{1}{2}l_2 C_{12}\left(\dot{\theta}_1 + \dot{\theta}_2\right)^2 \\ \left(l_1 C_1 + \frac{1}{2}l_2 C_{12}\right)\ddot{\theta}_1 + \frac{1}{2}l_2 C_{12}\ddot{\theta}_2 - l_1 S_1 \dot{\theta}_1^2 - \frac{1}{2}l_2 S_{12}\left(\dot{\theta}_1 + \dot{\theta}_2\right)^2 \end{bmatrix} \tag{14.22} $$

Equation (14.15) gives the torque at joint 2:

$$ \begin{aligned} {}^0\mathrm{T}_{1,2} = {}^0\mathrm{T}_{2,3} &+ m_2\left[\left({}^0\mathrm{d}_{1,2x}\hat{\mathbf{u}}_{x0} + {}^0\mathrm{d}_{1,2y}\hat{\mathbf{u}}_{y0}\right) \times \left({}^0\mathrm{a}_{2x}\hat{\mathbf{u}}_{x0} + {}^0\mathrm{a}_{2y}\hat{\mathbf{u}}_{y0}\right)\right]\cdot\hat{\mathbf{u}}_z \\ &+ \left[\left({}^0\mathrm{p}_{1,2x}\hat{\mathbf{u}}_{x0} + {}^0\mathrm{p}_{1,2y}\hat{\mathbf{u}}_{y0}\right) \times \left({}^0\mathrm{F}_{2,3x}\hat{\mathbf{u}}_{x0} + {}^0\mathrm{F}_{2,3y}\hat{\mathbf{u}}_{y0}\right)\right]\cdot\hat{\mathbf{u}}_z + \mathrm{I}_2{}^0\alpha_2 \end{aligned} $$

where $\hat{\mathbf{u}}_{x0}$, $\hat{\mathbf{u}}_{y0}$ are the unit vectors along the x_0, y_0 axes, respectively. Substituting and evaluating yields,

$$ \begin{aligned} {}^0\mathrm{T}_{1,2} = {}^0\mathrm{T}_{2,3} &+ \left(\tfrac{1}{2}m_2 l_1 l_2 C_2 + \tfrac{1}{4}m_2 l_2^2 + \mathrm{I}_2\right)\ddot{\theta}_1 + \left(\tfrac{1}{4}m_2 l_2^2 + \mathrm{I}_2\right)\ddot{\theta}_2 + \tfrac{1}{2}m_2 l_1 l_2 \mathrm{S}_1 \dot{\theta}_1^2 \\ &+ l_2 C_{12}{}^0\mathrm{F}_{2,3y} - l_2 \mathrm{S}_{12}{}^0\mathrm{F}_{2,3x} \end{aligned} \tag{14.23} $$

Similarly, Equation (14.16) gives the torque at joint 1:

$$ \begin{aligned} {}^0\mathrm{T}_{0,1} = {}^0\mathrm{T}_{1,2} &+ m_1\left[\left({}^0\mathrm{d}_{0,1x}\hat{\mathbf{u}}_{x0} + {}^0\mathrm{d}_{0,1y}\hat{\mathbf{u}}_{y0}\right) \times \left({}^0\mathrm{a}_{1x}\hat{\mathbf{u}}_{x0} + {}^0\mathrm{a}_{1y}\hat{\mathbf{u}}_{y0}\right)\right]\cdot\hat{\mathbf{u}}_z \\ &+ m_2\left[\left({}^0\mathrm{p}_{0,1x}\hat{\mathbf{u}}_{x0} + {}^0\mathrm{p}_{0,1y}\hat{\mathbf{u}}_{y0}\right) \times \left({}^0\mathrm{a}_{2x}\hat{\mathbf{u}}_{x0} + {}^0\mathrm{a}_{2y}\hat{\mathbf{u}}_{y0}\right)\right]\cdot\hat{\mathbf{u}}_z \\ &+ \left[\left({}^0\mathrm{p}_{0,1x}\hat{\mathbf{u}}_{x0} + {}^0\mathrm{p}_{0,1y}\hat{\mathbf{u}}_{y0}\right) \times \left({}^0\mathrm{F}_{2,3x}\hat{\mathbf{u}}_{x0} + {}^0\mathrm{F}_{2,3y}\hat{\mathbf{u}}_{y0}\right)\right]\cdot\hat{\mathbf{u}}_z + \mathrm{I}_1{}^0\alpha_1 \end{aligned} $$

which yields after substituting,

$$ \begin{aligned} {}^0\mathrm{T}_{0,1} = {}^0\mathrm{T}_{1,2} &+ \left(\tfrac{1}{4}m_1 l_1^2 + m_2 l_2^2 + \tfrac{1}{2}m_2 l_1 l_2 C_2 + \mathrm{I}_1\right)\ddot{\theta}_1 + \left(\tfrac{1}{2}m_2 l_1 l_2 C_2\right)\ddot{\theta}_2 \\ &- \tfrac{1}{2}m_2 l_1 l_2 \mathrm{S}_2 \dot{\theta}_1^2 - \tfrac{1}{2}m_2 l_1 l_2 \mathrm{S}_2 \dot{\theta}_2^2 - m_2 l_1 l_2 \mathrm{S}_2 \dot{\theta}_1 \dot{\theta}_2 \\ &+ l_1 C_1{}^0\mathrm{F}_{2,3y} - l_1 \mathrm{S}_1{}^0\mathrm{F}_{2,3x} \end{aligned} \tag{14.24} $$

The first term on the right-hand side of Equation (14.24) has been found in Equation (14.23) as the torque at joint 2. Substituting this torque into Equation (14.24) gives:

$$\begin{aligned}{}^{0}\mathrm{T}_{0,1} = {}^{0}\mathrm{T}_{2,3} &+ \left(\tfrac{1}{4}m_1 l_1^2 + \tfrac{1}{4}m_2 l_2^2 + m_2 l_1^2 + m_2 l_1 l_2 \mathrm{C}_2 + \mathrm{I}_1 + \mathrm{I}_2\right)\ddot{\theta}_1 \\ &+ \left(\tfrac{1}{4}m_2 l_2^2 + \tfrac{1}{2}m_2 l_1 l_2 \mathrm{C}_2 + \mathrm{I}_2\right)\ddot{\theta}_2 - \tfrac{1}{2}m_2 l_1 l_2 \mathrm{S}_2 \dot{\theta}_2^2 \\ &- m_2 l_1 l_2 \mathrm{S}_2 \dot{\theta}_1 \dot{\theta}_2 + (l_1 \mathrm{C}_1 + l_2 \mathrm{C}_{12})^0\mathrm{F}_{2,3y} - (l_1 \mathrm{S}_1 + l_2 \mathrm{S}_{12})^0\mathrm{F}_{2,3x}\end{aligned} \tag{14.25}$$

Closed-Form Equations. The N–E Equation (14.7) and Equation (14.9) can be expressed in an alternative form more suitable for use in controlling the motion of a robot. In this recast form they represent input–output relationships in terms of independent variables, sometimes referred to as the generalized coordinates, such as the joint-position variables. For direct dynamics, the inputs can be taken as the joint torques and the outputs as the joint-position variables, i.e., joint angles. For inverse dynamics, the inputs are the joint-position variables and the outputs are the joint torques. The inverse-dynamics form is well suited for robot control and programming, since it can be used to determine the appropriate inputs necessary to achieve the desired outputs.

The N–E equations can be written in scalar closed-form,

$$\mathrm{T}_{n-1,n} = \sum_{j=1}^{N} \mathrm{J}_{nj}\ddot{\theta}_j + \sum_{j=1}^{N}\sum_{k=1}^{N} \mathrm{D}_{njk}\dot{\theta}_j\dot{\theta}_k + \mathrm{T}_{ext,n}, \quad n = 1, \ldots, N \tag{14.26}$$

where J_{nj} is the mass moment of inertia of link j reflected to joint n assuming N total links, D_{njk} is a coefficient representing the centrifugal effect when $j = k$ and the Coriolis effect when $j \neq k$ at joint n, and $\mathrm{T}_{ext,n}$ is the torque arising from external forces and torques at joint n, including the effect of gravity and end-effector forces and torques.

For the two-link planar robot example, Equation (14.26) expands to:

$$\mathrm{T}_{0,1} = \mathrm{J}_{11}\ddot{\theta}_1 + \mathrm{J}_{12}\ddot{\theta}_2 + (\mathrm{D}_{112} + \mathrm{D}_{121})\dot{\theta}_1\dot{\theta}_2 + \mathrm{D}_{122}\dot{\theta}_2^2 + \mathrm{T}_{ext,1} \tag{14.27}$$

$$\mathrm{T}_{1,2} = \mathrm{J}_{21}\ddot{\theta}_1 + \mathrm{J}_{22}\ddot{\theta}_2 + \mathrm{D}_{211}\dot{\theta}_1^2 + \mathrm{T}_{ext,2} \tag{14.28}$$

Expressions for the coefficients in Equation (14.27) and Equation (14.28) can be found by comparison with Equation (14.23) and Equation (14.25). For example,

$$\mathrm{J}_{11} = \tfrac{1}{4}m_1 l_1^2 + \tfrac{1}{4}m_2 l_2^2 + m_2 l_1^2 + m_2 l_1 l_2 \mathrm{C}_2 + \mathrm{I}_1 + \mathrm{I}_2 \tag{14.29}$$

is the total effective moment of inertia of both links reflected to the axis of the first joint, where from the parallel-axis theorem the inertia of link 1 about joint 1 is $\tfrac{1}{4}m_1 l_1^2 + \mathrm{I}_1$ and the inertia of link 2 about joint 1 is $m_2\left(l_1^2 + \tfrac{1}{4}l_2^2 + l_1 l_2 \mathrm{C}_2\right) + \mathrm{I}_2$. Note that the inertia reflected from link 2 to joint 1 is a function of the configuration, being greatest when the arm is straight (when the cosine of θ_2 is 1).

The first term in Equation (14.27) is the torque due to the angular acceleration of link 1, whereas the second term is the torque at joint 1 due the angular acceleration of link 2. The latter arises from the contribution of the coupling force and torque across joint 2 on link 1. By comparison, the total effective moment of inertia of link 2 reflected to the first joint is

$$\mathrm{J}_{12} = \tfrac{1}{4}m_2 l_2^2 + m_2 l_1 l_2 \mathrm{C}_2 + \mathrm{I}_2$$

which is again configuration dependent.

The third term in Equation (14.27) is the torque due to the Coriolis force,

$$T_{0,1Coriolis} = (D_{112} + D_{121})\dot{\theta}_1\dot{\theta}_2 = -m_2 l_1 l_2 S_2 \dot{\theta}_1 \dot{\theta}_2$$

resulting from the interaction of the two rotating frames. The fourth term in Equation (14.27) is the torque due to the centrifugal effect on joint 1 of link 2 rotating at angular velocity $\dot{\theta}_2$,

$$T_{0,1centrifugal} = D_{122}\dot{\theta}_2^2 = -\tfrac{1}{2} m_2 l_1 l_2 S_2 \dot{\theta}_2^2$$

The Coriolis and centrifugal torque terms are also configuration dependent, and vanish when the links are co-linear. For moving links, these terms arise from Coriolis and centrifugal forces that can be viewed as acting through the center of mass of the second link, producing a torque around the second joint that is reflected to the first joint.

Similarly, physical meaning can be associated with the terms in Equation (14.28) for the torque at the second joint.

Robot Control

Control Overview

The subject of this section is robot control and how to design a controller to achieve the target dynamic performance for a robotic system. Nominally, the control of a robot takes the desired end-effector trajectory and converts it into desired arm-angle trajectories and eventually into signals that drive the arm actuators. The complex models developed in the previous section can be used for controller design and development. However, such controllers are complex in nature and require very specialized analysis and designs. The scope of this section is to present some of the considerations in controlling a single link of a robot. This type of control is called independent joint control. This approach suffices for slower moving robotic systems, but does not yield an optimal response for high-speed robots. In many robotic systems a set of single-link controllers are used that ignore the interlink dynamics. This approach often suffices and is an excellent starting point from which to develop more advanced controllers that take into consideration the higher-order and coupling dynamics discussed in the previous section.

Background

The fundamental control problem in robotics is to determine the actuator signals required to achieve the desired motion and specified performance criteria. If the robot is to perform a task while in contact with a surface, it is also necessary to control the contact force applied by the manipulator. Although the control problem can be stated simply, its solution may be quite complicated due to robot nonlinearities. For example, because of coupling among links, the robot dynamics in a serial-link robot design are described mathematically by a set of coupled nonlinear differential equations, making the controls problem challenging. The controls problem becomes even more difficult if the links exhibit flexibility, and hence cannot be modeled as rigid.

To achieve the desired motion and possibly contact-force characteristics, the planning of the manipulator trajectory is integrally linked to the control problem. The position of the robot can be described by a set of joint coordinates in **joint space** or by the position and orientation of the end-effector using coordinates along orthogonal axes in **Cartesian or task space**. The two representations are related, i.e., the Cartesian position and orientation can be computed from the joint positions via a mapping (or function) known as **forward kinematics.** The motion required to realize the desired task is generally specified in Cartesian space. The joint positions required to achieve the desired end-effector position and orientation can be found by a mapping known as the **inverse kinematics**. This inverse-kinematics problem may have more than one solution, and a closed-form solution may not be possible, depending on the geometric configuration of the robot. The desired motion may be specified as point-to-point, in which the end-effector moves from one point to another without regard to the path, or it may be specified as a continuous path, in which the end-effector follows a desired path between the points. A trajectory planner generally interpolates the desired path and generates a sequence of set points for the controller. The interpolation may be done in joint or Cartesian space.

Some of the robot control schemes in use today include independent joint control [Luh, 1983]; Cartesian-space control [Luh et al., 1980]; and force-control strategies such as hybrid position/force control [Raibert and Craig, 1981] and impedance control [Hogan, 1985] (Figure 14.15). In independent joint control, each joint is considered as a separate system and the coupling effects between the links are treated as disturbances to be rejected by the controller. The performance can be enhanced by compensating for the robot nonlinearities and interlink coupling using the method of computed torque or inverse dynamics. In Cartesian-space control the error signals are computed in Cartesian space and the inverse-kinematics problem need not be solved. In this chapter, independent joint control is analyzed in depth to provide insight into the use of various controllers for joint-position control. Information on the force-control schemes can be found in the references cited above as well as in [Bonitz, 1995].

Basic Joint-Position Dynamic Model

Many industrial robots operate at slow speeds, employing large drive reductions that significantly reduce the coupling effects between the links. For slowly varying command inputs, the drive system dominates the dynamics of each joint and each link can be considered rigid. Under these conditions each joint can be controlled as an independent system using linear system control techniques. Figure 14.16 is a model of a

FIGURE 14.15 Case packer cartesian robot. (Courtesy of CAMotion, Inc.)

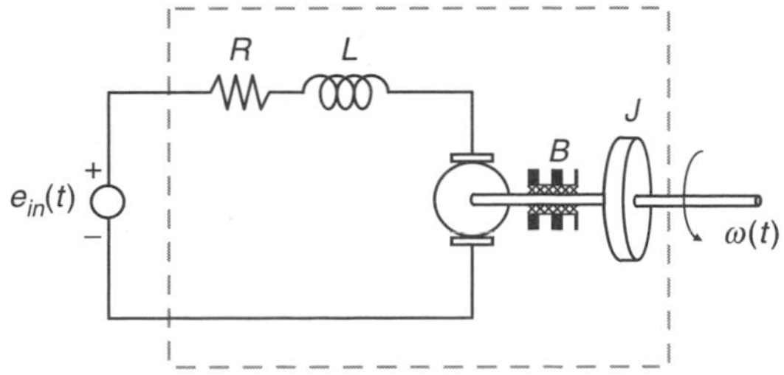

FIGURE 14.16 Simple actuator model for a single-link robot.

single-actuator-link model, representative of a single-link robot. The model includes an effective inertia, J,

$$J = J_m + r^2 J_l \tag{14.30}$$

where r is the gear ratio of the joint, J_m is the motor inertia, and J_l is the link inertia. The model also includes an effective damping coefficient, B, and a motor armature inductance and resistance, L and R, respectively. The relationship between the motor input voltage and the rotational speed is given by the second-order differential equation,

$$\ddot{\omega}(t) + (JR + BL)\dot{\omega}(t) + \left(\frac{RB}{JL} + \frac{K_t^2}{JL}\right)\omega(t) = \frac{K_t}{JL} e_{in}(t) \tag{14.31}$$

where $e_{in}(t)$ is the motor input voltage, $\omega(t)$ is the link angular velocity or rotational speed, and K_t, is a motor torque constant (torque per current ratio). Assuming zero initial conditions and defining the Laplace transformations,

$$\Im\{\omega(t)\} \equiv \Omega(s) \qquad \Im\{e_{in}(t)\} \equiv E_{in}(s),$$

the transfer function between the input voltage, $E_{in}(s)$, and the link rotational velocity, $\Omega(s)$, can be written as

$$\frac{\Omega(s)}{E_{in}(s)} = \frac{\frac{K_t}{JL}}{s^2 + (JR + BL)s + \left(\frac{RB}{JL} + \frac{K_t^2}{JL}\right)} \tag{14.32}$$

For most motor systems, the overall system dynamics are dominated by the mechanical dynamics rather than the electrical "dynamics," i.e., the effect of L is small in comparison with the other system parameters. Assuming negligible L in Equation (14.32), the following first-order transfer function can be developed

$$\lim_{L\to 0}\left[\frac{\Omega(s)}{E_{in}(s)}\right] = \lim_{L\to 0}\left[\frac{\frac{K_t}{JL}}{s^2 + (JR + BL)s + \left(\frac{RB}{JL} + \frac{K_t^2}{JL}\right)}\right] = \frac{K_t}{JRs + \left(RB + \frac{K_t^2}{JL}\right)} \tag{14.33}$$

Defining the open-loop joint time constant, T_m, as

$$T_m = \frac{JR}{RB + \left(\frac{K_t^2}{JL}\right)}, \tag{14.34}$$

and the open-loop link gain as

$$K_m \equiv \frac{K_t}{RB + \left(\frac{K_t^2}{JL}\right)}, \tag{14.35}$$

Equation (14.33) can be rewritten as

$$\frac{\Omega(s)}{E_{in}(s)} = \frac{K_m}{T_m s + 1}. \tag{14.36}$$

The assumption of the motor inductance being small, resulting in negligible electrical dynamics, is considered reasonable if the negative real part of the single pole from the electrical dynamics is approximately three times larger than the negative real part of the mechanical dynamics. This is known as the dominant-pole theory and is valid for most motors. However, for small motors, such as those used in microelectromechanical system (MEMS) devices, this assumption may not be valid.

Equation (14.36) provides a relationship between joint angular velocity and input motor voltage, and is a useful first-order model for velocity control. However, for most robot applications, position control rather than velocity control is desired. For the typical case of lower velocities, the dynamics of the individual joints can be considered to be decoupled. Recognizing that the position of the joint is the integral of the joint velocity, a transfer function between position $\Theta(s)$, which is the Laplace transform of $\theta(t)$, and the input voltage can be developed,

$$G(s) = \frac{\Theta(s)}{E_{\mathrm{in}}(s)} = \frac{\frac{1}{s}\Omega(s)}{E_{\mathrm{in}}(s)} = \left(\frac{1}{s}\right)\frac{K_{\mathrm{m}}}{T_{\mathrm{m}}s+1} = \frac{K_{\mathrm{m}}}{s(T_{\mathrm{m}}s+1)} \tag{14.37}$$

where $G(s)$ is the plant transfer function for the single link. Equation (14.9) is a model for position control, and the same simple dynamics can be used to model a variety of systems, including high-precision machine tools, which are essentially multiaxis Cartesian robots [Kurfess, 2000, 2002].

Independent Joint-Position Control

For many applications, the assumption of negligible link coupling is reasonable, and independent joint-position control can be a successful strategy. Two classical control methods — proportional derivative (PD) control and proportional integral derivative (PID) control — are typically implemented. This section develops design strategies for both types and, for completeness, considers the even simpler proportional (P) controller. Figure 14.17 represents the closed-loop feedback configuration for the single-link model, where $G(s)$ represents the link or plant dynamics given by Equation (14.37), $K(s)$ represents the controller that is to be designed, and $H(s)$ represents the feedback sensor dynamics. In general, the relationship between the desired angular position, $\Theta_d(s)$, and the actual angular position of the link, $\Theta(s)$, is given by Black's Law,

$$G_{\mathrm{CL}}(s) = \frac{\Theta(s)}{\Theta_{\mathrm{d}}(s)} = \frac{K(s)G(s)}{1+K(s)G(s)H(s)} \tag{14.38}$$

The angular error, $E(s)$, is given by the difference between the desired angle and the actual angle. The transfer function between $\Theta_{\mathrm{d}}(s)$ and $E(s)$ is given by

$$\frac{E(s)}{\Theta_{\mathrm{d}}(s)} = \frac{\Theta_{\mathrm{d}}(s)-\Theta(s)}{\Theta_{\mathrm{d}}(s)} = \frac{1}{1+K(s)G(s)H(s)} \tag{14.39}$$

The controller typically includes a power amplifier that provides the power to drive the system. In this example, the output of the controller is a voltage, i.e., the power amplifier's function is to provide sufficient current at the desired voltage to drive the system. Saturation occurs when the desired input voltage exceeds the maximum amplifier voltage output capacity, or when the product of the voltage and current exceeds the amplifier's rated power capacity. The effect of saturation is ignored in the examples to follow. Finally, the controller output voltage can be related to the input command signal via the following transfer function,

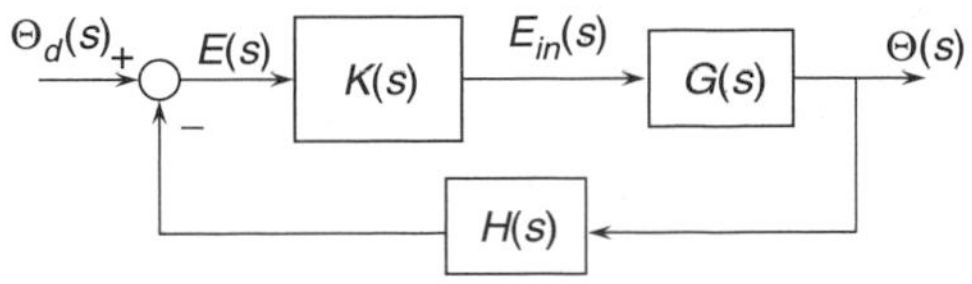

FIGURE 14.17 Generalized closed-loop system configuration.

$$\frac{E_{\text{in}}(s)}{\Theta_{\text{d}}(s)} = \frac{E(s)}{\Theta_{\text{d}}(s)}K(s) = \frac{\Theta_{\text{d}}(s) - \Theta(s)}{\Theta_{\text{d}}(s)}K(s) = \frac{K(s)}{1 + K(s)G(s)H(s)}. \tag{14.40}$$

Equation (14.40) can be used to determine the motor command voltage as a function of the desired angular trajectory.

Definition of Specifications. Specifications are generally provided to ensure that the desired system behavior is achieved. The specifications define the closed-loop system characteristics, and the usual practice is to tune the controller gains to achieve the desired performance. For this discussion, a 2% settling time, t_s, and a maximum percent overshoot, M_P, are specified. These two quantities (and others) are visualized in the typical second-order response shown in Figure 14.18. They define a desired dynamic response of the closed-loop system described by a second-order model of the form,

$$G_{\text{CL}}(s) = \frac{\omega_{\text{nd}}^2}{s^2 + 2\zeta_{\text{d}}\omega_{\text{nd}}s + \omega_{\text{nd}}^2}, \tag{14.41}$$

where ζ_{d} and ω_{nd} are the desired damping ratio and (undamped) natural frequency, respectively, and are related to the desired overshoot M_{Pd} and desired settling time t_{sd}, of the target closed-loop system by the expressions,

$$M_{\text{Pd}} = \mathrm{e}^{-\left(\zeta_{\text{d}}/\sqrt{1-\zeta_{\text{d}}^2}\right)\pi} \tag{14.42}$$

$$t_{\text{sd}} \approx \frac{4}{\zeta_{\text{d}}\omega_{\text{nd}}} \tag{14.43}$$

where the latter is a conservative approximation of the settling time. Solving for ζ_{d} in Equation (14.42) yields

$$\zeta_{\text{d}} = \sqrt{\frac{\ln(2M_{\text{Pd}})}{\pi^2 + \ln(2M_{\text{Pd}})}} = \frac{\ln(M_{\text{Pd}})}{\sqrt{\pi^2 + \ln(2M_{\text{Pd}})}} \tag{14.44}$$

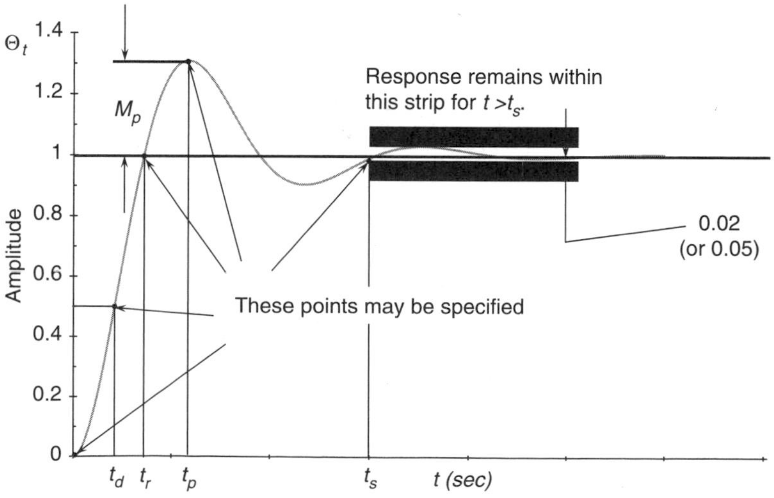

FIGURE 14.18 A typical second-order response.

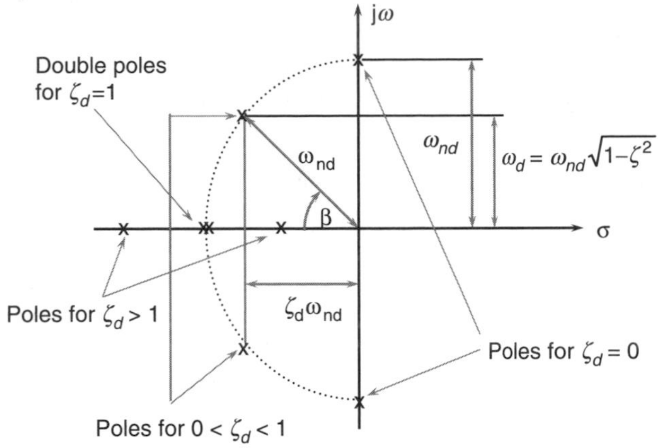

FIGURE 14.19 Pole locations for desired closed-loop system dynamics.

From Equation (14.43) and Equation (14.44), ω_{nd} can be expressed as

$$\omega_{\mathrm{nd}} \approx \frac{4\sqrt{\pi^2 + \ln(2M_{\mathrm{P}})}}{t_{\mathrm{sd}} \ln(M_{\mathrm{P}})} \tag{14.45}$$

From Equation (14.41), the characteristic equation that defines the system response is given by

$$s^2 + 2\zeta_{\mathrm{d}}\omega_{\mathrm{nd}}s + \omega_{\mathrm{nd}}^2 = 0. \tag{14.46}$$

This equation can be used for the design of various controllers. The poles of the closed-loop system are shown in Figure 14.19 for a variety of desired damping ratios, ζ_d. Nominally, for joint-position control a damping ratio of 1, corresponding to a critically damped system, is desired. This places both poles in the same location on the real axis, and yields the fastest system response without overshoot. For generality in the control design examples, the variables ζ_{d} and ω_{nd} are used rather than specific values for these variables [Nagurka and Kurfess, 1992].

Proportional (P) Control. In proportional (P) control, the simplest analog control algorithm, the control signal is proportionally related to the error signal by a constant gain, K_{P}. Thus, the transfer function for a proportional controller is a constant,

$$K(s) = K_P. \tag{14.47}$$

The closed-loop transfer function of the single actuator-link system with a proportional controller is

$$G_{\mathrm{CL}}(s) = \frac{K_{\mathrm{P}}K_{\mathrm{m}}/T_{\mathrm{m}}}{s^2 + (1/T_{\mathrm{m}})s + K_{\mathrm{P}}K_{\mathrm{m}}/T_{\mathrm{m}}} \tag{14.48}$$

from Equation (14.37), Equation (14.38), and Equation (14.47). The characteristic equation for the transfer function given in Equation (14.48) is

$$s^2 + (1/T_{\mathrm{m}})s + K_{\mathrm{P}}K_{\mathrm{m}}/T_{\mathrm{m}} = 0. \tag{14.49}$$

With one free design parameter, the proportional gain, K_{P}, the target closed-loop system dynamics will be virtually impossible to achieve. By equating the actual closed-loop characteristic Equation (14.49) to the desired closed-loop characteristic Equation (14.46),

$$s^2 + (1/T_m)s + K_P K_m/T_m = 0 = s^2 + 2\zeta_d \omega_{nd} s + \omega_{nd}^2, \tag{14.50}$$

only ω_{nd} can be specified directly, giving the value of K_P

$$K_P = \omega_{nd}^2 \frac{T_m}{K_m} \tag{14.51}$$

(found by equating the coefficients of s^0). Once ω_{nd} has been chosen, ζ_d is fixed to be

$$\zeta_d = \frac{1}{2\omega_{nd} T_m} \tag{14.52}$$

It is useful to investigate the location of the closed-loop poles as a function of K_P. The pole locations on the s-plane can be computed by solving for s in Equation (14.50) via the quadratic formula

$$s = \frac{-\frac{1}{T_m} \pm \sqrt{\left(\frac{1}{T_m}\right)^2 - 4\frac{K_P K_m}{T_m}}}{2} \tag{14.53}$$

Two values of K_P provide insight. When $K_P = 0$, the closed-loop poles are located at

$$s = \begin{cases} 0 \\ -\frac{1}{T_m} \end{cases} \tag{14.54}$$

corresponding to the open-loop pole locations. When K_P is selected such that the radical in Equation (14.53) vanishes,

$$K_P = \frac{1}{4K_m T_m} \tag{14.55}$$

and the two roots are located at

$$s = \begin{cases} -\frac{1}{2T_m} \\ -\frac{1}{2T_m} \end{cases} \tag{14.56}$$

When $0 \leqslant K_P \leqslant 1/(4K_m T_m)$, the closed-loop poles of the system are purely real. When $K_P > 1/(4K_m T_m)$, the closed-loop poles of the system are complex conjugates with a constant real part of $-1/(2T_m)$. Increasing the proportional gain beyond $1/(4K_m T_m)$ only increases the imaginary component of the poles; the real part does not change. A graphical portrayal of the pole locations of the system under proportional control is shown in the root locus plot of Figure 14.20. The poles start at their open-loop positions when $K_P = 0$ and transition to a break point at $-1/(2T_m)$ when $K_P = 1/(4K_m T_m)$. After the break point, they travel parallel to the imaginary axis, with only their imaginary parts increasing while their real parts remain constant at $-1/(2T_m)$. It is clear from Figure 14.20 that only specific combinations of ω_{nd} and ζ_d can be achieved with the proportional control configuration. Such relationships can be observed directly via a variety of control design tools for both single-variable and multivariable systems [Kurfess and Nagurka, 1993; Nagurka and Kurfess, 1993].

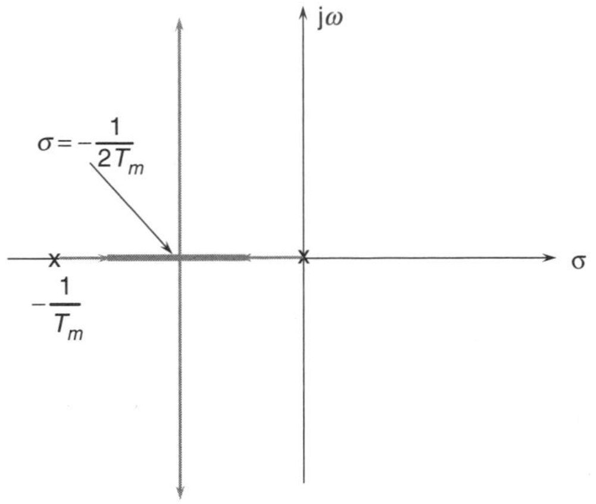

FIGURE 14.20 Root locus for proportional control on a single link.

Proportional Derivative (PD) Control. To improve performance beyond that available by proportional control, proportional-derivative or PD control can be employed. The dynamics for the PD controller are given by

$$K(s) = K_D s + K_P \tag{14.57}$$

where the constants K_P and K_D are the proportional and derivative gains, respectively. Employing Black's Law with the PD controller, the closed-loop transfer function for the single actuator-link model is

$$G_{CL}(s) = \frac{\frac{K_m}{T_m}(K_D s + K_P)}{s^2 + \left(\frac{K_m K_D + 1}{T_m}\right)s + K_m K_P / T_m} \tag{14.58}$$

There are now additional dynamics associated with the zero (root of the transfer function numerator) located at

$$s = -\frac{K_P}{K_D} \tag{14.59}$$

These dynamics can affect the final closed-loop response. Ideally, the zero given in Equation (14.59) will be far enough to the left of the system poles on the s-plane that the dominant-pole theory applies. (This would mean the dynamics corresponding to the zero are sufficiently fast in comparison to the closed-loop pole dynamics that their effect is negligible.)

Expressions for K_P and K_D can be found by equating the closed-loop characteristic equation of 14.58,

$$s^2 + \left(\frac{K_m K_D + 1}{T_m}\right)s + K_m K_P / T_m = 0 \tag{14.60}$$

to the target closed-loop characteristic equation that possesses the desired natural frequency and damping ratio, Equation (14.46). Equating the coefficients of s^0 and solving for K_P yields

$$K_P = \omega_{nd}^2 \frac{T_m}{K_m} \tag{14.61}$$

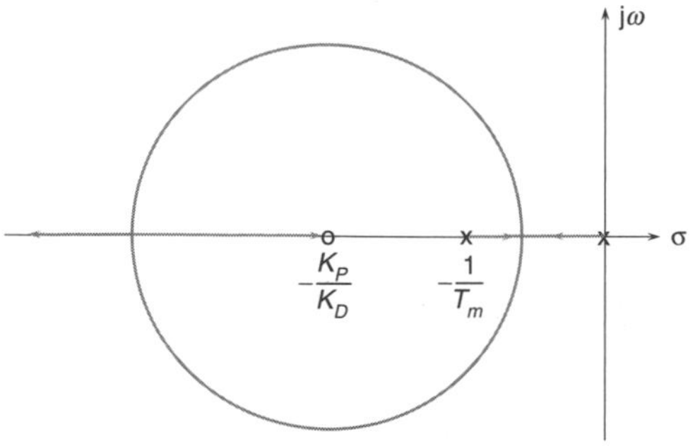

FIGURE 14.21 Root locus for PD control on a single link.

Similarly, equating the coefficients of s^1 and solving for K_D yields

$$K_D = \frac{2\zeta_d \omega_{nd} T_m - 1}{K_m} \tag{14.62}$$

This approach of solving for control gains is known as pole placement. As with all control design methods, the gains derived using Equation (14.61) and Equation (14.62) should be checked to ensure that they are realistic.

Figure 14.21 presents a typical root locus plot for a single-actuator-link model using a PD controller where the forward loop gain is varied. The location of the single zero ($s = -K_P/K_D$) can be seen on the real axis in the left of the s-plane. For this design, the zero dynamics will be insignificant in comparison to the pole dynamics. It is noted that the break point for the root locus presented in Figure 14.21 is slightly to the left of the break point in Figure 14.20 ($s = -1/2T_m$). Furthermore, the locus of the complex conjugates poles is circular and centered at the zero.

Classical control tools such as the root locus are critical in control design. Without such tools the designer could use the relationships given by Equation (14.61) and Equation (14.62) to place the poles at any location. However, the resulting location of the zero may have been too close to the closed-loop poles, resulting in system dynamics that do not behave as desired. This is described in detail in the following section.

Effect of Zero Dynamics. To demonstrate the effect of a zero, consider two systems that have the same damping ratios (ζ) and natural frequencies (ω_n). The first system is given by

$$G_1(s) = \frac{R_1(s)}{C(s)} = \frac{\omega_n^2}{s^2 + 2\zeta\omega_n s + \omega_n^2} \tag{14.63}$$

and the second system is

$$G_2(s) = \frac{R_2(s)}{C(s)} = \frac{s + \omega_n^2}{s^2 + 2\zeta\omega_n s + \omega_n^2} \tag{14.64}$$

where the input is $c(t)$ and the output is $r_i(t)\,(i = 1, 2)$. The difference between the two systems is the zero located at $s = -\omega_n^2$ for the system described by Equation (14.64). For a unit step input

$$C(s) = \frac{1}{s}, \tag{14.65}$$

the unit step responses of the two systems are

$$R_1(s) = \frac{\omega_n^2}{s^2 + 2\zeta\omega_n s + \omega_n^2}\left(\frac{1}{s}\right) \tag{14.66}$$

and

$$R_2(s) = \frac{s + \omega_n^2}{s^2 + 2\zeta\omega_n s + \omega_n^2}\left(\frac{1}{s}\right) = \frac{s}{s^2 + 2\zeta\omega_n s + \omega_n^2}\left(\frac{1}{s}\right) + \frac{\omega_n^2}{s^2 + 2\zeta\omega_n s + \omega_n^2}\left(\frac{1}{s}\right) \quad (14.67)$$

The difference between the two responses is the derivative term, corresponding to the first term on the right-hand side of Equation (14.67)

$$\frac{s}{s^2 + 2\zeta\omega_n s + \omega_n^2}\left(\frac{1}{s}\right)$$

The time domain responses for Equation (14.66) and Equation (14.67), respectively, are

$$r_1(t) = 1 - e^{-\zeta\omega_n t}\left[\left(\frac{\zeta}{\sqrt{1-\zeta^2}}\right)\sin(\omega_d t) + \cos(\omega_d t)\right] \quad (14.68)$$

and

$$\begin{aligned} r_2(t) = &\left\{1 - e^{-\zeta\omega_n t}\left[\left(\frac{\zeta}{\sqrt{1-\zeta^2}}\right)\sin(\omega_d t) + \cos(\omega_d t)\right]\right\} \\ &+ \left\{e^{-\zeta\omega_n t}\left(\frac{1}{\omega_n\sqrt{1-\zeta^2}}\right)\sin(\omega_d t)\right\} \end{aligned} \quad (14.69)$$

where ω_d, the damped natural frequency, is given by

$$\omega_d = \omega_n\sqrt{1-\zeta^2} \quad (14.70)$$

Several items are worth noting: (i) Equation (14.68) and Equation (14.69) are valid for $0 \leqslant \zeta < 1$. (ii) The difference between the two responses is

$$\left\{e^{-\zeta\omega_n t}\left(\frac{1}{\omega_n\sqrt{1-\zeta^2}}\right)\sin(\omega_d t)\right\}, \quad (14.71)$$

which is the impulse response of the system given by Equation (14.63). For the initial transient of the step response, the impulse response is positive and, therefore, adds to the overall system response. This indicates that the overshoot may be increased, which is the case for the example shown in Figure 14.22 for $\omega_n = 1$ rad/s and $\zeta = 0.707$. In Figure 14.22, the proportional term is the step response given by Equation (14.68); the derivative term is given by expression 14.71; and the total response is given by the sum of the two, Equation (14.69).

From this example, the addition of the zero results in increased overshoot. The real part of the poles is

$$s = \zeta\omega_n = 0.707 \quad (14.72)$$

and the location of the zero is

$$s = 1 \quad (14.73)$$

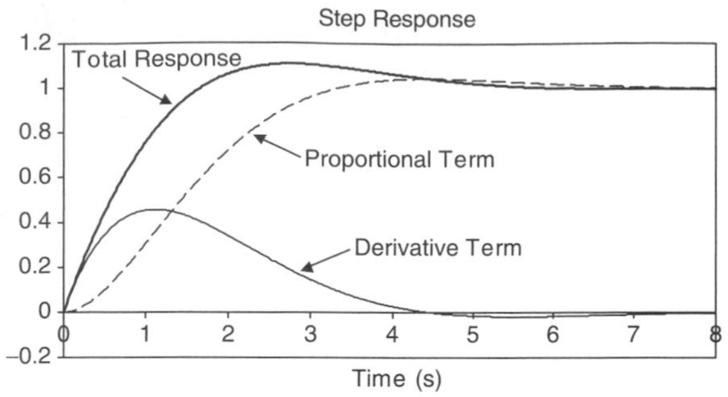

FIGURE 14.22 Step response of two systems.

Thus, the dominant-pole theorem does not hold and the dynamics of the zero cannot be ignored. Again, this is demonstrated by the increased overshoot illustrated in Figure 14.22.

Proportional Integral Derivative (PID) Control. A PID controller can be designed to provide better steady-state disturbance rejection capabilities compared with PD or P controllers. The form of the PID controller is given by

$$K(s) = K_{\mathrm{D}}s + K_{\mathrm{P}} + \frac{K_{\mathrm{I}}}{s} = \frac{K_{\mathrm{D}}s^2 + K_{\mathrm{P}}s + K_{\mathrm{I}}}{s} \tag{14.74}$$

For the single actuator-link model, the closed-loop transfer function using a PID controller is given by

$$G_{\mathrm{CL}}(s) = \frac{\frac{K_{\mathrm{m}}}{T_{\mathrm{m}}}(K_{\mathrm{D}}s^2 + K_{\mathrm{P}}s + K_{\mathrm{I}})}{s^3 + \left(\frac{K_{\mathrm{m}}K_{\mathrm{D}} + 1}{T_{\mathrm{m}}}\right)s^2 + \frac{K_{\mathrm{m}}K_{\mathrm{P}}}{T_{\mathrm{m}}}s + \frac{K_{\mathrm{m}}K_{\mathrm{I}}}{T_{\mathrm{m}}}} \tag{14.75}$$

The closed-loop dynamics are third-order and, in comparison to the second-order PD case, the control design has an extra gain that must be determined and an additional pole, located at $s = -d$, added to the target dynamics. The closed-loop characteristic equation,

$$s^3 + \left(\frac{K_{\mathrm{m}}K_{\mathrm{D}} + 1}{T_{\mathrm{m}}}\right)s^2 + \frac{K_{\mathrm{m}}K_{\mathrm{P}}}{T_{\mathrm{m}}}s + \frac{K_{\mathrm{m}}K_{\mathrm{I}}}{T_{\mathrm{m}}} = 0, \tag{14.76}$$

is equated to the new target closed-loop characteristic equation,

$$(s^2 + 2\zeta_{\mathrm{d}}\omega_{\mathrm{nd}}s + \omega_{\mathrm{nd}}^2)(s + d) = 0 \tag{14.77}$$

As with the PD controller, the location of the zeros in Equation (14.75) as well as the extra pole at $s = -d$, should have negative real parts that are at least three times as negative as the target closed-loop pole locations.

Figure 14.23 presents a possible root locus plot for the single actuator-link model with a PID controller where the forward loop gain is varied. The two zeros and extra pole can be seen on the real axis to the left in the s-plane. As with the PD control design, the gains used for achieving the target dynamics must be well understood. It is easy to pick gains that move both zeros and the extra pole far to the left in the s-plane. However, it should be noted that the further left the dynamics are shifted in the s-plane, the faster they become, influencing less the overall response of the system. However, to increase the response speed of these dynamics requires higher gains that ultimately result in higher power requirements (and the possibility of

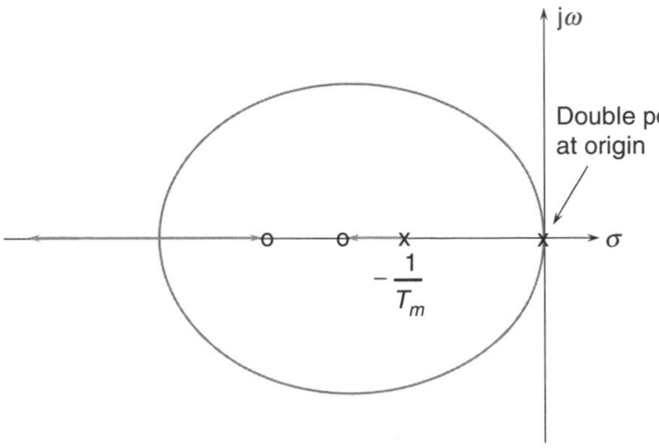

FIGURE 14.23 Root locus for PID control of a single link.

instability). As such, while designing for higher gains is rather straightforward mathematically, the implementation may be limited by the physical constraints of the system.

Unlike the locus of the complex conjugate poles for the model employing PD control (Figure 14.21), the locus here need not be circular. Also, given the increased number of closed-loop system parameters (including the control gains and design parameters such as *d*), the sensitivity of the closed-loop dynamics to changes of these parameters may be important and should be investigated. For example, if the proportional gain changes a small amount (which is bound to happen in a real system), the impact in the overall system performance must be understood. Mathematically, sensitivity can formally be shown to be a complex quantity (having both magnitude and angle), and a variety of techniques can be used to explore sensitivity to parameter changes [Kurfess and Nagurka, 1994].

Method of Computed Torque

If the robot is a direct-drive type without gear reduction or if the command inputs are not slowly varying, the control scheme of the previous section may exhibit poor performance characteristics, and instability may even result. One method of compensating for the effects of link coupling is to use feedforward disturbance cancellation. The disturbance torque is computed from the robot dynamic equation, summarized in vector-matrix form as

$$\tau = D(\theta)\ddot{\theta} + C(\theta, \dot{\theta}) + G(\theta) + F(\dot{\theta}) \tag{14.78}$$

where τ is the $n \times 1$ vector of joint torques (forces), $D(\theta)$ is the $n \times n$ inertia matrix, $C(\theta, \dot{\theta})$ is the $n \times 1$ vector of Coriolis and centrifugal torques (forces), $G(\theta)$ is the $n \times 1$ vector of torques (forces) due to gravity, and $F(\theta)$ is the $n \times 1$ vector of torques (forces) due to friction. In the feedforward disturbance cancellation scheme, the right-hand side of Equation (14.78) is computed using the desired value of the joint variables and injected as a compensating "disturbance." If the plant is minimum phase (has no right-half *s*-plane zeros), its inverse is also fed forward to achieve tracking of any reference trajectory.

Another version of computed-torque control, known as *inverse dynamics control*, involves setting the control torque to

$$\tau = D(\theta)v + C(\theta, \dot{\theta}) + G(\theta) + F(\dot{\theta}) \tag{14.79}$$

which results in $\ddot{\theta} = v$, a double integrator system. The value of v is now chosen as $v = \ddot{\theta}_d + K_D(\dot{\theta}_d - \dot{\theta}) + K_P(\theta_d - \theta)$, where subscript d denotes desired. This results in the tracking error, $e = \theta_d - \theta$, which satisfies

$$\ddot{e} + K_{\mathrm{D}}\dot{e} + K_{\mathrm{P}}e = 0 \tag{14.80}$$

The gains K_{P} and K_{D} can be chosen for the desired error dynamics (damping and natural frequency). In general, computed-torque control schemes are computationally intensive due to the complicated nature of Equation (14.78) and require accurate knowledge of the robot model [Bonitz, 1995].

Cartesian-Space Control

The basic concept of Cartesian-space control is that the error signals used in the control algorithm are computed in Cartesian space, obviating the solution of the inverse kinematics. The position and orientation of the robot end-effector can be described by a 3×1 position vector, p, and the three orthogonal axes of an imaginary frame attached to the end effector. (The axes are known as the normal (n), sliding (s), and approach (a) vectors.) The control torque is computed from

$$\tau = D(\theta)\ddot{\theta} + C(\theta,\dot{\theta}) + G(\theta) + F(\dot{\theta}) \tag{14.81}$$

$$\ddot{\theta} = J(\theta)^{-1}[\ddot{x}_{\mathrm{d}} + K_{\mathrm{D}}\dot{e} + K_{\mathrm{P}}e - \dot{J}(\theta,\dot{\theta})\dot{\theta}] \tag{14.82}$$

where $J(\theta)$ is the manipulator **Jacobian** that maps the joint velocity vector to the Cartesian velocity vector, $\ddot{x}_{\mathrm{d}}$ is the 6×1 desired acceleration vector, $e = [e_{\mathrm{p}}^T e_{\mathrm{o}}^T]^T$, e_{p} is the 3×1 position error vector, e_{o} is the 3×1 orientation error vector, K_{D} is the 6×6 positive-definite matrix of velocity gains, and K_{P} is the 6×6 positive-definite matrix of position gains. The actual position and orientation of the end-effector is computed from the joint positions via the forward kinematics. The position error is computed from $e_{\mathrm{p}} = \theta_{\mathrm{d}} - \theta$ and, for small error, the orientation error is computed from $e_{\mathrm{o}} = \frac{1}{2}[n\times n_{\mathrm{d}} + s\times s_{\mathrm{d}} + a\times a_{\mathrm{d}}]$. The control law of Equation (14.81) and Equation (14.82) results in the Cartesian error equation,

$$\ddot{e} + K_{\mathrm{D}}\dot{e} + K_{\mathrm{P}}e = 0 \tag{14.83}$$

The gain matrices K_{D} and K_{P} can be chosen to be diagonal to achieve the desired error dynamics along each Cartesian direction [Spong and Vidyasagar, 1989].

The Cartesian-space controller has the disadvantage that the inverse of the Jacobian is required, which does not exist at **singular configurations**. The planned trajectory must avoid singularities, or alternative methods such as the SR pseudoinverse [Nakamura, 1991] must be used to compute the Jacobian inverse [Bonitz, 1995].

Compliant Motion

While it is generally preferred for robots to move in rigid body modes, in many instances compliance (or flexibility) is desirable. Common practical applications include maintaining contact between a grasped object and a hard surface in the environment, such as in robotic grinding or deburring [Kurfess et al., 1988]. These applications involving contact are best performed by directly controlling the interaction force between the manipulator and its environment. The problem, commonly known as the compliant-motion control problem, is to exert a desired force profile in the constrained directions while following a reference position/velocity trajectory in the unconstrained degrees of freedom [Whitney, 1987].

The compliant-motion control problem has been investigated quite significantly, with techniques resulting in two basic solution approaches. One is a passive compliance (using the stiffness of the manipulator itself) for conversion of displacements normal to contact force. The second approach is an active compliance control of the joint torques via impedance control, compliance control, or one of several other approaches that follow a linear force vs. displacement or velocity relationships [Lanzon and Richards, 2000].

Passive compliance of a manipulator is often used to provide a controllable contact force via commanded position of joints in the force-constrained direction. The compliance of the robot should be designed such that the resulting manipulator with payload does not have excessive vibration or positional inaccuracies as a result of the compliance. In robotic applications of part insertion into another object, the use of a remote center of compliance (RCC) device (typically wrist mounted) is often required instead of compliant arm. The RCC allows an assembly robot or machine to compensate for positioning errors (misalignments) during insertion. This compliance can reduce or avoid the potential for part, robot, or end-effector damage, yet have suitable positional accuracy. It is noted that force-sensor feedback is still quite necessary to regulate or control the force. In many cases the force sensor itself provides sufficient compliance alone or it is integral to the RCC.

Adaptive Control

Previous methods presented for PID-type controllers work well when the dynamics of the robotic system is known and modeled correctly. However, in many instances the physical model may change with time. Common instances are the increase of damping resistance in a robot joint over time, or a varying payload weight. In these situations the real system may vary sufficiently from the modeled system, limiting the usefulness or stability of fixed-gain controllers. In these cases adaptive control may be applied to maintain stability and improve performance.

Many adaptive control schemes have been investigated. However, adaptive controllers can be divided into two basic types, direct and indirect methods [Astrom and Wittenmark, 1989]. Direct and indirect refer to the way controller gains are updated.

Model reference adaptive control (MRAC) is a direct control update approach. A functional diagram of this method is illustrated in Figure 14.24. While the technique has been known since the 1960s, Dubowsky and DesForges [1979] are noted as the first to use this method in the control of a robotic manipulator. In their approach, a linear second-order system was selected as the desired reference model for each joint of the manipulator. The critical damping ratio, ζ, and natural frequency, ω_n, were selected to achieve a desired performance level. This method works provided the manipulator changes configuration slowly relative to the adaptation rate, otherwise instability may occur. A good discussion on the stability of MRAC is given by Parks [1966].

Indirect methods of adaptive control determine controller parameters via a design procedure, not a direct error correction. A self-tuning regulator (STR) is an example of an indirect method. The general form of the functional diagram is provided in Figure 14.25. The self-tuning regulator uses model parameter estimation to determine current states for use in real-time design calculations to adjust controller parameters or gains. One of the key determinations of this approach is the desired form of the model for accurate estimation. Typically, the lowest-order sufficient model is preferred.

Many researchers have advanced the basic adaptive control methods for linear systems by employing the fuzzy logic and neural networks. However, two essential components of all indirect adaptive methods are system identification (model form or parameters) and a method using this information to update the applied control law.

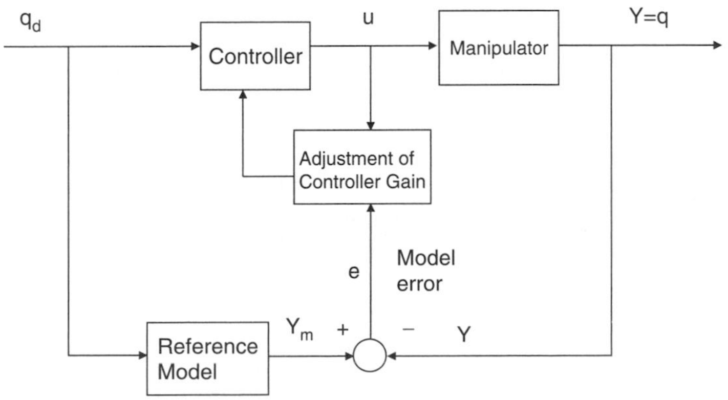

FIGURE 14.24 Function block diagram for model reference adaptive control (MRAC).

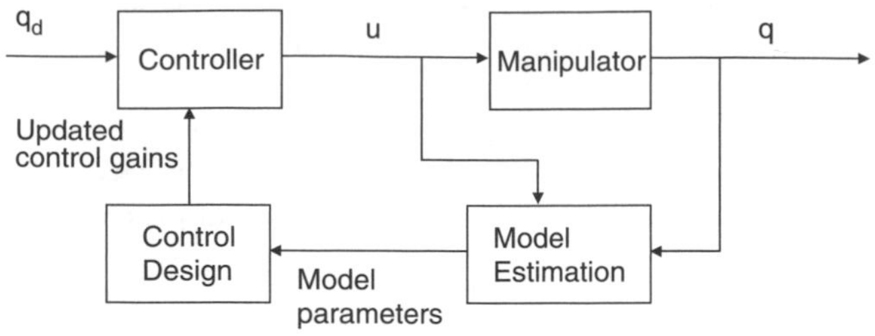

FIGURE 14.25 Function block diagram for self-tuning regulator.

Advanced adaptive techniques have employed recursive estimation and Kalman filtering, among other approaches, to estimate the current manipulator characteristics and apply control design directly. The Kalman filter is the most general of the multiple input, recursive least squares (RLS) methods for estimating time-varying parameters [Ljung, 1987]. Much of the current estimation research has focused on using Kalman filtering for state (parameter) estimation. It is an optimal estimator for time-varying parameters, minimizing the state error variance at each time. If appropriately applied, the Kalman filter has been demonstrated to yield better results than many of the other RLS-based methods (Danai and Ulsoy, 1987).

Resolved Motion Control

The methods previously presented provide a means for development of mathematical models of manipulator joints. In general, the prescribed motion of importance is that of the end-effector (or hand). (Motions of links are also significant for obstacle avoidance in the robot workspace.) Resolved motion control methods such as resolved motion rate control (RMRC) and resolved motion acceleration control (RMAC) have been used to control end-effector motion [Whitney, 1969, and Luh et al., 1980]. Using these approaches, the joint velocities (and displacements) are coordinated to achieve the desired end-effector motion. To coordinate joints and control the motion of a manipulator, RMRC can be employed. The RMRC relationship between the position vector (six-element column vector of displacements and rotations) of the end effector, $x_e(t)$, and the joint displacements, $q(t)$, is defined in Equation (14.84). J is the Jacobian matrix of the manipulator and is a function of $q(t)$.

$$\dot{x}_e(t) = J\dot{q}(t) \tag{14.84}$$

Multiplying by the inverse Jacobian yields the joint velocities for a desired end-effector motion.

$$\dot{q}(t) = J^{-1}\dot{x}_e(t) \tag{14.85}$$

For given joint velocities, $\dot{q}(t)$, the velocity of the end-effector can be derived from the kinematics of the manipulator. This method breaks down if the Jacobian is singular, when there are redundant degrees of freedom in the manipulator (x_e(t) and q(t) are not the same length).

Resolved motion acceleration control (RMAC) improves on the RMRC end-effector motions of Equation (14.84) to include end-effector acceleration effects. The derivative of Equation (14.84) yields the acceleration vector of the end-effector, $\ddot{x}_e(t)$.

$$\ddot{x}_e(t) = J\ddot{q}(t) + \dot{J}(q,\dot{q})\dot{q}(t) \tag{14.86}$$

To reduce both positional and velocity error, the end-effector acceleration can be determined as follows using proportional and derivative feedback, where proportional and derivative gains are K_p and K_d, respectively.

$$\ddot{x}_e(t) = \ddot{x}_{e,des}(t) + K_p(x_{e,des}(t) - x_e(t)) + K_d(\dot{x}_{e,des}(t) - \dot{x}_e(t)) \quad (14.87)$$

Equating the two right-hand sides of Equation (14.86) and Equation (14.87) yields the necessary joint accelerations, $\ddot{q}(t)$. This method has the same Jacobian issues as does the RMRC method.

Flexible Manipulators

It is desirable for robots and manipulators to move in rigid-body motion, with no dynamics other than at their joints. However, no robot is completely rigid. (In many instances compliance is built into a robot, by design, as discussed earlier.) Each robot has vibrational characteristics, by virtue of the robot linkage geometry, construction materials, and payloads.

Most commercial robots are designed to minimize vibration. To accomplish this, commercial robots maximize stiffness and are heavy. The large inertia of these heavy robots reduces their potential for speed. They are also limited in their payload, carrying capacity, typically less than 5% of their own weight. Linkage length (and workspace) is also minimized.

Flexible manipulators are designed to overcome the limits of speed, workspace, and payload of the standard commercial units. Flexible manipulators are lightweight in comparison and thus consume less power. However this comes at a cost. Fast movements initiate vibration that affect positional accuracy.

Several control strategies have been investigated to reduce or mitigate the problem of induced vibration in flexible manipulators. Most approaches have focused on open-loop or feed-forward control. Open-loop schemes are based on system models and may perform poorly from inaccurate model and trajectory definitions. One open-loop scheme involves modification of the command signal [Singer and Seering, 1990]. Here the commanded signal is filtered to remove resonant frequencies. While this method has been shown to reduce vibrations, a more effective method, input shaping [Meckl and Seering, 1990] has been developed. Input shaping is based on convolving a desired command signal with a sequence of impulses, or input shaper. The amplitude and duration of these impulses is based on the damped natural frequency of the system. Movement without vibration may be attained, but at the cost of a time delay.

Defining Terms

Cartesian or task space: The set of vectors describing the position and orientation of the end-effector using coordinates along orthogonal axes. The position is specified by a 3×1 vector of the coordinates of the end-effector frame origin. The orientation is specified by the 3×1 normal, sliding, and approach vectors describing the directions of the orthogonal axes of the frame. Alternatively, the orientation may be described by Euler angles, roll/pitch/yaw angles, or an axis/angle representation.

Compliant motion: Movement of a manipulator (robot) while maintaining contact with its "environment," such as writing on a chalkboard or precision grinding, applying a force profile normal to the surface of contact while moving across the surface.

Forward kinematics: The function that maps the position of the joints to the Cartesian position and orientation of the end-effector. It maps the joint space of the manipulator to Cartesian space.

Inverse kinematics: The function that maps the Cartesian position and orientation of the end-effector to the joint positions. It is generally a one-to-many mapping, and a closed-form solution may not always be possible.

Jacobian: The function that maps the joint-velocity vector to the Cartesian translational and angular-velocity vector of the end-effector: $\dot{x} = J(\theta)\dot{\theta}$.

Jacobian of a manipulator: A matrix of first-order partial derivatives that maps the joint velocities into end-effector velocities.

Model reference adaptive control (MRAC): Adaptive control algorithm to directly update controller gains based on error of model and actual output of a system.

Recursive least squares (RLS): A real-time estimation technique to determine a parameter by minimizing the cumulative square error.

Remote center of compliance (RCC): A device (typically wrist mounted) that allows a robot or assembly machine to compensate for positioning errors (misalignments) during insertion by lowering horizontal and rotational stiffnesses while retaining relatively higher axial insertion stiffness.

Self-tuning regulator (STR): Adaptive control algorithm that indirectly changes the controller gain of a system by model estimation and subsequent control design.

Singular configuration: A configuration of the manipulator in which the manipulator Jacobian loses full rank. It represents configurations from which certain directions of motion are not possible or when two or more joint axes line up and there is an infinity of solutions to the inverse kinematics problem.

References

K.J. Astrom and B. Wittenmark, *Adaptive Control*, Reading, MA: Addison-Wesley, 1989.

H.S. Black, "Stabilized feedback amplifiers," *Bell Syst. Tech. J.*, January 1934, and U.S. patent no. 2,102,671.

R.G. Bonitz, "Robots and control," *The Engineering Handbook*, 1st ed., R.D. Dorf., Ed., Boca Raton, FL: CRC Press, 1995.

J.J. Craig, *Introduction to Robotics: Mechanics and Control*, 2nd ed., Reading, MA: Addison-Wesley, 1989.

K. Danai and A.G. Ulsoy, "A dynamics state model for on-line tool wear estimation in turning," *ASME J. Eng. Ind.*, 109, 396, 1987.

J. Denavit and R.S. Hartenberg, "A kinematics notation for lower pair mechanisms based on matrices," *ASME J. Appl. Mech.*, 22, 215–221, 1955.

S. Dubowsky and D. Desforges, "The application of model-referenced adaptive control of robotic manipulators," *ASME J. Dyn. Syst. Meas. Control*, 101, no. 3, 193–200, 1979.

R. Featherstone and D.E. Orin, "Robot dynamics: equations and algorithms," *IEEE Int. Conf. Rob. Autom.*, San Francisco, CA, pp. 826–834, 2000.

R. Featherstone, *Robot Dynamics Algorithms*, Dordrecht: Kluwer Academic Publishers, 1987.

J. Feddema, G.R. Eisler, C. Dohrmann, G.G. Parker, D.G. Wilson, and D. Stokes, *Flexible Robot Dynamics and Controls*, R.D. Robinett, III, Ed., Dordrecht: Kluwer Academic Publishers, 2001.

N. Hogan, "Impedance control: an approach to manipulation, Parts I, II, and III," *ASME J. Dyn. Syst. Meas. Control*, 107–124, 1985.

T.R. Kurfess and M.L. Nagurka, "A geometric representation of root sensitivity," *ASME J. Dyn. Syst. Meas. Control*, 116, no. 4, 305–309, 1994.

T.R. Kurfess and M.L. Nagurka, "Foundations of classical control theory with reference to eigenvalue geometry," *J. Franklin Inst.*, 330, no. 2, 213–227, 1993.

T.R. Kurfess and H.E. Jenkins, "Ultra-high precision control," in *Control Systems Applications*, W. Levine, Ed., Boca Raton, FL: CRC Press, 2000, pp. 212–231.

T.R. Kurfess, "Precision manufacturing," in *Mechanical System Design Handbook*, O. Nwokah and Y. Hurmuzlu, Eds., Boca Raton, FL: CRC Press, 2002, pp. 151–179.

T.R. Kurfess, D.E. Whitney, and M.L. Brown, "Verification of a dynamic grinding model," *ASME J. Dyn. Syst. Meas. Control*, 110, 403–409, 1988.

A. Lanzon and R.J. Richards, "Compliant motion control for non-redundant rigid robotic manipulators," *Int. J. Control*, 73, no. 3, 225, 2000.

L. Ljung, *System Identification: Theory for the User*, Englewood Cliffs, NJ: Prentice Hall, 1987.

J.Y.S. Luh, "Conventional controller design for industrial robots—a tutorial," *IEEE Trans. Syst. Man Cybern.*, SMC-13, no. 3, 298–316, 1983.

J.Y.S. Luh, M.W. Walker, and R.P.C. Paul, "Resolved-acceleration control of mechanical manipulators," *IEEE Trans. Autom. Control*, AC-25, 464–474, 1980.

J.Y.S. Luh, M.W. Walker, and R.P. Paul, "On-line computational scheme for mechanical manipulators," *Trans. ASME, J. Dyn. Syst. Meas. Control*, 120, no. 180, 69–76.

M.T. Mason, *Mechanics of Robotic Manipulation*, Cambridge: MIT Press, 2001.

P.J. McKerrow, *Introduction to Robotics*, Reading, MA: Addison-Wesley, 1991.
P.H. Meckl and W.P. Seering, "Experimental evaluation of shaped inputs to reduce vibration of a Cartesian robot," *ASME J. Dyn. Syst. Meas. Control*, 112, no. 6, 159–165, 1990.
M.L. Nagurka and T.R. Kurfess, "An alternate geometric perspective on MIMO systems," *ASME J. Dyn. Syst. Meas. Control*, 115, no. 3, 538–543, 1993.
M.L. Nagurka and T.R. Kurfess, A Unified Classical/Modern Approach for Undergraduate Control Education, lecture notes from an NSF sponsored workshop, 1992.
Y. Nakamura, *Advanced Robotics, Redundancy, and Optimization*, Reading, MA: Addison-Wesley, 1991.
S. Niku, *Introduction to Robotics: Analysis, Systems, Applications*, Englewood Cliffs, NJ: Prentice-Hall, 2001.
D.E. Orin, R.B. McGhee, M. Vukobratovic, and G. Hartoch, "Kinematic and kinetic analysis of open-chain linkages utilizing Newton–Euler methods," *Math. Biosci.*, 43, 107–130, 1979.
P.C. Parks, "Lyanpunov redesign of a model reference adaptive control system," *IEEE Trans. Automat. Control*, AC-11, 362–367, 1966.
M.H. Raibert and J.J. Craig, "Hybrid position/force control of manipulators," *ASME J. Dyn. Syst. Meas. Control*, 103, 126–133, 1981.
L. Sciavicco and B. Siciliano, *Modelling and Control of Robot Manipulators*, 2nd ed., Berlin: Springer, 2000.
M. Shahinpoor, *A Robot Engineering Textbook*, New York: Harper and Row, 1986.
N.C. Singer and W.P. Seering, "Preshaping command inputs to reduce system vibration," *ASME J. Dyn. Syst. Meas. Control*, 112, no. 1, 76–82, 1990.
M.W. Spong and M. Vidyasagar, *Robot Dynamics and Control,* New York: Wiley, 1989.
A.K. Swain and A.S. Morris, "A Unified dynamic model formulation for robotic manipulator systems," *J. Rob. Syst.*, 20, no. 10, 601–620, 2003.
M. Vukobratovic, V. Potkonjak, and V. Matijevi, *Dynamics of Robots with Contact Tasks*, Dordrecht: Kluwer Academic Publishers, 2003.
M.W. Walker and D.E. Orin, "Efficient dynamic computer simulation of robotic mechanisms," *Trans. ASME, J. Syst. Meas. Control*, 104, 205–211, 1982.
D.E. Whitney, "Historical perspective and state of the art in robot force control," *Int. J. Rob. Res.*, 6, 3–13, 1987.
D.E. Whitney, "Resolved motion rate control of manipulators and human prostheses," *IEEE Trans. Man-Mach. Syst.*, MMS-10, no. 2, June 1969.
T. Yoshikawa, *Foundations of Robotics: Analysis and Control*, New York: MIT Press, 1990.

14.3 Applications

Nicholas G. Odrey

An important utilization of robotics has traditionally been in manufacturing operations. By their very design and reprogrammable features, robots have enhanced the capabilities for flexibility in automation. Robots have acquired mobility and enhanced "intelligence" over the past two decades. Robot applications initially focused on replacing repetitive, boring, and hazardous manual tasks. Such initial applications required minimal control, programming, or sensory capability. Modern robots have become more adaptive by incorporating more sensory capability and improved control structures. In potential applications, it is necessary to determine the degree of sophistication that one wishes to implement for the application. Implementation should be based on a detailed economic analysis. Aging equipment has also necessitated consideration of partial or full replacement strategies and the associated economics. A focus here is to present a practical implementation strategy for robots within a manufacturing environment, to review particular applications, and to discuss issues relevant to enhancing robot intelligence and applications on the manufacturing shop floor. Such issues include sensor placement strategies, sensor integration within an intelligent control system, the development of grippers for enhanced dexterity, and integration issues that pertain to hardware and software within a flexible manufacturing system.

Justification

Reprogrammable automated devices such as robots provide the flexible automation capability for modern production systems. Newman et al. [2000] note the multidisciplinary nature involved with building agile manufacturing systems where the system has the ability to be rapidly reconfigured within economic constraints. Their concept of agility refers to a design philosophy that considers hardware/software reuse and rapid system redesign. Our focus here is on the robots within such systems. To evaluate a potential robotic application within a manufacturing system environment, both technical and economic issues must be addressed. Typical technical issues include the choice of the number of degrees of freedom to perform a task, the level of controller and programming complexity, end-effector and sensor choices, and the degree of integration within the overall production system. Economic issues have typically been addressed from a traditional point of view, but it is important to note that other criteria should also be evaluated before a final decision is made to implement a robotic system. Such criteria may be both quantitative and qualitative.

Traditional economic approaches analyze investments and costs to compare alternative projects. Three methods are commonly used: (1) payback period method, (2) equivalent uniform annual costs (EUAC) method, and (3) return on investment (ROI). The payback method balances initial investment cost against net annual cash flow during the life of the project to determine the time required to recoup the investment. Many corporations today require relatively short (1- to 3-year) payback periods to justify an investment. In the current environment with the drive toward shortened product life cycles, it is not unusual to see payback period requirements of no greater than 1 year. The payback technique does not consider the time value of money and should be considered only as a first partial attempt at justification. The EUAC and ROI methods consider the time value of money (continuous or discrete compounding) and convert all investments, cash flows, salvage values, and any other revenues and costs into their equivalent uniform annual cash flow over the anticipated life of the project. In the EUAC method, the interest rate is known and set at a minimal acceptable rate of return, whereas the ROI method has the objective to determine the interest rate earned on the investment. Equipment replacement policies and economics also are an important issue as requirements and products change or equipment ages. Details to such techniques and replacement analysis are presented in various engineering economy texts such as those by DeGarmo et al. [1997] and Park [2004].

Other factors to be recognized in robotic justification are that robots are reusable from one project to the next and that there is a difference in production rates for a robotic implementation over a manual process. A changeover from a manual method to a robotic implementation would have the potential to affect revenues for any project. Many companies have also developed standard investment analysis forms for an economic evaluation of a proposed robot project. These forms are helpful in displaying costs and savings for a project. Groover et al. [1998] presents one such proposed form and gives several references to examples of forms specifically designed for projects devoted to robotics and related automation areas. Such economic techniques are important in performing an economic justification for a proposed robotic installation. Still, in general, there are other issues that should be included in the overall analysis. These issues are of particular importance if one is considering installing a more comprehensive system, i.e., a flexible manufacturing system that may include many robots and automated systems. As noted by Proth and Hillion [1990], these issues give rise to criteria that are both quantitative and qualitative. Quantitative criteria include not only reduced throughput time and work-in-process inventory, but also criteria related to increased productivity coupled with fewer resources. Another measurable criterion is the reduction in management and monitoring staff as a result of smaller quantities and automatic monitoring by sensors. Quality improvement can also be measured both quantitatively and qualitatively. Qualitative benefits from quality improvement can include increased customer satisfaction, increased competitiveness, simplified production management, and other factors. It should be noted that any benefits and cost reductions for installation of an automated system are difficult to evaluate and reflect a long-term commitment of the corporation. Strategic factors should be incorporated in the overall economic justification process, but they are difficult to access and incorporate due to their inherent complexity.

Implementation Strategies

A logical approach is a prerequisite to robotic implementation within a manufacturing firm. The following steps have been proposed by Groover et al. [1998] to implement a robotic system:

1. Initial familiarization with the technology
2. Plant survey to identify potential applications
3. Selection of an application(s)
4. Selection of a robot(s) for the application(s)
5. Detailed economic analysis and capital authorization
6. Plan and engineer the installation
7. Installation

It should be noted that a particular company may have nuances that could modify the above steps. Also of note is that the underlying issue is systems integration, and any robotic application should consider the total system impact. This must include the equipment, controllers, sensors, software, and other necessary hardware to have a fully functional and integrated system. In robot applications two categories may be distinguished: (1) a project for a new plant, or (2) placing a robot project in an existing facility. We focus here on the latter catcgory.

Critical factors for the introduction of robotics technology within a corporation are management support and production personnel acceptance of the technology. Companies such as General Electric have previously developed checklists to determine the degree of workforce acceptance. Given that the above two factors are met, a plant survey is conducted to determine suitability for automation or robotic implementation. General considerations for an industrial robot installation include hazardous, repetitive, or uncomfortable working conditions, difficult handling jobs, or multishift operations. High- and medium-volume production typically has many examples of repetitive operations. It can prove useful to investigate injury (particularly muscular injury) reports with medical personnel and ergonomics experts to identify potential manual operations that may be alleviated with the aid of robotics or automation. Multishift operations associated with high demand for a product are likely candidates for robot applications. As compared to manual work that typically has a high variable labor cost, a robot substitution would have a high fixed cost (which can be distributed over the number of shifts) plus a low variable cost. The overall effect of a robot application would then be to reduce the total operating cost.

Usually, a simple application that is easy to integrate into the overall system is a good initial choice. A fundamental rule is to implement any straightforward application to minimize the risk of failure. The General Electric Company has been successful in choosing robot applications by considering the following technical criteria:

- Operation is simple and repetitive
- Cycle time for the operation is greater than five seconds
- Parts can be delivered with the proper POSE (position and orientation)
- Part weight is suitable (typical upper weight limit is 1100 lb)
- No inspection is required for the operation
- One to two workers can be replaced in a 24-hour period
- Setups and changeovers are infrequent

A choice of a robot for a selected application can be a very difficult decision. Vendor information, expert opinion, and various Internet sources can aid in the selection. Selection needs to consider the appropriate combination of parameters suitable for the application. These parameters or technical features include the degrees of freedom, the type of drive and control system, sensory capability, programming features, accuracy and precision requirements, and load capacity of the selected robot. Various point or weighting schemes can be applied to rate different robot models.

The planning and engineering of a robot installation must address many issues, including the operational methods to be employed, workcell design and its control, the choice or design of end effectors, fixturing and tooling requirements, and sensory and programming requirements. In addition, one needs to focus on safety considerations for the workcell as well as overall systems integration. Computer-aided design (CAD) and simulation is very helpful to study potential machine interference and various layout problems as is useful in estimating various performance parameters. Analyzing the cycle time is basic to determining the production rate. An early approach developed by Nof and Lechtman in [1982], called robot time and motion (RTM), is useful for analyzing the cycle time of robots. More recent approaches have utilized computer-aided design and simulation packages in the overall design and analysis of robotic workcells, including motion planning, off-line programming, and animation. Examples include IGRIP (Deneb Robotics), ROBSCAD (ISRA), ROBEX (RWTH Aachen), KUSIM (Kuka robotics), and ROBOT-SIM (General Electric). One recent approach provided by Fanuc Robotics is the RoboGuide/SimPRO 3D simulation software, which allows users to test workcell layouts and robot motions in a virtual environment. The simulation can be optimized and downloaded into a robot controller [Knights, 2003]. Total workcell productivity can be also be addressed by optimizing robot placement and key performance parameters [Mahr, 2000]. Cycle-time issues, flexibility, and increased productivity are also being investigated for multiarmed, collaborative, and synchronous robots [Ranky, 2003].

Applications in Manufacturing

Robots have proven to be beneficial in many industrial and nonindustrial environments. Here, we focus on applications within a traditional manufacturing (shop floor) setting and, in particular, on applications that fall into the following three broad categories:

1. Material handling and machine loading/unloading
2. Processing
3. Assembly and inspection

Important to these applications is the choice of the proper end effector. End effectors can be divided into two categories: (1) grippers and (2) tools. Grippers grasp, manipulate, hold, and transport objects and are predominant in material handling. The standard gripper is modeled after the human hand, but other types of grippers include vacuum cups, magnetic grippers, adhesive grippers, and various devices such as hooks, ladles, etc. Tooling is designed to perform work on a part, e.g., spot welding electrodes attached to the end of the robot wrist. The discussion that follows is intended to present an overview of a small segment of industrial applications and a few of the more current topics that are impacting the shop floor, particularly as related to flexible manufacturing systems. In the latter case, such issues include developments in sensors, sensor integration, and sensory-interactive end effectors.

End Effectors

Robots are capable of interchanging end effectors by using quick-change wrists, or they may have a multiple spindle tool/gripper attached to the end of the arm. One such example is the automatic robotics tool-change system (ARTS), which is a product of Robotics and Automation Corporation of Minneapolis, Minnesota. ARTS was designed to meet the demand for multitask workcells for various manufacturing operations. In addition, ARTS was designed to work with a constant/controlled force device (CFD), which provides any necessary constant pressure for abrasive tooling. The CFD line includes three end-of–arm devices and two bench-mounted devices. ARTS-I is being used in industrial applications with six tool positions ranging from coarse sanding disks to cloth polishing wheels. Tool changing capability is possible with both ARTS-I and ARTS-II.

Sensors and Sensor Integration

With the increased emphasis on intelligence, the issues of sensor choice and their placement strategies have become more important. Sensors are used for safety monitoring, interlocks in workcell control, inspection, and determining position and orientation of objects. Ideally, a sensor should have high accuracy, precision, and response speed, a wide operating range, with high reliability and ease of calibration. Sensors in robotics can be divided into the following general categories: tactile sensors, proximity and range sensors, machine vision, and various miscellaneous sensors and sensor systems. In general, sensors have been developed for various applications in a manufacturing environment. A good reference source is Tonshoff and Inasaki [2002].

Multisensor fusion and integration techniques have been developed to provide more reliable and accurate information. Robots having multisensor integration and/or fusion capability can provide an enhanced flexibility for many industrial applications. Recent applications have investigated visual servoing, multi-robot systems, mobility issues and, in general, robot activity in an unstructured environment. Detailed explanations of techniques and applications can be found in Luo et al. [2002]. An example of sensor distribution strategies for a multistage assembly system have also been investigated to determine allocation of sensing stations and the minimal number of sensors for each station [Ding et al., 2003]. Sensor fusion techniques are important in applications other than manufacturing and will receive more emphasis in the future.

Material Handling and Machine Loading/Unloading

Robotic applications in material handling and machine loading pertain to the grasping and movement of a workpart or item from one location to another. General considerations for such applications pertain to the gripper design, distances moved, robot weight capacity, the POSE, and robot-dependent issues pertaining to the configuration, degrees of freedom, accuracy and precision, the controller, and programming features. POSE information is particularly important if there are no sensors (e.g., vision) to provide such information prior to pickup. A dexterous robotic gripper for use in unstructured environments has been developed for an outer-space environment that shows promise for industrial robotic applications [Biogiotti et al., 2003].

It should be noted that certain applications may require a high degree of accuracy and precision whereas others do not. Higher requirements result in more sophisticated drive mechanisms and controllers with associated increased costs. Typically, many industrial material-handling applications for robots are unsophisticated, with minimal control requirements. Two- to four-degrees-of-freedom robots may be sufficient in many tasks. More sophisticated operations such as palletizing may require up to six degrees of freedom with stricter control requirements and more programming features. Various criteria that have proved to contribute to the success of material handling and machine load/unload applications can be found in Groover et al. [1998]. In addition, excellent examples and case studies on robotic loading/unloading are given in the text by Asfahl [1992].

Processing

Robotic processing applications are considered here to be those applications in which a robot actually performs work on a part and requires that the end effectors serve as a tool. Examples include spot-welding electrodes, arc welding, and spray-painting nozzles. The most common robotic applications in manufacturing processes are listed in Table 14.3 [Odrey, 1992].

Spot welding and arc welding represent two major applications of industrial robots. Spot-welding robots are common in automotive assembly lines and have been found to improve weld quality and provide more consistent welds and better repeatability of weld locations. Continuous arc welding is a more difficult application than spot welding. Welding of dissimilar materials, variations in weld joints, dimensional

TABLE 14.3 Most Common Robotic Applications in Manufacturing Processes

Spot welding	Continuous arc welding	Spray coating
Drilling	Routing	Waterjet cutting
Grinding	Deburring	Polishing
Wire brushing	Riveting	Laser machining

variations from part to part, irregular edges, and gap variations are some of the difficulties encountered in the continuous arc welding processes. Typical arc welding processes include gas metal arc welding (GMAW), shielding metal arc welding (SMAW), i.e., the commonly known "stick" welding, and submerged arc welding (SAW). The most heavily employed robotic welding process is GMAW in which a current is passed through a consumable electrode and into a base metal, and a shielding gas (typically CO_2, argon, or helium) minimizes contamination during melting and solidification. It should be noted that arc welding, like many manufacturing processes, does not have an exact mathematical model to describe the process. One attempt to establish a mathematical model for predicting bead width compared neural-network approaches to regression techniques in an attempt to understand relations between GMAW parameters and top bead width [Kim et al., 2003]. In this study it was indicated that a neural-network model can serve as a better predictor than multiple regression methods for top bead width.

In welding, a worker can compensate automatically by varying welding parameters such as travel speed, deposition rate by current adjustment, weave patterns, and multiple welds where required. Duplicating human welding ability and skill requires that industrial robots have sensor capability and complex programming capability. Wide varieties of sensors for robotic arc welding are commercially available and are designed to track the welding seam and provide feedback information for the purpose of guiding the welding path.

Two basic categories of sensors exist to provide feedback information: noncontact sensors and contact sensors. Noncontact sensors include arc-sensing systems and machine vision systems. The former, also referred to as a through-the-arc system, uses feedback measurements via the arc itself. Specifically, measurements for feedback may be the current (constant-voltage welding) or the voltage (constant-current welding) obtained as a result of programming the robot to perform, for example, a weave pattern. The motion results in measurements that are interpreted as vertical and cross-seam position. Adaptive positioning is possible by regulating the arc length (constant-current systems) as irregularities in gaps or edge variations are encountered. Vision systems may also be used to track the weld seam, and any deviations from the programmed seam path are detected and fed back to the controller for automatic tracking. Single-pass systems detect variations and make corrections in one welding pass. Double-pass systems first do a high-speed scan of the joint to record in memory deviations from the programmed seam path, with actual welding corrections occurring on the second "arc-on" pass. Single-pass systems give the advantages of reduced cycle time and of being able to compensate for thermal distortions during the welding operation.

A robotic arc welding cell provides several advantages over manual welding operations. These advantages include not only higher productivity as measured by "arc-on" time, but also elimination of worker fatigue, decreased idle time, and improved safety. It is also important to correct upstream production operations to reduce variations. This is best accomplished during the design and installation phase of a robotic welding cell. During this phase, issues to consider include delivery of materials to the cell, fixtures and welding positioners, methods required for the processes, and any production and inventory control problems related to the efficient utilization and operation of the cell.

The National Institute for Standards and Technology (NIST) has an automated arc welding testbed. This automated and welding manufacturing system (AWMS) is focused on developing standards to contribute to increased implementation of automated welding technology by manufacturers. Various standards issues include hardware/robot interfaces, sensor interfaces, weld geometry and parameter data formats, and software application interfaces [Rippey and Falco, 1997]. The aim of the testbed is to work with manufacturers and technology suppliers to not only test systems and develop control systems, but also to develop and validate standards that contribute to improved welding technology. Open-architecture concepts will also be investigated by incorporating hardware and software components provided by others. Such open-architecture interface standards are focused on providing easy accessing and storing of real-time weld process data. Such data would enable the development of welding process models and in improved monitoring and control of welds. Another robotic testbed at NIST called RoboCrane is used to test concepts of portability and interoperability by moving welding components developed on the AWMS. The RoboCrane is a six degree-of-freedom work platform suspended by cables and driven by winches that effectively performs as a robot under computer control. A major advantage of this device is the large work volume. One version has a 20-meter^3 workspace.

A tandem gas metal arc welding (GMAW) dual-wire process with two power sources and two separate welding arcs is another recent robotic application. A lead arc provides penetration into the welded joint and the trailing arc controls bead shape and any reinforcement of the weld bead. Benefits of such an approach include increased productivity, less distortion, low spatter levels, high deposition rates, and improved weld quality [Nadzam, 2003].

Other processing applications for robot use include spray coating and various machining or cutting operations. Spray coating is a major application in the automotive industry where robots have proven suitable in overcoming various hazards such as fumes, mist, nozzle noise, fire, and possible carcinogenic ingredients. The advantages of robotic spray coating are lower energy consumption, improved consistency of finish, and reduced paint quantities used. To install a robotic painting application, one needs to consider certain manual requirements. These include continuous-path control to emulate the motion of a human operator, a hydraulic drive system to minimize electrical spark hazards, and manual lead-through programming with multiple program storage capability [Groover et al., 1998]. Schemes have considered geometric modeling, painting mechanics, and robot dynamics to output an optimal trajectory based on CAD data describing the objects [Suh et al., 1991]. The objective of such work is to plan an optimal robot trajectory that gives uniform coating thickness and minimizes coating time.

Machining operations utilizing robots typically employ end effectors that are powered spindles attached to the robot wrist. A tool is attached to the spindle to perform the processing operation. Examples of tools would be wire brushes or a grinding wheel. It should be noted that such applications are inherently flexible and have the disadvantage of being less accurate than a regular machine tool. Finishing operations, such as deburring, have provided excellent opportunities for robotic application. Force-control systems have proven particularly useful in regulating the contact force between the tool and the edge of the work to be deburred. In general, force-torque sensors mounted at the robot wrist have proven extremely useful in many applications in processing and assembly operations. The Lord Corporation and JR3 are two manufacturers of such commercially available sensors.

Assembly Applications

Automated assembly has become a major application for robotics. Assembly applications consider two basic categories: parts mating and parts joining. Part mating refers to peg-in-hole or hole-on-peg operations, whereas joining operations are concerned not only with mating but also a fastening procedure for the parts. Typical fastening procedures could include powered screwdrivers with self-tapping screws, glues, or similar adhesives.

In parts-mating applications, remote center compliance (RCC) devices have proven to be an excellent solution. In general, compliance is necessary for avoiding or minimizing impact forces, for correcting positioning error, and for allowing relaxation of part tolerances. In choosing an RCC device, the following parameters need to be determined prior to an application:

- Remote center distance (center of compliance). This is the point about which the active forces are at a minimum. The distance is chosen by considering the length of the part and the gripper.
- Axial force capacity: maximum designed axial force to function properly.
- Compressive stiffness: should be high enough to withstand any press-fitting requirements.
- Lateral stiffness: refers to force required to deflect RCC perpendicular to direction of insertion.
- Angular stiffness: relates to forces that rotate the part about the compliant center (also called the cocking stiffness).
- Torsion stiffness: relates to moments required to rotate a part about the axis of insertion.

Other parameters also include the maximum allowable lateral and angular errors as determined by the size of the part and by its design. These errors must be large enough to compensate for errors due to parts, robots, and fixturing. Passive and instrumented RCC devices have been developed for assembly applications. One such device that combines a passive compliance with active control is described by Xu and Paul [1992]. In addition, the SCARA (selective compliance articulated robot for assembly) class of robots is stiff vertically but relatively compliant laterally. Many opportunities exist for applying flexible assembly systems. The reader is referred to the book by Boothroyd, Dewhurst, and Knight [1994] for methods to evaluate a product's ease of assembly by robots.

Inspection

Inspection involves checking of parts, products, and assemblies as a verification of conformation to the specification of the engineering design. With the emphasis on product quality, there is a growing emphasis for 100% inspection. Machine vision systems, robot-manipulated active sensing for inspection, and automatic test equipment are being integrated into total inspection systems. Robot application of vision systems include part location, part identification, and bin picking. Machine vision systems for inspection typically perform tasks that include dimensional accuracy checks, flaw detection, and correctness and completeness of an assembled product. Vision inspection systems in industry are predominantly two-dimensional systems capable of extracting feature information, analyzing such information, and comparing to known patterns previously trained into the system.. Primary factors to be considered in the design or application of a vision system include the resolution and field of view of the camera, the type of camera, lighting requirements, and the required throughput of the vision system.

Machine vision application can be considered to have three levels of difficulty, namely, that (1) the object can be controlled in both appearance and position, (2) it can be controlled in either appearance or position, or (3) neither can be controlled. The ability to control both position and appearance requires advanced, potentially three-dimensional vision capabilities. The objective in an industrial setting is to lower the level of difficulty involved. It should be noted that inspection is but one category of robotic applications of machine vision. Two other broad categories are identification and visual servoing and navigation. In the latter case, the purpose of the vision system is to direct the motion of the robot based on visual input. Advances in machine vision technology have lowered the cost, increased the reliability, and provides enhanced "eyesight", interconnectivity, and interoperability of machine vision sensors. New vision sensors have been noted to have four times the resolution and eight times the speed than previously [Heil and Williams, 2001]

Emerging Issues

Robotics, by definition, is a highly multidisciplinary field. Applications are broad, and even those applications focused on the manufacturing shop floor are too numerous to cover in full here. The reader is referred to the various journals published by the IEEE and other societies and publishers, a few of which have been listed in the references. Still, it is worthwhile to note a few issues relevant to manufacturing shop-floor applications that could have an impact over the next decade. These issues include instrumented gripper development, machine vision integration, mobility, and the goal of intelligent robots. The objective of much of this work is the overall integration of a flexible manufacturing system. It has been noted that robot technology is moving in two different directions [Bloss, 2003]. On one hand there is a drive toward smaller, high-speed robots, whereas the other option is providing robots with greater reach and payload capacity. Higher autonomy for robots is also becoming a thrust area for manufacturers. Welding applications alone have benefited from increased use of sensor technology and enhanced control systems. The capability for simulation, testing, and programming a virtual model enables short lead times in preparing for production and nonphysical testing [Bolmsjo et al., 2002]. Recent developments on haptic robots shows promise on production lines. Haptic has its origin in the Greek word *hapien* meaning touch. Haptic robotic arms (teleoperated) have an integral force-feedback system that transmits forces to the user and provides the sensation of touching. Such robots are used in France (notably Renault using a Virtuose haptic robotic arm) to address the problem as to how to design automotive side windows for ease of assembly into the opening mechanism inside the car door [Kochan, 2003]. Advances in intelligent control, articulated hands, integrated joint-torque control, real-time sensory feed back including 3-D vision as well as advances in sensor fusion and control architectures have all occurred in the 1990s [Hirzinger et al., 1999]. The issues of intelligent systems, their architectures, and their implementation are a continuing source of investigation. One interesting reference pertaining to computational intelligence is the work of Meystel and Albus [2002]. This work had its origins in robotic control. Developments originating in the computer science (artificial intelligence) community have greatly influenced robotic development. The Internet now connects factory automation systems directly to customers and providers. The reader is referred to Brugali and Fayad [2002] for the effect of distributed computing on robotics and automation. Multiagent or holonic systems are one leading paradigm for future architectures and control of manufacturing systems.

Defining Terms

Cellular manufacturing: Grouping of parts by design and/or processing similarities such that the group (family) is manufactured on a subset of machines that constitute a cell necessary for the group's production.

Decision-tree analysis: Decomposing a problem into alternatives represented by branches where nodes (branch intersections) represent a decision point or chance event having probabilistic outcome. Analysis consists of calculating expected values associated with the chain of events leading to the various outcomes.

Degrees of freedom: The total number of individual motions typically associated with a machine tool or robot.

Intelligent control: A sensory-interactive control structure incorporating cognitive characteristics that can include artificial-intelligence techniques and contain knowledge-based constructs to emulate learning behavior with an overall capacity for performance and/or parameter adaptation.

Machine interference: The idle time experienced by any one machine in a multiple-machine system that is being serviced by an operator (or robot) and is typically measured as a percentage of the total idle time of all the machines in the system to the operator (or robot) cycle time.

Sensor fusion: Combining of multiple sources of sensory information into one representational format.

Sensor integration: The synergistic use of multiple sources of sensory information to assist in the accomplishment of a task.

References

C.R. Asfahl, *Robotics and Manufacturing Automation*, 2nd ed., New York: Wiley, 1992.

L. Biogiotti, C. Melchiorri, and G. Vassura, "A dexterous robotic gripper for autonomous grasping," *Ind. Rob.*, 30, no. 5, 449–458, 2003.

R. Bloss, "Innovations at IMTS" *Ind. Rob.*, 30, no. 2, 159–161, 2003.

G. Bolmsjo, M. Olsson, and P. Cederberg, "Robotic arc welding: trends and developments for higher autonomy," *Ind. Rob.*, 29, no. 2, 98–104, 2002.

G. Boothroyd, P. Dewhurst, and W. Knight, *Product Design and Manufacture for Assembly*, New York: Marcel Decker, 1994.

D. Brugali and M. Fayad, "Distributed computing in robotics and automation" *IEEE Trans. Rob. Autom.*, 18, no. 4, 409–420, 2002.

E.P. DeGarmo, W.G. Sullivan, J.A. Bontadelli, and E.M. Wicks, *Engineering Economy*, 10th ed., Englewood Cliffs, NJ: Prentice-Hall, 1997.

Y. Ding, P. Kim, D. Ceglarek, and J. Jin, "Optimal sensor distribution for variation diagnosis in multistage assembly systems," *IEEE Trans. Rob. Autom.*, 19, no. 4, 543–556, 2003.

R.C. Dorf, Ed., *International Encyclopedia of Robotics*, vols. 1–3, New York: Wiley, 1988.

M.P. Groover, M. Weiss, R.N. Nagel, N.G. Odrey, and Morris, *Industrial Automation and Robotics*, New York: McGraw-Hill, 1998.

P.J. Heil and M.E. Williams, *Advances in Machine Vision for Robot Control, Sensors*, June 2001, www.sensorsmag.com/articles/0601/68/main.shtml.

G. Hirzinger, M. Fischer, B. Brunner, R. Koeppe, M. Otter, M. Grebenstein, and I. Schafer, "Advances in robotics: the DLR experience," *The Int. J. Rob. Res.*, 18, no. 11, 1064–1087, 1999.

I.-S. Kim, J.-S. Son, and P.K.D.V. Yarlagadda, "A study on the quality improvement of robotic GMA welding process," *Rob. Comput. Integr. Manuf.*, 19, 567–572, 2003.

M. Knights, *Robots*, 44–45, October 2003, www.plastics.com.

A. Kochan, "Haptic robots help solve design issues," *Ind. Rob.*, 30, no. 6, 505–507, 2003.

R. Luo, C-C. Yih, and K.L. Su, "Multisensor fusion and integration: approaches, applications, and future research directions," *IEEE Sens. J.*, 2, no. 2, 107–119, 2002.

C. Mahr, "Beyond the standard cycle: increasing total work cell productivity by optimizing robot placement and other key performance factors," *Ind. Rob. Int. J.*, 27, no. 5, 334–337, 2000.

A. Meystel and J. Albus, *Intelligent Systems: Architecture, Design, and Control*, New York: Wiley-Interscience, 2002.

J. Nadzam, "Tandem GMAW offers quality weld deposits, high travel speeds," *Welding Design and Fabrication*, p. 28, November 2003.

W.S. Newman, A. Podgurski, R. Quinn, F. Merat, M. Branickey, N. Barendt, G. Causey, E. Haaser, Y. Kim, J. Swaminathan, and V. Velasco, Jr., "Design lessons for building agile manufacturing systems," *IEEE Trans. Rob. Autom.*, 16, no. 3, June 2000.

S.Y. Nof and H. Lechtman, "The RTM method of analyzing robot work," *Ind. Eng.*, April, 38–48, 1982.

S.Y. Nof, Ed., *The Handbook of Industrial Robotics,* 2nd ed., New York: Wiley, 1999.

N.G. Odrey, "Robotics and automation," in *Maynard's Industrial Engineering Handbook*, 4th ed., W.K. Hodson, Ed., New York: McGraw-Hill, 1992.

C.S. Park, *Fundamentals of Engineering Economics,* Englewood Cliffs, NJ: Pearson/Prentice-Hall, 2004.

J.M. Proth and H.P. Hillion, *Mathematical Tools in Production Management*, New York: Plenum Press, 1990.

P.G. Ranky, "Collaborative, synchronous robots serving machines and cells," *Ind. Rob. Int. J.*, 30, no. 3, 213–217, 2003.

W.G. Rippey and J.A. Falco, "The NIST automated arc welding testbed," *Proc. Seventh Int. Conf. Comput. Technol. Welding*, San Francisco, CA, July 8–11, 1997.

S.-H. Suh, I.-K. Wu, and S.-K. Noh, "Automatic trajectory planning system (ATPS) for spray painting robots," *J. Manuf. Syst.*, 10, no. 5, 396–406, 1991.

H.K. Tonshoff and I. Inasaki, Eds., *Sensors in Manufacturing*, Weinheim: Wiley-VCH GmbH, 31 May 2002.

Y. Xu and R.P. Paul, "Robotic instrumented compliant wrist," *ASME J. Eng. Ind.*, 114, 120–123, 1992.

Further Information

Various journals publish on topics pertaining to robots. Sources include the bimonthly *IEEE Journal of Robotics and Automation*, the quarterly journal *Robotics and Computer-Integrated Manufacturing* (published by Pergamon Press), *Robotics* (published by Cambridge University Press since 1983), and the *Journal of Robotic Systems* (published by Wiley).

The three-volume *International Encyclopedia of Robotics: Applications and Automation* (R.C. Dorf, Ed.), published by Wiley (1988), brings together the various interrelated fields constituting robotics and provides a comprehensive reference. Another good reference on various topics on robotics is the *Handbook of Industrial Robotics* edited by Nof [1999].

IEEE has sponsored since 1984 the annual "International Conference on Robotics and Automation." IEEE conference proceedings and journals are available from the IEEE Service Center, Piscataway, NJ. The Internet provides an immense access to researchers and robot manufacturers and research labs worldwide.

15

Aerospace Systems

Cary R. Spitzer
AvioniCon, Inc.

Daniel A. Martinec
Aeronautical Radio, Inc.

Cornelius T. Leondes
University of California, San Diego

Vyacheslav Tuzlukov
Ajou University

Won-Sik Yoon
Ajou University

Yong Deak Kim
Ajou University

15.1 Avionics Systems

Cary R. Spitzer, Daniel A. Martinec, and Cornelius T. Leondes

Avionics (aviation electronics) systems perform many functions: (1) for both military and civil aircraft, avionics are used for flight controls, guidance, navigation, communications, and surveillance; and (2) for military aircraft, avionics also may be used for electronic warfare, reconnaissance, fire control, and weapons guidance and control. These functions are achieved by the application of the principles presented in other chapters of this handbook, e.g., signal processing, electromagnetic, communications, etc. The reader is directed to these chapters for additional information on these topics. This section focuses on the system concepts and issues unique to avionics that provide the traditional functions listed in (1) above.

Development of an avionics system follows the familiar systems engineering flow from definition and analysis of the requirements and constraints at increasing level of detail, through detailed design, construction, **validation**, installation, and maintenance. Like some of the other aerospace electronic systems, avionics operate in real time and perform mission- and life-critical functions. These two aspects combine to make avionics system design and **verification** especially challenging.

Although avionics systems perform many functions, there are three elements common to most systems: data buses, controls and displays, and power. Data buses are the signal interfaces that lead to the high degree of integration found today in many modern avionics systems. Controls and displays are the primary form of crew interface with the aircraft and, in an indirect sense, through the display of synoptic information, also aid in the integration of systems. Power, of course, is the life blood of all electronics.

The generic processes in a typical avionics system are signal detection and preprocessing, signal fusion, computation, control/display information generation and transmission, and feedback of the response to the control/display information. (Of course, not every system will perform all of these functions.)

A Modern Example System

The B-777 Airplane Information Management System (AIMS) is the first civil transport aircraft application of the integrated, modular avionics concept, similar to that being used in the U.S. Air Force F-22. Figure 15.1 shows the AIMS cabinet with eight modules installed and three spaces for additional modules to be added as the AIMS functions are expanded. Figure 15.2 shows the AIMS architecture.

AIMS functions performed in both cabinets include flight management, electronic flight instrument system (EFIS) and engine indicating and crew alerting system (EICAS) display management, central maintenance, airplane condition monitoring, communications management, data conversion and gateway (ARINC 429 and ARINC 629), and engine data interface. AIMS does not control the engines nor flight controls, nor operate any internal or external voice or data link communications hardware, but it does select the data link path as part of the communications management function. Subsequent generations of AIMS may include some of these latter functions.

In each cabinet the line replaceable modules (LRMs) are interconnected by dual ARINC 659-like backplane data buses. The cabinets are connected to the quadruplex (not shown) or triplex redundant ARINC 629 system and fly-by-wire data buses and are also connected via the system buses to the three multifunction control display units (MCDU) used by flight crew and maintenance personnel to interact with AIMS. The cabinets transmit merged and processed data over quadruple-redundant custom-designed 100-MHz buses to the EFIS and EICAS displays.

In the AIMS the high degree of function integration requires levels of system availability and integrity not found in traditional distributed, federated architectures. These extraordinary levels of availability and integrity are achieved by the extensive use of **fault-tolerant** hardware and software maintenance diagnostics, which promise to reduce the chronic problem of unconfirmed removals and low mean time between unscheduled removals (MTBUR).

Figure 15.3 is a top-level view of the U.S. Air Force F-22 Advanced Tactical Fighter avionics. Like many other aircraft, the F-22 architecture is hybrid, part federated and part integrated. The left side of the figure is the highly integrated portion, dominated by the two common integrated processors (CIPs) that process, fuse,

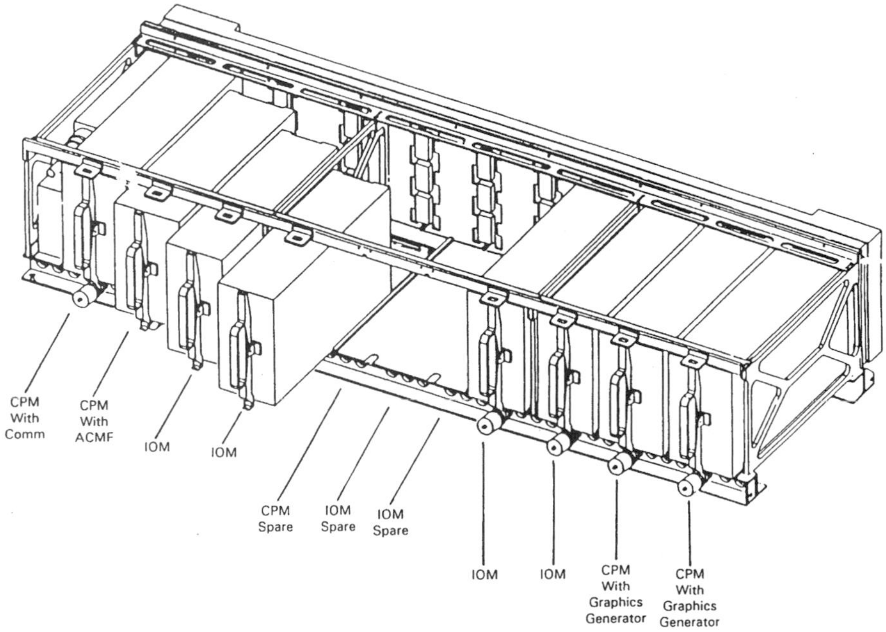

FIGURE 15.1 Cabinet assembly outline and installation (typical installation). (Courtesy of Honeywell, Inc.)

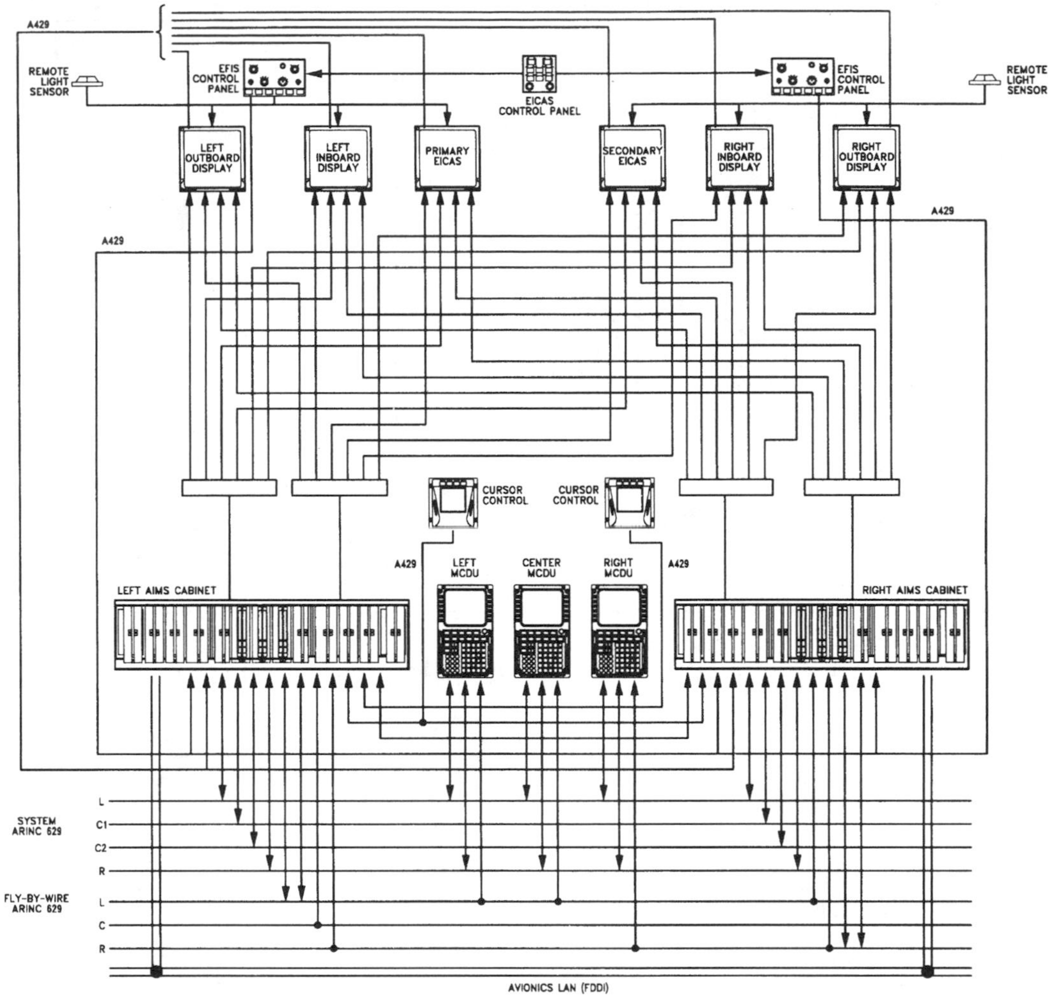

FIGURE 15.2 Architecture for AIMS baseline configuration. (Courtesy of Honeywell, Inc.)

and distribute signals received from the various sensors on the far left. The keys to this portion of the architecture are the processor interconnect (PI) buses within the CIPs and the high-speed data buses (HSDBs). (There are provisions for a third CIP as the F-22 avionics grow in capability.) The right side of the figure shows the federated systems, including the inertial reference, stores management, integrated flight and propulsion control, and vehicle management systems and the interface of the latter two to the integrated vehicle system control. The keys to this portion of the architecture are the triple- or quadruple-redundant AS 15531 (formerly MIL-STD-1553) command/response two-way data buses.

Data Buses

As noted earlier, data buses are the key to the emerging integrated avionics architectures. Table 15.1 summarizes the major features of the most commonly used system buses. MIL-STD-1553 and ARINC 429 were the first data buses to be used for general aircraft data communications. These are widely used today in military and civil avionics, respectively, and have demonstrated the significant potential of data buses. The others listed in the table build on their success.

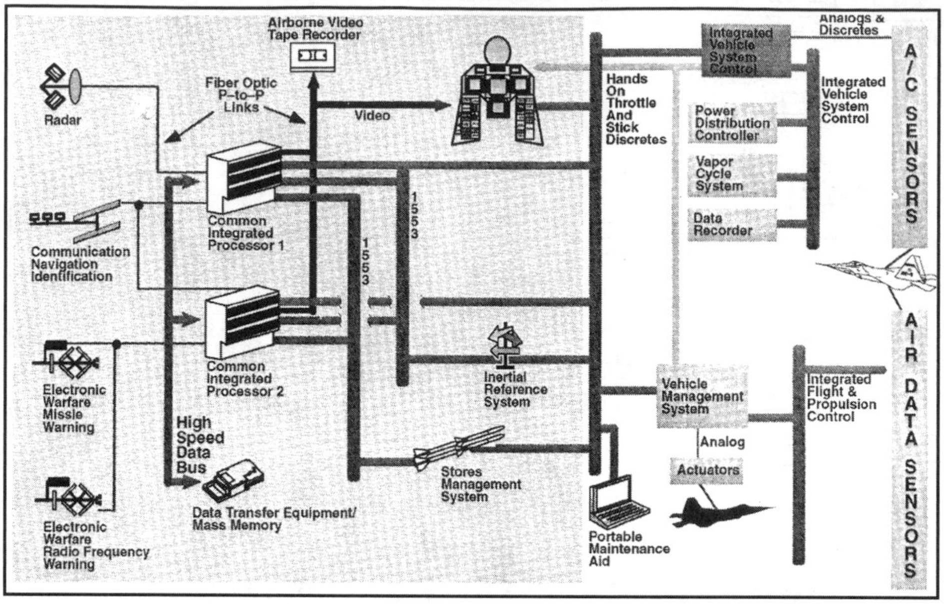

FIGURE 15.3 F-22 engineering and manufacturing development (EMD) architecture.

TABLE 15.1 Characteristics of Common Avionics Buses

Bus Name	Word Length	Bit Rate	Transmission Media
MIL-STD-1553	20 bits	1 Mb/s	Wire
DOD-STD-1773	20 bits	TBS	Fiber optic
High-speed data bus	32 bits	50 Mb/s	Wire or fiber optic
ARINC 429	32 bits	14.5/100 kb/s	Wire
ARINC 629	20 bits	2 Mb/s	Wire or fiber optic
ARINC 659	32 bits	100 Mb/s	Wire

Displays

All modern avionics systems use electronic displays, either CRTs or flat-panel LCDs, that offer exceptional flexibility in display format and significantly higher reliability than electromechanical displays. Because of the very bright ambient sunlight at flight altitudes, the principal challenge for an electronic display is adequate brightness. CRTs achieve the required brightness through the use of a shadow-mask design coupled with narrow bandpass optical filters. Flat-panel LCDs also use narrow bandpass optical filters and a bright backlight to achieve the necessary brightness.

Because of the intrinsic flexibility of electronic displays, a major issue is the design of display formats. Care must be taken not to place too much information in the display and to ensure that the information is comprehensible in high workload (aircraft emergency or combat) situations.

Power

Aircraft power is generally of two types: 28 vdc, and 115 vac, 400 Hz. Some 270 vdc is also used on military aircraft. Aircraft power is of poor quality when compared with power for most other electronics. Under normal conditions, there can be transients of up to 100% of the supply voltage and power interruptions of up to 1 second. This poor quality places severe design requirements on the avionics power supply, especially where the avionics are performing a full-time, flight-critical function. Back-up power sources include ram air turbines and batteries, although batteries require very rigorous maintenance practices to guarantee long-term reliable performance.

Software in Avionics

Most avionics currently being delivered are microprocessor controlled and are software intensive. The "power" achieved from software programs hosted on a sophisticated processor results in very complex avionics with many functions and a wide variety of options. The combination of sophistication and flexibility has resulted in lengthy procedures for validation and certification. The **brickwalling** of software modules in a system during the initial development process to ensure isolation between critical and noncritical modules has been helpful in easing the certification process.

There are no standard software programs or standard software certification procedures. RTCA has prepared document DO-178 to provide guidance (as opposed to strict rules) regarding development and certification of avionics civil software. The techniques for developing, categorizing, and documenting avionics civil software in DO-178 are widely used.

Military avionics software is "self certified," but generally follows, with varying degrees of rigor, the processes presented in DO-178.

The evolving definition of a standard for applications exchange (APEX) software promises to provide a common software platform whereby the specialized requirements of varying hardware (processor) requirements are minimized. APEX software is a hardware interface that provides a common link with the functional software within an avionics system. The ultimate benefit is the development of software independent of the hardware platform and the ability to reuse software in systems with advanced hardware while maintaining most, if not all, of the original software design.

CNS/ATM

The last 15 years have seen much attention focused on Communication/Navigation/Surveillance for Air Traffic Management (CNS/ATM), a satellite-based concept developed by the Future Air Navigation System (FANS) Committees of the International Civil Aviation Organization (ICAO), a special agency of the United Nations. Many studies have predicted significant economic and airspace capacity rewards of CNS/ATM for both aircraft operators and air traffic services providers.

The new CNS/ATM system should provide for:

- Global communications, navigation, and surveillance coverage at all altitudes and embrace remote, off-shore, and oceanic areas.
- Digital data exchange between air-ground systems (voice backup).
- Navigation/approach service for runways and other landing areas that need not be equipped with precision landing aids.

Navigation Equipment

A large portion of the avionics on an aircraft are dedicated to navigation. The following types of navigation and related sensors are commonly found on aircraft:

- Flight control computer (FCC)
- Flight management computer (FMC)
- Inertial navigation system (INS)
- Altitude heading and reference system (AHRS)
- Air data computer (ADC)
- Low range radio altimeter (LRRA)
- Radar

- Distance measuring equipment (DME)
- Instrument landing system (ILS)
- Microwave landing system (MLS)
- VHF omnirange (VOR) receiver
- Global navigation satellite system (GNSS)

Emphasis on Communications

An ever-increasing portion of avionics is dedicated to communications. Much of the increase comes in the form of digital communications for either data transfer or digitized voice. Military aircraft typically use digital communications for security. Civil aircraft use digital communications to transfer data for improved efficiency of operations and radio frequency (RF) spectrum utilization. Both types of aircraft are focusing more on enhanced communications to fulfill the requirements for better operational capability.

Various types of communications equipment are used on aircraft. The following list tabulates typical communications equipment:

- VHF transceiver (118–136 MHz)
- UHF transceiver (225–328 MHz/335–400 MHz for military)
- HF transceiver (2.8–24 MHz)
- Satellite (1530–1559/1626.5–1660.5 MHz, various frequencies for military)
- Aircraft communications addressing and reporting (ACARS)
- Joint tactical information distribution system (JTIDS)

In the military environment the need for communicating aircraft status and for aircraft reception of crucial information regarding mission objectives are primary drivers behind improved avionics. In the civil environment (particularly commercial transport), the desire for improved passenger services, more efficient aircraft routing and operation, safe operations, and reduced time for aircraft maintenance are the primary drivers for improving the communications capacity of the avionics.

The requirements for digital communications for civil aircraft have grown so significantly that the industry as a whole embarked on a virtually total upgrade of the communications system elements. The goal is to achieve a high level of flexibility in processing varying types of information as well as attaining compatibility between a wide variety of communication devices. The approach bases both ground system and avionics design on the ISO open-system interconnect (OSI) model. This seven-layer model separates the various factors of communications into clearly definable elements of physical media, protocols, addressing, and information identification.

The implementation of the OSI model requires a much higher level of complexity in the avionics as compared to avionics designed for simple dedicated point-to-point communications. The avionics interface to the physical medium will generally possess a higher bandwidth. The bandwidth is required to accommodate the overhead of the additional information on the communications link for the purpose of system management. The higher bandwidths pose a special problem for aircraft designers due to weight and electromagnetic interference (EMI) considerations. Additional avionics are required to perform the buffering and distribution of the information received by the aircraft. Generally a single unit, commonly identified as the communications management unit (CMU), will perform this function.

The CMU can receive information via RF transceivers operating in conjunction with terrestrial, airborne, or space-based transceivers. The capability also exists for transceiver pairs employing direct wire connections or very short-range optical links to the aircraft. The CMU also provides the routing function between the avionics, when applicable. Large on-board databases, such as an electronic library, may be accessed and provide information to other avionics via the CMU.

The increasing demand on data communication system capacity and flexibility is dictating the development of a system without the numerous limitations of current systems. Current communication systems require rather rigid protocols, message formatting, and addressing. The need for a more flexible and capable system has led to the initial work to develop an aeronautical telecommunications network (ATN).

The characteristics envisaged for the ATN are the initiation, transport, and application of virtually any type of digital message in an apparently seamless method between virtually any two end systems. The ATN is expected to be a continually evolving system.

Impact of "Free Flight"

"Free flight" is a term describing an airspace navigation system in which the "normal" air traffic controls are replaced by the regular transmission of position information from the airplane to the ground. The ground system, by projecting the aircraft position and time, can determine if the intended tracks of two aircraft would result in a cohabitation of the same point in the airspace. This is commonly called "conflict probe." If a potential conflict occurs, then a message is transmitted to one or more of the aircraft involved to make a change to course and/or speed.

"Free flight" dictates special requirements for the avionics suite. A highly accurate navigation system with high integrity is required. The communications and surveillance functions must exhibit an extremely high level of availability.

GNSS avionics performing the position determination functions will require augmentation to achieve the necessary accuracy. The augmentation will be provided by a data communications system and will be in the form of positional information correction. A data communications system will also be required to provide the frequent broadcast of position information to the ground. A modified Mode S transponder squitter is expected to provide that function.

The free-flight concept will require the equipage of virtually all aircraft operating within the designated free-flight airspace with a commensurate level of avionics capability. The early stages of the concept development uncovered the need to upgrade virtually all aircraft with enhanced CNS/ATM avionics. The air transport industry resolved this problem on older airplanes by developing improved and new avionics for retrofit applications. The new avionics design addresses the issues of increased accuracy of position and enhancement of navigation management in the form of the GNSS navigation and landing unit (GNLU) housed in a single unit and designed to be a physical and functional replacement for the ILS and/or MLS receivers. A built-in navigator provides enhanced navigation functionality for the airplane. The GNSS can provide ILS look-alike signals and perform landing guidance functions equivalent to Category I.

Avionics in the Cabin

Historically, the majority of avionics have been located in the electronics bay and the cockpit of commercial air transport airplanes. Cabin electronics had generally been limited to the cabin interphone and public address system, the sound and central video system, and the lighting control system. More recently the cabin has been updated with passenger telephones using both terrestrial and satellite systems. The terrestrial telephone system operates in the 900-MHz band in the United States and will operate near 1.6 GHz in Europe. The satellite system, when completely operational, will also operate near 1.6 GHz. Additional services available to the passengers are the ability to send facsimiles (FAXes) and to view virtually real-time in-flight position reporting via connection of the video system with the flight system. Private displays at each seat will allow personal viewing of various forms of entertainment including movies, games, casual reading, news programming, etc.

Avionics Standards

Standards play an important role in avionics. Military avionics are partially controlled by the various standards (MIL-STDs, DOD-STDs, etc.) for packaging, environmental performance, operating characteristics, electrical and data interfaces, and other design-related parameters. General aviation avionics are governed by fewer and less stringent standards. Technical standard orders (TSOs) released by the Federal Aviation Administration (FAA) are used as guidelines to ensure airworthiness of the avionics. TSOs are derived from and, in most cases, reference RTCA documents characterized as minimum operational performance standards and as minimum avionics system performance standards. EUROCAE is the European counterpart of RTCA.

The commercial air transport industry adheres to multiple standards at various levels. The International Civil Aviation Organization (ICAO) is commissioned by the United Nations to govern aviation systems including, but not limited to, data communications systems, on-board recorders, instrument landing systems, microwave landing systems, VHF omnirange systems, and distance measuring equipment. The ICAO standards and recommended practices (SARPS) control system performance, availability requirements, frequency utilization, etc. at the international level. The SARPS in general maintain alignment between the national avionics standards such as those published by EUROCAE and RTCA.

The commercial air transport industry also uses voluntary standards created by the Airlines Electronic Engineering Committee and published by Aeronautical Radio Inc. (ARINC). The ARINC "characteristics" define form, fit, and function of airline avionics.

Defining Terms

ACARS: A digital communications link using the VHF spectrum for two-way transmission of data between an aircraft and ground. It is used primarily in civil aviation applications.

Brickwalling: Used in software design in critical applications to ensure that changes in one area of software will not impact other areas of software or alter their desired function.

Distance measuring equipment: The combination of a receiver and a transponder for determining aircraft distance from a remote transmitter. The calculated distance is based on the time required for the return of an interrogating pulse set initiated by the aircraft transponder.

Fault tolerance: The built-in capability of a system to provide continued correct execution in the presence of a limited number of nonsimultaneously occurring hardware or software faults.

JTIDS: Joint Tactical Information Distribution System using spread-spectrum techniques for secure digital communication. It is used for military applications.

Validation: The process of evaluating a product at the end of the development process to ensure compliance with requirements.

Verification: (1) The process of determining whether the products of a given phase of the software development cycle fulfill the requirements established during the previous phase. (2) Formal proof of program correctness. (3) The act of reviewing, inspecting, testing, checking, auditing, or otherwise establishing and documenting whether items, processes, services, or documents conform to specified requirements (IEEE).

References

Airlines Electronic Engineering Committee Archives, Aeronautical Radio Inc.

FANS *Manual*, International Air Transport Association, Montreal, Version 1.1, May 1995.

Federal Radionavigation Plan, DOT-VNTSC-RSPA-90-3/DOD4650.4, Departments of Transportation and Defense, 1990.

M.J. Morgan, "Integrated modular avionics for next generation commercial airplanes," *IEEE/AES Systems Magazine*, pp. 9–12, August 1991.

C.R. Spitzer, *Digital Avionics Systems*, 2nd ed., (reprint) Caldwell, N.J., Blackburn Press, 2000.

C.R. Spitzer, Ed., *Avionics Handbook*, Boca Raton, Fla.: CRC Press, 2000.

Further Information

K. Feher, *Digital Communications*, Englewood Cliffs, N.J.: Prentice Hall, 1981.

J.L. Farrell, *Integrated Aircraft Navigation*, New York: Academic Press, 1976.

L.E. Tannas, Jr., *Flat Panel Displays and CRTs*, New York: Van Nostrand Reinhold, 1985.

M. Kayton and W.R. Fried, *Avionics Navigation Systems*, 2nd ed., New York: John Wiley and Sons, 1997.

15.2 Satellite Communications Systems: Applications

Vyacheslav Tuzlukov, Won-Sik Yoon, and Yong Deak Kim

Background of Satellite Communications Systems

Satellite communications are the outcome of research in the area of communications whose objective is to achieve ever-increasing ranges and capacities with the lowest possible cost. Satellite communications now form an integral part of our new "wired" world. Since their introduction in about 1965, satellite communications have generated a multiplicity of new telecommunication or broadcasting services on a global or regional basis. In particular, they have enabled a global, automatically switched telephone network to be created. Although, since 1956, submarine cables began to operationally connect the continents of the world with readily available telephone circuits, only satellite communications have enabled completely reliable communication links for telephone, television, and data transmission to be provided over all types of terrestrial obstacles, regardless of the distance or remoteness of the locations that have to be connected.

Satellite communications systems were originally developed to provide long-distance telephone services. In the late 1960s, launch vehicles had been developed that could place a 500-kg satellite in geostationary Earth orbit (GEO), with a capacity of 5000 telephone circuits, marking the start of an era of expansion for telecommunication satellites. Geostationary satellites were soon carrying transoceanic and transcontinental telephone calls. For the first time, live television links could be established across the Atlantic and Pacific Oceans to carry news and sporting events.

The geostationary orbit is preferred for all high-capacity communication satellite systems because a satellite in GEO appears to be stationary over a fixed point on the ground. It can establish links to one third of the Earth's surface using fixed antennas at the Earth stations. This is particularly valuable for broadcasting, as a single satellite can serve an entire continent. Direct broadcast satellite television (DBS-TV) and the distribution of video signals for cable television networks are the largest single revenue source for geostationary satellites, accounting for $17 billion in revenues in 1998. By the year 2001, nearly 200 GEO communication satellites were in orbit, serving every part of the globe. Although television accounts for much of the traffic carried by these satellites, international and regional telephony, data transmission, and Internet access are also important. In the populated parts of the world, the geostationary orbit is filled with satellites every 2° or 3° operating in almost every available frequency band.

GEO satellites have grown steadily in weight, size, lifetime, and cost over the years. Some of the largest satellites launched to date are the *KH* and *Lacrosse* surveillance satellites of the U.S. National Reconnaissance Office, weighing an estimated 13,600 kg [*Aviation Week and Space Technology*, 2000]. By 2000, commercial telecommunications satellites weighing 6000 kg with lifetimes of 15 years were being launched into geostationary orbit at a typical cost of around $125 million for the satellite and launch. The revenue-earning capacity of these satellites must exceed $20 million per year for the venture to be profitable, and they must compete with optical fibers in carrying voice, data, and video signals. A single optical fiber can carry 4.5 Gbps, a capacity similar to that of the largest GEO satellites, and optical fibers are never laid singly but always in bundles. GEO satellites can compete effectively on flexibility of delivery point. Any place within the satellite coverage can be served by simply installing the Earth terminal. To do the same with the fiber-optic link requires it to be laid. Fiber-optic transmission systems compete effectively with satellites where there is a requirement for high capacity or, equivalently, when the user density exceeds the required economic threshold.

GEO satellites have been supplemented by low and medium Earth orbit satellites for special applications. Low Earth orbit (LEO) satellites can provide satellite telephone and data services over continents or over the entire world, and by 2000 three systems were in orbit or nearing completion, with a total of 138 LEO satellites. LEO satellites are also used for Earth imaging and surveillance. Although not strictly a satellite communications system, the Global Positioning System (GPS), which uses 24 medium Earth orbit (MEO) satellites, has revolutionized navigation. GPS receivers have become a consumer product. Eventually every car and cellular telephone will have a GPS receiver built into it so that drivers will not get lost and emergency calls from cellular phones will automatically carry information about the phone's location.

At present, about 140,000 equivalent channels are in operation, accounting only for the International Telecommunications Satellite Consortium (INTELSAT) system (the term "equivalent channel" is used in lieu of "telephone circuit" to indicate that this channel may carry any type of multimedia traffic — voice, data, video, etc.). Although the growth of the satellite part on intercontinental telephone highways has been somewhat slowed by the installation of new, advanced, optical-fiber submarine cables, satellites continue to carry out most of the public switched telephone traffic between the developed and the developing world and between developing regions, and this situation should continue, thanks to the advantages of satellite communications systems and, in particular, to the simplicity and the flexibility of the required infrastructure.

It can be predicted that the future of satellite communications systems will rely more and more on the effective use of their specific characteristics:

- Multiple access capability, i.e., point-to-point, point-to-multipoint, or multipoint-to-multipoint connectivity, in particular, for small- or medium-density scattered data or voice/data traffic between small, low-cost Earth stations (business or private communications networks, rural communications, etc.)
- Distribution capability (a particular case of point-to-multipoint transmission), including:
    - TV program broadcasting and other video and multimedia applications (these services are currently in full expansion)
    - Data distribution, e.g., for business services, Internet broadband services, etc.
- Flexibility for changes in traffic and in network architecture and also ease of orientation and commission

A major factor in the recent revolution of satellite communications is the nearly complete replacement by digital modulation and transmission of the previous conventional analogue technique, bringing the full advantage of digital techniques in terms of signal processing, hardware and software implementation, etc. It should also be emphasized that satellite communications will no longer be based, as was the case up till now, only on GEO satellites, but also on the utilization of LEO and MEO satellite communications. These new satellite communications systems are opening the way to new applications, such as personal communications, fixed or mobile, innovative wideband data services, etc.

History of Satellite Communications Systems

The main events that have contributed to the creation of a new era in worldwide communications are summarized below. It is of interest to note that only 11 years passed between the launching of the first artificial satellite *Sputnik-1* by the U.S.S.R. and the actual implementation of a fully operational global satellite communications system (*Intelsat-III*) in 1968.

1903: Russian scientist Konstantin Tsyolkovsky describes the Theory of Rocket Flight in Cosmic Space.

1927: Russian scientist Konstantin Tsyolkovsky introduces the statement "artificial satellite of the Earth" and describes the Theory of Cosmic Orbital Station Construction.

1929: Hermann Noordung describes the concept of the geostationary orbit.

1945 (May): Arthur C. Clark describes a world communication and broadcasting system based on geosynchronous stations.

1957 (4 Oct.): Launching of the *Sputnik-1* artificial satellite (U.S.S.R.) and detection of the first satellite-transmitted signals.

1959 (March): Pierce's basic paper on satellite communications possibilities.

1960 (Aug.): Launching of the *Echo-1* balloon satellite (U.S./NASA). Earth-station to Earth-station passive relaying of telephone and television signals at 1 and 2.5 GHz by reflection on the metalized surface of this 30-m balloon placed in a circular orbit at 1600-km altitude.

1960 (Oct.): First experiment of active relaying communications using a space-borne amplifier at 2 GHz (delayed relaying communications) by the *Courier-1B* satellite (U.S.) at about 1000-km altitude.

1962: Foundation of the COMSAT Corporation (U.S.), the first company specifically devoted to domestic and international satellite communications.

1962: Launching of the *Telstar-1* satellite (U.S./AT&T) (July) and of the *Relay-1* satellite (U.S./NASA) (December). Both were nongeostationary, low-altitude satellites operating in the 6/4-GHz bands.

1962: First experimental transatlantic communications (television and multiplexed telephony) between the first large-scale, preoperational Earth stations (Andover, Maine, U.S., Pleumeu-Bodou, France, and Goonhilly, U.K.).

1963: First international regulations of satellite communications (ITU Extraordinary Radio Conference). Initiation of sharing between space and terrestrial services.

1963 (July): Launching of *Syncom-2* (U.S./NASA), the first geostationary satellite (300 telephone circuits or one TV channel).

1964 (Aug.): Establishment of the INTELSAT organization (19 national administrations as initial signatories).

1965 (April): Launching of the *Early Bird* (*Intelsat-1*) satellite, first commercial geostationary communication satellite (240 telephone circuits or one TV channel). First operational communications (U.S., France, Federal Republic of Germany, U.K.).

1965: Launching of *Molniya-1* (U.S.S.R.), a nongeostationary satellite (elliptical orbit, 12 hours revolution). Beginning of television transmission to small-size receive Earth stations in the U.S.S.R. (29 *Molniyas* were launched between 1965 and 1975).

1967: *Intelsat-II* satellites (240 telephone circuits in multiple access mode or one TV channel) over Atlantic and Pacific Ocean regions.

1968–1970: *Intelsat-III* satellites (1500 telephone circuits, four TV channels or combinations thereof). INTELSAT worldwide operation.

1969: Launch of *ATS-5* (U.S./NASA). First geosynchronous satellite with a 15.3- and 3.6-GHz bands propagation experiment.

1971 (Jan): First *Intelsat-IV* satellite (4000 circuits + two TV channels).

1971 (Nov.): Establishment of the INTERSPUTNIK Organization (U.S.S.R. and nine initial signatories).

1972 (Nov.): Launching of the *Anik-1* satellite and first implementation of a national (domestic) satellite communications system outside the U.S.S.R. (Canada/TELESAT).

1974 (April): *Westar-1* satellite. Beginning of national satellite communications in the U.S.

1974 (Dec.): Launching of the *Symphonie-1* satellite (France, Federal Republic of Germany): the first three-axis stabilized geostationary communications satellite.

1975 (Jan.): Algerian satellite communications system: First operational national system (14 Earth stations) using a leased INTELSAT transponder.

1975 (Sept.): First *Intelsat-IVA* satellite (20 transponders: more than 6000 circuits + two TV channels, Frequency reuse by beam separation).

1975 (Dec.): Launching of the first U.S.S.R. geostationary *Stationar* satellite.

1976 (Jan.): Launching of the *CTS* (or *Hemes*) satellite (Canada), first experimental high-power broadcasting satellite (14/12 GHz).

1976 (Feb.): Launching of the *Marisat* satellite (U.S.), first maritime communications satellite.

1976 (July): Launching of the *Palapa-1* satellite. First national system (40 Earth stations) operating with a dedicated satellite in a developing country (Indonesia).

1976 (Oct.): Launching of the first *Ekran* satellite (U.S.S.R.). Beginning of the implementation of the first operational broadcasting satellite system (6.2/0.7 GHz).

1977 (June): Establishment of the EUTELSAT organization with 17 administrations as initial signatories.

1977 (Aug.): Launching of the *Siro* satellite (Italy). First experimental communications satellite using frequencies above 15 GHz (17/11 GHz).

1977: ITU world broadcasting satellite Administrative Radio Conference (Geneva, 1977) (WARC SAT-77).

1978 (Feb.): Launching of the *BSE* experimental broadcasting satellite for Japan (14/12 GHz).

1978 (May): Launching of the *OTS* satellite, first communication satellite in the 14/11-GHz band and first experimental regional communication satellite for Europe (ESA: European Space Agency).

1979 (June): Establishment of the INMARSAT organization for global maritime satellite communications (26 initial signatories).

1980 (Dec.): First *Intelsat-V* satellite (12,000 circuits, FDMA + TDMA operation, 6/4-GHz and 14/11-GHz wideband transponders, frequency reuse by beam separation + dual polarization).

1981: Beginning of operation in the U.S. of satellite business systems based on very small data receive Earth stations (using VSATs).

1983: ITU Regional Administrative Conference for the Planning of the Broadcasting-Satellite Service in Region 2.

1983 (Feb.): Launching of the *CS-2* satellite (Japan). First domestic operational communication satellite in the 30/20-GHz band.

1983 (June): First launch of the *ECS* (EUTELSAT) satellite, (nine wideband transponders at 14/11 GHz: 12,000 circuits with full TDMA operation + TV, frequency reuse by beam separation and by dual polarization).

1984: Beginning of operation of satellite business systems (using VSATs) with full transmit/receive operation.

1984 (April): Launching of *STW-1*, the first communication satellite of China, providing TV, telephone, and data transmission services.

1984 (Aug.): Launching of the first French domestic *Telecom-1* multimission satellite: 6/4 GHz, telephony and TV distribution; 8/7 GHz, military communications; 14/12 GHz, TVRO and business communications in TDMA/DA.

1984 (Nov.): First retrieval of communication satellites from space, using the space shuttle (U.S.).

1985 (Aug.): ITU World Administrative Radio Conference (WARC Orb-85) (first session on utilization of the geostationary orbit).

1988 (Oct.): ITU World Administrative Radio Conference (WARC Orb-88) (second session on utilization of the geostationary orbit).

1989: *Intelsat-VI* satellite (satellite-switched TDMA, up to 120 circuits [with DCME], etc.).

1992 (Feb.): ITU World Administrative Radio Conference.

1992 (Feb.): Launching of the first Spanish *Hispasat-1* multimission satellite: 14/11–12-GHz distribution, contribution, SNG, TVRO, VSAT, business services, TV America, etc.; 17/12-GHz, DBS analogue and digital television; 8/7-GHz governmental communications.

Up to 1996: Nine *Intelsat VII-VIIA* satellites.

1997–1998: *Intelsat VIII* satellites.

1998 onwards: Launching of various nongeostationary satellites and implementation of the corresponding MSS systems (Iridium, Globalstar, etc.) and FSS systems (Teledisc, Skybridge, etc.).

1999: First *Intelsat K-TV* satellite (30 14/11–12-GHz transponders for up to 210 TV programs with possible direct to home (DTH) broadcast and VSAT services).

2000: *Intelsat IX* satellites (up to 160,000 circuits [with DCME]).

Development and Overview of Satellite Communications Systems

Satellite communications began in October 1957 with the launch by the U.S.S.R. of a small satellite called *Sputnik-1*. This was the first artificial Earth satellite, and it sparked the space race between the United States and the U.S.S.R. *Sputnik-1* carried only a beacon transmitter and did not have communications capability, but it demonstrated that satellites could be placed in orbit by powerful rockets. The first voice heard from space was that of president Eisenhower, who recorded a brief Christmas message that was transmitted back to the

Earth from the *Score* satellite in December 1958. The *Score* satellite was essentially the core of the Atlas intercontinental ballistic missile (ICBM) booster with a small payload in the nose. A tape recorder on *Score* had a storage capacity that allowed a 4-minute message received from the Earth station to be retransmitted.

After some early attempts to use large balloons (*Echo I* and *II*) as passive reflectors for communication signals, and some small experimental satellite launches, the first true communications satellites, *Telstar I* and *II*, were launched in July 1962 and May 1963. The *Telstar* satellites were built by Bell Telephone Laboratories and used C-band transponders adapted from terrestrial microwave link equipment. The uplink was at 6389 MHz and the downlink was at 4169 MHz, with 50-MHz bandwidth. The satellites carried solar cells and batteries that allowed continuous use of the single transponder, and demonstrations of live television links and multiplexed telephone circuits were made across the Atlantic Ocean, emphatically demonstrating the feasibility of satellite communications. The *Telstar* satellites were MEO satellites with periods of 158 and 225 min. This allowed transatlantic links to operate for about 20 min while the satellite was mutually visible. The orbits chosen for the *Telstar* satellites took them through several bands of high-energy radiation that caused early failure of the electronics on board. However, the value of communication satellites had been demonstrated and work was begun to develop launch vehicles that could deliver a payload to GEO, and to develop satellites that could provide useful communication capacity.

On December 20, 1961, the U.S. Congress recommended that the International Telecommunications Union (ITU) should examine the aspects of space communications for which international cooperation would be necessary. The most critical step was in August 1962, when the U.S. Congress passed the *Communications Satellite Act.* This set the stage for commercial investment in an international satellite organization and, on July 19, 1964, representatives of the first 12 countries to invest in what became INTELSAT signed an initial agreement. The company that represented the United States at this initial signing ceremony was Communicating Satellite Corporation (COMSAT), an entity specifically created to act for the United States within INTELSAT. It should be remembered that, at this point, Bell Systems had a complete monopoly of all long-distance telephone communications within the United States. When Congress passed the *Communications Satellite Act,* Bell Systems was specifically barred from directly participating in satellite communications, although it was permitted to invest in COMSAT. COMSAT essentially managed INTELSAT in the formative years and should be credited with remarkable success of the international venture. The first five *Intelsat* series (*Intelsat I* through *V*) were selected, and their procurement managed, by teams put in place under COMSAT leadership. Over this same phase, though, large portions of the COMSAT engineering and operations groups transferred over to INTELSAT so that, when Permanent Management Arrangements came into force in 1979, many former COMSAT groups were now part of INTELSAT.

In mid-1963, 99% of all satellites had been launched into LEO. LEO and the slightly higher MEO were much easier to reach than GEO with the small launchers available at that time. The intense debate was eventually settled on launcher reliability issues rather than on payload capabilities. The first six years of the so-called space age was a period of both payload and launcher development. The new frontier was very risky, with about one launch in four being fully successful. The system architecture of the first proposed commercial satellite communications system employed 12 satellites in an equatorial MEO constellation. Thus, with the launch failure rate at the time, 48 launches were envisioned to guarantee 12 operational satellites in orbit. Without 12 satellites in orbit, continuous 24-h coverage could not be offered. Twenty-four hours a day, seven days a week — referred to as 24/7 operation — is a requirement for any successful communications service. GEO systems architecture requires only one satellite to provide 24/7 operation over essentially one third of the inhabited world. On this basis, four launches would be required to achieve coverage of the one third of the Earth, 12 for the entire inhabited world. Despite its unproven technological approach, the geostationary orbit was selected by the entities that became INTELSAT.

The first INTELSAT satellite, *Intelsat* (formerly *Early Bird*) was launched on April 16, 1965. The satellite weighed a mere 36 kg and incorporated two 6/4-GHz transponders, each with 25-MHz bandwidth. Commercial operations commenced between Europe and the United States on June 28, 1965. INTELSAT was highly successful and grew rapidly as many countries saw the value of improved telecommunications, not just internationally but for national systems that provided high-quality satellite communications within the borders of large countries.

Canada was the first country to build a national telecommunication system using GEO satellites. *Anik 1A* was launched in May 1974, just two months before the first U.S. domestic satellite, *Westar I*. The honor of the first regional satellite system, however, goes to the U.S.S.R's. *Molniya* system of highly elliptic orbit (HEO) satellites, the first of which was launched in April 1965 (the same month as *Intelsat I*). Countries that are geographically spread like the U.S.S.R., which covers 11 time zones, have used regional satellite systems very effectively. Another country that benefited greatly from a GEO regional system was Indonesia, which consists of more than 3000 islands spread out over more than 1000 miles. A terrestrially based telecommunication system was not economically feasible for these countries, while a single GEO satellite allowed instant communications region wide. Such ease of communications via GEO satellites proved to be very profitable. Within less than ten years, INTELSAT was self-supporting and, since it was not allowed to make a profit, it began returning substantial revenues to what were known as its *signatories*. Within 25 years, INTELSAT had more than 100 signatories [Ress, 1989] and, in early 2000, there were 143 member countries and signatories that formed part of the international INTELSAT community.

The astonishing commercial success of INTELSAT led many nations to invest in their satellite communications systems. This was particularly true in the United States. By the end of 1983, telephone traffic carried by the U.S. domestic satellite communications systems earned more revenue than the INTELSAT system. Many of the original INTELSAT signatories had been privatized by the early 1990s and were, in effect, competing not only with each other in space communications, but with INTELSAT. It was clear that some mechanism had to be found whereby INTELSAT could be turned into a for-profit, private entity, which could then compete with other commercial organizations while still safeguarding the interests of the smaller nations that had come to depend on the remarkably low communications cost that INTELSAT offered. The first step in the move to privatizing INTELSAT was the establishment of a commercial company called *New Skies* and the transfer of a number of INTELSAT satellites to *New Skies*.

In the 1970s and 1980s there was rapid development of GEO satellite communications systems for international, regional, and domestic telephone traffic and video distribution. In the United States, the expansion of fiber-optic links with very high capacity and low delay caused virtually all telephone traffic to move to terrestrial circuits by 1985. However, the demand for satellite communications systems grew steadily through this period, and the available spectrum in C-band was quickly occupied, leading to expansion into Ku-band. In the United States, most of the expansion after 1985 was in the areas of video distribution and very small aperture terminal (VSAT) networks. By 1995 it was clear that the GEO orbit capacity at Ku-band would soon be filled, and Ka-band satellite communications systems would be needed to handle the expansion of digital traffic, especially wideband delivery of high-speed Internet data. The Societe Europeén de Satellites (SES), based in Luxemburg, began two-way multimedia and Internet access service in western and central Europe at Ka-band using the *Astra 1H* satellite in 2001 [http://www.astra.ln]. Several Ka-band satellite communications systems are operational starting from 2003 [http://www.astrolink.com, http://www.hns.com.spaceway].

The ability of satellite systems to provide communication with mobile users had long been recognized, and the International Maritime Satellite Organization (INMARSAT) has provided service to ships and aircraft for several decades, although at a high price. LEO satellites were seen as one way to create a satellite telephone system with worldwide coverage; numerous proposals were floated in the 1990s, with three LEO satellite communications systems eventually reaching completion by 2000 (*Iridium, Globalstar,* and *Orbcomm*). The implementation of a LEO and MEO satellite system for mobile communication has proved much more costly than anticipated, and the capacity of the systems is relatively small compared to GEO satellite communications systems, leading to a higher cost per transmitted bit. Satellite telephone systems were unable to compete with cellular telephone systems because of the high cost and relatively low capacity of the space segment. The *Iridium* satellite communications system, for example, cost over $5 billion to implement, but provided a total capacity for the United States of less than 10,000 telephone circuits [Pratt et al., 2003]. *Iridium Inc.* declared bankruptcy in early 2000, having failed to establish a sufficiently large customer base to make the venture viable. The entire *Iridium* satellite communications system was sold to *Iridium Satellite LLC* for a reported $25 million, approximately 0.5% of the system's construction cost. The future of the other LEO and MEO satellite communications systems also seem uncertain at the present time.

Satellite navigation systems, notably the GPS, have revolutionized navigation and surveying. The GPS took almost 20 years to design and fully implement, at a cost of $12 billion. By 2000, GPS receivers could be built in original equipment manufacturer (OEM) form for less than $25, and the worldwide GPS industry was earning billions of dollars from equipment sales and services. In the United States, aircraft navigation will depend almost entirely on GPS by 2010, and blind landing systems using GPS will also be available. Accurate navigation of ships, especially in coastal waters and bad weather, is also heavily reliant on GPS. Europe is building a comparable satellite navigation system called *Galileo.*

Main Functioning Principles of Satellite Communications Systems

Satellite communications systems exist because the Earth is a sphere. Radio waves travel in straight lines at the microwave frequencies used for wideband communications, so a *repeater* is needed to convey signals over long distances. A satellite, because it can link places on the Earth that are thousands of miles apart, is a good place to locate a repeater, and a GEO satellite is the best place of all. A repeater is simply a receiver linked to a transmitter, always using different radio frequencies, that can receive a signal from one Earth station, amplify it, and retransmit it to another Earth station. The repeater derives its name from 19th-century telegraph links, which had a maximum length of about 50 miles. Telegraph repeater stations were required every 50 miles in a long-distance link so that the Morse code signals could be resent before they became too weak to read.

The majority of communication satellites are in GEO, at an altitude of 35,786 km. Typical path length from the Earth station to a GEO satellite is 38,500 km. Radio signals get weaker in proportion to the square of the distance traveled, so signals reaching a satellite are always very weak. Similarly, signals received on the Earth from a satellite 38,500 km away are also very weak, because of limits on the weight of GEO satellites and the electrical power they can generate using solar cells. It costs roughly $25,000 per kilogram to get a geostationary satellite in orbit. This obviously places severe restrictions on the size and weight of GEO satellites, since the high cost of building and launching a satellite must be recovered over a 10- to 15-year lifetime by selling communications capacity.

Satellite communications systems are dominated by the need to receive very weak signals. In the early days, very large receiving antennas, with diameters up to 30 m, were needed to collect sufficient signal power to drive video signals or multiplexed telephone channels. As satellites have become larger, heavier, and more powerful, smaller Earth station antennas have become feasible, and DBS-TV receiving systems can use dish antennas as small as 0.5 m in diameter. Satellite communications systems operate in the microwave and millimeter wave frequency bands, using frequencies between 1 and 50 GHz. Above 10 GHz, rain causes significant attenuation of the signal, and the probability that rain will occur in the path between the satellite and the Earth station must be factored into the satellite communications system design. Above 20 GHz, attenuation in heavy rain (usually associated with thunderstorms) can cause sufficient attenuation that the link will fail.

For the first 20 years of satellite communications systems, analog signals were widely used, with most links employing frequency modulation (FM). Wideband FM can operate at low carrier-to-noise ratios (C/N), in the 5- to 15-dB range, but adds a signal-to-noise ratio (SNR) improvement so that video and telephone signals can be delivered with SNR of 50 dB. The penalty for the improvement is that the radio frequency signal occupies a much larger bandwidth than the baseband signal. In satellite links, that penalty results because signals are always weak and the improvement in SNR is essential.

The move toward digital communications in terrestrial telephone and data transmission has been mirrored by a similar move toward digital transmission over satellite links. In the United States, only TV distribution at C-band remains as the major analog satellite communications system. Even this last bastion of analog signaling seems destined to disappear as cable TV stations switch over to digital receivers that allow six TV signals to be sent through a single Ku-band transponder. More importantly, dual standards permitting the transmission of not only digital TV, but also high-definition TV (HDTV), will eventually remove analog TV from consideration. Almost all other signals are digital — telephone, data, DBS-TV, radio broadcasting, and navigation with GPS, all use digital signaling techniques. All of the LEO and MEO mobile communications systems are digital, taking advantage of voice compression techniques that allow a digital voice signal to be compressed into a bit stream at 4.8 kbps. Similarly, MPEG 2 (Moving Picture Expert Group) and other video compression techniques allow video signals to be transmitted in full fidelity at rates less then 6.2 Mbps.

The Structure of a Satellite Communications System

Figure 15.4 gives an overview of a satellite communications system and illustrates its interfacing with terrestrial entities. The satellite communications system consists of a space segment, a control segment, and a ground segment. The *space segment* contains one or several active and spare satellites organized in a constellation. The *control segment* consists of all ground facilities for the control and monitoring of the satellites, also named TTC (tracking, telemetry, and command) stations, and for the management of the traffic and the associated resources onboard the satellite. The *ground segment* consists of all the traffic Earth stations; depending on the type of service considered, these stations can be of different size, from a few centimeters to tens of meters. Table 15.2 gives examples of traffic Earth stations in connection with the types of service. Earth stations come in three classes (see Figure 15.4): *user stations*, such as handsets, portables, mobile stations, and VSATs, which allow the customer a direct access to the space segment; *interface stations*, known as gateways, which interconnect the space segment to a terrestrial network; and *service stations*, such as hub/feeder stations, which collect/distribute information from/to user stations via the space segment.

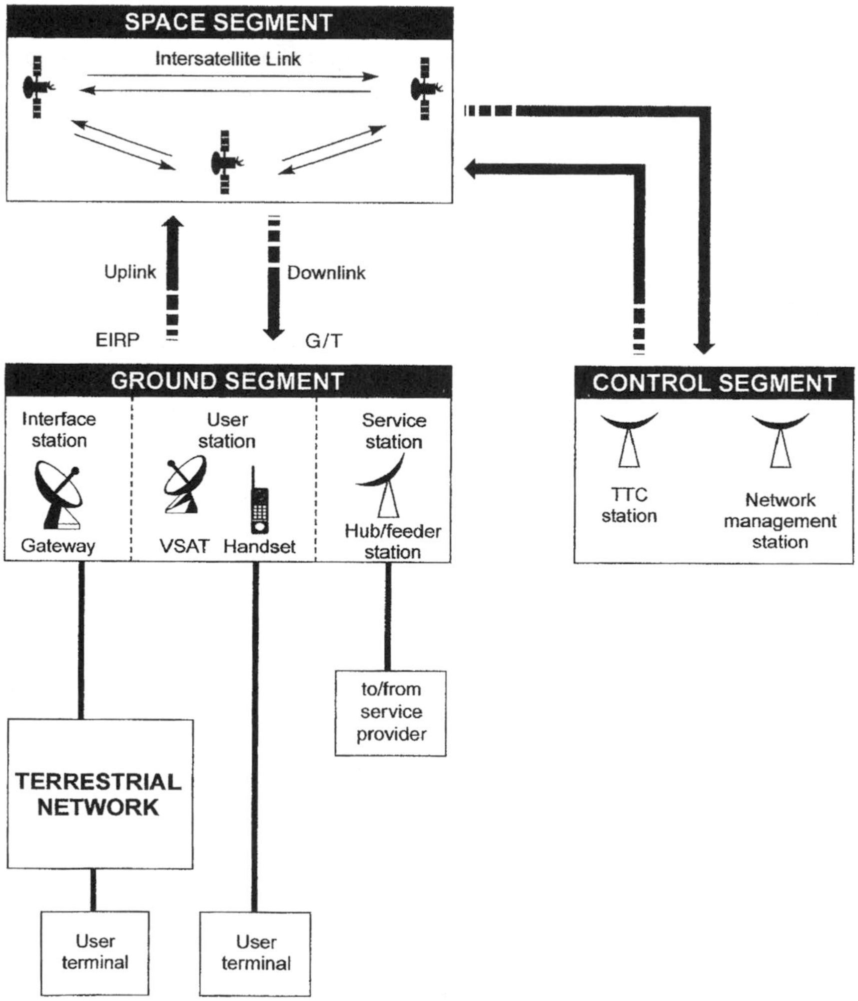

FIGURE 15.4 Satellite communications system and interfacing with terrestrial entities.

TABLE 15.2 Services from Different Types of Traffic Earth Station

Type of Service	Type of Earth Station	Typical Size
Point-to-point	Gateway, hub	2–10 m
	VSAT	1–2 m
Broadcast/multicast	Feeder station	1–5 m
	VSAT	0.5–1.0 m
Collect	VSAT	0.1–10 m
	Hub	2–10 m
Mobile	Handset, portable, mobile	0.1–0.8 m
	Gateway	2–10 m

Communications between users are set up through *user terminals* that consist of equipment such as telephone sets, fax machines, and computers that are connected to the terrestrial network or to the user stations (e.g., a VSAT), or are part of the user station (e.g., if the terminal is mobile). The path from a source user terminal to a destination user terminal is named a simplex connection. There are two basic schemes: *single connection per carrier* (SCPC), where the modulated carrier supports one connection only, and *multiple connections per carrier* (MCPC), where the modulated carrier supports several time or frequency multiplexed connections. Interactivity between two users requires a duplex connection between their respective terminals, i.e., two simplex connections, each along one direction. Each user terminal should then be capable of sending and receiving information. A connection between a service provider and a user goes through a hub (for collecting services) or a feeder station (e.g., for broadcasting services). A connection from a gateway, hub, or feeder station to a user terminal is called a *forward* connection. The reverse connection is the *return* connection. Both forward and return connections entail an uplink and a downlink, and possibly one or more intersatellite links.

Communications Links

A link between a transmitting terminal and a receiving terminal consists of a radio or optical modulated carrier. The transmit performance of the terminal is measured by its *effective isotropic radiated power* (EIRP), which is the power fed to the antenna multiplied by the gain of the antenna in the considered direction. The receive performance of the terminal is measured by G/T, the ratio of the antenna receive gain, G, in the considered direction and the system noise temperature, T; G/T is called the receiver's *figure of merit*. The types of link shown in Figure 15.4 are: the *uplinks* from the Earth stations to the satellites; the *downlinks* from the satellites to the Earth stations; and the *intersatellite links*, between the satellites. Uplinks and downlinks consist of radio-frequency-modulated carriers, while intersatellite links can be either radio frequency or optical. Carriers are modulated by baseband signals conveying information for communications purposes.

The link performance can be measured by the ratio of the received carrier power, C, to the noise power spectral density, N_0, and is quoted as the C/N_0 ratio, expressed in hertz (Hz). The values of C/N_0, for the links that participate in the connection between the end terminals, determine the quality of service, specified in terms of either baseband SNR or *bit error rate* (BER) for analog and digital communications, respectively.

Another parameter of importance for the design of a link is the bandwidth, B, occupied by the carrier. This bandwidth depends on the type of baseband signal and the type of modulation used to modulate the carrier. For satellite links, the trade-off between required carrier power and occupied bandwidth is paramount to the cost-effective design of the link. This is an important aspect of satellite communications systems, as power impacts both satellite mass and Earth station size, and bandwidth is constrained by regulations. Moreover, a service provider who rents satellite bandwidth from the satellite operator is charged according to the highest share of either power or bandwidth resource available from the satellite. The service provider's revenue is based on the number of established connections, so the objective is to maximize the capacity of the considered link while keeping a balanced share of power and bandwidth usage.

In a satellite communications system several stations transmit their carriers to a given satellite; therefore, the satellite acts as a network node. The techniques used to organize the access to the satellite by the carriers are called multiple-access techniques.

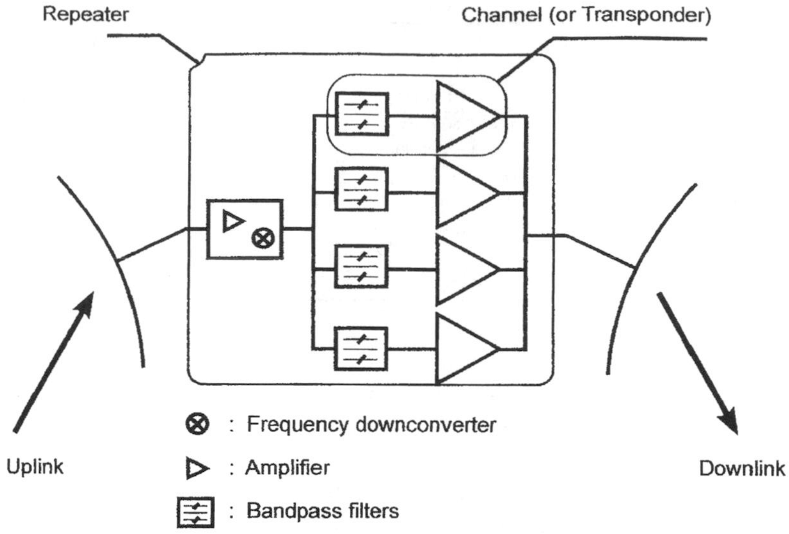

FIGURE 15.5 Payload organization: transparent.

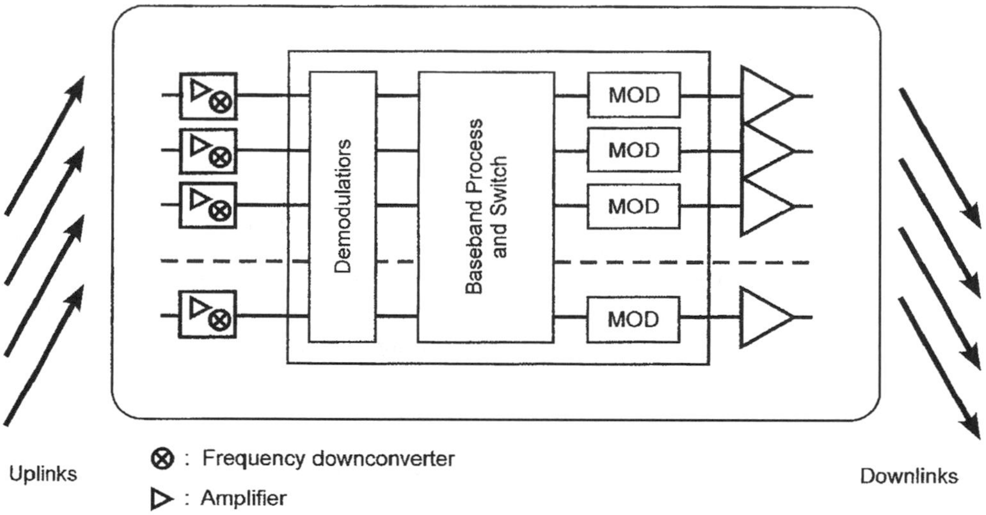

FIGURE 15.6 Payload organization: regenerative.

The Space Segment

The satellite consists of the *payload* and the *platform*. The payload consists of the receiving and transmitting antennas and all the electronic equipment, which support the transmission of the carriers. The two types of payload organization are illustrated in Figure 15.5 and Figure 15.6.

Figure 15.5 shows a *transparent* payload where carriers are power amplified and frequency downconverted. Power gain is of the order of 100 to 130 dB, required to raise the power level of the received carrier from a few hundred picowatts to the power level of the carrier fed to the transmit antenna of a few watts to a few tens of watts. Frequency conversion is required to increase isolation between the receiving input and the transmitting output. Due to technology power limitations, the overall bandwidth is split into several subbands, the carriers in each subband being amplified by a dedicated power amplifier. The amplifying chain associated with each subband is called a *satellite channel*, or transponder. The bandwidth splitting is achieved using a set of filters called the *input multiplexer* (IMUX). Power-amplified carriers are recombined in the *output multiplexer* (OMUX). The above transparent payload belongs to a single-beam satellite, where each transmit and receive antenna generates one beam only. One could also consider multiple-beam antennas. The payload would then have as many inputs/outputs as upbeams/downbeams. Routing of carriers from one upbeam to a given downbeam implies either routing through different satellite channels (channel hopping) or *onboard switching at radio frequency.*

Figure 15.6 shows a multiple-beam *regenerative* payload where the uplink carriers are demodulated. The availability of the baseband signals allows *onboard processing* and routing of information from up beam to down beam through *onboard switching at baseband.* The frequency conversion is achieved by modulating onboard-generated carriers at downlink frequency. The modulated carriers are then power amplified and delivered to the destination downbeam.

Figure 15.7 illustrates a multiple-beam satellite antenna and its associated coverage areas. Each beam defines a *beam coverage area* also called a *footprint* on the Earth's surface. The aggregate beam coverage areas define the *multibeam antenna coverage area.* A given satellite may have several multiple-beam antennas, and their combined coverage defines the *satellite coverage area.* Figure 15.8 illustrates the concept of instantaneous system coverage and long-term coverage. The *instantaneous system coverage* consists of the aggregation at a given time of the coverage areas of the individual satellites participating in the constellation. The *long-term coverage* is the area on the Earth scanned over time by the antennas of the satellites in the constellation. The coverage area should encompass the *service zone,* which corresponds to the geographical region where stations are installed. For real-time services, the instantaneous system coverage should at any time encompass the service zone, while for delayed (store-and-forward) services, it is the long-term coverage that should encompass the service zone. The platform consists of all the subsystems that permit the payload to operate. Table 15.3 lists these subsystems and indicates their respective main functions and characteristics.

To ensure a service with a specified availability, a satellite communications system must make use of several satellites in order to ensure redundancy. A satellite can cease to be available due to a failure or because it has reached the end of its lifetime. In this respect, it is necessary to distinguish between the reliability and the lifetime of a satellite. *Reliability* is a measure of the probability of breakdown and depends on the reliability of the equipment and many schemes to provide redundancy. The *lifetime* is conditioned by the ability to maintain the satellite on station in the nominal attitude, and depends on the quantity of fuel available for the propulsion system and attitude and orbit control. In a system, provision is generally made for an operations

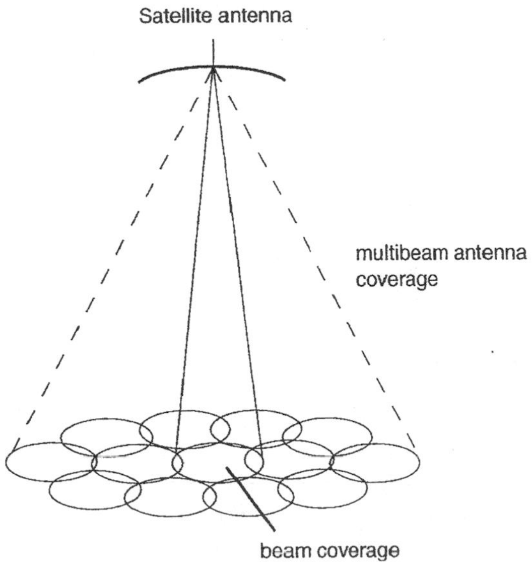

FIGURE 15.7 Multibeam satellite antenna and associated coverage area.

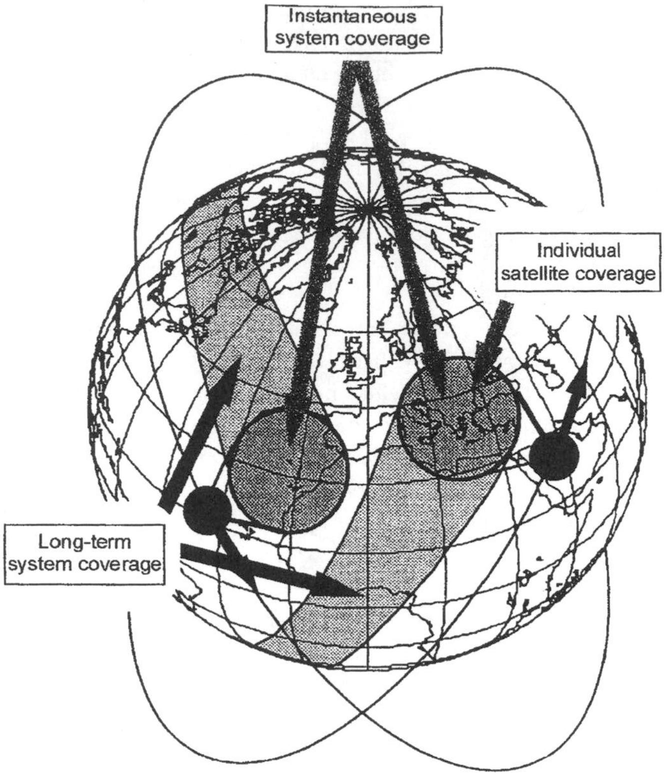

FIGURE 15.8 Types of coverage.

TABLE 15.3 Platform Subsystem

Subsystem	Principal Functions	Characteristics
Attitude and orbit control (AOCS)	Attitude stabilization Orbit determination	Accuracy
Propulsion	Provision of velocity increments	Specific impulse, mass of propellant
Electric power supply	Provision of electrical energy	Power, voltage stability
Telemetry, tracking and command (TTC)	Exchange of housekeeping information	Number of channels, security of communications
Thermal control	Temperature maintenance	Dissipation capability
Structure	Equipment support	Rigidity, lightness

satellite, a backup satellite in orbit, and a backup satellite on the ground. The reliability of the system will involve not only the reliability of each of the satellites, but also the reliability of launching.

The Ground Segment

The ground segment consists of all Earth stations; these are most often connected to the end-user's terminal by a terrestrial network or, in the case of small stations (VSAT), directly connected to the end-user's terminal. Stations are distinguished by their size, which varies according to the volume of traffic to be carried on the satellite link and the type of traffic (telephone, television, or data). The largest are equipped with antennas of 30-m diameter (*Standard A* of the INTELSAT network [Pratt et al., 2003]). The smallest have 0.6-m antennas (receiving stations from direct broadcasting satellites) or even smaller antennas (mobile stations, portable stations, or handsets). Some stations both transmit and receive. Others are receive-only stations; this is the case, for example, with

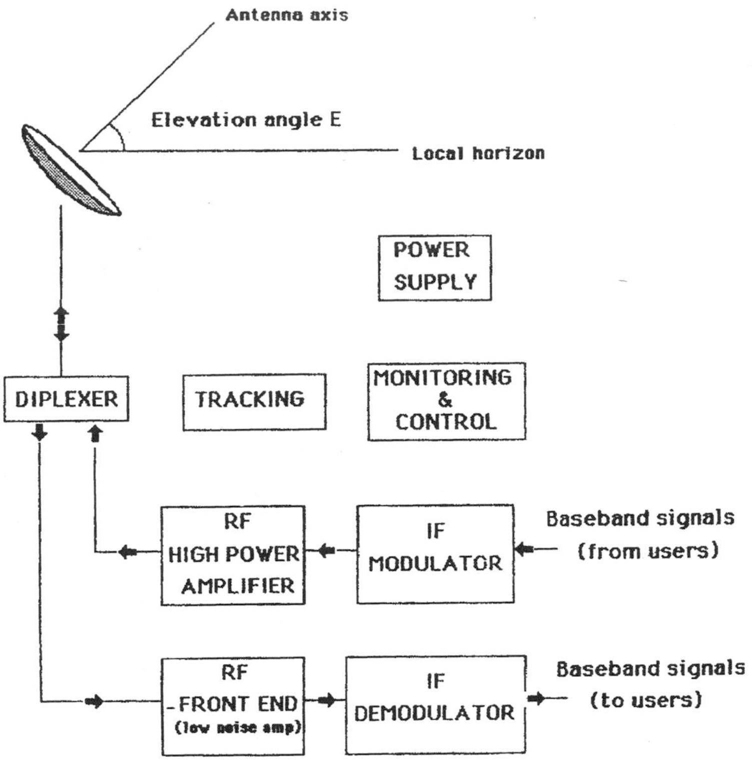

FIGURE 15.9 The organization of the Earth station. RF = radio frequency, IF = intermediate frequency.

receiving stations for a broadcasting satellite system or a distribution system for television or data signals. Figure 15.9 shows the typical architecture of the Earth station for both transmission and reception.

Type of Orbits

The orbit is the trajectory followed by the satellite. The trajectory is within a plane and shaped as an ellipse with a maximum extension at the apogee and a minimum at the perigee. The satellite moves more slowly in its trajectory as the distance from the Earth increases. The most favorable orbits are as follows.

Elliptical orbits inclined at an angle 64° with respect to the equatorial plane. This type of orbit is particularly stable with respect to irregularities in terrestrial gravitational potential and, owing to its inclination, enables the satellite to cover regions of high latitude for a large fraction of the orbital period as it passes to the apogee. This type of orbit has been adopted by the U.S.S.R. for the satellites of the *Molniya* system, with period of 12 hours. Figure 15.10 shows the geometry of the orbit. The satellite remains above the regions located under the apogee for a time interval of the order of eight hours. Continuous coverage can be ensured with three phased satellites on different orbits. Several studies relate to elliptical orbits with a period of 24 h (*Tundra* orbits) or a multiple of 24 h. These orbits are particularly useful for satellite systems for communications with mobiles, where the masking effects caused by surrounding obstacles such as buildings and trees and multiple path effects are pronounced at low elevation angles (say, less than 30°). In fact, inclined elliptic orbits can provide the possibility of links at medium latitudes when the satellite is close to the apogee with elevation angles close to 90°; these favorable conditions cannot be provided at the same latitudes by geostationary satellites. The European Space Agency (ESA) has been studying the use of HEO for digital audio broadcasting and mobile communications in the framework of its *Archimedes* program. An optimized system of six satellites in eight-hour orbits could provide simultaneous uninterrupted coverage of the world's most important markets: Europe, North America, and East Asia.

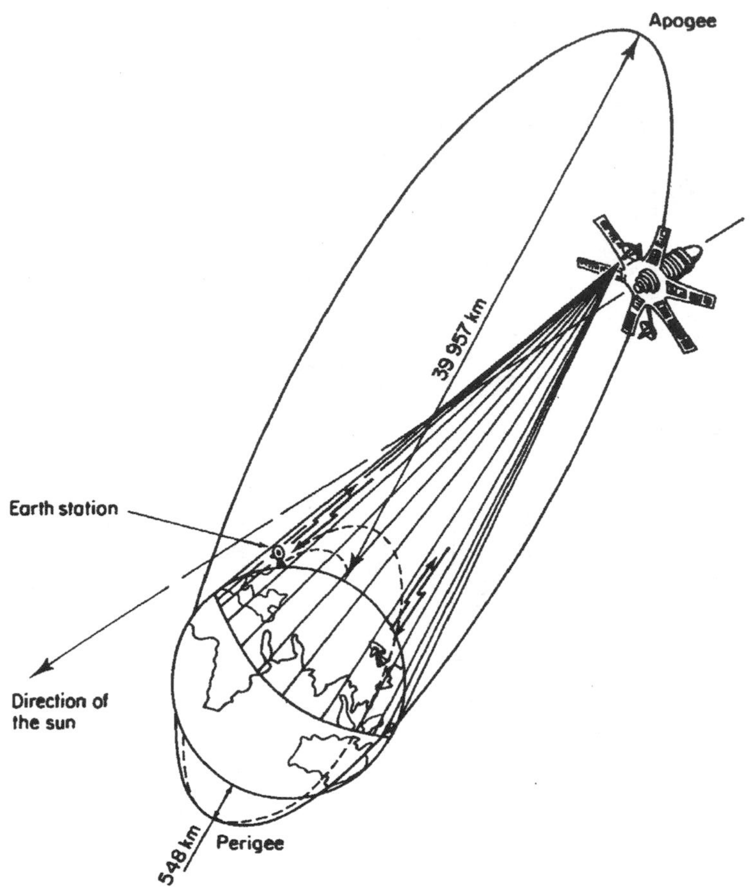

FIGURE 15.10 The orbit of a *Molniya* satellite.

Circular LEO. The altitude of the satellite is constant and equal to several hundreds of kilometers. The period is of the order of 90 min. With near 90° inclination, this type of orbit guarantees a worldwide long-term coverage as a result of the combined motion of the satellite and Earth rotation, as shown in Figure 15.11. This is the reason for choosing this type of orbit for observation satellites (for example, the *Spot* satellite; altitude 830 km, orbit inclination 98.7°, period 101 minutes). One can envisage the establishment of store-and-forward communications if the satellite is equipped with a means of storing information. A constellation of several tens of satellites in low-altitude (e.g., *Iridium* with 66 satellites at 780 km) circular orbits can provide worldwide real-time communication. Nonpolar orbits with less than 90° inclination can also be envisaged. Several such systems have been proposed (*Globalstar, Echo,* etc.).

Circular MEO. These orbits are also named intermediate circular orbits (ICO). Such orbits have an altitude of about 10,000 km and an inclination of about 50°. The period is six hours. With constellations of about 10 to 15 satellites, a continuous coverage of the world is guaranteed, allowing worldwide real-time communications. A planned satellite communications system of this kind is the *ICO* system (which has emerged from project 21 of INMARSAT) with a constellation of ten satellites in two planes, at 45° inclination.

Equatorial orbits. These are circular orbits with zero inclination. The most popular is the geostationary satellite orbit; the satellite orbits around the Earth in the equatorial plane according to the Earth rotation at an altitude of 35,786 km. The period is equal to that of the rotation of the Earth. The satellite thus appears as a point fixed in the sky and ensures continuous operations as a radio relay in real time for the area of visibility of the satellite (43% of the Earth's surface).

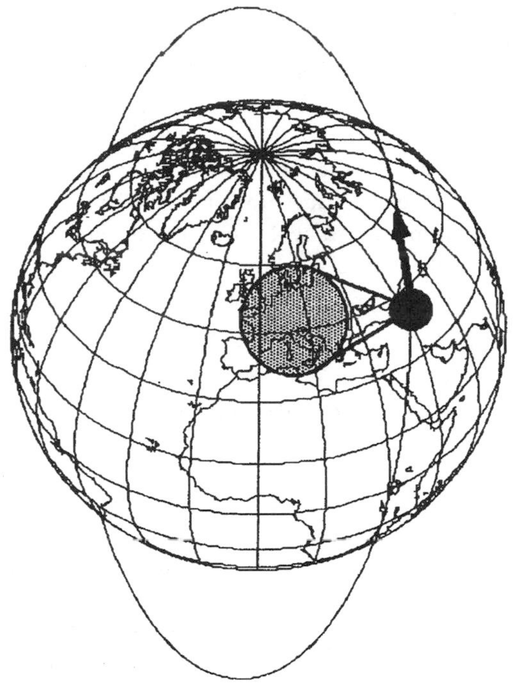

FIGURE 15.11 Circular polar low Earth orbit (LEO).

Hybrid systems. Some satellite communications systems may include combinations of orbits with circular and elliptical orbits. An example is the *Ellipso* satellite communications system. The choice of orbit depends on the nature of the mission, the acceptable interference, and the performance of the launchers.

The extent and latitude of the area to be covered. Contrary to widespread opinion, the altitude of the satellite is not a determining factor in the link budget for a given Earth coverage. The propagation attenuation varies as the inverse square of the distance, and this favors a satellite following a low orbit on account of its low altitude [Maral and Bousquet, 2003]; however, this disregards the fact that the area to be covered is then seen through a larger solid angle. The result is a reduction in the gain of the satellite antenna, which offsets the distance advantage. Now a satellite following a low orbit provides only limited coverage of the Earth at a given time and limited time at a given location. Unless low-gain antennas (of the order of a few dB) that provide low directivity and hence almost omnidirectional radiation are installed, Earth stations must be equipped with satellite-tracking devices, which increases the cost. The geostationary satellite thus appears to be particularly useful for continuous coverage of extensive regions. However, it does not permit coverage of the polar regions, which are accessible by satellites in inclined elliptical orbits or polar orbits.

The elevation angle. A satellite in an inclined or polar elliptical orbit can appear overhead at certain times, which enables communication to be established in urban areas without encountering the obstacles that large buildings constitute for elevation angles between 0° and approximately 70°. With a geostationary satellite, the angle of elevation decreases as the difference in latitude or longitude between the Earth station and the satellite increases.

Transmission duration and delay. The geostationary satellite provides a continuous relay for stations within visibility, but the propagation time of the waves from one station to the other is of the order of 0.25 sec. This requires the use of echo-control devices on telephone channels or special protocols for data transmission. A satellite moving in a low orbit confers a reduced propagation time. The transmission time is thus low between stations that are close and simultaneously visible to the satellite, but it can become long (several hours) for distant stations if only store-and-forward transmission is considered.

Interference. Geostationary satellites occupy fixed positions in the sky with respect to the stations with which they communicate. Protection against interference between satellite communications systems is ensured by planning the frequency bands and orbital positions. The small orbital spacing between adjacent satellites operating at the same frequencies leads to an increase in the level of interference, and this impedes the installation of new satellites. Different systems could see different frequencies, but this is restricted by the limited number of frequency bands assigned for space radio communication by the radio communication regulations. In this context one can refer to an "orbit-spectrum" resource that is limited. With orbiting satellites, the geometry of each system changes with time, and the relative geometries of one system with respect to another are variable and difficult to synchronize. The probability of interference is thus high.

The performance of launchers. The mass, which can be launched, decreases as the altitude increases.

The geostationary satellite communications systems are certainly the most popular. At the present time, there are around 200 geostationary satellites in operation within 360° of the whole orbital arc. Some parts of this orbital arc, however, tend to be highly congested (for example, above the American continent and the Atlantic).

Radio Regulations

The International Telecommunication Union (ITU). Radio regulations are necessary to ensure an efficient and economical use of the radio-frequency spectrum by all communications systems, both terrestrial and satellite. While so doing, the sovereign right of each state to regulate its telecommunications must be preserved. It is the role of the ITU to promote, coordinate, and harmonize the efforts of its members to fulfill these possibly conflicting objectives. The ITU, a United Nations organ, operates under a convention adopted by its member administrations. The ITU publishes the Radio Regulations, which are reviewed by the delegates from ITU member administrations at periodic World/Regional Radio Conferences.

Space radio communications services. The Radio Communications Regulations refer to the following space radio communications services, defined as transmission and reception of radio waves for specific telecommunication applications: Fixed Satellite Service (FSS); Mobile Satellite Service (MSS); Broadcasting Satellite Service (BSS); Earth Exploration Satellite Service (EESS); Space Research Service (SRS); Space Operation Service (SOS); Radio Determination Satellite Service (RDSS); Inter-Satellite Service (ISS); Amateur Satellite Service (ASS).

Frequency allocation. Frequency bands are allocated to the above radio communications services to allow compatible use. The allocated bands can be either exclusive for a given service or shared among several services. Allocations refer to the following division of the world into three regions. Region 1 is Europe, Africa, the Middle East, and the former U.S.S.R.; Region 2 is the Americas; and Region 3 is Asia and Oceania, except the Middle East and the former U.S.S.R. Table 15.4 summarizes the main frequency allocation and indicates some usual terminology for FSS, MSS, and BSS.

TABLE 15.4 Frequency Allocations

Radio Communications Service	Typical Frequency Bands for Uplink/Downlink	Usual Terminolgy
Fixed Satellite Service (FSS)	6/4 GHz	C-band
	8/7 GHz	X-band
	14/12–11 GHz	Ku-band
	30/20 GHz	Ka-band
	50/40 GHz	V-band
Mobile Satellite Service (MSS)	1.6/1.5 GHz	L-band
	30/20 GHz	Ka-band
Broadcasting Satellite Service (BSS)	2/2.2 GHz	S-band
	12 GHz	Ku-band
	2.6/2.5 GHz	S-band

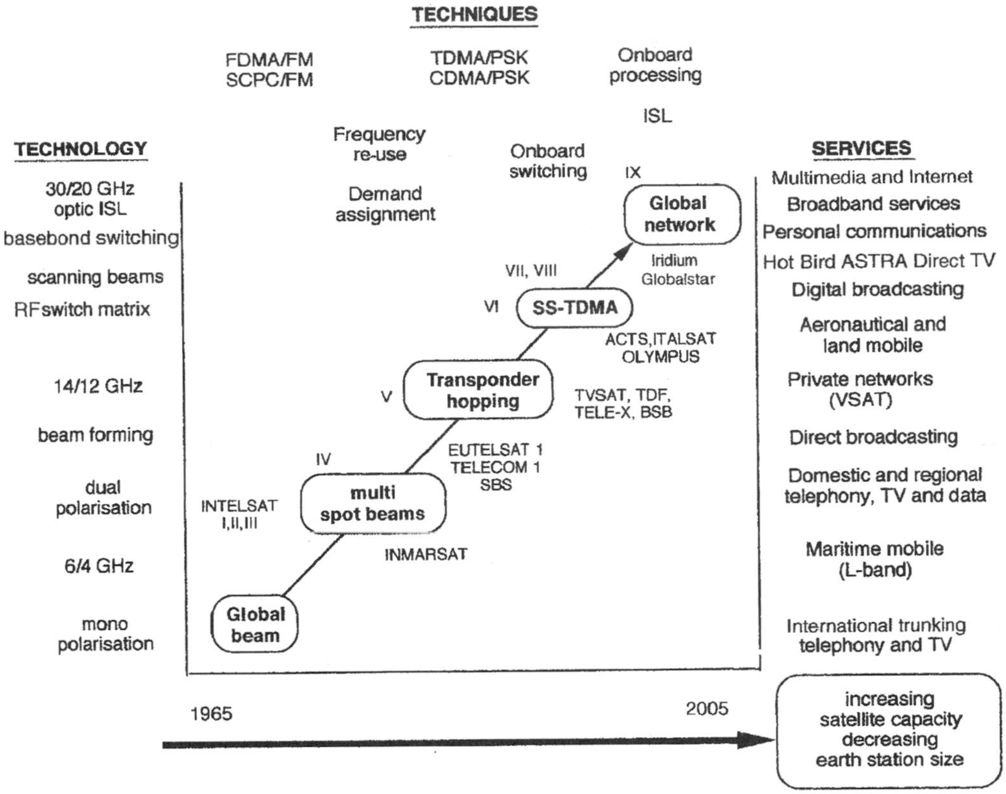

FIGURE 15.12 The evolution of satellite communication.

Progress in Technology

Figure 15.12 shows the progress in technology since start of the satellite communications era. The start of commercial satellite telecommunications can be traced back to the commissioning of *Intelsat I* (*Early Bird*) in 1965. Until the beginning of the 1970s, the services provided were telephone and television (TV) signal transmission between continents. The satellite communications system was designed to complement the submarine cable and played essentially the role of a telephone trunk connection. The goal of increased capacity has led rapidly to the institution of multibeam satellites and the reuse of frequencies, first by orthogonal polarization and subsequently by spatial separation. The transmission techniques were analogous, and each carrier conveyed either a TV signal or frequency division multiplexed telephone channels. Multiple access to the satellite was resolved by frequency division multiple access (FDMA). The increasing demand for a large number of low-capacity links, for example, for national requirements or for communication with ships, led in 1980 to the introduction of demand assignment, first using FDMA with single-channel-per-carrier/frequency modulation or phase-shift keying and subsequently using time-division-multiple-access/phase-shift keying in order to profit from the flexibility of digital techniques. Simultaneously, the progress in antenna technology enabled the beams to conform to the coverage of the service area; in this way the performance of the link was improved while reducing the interference between systems.

Multibeam satellites emerged, with interconnection between beams achieved by transponder hopping or onboard switching using satellite-switched time division multiple access (SSTDMA). Scanning or hopping beams have been implemented in connection with onboard processing on some experimental satellites such as *Advanced Communications Technology Satellite* (ACTS). Onboard processing is a technique that takes advantage of the availability of base-band signal onboard the satellite thanks to carrier demodulation.

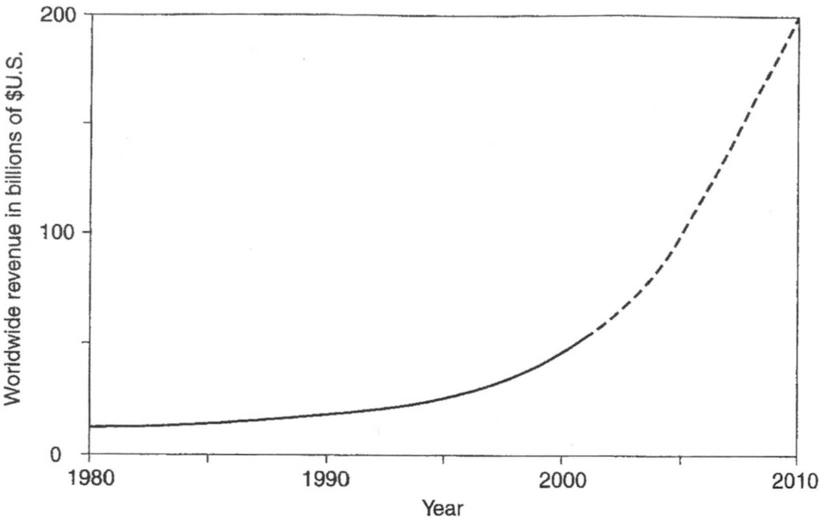

FIGURE 15.13 Growth of worldwide revenues from satellite communications systems 1980 through 2010. Beyond 2000, the curve is a projection.

Intersatellite links were developed for civilian applications in the framework of multisatellite constellations, such as *Iridium* for mobile applications, and eventually will develop for geostationary satellite. GEO satellites have always been the backbone of the commercial satellite communications industry. Large GEO satellites can serve one-third of the Earth's surface, and they can carry up to 4 Gbps of data or transmit up to 16 high DBS-TV signals, each of which can deliver several video channels. The weight and power of GEO satellites have also increased. In 2000 a large GEO satellite could weigh 10,000 kg (10 tons), might generate 12 kW of power, and carry 60 transponders, with a trend toward even higher powers but lower weight. For example, in 2001 *Space System/Load* contracted with *APT Satellite Company Ltd.* in Hong Kong to build the *Apstar-V* satellite, a GEO satellite serving Asia with a mass of 4845 kg when injected into geostationary orbit and an expected lifetime of 13 years. *Apstar-V* generates an initial power of 10.6 kW and carries 38 C-band transponders with 60-W output power and 16 Ku-band transponders at 141 W each [*Aviation Week and Space Technology*, 2000]. Satellites generating 25 kW and carrying antennas with hundreds of beams are planned for the time frame 2005 to 2010 [Pratt et al., 2003].

Television program distribution and DBS-TV have become the major source of revenue for commercial satellite communications system operators, earning more than half of the industry's $30 billion revenues for 1998. By the end of 2000 there were over 14 million DBS-TV customers in the United States. The high capacity of GEO satellites results from the use of high-power terrestrial transmitters and relatively high-gain Earth station antennas. Earth station antenna gain translates directly into communication capacity, and therefore into revenue. Increased capacity lowers the delivery cost per bit for a customer. Satellite communications systems with fixed directional antennas can deliver bits at a significantly lower cost than those using low-gain antennas, such as satellite communications systems designed for use by mobile users. Consequently, GEO satellites look set to be the largest revenue earners in space for the foreseeable future. Figure 15.13 shows the estimated growth in revenue from all satellite communication services, projected to 2010.

All radio systems require frequency spectrum, and the delivery of high-speed data requires a wide bandwidth. Satellite communications systems started at C-band, with an allocation of 500 MHz, shared with terrestrial microwave links. As the GEO orbit filled up with satellites operating at C-band, satellites were built for the next available frequency band, Ku-band. There is a continuing demand for ever-more spectrum to allow satellites to provide new services, with high-speed access to the Internet forcing a move to Ka-band and even higher frequencies. But the use of higher frequencies (Ka-band at 30/20 GHz) will enable the emergence of broadband services, thanks to the large amount of unused bandwidth currently available, in spite of the propagation problems caused by rain effects. Access to the Internet from small transmitting Ka-band Earth stations located at the home offers a way to bypass the terrestrial telephone network and achieve much higher bit rates. SES began two-way Ka-band Internet access in Europe in 1998 with the *Astra-K* satellite, and the next generation of Ka-band satellites in the United States will offer similar services.

Successive World Radio Conferences have allocated new frequency bands for commercial satellite services that now include L-, S-, C-, Ku-, Ka-, V-, and Q-bands. Mobile satellite systems use VHF-, UHF-, L-, and S-bands with carrier frequencies from 137 to 2500 MHz, and GEO satellites use frequency bands extending from 3.2 to 50 GHz [*Handbook on Satellite Communications*, 2002]. Despite the growth of fiber-optic links with very high capacity, the demand for the satellite communications systems continues to increase. Satellites have also become integrated into complex communications architectures that use each element of the network to its best advantage. Examples are VSAT/WLL (very small aperture terminals/wireless local loop) in countries where the communications infrastructure is not yet mature and GEO/LMDS (local multipoint distribution systems) for the urban fringes of developed nations where the build-out of fiber has yet to be an economic proposition.

Services

Initially designed as "trunks" that duplicate long-distance terrestrial links, satellite links have rapidly conquered specific markets. Satellite communications systems have three properties that are not, or only to a lesser extent, found in terrestrial networks. These are (1) the possibility of broadcasting, (2) a wide bandwidth, and (3) rapid setup and ease of reconfiguration. Initially, satellite communications systems contained a small number of Earth stations (several stations per country equipped with antennas of 15- to 30-m diameter collecting the traffic from an extensive area by means of ground networks). Subsequently, the number of Earth stations has increased with a reduction in size (antennas of 1 to 4 m) and a greater geographical dispersion. The Earth stations have become closer to the user, possibly being transportable or mobile. The potential of the services offered by satellite communications systems has thus diversified.

Trunking telephony and television program exchange. This is a continuation of the original service. The traffic concerned is part of a country's international traffic. It is collected and distributed by the ground network on a scale appropriate to the country concerned. Examples are INTELSAT and EUTELSAT (TDMA network). The Earth stations are equipped with 15- to 30-m diameter antennas.

"Multiservice" systems. These are telephone and data for user groups who are geographically dispersed. Each group shares the Earth station and accesses it through a ground network whose extent is limited to one district of a town or an industrial area. Examples are TELECOM 2, EUTELSAT, SMS, and INTELSAT — International Business Service (IBS) network. The Earth stations are equipped with antennas of 3 to 10 m in diameter.

VSAT systems. Low-capacity data transmission (uni- or bidirectional), television, or digital sound program broadcasting [Maral, 1995]. Most often, the user is directly connected to the station. VSATs are equipped with antennas of 0.6 to 1.2 m in diameter. The introduction of Ka-band will allow even smaller antennas (USAT, ultra small aperture terminals) to provide even larger capacity for data transmission, allowing multimedia interactivity, data-intensive business applications, residential and commercial Internet connections, two-way videoconferencing, distance learning, and telemedicine.

Mobile and personal communications. Despite the proliferation in cellular and terrestrial personal communication services around the world, there will still be vast geographic areas not covered by any wireless terrestrial communications. These areas are open fields for mobile and personal satellite communications systems, and they are key markets for the operators of geostationary satellites like *Inmarsat* and of nongeostationary satellite constellations like *Iridium*, *Globalstar*, *Ico*, *Ellipso*, and *Echo*.

Digital sound TV and data broadcasting. The emergence of standards for compression, such as MPEG standard for video, has triggered the implementation of digital services to small Earth stations installed at the user's premises with antennas of the order of a few tens of centimeters. For *television*, such services using the S-DVB (satellite digital video broadcasting) standard are progressively replacing the former broadcasting of analog programs. Examples of satellite communications systems broadcasting digital television are *Astra*,

Hot Bird, *Direct TV*, *Asiasat*, etc. For *sound*, several satellite communications systems incorporating onboard processing have been launched in such a way as to allow FDMA access by several broadcasters on the uplink and time division multiplexing (TDM) on a single downlink carrier of the sound programs. It avoids a delivery of the programs to a single-feeder Earth station and allows operation of the satellite payload at full power. This approach combines flexibility and efficient use of the satellite. Examples of such satellite communications systems are World-Space, Sirius-Radio, and XM-Radio. The ability of the user terminal to process digital data paves the way for satellite distribution of files on demand through the Internet, with a terrestrial request channel or even a satellite-based channel. This anticipates the broadband multimedia satellite services.

Multimedia services. These services aggregate different media, such as text data, audio, graphics, fixed- or slow-scan pictures, and video, in a common digital format, so as to offer excess potential for online services, teleworking, distance learning, interactive television, telemedicine, etc. Interactivity is therefore an imbedded feature. They require an increased bandwidth compared to conventional services such as telephony. This has triggered the concept of an information superhighway. Satellites will complement terrestrial high-capacity fiber cable-based networks. Several satellite-based projects have been proposed. Examples are: Space Way, Wild Blue, Astrolink, using geostationary satellites, and Skybridge, Teledisc, based on constellations of LEO satellites. Hybrid systems incorporating geostationary and nongeostationary satellites are also planned, such as *West*. These systems incorporate the following characteristics: use of Ka-band, multibeam antennas, wideband transponders (typically 125 MHz), onboard processing and switching, a large range of service rates (from 16 kb/s to 10 Mb/s) and quasi error-free transmission (typically, BER is equal to 10^{-10}).

The Way to the Next Generation

In the last 30 years, the satellite communications system landscape has changed significantly. Advances in satellite technology have enabled satellite communications system providers to expand their service offerings. The mix of satellite communications systems is continuously evolving. Point-to-point trunking for analog voice and television was the sole service initially provided by satellite communications systems 30 years ago. In addition, satellite communications systems are today providing digital sound and video broadcasting, mobile communications, on-demand narrowband data services, and broadband multimedia and Internet services. The mix of service offering will continue to change significantly in the next ten years.

Satellite communications system services can be characterized as either satellite-relay applications or end-user applications (fixed or mobile). For *satellite relay applications*, a content provider or carrier will lease capacity from a satellite operator, or will use its own satellite communications system to transmit content to and from terrestrial stations, where the content is routed to the end user. Relay applications accounted for around $10 billion in 2000. *End-user satellite applications* provide information directly to individual customers via consumer devices such as small-antenna (less than Earth station) and handheld satellite phones. End-user applications accounted for about $25 billion in 2000.

Numerous new technologies are under development, in response to the tremendous demand for emerging global satellite communications system applications. Improved technology leads to the production of individual satellites that are more powerful and capable than earlier models. With larger satellites (up to 10,000 kg) able to carry additional transponders and more powerful solar arrays and batteries, these designs will provide a higher power supply (up to 20 kW) to support a greater number of transponders (up to 150). New platform designs allowing additional capacity for station-keeping propellant and the adoption of new types of thrusters are contributing to increased service life for GEO satellites up to 20 years. This translates into an increased capacity from satellites with more transponders, longer lives, and the ability to transmit more data through increasing rates of data compression.

Nomenclature

Symbol	Quantity
ACTS	Advanced communications technology satellite
ASS	Amateur Satellite Service
BER	Bit error rate
BSS	Broadcasting Satellite Service
COMSAT	Communication Satellite Corporation
C/N	Carrier-to-noise ratio
DBS-TV	Direct broadcast satellite television
DTH	Direct-to-home television distribution
OEM	Original equipment manufacturer
EESS	Earth Exploration Satellite Service
EIRP	Effective isotropic radiated power
ESA	European Space Agency
FDMA	Frequency division multiple access
FM	Frequency modulation
FSS	Fixed Satellite Service
GPS	Global Positioning System
GEO	Geostationary Earth orbit
HEO	Highly elliptic orbit
HDTV	High-definition TV
ICO	Intermediate circular orbits
IMUX	Input multiplexer
INMARSAT	International Maritime Satellite Organization
INTELSAT	International Telecommunications Satellite Consortium
ISS	Inter-Satellite Service
ITU	International Telecommunication Union
LEO	Low Earth orbit
LMDS	Local multipoint distribution system

Symbol	Quantity
MCPC	Multiple connections per carrier
MEO	Medium Earth orbit
MPEG	Moving Picture Expert Group
MSS	Mobile Satellite Service
NASA	National Aeronautics and Space Administration
OMUX	Output multiplexer
RDSS	Radio Determination Satellite Service
SCPC	Single connection per carrier
S-DVB	Satellite digital video broadcasting
SES	Societe Europeén de Satellites
SNG	Satellite newsgathering
SNR	Signal-to-noise ratio
SOS	Space Operation Service
SRS	Space Research Service
SSTDMA	Satellite-switched time division multiple access
TDM	Time division multiplexing
TDMA	Time division multiple access
TTC	Tracking, telemetry, and command
TV	Television
TVRO	Television receive only (Earth stations)
USAT	Ultra small aperture terminals
VSAT	Very small aperture terminal. This term designates small Earth stations, generally directly installed on the user premises. It also designates satellite communications systems and network implementing VSAT Earth stations.
WLL	Wireless local loop.

References

http://www.astra.ln.

http://www.astrolink.com.

Aviation Week and Space Technology, Aerospace Source Book, vol. 153, no. 3, New York: McGraw-Hill, January 17, 2000.

Handbook on Satellite Communications, 3rd ed., New York: Wiley, ITU, 2002.

http://www.hns.com.spaceway.

Maral, G., *VSAT Networks*, New York: Wiley, 1995.

Maral, G. and Bousquet, M., *Satellite Communications System*, 4th ed., New York: Wiley, 2003.

Pratt, T., Bostian, Ch., and Allnutt, J., *Satellite Communications*, 2nd ed., New York: Wiley, 2003.

Ress, D., *Satellite Communications: The First Quarter Century of Service*, New York: Wiley, 1989.

16

Embedded Systems

Grant Martin
Tensilica Inc.

Luciano Lavagno
Cadence Berkeley Laboratory

Hans Hansson
Mälardalen University

Mikael Nolin
Mälardalen University

Thomas Nolte
Mälardalen University

Kleanthis Thramboulidis
University of Patras

16.1 Embedded Systems — An Overview

Grant Martin and Luciano Lavagno

Embedded systems span a number of scientific and engineering disciplines, which evolve continuously. So does the field of embedded systems. To write on this topic requires a careful choice of the material to be covered, to make a presentation of a reasonable length, and provide a fairly up-to-date material. This chapter focuses on two main areas of embedded systems, which have evolved only in recent years, namely, systems on a chip and networked embedded systems. The section on System-on-Chip surveys a large number of the issues involved in its design. The section on networked embedded systems presents an overview of trends for networking of embedded systems, their design, and their application for in-car controls and automation. This material is preceded by a round up of embedded systems in general, and design trends.

The Embedded System Revolution

The world of electronics has witnessed a dramatic growth of its applications in the last few decades. From telecommunications to entertainment, from automotive to banking, almost any aspect of our everyday life employs some kind of electronic component. In most cases, these components are computer based systems, which are not however used or perceived as a computers. For instance, they often do not have a keyboard or a display to interact with the user, and they do not run standard operating systems and applications. Sometimes, these systems constitute a self-contained product themselves (e.g., a mobile phone), but they are frequently embedded inside another system, for which they provide better functionalities and performance (e.g., the engine control unit of a motor vehicle). We call these computer based systems *embedded systems*.

The huge success of embedded electronics has several causes. The main one in our opinion is that, embedded systems bring the advantages of Moore's Law into everyday life, that is an exponential increase in performance and functionality at an ever decreasing cost. This is possible because of the capabilities of integrated circuit technology and manufacturing, which allow one to build more and more complex devices, and because of the development of new design methodologies, which allows one to efficiently and cleverly use those devices. Traditional steel-based mechanical development, on the other hand, had reached a plateau near the middle of the 20th century, and thus it is not a significant source of innovation any longer, unless coupled to electronic manufacturing technologies (MEMS) or embedded systems, as argued above.

There are many examples of embedded systems in the real world. For instance, a modern car contains tens of electronic components (control units, sensors, and actuators) that perform very different tasks. The first embedded system that appeared in a car was related to the control of mechanical aspects, such as the control of the engine, the antilock brake system, and the control of suspension and transmission. However, nowadays cars also have a number of components which are not directly related to mechanical aspects, but are mostly related to the use of the car as a vehicle for moving around, or the communication and entertainment needs of the passengers; navigation systems, digital audio and video players, and phones are just a few examples. Moreover, many of these embedded systems are connected together using a network, because they need to share information regarding the state of the car.

Other examples come from the communication industry. A cellular phone is an embedded system whose environment is the mobile network. These are very sophisticated computers whose main task is to send and receive voice, but are also currently used as personal digital assistants, for games, to send and receive images and multimedia messages, and to wirelessly browse the Internet. They have been so successful and pervasive that in just a decade they became essential in our life. Other kinds of embedded systems significantly changed our life as well, for instance, ATM and point-of-sale (POS) machines modified the way we do payments, and multimedia digital players changed how we listen to music and watch videos.

We are just at the beginning of a revolution that will have an impact on every other industrial sector. Special purpose embedded systems will proliferate and will be found in almost any object that we use. They will be optimized for their application and show a natural user interface. They will be flexible, in order to adapt to a changing environment. Most of them will also be wireless, in order to follow us wherever we go and keep us constantly connected with the information we need and the people we care about. Even the role of computers will have to be reconsidered, as many of the applications for which they are used today will be performed by specially designed embedded systems.

What are the consequences of this revolution in the industry? Modern car manufacturers today need to acquire a significant amount of skills in hardware and software design, in addition to the mechanical skills that they already had in-house, or they should outsource the requirements, they have, to an external supplier. In either case, a broad variety of skills needs to be mastered, from the design of software architectures for implementing the functionality to being able to model the performance, because real-time aspects are extremely important in embedded systems, especially those related to safety-critical applications. Embedded system designers must also be able to architect and analyze the performance of networks, as well as validate the functionality that has been implemented over a particular architecture and the communication protocols that are used.

A similar revolution has happened, or is about to happen, to other industrial and socioeconomical areas as well, such as entertainment, tourism, education, agriculture, government, and so on. It is therefore clear that new, more efficient, and easy-to-use embedded electronics design methodologies need to be developed, in order to enable the industry to make use of the available technology.

Embedded systems are informally defined as a collection of programmable parts surrounded by application specific integrated circuits (ASICs) and other standard components (application specific standard parts, ASSPs), that interact continuously with an environment through sensors and actuators. The collection can

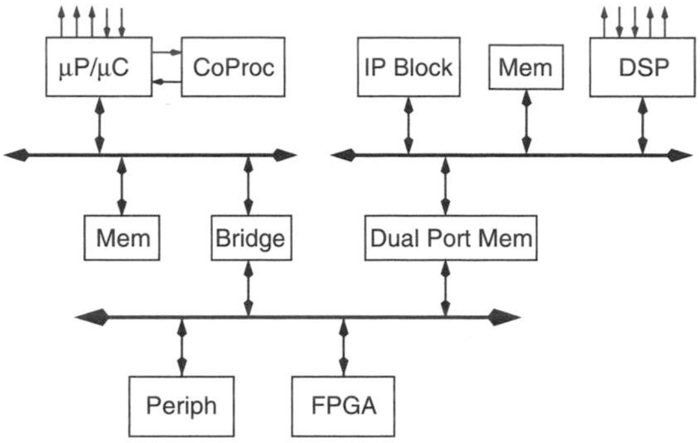

FIGURE 16.1 A reactive real-time embedded system architecture. (*Source*: "Design of Embedded Systems," by Luciano Lavagno and Claudio Passerone in *The Embedded Systems Handbook*, CRC Press, 2005.)

physically be a set of chips-on-board, or a set of modules on an integrated circuit. Software is used for features and flexibility, while dedicated hardware is used for increased performance and reduced power consumption. An example of an architecture of an embedded system is shown in Figure 16.1. The main programmable components are microprocessors and DSPs that implement the software partition of the system. One can view reconfigurable components, especially if they can be reconfigured at runtime, as programmable components in this respect. They exhibit area, cost, performance, and power characteristics that are intermediate between dedicated hardware and processors. Custom and programmable hardware components, on the other hand, implement application specific blocks and peripherals. All components are connected through standard and dedicated buses, and networks, and data is stored on a set of memories. Often several smaller subsystems are networked together to control, e.g., an entire car, or to constitute a cellular or wireless network.

We can identify a set of typical characteristics that are commonly found in embedded systems. For instance, they are usually not very flexible and are designed to always perform the same task. If you buy an engine control embedded system, you cannot use it to control the brakes of your car or to play games. A PC, on the other hand, is much more flexible because it can perform several very different tasks. An embedded system is often part of a larger control system. Moreover, cost, reliability, and safety are often more important criteria than performance, because the customer may not even be aware of the presence of the embedded system, and so he looks at other characteristics, such as the cost, the ease of use, or the lifetime of a product.

Another common characteristic of many embedded systems is that they need to be designed in an extremely short time to meet their time-to-market. Only a few months should elapse from conception of a consumer product to the first working prototypes. If these deadlines are not met, the result is a concurrent increase in design costs and decrease of profits, because fewer items will be sold. So delays in the design cycle may make a huge difference between a successful product and an unsuccessful one.

In the current state-of-the-art, embedded systems are designed with an *ad hoc* approach that is heavily based on earlier experience with similar products and on manual design. Often the design process requires several iterations to obtain convergence, because the system is not specified in a rigorous and unambiguous fashion, and the level of abstraction, details and design style in various parts are likely to be different. As the complexity of embedded systems scales up, this approach is showing its limits, especially regarding design and verification time.

New methodologies are being developed to cope with the increased complexity and enhance designers' productivity. In the past, a sequence of two steps has always been used to reach this goal: *abstraction* and *clustering*. Abstraction means describing an object (i.e., a logic gate made of MOS transistors) using a model where some of the low-level details are ignored (i.e., the Boolean expression representing that logic gate).

Clustering means connecting a set of models at the same level of abstraction, to get a new object, which usually shows new properties that are not part of the isolated models that constitute it. By successively applying these two steps, digital electronic design went from drawing layouts to transistor schematics, to logic gate netlists, to register transfer level descriptions, as shown in Figure 16.2.

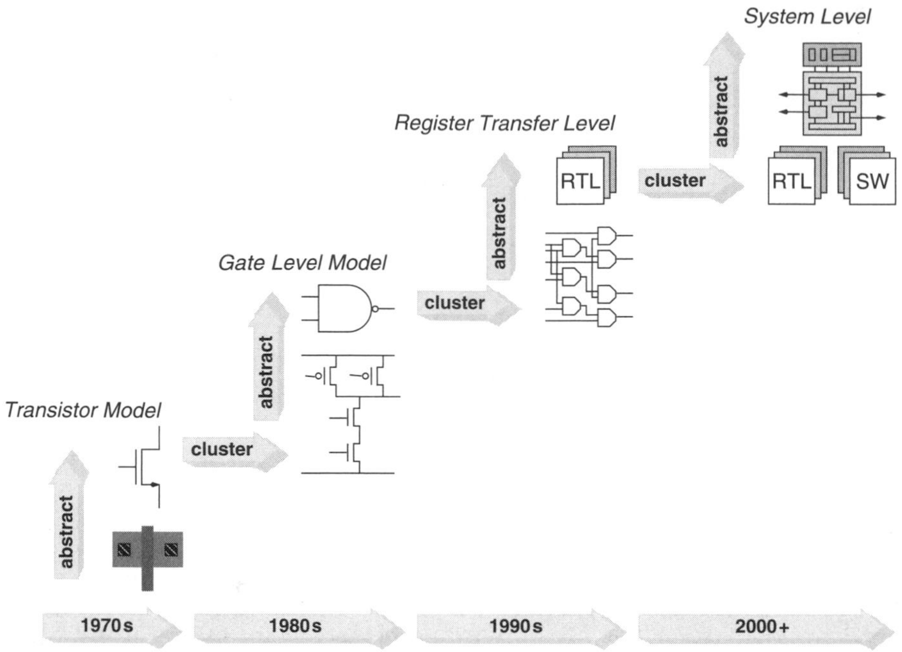

FIGURE 16.2 Abstraction and clustering levels in hardware design. (*Source*: "Design of Embedded Systems," by Luciano Lavagno and Claudio Passerone in *The Embedded Systems Handbook*, CRC Press, 2005.)

The notion of *platform* is key to the efficient use of abstraction and clustering. A platform is a single abstract model that hides the details of a set of different possible implementations as clusters of lower level components. The platform, e.g., a family of microprocessors, peripherals, and a bus protocol, allows developers of designs at the higher level to operate without detailed knowledge of the implementation (e.g., the pipelining of the processor or the internal implementation of the UART). At the same time, it allows platform implementers to share design and fabrication costs among a broad range of potential users, broader than if each design was a one-of-a-kind type.

Today we are witnessing the appearance of a new higher level of abstraction, as a response to the growing complexity of integrated circuits. Objects can be functional descriptions of complex behaviors or architectural specifications of complete hardware platforms. They make use of formal high level models that can be used to perform an early and fast validation of the final system implementation, although with reduced detail with respect to a lower level description.

The design methodology for embedded systems put in the context of "System-on-Chip" is presented in the next section.

The System-on-Chip Revolution

"System-on-Chip" is a phrase that has been much bandied about in recent years (IEEE Spectrum, November 1996). It is more than a design style, more than an approach to the design of application-specific integrated circuits (ASICs), more than a methodology. Rather, System-on-Chip (SoC) represents a major revolution in IC design, a revolution enabled by the advances in process technology allowing the integration of all or most of the major components and subsystems of an electronic product onto a single chip, or integrated chipset [2]. This revolution in design has been embraced by many designers of complex chips, as the performance, power consumption, cost, and size advantages of using the highest level of integration made available have proven to be extremely important for many designs. In fact, the design and use of SoCs is arguably one of the key problems in designing real-time embedded systems.

The move to SoC began sometime in the mid 1990s. At this point, the leading CMOS-based semiconductor process technologies of 0.35 and 0.25 microns were sufficiently capable of allowing the integration of many of the major components of a second generation wireless handset or a digital set-top box on to a single chip. The digital baseband functions of a cellphone; a Digital Signal Processor (DSP), hardware support for voice encoding and decoding, and a RISC processor, could all be placed on to a single die. Although such a baseband SoC was far from the complete cellphone electronics, there were major components such as the RF transceiver, the analogue power control, analogue baseband, and passives which were not integrated — the evolutionary path with each new process generation, to integrate more and more on to a single die, was clear. Today's chipset would become tomorrow's chip. The problems of integrating hybrid technologies involved in making up a complete electronic system would be solved. Thus, eventually, SoC could encompass design components drawn from the standard and more adventurous domains of digital, analogue, RF, reconfigurable logic, sensors, actuators, optical, chemical, microelectronic mechanical systems, and even biological and nanotechnology.

With this viewpoint of continued process evolution leading to ever-increasing levels of integration into ever-more-complex SoC devices, the issue of an SoC being a single *chip* at any particular point in time is somewhat moot. Rather, the word "system" in System-on-Chip is more important than "chip." What is most important about an SoC whether packaged as a single chip or integrated chipset, or System-in-Package (SiP) or System-on-Package (SoP) is that it is designed as an integrated *system*, making design tradeoffs across the processing domains and across the individual chip, and package boundaries.

System-on-Chip

Let us define a System-on-Chip (SoC) as a complex integrated circuit or integrated chipset, which combines the major functional elements or subsystems of a complete end-product into a single entity. These days, all interesting SoC designs include at least one programmable processor, and very often a combination of at least one RISC control processor and one DSP. They also include on-chip communications structures, such as processor bus(es), peripheral bus(es), and perhaps a high-speed system bus. A hierarchy of on-chip memory units and links to off-chip memory are important especially for SoC processors (cache, main memories, very often separate instruction and data caches are included). For most signal processing applications, some degree of hardware-based accelerating functional units are provided, offering higher performance and lower energy consumption. For interfacing to the external, real world, SoCs include a number of peripheral processing blocks, and due to the analogue nature of the real world, this may include analogue components as well as digital interfaces (for example, to system buses at a higher packaging level). Although there is much interesting research in incorporating MEMS-based sensors and actuators, and in SoC applications incorporating chemical processing (lab-on-a-chip), these are, with rare exceptions, research topics only. However, future SoCs of a commercial nature may include such subsystems, as well as optical communications interfaces.

One key point about SoC which is often forgotten for those approaching them from a hardware-oriented perspective is that all interesting SoC designs encompass hardware and software components, i.e., programmable processors, real-time operating systems, and other aspects of hardware-dependent software such as peripheral device drivers, as well as middleware stacks for particular application domains, and possibly optimized assembly code for DSPs. Thus, the design and use of SoCs cannot remain a hardware-only concern, it involves aspects of system-level design and engineering, hardware-software tradeoff and partitioning decisions, and software architecture, design and implementation.

System-on-Programmable-Chip

Recently, attention has begun to expand in the SoC world from SoC implementations using custom, ASIC or application-specific standard part (ASSP) design approaches, to include the design and use of complex reconfigurable logic parts with embedded processors and other application oriented blocks of intellectual property. These complex FPGAs (Field-Programmable Gate Arrays) are offered by several vendors, including Xilinx (Virtex-II PRO Platform FPGA, Virtex-IV) and Altera (SOPC), but are referred to by different names such as highly programmable SoCs, System-on-a-programmable-chip, embedded FPGAs. The key idea behind this approach to SoC is to combine large amounts of reconfigurable logic with embedded RISC processors (either custom laid-out, "hardened" blocks, or synthesizable processor cores), in order to allow very flexible and tailorable combinations of hardware and software processing to be applied to a particular design problem. Algorithms that consist of significant amounts of control logic plus significant quantities of dataflow processing, can be partitioned into the control RISC processor (for example, in Xilinx Virtex-II PRO, a PowerPC processor) and reconfigurable logic offering hardware acceleration. Although the resulting combination does not offer the highest performance, lowest energy consumption, or lowest cost, in comparison with custom IC or ASIC/ASSP implementations of the same functionality, it does offer tremendous flexibility in modifying the design in the field and avoiding expensive nonrecurring engineering (NRE) charges in the design. Thus, new applications, interfaces and improved algorithms can be downloaded to products working in the field using this approach.

Products in this area also include other processing and interface cores, such as multiply-accumulate (MAC) blocks that are specifically aimed at DSP-type dataflow signal and image processing applications; and high speed serial interfaces for wired communications such as SERDES (serializer/deserializer) blocks. In this sense, system-on-a-programmable-chip SoCs are not exactly application-specific, but not completely generic either.

It remains to be seen whether system-on-a-programmable chip SoCs are going to be a successful way of delivering high-volume consumer applications or will end up restricted to the two main applications for high-end FPGAs; rapid prototyping of designs which will be retargeted to ASIC or ASSP implementations, and used in high-end, relatively expensive parts of the communications infrastructure which require in-field flexibility and can tolerate the tradeoffs in cost, energy consumption, and performance. Certainly, the use of synthesizable processors on more moderate FPGAs to realize SoC-style designs is one alternative to the cost issue. Intermediate forms, such as the use of metal-programmable gate-array style logic fabrics together with hard-core processor subsystems and other cores, such as is offered in the "Structured ASIC" offerings of LSI Logic (RapidChip) and NEC (Instant Silicon Solutions Platform) represents an intermediate form of SoC between the full-mask ASIC and ASSP approach, and the field-programmable gate array approach. Here the tradeoffs are much slower design creation (a few weeks rather than a day or so), higher NRE than FPGA (but much lower than a full set of masks), and better cost, performance and energy consumption than FPGA (perhaps 15 to 30% worse than an ASIC approach). Further interesting compromise or hybrid style approaches, such as ASIC/ASSP with on-chip FPGA regions, are also emerging to give design teams more choices. A final interesting variation is a combination of a configurable processor which is implemented partly in fixed silicon, together with an FPGA region which is used for instruction extensions and other hardware implementations in the field. Stretch, for example, uses the Tensilica configurable processor to implement such a platform SoC.

IP Cores

The design of SoC would not be possible if every design started from scratch. In fact, the design of SoC depends heavily on the reuse of Intellectual Property blocks – what are called "IP Cores." IP reuse has emerged as a strong trend over the last eight to nine years [3] and has been one key element in closing what the International Technology Roadmap for Semiconductors [4] calls the "design productivity gap" – the difference between the rate of increase of complexity offered by advancing semiconductor process technology, and the rate of increase in designer productivity offered by advances in design tools and methodologies.

Reuse is not just important to offer ways of enhancing designer productivity – although it has dramatic impacts on that. It also provides a mechanism for design teams to create SoC products that span multiple design disciplines and domains. The availability of hard (laid-out and characterized) and soft (synthesizable) processor cores from a number of processor IP vendors allows design teams who would not be able to design their own processor from scratch to drop them into their designs and thus add RISC control and DSP functionality to an integrated SoC without having to master the art of processor design within the team. In this sense, the advantages of IP reuse go beyond productivity – it offers a large reduction in design risk, and also a way for SoC designs to be done that would otherwise be infeasible due to the length of time it would take to acquire expertise and design IP from scratch.

This ability when acquiring and reusing IP cores – to acquire, in a *prepackaged* form, design domain expertise outside one's own design team's set of core competencies, is a key requirement for the evolution of

SoC design going forward. SoC up to this point has concentrated to a large part on integrating digital components together, perhaps with some analogue interface blocks that are treated as black boxes. The hybrid SoCs of the future, incorporating domains unfamiliar to the integration team, such as RF or MEMS, requires the concept of "drop-in" IP to be extended to these new domains. We are not yet at that state – considerable evolution in the IP business and the methodologies of IP creation, qualification, evaluation, integration and verification are required before we will be able to easily specify and integrate truly heterogeneous sets of disparate IP blocks into a complete hybrid SoC.

However, the same issues existed at the beginning of the SoC revolution in the digital domain. They have been solved to a large extent, through the creation of standards for IP creation, evaluation, exchange and integration – primarily for digital IP blocks but extending also to analogue/mixed-signal (AMS) cores. Among the leading organizations in the identification and creation of such standards has been the Virtual Socket Interface Alliance (VSIA) [5], formed in 1996 and having at its peak a membership more than 200 IP, systems, semiconductor, and Electronic Design Automation (EDA) corporate members. Although often criticized over the years for a lack of formal and acknowledged adoption of its IP standards, VSIA has had a more subtle influence on the electronics industry. Many companies instituting reuse programmes internally; many IP, systems and semiconductor companies engaging in IP creation and exchange; and many design groups have used VSIA IP standards as a key starting point for developing their own standards and methods for IP based design. In this sense, use of VSIA outputs has enabled a kind of IP reuse in the IP business.

VSIA, for example, in its early architectural documents of 1996 to1997, helped define the strong industry-adopted understanding of what it meant for an IP block to be considered to be in "hard" or "soft" form. Other important contributions to design included the widely-read system-level design-model taxonomy created by one of its working groups. Its standards, specifications and documents thus represent a very useful resource for the industry [6].

Other important issues for the rise of IP-based design and the emergence of a third party industry in this area (which has taken much longer to emerge than originally hoped in the mid-1990s) are the business issues surrounding IP evaluation, purchase, delivery and use. Organizations such as the Virtual Component Exchange (VCX) [7] emerged to look at these issues and provide solutions. Although still in existence, it is clear that the vast majority of IP business relationships between firms occur within a more *ad hoc* supplier to customer business framework.

Virtual Components

The VSIA has had a strong influence on the nomenclature of the SoC and IP-based design industry. The concept of the "virtual socket" – a description of the all the design interfaces which an IP core must satisfy, and design models and integration information which must be provided with the IP core – required to allow it to be more easily integrated or "dropped into" an SoC design – comes from the concept of Printed Circuit Board (PCB) design where components are sourced and purchased in prepackaged form and can be dropped into a board design in a standardized way.

The dual of the "virtual socket" then becomes the "virtual component." Specifically in the VSIA context, but also more generally in the interface, an IP core represents a design block that *might* be reusable. A virtual component represents a design block which is *intended* for reuse, and which has been *developed* and *qualified* to be highly reusable. The things that separate IP cores from virtual components are in general:

- Virtual components conform in their development and verification processes to well-established design processes and quality standards.
- Virtual components come with design data, models, associated design files, scripts, characterization information and other deliverables which conform to one or other well-accepted standards for IP reuse – for example, the VSIA deliverables, or another internal or external set of standards.
- Virtual components in general should have been fabricated at least once, and characterized post fabrication to ensure that they have validated claims.
- Virtual components should have been reused at least once by an external design team, and usage reports and feedback should be available.

- Virtual components should have been rated for quality using an industry standard quality metric such as OpenMORE (originated by Synopsys and Mentor Graphics) or the VSI Quality standard (which has OpenMORE as one of its inputs).

To a large extent, the developments over the last decade in IP reuse have been focused on defining the standards and processes to turn the *ad hoc* reuse of IP cores into a well-understood and reliable process for acquiring and reusing virtual components – thus enhancing the analogy with PCB design.

Platforms and Programmable Platforms

The emphasis in the preceding sections has been on IP (or virtual component) reuse on a somewhat *ad hoc* block by block basis in SoC design. Over the past several years, however, there has arisen a more integrated approach to the design of complex SoCs and the reuse of virtual components – what has been called "Platform based design." Much more information is available in Ref. 8 to Ref. 11. Suffice it here to define platform-based design in the SoC context from one perspective.

We can define platform-based design as a planned design methodology which reduces the time and effort required, and risk involved, in designing and verifying a complex SoC. This is accomplished by extensive reuse of combinations of hardware and software IP. As an alternative to IP reuse in a block by block manner, platform-based design assembles groups of components into a reusable platform architecture. This reusable architecture, together with libraries of preverified and precharacterized, application-oriented HW and SW virtual components, is an SoC integration platform.

There are several reasons for the growing popularity of the platform approach in industrial design. These include the increase in design productivity, the reduction in risk, the ability to utilize preintegrated virtual components from other design domains more easily, and the ability to reuse SoC architectures created by experts. Industrial platforms include full application platforms, reconfigurable platforms, and processor-centric platforms [12]. Full application platforms, such as Philips Nexperia and TI OMAP provide a complete implementation vehicle for specific product domains [13]. Processor-centric platforms, such as ARM PrimeXsys concentrate on the processor, its required bus architecture, and basic sets of peripherals along with RTOS and basic software drivers. Reconfigurable, or "highly programmable" platforms such as the Xilinx Platform FPGA and Altera's SOPC deliver hardcore processors plus reconfigurable logic along with associated IP libraries and design tool flows.

Integration Platforms and SoC Design

The use of SoC integration platforms changes the SoC design process in two fundamental ways:

1. The basic platform must be designed, using whatever *ad hoc* or formalized design process for SoC that the platform creators decide on. The next section outlines some of the basic steps required to build an SoC, whether building a platform or using a block-based more *ad hoc* integration process. However, when constructing an SoC platform for reuse in derivative design, it is important to remember that it may not be necessary to take the whole platform and its associated HW and SW component libraries through complete implementation. Enough implementation must be done to allow the platform and its constituent libraries to be fully characterized, and modeled for reuse. It is also essential that the platform creation phase produces all the design files required for the platform and its libraries to be reused in a derivative design process in an archivable and retrievable form. This must also include the setup of the appropriate configuration programs or scripts to allow automatic creation of a configured platform during derivative design.
2. A design process must be created and qualified for all the *derivative* designs that will be created based on the SoC integration platform. This must include processes for retrieving the platform from its archive, for entering the derivative design configuration into a platform configurator, the generation of the design files for the derivative, the generation of the appropriate verification environment(s) for the derivative, the ability for derivative design teams to select components from libraries, to modify these components and validate them within the overall platform context, and, to the extent supported by the platform, to create new components for their particular application.

Reconfigurable or highly programmable platforms introduce an interesting addition to the platform based SoC design process [14]. Platform FPGAs and SOPC devices can be thought of as a "meta-platform," a platform for creating platforms. Design teams can obtain these devices from companies such as Xilinx and Altera, containing a basic set of more generic capabilities and IP embedded processors, on-chip buses, special IP blocks such as MACs and SERDES, and a variety of other prequalified IP blocks. They can then customize the meta-platform to their own application space by adding application domain-specific IP libraries. The combined platform can be provided to derivative design teams, who can select the basic meta-platform and configure it within the scope intended by the intermediate platform creation team, selecting the IP blocks needed for their exact derivative application.

Overview of the SoC Design Process

The most important thing to remember about SoC design is that it is a multi-disciplinary design process, which needs to exercise design tasks from across the spectrum of electronics. Design teams must gain some fluency with all these multiple disciplines, but the integrative and reusable nature of SoC design means that they may not need to become deep experts in all of them. Indeed, avoiding the need for designers to understand all methodologies, flows and domain-specific design techniques is one of the key reasons for reuse and enablers of productivity. Nevertheless, from DFT through digital and analogue HW design, from verification through system level design, from embedded SW through IP procurement and integration, from SoC architecture through IC analysis, a wide variety of knowledge is required by the team, if not by every designer.

Figure 16.3 Some of the basic constituents of the SoC design process.

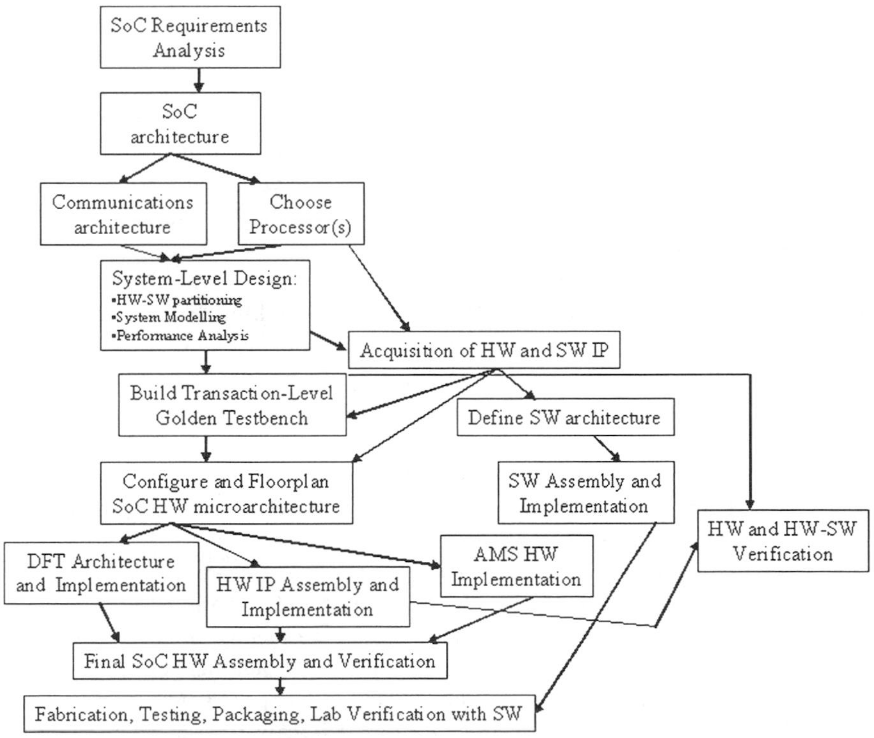

FIGURE 16.3 Steps in the SoC Design Process. (*Source*: "System-on-Chip Design," Grant Martin in *The Embedded Systems Handbook*, CRC Press, 2005.)

We will now define each of these steps as illustrated:

- *SoC requirements analysis* is the basic step of defining and specifying a complex SoC, based on the needs of the end product into which it will be integrated. The primary input into this step is the marketing definition of the end product and the resulting characteristics of what the SoC should be, functional and nonfunctional (e.g., cost, size, energy consumption, performance; latency and throughput, package selection). This process of requirements analysis must ultimately answer the question — is the product feasible? Is the desired SoC feasible to design and with what effort, and in what timeframe? How much reuse will be possible? Is the SoC design based on legacy designs of previous generation products (or, in the case of platform-based design, to be built based on an existing platform offering)?
- *SoC architecture*: In this phase the basic structure of the desired SoC is defined. Vitally important is to decide on the *communications architecture* that will be used as the backbone of the SoC on-chip communications network. Inadequate communications architecture will cripple the SoC and have as big an impact as the use of an inappropriate processor subsystem. Of course, the choice of communications architecture is impossible to divorce from making the basic *processor(s) choice*, e.g., Do I use a RISC control processor? Do I have an onboard DSP? How many of each? What are the processing demands of my SoC application? Do I integrate the bare processor core, or use a whole processor subsystem provided by an IP company (most processor IP companies have moved from offering just processor cores, to whole processor subsystems including hierarchical bus fabrics tuned to their particular processor needs)? Do I use a configurable and extensible processor, or more than one, tailored very closely to the specific application domain? Do I have some ideas, based on legacy SoC design in this space, as to how SW and HW should be partitioned? What memory hierarchy is appropriate? What are the sizes, levels, performance requirements and configurations of the embedded memories most appropriate to the application domain for the SoC?
- *System-level design* is an important phase of the SoC process, but one that is often done in a relatively *ad hoc* way. The whiteboard and the spreadsheet are as much used by the SoC architects as more capable toolsets. However, there has long been use of *ad hoc* C/C++ based models for the system design phase to validate basic architectural choices. Designers of complex signal processing algorithms for voice and image processing have long adopted dataflow models and associated tools to define their algorithms, define optimal bit-widths and validate performance whether destined for hardware or software implementation. A flurry of activity in the last few years on different C/C++ modeling standards for system architects has consolidated on SystemC [15]. The *system* nature of SoC demands a growing use of system-level design modeling and analysis as these devices grow more complex. The basic processes carried out in this phase include HW-SW partitioning (the allocation of functions to be implemented in dedicated HW blocks, in SW on processors (and the decision of RISC vs. DSP), or a combination of both, together with decisions on the communications mechanisms to be used to interface HW and SW, or HW–HW and SW–SW). In addition, the construction of system-level models, and the analysis of correct functioning, performance, and other nonfunctional attributes of the intended SoC through simulation and other analytical tools, are necessary. Finally, all additional IP blocks required which can be sourced outside, or reused from the design group's legacy, must be identified, HW and SW. The remaining new functions will need to be implemented as part of the overall SoC design process.
- After system level design and the identification of the processors and communications architecture, and other HW or SW IP required for the design, the group must undertake an *IP acquisition* stage. This can to a large extent be done at least in part in parallel with other work such as system-level design (assuming early identification of major external IP is made) or building golden transaction-level test-bench models. Fortunate design groups will be working in companies with a large legacy of existing well-crafted IP (rather, "virtual components") organized in easy to search databases; or those with access via supplier agreements to large external IP libraries; or at least those with experience at IP search, evaluation, purchase, and integration. For these lucky groups, the problems at this stage are greatly ameliorated. Others with less experience or infrastructure will need to explore these processes

for the first time, hopefully making use of IP suppliers' experience with the legal and other standards required. Here the external standards bodies such as VSIA and VCX have done much useful work that will smooth the path, at least a little. One key issue in IP acquisition is to conduct rigorous and thorough incoming inspection of IP to ensure its completeness and correctness to the greatest extent possible prior to use, and to resolve any problems with quality early with suppliers, long before SoC integration. Every hour spent on this at this stage, will pay back in avoiding much longer schedule slip later. The IP quality guidelines discussed earlier are a foundation level for a quality process at this point.

- *Build a transaction-level golden test bench*: The system model built up during the system level design stage can form the basis for a more elaborated design model, using "transaction-level" abstractions [16], which represent the underlying HW–SW architecture and components in more detail, sufficient detail to act as a functional virtual prototype for the SoC design. This golden model can be used at this stage to verify the microarchitecture of the design and to verify detailed design models for HW IP at the Hardware Description Language (HDL) level within the overall system context. It thus can be reused all the way down the SoC design and implementation cycle.
- *Define the SoC SW Architecture*: SoC is of course not just about HW [17]. As well as often defining the right on-chip communications architecture, the choice of processor(s), and the nature of the application domain have a very heavy influence on the SW architecture. For example, real-time operating system (RTOS) choice is limited by the processor ports, which have been done and by the application domain (OSEK is an RTOS for automotive systems; Symbian OS for portable wireless devices; PalmOS for Personal Digital Assistants, etc.). As well as the basic RTOS, every SoC peripheral device will need a device driver, hopefully based on reuse and configuration of templates; various middleware application stacks (e.g., telephony, multimedia image processing) are important parts of the SW architecture; voice and image encoding and decoding on portable devices often are based on assembly code IP for DSPs. There is thus a strong need in defining the SoC to fully elaborate the SW architecture to allow reuse, easy customization, and effective verification of the overall HW–SW device.
- *Configure and Floorplan SoC Microarchitecture*: At this point we are beginning to deal with the SoC on a more physical and detailed logical basis. Of course, during high-level architecture and system-level design, the team has been looking at physical implementation issues (although our design process diagram shows everything as a waterfall kind of flow, in reality SoC design like all electronics design is more of an iterative, incremental process, i.e., more akin to the famous "spiral" model for SW). But before beginning detailed HW design and integration, it is important that there is agreement among the teams on the basic physical floorplan; that all the IP blocks are properly and fully configured; that the basic microarchitectures (test, power, clocking, bus, timing) have been fully defined and configured, and that HW implementation can proceed. In addition, this process should also generate the downstream verification environments that will be used throughout the implementation processes, whether SW simulation based, emulation based, using rapid prototypes, or other hybrid verification approaches.
- *DFT architecture and implementation:* The test architecture is only one of the key microarchitectures which must be implemented; it is complicated by IP legacy and the fact that it is often impossible to impose one design-for-test style (such as BIST or SCAN) on all IP blocks. Rather, wrappers or adaptations of standard test interfaces (such as JTAG ports) may be necessary to fit all IP blocks together into a coherent test architecture and plan.
- *AMS HW Implementation:* Most SoCs incorporating analogue/mixed-signal (AMS) blocks use them to interface to the external world. VSIA, among other groups, has done considerable work in defining how AMS IP blocks should be created to allow them to be more easily integrated into mainly digital SoCs (the "Big D/little a" SoC); and guidelines and rules for such integration. Experiences with these rules and guidelines and extra deliverables has been on the whole promising, but they have more impact between internal design groups today than on the industry as a whole. The "Big A/Big D" mixed-signal SoC is still relatively rare.
- *HW IP Assembly and Integration:* This design step is in many ways the most traditional. Many design groups have experience in assembling design blocks done by various designers or subgroups in an

incremental fashion, into the agreed on architectures for communications, bussing, clocking, power, etc. The main difference with SoC is that many of the design blocks may be externally sourced IP. To avoid difficulties at this stage, the importance of rigorous qualification of incoming IP and the early definition of the SoC microarchitecture, to which all blocks must conform, cannot be overstated.

- *SW Assembly and Implementation:* Just as with HW, the SW IP, together with new or modified SW tasks created for the particular SoC under design, must be assembled together and validated as to conformance to interfaces and expected operational quality. It is important to verify as much of the SW in its normal system operating context as possible (see below).
- *HW and HW-SW Verification:* Although represented as a single box on the diagram, this is perhaps one of the largest consumers of design time and effort and the major determinant of final SoC quality. Vital to effective verification is the setup of a targeted SoC verification environment, reusing the golden test bench models created at higher levels of the design process. In addition, highly capable, multi-language, mixed simulation environments are important (for example, SystemC models and HDL implementation models need to be mixed in the verification process and effective links between them are crucial). There are a large number of different verification tools and techniques [18], ranging from SW based simulation environments, to HW emulators, HW accelerators, and FPGA and bonded-core-based rapid prototyping approaches. In addition, formal techniques such as equivalence checking, and model/property checking have enjoyed some successful usage in verifying parts of SoC designs, or the design at multiple stages in the process. Mixed approaches to HW–SW verification range from incorporating instruction set simulators (ISSs) of processors in SW-based simulation to linking HW emulation of the HW blocks (compiled from the HDL code) to SW running natively on a host workstation, linked in an *ad hoc* fashion by design teams or using a commercial mixed verification environment. Alternatively, HDL models of new HW blocks running in a SW simulator can be linked to emulation of the rest of the system running in HW, a mix of emulation and use of bonded-out processor cores for executing SW. It is important that as much of the system SW be exercised in the context of the whole system as possible, using the most appropriate verification technology that can get the design team close to realtime execution speed (no more than 100 times slower is the minimum to run significant amounts of software). The trend to transaction-based modeling of systems, where transactions range in abstraction from untimed functional communications via message calls, through abstract bus communications models, through cycle-accurate bus functional models, and finally to cycle- and pin-accurate transformations of transactions to the fully detailed interfaces, allows verification to occur at several levels or with mixed levels of design description. Finally, a new trend in verification is assertion-based verification, using a variety of input languages (PSL/Sugar, e, Vera, or regular Verilog and VHDL, and emerging SystemVerilog) to model design properties, which can then be monitored during simulation, either to ensure that certain properties will be satisfied, or certain error conditions never occur. Combinations of formal property checking and simulation-based assertion checking have been created – "semiformal verification." The most important thing to remember about verification is that armed with a host of techniques and tools, it is essential for design teams to craft a well-ordered verification process which allows them to definitively answer the question "how do we know that verification is done?" and thus allow the SoC to be fabricated.
- *Final SoC HW Assembly and Verification:* Often done in parallel or overlapping "those final few simulation runs" in the verification stage, the final SoC HW assembly and verification phase includes final place and route of the chip, any hand-modifications required, and final physical verification using design rule checking and layout vs. schematic (netlist) tools. In addition, this includes important analysis steps for issues that occur in advanced semiconductor processes such as IR-drop, signal integrity, power-network integrity, as well as satisfaction and design transformation for manufacturabilty and yield (OPC, reticle enhancement, etc.).
- *Fabrication, Testing, Packaging, and Lab Verification:* When an SoC has been shipped to fabrication, it would seem time for the design team to relax. Instead, this is an opportunity for additional verification to be carried out, especially more verification of system software running in context of the hardware design and for fixes, either of software, or of the SoC HW on hopefully no more than one expensive

iteration of the design, to be determined and planned. When the tested packaged parts arrive back for verification in the lab, the ideal scenario is to load the software into the system and have the SoC and its system booted up and running software within a few hours. Interestingly, the most advanced SoC design teams, with well-ordered design methodologies and processes, are able to achieve this quite regularly.

System-Level Design

As we touched on earlier, when describing the overall SoC design flow, system-level design and SoC are essentially made for each other. A key aim of IP reuse and of SoC techniques such as platform-based design is to make the "back end" (RTL to GDS II) design implementation processes easier, fast and with low risk; and to shift the major design phase for SoC up in time and in abstraction level to the system level. This also means that the back-end tools and flows for SoC designs do not necessarily differ from those used for complex ASIC, ASSP, and custom IC design, it is the methodology of how they are used, and how blocks are sourced and integrated, overlaying the underlying design tools and flows, that may differ for SoC. However, the fundamental nature of IP based design of SoC has a stronger influence on the system level.

It is at the system level that the vital tasks of deciding on and validating the basic system architecture and choice of IP blocks are carried out. In general, this is known as "design space exploration." As part of this exploration, SoC platform customization for a particular derivative is carried out, should the SoC platform approach be used. Essentially one can think of platform design space exploration (DSE) as being a similar task to general DSE, except that the scope and boundaries of the exploration are much more tightly constrained, the basic communications architecture and platform processor choices may be fixed, and the design team may be restricted to choosing certain customization parameters and choosing optional IP from a library. Other tasks include HW–SW partitioning, usually restricted to decisions about key processing tasks which might be mapped into either HW or SW form and which have a big impact on system performance, energy consumption, on-chip communications bandwidth consumption, or other key attributes. Of course, in multi-processor systems, there are "SW–SW" partitioning or codesign issues as well, deciding on the assignment of SW tasks to various processor options. Again, perhaps 80 to 95% of these decisions can or are made *a priori*, especially if an SoC is either based on a platform or an evolution of an existing system; such codesign decisions are usually made on a small number of functions which have critical impact.

Because partitioning, codesign and DSE tasks at the system level involve much more than HW–SW issues, a more appropriate term for this is "function-architecture codesign" [19,20]. In this codesign model, systems are described on two equivalent levels:

- The functional intent of the system; e.g., a network of applications, decomposed into individual sets of functional tasks which may be modeled using a variety of models of computation, such as discrete event, finite state machine or dataflow.
- The architectural structure of the system; the communications architecture, major IP blocks such as processors, memories, and HW blocks, captured or modeled for example using some kind of IP or platform configurator.

The methodology implied in this approach is then to build explicit mappings between the functional view of the system and the architectural view, that carry within them the implicit partitioning that is made for computation and communications. This hybrid model can then be simulated, the results analyzed, and a variety of ancillary models (for example, cost, power, performance, communications bandwidth consumption, etc.) can be utilized in order to examine the suitability of the system architecture as a vehicle for realizing or implementing the end product functionality.

The function-architecture codesign approach has been implemented and used in research and commercial tools [21] and forms the foundation of many system-level codesign approaches going forward. In addition, it has been found extremely suitable as the best system-level design approach for platform-based design of SoC [22].

Interconnection and Communication Architectures for Systems on-Chip

This topic is a large topic and beyond the limited space of this chapter. Suffice it to say here that current SoC architectures deal in fairly traditional hierarchies of standard on-chip buses, for example, processor-specific buses, high-speed system buses, and lower-speed peripheral buses, using standards such as ARM's AMBA and IBM's CoreConnect [13], and traditional master-slave bus approaches. Recently there has been a lot of interest in network-on-chip communications architectures, based on packet-sw and a number of approaches have been reported in the literature but this remains primarily a research topic in universities and industrial research labs [23,24].

Computation and Memory Architectures for System-on-Chip

The primary processors used in SoC are embedded RISCs such as ARM processors, PowerPCs, MIPS architecture processors, and some of the configurable, extensible processors designed specifically for SoC such as Tensilica and ARC. In addition, embedded DSPs from traditional suppliers as TI, Motorola, ParthusCeva, and others are also quite common in many consumer applications, for embedded signal processing for voice and image data. Research groups have looked at compiling or synthesizing application-specific processors or coprocessors [25,26] and these have interesting potential in future SoCs which may incorporate networks of heterogeneous configurable processors collaborating to offer large amounts of computational parallelism. This is an especially interesting prospect given wider use of reconfigurable logic which opens up the prospect of dynamic adaptation of SoC to application needs. However, most multi-processor SoCs today involve at most two to four processors of conventional design; the larger networks are more often found today in the industrial or university lab. There are of course outliers, one Cisco chip for a high-end router includes 192 Tensilica processors configured for the routing application.

Although several years ago most embedded processors in early SoCs did not use cache memory-based hierarchies, this has changed significantly over the years, and most RISC and DSP processors now involve significant amounts of Level 1 Cache memory, as well as higher level memory units on and off-chip (off-chip flash memory is often used for embedded software tasks which may be only infrequently required). System design tasks and tools must consider the structure, size, and configuration of the memory hierarchy as one of the key SoC configuration decisions that must be made.

IP Integration Quality and Certification Methods and Standards

We have emphasized the design reuse aspects of SoC and the need for reuse of internal and externally sourced IP blocks by design teams creating SoCs. In the discussion of the design process above we mentioned issues such as IP quality standards and the need for incoming inspection and qualification of IP. The issue of IP quality remains one of the biggest impediments to the use of IP based design for SoC [27]. The quality standards and metrics available from VSIA and OpenMORE, and their further enhancement help, but only to a limited extent. The industry could clearly use a formal certification body or lab for IP quality that would ensure conformance to IP transfer requirements and the integration quality of the blocks. Such a certification process would be of necessity quite complex due to the large number of configurations possible for many IP blocks and the almost infinite variety of SoC contexts into which they might be integrated. Certified IP would begin the deliver the "virtual components" of the VSIA vision.

In the absence of formal external certification (and such third-party labs seem a long way off, if they ever emerge), design groups must provide their own certification processes and real reuse quality metrics, based on their internal design experiences. Platform-based design methods help due to the advantages of prequalifying and characterizing groups of IP blocks and libraries of compatible domain-specific components. Short of independent evaluation and qualification, this is the best that design groups can do currently. Of course, the most reliable IP suppliers in the industry have built up their own reputations for delivering quality IP together with the advanced tools required to design this IP into the overall SoC. Embedded processor vendors in general have the highest reputations for IP quality.

One key issue to remember is that IP not created for reuse, with all the deliverables created and validated according to a well-defined set of standards, is inherently not reusable. The effort required to make a reusable IP block has been estimated to be 50 to 200% more effort than that required to use it once; however, assuming the most conservative extra cost involved implies positive payback with three uses of the IP block. Planned and systematic IP reuse and investment in those blocks with greatest SoC use potential gives a high chance of achieving significant productivity soon after starting a reuse programme. But *ad hoc* attempts to reuse existing design blocks not designed to reuse standards have failed in the past and are unlikely to provide the quality and productivity desired.

The Networked Embedded Systems Revolution

Another phase in the revolution and evolution of embedded systems, already in progress, is the emergence of distributed embedded systems; frequently termed networked embedded systems, where the word "networked" signifies the importance of the networking infrastructure and communication protocol. A networked embedded system is a collection of spatially and functionally distributed embedded nodes interconnected by means of wireline or wireless communication infrastructure and protocols, interacting with the environment (via a sensor/actuator elements) and each other, and, possibly, a master node performing some control and coordination functions, to coordinate computing and communication in order to achieve certain goals. The networked embedded systems appear in a variety of application domains to mention automotive, train, aircraft, office building, and industrial, primarily for monitoring and control, environment monitoring, and also in the future, control.

There have been various reasons for the emergence of networked embedded systems, influenced largely by their application domains. The benefits of using distributed systems and an evolutionary need to replace point-to-point wiring connections in these systems by a single bus are some of the most important ones.

The advances in design of embedded systems, tools availability, and falling fabrication costs of semiconductor devices and systems have allowed for infusion of intelligence into field devices such as sensors and actuators. The controllers used with these devices provide typically on-chip signal conversion, data processing, and communication functions. The increased functionality, processing, and communication capabilities of controllers have been largely instrumental in the emergence of a wide-spread trend for networking of field devices around specialized networks, frequently referred to as field area networks.

The field area networks [28] (a field area network is, in general, a digital, two-way, multi-drop communication link) as commonly referred to, are, in general, networks connecting field devices such as sensors and actuators with field controllers, such as programmable logic controllers (PLCs) in industrial automation, or electronic control units (ECUs) in automotive applications, as well as man-machine interfaces, for instance, dashboard displays in cars.

In general, the benefits of using those specialized networks are numerous, including increased flexibility attained through combination of embedded hardware and software, improved system performance, and ease of system installation, upgrade, and maintenance. Specifically, in automotive and aircraft applications, for instance, they allow for a replacement of mechanical, hydraulic, and pneumatic systems by mechatronic systems, where mechanical or hydraulic components are typically confined to the end-effectors; just to mention these two different application areas.

Unlike LANs, due to the nature of communication requirements imposed by applications, field area networks, by contrast, tend to have low data rates, small size of data packets, and typically require real-time capabilities which mandate determinism of data transfer. However, data rates above 10 Mbit/s, typical of LANs, have become a commonplace in field area networks.

The specialized networks tend to support various communication media like twisted-pair cables, fiber-optic channels, power-line communication, radio frequency channels, infrared connections, etc. Based on the physical media employed by the networks, they can be in general divided into three main groups, namely, wireline based networks using media such as twisted-pair cables, fiber-optic channels (in hazardous environments like chemical and petrochemical plants), and power lines (in building automation); wireless networks supporting radio frequency channels, and infrared connections; and hybrid networks composed of wireline and wireless networks.

Although the use of wireline based field area networks is dominant, the wireless technology offers a range of incentives in a number of application areas. In industrial automation, for instance, wireless device (sensor/actuator) networks can provide a support for mobile operation required in case of mobile robots, monitoring and control of equipment in hazardous and difficult to access environments, etc. In a wireless sensor/actuator network, stations may interact with each other on a peer-to-peer basis, and with a base station. The base station may have its transceiver attached to a cable of a (wireline) field area network, giving rise to a hybrid wireless–wireline system [29]. A separate category is the wireless sensor networks, mainly envisaged to be used for monitoring purposes, which is discussed in detail in this book.

The variety of application domains impose different functional and nonfunctional requirements on to the operation of networked embedded systems. Most of them are required to operate in a reactive way; for instance, systems used for control purposes. With that comes the requirement for real-time operation, in which systems are required to respond within a predefined period of time, mandated by the dynamics of the process under control. A response, in general, may be periodic to control a specific physical quantity by regulating dedicated end effector(s), or aperiodic arising from unscheduled events such as out-of-bounds state of a physical parameter, or any other kind of abnormal conditions. Broadly speaking, systems that can tolerate a delay in response are called soft real-time systems; in contrast, hard real-time systems require deterministic responses to avoid changes in the system dynamics which potentially may have negative impact on the process under control, and as a result may lead to economic losses or cause injury to human operators. Representative examples of systems imposing hard real-time requirements on their operation are fly-by-wire in aircraft control, and steer-by-wire in automotive applications, to mention some.

The need to guarantee a deterministic response mandates using appropriate scheduling schemes, which are frequently implemented in application domain specific real-time operating systems or frequently custom designed "bare-bones" real-time executives.

The networked embedded systems used in safety critical applications such as fly-by-wire and steer-by-wire require a high level of dependability to ensure that a system failure does not lead to a state in which human life, property, or environment are endangered. The dependability issue is critical for technology deployment; various solutions are discussed in this chapter in the context of automotive applications. One of the main bottlenecks in the development of safety-critical systems is the software development process. This issue is also briefly discussed in this chapter in the context of the automotive application domain.

As opposed to applications mandating hard real-time operation, such as the majority of industrial automation controls or safety critical automotive control applications, building automation control systems, for instance, seldom have a need for hard real-time communication; the timing requirements are much more relaxed. The building automation systems tend to have a hierarchical network structure and typically implement all seven layers of the ISO/OSI reference model [30]. In the case of field area networks employed in industrial automation, for instance, there is little need for the routing functionality and end-to-end control. As a consequence, typically, only the layers one (physical layer), two (datalink layer, including implicitly the medium access control layer), and seven (application layer, which covers also user layer) are used in those networks.

This diversity of requirements imposed by different application domains (soft/hard real-time, safety critical, network topology, etc.) necessitated different solutions, and using different protocols based on different operation principles. This has resulted in plethora of networks developed for different application domains.

With the growing trend for networking of embedded system and their internetworking with LAN, WAN, and the Internet (for instance, there is a growing demand for remote access to process data at the factory floor), many of those systems may become exposed to potential security attacks, which may compromise their integrity and cause damage as a result. The limited resources of embedded nodes pose considerable challenge for the implementation of effective security policies which, in general, are resource-demanding. These restrictions necessitate a deployment of lightweight security mechanisms. Vendor tailored versions of standard security protocol suites such as Secure Sockets Layer (SSL) and IP Security Protocol (IPSec) may still not be suitable due to excessive demand for resources. Potential security solutions for this kind of systems depend heavily on the specific device or system protected, application domain, and extent of internetworking and its architecture.

Design Methods for Networked Embedded Systems

Design methods for networked embedded systems fall into the general category of system-level design. They include two separate aspects, which will be discussed briefly in the following. A first aspect is the network architecture design, in which communication protocols, interfaces, drivers and computation nodes are selected and assembled. A second aspect is the system-on-chip design, in which the best HW/SW partition is selected, and an existing platform is customized, or a new chip is created for the implementation of a computation or a communication node. Both aspects share several similarities, but so far have generally been solved using *ad hoc* methodologies and tools, since the attempt to create a unified electronic system-level design methodology so far has failed.

When one considers the complete networked system, including several digital and analog parts, many more trade-offs can be played at the global level. However, it also means that the interaction between the digital portion of the design activity and the rest is much more complicated, especially in terms of tools, formats, and standards with which one must interoperate and interface.

In the case of network architecture design, tools such as OpNet and NS are used to identify communication bottlenecks, investigate the effect of parameters such as channel bit-error rate, and analyze the choice of coding, medium access and error correction mechanisms on the overall system performance. For wireless networks, tools such as MATLAB and SIMULINK are also used, in order to analyze the impact of detailed channel models, thanks to their ability to model digital and analogue components, as well as physical elements, at a high level of abstraction. In all cases, the analysis is essentially *functional*, i.e., it takes into account only in a very limited manner effects such as power consumption, computation time, and cost. This is the main limitation that will need to be addressed in the future, if one wants to be able to model and design in an optimal manner low power networked embedded systems, such as those that are envisioned for wireless sensor network applications.

At the system-on-chip architecture level, the first decision to be made is whether to use a *platform instance* or design an application-specific integrated circuit from scratch. The first option builds on the availability of large libraries of intellectual properties (IP), in the form of processors, memories, and peripherals, from major silicon vendors. These IP libraries are guaranteed to work together. Processors (and the software executing on them) provide flexibility to adapt to different applications and customizations (e.g., localization and adherence to regional standards), while hardware IPs provide efficient implementation of commonly used functions. Configurable processors can be adapted to the requirements of specific applications and via instruction extensions, offer considerable performance and power advantages over fixed instruction set architectures. Design methods that exploit the notion of platform generally start from a functional specification, which is then *mapped* on to an architecture (a platform instance) in order to derive performance information and explore the design space.

Full exploitation of the notion of platform results in better reuse, by decoupling independent aspects, that would otherwise tie, e.g., a given functional specification to low-level implementation details. The guiding principle of separation of concerns distinguishes between:

- Computation and communication. This separation is important because refinement of computation is generally done by hand, or by compilation and scheduling, while communication makes use of patterns.
- Application and platform implementation, because they are often defined and designed independently by different groups or companies.
- Behavior and performance, which should be kept separate because performance information can either represent nonfunctional requirements (e.g., maximum response time of an embedded controller), or the result of an implementation choice (e.g., the worst-case execution time of a task). Nonfunctional constraint verification can be performed traditionally, by simulation and prototyping, or with static formal checks, such as schedulability analysis.

Tool support for system-on-chip architectural design is so far mostly limited to simulation and interface generation. The first category includes tools such as NC SystemC from Cadence, ConvergenSC from CoWare, and SystemStudio from Synopsys. Simulators at the SOC level provide abstractions for the main architectural components (processors, memories, busses, HW blocks) and permit quick instantiation of complete platform instances from template skeletons. Interface synthesis can take various forms, from the automated instantiation of templates offered by N2C from CoWare, to the automated consistent file generation for SW and HW offered by Beach Solutions.

A key aspect of design problems in this space is compatibility with respect to specifications, at the interface level (bus and networking standards), instruction-set architecture level, and application procedural interface level. Assertion-based verification techniques can be used to ease the problem of verifying compliance with a digital protocol standard (e.g., for a bus).

Let us consider in the following an example of a design flow in the automotive domain, which can be considered as a paradigm of any networked embedded system. Automotive electronic design starts, usually five to ten years before the actual introduction of a product, when a car manufacturer defines the specifications for its future line of vehicles.

It is now accepted practice to use the notion of platform also in this domain, so that the electronic portion (as well as the mechanical one, which is outside the scope of this discussion) is modularized and componentized, enabling sharing across different models. An electronic control unit (ECU) generally includes a microcontroller (8, 16, and 32 bits), memory (SRAM, DRAM, and Flash), some ASIC or FPGA for interfacing, one or more in-vehicle network interfaces (e.g., CAN or FlexRay), and several sensor, and actuator interfaces (analog/digital and digital/analog converters, pulse-width modulators, power transistors, display drivers, and so on).

The system-level design activity is performed by a relatively small team of architects, who know the domain fairly well (mechanics, electronics and business), define the specifications for the electronic component suppliers, and interface with the teams that specify the mechanical portions (body and engine). These teams essentially use past experience to perform their job, and currently have serious problems forecasting the state of electronics ten years in advance.

Control algorithms are defined in the next design phase, when the first engine models (generally described using Simulink, Matlab, and StateFlow) become available, as a specification for the electronic design and the engine design. An important aspect of the overall flow is that these models are not frozen until much later, and hence algorithm design and (often) ECU software design must cope with their changes. Another characteristic is that they are parametric models, sometimes reused across multiple engine generations and classes, whose exact parameter values will be determined only when prototypes or actual products will be available. Thus control algorithms must consider allowable ranges and combinations of values for these parameters, and the capability to measure directly or indirectly their values from the behavior of engine and vehicle. Finally, algorithms are often distributed over a network of cooperating ECUs, thus deadlines and constraints generally span a number of electronic modules.

While control design progresses, ECU hardware design can start, because rough computational and memory requirement, as well as interfacing standards, sensors and actuators, are already known. At the end of control design and hardware design, software implementation can start. As mentioned above, most of the software running on modern ECUs is automatically generated (model-based design).

The electronic subsystem supplier in the HW implementation phase can use off-the-shelf components (such as memories), ASSPs (such as microcontrollers and standard bus interfaces), and even ASICs and FPGAs (typically for sensor and actuator signal conditioning and conversion).

The final phase, called system integration, is generally performed by the car manufacturer again. It can be an extremely lengthy and costly phase, because it requires the use of expensive detailed models of the controlled system (e.g., the engine, modeled with DSP based multiprocessors) or even of actual car prototypes. The goal of integration is to ensure smooth subsystem communication (e.g., checking that there are no duplicate module identifiers and that there is enough bandwidth in every in-vehicle bus). Simulation support in this domain is provided by companies such as Vast and Axys (now part of ARM), who sell fast instruction set simulators for the most commonly used processors in the networked embedded system domain, and network simulation models exploiting either proprietary simulation engines, e.g., in the case of Virtio, or standard simulators (HDL or SystemC).

Automotive Networked Embedded Systems

In the automotive electronic systems, the electronic control units are networked by means of one of automotive specific communication protocols for the purpose of controlling one of the vehicle functions; for instance, electronic engine control, antilocking brake system, active suspension, telematics, to mention a few. In Ref. [31] a number of functional domains have been identified for the deployment of automotive-networked embedded systems. They include the powertrain domain, involving, in general, control of engine and transmission; the chassis domain involving control of suspension, steering and braking, etc.; the body domain involving control of wipers, lights, doors, windows, seats, mirrors, etc.; the telematics domain involving mostly the integration of wireless communications, vehicle monitoring systems, and vehicle location systems; and the multimedia and human–machine interface domains. The different domains impose varying constraints on the networked embedded systems in terms of performance, safety requirements, and quality of service (QoS). For instance, the powertrain and chassis domains will mandate real-time control; typically bounded delay is required, as well as fault-tolerant services.

There are a number of reasons for the interest of the automotive industry in adopting mechatronic solutions, known by their generic name as X-by-Wire, aiming to replace mechanical, hydraulic, and pneumatic systems by electrical/electronic systems. The main factors seem to be economic in nature, improved reliability of components, and increased functionality to be achieved with a combination of embedded hardware and software. Steer-by-Wire, Brake-by-Wire, or Throttle-by-Wire systems are representative examples of those systems. An example of a possible architecture of a Steer-by-Wire is shown in Figure 16.4. But, it seems that certain safety critical systems such as Steer-by-Wire and Brake-by-Wire will be complemented with traditional

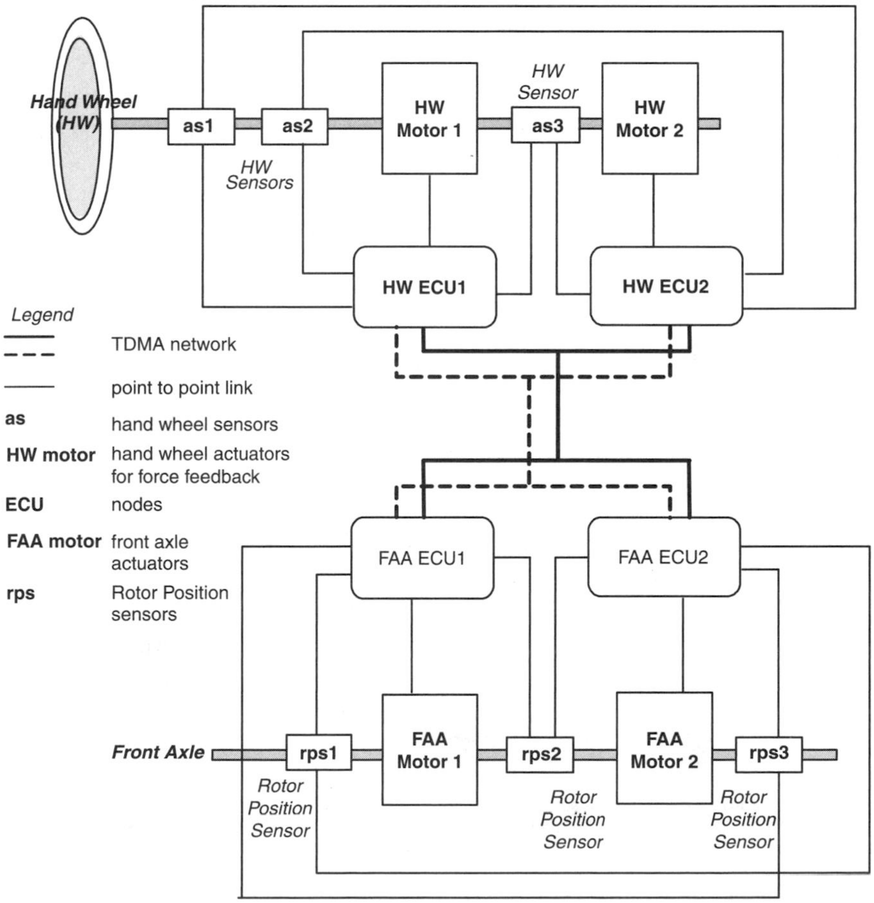

FIGURE 16.4 Steer-by-Wire architecture. (*Source*: "Design of automotive X-by-Wire systems," Cédric Wilwert et al., *The Industrial Communication Technology Handbook*, CRC Press, 2005.)

mechanical/hydraulic backups, for safety reasons. The dependability of X-by-Wire systems is one of the main requirements, as well as constraints on the adoption of this kind of systems. In this context, a safety critical X-by-Wire system has to ensure that a system failure does not lead to a state in which human life, property, or environment are endangered; and a single failure of one component does not lead to a failure of the whole X-by-Wire system [32]. When using Safety Integrity Level (SIL) scale, it is required for X-by-Wire systems that the probability of a failure of a safety critical system does not exceed the figure of 10^{-9} per hour/system. This figure corresponds to the SIL4 level. Another equally important requirement for the X-by-Wire systems is to observe hard real time constraints imposed by the system dynamics; the end-to-end response times must be bounded for safety critical systems. A violation of this requirement may lead to performance degradation of the control system, and other consequences as a result. Not all automotive electronic systems are safety critical. For instance, systems to control seats, door locks, internal lights, etc., are not. Different performance, safety, and QoS requirements dictated by various in-car application domains necessitate adoption of different solutions, which, in turn, gave rise to a significant number of communication protocols for automotive applications. Time-triggered protocols [35] based on TDMA (Time Division Multiple Access) medium access control technology are particularly well suited for the safety critical solutions, as they provide deterministic access to the medium. (In timed triggered protocols, activities occur at predefined instances of time, as opposed to an event triggered in which actions are triggered by occurrences of events.) In this category, there are two protocols, which, in principle, meet the requirements of X-by-Wire applications, namely TTP/C [33] and FlexRay [34] (FlexRay can support a combination of time-triggered and event-triggered transmissions). The following discussion will focus mostly on TTP/C and FlexRay.

The TTP/C (Time Triggered Protocol) is a fault-tolerant time triggered protocol; one of two protocols in the Time Triggered Architecture (TTA) [35]. The other one is a low cost fieldbus protocol TTP/A [36] (fieldbus is another term for field area network). In TTA, the nodes are connected by two replicated communication channels forming a cluster. In TTA, a network may have two different interconnection topologies, namely bus and star. In the bus configuration, Figure 16.5, each node is connected to two replicated passive buses via bus guardians. The bus guardians are independent units preventing associated nodes from transmitting outside predetermined time slots, by blocking the transmission path; a good example may be a case of a controller with a faulty clock oscillator that attempts to transmit continuously. In the star topology, Figure 16.6, the guardians are integrated into two replicated central star couplers. The guardians are required to be equipped with their own clocks, distributed clock synchronization mechanism, and power supply. In addition, they should be located at a distance from the protected node to increase immunity to spatial proximity faults. To cope with internal physical faults, TTA employs partitioning of nodes into so-called fault-tolerant units (FTUs), each of which is a collection of several stations performing the same computational functions. As each node is (statically) allocated a transmission slot in a TDMA round, failure of any node or a frame corruption is not going to cause degradation of the service. In addition, data redundancy allows, by voting process, to ascertain the correct data value.

TTP/C employs a synchronous TDMA medium access control scheme on replicated channels, which ensures fault-tolerant transmission with known delay and bounded jitter between the nodes of a cluster. The use of replicated channels, and redundant transmission, allows for the masking of a temporary fault on one of channels. The payload section of the message frame contains up to 240 bytes of data protected by a

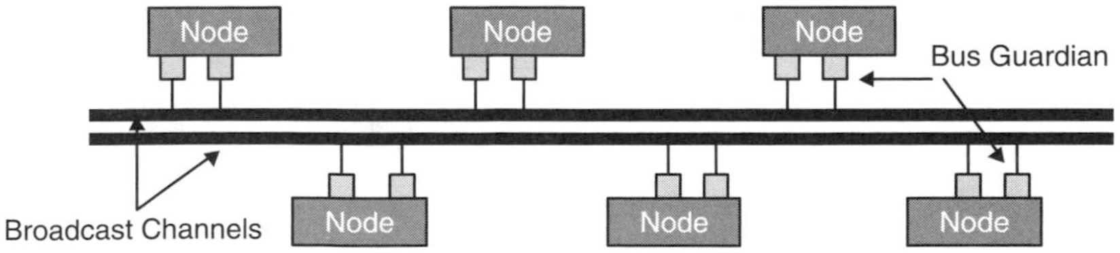

FIGURE 16.5 TTA–bus topology. (*Source*: "Dependable Time-Triggered Communication," Hermann Kopetz, Günther Bauer, and Wilfried Steiner in *The Industrial Communication Technology Handbook*, CRC Press, 2005.)

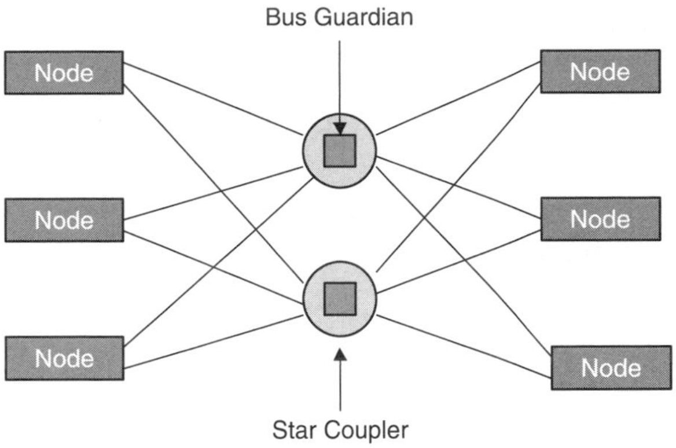

FIGURE 16.6 TTA–Star topology. (*Source*: "Dependable Time-Triggered Communication," Hermann Kopetz, Günther Bauer, and Wilfried Steiner in *The Industrial Communication Technology Handbook*, CRC Press, 2005.)

24-bit CRC checksum. In TTP/C, the communication is organized into rounds. In a round, different slot sizes may be allocated to different stations. However, slots belonging to the same station are of the same size in successive rounds. Every node must send a message in every round. Another feature of TTP/C is fault-tolerant clock synchronization that establishes global time base without a need for a central time provider. In the cluster, each node contains the message schedule. Based on that information, a node computes the difference between the predetermined and actual arrival time of a correct message. Those differences are averaged by a fault-tolerant algorithm, which allows for the adjustment of the local clock to keep it in synchrony with clocks of other nodes in the cluster. TTP/C provides so-called membership services to inform every node about the state of every other node in the cluster; it is also used to implement the fault-tolerant clock synchronization mechanism. This service is based on a distributed agreement mechanism, which identifies nodes with failed links. A node with a transmission fault is excluded from the membership until restarted with a proper state of the protocol. Another important feature of TTP/C is a clique avoidance algorithm to detect and eliminate formation of cliques in case the fault hypothesis is violated. In general, the fault-tolerant operation based on FTUs cannot be maintained if the fault hypothesis is violated. In such a situation, TTA activates never-give-up (NGU) strategy [35]. The NGU strategy, specific to the application, is initiated by TTP/C in combination with the application with an aim to continue operation in a degraded mode.

The TTA infrastructure, and the TTP/A and TTP/C protocols have a long history dating back to 1979 when the Maintainable Architecture for Real Time Systems (MARS) project started at the Technical University of Berlin. Subsequently, the work was carried out at the Vienna University of Technology. TTP/C protocols have been experimented with and considered for deployment for quite some time. However, to date, there have been no actual implementations of that protocol involving safety critical systems in commercial automobiles, or trucks. In 1995, a "proof of concept," organized jointly by Vienna University of Technology and DaimlerChrysler, demonstrated a car equipped with a "Brake-by-Wire" system based on time-triggered protocol. The TTA design methodology, which distinguishes between the node design and the architecture design, is supported by a comprehensive set of integrated tools from TTTech. A range of development and prototyping hardware is available from TTTech, as well. Austriamicrosystems offers automotive certified TTP-C2 Communication Controller (AS8202NF).

FlexRay, which appears to be the frontrunner for future automotive safety-critical control applications, employs a modified TDMA medium access control scheme on a single or replicated channel. A FlexRay node is shown in Figure 16.7. The payload section of a frame contains up to 254 bytes of data protected by a 24-bit CRC checksum. To cope with transient faults, FlexRay also allows for a redundant data transmission over the same channel(s) with a time delay between transmissions. The FlexRay communication cycle comprises a network communication time, and network idle time, Figure 16.8. Two or more

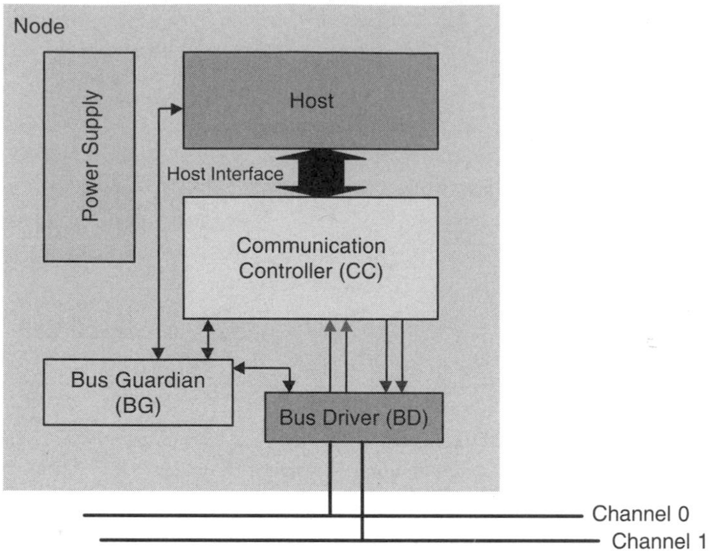

FIGURE 16.7 FlexRay node. (*Source*: "FlexRay Communication Technology," Dietmar Millinger and Roman Nossal in *The Industrial Communication Technology Handbook*, CRC Press, 2005.)

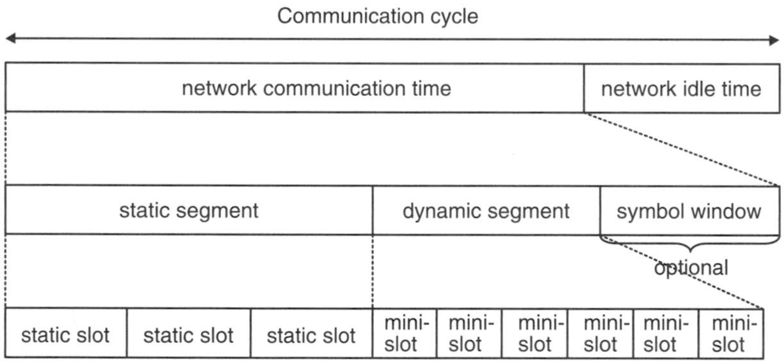

FIGURE 16.8 FlexRay communication cycle. (*Source*: "FlexRay Communication Technology," Dietmar Millinger and Roman Nossal in *The Industrial Communication Technology Handbook*, CRC Press, 2005.)

communication cycles can form an application cycle. The network communication time is a sequence of static segment, dynamic segment, and symbol window. The static segment uses a TDMA MAC protocol. The static segment comprises static slots of fixed duration. Unlike in TTP/C, the static allocation of slots to a node (communication controller) applies to one channel only. The same slot may be used by another node on the other channel. Also, a node may possess several slots in a static segment. The dynamic segment uses a FTDMA (flexible time division multiple access) MAC protocol, which allows for a priority and demand driven access pattern. The dynamic segment comprises so-called mini-slots with each node allocated a certain number of mini-slots, which do not have to be consecutive. The mini-slots are of a fixed length, and much shorter than static slots. As the length of a mini-slot is not sufficient to accommodate a frame (a mini-slot only defines a potential start time of a transmission in the dynamic segment), it has to be enlarged to accommodate transmission of a frame. This in turn reduces the number of mini-slots in the reminder of the dynamic segment. A mini-slot remains silent if there is nothing to transmit. The nodes allocated mini-slots towards the end of the dynamic segment are less likely to get transmission time. This in turn enforces a priority scheme. The symbol window is a time slot of fixed duration used for network management purposes. The network idle time is a protocol specific time window, in which no traffic is scheduled on the communication channel. It is used by the communication controllers for the clock synchronization activity; in principle, similar to the one described for TTP/C. If the dynamic segment and idle window are optional, the idle time and minimal static segment are mandatory parts of a communication cycle; minimum two static slots (degraded static segment), or four static slots for fault-tolerant clock synchronization are required. With all that, FlexRay allows for three configurations, pure static; mixed, with static and dynamic, bandwidth ratio depends on the application; and pure dynamic, where all bandwidth is allocated to the dynamic communication.

FlexRay supports a range of network topologies offering a maximum of scalability and a considerable flexibility in the arrangement of embedded electronic architectures in automotive applications. The supported configurations include bus, active star, active cascaded stars, and active stars with bus extension. FlexRay also uses the bus guardians in the same way as TTP/C.

The existing FlexRay communication controllers support communication bit rates of up to 10 Mbps on two channels. The transceiver component of the communication controller also provides a set of automotive network specific services. Two major services are alarm handling and wakeup control. In addition to the alarm information received in a frame, an ECU also receives the alarm symbol from the communication controller. This redundancy can be used to validate critical signals; for instance, an air-bag fire command. The wakeup service is required where electronic components have a sleep mode to reduce power consumption.

FlexRay is a joint effort of a consortium involving some of the leading car makers and technology providers: BMW, Bosch, DaimlerChrysler, General Motors, Motorola, Philips, and Volkswagen, as well as Hyundai Kia Motors as a premium associate member with voting rights. DECOMSYS offers Designer Pro, a comprehensive set of tools to support the development process of FlexRay based applications. The FlexRay protocol specification version 2.0 was released in 2004. The controllers are currently available from Freescale, and in future from NEC. The latest controller version, MFR4200, implements the protocol specification versions 1.0 and 1.1. Austriamicrosystems offers high-speed automotive bus transceiver for FlexRay (AS8221). The special physical layer for FlexRay is provided by Phillips. It supports the topologies described above, and a data rate of 10 MBPS on one channel. Two versions of the bus driver will be available.

TTCAN (Time Triggered Controller Area Network) [37], that can support a combination of time-triggered and event-triggered transmissions, utilize physical and Data-Link Layer of the CAN protocol. Since this protocol, as in the standard, does not provide the necessary dependability services, it is unlikely to play any role in fault-tolerant communication in automotive applications.

TTP/C and FlexRay protocols belong to class D networks in the classification published by the Society for Automotive Engineers [38,39]. Although the classification dates back to 1994, it is still a reasonable guideline for distinction of different protocols based on data transmission speed and functions distributed over the network. The classification comprises four classes. Class A includes networks with a data rate less than 10 Kbit/s. Some of the representative protocols are LIN (Local Interconnect Network) [40] and TTP/A [36]. Class A networks are employed largely to implement the body domain functions. Class B networks operate within the range of 10 Kbit/s to 125 Kbit/s. Some of the representative protocols are J1850 [41], low-speed CAN (Controller Area Network) [42], and VAN (Vehicle Area Network) [43]. Class C networks operate within the range of 125Kbit/s to 1Mbit/s. Examples of this class networks are high-speed CAN [44] and J1939 [45]. Network in this class are used for the control of powertrain and chassis domains. High-speed CAN, although used in the control of powertrain and chassis domains, is not suitable for safety-critical applications as it lacks the necessary fault-tolerant services. Class D networks (not formally defined as yet) includes networks with a data rate over 1Mb/s. Networks to support the X-by-Wire solutions fall in to this class, to include TTP/C and FlexRay. Also, MOST (Media Oriented System Transport [46] and IDB-1394 [47] for multimedia applications belong to this class.

The cooperative development process of networked embedded automotive applications brings with itself, heterogeneity of software and hardware components. Even with the inevitable standardization of those components, interfaces, and even complete system architectures, the support for reuse of hardware and software components is limited. This potentially makes the design of networked embedded automotive applications labor-intensive, error-prone, and expensive. This necessitates the development of component-based design integration methodologies. An interesting approach is based on platform-based design [8,48], with a view for automotive applications. Some industry standardization initiatives include: OSEK/VDX with its OSEKTime OS (OSEK/VDX Time-Triggered Operating Systems) [49]; OSEK/VDX Communication [50] which specifies a communication layer that defines common software interfaces and common behavior for internal and external communications among application processes; and OSEK/VDX FTCom (Fault-Tolerant Communication) [51] – a proposal for a software layer to provide services to facilitate development of fault-tolerant applications on top of time-triggered networks.

One of the main bottlenecks in the development of safety-critical systems is the software development process. The automotive industry clearly needs a software development process model and supporting tools suitable for the development of safety-critical software. At present, there are two potential candidates, MISRA (Motor Industry Software Reliability Association) [52], which published recommended practices for safe automotive software. The recommended practices, although automotive specific, do not support X-by-Wire. IEC 61508 [53] is an international standard for electrical, electronic, and programmable electronic safety related systems. IEC 61508 is not automotive specific, but broadly accepted in other industries.

Summary

This section introduced embedded systems, their design in the context of system-on-chip, as well as networked embedded systems and applications for in-car controls and automation.

In this section we have defined system-on-chip and surveyed a large number of the issues involved in its design. An outline of the important methods and processes involved in SoC design define a methodology that can be adopted by design groups and adapted to their specific requirements. Productivity in SoC design demands high levels of design reuse and the existence of third-party, and internal IP groups, and the chance to create a library of reusable IP blocks (true virtual components) are all possible for most design groups today. The wide variety of design disciplines involved in SoC mean that unprecedented collaboration between designers of all backgrounds, from systems experts through embedded software designers through architects through HW designers, is required. But the rewards of SoC justify the effort required to succeed.

This chapter has also presented an overview of trends for networking of embedded systems, their design, and selected application domain specific network technologies. The networked embedded systems appear in a variety of application domains to mention automotive, train, aircraft, office building, and industrial automation. The networked embedded systems pose a multitude of challenges in their design, particularly for safety critical applications, deployment, and maintenance. The majority of the development environments and tools for specific networking technologies do not have firm foundations in computer science, and software engineering models and practices making the development process labor-intensive, error-prone, and expensive.

References

1. M. Hunt and J. Rowson, "Blocking in a system on a chip," *IEEE Spectrum*, vol. 33, no. 11, pp. 35–41, 1996.
2. R. Rajsuman, *System-on-a-Chip Design and Test*, Norwood, MA: Artech House, 2000.
3. M. Keating and P. Bricaud, *Reuse Methodology Manual for System-on-a-Chip Designs*, Dordrecht: Kluwer Academic Publishers, 3rd ed., 2002, (1st ed., 1998; 2nd ed., 1999).
4. International Technology Roadmap for Semiconductors (ITRS), 2001 ed., URL: http://public.itrs.net/.

5. Virtual Socket Interface Alliance, on the web at URL: http://www.vsia.org. This includes access to its various public documents, including the original Reuse Architecture document of 1997, as well as more recent documents supporting IP reuse released to the public domain.
6. B. Bailey, G. Martin, and T. Anderson, Eds., *Taxonomies for the Development and Verification of Digital Systems*, New York: Springer, 2005.
7. The Virtual Component Exchange (VCX), Web URL: http://www.thevcx.com/.
8. H. Chang, L. Cooke, M. Hunt, G. Martin, A. McNelly, and L. Todd, *Surviving the SOC Revolution: A Guide to Platform-Based Design*, Dordrecht: Kluwer Academic Publishers, 1999.
9. K. Keutzer, S. Malik, A. R. Newton, J. Rabaey, and A. Sangiovanni-Vincentelli, "System-level design: orthogonalization of concerns and platform-based design," *IEEE Trans. CAD ICs Syst.*, vol. 19, no. 12, p. 1523, 2000.
10. A. Sangiovanni-Vincentelli and G. Martin, "Platform-based design and software design methodology for embedded systems," *IEEE Des. Test Comput.*, vol. 18, no. 6, pp. 23–33, 2001.
11. *IEEE Design and Test Special Issue on Platform-Based Design of SoCs*, vol. 19, no. 6, pp. 4–63, 2002.
12. G. Martin and F. Schirrmeister, "A design chain for embedded systems," *IEEE Comput. Embed. Syst. Column*, vol. 35, no. 3, pp. 100–103, 2002.
13. G. Martin and H. Chang, Eds., *Winning the SOC Revolution: Experiences in Real Design*, Dordrecht: Kluwer Academic Publishers, 2003.
14. P. Lysaght, "FPGAs as Meta-platforms for Embedded Systems," *Proc. IEEE Conf. Field Program. Technol.*, Hong Kong, December 2002.
15. T. Groetker, S. Liao, G. Martin, and S. Swan, *System Design with System C*, Dordrecht: Kluwer Academic Publishers, 2002.
16. J. Bergeron, *Writing Testbenches*, 3rd. ed., Dordrecht: Kluwer Academic Publishers, 2003.
17. G. Martin, and C. Lennard, Invited CICC paper, "Improving embedded SW design and integration for SOCs," *Custom Integrated Circuits Conference*, May 2000, pp. 101–108.
18. P. Rashinkar, P. Paterson, and L. Singh, *System-on-a-Chip Verification: Methodology and Techniques*, Dordrecht: Kluwer Academic Publishers, 2001.
19. F. Balarin, M. Chiodo, P. Giusto, H. Hsieh, A. Jurecska, L. Lavagno, C. Passerone, A. Sangiovanni-Vincentelli, E. Sentovich, K. Suzuki, and B. Tabbara, *Hardware-Software Co-Design of Embedded Systems: The POLIS Approach*, Dordrecht, The Netherlands: Kluwer Academic Publishers, 1997.
20. S. Krolikoski, F. Schirrmeister, B. Salefski, J. Rowson, and G. Martin, *Methodology and Technology for Virtual Component Driven Hardware/Software Co-Design on the System Level*, paper 94.1, ISCAS 99, Orlando, Florida, May 30–June 2, 1999.
21. G. Martin and B. Salefski, "System level design for soc's: a progress report-two years on," in *System-on-Chip Methodologies and Design Languages*, J. Mermet, Ed., Dordrecht: Kluwer Academic Publishers, 2001, pp. 297–306, chap. 25.
22. G. Martin, "Productivity in VC Reuse: Linking SOC platforms to abstract systems design methodology," in *Virtual Component Design and Reuse*, edited by Ralf Seepold and Natividad Martinez Madrid, Dordrecht: Kluwer Academic Publishers, 2001, pp. 33–46, chap. 3.
23. J. Nurmi, H. Tenhunen, J. Isoaho and A. Jantsch Eds., *Interconnect-Centric Design for Advanced SoC and NoC*, Dordrecht: Springer, 2004.
24. A. Jantsch and H. Tenhunen, Eds., *Networks on Chip*, Dordrecht: Kluwer Academic Publishers, 2003.
25. V. Kithail, S. Aditya, R. Schreiber, B. Ramakrishna Rau, and D.C. Cronquist and M. Sivaraman, "PICO: automatically designing custom computers," *IEEE Comput.*, vol. 35, no. 9, pp. 39–47, 2002.
26. T.J. Callahan, J.R. Hauser, and J. Wawrzynek, "The Garp architecture and C compiler," *IEEE Comput.*, vol. 33, no. 4, pp. 62–69, 2000.
27. DATE 2002 Proceedings, Session 1A: "How to Choose Semiconductor IP?: Embedded Processors, Memory, Software, Hardware," in *Proc. DATE 2002*, Paris, March 2002, pp. 14–17.
28. "The industrial communication systems, special issue," in *Proc. IEEE*, R. Zurawski, Ed., vol. 93, no. 6, June 2005.

29. J.-D. Decotignie, P. Dallemagne, and A. El-Hoiydi, "Architectures for the interconnection of wireless and wireline fieldbusses," in *Proc. Fourth IFAC Conf. Fieldbus Syst. Appl.*, (FET 2001), Nancy, France, 2001.
30. H. Zimmermann, "OSI reference model: the ISO model of architecture for open system interconnection," *IEEE Trans. Commun.*, vol. 28, no. 4, pp. 425–432, 1980.
31. F. Simonot-Lion, "In-car embedded electronic architectures: how to ensure their safety," in *Proc. fifth IFAC Int. Conf. Fieldbus Syst. Appl. — FeT'2003*, Aveiro, Portugal, July 2003.
32. X-by-Wire Project, Brite-EuRam 111 Program, X-By-Wire — Safety Related Fault Tolerant Systems in Vehicles, Final Report, 1998.
33. TTTech Computertechnik GmbH, *Time-Triggered Protocol TTP/C, High-Level Specification Document, Protocol Version 1.1*, November 2003, www.tttech.com.
34. FlexRay Consortium, *FlexRay Communication System, Protocol Specification, Version 2.0*, June 2004, www.flexray.com.
35. H. Kopetz and G. Bauer, "The time triggered architecture," *Proc. IEEE*, vol. 91, no. 1, pp. 112–126, 2003.
36. H. Kopetz et al., *Specification of the TTP/A Protocol*, University of Technology Vienna, September 2002.
37. International Standard Organization, 11898-4, *Road Vehicles — Controller Area Network (CAN) — Part 4: Time-Triggered Communication*, ISO, 2000.
38. Society of Automotive Engineers, "J2056/1 class C application requirements classifications," *SAE Handbook*, 1994.
39. Society of Automotive Engineers, "J2056/2 survey of known protocols," *SAE Handbook*, 2, 1994.
40. A. Rajnak, "The LIN standard," *The Industrial Communication Technology Handbook*, Boca Raton, FL: CRC Press, 2005.
41. Society of Automotive Engineers, *Class B data communications network interface — SAE J1850 Standard — rev. November 96*, 1996.
42. International Standard Organization, *ISO 11519-2, Road Vehicles — Low Speed Serial Data Communication — Part 2: Low Speed Controller Area Network, ISO*, 1994.
43. International Standard Organization, *ISO 11519-3, Road Vehicles — Low Speed Serial Data Communication — Part 3: Vehicle Area Network (VAN), ISO*, 1994.
44. International Standard Organization, *ISO 11898, Road Vehicles — Interchange of Digital Information — Controller Area Network for High-Speed Communication, ISO*, 1994.
45. SAE J1939 Standards Collection, www.sae.org.
46. MOST Cooperation, *MOST Specification Revision 2.3*, August 2004, www.mostnet.de.
47. www.idbforum.org.
48. K. Keutzer, S. Malik, A. R. Newton, J. Rabaey, and A. Sangiovanni-Vincentelli, "System level design: orthogonalization of concerns and platform-based design," *IEEE Trans. Comput. Aided Des. Integr. Circ. Syst.*, vol. 19, no. 12, December 2000.
49. OSEK Consortium, *OSEK/VDX Operating System, Version 2.2.2*, July 2004, www.osek-vdx.org.
50. OSEK Consortium, *OSEK/VDX Communication, Version 3.0.3*, July 2004, www.osek-vdx.org.
51. OSEK Consortium, *OSEK/VDX Fault-Tolerant Communication, Version 1.0*, July 2001, www.osek-vdx.org.
52. www.misra.org.uk.
53. International Electrotechnical Commission, *IEC 61508:2000, Parts 1–7, Functional Safety of Electrical/Electronic/Programmable Electronic Safety-Related Systems*, 2000.

16.2 Real-Time in Embedded Systems

Hans Hansson, Mikael Nolin, and Thomas Nolte

This chapter will provide an introduction to issues, techniques and trends in real-time systems. We will discuss the design of real-time systems, real-time operating systems, real-time scheduling, real-time communication, real-time analysis and both testing and debugging of real-time systems. For each area, state-of-the-art tools and standards are presented.

Introduction

Consider the *airbag* in the steering wheel of your car. After the detection of a crash and and only then, it should inflate just in time to softly catch your head to prevent it from hitting the steering wheel. It should not inflate too early, since this would make the airbag deflate before it can catch you; nor should it inflate too late, since the exploding airbag then could injure you by blowing up in your face or catch you too late to prevent your head from banging into the steering wheel.

The computer-controlled airbag system is an example of a real-time system (RTS). RTSs come in many different varieties, including vehicles, telecommunication systems, industrial automation systems and household appliances.

There is no commonly agreed upon definition of an RTS, but the following characterization is almost universally accepted:

- RTSs are computer systems that physically interact with the real world.
- RTSs have requirements on the timing of these interactions.

Typically, those real-world interactions occur by sensors and actuators rather than through the keyboard and screen of a standard personal computer.

Real-time requirements typically demand that an interaction occur within a specified time. This is quite different from requiring the interaction to occur as quickly as possible.

Essentially, all RTSs are embedded in products, and the vast majority of *embedded computer systems* use RTSs. RTSs are the dominant application for computer technology. More than 99 percent of manufactured processors (more than 8 billion in 2000) are used in embedded systems [Hal00].

Returning to the airbag system, in addition to being a RTS it is a safety-critical system. That means it is a system that, due to severe risks of damage, has strict *Quality of Service* (QoS) requirements, including requirements on its functional behavior, robustness, reliability and timeliness.

A certain response to an interaction always must occur within some prescribed time. For instance, the airbag charge must detonate between 10 and 20 ms after detection of a crash. Violating this must be avoided, since it would lead to something unacceptable, like you having to spend a couple of months in the hospital. A system designed to meet strict timing requirements is often referred to as a *hard real-time system*. In contrast, systems for which occasional timing failures are acceptable –possibly because this will not lead to something terrible – are termed *soft real-time systems*.

A comparison between hard and soft RTSs that highlights the difference between these extremes is shown in Table 16.1. A typical hard real-time system could, in this context, be an engine control system. This system must operate with μs-precision and will severely damage the engine if timing requirements fail by more than a few *ms*. A typical soft real-time system could be a banking system, where timing is important but no strict deadlines and some timing variations are acceptable.

TABLE 16.1 Typical characteristics of Hard and Soft Real-Time Systems [Kop03]

Characteristic	Hard Real-Time	Soft Real-Time
Timing requirements	hard	soft
Pacing	environment	computer
Peak-load performance	predictable	degraded
Error detection	system	user
Safety	critical	non-critical
Redundancy	active	standby
Time granularity	millisecond	second
Data files	small	large
Data integrity	short term	long term

Unfortunately, it is impossible to build real systems that satisfy hard real-time requirements. Due to the imperfection of hardware (and of designers), any system may break. The best that can be achieved is a system that has a high probability to provide the intended behavior during a finite time interval.

On the conceptual level, hard real-time makes sense since it implies a certain amount of rigor in the way the system is designed. This implies an obligation to prove that strict timing requirements are met, at least under some simplified but realistic assumptions.

Since the early 1980s, a substantial research effort has provided a sound theoretical foundation (RTS, Klu) and useful results for the design of hard real-time systems. Most notably, hard real-time system scheduling has evolved into a mature discipline. It uses abstract, but realistic, models of tasks executed on a single CPU, a multiprocessor or distributed computer systems, together with associated methods for timing analysis. Such *schedulability analysis*, including the well-known rate-monotonic analysis (LL73, KRP$^+$98, ABD$^+$95), have found significant uses in some industrial segments.

However, hard real-time scheduling is not the cure for all RTSs. Its main weakness is that it is based on analysis of the worst possible scenario. For safety-critical systems, this is of course a must, but for other systems, where general customer satisfaction is the main criterion, it may be too costly to design the system for a worst-case scenario that may not occur during the system's lifetime.

If we look at the other end of the spectrum, we find the *best-effort approach*, which is still the dominant industry approach. The essence of this approach is to implement the system using some best practice and then use measurements, testing and tuning to make sure that the system is of sufficient quality. Such a system should satisfy some soft real-time requirement; the weakness is that we do not know which requirement. However, compared to the hard real-time approach, the system can be better optimized for available resources. A further difference is that hard real-time system methods essentially are applicable to static configurations only. It is less problematic to handle dynamic task creation in best-effort systems.

Having identified the weaknesses of the hard real-time and best-effort approaches, efforts are being put into more flexible techniques for soft real-time systems. These techniques provide the ability to analyze (similar to hard real-time), together with flexibility and resource efficiency (like best-effort). The basis for these flexible techniques are often quantified, *Quality-of-Service* (QoS) characteristics. These are typically related to non-functional aspects, such as timeliness, robustness, dependability and performance. To provide a specified QoS, resource management is needed. Such QoS-management is either handled by the application, by the operating system, by some middleware or by a mix of the above. QoS management often is a flexible online mechanism that dynamically adapts the resource allocation to balance conflicting QoS-requirements.

Design of Real-Time Systems

The main issue in designing RTSs is timeliness, ensuring that the system performs its operations at proper instances in time. When timeliness is not considered at the design phase, it is virtually impossible to analyze and predict the timely behavior of the RTS. This section presents some important architectural issues for embedded RTSs, together with some supporting commercial tools.

Reference Architecture

A generic system architecture for a RTS is depicted in Figure 16.9. This architecture is a model of any computer-based system interacting with an external environment by sensors and actuators.

Since our focus is on the RTS, we will look more into different organizations of that part of the generic architecture in Figure 16.9. The simplest RTS is a single processor, but, in many cases, the RTS is a distributed computer system consisting of a set of processors interconnected by a communications network. There could be several reasons for making an RTS distributed, including

- the physical distribution of the application,
- the computational requirements, which may not be conveniently provided by a single CPU,
- the need for redundancy to meet availability, reliability, or other safety requirements, or
- to reduce the system's cabling.

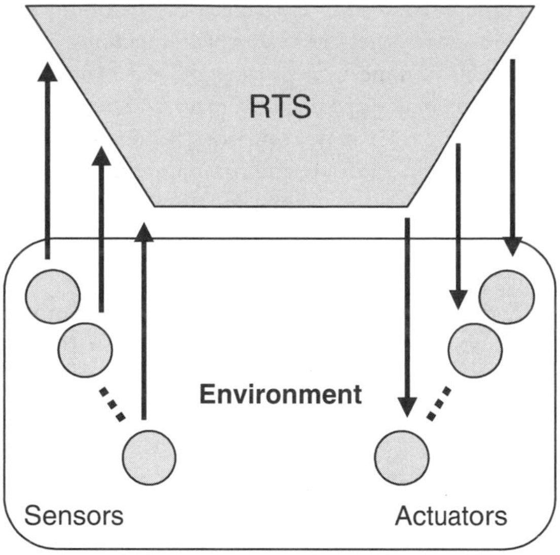

FIGURE 16.9 A generic real-time system architecture.

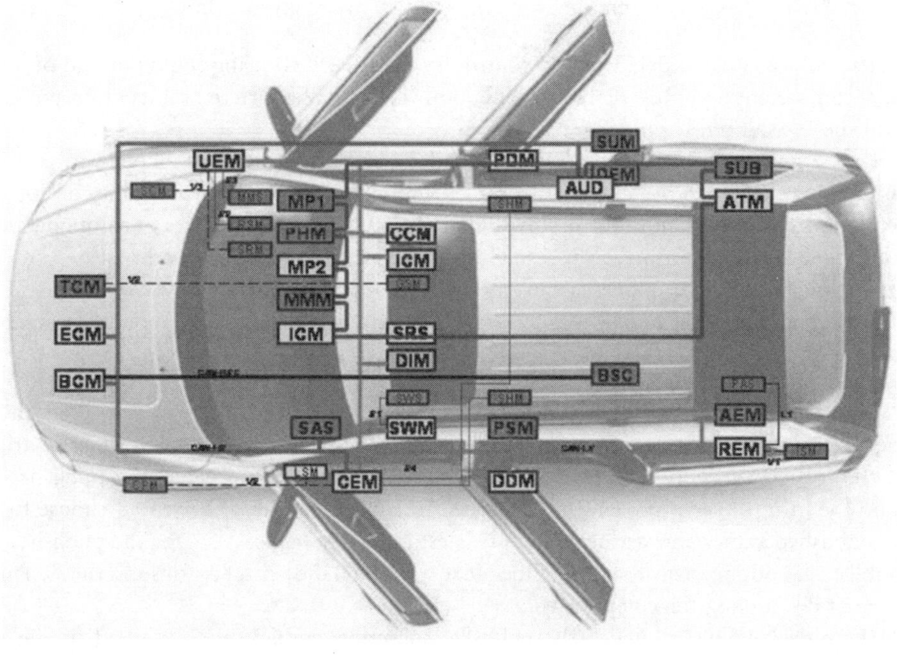

FIGURE 16.10 Network infrastructure of the Volvo XC90.

Figure 16.10 shows an example of a distributed RTS. In a modern car, like the one depicted in the figure, there are some 20 to 100 computer nodes (in the automotive industry, they are called electronic control units or ECUs) interconnected with one or more communication networks. The initial motivation for this type of electronic architecture in cars was the need to reduce the amount of cabling. However, the electronic architecture has led to other significant improvements, including substantial pollution reduction and new safety mechanisms such as computer-controlled electronic stabilization programs (ESPs). The current development is towards making the most safety-critical vehicle functions, including braking and steering, completely computer-controlled. This is done by removing the mechanical connections between steering wheel and front wheels and between brake pedals and brakes and replacing them with computers and computer networks. Meeting the stringent safety requirements for these functions will require careful introduction of redundancy mechanisms in hardware and communications, as well as software. For the latter, a safety-critical system architecture is needed (an example of such an architecture is TTA [KB03]).

Models of Interaction

Above, we presented the physical organization of a RTS, but for an application programmer, this is not the most important aspect of the system architecture. From an application programmer's perspective, system architecture is given by the *execution paradigm* (execution strategy) and the *interaction model* in the system. In this section, we describe what an interaction model is and how it affects the real-time properties of a system. In Section "Execution Strategies," we discuss the execution strategies in RTSs.

An ineraction model describes the rules by which components interact with each other (in this section we will use the term component to denote any type of software unit, such as a task or a module). The interaction model can govern control flow and data flow between system components. One of the most important design decisions for all system types is deciding which interaction models to use (sadly, this decision is often implicit and hidden in the system's architectural description).

When designing RTSs, attention should be paid to the timing properties of the interaction models chosen. Some models have a more predictable and robust timing behavior than do other models. Examples of some of the more predictable models also commonly used in RTS design are given below.

Pipes-and-Filters. In this model, data and control flow are specified using the input and output ports of components. A component becomes eligible for execution when data has arrived on its input ports. When the component finishes execution, it produces output on its output ports.

This model fits many types of control programs, and control laws are easily mapped to this interaction model. It has gained widespread use in the real-time community. The real-time properties of this model also are quite useful. Since both data and control flows unidirectionally through a series of components, the order of execution and end-to-end timing delay usually become predictable. The model also provides a high degree of decoupling in time; components can often execute without having to worry about timing delays caused by other components. Hence, it is usually straightforward to specify the compound timing behavior of a set of components.

Publisher-Subscriber. The publisher-subscriber model is similar to the pipes-and-filters model but it usually decouples data and control flow. A subscriber can usually choose different forms for triggering its execution. If the subscriber chooses to be triggered on each new published value, the publisher-subscriber model takes on the form of the pipes-and-filters model. In addition, a subscriber could choose to ignore the timing of the published values and decide to use the latest published value. Also, for the publisher-subscriber model, the publisher is not necessarily aware of the identity, or even the existence, of subscribers. This provides a higher degree of decoupling for components.

Similar to the pipes-and-filters model, the publisher-subscriber model provides good timing properties. However, a prerequisite for systems analysis using this model is that subscriber-components make explicit the values they will subscribe to, even though this is not mandated by the model itself. However, when using the publisher-subscriber model for embedded systems, subscription information commonly is available (this information is used, for instance, to decide which values are to be published over a communications network and to decide the receiving nodes of those values).

Blackboard. The blackboard model allows variables to be published on a globally available blackboard area. It resembles the use of global variables. The model allows any component to read or write values to blackboard variables. Hence, the software engineering qualities of the blackboard model are questionable.

Nevertheless, it is a model that is commonly used, and in some situations it provides a pragmatic solution to problems difficult to address with more stringent interaction models.

Software engineering aspects aside, the blackboard model does not introduce extra elements of unpredictable timing. On the other hand, the model's flexibility does not help engineers to achieve predictable systems. Since the model does not address the control flow, components can execute relatively undisturbed and decoupled from other components.

At the other end of the spectrum are interaction models that increase the unpredictable timing of the system. These models should, if possible, be avoided when designing RTSs. The two most notable, and commonly used, are:

Client-Server. In the client-server model, a client asynchronously invokes a service of a server. The service invocation passes the control flow (plus any input data) to the server, and control stays at the server until it has completed the service. When the server is done, the control flow (and any return data) is returned to the client, which in turn resumes execution.

The client-server model has inherently unpredictable timing. Since services are invoked asynchronously, it is difficult to *a priori* assess the load on the server for a certain service invocation. Thus, it is difficult to estimate the delay of the service invocation, and it is difficult to estimate the response time of the client. This is furthermore complicated by the fact that most components often behave both as clients and servers (a server often uses other servers to implement its own services); leading to complex and unanalyzable control flow paths.

Message-Boxes. A component can have a set of message-boxes, where components communicate by posting messages in each other's message-boxes. Messages are typically handled in FIFO order or in priority order, where the sender specifies a priority. Message passing does not change the flow of control for the sender. However, a component that tries to receive a message from an empty message-box blocks on that message-box until a message arrives (often the receiver can specify a timeout to prevent indefinite blocking).

From a sender's point of view, the message-box model has similar problems as the client-server model. The data sent by the sender (and the action that the sender expects the receiver to perform) may be delayed in an unpredictable way when the receiver is highly loaded. Also, the asynchronous nature of the message-passing makes it difficult to foresee the receiver's load at any particular moment.

From the receiver's point of view, the reading of message-boxes also is unpredictable because the receiver may block on the message-box. Since message boxes often are of limited size, a highly loaded receiver also risk losing a message. Which messages are lost is another source of unpredictability.

Execution Strategies

There are two main execution paradigms for RTSs: *time-triggered* and *event-triggered*. With time-triggered execution, activities occur at predefined instances of time. For instance,, a specific sensor value is read exactly every 10 ms. Exactly 2 ms later, the corresponding actuator receives an updated control parameter. In an event-triggered execution, actions are triggered by event occurrences. For instance, when the level of toxic fluid in a tank reaches a certain level, an alarm will go off. It should be noted that the same functionality typically can be implemented in both examples. A time-triggered implementation of the above alarm would be to periodically read the level-measuring sensor and set of the alarm when the read level exceeds the maximum allowed. If alarms are rare, the time-triggered version will have much higher computational overhead than the event-triggered ones. On the other hand, the periodic sensor readings will facilitate detection of a malfunctioning sensor.

Time-triggered executions are used in many safety-critical systems with high dependability requirements (such as avionic control systems), whereas the majority of other systems are event-triggered. Dependability can be guaranteed in the event-triggered paradigm. Due to the observability provided by the exact timing of time-triggered executions, most experts argue for using time-triggered systems in ultra-dependable systems. The main argument against time-triggered executions is theit lack of flexibility and the requirement of preruntime schedule generation (which is a nontrivial and possibly time-consuming task).

Time-triggered systems are mostly implemented by simple, proprietary, table-driven dispatchers (XP90), but complete commercial systems that include design tools are also available (TTT, KG94). For an event-triggered paradigm, a large number of commercial tools and operating systems are available. Systems also can integrate the two execution paradigms, attempting to gain the best of both worlds: time-triggered dependability and event-triggered flexibility. One example is the Basement system (HLS96) and its associated real-time kernel Rubus (Arc).

Since computation in time-triggered systems are statically allocated both in space (to a specific processor) and time, some configuration tool is frequently used. This tool assumes that the computations are packaged into schedulable units (corresponding to tasks or threads in an event-triggered system). Typically, basement, computations are control-flow based and defined by sequences of schedulable units. Each unit performs a computation based on its inputs and produces outputs to the next unit in sequence. The system is configured by defining the sequences and their timing requirements. The configuration tool will then automatically (if possible[1]) generate a schedule guaranteeing that all timing requirements are met.

Event-triggered systems typically have richer and more complex APIs (application programmer interfaces), defined by the user operating system and middleware.

Component-Based Design of Real-Time Systems

Component-based design (CBD) of software systems is an interesting idea for software engineering in general and for the engineering of RTSs in particular. In CBD, a *software component* encapsulates some functionality. That functionality is only accessed though the component's *interface*. A system is composed by assembling a set of components and connecting their interfaces.

The reason CBD could prove more useful for real-time systems is the possibility to extend components with *introspective interfaces*. An introspective interface does not provide any functionality *per se*. Instead, the interface can retrieve information about *extra-functional* properties of the component. Extra-functional properties can include attributes such as memory consumption, execution times and task periods. For RTS, timing properties are of particular interest.

Unlike the functional interfaces of components, introspective interfaces can be available offline, during the component's assembly phase. This way, the timing attributes of the system components can be obtained at design time and tools to analyze the system's timing behavior can be used. If the introspective interfaces are also available online, they could be used in admission control algorithms, for instance. An admission control could query new components for their timing behavior and resource consumption before deciding to accept new components to the system.

Unfortunately, many industry standard software techniques are based on the client-server or the message-box models of interaction, which we deemed unfit for RTSs in Section "Models of Interaction" This is especially true for the most commonly used component models. For instance, the Corba Component Model (CCM) (OMG02), Microsoft's COM (Mica) and .NET (Micb) models, and Java Beans (SUN) all have the client-server model as their core model. None of these component technologies allow the specification of extra-functional properties through introspective interfaces. From the real-time perspective, the biggest advantage of CBD is void for these technologies.

However, there are numerous research projects addressing CBD for real-time and embedded systems (Sta01, vO02, MSZ02, SVK97). These projects address the issues left behind by the existing commercial technologies, such as timing predictability (using suitable computational models), support for offline analysis of component assemblies and better support for resource-constrained systems. Often, these projects strive to remove the considerable runtime flexibility provided by existing technologies. This runtime flexibility is judged to be the foremost contributor to unpredictability (the flexibility adds to the runtime complexity and prevents CBD for resource-constrained systems).

---

[1]This scheduling problem is theoretically intractable, so the configuration tool will have to rely on some heuristics which works well in practice, but do not guarantee to find a solution in all cases when there is a solution.

Tools for Design of Real-Time Systems

In the industry, the term *real-time system* is highly overloaded and can mean anything from interactive systems to super-fast, or embedded, systems. Consequently, it is not easy to judge what tools are suitable for developing RTSs (as we define real-time in this chapter).

For instance, UML (OMG03b) is commonly used for software design. However, UML's focus is mainly on client-server solutions and it has proven unacceptable for RTSs design. As a consequence, UML-based tools that extend UML with constructions suitable for real-time programs have emerged. The two most known such products are Rational's Rose RealTime (Rat) and i-Logix' Rhapsody (IL). These tools provide UML-support with the extension of *real-time profiles.* While giving real-time engineers access to suitable abstractions and computational models, these tools do not provide means to describe timing properties or requirements in a formal way. Thus, they do not allow automatic verification of timing requirements.

TeleLogic provides programming and design support using the language SDL (Tel). SDL was originally developed as a specification language for the telecom industry and is suitable for describing complex reactive systems. However, the fundamental model of computation is the message-box model, which has an inherently unpredictable timing behavior. For soft embedded RTSs, SDL can offer time- and space-efficient implementations.

For resource constrained hard RTSs, design tools are provided by Arcticus Systems (Arc), TTTech (TTT) and Vector (Veca). These tools are instrumental during system design and implementation and provide timing analysis techniques allowing verification of the system or parts of it. However, these tools are based on proprietary formats and processes and have reached a limited customer base, mainly within the automotive industry.

In the near future, UML2 will become an adopted standard (OMG03c). UML2 has support for computational models suitable for RTSs. This support comes mainly in the form of *ports* that have *protocols* associated to them. Ports are either *provided* or *required*, allowing type-matching of connections between components. UML2 also includes much of the concepts from Rose RealTime, Rhapsody and SDL. Other, future design techniques that are expected to have an impact on the design of RTSs include, the EAST/EEA *Architecture Description Language* (EAST-ADL) (ITE). The EAST-ADL is being developed by the automotive industry and is a description language covering the complete development cycle of distributed, resource constrained, safety critical RTSs. Tools to support development with EAST-ADL (which is a UML2 compliant language) are expected to be provided by automotive tool vendors such as ETAS (ETA), Vector (Vecb) and Siemens (Sie).

Real-Time Operating-Systems

A *Real-Time Operating System* (RTOS) provides services for resource access and resource sharing that are similar to a general-purpose operating-system. An RTOS, provides additional services suited for real-time development and also supports the development process for embedded-systems development. Using a general-purpose operating-system when developing RTSs have several drawbacks:

- High resource utilization, such as large RAM and ROM footprints, and high internal CPU demand.
- Difficulty in accessing hardware and devices in a timely manner, with no application-level control over interrupts.
- The lack of services to allow timing-sensitive interactions between different processes.

Typical Properties of RTOSs

The state of practice in real-time operating systems (RTOS) is reflected in (FAQ). Not all operating systems are RTOS. An RTOS is typically multi-threaded and preemptible, there has to be thread priority, predictable thread synchronization must be supported, priority inheritance should be supported, and the OS behavior should be known (Art03). The interrupt latency, worst-case execution time of system calls and maximum time during which interrupts are masked must be known. A commercial RTOS is usually marketed as the runtime component of an embedded development platform.

As a general rule of thumb, RTOSs are:

Suitable for resource-constrained environments. RTSs typically operate in such environments.

Most RTOSs can be configured preruntime (at compile time) to include only a subset of the total functionality. The application developer can choose to leave out unused portions of the RTOS to save resources.

RTOSs typically store much of their configuration in ROM. This is done for two purposes: (1) to minimize use of expensive RAM memory, and (2) to minimize the risk that critical data will be overwritten by an erroneous application.

Giving the application programmer easy access to hardware features. These include interrupts and devices.

Most often the RTOSs give the application programmer a means to install interrupt service routines (ISRs) during compile time run time. The RTOS leaves all interrupt handing to the application programmer, allowing fast, efficient and predictable handling of interrupts.

In general-purpose operating systems, memory-mapped devices normally are protected from direct access using the MMU (memory management unit) of the CPU. This forces all device accesses to go through the operating system. RTOSs typically do not protect such devices and allow the application to directly manipulate the devices. This allows faster and more efficient access to the devices. (However, this efficiency comes at the price of an increased risk of erroneous use of the device.)

Providing services that allow implementation of timing sensitive code. An RTOS typically has many mechanisms to control the relative timing between different system processes. Notably, an RTOS has a real-time process scheduler whose function is to make sure that the processes execute in the way the application programmer intended. We will elaborate more below on the issues of scheduling.

An RTOS also provides mechanisms to control the processes' relative performance when accessing shared resources. This can be done, for instance, by priority queues instead of plain FIFO-queues used in general-purpose operating systems. Typically, an RTOS supports one or more real-time resource-locking protocols, such as priority inheritance or priority ceiling.

Tailored to fit the embedded systems development process. RTSs are usually constructed in a *host environment* different from the *target environment*, or *cross-platform* development. It is typical that the whole memory image, including both RTOS and one or more applications, is created at the host platform and downloaded to the target platform. Hence, most RTOSs are delivered as source code modules or pre-compiled libraries statically linked with the applications at compile time.

Mechanisms for Real-Time

One of the most important functions of an RTOS is to arbitrate access to shared resources so that the timing behavior of the system becomes predictable. The two most obvious resource that the RTOS manages access to are:

- The CPU. The RTOS should allow processes to execute in a predictable manner.
- Shared memory areas. The RTOS should resolve contention to shared memory in a way that offers predictable timing.

The CPU access is arbitrated with a *real-time scheduling policy*. Section "Real-Time Scheduling" will describe real-time scheduling policies. Examples of scheduling policies that can be used in real-time systems are priority scheduling, deadline scheduling and rate scheduling. Some of these policies directly use timing attributes (like deadlines) to perform scheduling decisions, while other policies uses scheduling parameters (like priority, rate or bandwidth) that indirectly affect the timing of the tasks.

A special form of scheduling that is also useful for real-time systems is table driven, or static, scheduling. Table-driven scheduling is described further in Section "Offline Schedulers". In table-driven scheduling, arbitration decisions are made offline and the RTOS scheduler just follows a simple table. This gives good timing predictability, albeit at the expense of system flexibility.

The most important aspect of a real-time scheduling policy is that is should provide a means to analyze the timing behavior of the system. Scheduling in general-purpose operating systems normally emphasizes properties such as fairness, throughput and guaranteed progress; these properties may be adequate in their own respect, but they usually conflict with the requirement that an RTOS should provide timing predictability.

Shared resources (such as memory areas, semaphores and mutexes) also are arbitrated by the RTOS. When a task locks a shared resource, it will *block* all other tasks that subsequently try to lock the resource. To achieve predictable blocking times, special real-time resource-locking protocols have been proposed [But97, BW96].

Priority Inheritance Protocol (PIP). PIP makes a low priority task inherit the priority of any higher priority task that becomes blocked on a resource locked by the lower priority task.

This is a simple and straightforward method to lower blocking time. However, it is computationally intractable to calculate the worst-case blocking (which may be infinite since the protocol does not prevent deadlocks). For hard RTSs or when timing performance needs to be calculated *a priori*, the PIP is not adequate.

Priority Ceiling inheritance Protocol (PCP). PCP associates to each resource a *ceiling value* equal to the highest priority of any task that may lock the resource. By clever use of the ceiling values of each resource, the RTOS scheduler will manipulate task priorities to avoid the problems of PIP.

PCP guarantees freedom from deadlocks, and the worst case blocking is relatively easy to calculate. However, the computational complexity of keeping track of ceiling values and task priorities gives PCP high runtime overhead.

Immediate ceiling priority Inheritance Protocol (IIP). IIP, also associates to each resource a *ceiling value* equal to the highest priority of the task that may lock the resource. However, compared to PCP, in IIP a task is immediately assigned the ceiling priority of the resource it is locking.

IIP has the same real-time properties as the PCP (including the same worst-case blocking time[1]). However, IIP is significantly easier to implement. In fact, for single node systems it is easier to implement IIP than any other resource-locking protocol (including non real-time protocols). In IIP no actual locks need to be implemented. It is enough for the RTOS to adjust the priority of the task that locks or releases a resource. IIP has other operational benefits. Notably, it paves the way for letting multiple tasks use the same stack area. Operating systems based on IIP can build systems with footprints that are extremely small [Ast, Liv99].

Commercial RTOSs

There are an abundance of commercial RTOSs. Most provide adequate mechanisms to enable development of RTSs. Some examples are Tornado/VxWorks (Win), LYNX (LYN), OSE (Sys), QNX (QNX), RT-Linux (RTL), and ThreadX (Log). The major problem with these tools is the rich set of primitives provided. These systems provide both primitives that are suitable for RTSs and primitives unfit for real-time systems or that should be used with great care. For instance, they usually provide multiple resource locking protocols, some of which are suitable and some of which are not suitable for real-time.

This richness becomes a problem when these tools are used by inexperienced engineers, when projects are large and when project management does not provide clear design guidelines or rules. In these situations, it is easy to use primitives that will contribute to timing unpredictability of the developed system. Rather, an RTOS should help the engineers and project managers by providing mechanisms that only help in designing predictable systems. There is an obvious conflict between the desire or need of RTOS manufacturers to provide rich interfaces and the stringency needed by designers of RTSs.

There is a smaller set of RTOSs designed to resolve these problems and, at the same time, allow extreme, lightweight implementations of predictable RTSs. The driving idea is to provide a small set of primitives that guides the engineers towards good design of their systems. Typical examples are the research RTOS Asterix (Ast) and the commercial RTOS SSX5 [Liv99]. These systems provide a simplified task model, where tasks

[1]However, the average blocking time will be higher in IIP than in PCP.

cannot suspend themselves (such as no sleep primitive) and tasks are restarted from their entry point on each invocation. The only resource locking protocol that is supported is IIP, and the scheduling policy is fixed priority scheduling. These limitations makes it possible to build an RTOS that is able to run 10 tasks using less that 200 bytes of RAM and, at the same time, gives predictable timing behavior [App98]. Other commercial systems that follow a similar principle of reducing the degree of freedom and promote stringent design of predictable RTSs includes Arcticus Systems' Rubus OS (Arc).

Many commercial RTOSs provide standard APIs. The most important RTOS-standards are RT-POSIX (IEE98), OSEK (Gro), and APEX (Air96). We will here only discuss POSIX, since it is the most widely adopted RTOS standard, but those interested in automotive and avionic systems should take a closer look at OSEK and APEX, respectively.

The POSIX standard is based on Unix, and its goal is portability of applications at the source code level. The basic POSIX services include task and thread management, file system management, input and output and event notification by signals. The POSIX real-time interface defines services facilitating concurrent programming and providing predictable timing behavior. Concurrent programming is supported by synchronization and communication mechanisms that allow predictability. Predictable timing behavior is supported by preemptive, fixed-priority scheduling, time management with high resolution and virtual memory management. Several restricted subsets of the standard, intended for different types of systems, have been defined, as have specific language bindings, such as for Ada (ISO95).

Real-Time Scheduling

Traditionally, real-time schedulers are divided into *offline* and *online* schedulers. Offline schedulers make all scheduling decisions before the system is executed. At runtime, a simple dispatcher is used to activate tasks according to an offline-generated schedule. Online schedulers, on the other hand, decide during execution, based on various parameters, which tasks should execute at any given time.

As there are many different schedulers developed in the research community, we have in this section focused on highlighting the main categories of schedulers readily available in existing RTOSs.

Introduction to Scheduling

A real-time system consists of a set of real-time programs, which in turn consists of a set of *tasks*. These tasks are sequential pieces of code, executing on a platform with limited resources. The tasks have different timing properties, including *execution times*, *periods* and *deadlines*. Several tasks can be allocated to a single processor. The scheduler decides, at each moment, which task to execute.

A real-time system can be *preemptive* or *nonpreemptive*. In a preemptive system, tasks can preempt each other, letting the task with the highest priority execute. In a non-preemptive system a task allowed to start will execute until its completion.

Tasks can be categorized into *periodic*, *sporadic* or *aperiodic*. Periodic tasks execute with a specified time period between task releases. Aperiodic tasks have no information saying when the task is to release. Usually, aperiodics are triggered by interrupts. Similarly, sporadic tasks have no time period, but in contrast with aperiodics, sporadic tasks have a known minimum time between releases. Typically, tasks that perform measurements are periodic, collecting some value every *n*th time unit. A sporadic task typically reacts to an event or interrupt that has a minimum inter-arrival time, such as an alarm or the emergency shutdown of a production robot. The minimum inter-arrival time can be constrained by physical laws or can be enforced by a hardware mechanism. If the minimum time between two consecutive events is not known, the event-handling task is classified as aperiodic.

A real-time *scheduler* schedules the real-time tasks sharing the same resource (e.g., a CPU or a network link). The goal of the scheduler is to make certain the timing requirements of these tasks are satisfied. The scheduler decides, based on the task timing properties, which task is to execute or use the resource.

Offline Schedulers

Offline schedulers, or table-driven schedulers, work in the following way: The schedulers create a schedule, or table, before the system is started offline. At runtime, a *dispatcher* follows the schedule and ensures that tasks are only executing at their predetermined time slots, according to the schedule.

By creating a schedule offline, complex timing constraints can be handled in a way that would be difficult to do online. The schedule will be used at runtime. Therefore, the online behavior of table-driven schedulers is deterministic. Because of this determinism, table-driven schedulers are commonly used in applications that have high safety-critical demands. Since the schedule is created offline, flexibility is limited, however, As soon as the system changes (due to the adding of functionality or a change of hardware), a new schedule has to be created and given to the dispatcher. To create new schedules is difficult and sometimes time-consuming.

There also are combinations of the predictable, table-driven scheduling and the more flexible, priority-based schedulers, and there are methods to convert one policy to another [FLD03, Arc, MTS02].

Online Schedulers

Scheduling policies that make their decisions during runtime are classified as *online schedulers*. These schedulers make their scheduling decisions based on some task properties or task priorities. Schedulers that base their scheduling decisions based on task priorities are called *priority-based schedulers*.

Priority-Based Schedulers. Using priority-based schedulers increases flexibility, compared to table-driven schedulers, since the schedule is created online and based on the currently active task's constraints. Hence, priority-based schedulers can cope with changes in workload and added functions, as long as a task set can be scheduled. However, the exact behavior of priority-based schedulers is harder to predict. Therefore, these schedulers are not used often in the most safety-critical applications.

Two common priority-based scheduling policies are *Fixed Priority Scheduling (FPS)* and *Earliest Deadline First (EDF)*. The difference between these scheduling policies is whether the priorities of the real-time tasks are fixed or can change during execution (or whether they are dynamic).

In FPS, priorities are assigned to the tasks offline before execution. The task with the highest priority among all tasks available for execution is scheduled. Some priority assignments are better than others. For instance, for a simple model with strictly periodic, non-interfering tasks that has deadlines equal to the task period, a Rate Monotonic (RM) priority assignment has been shown by Liu and Layland [LL73] to be optimal. In RM, the priority is assigned based on the task period. The shorter the period, the higher the priority assigned to the task.

Using EDF, the task with the nearest (or earliest) deadline among all available tasks is selected for execution. The priority is not fixed and changes with time. It has been shown that EDF is an optimal dynamic priority scheme [LL73].

Scheduling with Aperiodics. In order for the priority-based schedulers to cope with aperiodic tasks, different service methods have been presented. The objective of these service methods is to give a good average-response time for aperiodic requests while preserving the timing properties of periodic and sporadic tasks. These services are implemented using special *server* tasks. In the scheduling literature, many types of servers are described. Using FPS, for instance, the Sporadic Server (SS) is presented by Sprunt *et al.* [SSL89]. SS has a fixed priority chosen according to the RM policy. Using EDF, Dynamic Sporadic Server (DSS) [SB94, SB96] extends SS. Other EDF-based schedulers are the Constant Bandwidth Server (CBS), presented by Abeni (AB98), and the Total Bandwidth Server (TBS) by Spuri and Buttazzo [SB94, SBS95]. Each server is characterized partly by its unique mechanism for assigning deadlines and partly by a set of variables to configure the server. Examples of such variables are bandwidth, period and capacity.

Real-Time Communications

Real-time communication aims at providing timely and deterministic communication of data between distributed devices. In many cases, there are requirements to provide guarantees of the real-time properties of these transmissions. There are real-time communication networks of different types, ranging from small fieldbus-based control systems to large Ethernet/Internet-distributed applications. There is also a growing interest in wireless solutions.

In this section we give a brief introduction to communications in general and real-time communications in particular. We then provide an overview of the most-popular real-time communication systems and protocols, both in industry and academia.

Communications Techniques

Common access mechanisms in communication networks are CSMA/CD (carrier sense multiple access/collision detection), CSMA/CA (carrier sense multiple access/collision avoidance), TDMA (time division multiple access), tokens, central maste, and mini-slotting. These techniques are all used both in real-time and non real-time communication, and each technique has different timing characteristics.

In CSMA/CD, collisions between messages are detected, causing the messages involved to be retransmitted. CSMA/CD is used in Ethernet. CSMA/CA, on the other hand, avoids collisions and is more deterministic in its behavior compared to CSMA/CD. CSMA/CA is more suitable for hard real-time guarantees, while CSMA/CD can provide soft real-time guarantees. Examples of networks that implement CSMA/CA are CAN and ARINC 629.

TDMA uses time to achieve exclusive usage of the network. Messages are sent at predetermined instances in time. Hence, the behavior of TDMA-based networks is deterministic and suitable to provide real-time guarantees. One example of a TDMA-based real-time network is TTP.

An alternative way of eliminating collisions on the network is to use tokens. In token-based networks, only the owner (unique within the network) of the token is allowed to send messages on the network. Once the token holder is done or has used its allotted time, the token is passed to another node. Tokens are used in Profibus.

It is also possible to eliminate collisions by letting one node in the network be the master node. The master node controls traffic on the network, and it decides which messages are allowed to be sent and when they are sent. This approach is used in LIN and TTP/A.

Finally, mini-slotting can be used to eliminate collisions. When using mini-slotting, as soon as the network is idle and some node would like to transmit a message, the node has to wait for a unique (for each node) time before sending messages. If there are several competing nodes wanting to send messages, the node with the longer waiting time will see that there is another node that has started transmission. In that situation the node must wait until the next time the network will become idle. Hence, collisions are avoided. Mini-slotting can be found in FlexRay and ARINC 629.

Fieldbuses

Fieldbuses are a family of factory communication networks that have evolved in response to the demand for reduced cabling costs in factory automation systems. By moving from a situation where every controller has its own cables connecting the sensors to the controller (in a parallel interface) to a system with a set of controllers sharing a bus (a serial interface), costs could be cut and flexibility increased. Driving this evolution in technology was the fact that the number of cables in the system increases as the number of sensors and actuators grew. Also, controllers have moved from being specialized with their own microchip to sharing a microprocessor with other controllers. Fieldbuses soon were ready to handle the most demanding applications on the factory floor.

Several fieldbus technologies, usually specialized, were developed by different companies to meet application demands. Fieldbuses in the automotive industry include CAN, TT-CAN, TTP, LIN and FlexRay. In avionics, ARINC 629 is one of the more-used communications standards. Profibus is widely used in automation and robotics, while in trains, TCN and WorldFIP are popular communication technologies. We will present each fieldbuses in more detail, outlining key features and specific properties.

Controller Area Network (CAN). The controller area network (CAN) [CAN02] was standardized by the International Standardization Organisation (ISO) [CAN92] in 1993. Today CAN is a widely used fieldbus, mainly in automotive systems but in other real-time applications, such as medical equipment. CAN is an event-triggered broadcast bus designed to operate at speeds of up to 1 Mbps. CAN uses a fixed-priority-based arbitration mechanism that can provide timing guarantees using fixed-priority scheduling (FPS) analysis (Tindell et al. [TBW95, THW94]). An example will be provided in Section "Example of Analysis."

CAN is a collision-avoidance broadcast bus using deterministic collision resolution to control access to the bus (called CSMA/CA). The basis for the access mechanism is the electrical characteristics of a CAN bus, allowing sending nodes to detect collisions in a nondestructive way. By monitoring the resulting bus value during message arbitration, a node detects if there are higher priority messages competing for bus access. If this is the case, the node will stop the message transmission and attempt to retransmit the message as soon as the bus becomes idle again. Hence, the bus is behaving like a priority-based queue.

Time-Triggered CAN (TT-CAN). Time-triggered communication on CAN (TT-CAN) (TTC) is a standardized session layer extension to the original CAN. In TT-CAN, the exchange of messages is controlled by the temporal progression of time, and all nodes follow a predefined static schedule. It is possible to support original, event-triggered CAN traffic together with time-triggered traffic. This traffic is sent in dedicated arbitration windows, using the same arbitration mechanism as native CAN.

The static schedule is based on a time division (TDMA) scheme, where message exchanges only may occur during specific time slots or in time windows. Synchronization of the nodes is done using a clock synchronization algorithm or by periodic messages from a master node. In the latter case, all system nodes synchronize with this message, allowing a reference point in the temporal domain for the static schedule of the message transactions. The master's view of time is referred to as the network's global time.

TT-CAN appends a set of new features to the original CAN and several semiconductor vendors are manufacturing the standardized, TT-CAN-compliant devices.

Flexible Time-Triggered CAN (FTT-CAN). Flexible time-triggered communication on CAN (FTT-CAN) [AFF98, AFF99] provides a way to schedule CAN in a time-triggered fashion with support for event-triggered traffic. In FTT-CAN, time is partitioned into Elementary Cycles (ECs) initiated by a special message, the trigger message (TM). This message triggers the start of the EC and contains the schedule for the time-triggered traffic sent within this EC. The schedule is calculated and sent by a master node. FTT-CAN supports both periodic and aperiodic traffic by dividing the EC into two parts. In the first part, called the asynchronous window, aperiodic messages are sent. In the second part, called the synchronous window, traffic is sent in a time-triggered fashion according to the schedule delivered by the TM. FTT-CAN is still mainly an academic communication protocol.

Time-Triggered Protocol (TTP). TTP/C (TTT, TTT99), or the time-triggered protocol Class C, is a TDMA-based communication network intended for truly hard real-time communication. TTP/C is available for network speeds of as much as 25 Mbps. TTP/C is part of the time-triggered architecture (TTA) by Kopetz *et. al* [TTT, Kop98], designed for safety-critical applications. TTP/C has support for fault tolerance, clock synchronization, membership services, fast error detection and consistency checks. Several major automotive companies are supporting this protocol, including Audi, Volkswagen and Renault.

For the less hard real-time systems (or for soft real-time systems), there exists a scaled-down version of TTP/C called TTP/A (TTT).

Local Interconnect Network (LIN). LIN, or the Local Interconnect Network, was developed by the LIN Consortium (including Audi, BMW, DaimlerChrysler, Motorola, Volvo and VW) as a low-cost alternative for small networks. LIN is cheaper than CAN. LIN uses the UART/SCI interface hardware, and transmission speeds are possible up to 20 Kbps. Among the nodes in the network, one node is the master node, responsible for synchronization of the bus. The traffic is sent in a time-triggered fashion.

FlexRay. FlexRay (BBE + 02) was proposed in 1999 by several major automotive manufacturers, including DaimlerChrysler and BMW, as a competitive next-generation fieldbus replacing CAN. FlexRay is a real-time communication network that provides both synchronous and asynchronous transmissions with network speeds up to 10 Mbps. For synchronous traffic, FlexRay uses TDMA, providing deterministic data transmissions with a bounded delay. For asynchronous traffic, mini-slotting is used. Compared with CAN,

FlexRay is more suited for the dependable application domain by including support for redundant transmission channels, bus guardians and fast error detection and signaling.

ARINC 629. For avionic and aerospace communication systems, the ARINC 429 [ARI99] standard and its newer ARINC 629 [ARI99] successor are the most commonly used communication systems today. ARINC 629 supports periodic and sporadic communication. The bus is scheduled in cycles, divided into two parts. In the first part, periodic traffic is sent and, in the second part, sporadic traffic is sent. The arbitration of messages is based on collision avoidance (i.e. CSMA/CA) using mini-slotting. Network speeds are as high as 2 Mbps.

Profibus. Profibus (PRO) is used in process automation and robotics. There are three versions of Profibus: (1) Profibus - DP is optimized for speed and low cost. (2) Profibus - PA is designed for process automation. (3) Profibus - FMS is a general-purpose version of Profibus. Profibus provides master-slave communication together with token mechanisms. Profibus is available with data rates up to 12 Mbps.

Train Communication Network (TCN). The train communication network (TCN) [KZ01] is widely used in trains, and it implements the IEC 61275 standard as well as the IEEE 1473 standard. TCN is composed of two networks: the wire train bus (WTB) and the multifunction vehicle bus (MVB). The WTB is the network used to connect all vehicles of the train. The network data rate is up to 1 Mbps. The MVB is the network used within one vehicle. Here the maximum data rate is 1.5 Mbps.

Both the WTB and the MVB are scheduled in cycles called basic periods. Each basic period consists of a periodic phase and a sporadic phase. There is a support for both periodic and sporadic types of traffic. The difference between the WTB and the MVB (apart from the data rate) is the length of the basic periods (1 or 2 ms for the MVB and 25 ms for the WTB).

WorldFIP. The WorldFIP (Wor) is a popular communication network in train control systems. WorldFIP is based on the producer-distributor-consumers (PDC) communication model. Currently, network speeds are as high as 5 Mbps. The WorldFIP protocol defines an application layer that includes PDC and messaging services.

Ethernet for Real-Time Communication

In parallel with the search for the holy grail of real-time communication, Ethernet has established itself as the de facto standard for non real-time communication. Comparing networking solutions for automation networks and office networks, fieldbuses was the choice for the former. At the same time, Ethernet developed as the standard for office automation. Due to its popularity, prices on networking solutions dropped. Ethernet was not originally developed for real-time communication, since the original intention with Ethernet is to maximize throughput (bandwidth). However, nowadays, due to its popularity, an effort is being made to provide real-time communication using Ethernet. The biggest challenge is to provide real-time guarantees using standard Ethernet components.

The reason that Ethernet is not suitable for real-time communication is its handling of collisions on the network. Several proposals to minimize or eliminate the occurrence of collisions on Ethernet have been proposed. Below, we present some of these proposals.

TDMA. A simple solution would be to eliminate the occurrence of collisions on the network. This has been explored by, e.g., Kopetz et al. [KDKM89], using a TDMA protocol on top of Ethernet.

Use of Tokens. Another solution to eliminate the occurrence of collisions is the use of tokens. Token-based solutions [VC94, PMBL95] on the Ethernet also eliminates collisions but is not compatible with standard hardware.

A token-based communication protocol provides real-time guarantees on most types of networks, since they are deterministic in their behavior. However, a dedicated network is required. All nodes sharing the network must obey the token protocol. Examples of token-based protocols are the Timed-Token Protocol (TTP) [MZ94] and the IEEE 802.5 Token Ring Protocol.

Modified Collision Resolution Algorithm. A different approach is to modify the collision resolution algorithm [RY94, Mol94]. Using standard Ethernet controllers, the modified collision resolution algorithm is nondeterministic. To make a deterministic modified collision resolution algorithm, a major modification of Ethernet controllers is required [LR93].

Virtual Time and Window Protocols. Another solution to real-time communication using Ethernet is the use of the Virtual Time CSMA (VTCSMA) (MK85,ZR86,EDES90) Protocol, where packets are delayed in a deterministic way to eliminate the occurrence of collisions. Window protocols [ZSR90] are using a global window (a synchronized time interval) that also removes collisions. The Windows protocol is more dynamic and somewhat more efficient in its behavior than the VTCSMA approach.

Master/Slave. A fairly straightforward way of providing real-time traffic on the Ethernet is to use a master/slave approach. As part of the flexible time-triggered (FTT) framework [APF02], FTT-Ethernet [PAG02] is proposed as a master/multi-slave protocol. At the cost of some computational overhead at each node in the system, the timely delivery of messages on Ethernet is provided.

Traffic Smoothing. The most recent work, without modification to hardware or networking topology (or infrastructure), is the use of traffic smoothing. Traffic smoothing can eliminate bursts of traffic [KSW00, CCLM02] that have severe impact on the timely delivery of message packets on the Ethernet. By keeping the network load below a given threshold, a probable guarantee of message delivery can be provided. Traffic smoothing could be a solution for soft real-time systems.

Black Bursts. Black Bursts (SK98) use the fact that if a transmitting station detects a collision on the bus, the station will wait some time before it will re-transmit its message (based on the collision resolution algorithm). However, suppose astation is jamming the network, causing stations to wait for retransmission. What the jamming station does is send its message right after it has stopped the jamming. If all stations are using unique-length jamming signals, there will always be a unique winner. This is what the black-burst approach uses.

Switches. Finally, a completely different approach to achieve real-time communication using the Ethernet is to change the infrastructure. One way of doing this is to construct the Ethernet using switches to separate collision domains. By using these switches, a collision-free network is provided. However, this requires new hardware supporting the IEEE 802.1p standard. Therefore it is not as attractive a solution for existing networks as traffic smoothing.

Wireless Communication

There are no commercially available wireless communication protocols providing real-time guarantees.[1] Two of the more common wireless protocols today are the IEEE 802.11 (WLAN) and Bluetooth. However, these protocols do not provide the temporal guarantees needed for hard real-time communication. Today, a big effort is being made (as with Ethernet) to provide real-time guarantees for wireless communication, possibly by using either WLAN or Bluetooth.

Analysis of Real-Time Systems

The most important property to analyze in a RTS is its temporal behavior, for the timeliness of the system. The analysis should provide strong evidence that the system performs as intended at the correct time. This section will give an overview of the basic properties analyzed in a RTS. The section concludes with a presentation of trends and tools in RTS analysis.

[1]Bluetooth provides real-time guarantees limited to streaming voice traffic.

Timing Properties

Timing analysis is a complex problem. Not only are the techniques sometimes complicated but the problem itself is elusive. For instance, what is the meaning of the term "program execution time"? Is it the average time to execute the program, or the worst possible time? Or does it mean some form of "normal" execution time? Under what conditions does a statement regarding program execution times apply? Is the program delayed by interrupts or higher-priority tasks? Does the time include waiting for shared resources?.

To straighten out some of these questions and to study some existing techniques for timing analysis, we have structured timing analysis into three major types. Each type has its own purpose, benefits and limitations. The types are listed below.

Execution Time. This refers to the execution time of a single task (or program, function or any other unit of single-threaded sequential code). The result of an execution-time analysis is the time (or the number of clock cycles) the task takes to execute when undisturbed on a single CPU. The result should not account for interrupts, preemption, background DMA transfers, DRAM refresh delays or any other types of interfering background activities.

At first glance, leaving out all types of interference from execution-time analysis would give unrealistic results, but the purpose of execution-time analysis is *not* to deliver estimates on "real-world" timing when executing the task. Instead, the role of execution-time analysis is to find out how many computing resources are needed to execute the task. (Hence, background activities unrelated to the task should not be considered.)

There are several different types of execution times that can be of interest:

- Worst-Case Execution-Time (WCET) – This is the worst possible execution time a task could exhibit, or equivalently, the maximum amount of computing resources to execute the task. The WCET should include any possible atypical task execution, such as exception handling or cleanup after abnormal task termination.
- Best-Case Execution-Time (BCET) – During some types of real-time analysis, having knowledge about a task's BCET is useful.
- Average Execution-Time (AET) – The AET can be useful in calculating throughput figures for a system. However, for most RTS analysis, the AET is of less importance, since a reasonable approximation of the average case is easy to obtain during testing (where typically, the average system behavior is studied). Also, only knowing the average and not knowing any other statistical parameters, such as standard deviation or distribution function, makes statistical analysis difficult. For analysis purposes, a more pessimistic metric such as the 95% quartile would be more useful. However, analytical techniques using statistical metrics of execution time are scarce and not well developed.

Response Time. The response time of a task is the time it takes from the *invocation* of the task to the task's *completion*. In other words, this is the time from when the task is placed in the operating system's ready position to the time when it is removed from the running state and placed in the idle or sleeping state.

For analysis purposes, it is assumed that a task does not voluntarily suspend itself during execution. The task may not call primitives such as sleep(...) or delay(...). However, involuntarily suspension, such as blocking on shared resources, is allowed. Primitives such as get_semaphore(...) and lock_database_tuple(...) are allowed. When a program voluntarily suspends itself, that program should be broken into two or more analysis tasks.

Response-time is typically a *system level property*, including interference from other, unrelated, tasks and parts of the system. The response-time also includes delays caused by contention on shared resources. The response-time is only meaningful when considering a complete system, or in distributed systems, when evaluating a complete node.

End-to-End Delay. The above described "execution time" and "response time" are useful concepts, since they are easy to understand and have well-defined scopes. However, when trying to establish the temporal correctness of a system, knowing the WCET or the response-times of tasks often is not enough. Typically, the correctness criterion is stated using *end-to-end* latency timing requirements. For instance, an upper bound on the delay between the input of a signal and the output of a response could be used.

In a given implementation, there may be a chain of events between the input of a signal and the output of a response. For instance, one task may be in charge of reading the input and another task for generating the response. The two tasks may have to exchange messages on a communications link before the response can be generated. The end-to-end timing denotes timing of externally visible events.

Jitter. The term *jitter* is a metric for variability in time. For instance, the jitter in the execution time of a task is the difference between the task's BCET and WCET. Similarly, the response-time jitter of a task is the difference between its best-case response time and its worst-case response time. Often, control algorithms have requirements that limit the jitter of the output. Hence, the jitter is sometimes a metric equally important as the end-to-end delay.

Input to the system can have jitter. For instance, an interrupt which is expected to be periodic may have a jitter (due to some imperfection in the process generating the interrupt). In this case, the jitter value is bound on the maximum deviation from the ideal period of the interrupt. Figure 16.11 illustrates the relation between the period and the jitter for this example.

Note that jitter should not accumulate over time. For our example, even though two successive interrupts could arrive closer than one period, in the long-run, the average interrupt inter-arrival time will be that of the period.

In the above list of types of time, we only mentioned time to execute programs. However, in many RTSs, other timing properties may exist. Most typical are delays on a communications network, but other resources such as hard-disk drives may cause delays and need to be analyzed. The above introduced times can be mapped to different types of resources. For instance, the WCET of a task corresponds to the maximum size of a message to be transmitted, and the response time of a message is defined analogous to the response time of a task.

Methods for Timing Analysis

When analyzing hard RTSs, it is essential that the estimates obtained during timing analysis are *safe*. An estimate is considered safe if it is guaranteed not to underestimate the actual worst-case time. It is also important that the estimate is *tight*, meaning that the estimated time is close to the actual worst-case time.

For the previously defined types of timings, there are different methods available:

Execution-Time Estimation. For real-time tasks, the WCET is the most important execution-time measure to obtain. However, it is also often the most difficult measure to obtain.

Methods to obtain the WCET of a task can be divided into two categories: (1) static analysis, and (2) dynamic analysis. Dynamic analysis is equivalent to testing (or executing the task on the target hardware) and has all the drawbacks and problems that testing exhibits (such as being tedious and error-prone). One major

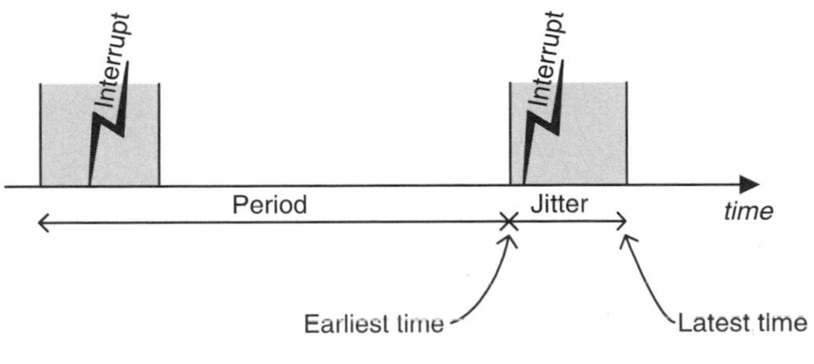

FIGURE 16.11 Jitter used as a bound on variability in periodicity.

problem with dynamic analysis is that it does not produce safe results. The result can never exceed the true WCET, and it is difficult to be certain that the estimated WCET is really the true WCET.

Static analysis, on the other hand, can provide guaranteed safe results. Static analysis is performed by analyzing the code (source or object code is used) and basically counts the number of clock cycles that the task may use to execute in the worst possible case. Static analysis uses hardware models to predict execution times for each instruction. For modern hardware, it may be difficult to produce static analyzers that give good results. One source of pessimism in the analysis is hardware caches; whenever an instruction or data item cannot be guaranteed to reside in the cache, a static analyzer must assume a cache miss. And since modeling the exact state of caches (sometimes of multiple levels), branch predictors are difficult and time-consuming. Few tools exist that give adequate results for advanced architectures. It also is difficult to perform a program flow and data analysis that exactly calculates the number of times a loop iterates or the input parameters for procedures.

Methods for good hardware and software modeling do exist in the research community. However, combining these methods into good quality tools has proven tedious.

Schedulability Analysis. The goal of schedulability analysis is to determine whether a system is *schedulable.* A system is deemed schedulable if it is guaranteed that all task deadlines always will be met. For statically scheduled (or table-driven) systems, calculation of response times are trivially given from the static schedule. For dynamically scheduled systems (such as fixed-priority or deadline scheduling) more advanced techniques must be used.

There are two classes of schedulability analysis techniques: (1) response-time analysis, and (2) utilization analysis. As the name suggests, a response-time analysis calculates a (safe) estimate of the worst-case response time of a task. That estimate can be compared to the task deadline. If it does not exceed the deadline, then the task is schedulable. Utilization analysis, in contrast, does not directly derive the response times for tasks. Rather, they give a Boolean result for each task, forecasting whether the task is schedulable. This result is based on the fraction of utilization of the CPU for a relevant subset of the tasks, leading to the term utilization analysis.

Both types of analysis are based on similar types of task models. Typically,the task models used for analysis are not the task models provided by commercial RTOSs. This problem can be resolved by mapping one or more OS tasks to one or more analysis tasks. This mapping has to be performed manually and requires an understanding of the limitations of the analysis task model and the analysis technique used.

End-to-End Delay Estimation. The typical way to obtain end-to-end delay estimations is to calculate the response time for each task or message in the end-to-end chain and to summarize these response times to obtain and end-to-end estimate. When using a utilization-based analysis technique (in which no response time is calculated), one has to resort to using the task or message deadlines as safe upper bounds on the response times.

However, when analyzing distributed RTSs, it may not be possible to calculate all response times in one pass. The reasons is that delays on one node will lead to jitter on another node, and that this jitter may affect the response times on that node. Since jitter can propagate in several steps between nodes in both directions, there may not exist a right order to analyze the nodes. (If A sends a message to B and B sends a message to A, which node should one analyze first?) Solutions to this type of problems are called *holistic* schedulability analysis methods (since they consider the whole system). The standard method for holistic response-time analysis is to repeatedly calculate response times for each node (and update jitter values in the nodes affected by the node just analyzed) until response times do not change (or a fixed point is reached).

Jitter Estimation. To calculate the jitter, one need not only perform a worst-case analysis (of, for instance, response time or end-to-end delay). It also is necessary to perform a best-case analysis.

Even though best-case analysis techniques often are conceptually similar to worst-case analysis techniques, there has been little attention paid to best-case analysis. One reason for not spending too much time on best-case analysis is that it is quite easy to make a conservative estimate of the best case: the best-case time is never less than zero (0). In many tools, it is simply assumed that the BCET, for instance. is zero, where greater efforts can be spent analyzing the WCET.

However, it is important to record tight estimates of the jitter and to keep jitter as low as possible. It has been shown that the number of *execution paths* a multitasking RTS can take dramatically increases if jitter increases [TH99]. Unless the number of possible execution paths is kept as low as possible, it becomes difficult to achieve good coverage during testing.

Example of Analysis

In this section, we give simple examples of schedulability analysis. We show a simple example of how a set of tasks running on a single CPU can be analyzed, and we also give an example of how the response times for a set of messages sent on a CAN bus can be calculated.

Analysis of Tasks. This example is based on 30-year-old task models and is intended to give the reader a feeling for how these types of analyses work. Today's methods allow for richer and more realistic task models, with a resulting increase in complexity of the equations used. Hence, they are not suitable for use in our simple example.

In the first example, we will analyse a small task set described in Table 16.2, where T, C, and D denote the task period, WCET and deadline, respectively. In this example, $T = D$ for all tasks, and priorities have been assigned in *rate monotonic* order, with the highest rate prividing the highest priority.

For the task set in Table 16.2, original analysis techniques of Liu and Layland [LL73], and Joseph and Pandya [JP86] are applicable, and we can perform both utilization-based and response-time based schedulability analysis.

We start with the utilization-based analysis. For this task model, Liu and Layland conclude that a task set of n tasks is schedulable if its total utilization, U_{tot}, is bounded by the following equation

$$U_{\text{tot}} \leqslant n(2^{1/n} - 1)$$

Table 16.3 shows the utilization calculations performed for the schedulability analysis. For our example task, set $n = 3$ and the bound is approximately 0.78. However, the utilization $\left(U_{\text{tot}} = \sum_{i=1}^{n} \frac{C_i}{T_i}\right)$ for our task set is 0.81, exceeding the bound. Hence, the task set fails the *rate monotonic test* and cannot be deemed schedulable.

Joseph and Pandya's response-time analysis allows us to calculate worst-case response-time, R_i, for each task i in our example (Table 16.2). This is done using the following formula

TABLE 16.2 Example Task Set for Analysis

Task	T	C	D	Prio
X	30	10	30	High
Y	40	10	40	Medium
Z	52	10	52	Low

TABLE 16.3 Result of Rate Monotonic Test

Task	T	C	D	Prio	U
X	30	10	30	High	0.33
Y	40	10	40	Medium	0.25
Z	52	10	52	Low	0.23
				Total:	0.81
				Bound:	0.78

$$R_i = C_i + \sum_{j \in hp(i)} \left\lceil \frac{R_i}{T_j} \right\rceil C_j \tag{16.1}$$

where hp(i) denotes the set of tasks with priority higher than i.

The observant reader may have noticed that Equation (16.1) is not on closed form, in that R_i is not isolated on the left-hand side of the equality. R_i cannot be isolated on the left-hand side of the equality; instead Equation (16.1) has to be solved using *fix-point iteration*. This is done with the recursive formula in Equation (16.1), starting with $R_i^0 = 0$ and terminating when a fix-point has been reached (i.e., when $R_i^{m+1} = R_i^m$).

$$R_i^{m+1} = C_i + \sum_{j \in \text{hp}(i)} \left\lceil \frac{R_i^m}{T_j} \right\rceil C_j \tag{16.2}$$

For our example task set, Table 16.4 show the results of calculating Equation (16.1). From the table, we can conclude that no deadlines will be missed and that the system is schedulable.

Remarks. As we see for our example task set in Table 16.2, the utilization-based test could not deem the task set as schedulable while the response-time-based test could. This situation is symptomatic for the relation between utilization-based and response-time-based schedulability tests. The response-time based tests find more tasksets schedulable than do the utilization-based tests.

However, as shown by the example, the response-time-based test needs to perform more calculations than do the utilization-based tests. For this simple example, the extra computational complexity of the response-time test is insignificant. However, when using modern task models that are capable of modeling realistic systems, the computational complexity of response time-based tests is significant. Unfortunately, for these advanced models, utilization-based tests are not always available.

Analysis of Messages. In our second example, we show how to calculate the worst-case response times for a set of periodic messages sent over the CAN bus. We use a response-time analysis technique similar to the one we used when we analyzed the task set in Table 16.2. In this example, our message set is given in Table 16.5, where T, S, D, and Id denote the messages' period, data size (in bytes), deadline and CAN-identifier, respectively. (The time-unit used in this example is "bit time", or. the time it takes to send one bit. For a 1Mbit CAN, this means that 1 time unit is 10^{-6} seconds.)

Before we attack the problem of calculating response times we extend Table 16.5 with two columns. First, we need the priority of each message; in CAN this is given by the identifier. The lower numerical value. the higher the priority. Second, we need to know the worst-case transmission time of each message. The transmission

TABLE 16.4 Result of Response-Time Analysis for Tasks

Task	T	C	D	Prio	R	$R \leq D$
X	30	10	30	High	10	Yes
Y	40	10	40	Medium	20	Yes
Z	52	10	52	Low	52	Yes

TABLE 16.5 Example CAN-Message Set

Message	T	S	D	Id
X	350	8	300	00010
Y	500	6	400	00100
Z	1000	5	800	00110

TABLE 16.6 Result of Response-Time Analysis for CAN

Message	T	S	D	Id	Prio	C	w	R	$R \leqslant D$
X	350	8	300	00010	High	130	130	260	Yes
Y	500	6	400	00100	Medium	111	260	371	Yes
Z	1000	5	800	00110	Low	102	612	714	Yes

time is given partly by the message data size, but we also need to add time for the frame-header and for any *stuff bits*.[1] The formula to calculate the transmission time, C_i, for a message i is given below

$$C_i = S_i + 47 + \left\lfloor \frac{34 + 8S_i + 1}{5} \right\rfloor$$

In Table 16.6, the two columns Prio and C shows the priority assignment and the transmission times for our example message set.

Now we have all the data needed to perform the response-time analysis. However, since CAN is a nonpreemptive resource, the equation's structure is slightly different from Equation (16.1), which we used for task analysis. The response time equation for CAN is given in Equation (16.3).

$$\begin{aligned} R_i &= w_i + C_i \\ w_i &= 130 + \sum_{\forall j \in \mathrm{hp}(i)} \left\lceil \frac{w_i + 1}{T_j} \right\rceil C_j \end{aligned} \qquad (16.3)$$

In Equation (16.3), hp(i) denotes the set of messages with higher priority than message i. Note, that, similar to Equation (16.1), w_i is not isolated on the left-hand side of the equation, and its value has to be calculated using fix-point iteration (compare this to Equation (16.2)).

Applying Equation (16.3), we can calculate the worst-case response time for our example messages. In Table 16.6, the two columns w and R show the results of the calculations, and the final column shows the schedulablilty verdict for each message.

As we can see from Table 16.6, our example message set is schedulable, meaning that the messages will always be transmitted before their deadlines. Note that this analysis was made assuming that there will not be any retransmissions of broken messages. CAN normally retransmits any message automatically that has been broken due to interference on the bus. To account for such automatic retransmissions, an *error model* needs to be adopted and the response-time equation adjusted [THW94].

Trends and Tools

As pointed out earlier, and illustrated by our example in Table 16.2, there is a mismatch between the analytical task models and the task models provided by commonly used RTOSes. Unfortunately,there is no one-to-one mapping between analysis tasks and RTOS tasks. For many systems, there is a N-to-N mapping between the task types. For instance, an interrupt handler may have to be modeled as several different analysis tasks (one analysis task for each type of interrupt it handles), and one OS task may have to be modeled as several analysis tasks (for instance, one analysis task per call to sleep(...) primitives).

Current schedulability analysis techniques cannot adequately model other types of task synchronization than locking/blocking on shared resources. Abstractions such as message queues are difficult to include in the

[1]CAN adds stuff bits, if necessary, to avoid the two reserved bit patterns 000000 and 111111. These stuff bits are never seen by the CAN user but have to be accounted for in the timing analysis.

schedulability analysis.[1] Furthermore, tools to estimate the WCET are also scarce. Currently, only two tools that give safe WCET estimates are commercially available (Abs, Bou).

These problems have led to low penetration of schedulability analysis in industrial software-development processes. However, in isolated domains, such as real-time networks, some commercial tools based on real-time analysis do exists. For instance, Volcano [CRTM98, Vol] provides tools for the CAN bus that allow system designers to specify signals on and abstract level (giving signal attributes such as size, period and deadline) and automatically derives a mapping of signals to CAN messages, where all deadlines are guaranteed to be meet.

On the software side, tools provided by TimeSys (Timb), Arcticus Systems (Arc), and TTTech (TTT) can provide system development environments with timing analysis as an integrated part of the tool suite. However, these tools require that software development processes are under complete control of the respective tool. This requirement has limited the widespread use of these tools.

The widespread use of UML (OMG03b) in software design has led to some specialized UML-products for real-time engineering (Rat, IL). However, these products, as of today, do support timing analysis of the designed systems. There is recent work within the OMG that specifies a *profile* "Schedulability, Performance and Time" (SPT) (OMG03a), allowing specification of both timing properties and requirements in a standardized way. This will lead to products that can analysz UML models conforming to the SPT profile.

The SPT profile has not been received without criticism. That critique has mainly come from researchers active in the timing analysis field who claim both that the profile is not precise enough and that some important concepts are missing. For instance, the Universidad de Cantabria has developed the MAST-UML profile and an associated MAST tool for analyzing MAST-UML models (MHD01, MAS). MAST allows modeling of advanced timing properties and requirements, and the tool also provides state-of-the-art timing analysis techniques.

Component-Based Design of RTS

Component-Based Design (CBD) is a current trend in software engineering. In the desktop area, component technologies like COM (Mica), .NET (Micb) and Java Beans (SUN) have gained widespread support. These technologies provide substantial benefits, in reduced development time and software complexity when designing complex or distributed systems. However, for real-time systems. these and other, desktop-oriented component technologies do not suffice.

As stated before, the main challenge of designing real-time systems is the need to consider issues that do not typically apply to general purpose computing systems. These issues include:

- constraints on extra-functional properties, such as timing, QoS and dependability,
- the need to statically predict and verify these extra-functional properties,
- managing scarce resources, including processing power, memory and communication bandwidth.

In the commercially available component technologies today, there is little or no support for these issues. On the academic scene, there are no readily available solutions to satisfactory handle these issues.

In the remainder of this chapter, we will discuss how these issues can be addressed in the context of CBD. We also will highlight the challenges in designing a CBD process and component technology for RTS development.

Timing Properties and CBD

In general, for systems where timing is crucial, there will be at least some global timing requirements that have to be met. If the system is built from components, this will imply the need for timing parameters and component properties and some proof that global timing requirements are met.

[1]Techniques to handle more advanced models include timed logic and model checking. However, the computational and conceptual complexity of these techniques has limited their industrial impact. There are examples of commercial tools for this type of verification, e.g. (Tima).

In Section "Analysis of Real-Time Systems", we introduced the following four types of timing properties:

- execution time,
- response time,
- end-to-end delay and
- jitter.

How are these related to the use of a component-based design methodology?

Execution Time. For a component used in a real-time context, an execution time measure will have to be derived. Since execution time is inherently dependent on target hardware, and since reuse is the primary motivation for CBD, it would be highly desirable if the execution time for several targets were available (or for the execution time for new hardware platforms to be automatically derivable.)

The nature of the applied component model may also make execution-time estimation more or less complex. Consider, for instance, a client-server-oriented component model, with a server component that provides services of different types, as illustrated in Figure 16.12(a). What does "execution time" mean for such a component? Clearly, a single execution time is not appropriate. Rather, the analysis will require a set of execution times related to servicing different requests. On the other hand, for a simple port-based-object component model [SVK97] where components are connected in sequence to form periodically executing transactions (illustrated in Figure 16.12(b)), it could be possible to use a single execution time measure. That measure would, correspond to the execution time required for reading the values at the input ports, performing the computation and writing values to the output ports.

Response Time. Response times denote the time from invocation to completion of tasks, and response-time analysis is the activity to statically derive response-time estimates.

The first question to ask from a CBD perspective is: What is the relation between a "task" and a "component?"

This is obviously related to the component model used. As illustrated in Figure 16.13(a), there could be a one-to-one mapping between components and tasks, but in general, several components could be implemented in one task (Figure 16.13(b)) or one component could be implemented by several tasks (Figure 16.13(c)). There is a many-to-many relation between components and tasks. In principle, there could even be more irregular correspondence between components and tasks, as illustrated in Figure 16.13(d). In a distributed system, there could be a many-to-many relation between components and processing nodes, making the situation more complicated.

Once we have sorted out the relation between tasks and components, we can calculate the response times of tasks. We must have an appropriate analysis method for the used execution paradigm and relevant execution-time measures. However, how to relate these response times to components and the

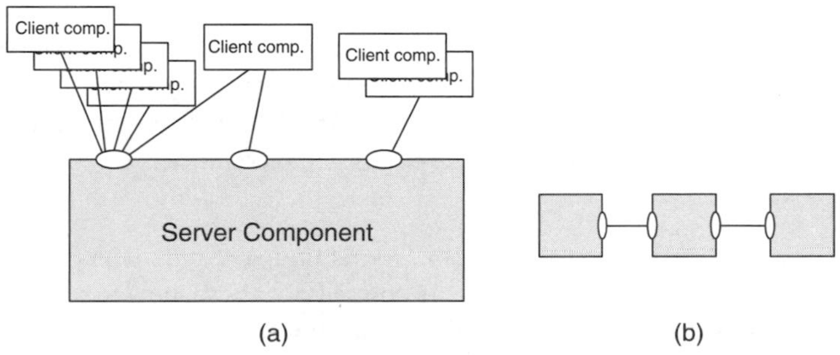

FIGURE 16.12 (a) A complex server component, providing multiple services to multiple users, and (b) a simple chain of components implementing a single thread of control.

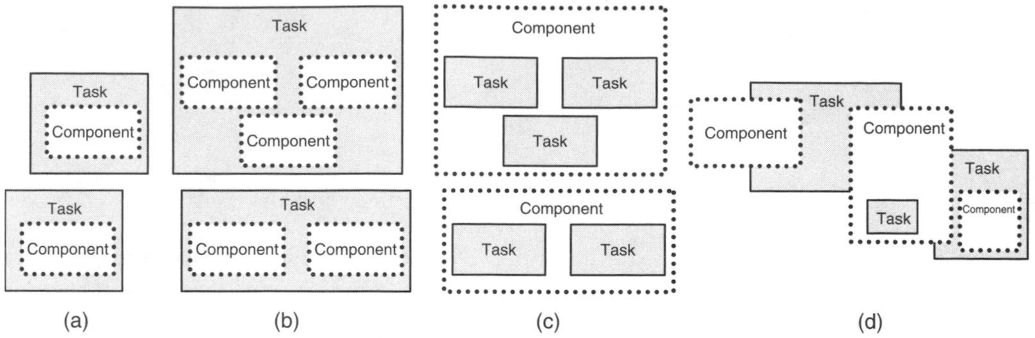

FIGURE 16.13 Tasks and components: (a) one-to-one correspondence, (b) one-to-many correspondence, (c) many-to-one correspondence, (b + c) many-to-many correspondence, and (d) irregular correspondence.

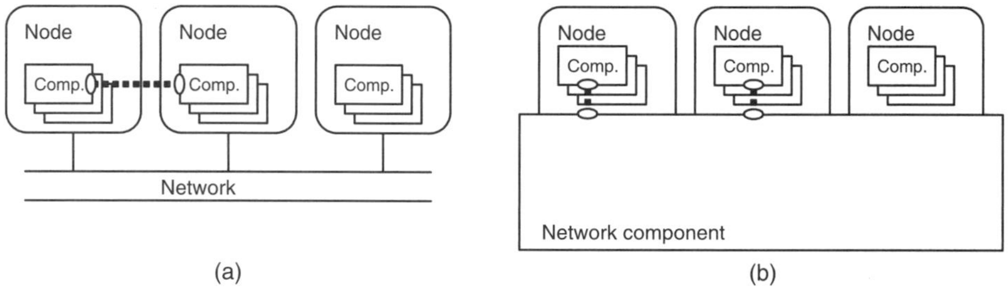

FIGURE 16.14 Components and communication delays: (a) communication delays can be part of the inter-component communication properties, and (b) communication delays can be timing properties of components.

application-level timing requirements may not be straightforward, but this is an issue for the subsequent end-to-end analysis.

Another issue with respect to response times is how to handle communication delays in distributed systems. In essence, there are two ways to model the communication, as depicted in Figure 16.14. In Figure 16.14(a), the network is abstracted and the inter-component communication is handled by the framework. In this case, response-time analysis is made more complicated, since it must account for delays in inter-component communication, depending on the physical location of components. In Figure 16.14(b), on the other hand, the network is modeled as a component itself, and network delays can be modeled as delays in any other component (and inter-component communication can be considered instantaneous).

The choice of how to model network delays has an impact on the software engineering aspects of the component model. In Figure 16.14(a), communication is completely hidden from the components (and the software engineers), giving optimizing tools many degrees of freedom with respect to component allocation, signal mapping and scheduling parameter selection. In Figure 16.14(b), the communication is explicitly visible to the components (and the software engineers), putting a larger burden on the software engineers to manually optimize the system.

End-to-End Delay. End-to-end delays are application-level timing requirements relating the occurrence in time of one event to the occurrence of another event. As pointed out above, how to relate such requirements to the lower-level timing properties of components discussed above is highly dependent on the component model and the timing analysis model.

When designing RTS using CBD, the component structure gives excellent information about the points of interaction between the RTS and its environment. Since end-to-end delays concern timing estimates and timing requirements on such interactions, CBD gives a natural way of stating timing requirements in terms of signals received or generated. (In traditional RTS development, the reception and generation of signals is embedded into the code of tasks and are not externally visible. This makes it difficult to relate response times of tasks to end-to-end requirements.)

Jitter. Jitter is an important timing parameter related to execution time, and that affects response times and end-to-end delays. There may be specific jitter requirements. Jitter has the same relation to CBD as does end-to-end delay.

Summary of Timing and CBD. As described above, there is no single solution on how to apply CBD for RTS. In some cases, timing analysis is made more complicated when using CBD or, when using client-server-oriented component models. In other cases, CBD helps timing analysis in, identifying interfaces and events associated with end-to-end requirements when using CBD.

Further, the characteristics of the component model has great impact on the analysisy of CBD and RTS. For instance, interaction patterns like client servers does not map well to established analysis methods and makes analysis difficult. However, pipes-and-filter based patterns, such as the port-based objects component model [SVK97] map well to existing analysis methods and allow for tight analysis of timing behavior. The execution semantics of the component model has an impact on analyzability. The execution semantics give restrictions on how to map components to tasks. In the Corba Component Model [OMG02], each component is assumed to have its own thread of execution, making it difficult to map multiple components to a single thread. On the other hand, the simple execution semantics of pipes-and-filter-based models allow for automatic mapping of multiple components to a single task, simplifying timing analysis and making better use of system resources.

Real-Time Operating Systems

There are two important aspects regarding CBD and real-time operating systems (RTOSs): (1) the RTOS may itself be component-based, and (2) the RTOS may support or provide a framework for CBD.

Component-Based RTOSs. Most RTOSs allow for offline configurations where the engineer can choose to include or exclude large parts of functionality. For instance, which communications protocols to include is typically configurable. However, this configurability is not the same as the RTOS being component-based (even though the unit of configuration is often referred to as components in marketing materials). For an RTOS to be component-based, it is required that the parts conform to a component model, which is typically not the case in most configurable RTOSs.

There has been some research on component based RTOSs. For instance, the research RTOS VEST [Sta01]. In VEST, schedulers, queue managers and memory management is built out of components. Furthermore, special emphasis has been put on predictability and analyzability. However, VEST is still on the research stage and has not been released to the public. Publicly available sources, however, include the eCos RTOS (Mas02, eCo) providing a component-based configuration tool. Using eCos components, the RTOS can be configured by the user and third party extension can be provided.

RTOSs that Support CBD. Looking at component models in general and those intended for embedded systems in particular, we observe that they are all supported by some runtime executive or simple RTOS. Many component technologies provide frameworks independent of the underlying RTOS. Hence, RTOS can support CBD using such an RTOS-independent framework. Examples include Corba's ORB (OMG) and the framework for PECOS (MSZ02, PEC).

Other component technologies have a tighter coupling between the RTOS and component framework, in that the RTOS explicitly supports the component model by providing the framework (or part of the framework). Such technologies include:

- Koala [vO02] is a component model and architectural description language from Philips, introduced in Section "Offline Schedulers" As mentioned in that section, Koala provides high-level APIs to the computing and audio/video hardware. The computing layer provides a simple, proprietary, real-time kernel with priority-driven preemptive scheduling. Special techniques for thread-sharing are used to limit the number of concurrent threads.

- The Chimera RTOS provides an execution framework for the port-based-object component model [SVK97]. These are intended for development of sensor-based control systems, specifically reconfigurable robotics applications. Chimera has multiprocessor support and handles both static and dynamic scheduling, the latter EDF-based.
- Rubus is a RTOS introduced in Section "Online Schedulers" Rubus supports a component model where behaviors are defined by sequences of port-based objects. The Rubus kernel supports predictable execution of statically scheduled periodic tasks (termed red tasks in Rubus) and dynamically fixed-priority, preemptive scheduled tasks (termed Blue). In addition, support for handling interrupts is provided. In Rubus, support is provided for transforming sets of components into sequential chains of executable code. Each such chain is implemented as a single task. Support is also provided for analysis of response times and end-to-end deadlines, based on execution-time measures that have to be provided. Execution time analysis is not provided by the framework.
- The time-triggered operating system (TTOS) is an adapted and extended version of the MARS OS [KDKM89]. Task scheduling in TTOS is based on an offline-generated scheduling table and relies on the global time base provided by the TTP/C communication system. All synchronization is handled by offline scheduling. TTOS, and in general the entire time-triggered architecture (TTA) is (just as is IEC61131–3) well suited for the synchronous execution paradigm.

In a synchronous execution the system is considered sequential, computing in each step (or cycle) a global output based on a global input. The effect of each step is defined by a set of transformation rules. Scheduling is done statically by compiling the set of rules into a sequential program implementing these rules and executing them in some statically defined order. A uniform timing bound for the execution of global steps is assumed. In this context, a component is a design level entity.

TTA defines a protocol for extending the synchronous language paradigm to distributed platforms, allowing distributed components to interoperate, as long as they conform to imposed timing requirements.

Real-Time Scheduling

Ideally, from a CBD perspective, a component's response time should be independent of the environment in which it is executing (since this would facilitate reuse of the component). However, this is normally unrealistic since

1. the execution time of the task will be different in different target environments and
2. the response time is dependent on the other tasks competing for the same resources (CPU etc.) and the scheduling method to resolve the resource contention.

Rather than aiming for the nonachievable ideal, a realistic ambition could be to have a component model and framework which allows for analysis of response times based on abstract models of components and their compositions. Time-triggered systems go one step towards this ideal solution, in that components can be isolated from each other. While not having a major impact on the component model, time-triggered systems simplify implementation of the component framework since all synchronization between components is resolved off-line. Also, from a safety perspective, the time-triggered paradigm reduces the number of possible execution scenarios (due to the static order of execution of components and the lack of preemption).

In time-triggered component models, it is possible to use the structure given by the component composition to synthesize scheduling parameters. For instance, in Rubus (Arc) and TTA [KB03], this is already done today by generating the static schedule using the components as schedulable entities.

In theory, a similar approach could be used for dynamically scheduled systems. Ths method uses a scheduler or task configuration tool to automatically derive mappings of components for tasks and scheduling parameters (such as priorities or deadlines). However, this approach is still in the research stage.

Testing and Debugging of Real-Time Systems

According to a recent study by NIST (U.S02) up to 80% of the life-cycle cost for software is spent on testing and debugging. Despite the importance, there are few results on RTS testing and debugging.

It is quite difficult to test and debug RTS. RTSs are timing-critical and they interact with the real world. Since testing and debugging typically involves some code instrumentation, the timing behavior of the system will be different when testing and debugging, compared to the execution of the deployed system. Hence, the test cases passed during testing may lead to failures in the deployed system, and tests that failed may not cause any problem in the deployed system. For debugging, the situation is possibly worse. In addition to a similar effect when running the system in a debugger, entering a break point will stop the execution for an unspecified time. The problem with this is that the controlled external process will continue to evolve (e.g., a car will not momentarily stop by stopping the execution of the controlling software). The result is that we get a behavior of the debugged system not possible in the real system. Also, it is often the case that the external process cannot be completely controlled, which means that we cannot reproduced the observed behavior. It will be difficult to use cyclic debugging to track down an error that cased a failure.

The following are two possible solutions to the above problems:

- To build a simulator that faithfully captures the functional and timing behavior of both the RTS and the environment it is controlling. Since this is both time consuming and costly, this approach is only feasible in special situations. Since such situations are rare, we will not consider this alternative here.
- To record the RTS's behavior during testing or execution. Then, if a failure is detected, replay the execution in a controlled way. For this to work it is essential that timing behavior is the same during testing as in the deployed system. This can be achieved by using non-intrusive hardware recorders or by leaving the software for instrumentation in the deployed system. The latter comes at a cost in memory space and execution time but gives the added benefit of debugging the deployed system in case of a failure [RDBC + 03].

An additional problem for most RTSs is that the system consists of several, concurrently executing threads. This is the case for the majority of non-real-time systems. This concurrency will lead to a problematic nondeterminism. Due to race conditions caused by slight variations in execution time, the exact preemption points will vary. This causes unpredictability, both in the number of scenarios and in being able to predict which scenario will actually be executed in a specific situation.

In conclusion, testing and debugging of RTSs are difficult and challenging tasks.

The following is a brief account of some of the few results on testing of RTSs reported in the literature:

- Thane et al. [TH99] propose a method for deterministic testing of distributed RTSs. The key element is identifying the different execution orderings (serialisations of the concurrent system) and treating each as a sequential program. The main weakness of this approach is the potentially exponential blow-up of the number of execution orderings.
- For testing of temporal correctness, Tsai et al. [TFB90] provide a monitoring technique that records runtime information. This information is then used to analyze if temporal constraints are violated.
- Schütz [Sch94] have proposed a testing strategy fordistributed RTSs. The strategy is tailored for the time-triggered MARS system [KDKM89].
- Zhu *et. al* [ZHM97] have proposed a framework for regression testing of real-time software in distributed systems. The framework is based on the Onomas [OTPS98] regression testing process.

When it comes to RTS debugging, the most promising approach is record/replay [CAN + 01, MCL89, TCO91, TH00, ZN99], as mentioned above. Using record/replay,, a reference execution of the system is first executed and observed. A replay execution is then performed, based on the observations made during the reference execution. Observations are performed by instrumenting the system, extracting information about the execution.

The industrial practice for testing and debugging of multitasking RTS is a time-consuming activity. At best, hardware emulators such Lau are used to get some level of observability without interfering with the observed system. More often, it is an ad-hoc activity, using intrusive code instrumentations to observe test results or attempt to track down intricate timing errors. Some tools using the above record/replay method are now emerging on the market (Zea).

Summary

This section has presented the most important issues, methods and trends in the area of embedded real-time systems (embedded RTSs). A wide range of topics have been covered, from the initial design of embedded RTSs to analysis and testing. Important issues discussed and presented were design tools, operating systems and major underlying mechanisms such as architectures, models of interactions, real-time mechanisms, executions strategies and scheduling. Moreover, communications, analysis and testing techniques were presented.

Over the years, the academic community has put an effort into enhancing the techniques to compose and design complex embedded RTSs. Industry is following a slower pace while also adopting and developing area-specific techniques. Today, we can see a diversity of techniques in different application domains, such as automotive, aerospace and trains. In communications, an effort has been made in the academic community, and also in some parts of industry, towards using Ethernet. This is a step towards a common technique for several application domains.

Different real-time demands have led to domain-specific operating systems, architectures and models of interaction. As many of these have several commonalities, there is a potential for standardization across several domains. However, this takes time, and we will certainly stay with application-specific techniques for a while. For specific domains with extreme demands on safety or low cost, specialized solutions will most likely be used in the future. Knowledge of the techniques suitable for the various domains will remain important.

References

[AB98] L. Abeni and G. Buttazzo, "Integrating multimedia applications in hard real-time systems," in *Proc. 19th IEEE Real-Time Syst. Symp. (RTSS '98)*, Madrid, Spain, December 1998, IEEE Computer Society.

[ABD+95] N.C. Audsley, A. Burns, R.I. Davis, K. Tindell, and A.J. Wellings, "Fixed priority pre-emptive scheduling: an historical perspective," *Real-Time Systems*, vol. 8, no. 2/3, pp. 129–154, 1995.

[Abs] AbsInt. http://www.absint.com.

[AFF98] L. Almeida, J. A. Fonseca, and P. Fonseca, "Flexible time-triggered communication on a controller area network," in *Proc. Work Prog. Session 19th IEEE Real Time Syst. Symp. (RTSS'98)*, Madrid, Spain, December 1998, IEEE Computer Society.

[AFF99] L. Almeida, J.A. Fonseca, and P. Fonseca, "A flexible time-triggered communication system based on the controller area network: experimental results," in *Proc. IFAC Int. Conf. Filedbus Technol.*, 1999.

[Air96] Airlines Electronic Engineering Committee (AEEC), ARINC 653:Avionics Application Software Standard Interface (Draft 15), June 1996.

[APF02] L. Almeida, P. Pedreiras, and J. A. Fonseca, "The FTT-CAN protocol: why and how," *IEEE Trans. Ind. Electron.*, vol. 49, no. 6, December 2002.

[App98] Northern Real-Time Applications, Total time predictability, 1998, Whitepaper on SSX5.

[Arc] Arcticus Systems, *The Rubus Operating System*, http://www.arcticus.se.

[ARI99] ARINC/RTCA-SC-182/EUROCAE-WG-48, Minimal Operational Performance Standard for Avionics Computer Resources, 1999.

[Art03] Roadmap — Adaptive Real-Time Systems for Quality of Service Management, ARTIST — Project IST-2001–34820, May 2003, http://www.artistembedded.org/Roadmaps/.

[Ast] The Asterix Real-Time Kernel, http://www.mrtc.mdh.se/projects/asterix/.

[BBE+02] R. Belschner, J. Berwanger, C. Ebner, H. Eisele, S. Fluhrer, T. Forest, T. Führer, F. Hartwich, B. Hedenetz, R. Hugel, A. Knapp, J. Krammer, A. Millsap, B. Müller, M. Peller, and A. Schedl, *FlexRay — Requirements Specification*, April 2002, http://www.flexray-group.com.

[Bou] Bound-T Execution Time Analyzer, http://www.bound-t.com.

[But97] G.C. Buttazzo, *Hard Real-Time Computing Systems*, Dordrecht: Kluwer Academic Publishers, 1997, ISBN 0-7923-9994-3.

[BW96] A. Burns and A. Wellings, *Real-Time Systems and Programming Languages*, 2nd ed., Reading, MA: Addison-Wesley, 1996, ISBN 0-201-40365-X.

[CAN92] Road Vehicles — Interchange of Digital Information — Controller Area Network (CAN) for High Speed Communications, February 1992, ISO/DIS 11898.

[CAN+01] J.D. Choi, B. Alpern, T. Ngo, M. Sridharan, and J. Vlissides, "A pertrubation-free replay platform for cross-optimized multithreaded applications," in *Proc. 15th Int. Parallel Distrib. Process. Symp.*, April 2001, IEEE Computer Society.

[CAN02] CAN Specification 2.0, Part-A and Part-B, CAN in Automation (CiA), Am Weichselgarten 26, D-91058 Erlangen, 2002, http://www.can-cia.de.

[CCLM02] A. Carpenzano, R. Caponetto, L. LoBello, and O. Mirabella, "Fuzzy Traffic Smoothing: an Approach for Real-Time Communication over Ethernet Networks," in *Proc. fourth IEEE Int. Workshop Fact. Commun. Syst. (WFCS'02)*, Västerås, Sweden, August 2002, pp. 241–248, IEEE Industrial Electronics Society.

[CRTM98] L. Casparsson, A. Rajnak, K. Tindell, and P. Malmberg, "Volcano - a revolution in on board communications," *Volvo Technology Report*, 1, 9–19, 1998.

[eCo] eCos Home Page, http://sources.redhat.com/ecos.

[EDES90] M. El-Derini and M. El-Sakka, "A novel protocol under a priority time constraint for real-time communication systems," in *Proc. Second IEEE Workshop Future Trends Distrib. Comput. Syst. (FTDCS'90)*, Cairo, Egypt, September 1990, pp. 128–134, IEEE Computer Society.

[ETA] ETAS, http://en.etasgroup.com.

[FAQ] Comp.realtime FAQ, available at http://www.faqs.org/faqs/realtime-computing/-faq/.

[FLD03] G. Fohler, T. Lennvall, and R. Dobrin, "A component based real-time scheduling architecture," in *Architecting Dependable Systems*, R. de Lemos, C. Gacek, and A. Romanovsky, Eds., vol. LNCS-2677, Berlin: Springer, 2003.

[Gro] OSEK Group, OSEK/VDX Operating System Specification 2.2.1, http://www.osek-vdx.org/.

[Hal00] Tom R. Halfhill, "Embedded markets breaks new ground," *Microprocessor Report*, 17, January 2000.

[HLS96] H. Hansson, H. Lawson, and M. Strömberg, "BASEMENT a distributed real-time architecture for vehicle applications," *Real-Time Systems*, vol. 3, no. 11, pp. 223–244, 1996.

[IEE98] IEEE, Standard for Information Technology — Standardized Application Environment Profile — POSIX Realtime Application Support (AEP), 1998, IEEE Standard P1003.13.

[IL] I-Logix, Rhapsody, http://www.ilogix.com/products/rhapsody.

[ISO95] ISO, Ada95 Reference Manual, 1995, ISO/IEC 8652:1995(E).

[ITE] ITEA, EAST/EEA Project Site. http://www.east-eea.net.

[JP86] M. Joseph and P. Pandya, "Finding response times in a real-time system," *Comput. J.*, vol. 29, no. 5, pp. 390–395, 1986.

[KB03] H. Kopetz and G. Bauer, "The time-triggered architecture,"*Proc. IEEE Special Issue Model. Des. Embedded Software*, vol. 91, no. 1, pp. 112–126, 2003.

[KDKM89] H. Kopets, A. Damm, C. Koza, and M. Mullozzani, "Distributed fault tolerant real-time systems: the MARS approach," *IEEE Micro*, vol. 9, no. 1, 1989.

[KG94] H. Kopetz and G. Grünsteidl, "TTP-A protocol for fault-tolerant real-time systems," *IEEE Comput.*, 14–23, January 1994.

[Klu] Kluwer, *Real-Time Syst.*, http://www.wkap.nl/kapis/CGI-BIN/WORLD/journalhome.htm?0922–6443.

[Kop98] H. Kopetz, "The time-triggered model of computation," in *Proc. 19th IEEE Real-Time Syst. Symp. (RTSS'98)*, Madrid, Spain, December 1998, pp. 168–177, IEEE Computer Society.

[Kop03] H. Kopetz, Introduction in Real-Time Systems: introduction and overview, Part XVIII of Lectures Notes from ESSES 2003 — European Summer School on Embedded Systems, Västerås, Sweden, September 2003.

[KRP+98] M.H. Klein, T. Ralya, B. Pollak, R. Obenza, and M.G. Harbour, *A Practitioners Handbookfor Rate-Monotonic Analysis*, Dordrecht: Kluwer, 1998.

[KSW00] S.K. Kweon, K. G. Shin, and G. Workman. "Achieving real-time communication over ethernet with adaptive traffic smoothing," in *Proc. Sixth IEEE Real Time Technol. Appl. Symp. (RTAS'00)*, Washington DC, USA, June 2000, pp. 90–100, IEEE Computer Society.

[KZ01] H. Kirrmann and P.A. Zuber, "The IEC/IEEE train communication network," *IEEE Micro*, vol. 21, no. 2, pp. 81–92, 2001.

[Lau] Lauterbach, http://www.laterbach.com.

[LIN] LIN — Local Interconnect Network, http://www.lin-subbus.de.

[Liv99] LiveDevices, Realogy Real-Time Architect, SSX5 Operating System, 1999, http://www.livedevices.com/realtime.shtml.

[LL73] C. Liu and J. Layland, "Scheduling algorithms for multiprogramming in a hard-real-time environment," *J. ACM*, vol. 20, no. 1, pp. 46–61, 1973.

[Log] Express Logic, Threadx, http://www.expresslogic.com.

[LR93] G. Lann and N. Riviere, Real-Time Communications over Broadcast Networks: the CSMA/DCR and the DOD-CSMA/CD Protocols, Technical report, TR 1863, INRIA, 1993.

[LYN] Lynuxworks, http://www.lynuxworks.com.

[MAS] MAST home-page, http://mast.unican.es/.

[Mas02] A. Massa, *Embedded Software Development with eCos*, Englewood Cliffs, NJ: Prentice Hall, November 2002.

[MCL89] J. Mellor-Crummey and T. LeBlanc, "A software instruction counter," in *Proc. Third Int. Conf. Archit. Support Program. Lang. Operating Syst.*, ACM, April 1989, pp. 78–86.

[MHD01] J.L. Medina, M. González Harbour, and J.M. Drake, "MAST real-time view: a graphic uml tool for modeling object-oriented real-time systems," in *Proc. 22th IEEE Real-Time Syst. Symp. (RTSS)*, December 2001.

[Mica] Microsoft, Microsoft COM Technologies, http://www.microsoft.com/com/.

[Micb] Microsoft.NET Home Page, http://www.microsoft.com/net/.

[MK85] M. Molle and L. Kleinrock, "Virtual time CSMA: why two clocks are better than one," *IEEE Trans. Commun.*, vol. 33, no. 9, pp. 919–933, 1985.

[Mol94] M. Molle, "A new binary logarithmic arbitration method for Ethernet," *Technical report, TR CSRI-298, CRI*, University of Toronto, Canada, 1994.

[MSZ02] P.O. Müller, C.M. Stich, and C. Zeidler, "Component based embedded systems," *Building Reliable Component-Based Software Systems*, Norwood, MA: Artech House publisher, 2002, pp. 303–323, ISBN 1-58053-327-2.

[MTS02] J. Mäki-Turja and M. Sjödin, "Combining dynamic and static scheduling in hard real-time systems," *Technical Report MRTC no. 71*, Mälardalen Real-Time Research Centre (MRTC), October 2002.

[MZ94] N. Malcolm and W. Zhao, "The timed token protocol for real-time communication," *IEEE Comput.*, vol. 27, no. 1, pp. 35–41, 1994.

[OMG] OMG, CORBA Home Page, http://www.omg.org/corba/.

[OMG02] OMG, CORBA Component Model 3.0, June 2002, http://www.omg.org/technology/documents/formal/components.htm.

[OMG03a] OMG, UML Profile for Schedulability, Performance and Time Specification, September 2003, OMG document formal/2003-09-01.

[OMG03b] OMG, Unified Modeling Language (UML), Version 1.5, 2003, http://www.omg.org/technology/documents/formal/uml.htm.

[OMG03c] OMG, Unified Modeling Language (UML), Version 2.0 (draft), September 2003, OMG document ptc/03-09-15.

[OTPS98] K. Onoma, W.-T. Tsai, M. Poonawala, and H. Suganuma, "Regression testing in an industrial environment," *Commun. ACM*, vol. 41, no. 5, pp. 81—86, 1998.

[PAG02] P. Pedreiras, L. Almeida, and P. Gai, "The FTT-ethernet protocol: merging flexibility, timeliness and efficiency," in *Proc. 14th Euromicro Conf. Real-Time Syst. (ECRTS'02)*, Vienna, Austria, June 2002, pp. 152–160, IEEE Computer Society.

[PEC] PECOS Project web site, http://www.pecos-project.org.

[PMBL95] D.W. Pritty, J.R. Malone, S.K. Banerjee, and N. L. Lawrie, "A real-time upgrade for ethernet based factory networking," in *Proc. IECON'95*, 1995, pp. 1631–1637.

[PRO] PROFIBUS International, http://www.profibus.com.

[QNX] QNX Software Systems, QNX realtime OS, http://www.qnx.com.

[Rat] Rational Rose RealTime, http://www.rational.com/products/rosert.

[RDBC+03] M. Ronsse, K. De Bosschere, M. Christiaens, J. Chassin de Kergommeaux, and D. Kranzlmüller, "Record/replay for nondeterministic program executions," *Communications of the ACM*, vol. 46, no. 9, pp. 62–67, 2003.

[RTL] List of real-time Linux variants, http://www.realtimelinuxfoundation.org/variants/variants.html.

[RTS] IEEE Computer Society, Technical Committee on Real-Time Systems Home Page, http://www.cs.bu.edu/pub/ieee-rts/.

[RY94] K.K. Ramakrishnan and H. Yang, "The ethernet capture effect: analysis and solution," in *Proc. 19th IEEE Local Comput. Networks Conf. (LCNC'94)*, October 1994, pp. 228–240.

[SB94] M. Spuri and G.C. Buttazzo, "Efficient aperiodic service under earliest deadline scheduling," in *Proc. 15th IEEE Real-Time Syst. Symp. (RTSS'94)*, San Juan, Puerto Rico, December 1994, pp. 2–11, IEEE Computer Society.

[SB96] M. Spuri and G.C. Buttazzo, "Scheduling aperiodic tasks in dynamic priority systems," *Real Time Syst.*, vol. 10, no. 2, pp. 179–210, 1996.

[SBS95] M. Spuri, G.C. Buttazzo, and F. Sensini, "Robust aperiodic scheduling under dynamic priority systems," in *Proc. 16th IEEE Real Time Syst. Symp. (RTSS'95)*, Pisa, Italy, December 1995, pp. 210–219, IEEE Computer Society.

[Sch94] W. Schütz, "Fundamental issues in testing distributed real-time systems," in *Real Time Systems*, vol. 7, Dordrecht: Kluwer, 1994, pp. 129–157.

[Sie] Siemens, http://www.siemensvdo.com.

[SK98] J.L. Sobrinho and A.S. Krishnakumar, "EQuB-ethernet quality of service using black bursts," in *Proc. 23rd IEEE annu. Conf. Local Comput. Networks (LCN'98)*, Lowell, MA, USA, October 1998, pp. 286–296, IEEE Computer Society.

[SSL89] B. Sprunt, L. Sha, and J.P. Lehoczky, "Aperiodic task scheduling for hard realtime systems," *Real Time Syst.*, vol. 1, no. 1, pp. 27–60, 1989.

[Sta01] J.A. Stankovic, "VEST — A toolset for constructing and analyzing component based embedded systems," *Lect. Notes Comput. Sci.*, 2211, 2001.

[SUN] SUN Microsystems, Introducing Java Beans, http://developer.java.sun.com/-developer/onlineTraining/Beans/Beans1/index.html.

[SVK97] D.B. Stewart, R.A. Volpe, and P.K. Khosla, "Design of dynamically recon-figurable real-time software using port-based objects," *IEEE Trans. Software Eng.*, vol. 23, no. 12, 1997.

[Sys] Enea OSE Systems, Ose, http://www.ose.com.

[TBW95] K.W. Tindell, A. Burns, and A.J. Wellings, "Calculating controller area network (CAN) message response times," *Control Eng. Pract.*, vol. 3, no. 8, pp. 1163–1169, 1995.

[TCO91] K.C. Tai, R. Carver, and E. Obaid, "Debugging concurrent ADA programs by deterministic execution," *IEEE Trans. Software Eng.*, vol. 17, no. 1, pp. 280–287, 1991.

[Tel] Telelogic tau, http://www.telelogic.com/products/tau.

[TFB90] J.J.P. Tsai, K.Y. Fang, and Y.D. Bi, "On realtime software testing and debugging," in *Proc. 14th Annu. Int. Comput. Software Appl. Conf.*, 1990, pp. 512–518.

[TH99] H. Thane and H. Hansson, "Towards systematic testing of distributed real-time systems," in *Proc. 20th IEEE Real Time Syst. Symp. (RTSS)*, December 1999, pp. 360–369.

[TH00] H. Thane and H. Hansson, "Using deterministic replay for debugging of distributed real-time systems," in *Proc. 12th Euromicro Conf. Real Time Syst.*, June 2000, pp. 265–272, IEEE Computer Society.

[THW94] K. Tindell, H. Hansson, and A. Wellings, "Analysing real-time communications: controller area network (CAN)," in *Proc. 15th IEEE Real Time Syst. Symp. (RTSS)*, Los Alamitos, CA: IEEE Computer Society Press, December 1994, pp. 259–263.

[Tima] The Times Tool, http://www.docs.uu.se/docs/rtmv/times.

[Timb] TimeSys, Timewiz — a modeling and simulation tool, http://www.timesys.com/.

[TTC] Road vehicles — controller area network (CAN) — Part 4: Time triggered communication, ISO/CD 11898-4.

[TTT] Time Triggered Technologies, http://www.tttech.com.

[TTT99] TTTech Computertechnik AG, Specification of the TTP/C Protocol v0.5, July 1999.

[U.S02] U.S. Department of Commerce, The Economic Impacts of Inadequate Infrastructure for Software Testing, May 2002, NIST Report.

[VC94] C. Venkatramani and T. Chiueh, "Supporting Real-Time Traffic on Ethernet," In *Proc. 15th IEEE Real Time Syst. Symp. (RTSS'94)*, San Juan, Puerto Rico, December 1994, pp. 282–286, IEEE Computer Society.

[Veca] Vector, DaVinci Tool Suite, http://www.vector-informatik.de/.

[Vecb] Vector, http://www.vector-informatik.com.

[vO02] Rob van Ommering, "The koala component model," in *Building Reliable Component-Based Software Systems*, Norwood, MA: Artech House Publishers, July 2002, pp. 223–236, ISBN 1-58053-327-2.

[Vol] Volcano automotive group, http://www.volcanoautomotive.com.

[Win] Wind River Systems Inc, VxWorks Programmer's Guide, http://www.windriver.com/.

[Wor] WorldFIP Fieldbus, http://www.worldfip.org.

[XP90] J. Xu and D.L. Parnas, "Scheduling processes with release times, deadlines, precedence, and exclusion relations," *IEEE Trans. Software Eng.*, vol. 16, no. 3, pp. 360–369, 1990.

[Zea] ZealCore Embedded Solutions AB, http://www.zealcore.com.

[ZHM97] H. Zhu, P. Hall, and J. May, "Software unit test coverage and adequacy," *ACM Comput. Surv.*, vol. 29, no. 4, 1997.

[ZN99] F. Zambonelli and R. Netzer, "An efficient logging algorithm for incremental replay of message-passing applications," in *Proc. 13th Int. 10th Symp. Parallel Distrib. Process.*, IEEE, April 1999, pp. 392–398.

[ZR86] W. Zhao and K. Ramamritham, "A virtual time CSMA/CD protocol for hard real-time communication," in *Proc. 7th IEEE Real Time Syst. Symp. (RTSS'86)*, New Orleans, Louisiana, USA, December 1986, pp. 120–127, IEEE Computer Society.

[ZSR90] W. Zhao, J.A. Stankovic, and K. Ramamritham, "A window protocol for transmission of time-constrained messages," *IEEE Trans. Comput.*, vol. 39, no. 9, pp. 1186–1203, 1990.

16.3 Using UML for Embedded Software and System Modeling

Kleanthis Thramboulidis

Introduction

Embedded Systems (ESs) and mainly distributed embedded systems with multiple processing elements are becoming common in various application areas including telecommunications, robotics, industrial control, automotive, medicine, aerospace, military, etc. The ES market is growing very fast. It has been predicted that, by the year 2010, about 90% of all program code will be implemented for embedded systems. Developers of such systems face additional challenges compared to the developers of traditional software. The systems they have to develop are highly event driven, concurrent, and usually distributed, and they must satisfy stringent requirements for latency, throughput, and dependability. Furthermore, a reasonable behavior is expected from ESs, even under faults of their hardware or software components. Timing, reliability, robustness, and power consumption should be considered as primary-level aspects and not as secondary, as is the case for conventional software development notations and methodologies [Karsai et al., 2003]. These physical characteristics of computation must be included and appropriately handled in the design phase.

To capture the requirements, proceed to the design, and effectively communicate the produced artifacts for embedded systems is a daunting task and demands new approaches that go beyond the traditional hardware-based notations such as HDLs (hardware description languages) and the traditional software-based approaches. Martin [2002] claims that the growing dominance of software in embedded systems design requires a careful look at the latest methods for software specification and analysis. However embedded systems are different in several ways from traditional software systems. The hardware on which they run is often resource-constrained in terms of memory, processor cycles, power, etc., but they must still respond in real time. It is also accepted that a better temporal software model is required. The use of existing proven software engineering concepts and methodologies without extensions and modifications is not a solution to the problem. Edward [2000] argues that "the most pressing problem is how to adapt existing software techniques to meet the challenges of the physical world.." Karsai et al. [2003] claim that "current design and implementation strategies used in the practice do not provide enough support" for today's complex embedded systems. Lee [2002] presents an evaluation of the best-known specification languages for real-time systems with interesting results. In the above context, the unified modeling language (UML) has attracted the interest of many researchers. A number of extensions and modifications to the UML notation have been proposed to address the requirements of embedded systems.

UML was conceived as a general-purpose language for modeling object-oriented software applications and has already been accepted as the new industry standard for modeling software-intensive systems. The language represents a further abstraction step away from the one provided by high-level programming languages, which are close to the underlying implementation technology. UML is methodology-independent, which means that UML can be used as a notation to represent the deliverables defined by the methodology regardless of the type of methodology used to perform the system's development, i.e., whether it is rigorous such as the Unified Process [Jacobson et al., 1999a] or lightweight such as Extreme Programming [Paulk, 2001]. In any case, the methodology will guide the engineer in deciding what artifacts to produce, what activities and what workers to use to create and manage them, and how to use those artifacts to measure and control the project as a whole [Booch et al., 1999].

The 12 diagram types, which can be used to model every aspect of the system, make UML extremely expressive, allowing multiple aspects of a system to be modeled at the same time. However, this expressiveness comes at a price. UML is extremely large with many notational possibilities [LeBlanc and Stiller, 2000]. In this section a subset of the UML that represents the core of the language, is presented. Special emphasis is given to diagrams and constructs that are greatly utilized for the development of ESs. This subset will allow the reader to understand the basic concepts of the language and communicate, through UML diagrams, the knowledge of the ES that is under development. The UML-RT notation, which is utilized by many approaches that support the development of embedded real-time systems, is also considered, and a brief description of the basic constructs of this notation is given. The evolution of the language through the new version 2.0 and its profiles for use in embedded systems as well as in the real-time domain are also considered. Reference is also made to the widely used, by the industry, development methodologies and the most important CASE (computer-aided software engineering) tools that support the engineer in analyzing the application's requirements and designing a solution to meet requirements. Throughout the section, a simplified embedded control system, namely, the teabag boxing control system (TBCS), is used as a running example.

The remainder of this section is organized as follows. In the next subsection the basics of the object-oriented approach and the UML for system modeling are given. In the section "UML Models and Views," the models or views provided by UML to create system models are presented. In the section "The Teabag Boxing System Case Study," the teabag boxing system that is used as a running example throughout the section is introduced. In the section "The Basic Diagrams of the Unified Modeling Language," the most important UML diagrams are presented with examples that highlight their basic constructs. The mechanisms for extending UML in different application domains are discussed in the section "Extending UML." In the section "The UML-RT notation," the UML-RT notation is presented and a brief description of its basic constructs is given. In the section "UML Profile for Schedulability, Performance, and Time," the UML profile for schedulability, performance and time is briefly presented. The most known notations and methodologies for embedded systems are referenced in the section "UML-Based Notations and Methodologies for Embedded Applications and Systems." In the section "The UML 2.0 Specification," UML 2.0, which is the upcoming specification of the UML, is presented, while in the section "UML-based CASE Tools for Embedded Systems," the most known CASE tools that support the development of embedded systems are briefly referenced. Finally, the last section gives a summary of the chapter.

Object-Oriented Modeling and UML

System Modeling and the Object-Oriented Approach

Modeling, which is as old as engineering, is the designing of software applications before coding. It is an essential part of large software projects, and helpful to medium and even small projects as well [OMG, 2003c]. A large number of methods have been proposed over the last few years to support the creation of models for software systems. Until 1989 the proposed methods followed the structured approach with the structured analysis and structured design (SA/SD) created by Demarco [1979] and Yourdon and Constantine [1979], considered to be the most widely used method. An updated version of this method, called modern structured analysis, was published in 1989 by Yourdon [1989]. However, this method did not attain wide acceptance, since (a) the complexity of today's systems is almost impossible to handle with the traditional procedural-like approaches, and (b) many new proposals following the more promising object-oriented (OO) approach appeared at the same time. The great advantage of the OO paradigm, which has been gradually accepted during the last decade, is the conceptual continuity across all phases of the software development process [Capretz, 2003].

Page-Jones and Weiss [1989] state that "the object-oriented approach is a refinement of some of the best software engineering ideas of the past." Capretz [2003] claims that the background of the object-oriented paradigm stems from many different research fields. The most important of them are: system simulation with classes and objects, operating systems with monitors, data abstraction with types and encapsulation, and artificial intelligence with frames. Pressman and Ince [2000] in their introduction to object-oriented concepts state:

> We live in a world of objects. These objects exist in nature, in manmade entities, in business, and in the products that we use. They can be categorized, described, organized, combined, manipulated and created. Therefore, it is no surprise that an object-oriented view would be proposed for the creation of computer software—an abstraction that enables us to model the world in ways that help us to better understand and navigate it.

Therefore, the *object* is the basic concept of the new approach instead of the *process*, which was the basic concept in early structured methodologies. An object, as defined by Booch [1994], "represents an individual, identifiable item, unit, or entity, either real or abstract, with a well-defined role in the problem domain." The object is the atomic unit of encapsulation and is used for the decomposition of the system. An object has:

- State
- Behavior
- Identity

The structure and behavior of similar objects are defined in their common *class* [Booch, 1994]. An object contains attributes that represent the object's state, and operations, called methods, which define the behavior of the object. Methods specify the response of objects to received messages. They accept parameters and have access to an object's attributes; they usually result in a state change. Each object is composed of two parts: the interface and the implementation. The implementation is the inner workings of the object and is hidden from the outside (information hiding). The interface is visible to other objects that use it to get the services provided by the object.

A system is considered as an aggregation of discrete objects that collaborate in order to perform work that ultimately benefits an outside user. The collaboration between objects that compose the system is obtained through the exchange of messages. *Message passing* is a means of communication among objects within an application.

In the 1990s, at least 20 object-oriented methods were proposed in books, and many more were proposed in conference and journal papers. A survey and a classification scheme for object-oriented methodologies can be found in [Capretz, 2003]. In the same article, the author states that "after more than thirty years since the first OO programming language was introduced, the debate over the claimed benefits of the object-oriented paradigm still goes on. But there is no doubt that most new software systems will be object-oriented; that no-body disputes." Wieringa [1998] surveyed state-of-the-art structured and object-oriented development methods. He identified the underlying composition of structured and object-oriented software specifications and investigated in which respect OO specifications differ essentially from structured ones. The last specification method in his catalogue of OO methods is the unified modeling language (UML) version 1.0.

What Is UML

The easiest answer to the question "what is UML?" is, according to Quatrani [2001], the following: "UML is the standard language for specifying, visualizing, constructing, and documenting all the artifacts of a software system." The above terms are explained in [Holt, 2001] as follows: "To *specify* means to refer to or state specific needs, requirements or instructions concerned with a complex system. To *visualize* means to realize in a visual fashion or, in other words, to represent information using diagrams. To *construct* means to create a system based on specified components of that system. To *document* means to record the knowledge and experience of the software in order to show how the software was conceived, designed, developed verified and validated." However, UML can be used to communicate any form of information and should not be limited to software. Holt [2001], for example, applies UML to systems engineering rather than simply to software engineering.

The Evolution of UML

The UML started out as a collaboration between three outstanding methodologists: Grady Booch, Ivar Jacobson, and James Rumbaugh. As a first step Booch and Rumbaugh collaborated to combine the best features of their individual object-oriented analysis and design methods [Booch, 1994; Rumbaugh et al., 1991] and presented the Unified Method ver. 0.8 at OOPSLA in 1995. At that time, the unified method was both a language and a process. Later Jacobson joined the group and contributed the best features of the OOSE methodology [Jacobson, 1992]. The result of this collaboration was the separation of the language from the process, which was later described in [Jacobson et al., 1999b]. The language was defined and was presented in 1996 as UML 0.9. Later, in January 1997, they submitted their initial proposal as UML 1.0. Since then the standardization odyssey of UML has been in evolution.

Kobryn [1999] placed the end of this odyssey in 2001, but UML 2.0 is still under discussion. Crawford [2002] notes that "proposals of the UML2 are now under consideration, and what's clear is the future is unclear." As Miller [2002] argues, UML1 unified several of the competing schools of modeling, however some equally important ideas for good OO design did not influence the language. Furthermore, a lot of problems have been cited [e.g., Kobryn, 1999; Dori, 2002; Kobryn, 2002] and are waiting for a solution in UML2. A spirited debate on the future of UML is in evolution. Five groups submitted proposals in response to the OMG RFPs, but hopefully there is a desire for a consensus version [Miller, 2002]. They all agree that UML2 should be in the context of OMG's model driven architecture[1] (MDA) initiative [Miller and Mukerji, 2001], which considers as primary artifacts of software development not programs, but models created by modeling languages. UML artifacts should be used to create the platform-independent model (PIM), which should then be mapped by a model compiler into a platform-specific model (PSM) as Mellor [2002] states.

[1]According to the OMG, "MDA provides an approach and tools for: a) specifying a system independently of the platform that supports it, b) specifying the platforms, c) choosing a particular platform for the system, and d) transforming the system specification into one for a particular platform."

The large number of practitioners and researchers that have adopted UML at a rate exceeding even OMG's most optimistic predictions is a strong argument for UML 2 to follow an evolutionary rather than revolutionary approach. However, as Selic et al. [2002] state, "the same forces have also created a strong pressure to improve the effectiveness of UML and provide it with a multitude of new features." The one thing that is accepted by all parties is that the language's popularity has confirmed the urgency for a standard communication medium for humans and software tools. It is already widely accepted that UML 2 will drive the software development industry for the next decades.

UML Models and Views

As has already been noted, UML is a language for visualizing, specifying, constructing, and documenting the artifacts of a software-intensive system. A language provides a vocabulary and the rules for combining words in that vocabulary for the purpose of communication. The vocabulary and rules of the UML focus on the conceptual and physical representation of a system. They define how to create and read system models, but they do not define what models should be created and when they should be created. This is why a methodology should be used throughout the development process. Erikson and Penker [1998] discriminated the following three kinds of rules: syntactic, semantic, and pragmatic rules. According to them: (a) the syntax defines how the symbols should look and how they are combined, (b) the semantic rules define what each symbol means, and how it should be interpreted by itself and in the context of other symbols, and (c) the pragmatic rules define the intentions of the symbols through which the purpose of the model is achieved and becomes understandable for others.

A system is represented in UML using multiple models. Each model describes the system from a distinctly different perspective. The following three kinds of views are defined at the top level [Rumbaugh et al., 1999].

Structural Classification

Views of this category consider the things of the system and their relationships with each other. Class, use case, component, and node are the UML classifiers used to represent things. The static view is expressed through class diagrams, while the use case view is expressed through use case diagrams, and the implementation view is expressed through component and deployment diagrams that constitute the classification views of this category.

Dynamic Behavior

Views of this category describe the different aspects of the application's dynamic behavior, i.e., its behavior over time. Views of this category include: (a) the state machine view, which is expressed through statechart diagrams, (b) the activity view, expressed through activity diagrams, and (c) the interaction view, expressed through sequence or collaboration diagrams.

Model Management

Views of this category describe the organization of the system models into hierarchical units. This organization is expressed using class diagrams. The model management view, which uses the package as its generic construct, crosses the other views and organizes the application's models during development and configuration control.

Alternatively, Holt [2001] describes two types of models or views of the system in order for this to be fully defined. These models are the static and the behavioral one. The static model or view of the system shows the "things" or entities in a system and the relationships between them. Four UML diagrams are used to construct the static model of the system: class diagrams, object diagrams, deployment diagrams, and component diagrams. Behavioral models demonstrate how a system behaves over time by showing the order in which things happen, the conditions under which they happen, and the interactions between things. Five UML diagrams are used to construct the behavior model of the system: statecharts, use cases, sequence diagrams, collaboration diagrams, and activity diagrams.

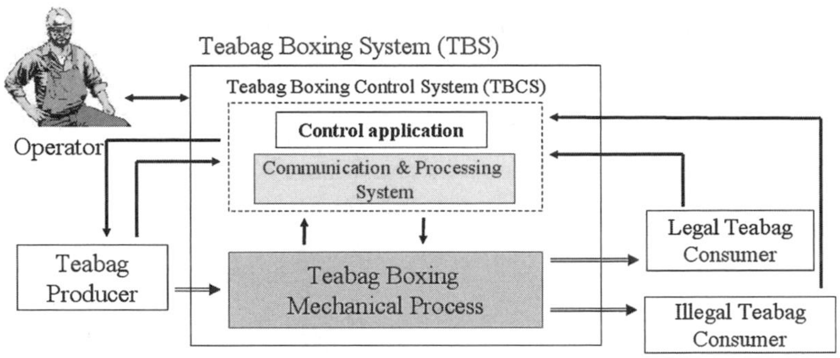

FIGURE 16.15 The teabag boxing system.

The Teabag Boxing System Case Study

We are using as a running example in this section a simplified version of a real system used in the production chain of packed teabags, namely the teabag boxing system (TBS). The TBS receives teabags from a teabag producer, checks their weight, and forwards them either for packaging to a valid-teabag consumer, or for recycling to an invalid-teabag consumer. TBS is composed of: (a) a scale, for weighing teabags, (b) a feeder, for rejecting out-of-range teabags, and (c) three conveyor belts for moving teabags.

Figure 16.15 presents the environment model of the teabag boxing system and its main components: (a) the teabag boxing mechanical process (TBMP) that is the controlled system, and (b) the teabag boxing control system (TBCS) that is the controlling system. Sensors and actuators are embedded in the TBMP.

We consider the development of the mechanical process, namely the TBMP, to be carried out independently of the development process of the controlling system. Figure 16.16 presents an abstract model of the TBMP with the appropriate sensors and actuators.

We assume that each one of the mechanical process subsystems (entities), i.e., the three belts, the scale, and the feeder, has its own processing and memory unit that we call node. Nodes are interconnected with a communication system and constitute the infrastructure on which the embedded software will be executed.

The Basic Diagrams of the Unified Modeling Language

In this section the most commonly used UML diagrams are presented using as a running example the TBCS. These diagrams are: use case, class, statecharts, and sequence and collaboration diagrams. Component and deployment diagrams are also presented, since they are important in distributed embedded systems modeling.

Use-Case Diagram

The use-case diagram was introduced by Jacobson [1992] as the basic artifact for requirements modeling. The diagram, which models the functionality of the system as perceived by outside users, contains a set of actors, use cases, and their relationships. The construct of actor is used to model what exists outside of the system, while the construct of use case is used to model what should be performed by the system. So the

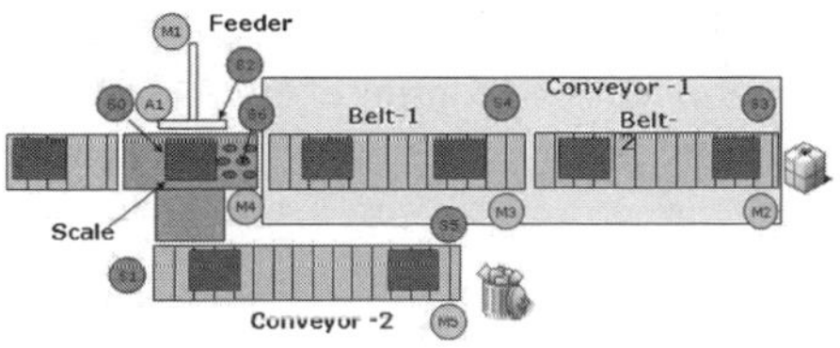

FIGURE 16.16 Abstract model of the teabag boxing mechanical process (TBMP).

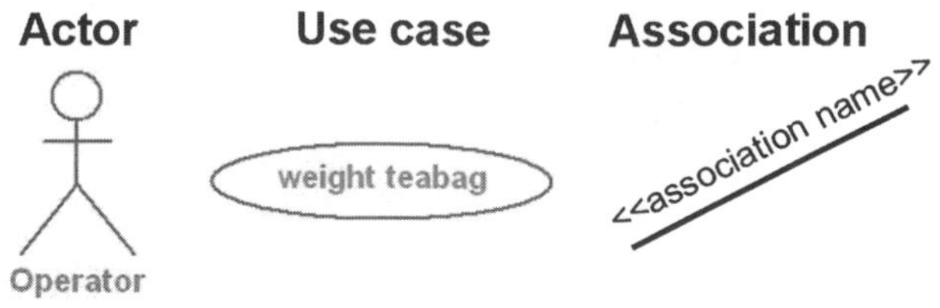

FIGURE 16.17 Basic constructs of use-case diagrams.

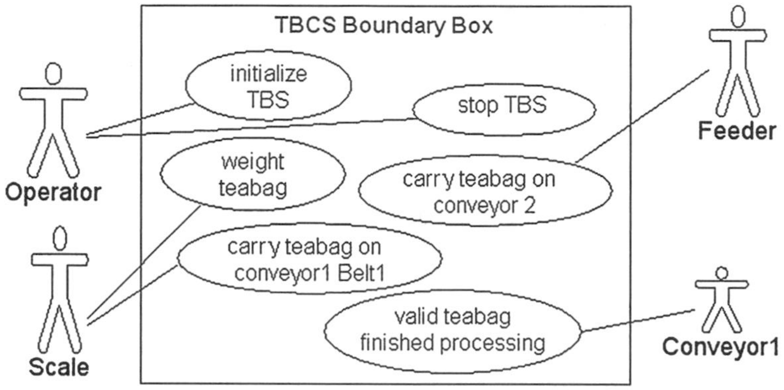

FIGURE 16.18 Example use-case diagram of the teabag boxing control system.

use-case diagram is very close to the set of context diagram and event list as they are defined by the modern structured analysis methodology. Figure 16.17 shows the notational elements of the above constructs of the use-case diagram.

An ***actor*** characterizes and abstracts an outside user (human or system) or related set of users that interact with the system. Actors are the only external entities that interact with the system. A ***use case*** defines a specific way of using the system. It specifies the interaction that takes place between one or more actors and the system. Each use case constitutes a complete course of events initiated by an actor. It defines the behavior of some aspect of the system without revealing its internal structure. However, later, during subsequent analysis modeling, the objects that participate in each use case are determined [Thramboulidis, 2003]. A ***relationship*** represents an association between use cases. *Dependency*, *generalization*, and *association* are the most commonly used relationships.

Figure 16.18 shows part of the TBCS use-case diagram where the "initialize TBS," "stop TBS," "weight teabag," "carry teabag on conveyor 2," "carry teabag on conveyor-1 belt-1," and "valid teabag finished processing" use cases are shown. We must note that the actors and use cases of the TBCS are different from the actors and use cases of the whole system, i.e., the teabag boxing system. From Figure 16.15 it is clear that the actors for the TBS are: operator, teabag producer, legal teabag consumer, and illegal teabag consumer. The following are use cases of the TBS: switch on, switch off, start, reset, box teabag, legal teabag consumer unable to consume, illegal teabag consumer unable to consume. In the rest of this section we consider the development of the TBCS assuming the existence of the teabag boxing mechanical process. An approach for the integrated development of software, electrical, and mechanical subsystems is given in [Thramboulidis, 2005].

For each use case of the use-case diagram, a detailed description is required. Figure 16.19 provides, as an example, the description of the "carry teabag on conveyor-1 belt-2" use case of the TBCS. The diagrams presented in this subsection are drawn using the Rhapsody CASE tool by i-Logix.

Use case: Carry Teabag to conveyor-1 belt-2
Actor (Initiating IPT)**:** Conveyor-1 belt-1 (IPP: Teabag leaving Conveyor-1 Belt-1 (S4))
Involved IPTs: Conveyor-1 Belt-1 M3 (out), Conveyor-1-Belt-2 M2 (out)
Precondition: Conveyor1-Belt-1 is in the MOVING state.
Description
The Conveyor-1 Belt-1 notifies the system (through S4) that a teabag has moved out from conveyor-1 belt-1. The system checks the number of teabags (internal representation) on conveyor-1 belt-2. If the number of teabags is zero it sends a MOVE FORWARD message to conveyor-1 belt-2 (M2) and increments the number of teabags. The system decrements the number of teabags (internal representation) on conveyor-1 belt-1. If this number is zero the system sends a STOP message to conveyor-1 belt-1 (M3). The use case is terminated.
Postcondition: Conveyor-1 belt-2 is in the MOVING state.
Exceptions: [tbd]

FIGURE 16.19 Description of the "carry teabag on conveyor-1 belt-2" use case.

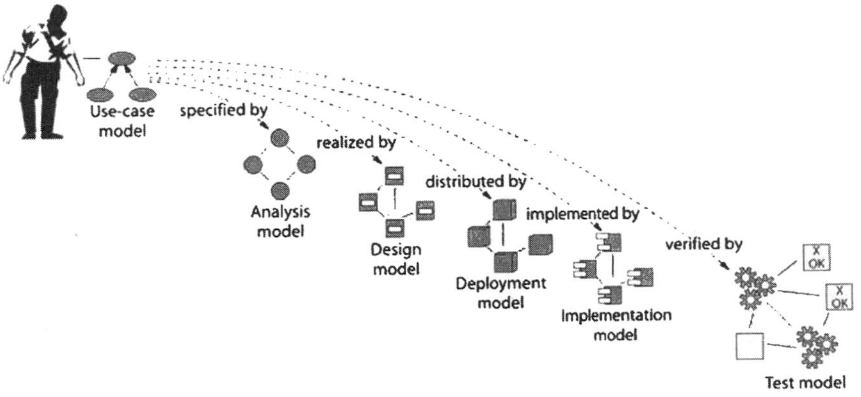

FIGURE 16.20 Use-case model dependencies with the other models of the development process.

Use-case diagrams were adopted by the majority of OO methodologies as a means for capturing and documenting requirements. Lee and Xue [1999] note this when they state:

> Use case approaches are increasingly attracting attention in requirements engineering because the user-centered concept is valuable in eliciting, analyzing and documenting requirements. One of the main goals of requirements engineering process is to get agreement on the views of the involved users, and use cases are a good way to elicit requirements from a user's point of view. An important advantage of use case driven analysis is that it helps manage complexity, since it focuses on one specific usage aspect at a time.

Use-case models play a key role in the development process. They are used as input to several activities in the software development process. Jacobson et al. [1999b] note this by capturing the dependencies between the use-case model and the other models in the context of the Unified Process. Figure 16.20, taken from [Jacobson et al., 1999a], highlights the core role of use-case models in the development process.

Although use cases were first introduced as a means to capture functional requirements, there is a trend to expand the whole concept to cover nonfunctional requirements too. Use-case models can be used, for example, to capture the performance requirements in the case of real-time systems. Moore and Cooling [2000] characterize use case as either periodic or aperiodic. In the first case in which the use case happens regularly at constant time intervals, performance is expressed as the period of the use case and optionally as the proposed maximum processor load. In the later case, in which the use case happens unpredictably at random time intervals, performance is expressed normally in terms of a maximum response time and sometimes a minimum response time and often a specified tolerance. The requirement is normally expressed as an end-to-end time, but it can also be expressed as a message transmission time, or the time to execute a particular operation. Alexander [2003] presents a way of using use and misuse cases to elicit security requirements. A misuse case is a use case from the point of view of an actor hostile to the system under development. Misuse cases have many interesting applications and interact with use cases in interesting and helpful ways. A use/misuse analysis can complement existing analysis, design, and verification practices.

It is widely accepted that the quality of the use-case model has an important impact on the quality of the resulting software product. Software inspection, one of the most efficient methods for verifying software, may be used to ensure the quality of the use-case model. Anda and Sjoberg [2002] present a taxonomy of typical defects in use-case models and propose a check list-based inspection technique for detecting such defects.

UML 2.0 introduces minor changes to the use-case constructs defined in UML 1.x. In fact, there are no changes regarding use-case diagrams from UML 1.x; only some aspects of notation to model element mapping have been clarified as is stated in [OMG, 2003a].

Class Diagram

The class diagram can be considered as an evolution of the entity-relationship diagram (ERD), the main diagram used for many years in information modeling. The big difference between the two diagrams is that the class, which is the basic construct of the class diagram, has, in addition to the structure of the entity, behavior that is described by a set of operations. A class diagram is composed of classes and their relationships.

Berard [1993] defines the class in the following way: "A class is often defined as a template, description, pattern or blueprint for a category of very similar items. Classes are used as templates or 'factories' for the creation of specific items that meet the criteria defined in the class." An implementation-influenced definition is given by Rumbaugh [1991]: "A class describes a group of objects with similar properties (attributes), common behavior (operations), common relationships to other objects, and common semantics."

A class is represented in UML by a rectangle that is divided into three compartments, as shown in Figure 16.21. This figure represents the `BeltSR` class that is the software representation of the corresponding real-world entity of the mechanical process. The fact that the class Belt represents an entity of the mechanical process is denoted in the figure with the name of the <<MP EntityImage>> stereotype. This stereotype was defined to represent the images of the mechanical process entities in the model. Stereotypes are an extension mechanism of UML and will be described in subsection "Extending UML." The part in the top of Figure 16.21 contains the name of the class (`BeltSR`), while the specific icon in the upper corner means that the dynamic behavior of the class is defined by a statechart. The second part contains the attributes, while the

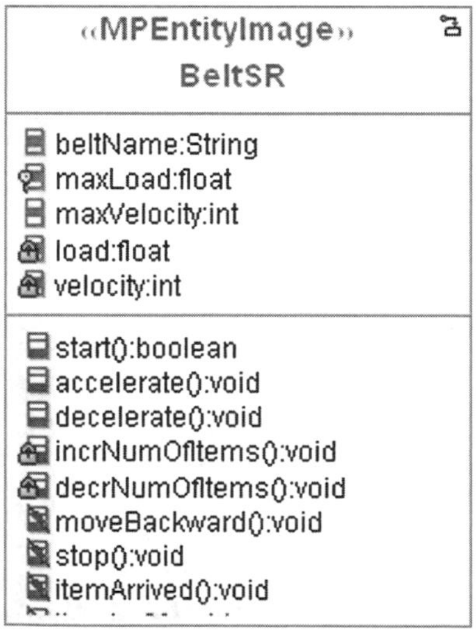

FIGURE 16.21 The `BeltSR` class.

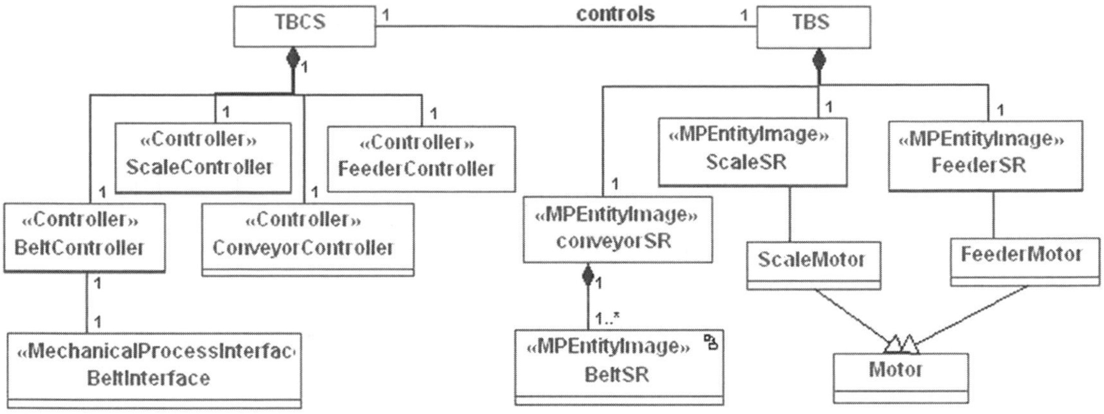

FIGURE 16.22 Part of the class diagram of the teabag boxing control system.

last part contains the operations. For each attribute, the name, the type, and the visibility are shown. Specific symbols are used to represent the visibility of each attribute, i.e., public, protected, and private. `beltName`, `maxLoad`, `maxVelocity`, `load`, and `velocity` are the only attributes of the `BeltSR` in this representation; `moveBackward`, `stop`, `itemArrived`, `itemLeft`, `accelerate`, `decelerate`, etc. are operations that define the behavior of any instance of this class. `IncrNumOfInstances` and `decrNumOfInstances` are private operations, while the rest are public. The accelerate operation specifies the behavior of the belt to the message, "accelerate," that instructs the belt instance to increase its speed with a predefined value. The `itemArrived` operation specifies the behavior of `beltSR` instances to the message `itemArrived` that informs the corresponding instance that an item has left the belt.

Relationships between classes include: associations, composition and aggregation relationships, generalization/specialization relationships, and the various kinds of dependency such as realization and usage.

An association defines a relationship between two or more classes, denoting a static, structural relationship between them. The association `Controls` in Figure 16.22 defines a relationship between `TeabagBoxingControlSystem` (TBCS) and TBS. An association is inherently bidirectional. Multiplicity must be specified for both sides of the association to define the number of instances of one class that may relate to a single instance of the other class. For the Controls association, for example, the multiplicity is "*one-to-one*." This means that a TBCS controls only one TBS, and a TBS has to be controlled by only one TBCS.

Composition and aggregation are used to represent the relationships between the parts and the whole. Composition is stronger than aggregation, which is stronger than association. The composition represents a physical relationship between the whole and its parts. Aggregation is likely to be used to model conceptual classes rather than physical classes. A part may belong to more than one aggregation. An example of composition relationship is the one connecting the TBCS in Figure 16.22 with its physical components `BeltController`, `ScaleController`, `ConveyorController`, and `FeederController`.

The generalization/specialization (gen/spec) relationship is used to represent the special kind of association between two classes in which one class represents a general concept and the other a specialization of this general concept. The class `Motor` in Figure 16.22, which is referred to as a *superclass* or *ancestor* class, represents the abstract concept Motor. It is used to abstract common structure and behavior of both scale and feeder motors. The class `ScaleMotor`, which is referred to as a *subclass or descendent* class, represents the specific type of motor used in the Scale. Each subclass inherits the properties of the superclass but then extends these properties in different ways. Some authors call this relationship inheritance. However, inheritance is the mechanism of the implementation environment used to implement the gen/spec relationship.

Statechart Diagram

Statecharts were defined by Harel [1987] as a notation that extends finite-state machines. Statecharts, overcoming the limitations of traditional finite state machines (FSMs) while retaining the benefits of

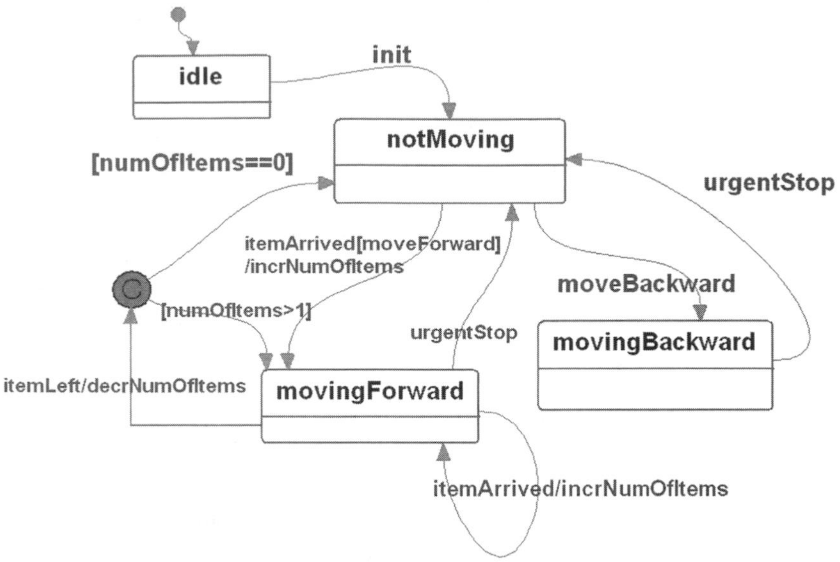

FIGURE 16.23 A simplified statechart diagram for the `BeltSR` class.

finite-state modeling, became popular for modeling system behavior in the traditional structured analysis paradigm. Coleman et al. [1992] extended statecharts with three basic extensions, i.e., hierarchy, concurrency, and broadcast communication, to produce modular, hierarchical, and structured descriptions in the context of the object-oriented design. Harel and Gery [1997] claim that statecharts form the core of the emerging unified modeling language to produce a fully executable language set for modeling object-oriented systems. Modern CASE tools exploit statecharts: (a) to allow the developer to produce fully executable models, and (b) to obtain automatic code generation.

The statechart diagram is used to capture the behavior of the system. To avoid the "state explosion" problem, a statechart in UML specifies all behavioral aspects of the instances of a class. A statechart contains states connected by transitions. States are represented by boxes with rounded corners, while transitions are shown as arrowed lines. Transitions are fired by events; they may cause the execution of actions attached to the transition, and they usually take the object (class or instance) to a new state. A transition usually has: (a) a source and a target state, (b) a guard condition that is optional, (c) an event trigger that fires the transition if the guard condition is true, and (d) optionally an action that is an executable atomic computation that may act on the same or other objects.

Figure 16.23 shows a simplified version of the statechart of the `BeltSR` class of the TBCS. When the TBCS is initialized, each `BeltSR` instance is in the `idle` state, which is called initial state. When a `BeltSR` instance accepts the `itemArrived` message and the belt is in the `moveForward` mode (i.e. the guard of the transition is true), the transition is fired, the defined action is executed (i.e., the `incrNunOfItems` operation is executed), and the `BeltSR` instance changes its state to `movingForward`. An `urgentStop` message fires the transition from the `movingForward` state to the `idle` state.

A state may have entry and exit actions, i.e., actions that are executed on entering and exiting the state, respectively. Internal transitions, i.e., transitions that are handled without changing the state, as the one fired by the event `itemArrived` when the `BeltSR` is in the `movingForward` state, are also allowed. A state may have substates. This feature supports the construction of statecharts of different levels of abstraction and allows concurrency to be handled by using substates [Coleman et al., 1992].

Sequence Diagram

Class diagrams model the static perspective of an object-oriented system. During the execution of such a system, objects that constitute the system interact with each other to provide a higher-level behavior.

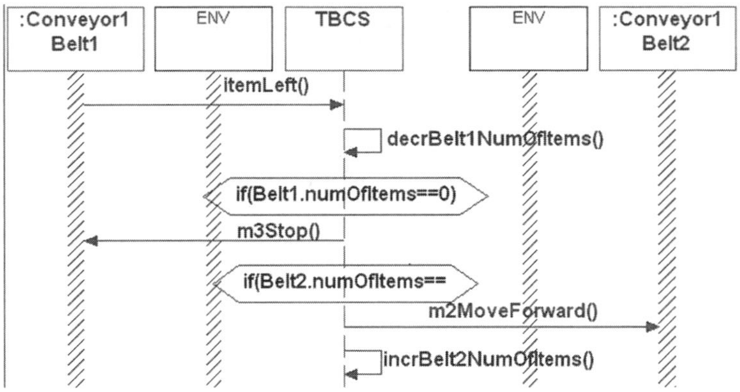

FIGURE 16.24 First-level sequence diagram for the "carry teabag on conveyor-1 belt-2" use case.

Interaction diagrams in the form of sequence or collaboration diagrams are used to model these interactions, which capture the dynamic aspects of the system.

The sequence diagram shows the sequence of messages, which are exchanged among roles that implement the behavior of the system, arranged in time. It shows the flow of control across many objects that collaborate in the context of a scenario, i.e., the individual history of a transaction. It is clear that the sequence diagram "does not tell the complete story" of the scenario. There are usually other legal and possible traces in the context of the same scenario that must be documented.

The sequence diagram is used in early development phases for the realization of use cases. Thramboulidis in [2004] describes the use of sequence diagrams for the use-case realization in two levels of abstraction: (a) the first-level component interaction diagram (CID), which captures the interactions of the application, represented as a component, with external entities, and (b) the second-level CID (detailed-CID), in which the components required to compose the system for the described behavior to be obtained should be identified and their collaboration defined. Figure 16.24 presents the first-level CID of the "carry teabag on conveyor-1 belt-2" use case, while Figure 16.25 presents the detailed CID of the same use case. This CID is considered as the first formal realization of the corresponding use case.

Time is represented vertically in the diagram, so the first message in the sequence diagram of Figure 16.25 is the `itemLeft` that is sent from the corresponding sensor of belt-1 of conveyor-1 to the `conv1Belt1` instance of the `BeltInterface` class.[1] The instance of the `Conv1Belt2Controller` class of the sequence diagram sends the message `itemArrived` to the `Belt2` instance. The `Belt2` instance checks the number of items on the belt and, if these are 0, sends a `moveForward` message to `Conv1Belt2 Controller` and then executes its `incNumOfItems` operation to update the number of items. The `Conv1Belt2Controller` passes a `moveForward` message to the M2 motor of Belt-2 of conveyor-1.

Sequence diagrams can be used to express the performance requirements that have been captured in the corresponding use-cases. These high-level response requirements are split either top-down or bottom-up among the components that constitute the system and are expressed as end-to-end times, message latency, or operation duration in sequence diagrams, as is shown in Figure 16.26 taken from [Moore and Cooling, 2000].

Sequence diagrams differ from collaboration diagrams in the following features [Booch et al., 1999]:

1. The sequence diagram uses the lifeline to highlight the existence of the object for the duration of the interaction. Usually the objects that appear in an interaction are in existence for the whole duration of

[1]The notation used to identify the objects in the sequence diagram is <instance name>:<class name>, with at least one of the constituent parts to be present for a valid identification. It must be noted that <instance name> is a unique identifier and not the actual name of the object represented by the instance.

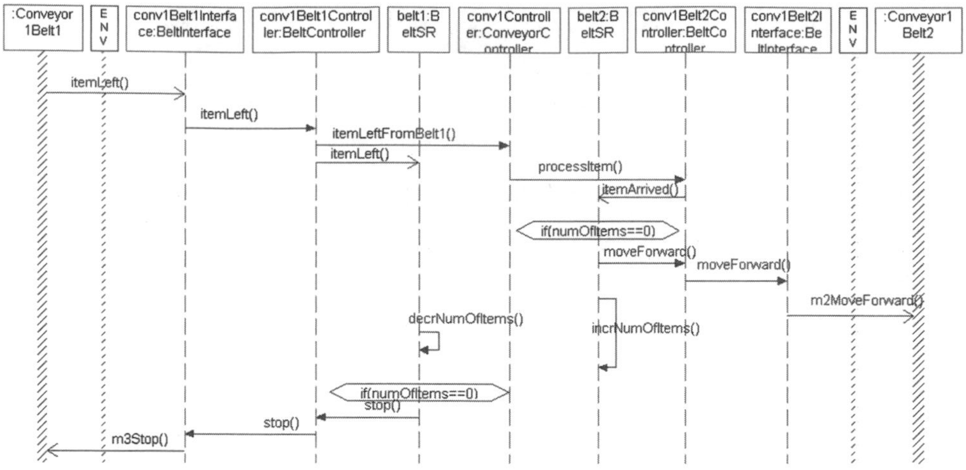

FIGURE 16.25 Detailed sequence diagram for the "carry teabag on conveyor-1 belt-2" use case.

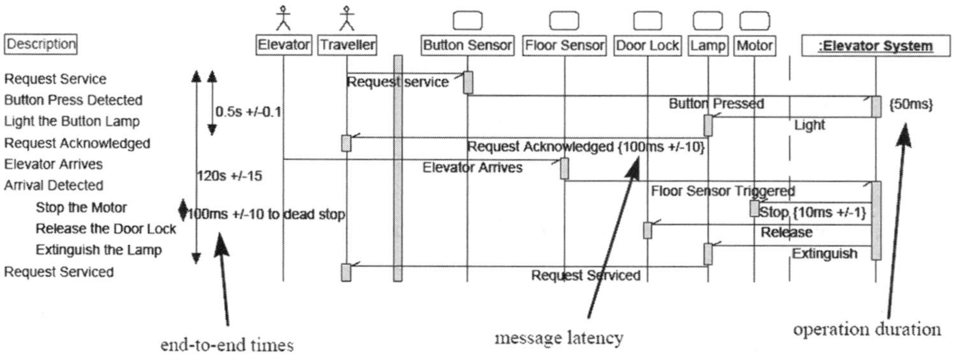

FIGURE 16.26 Performance requirements expressed in sequence diagrams. [From Moore and Cooling, 2000.]

the interaction. The lifelines of these objects are drawn from the top of the diagram to the bottom. However, an object may be created or destroyed during the interaction, and its lifeline starts after the corresponding message for creation (<<create>> stereotype) and ends with the corresponding message for destruction (<<destroy>> stereotype).

2. The sequence diagram shows the focus of control represented by a thin rectangle that shows the period of time during which the object is active, i.e., performs an action as a response to a message that it has accepted.

Interaction diagrams can be used, except from realizing use cases and the precise definition of interobject communication, in the detailed design phase, when the testing of the system is performed. During testing, the traces of the system are described as interactions, and then they are compared with those of the earlier phases of development.

Collaboration Diagram

A collaboration diagram represents the objects that collaborate for the specific behavior to occur and the meaningful links between them. It captures some of the dynamic aspects of message passing required for the realization of a particular behavior. Objects are represented by rectangles, while links representing associations in the context of the particular collaboration are shown by lines connecting rectangles. Sequence numbers preceding to message descriptions indicate the sequence of messages in time. Collaboration diagrams are used during the design phase to model use cases as well as to show the implementation of an operation provided by an object in collaboration with other objects. An example of a collaboration diagram from the TBCS is given in Figure 16.27. From this diagram, which was drawn with Rational ROSE Enterprise edition 2000, it is clear that specific notation is required to express concurrency in the model. UML 2.0 provides solutions to this problem.

Collaboration diagrams help in determining the operations of objects, because the arrival of a message to an object usually invokes an operation. However, during analysis, emphasis is on capturing the information passed between objects rather than the operations invoked [Gomaa 2000]. Later during design, it is decided if two different messages arriving at an object invoke different operations or the same one. In the latter case, the name of the message is handled as a parameter of the operation. The decision for the kind of message, i.e., synchronous or asynchronous, is postponed to the design phase.

Collaboration diagrams have two features that distinguish them from sequence diagrams [Booch et al., 1999]. The first one can be used to show how an object is linked to another. A path stereotype such as <<local>> is attached to the receiver's object end of a link to show that the receiving object is local to the sender. The second one is the sequence number that is used to indicate the time order of a message. The sequence number prefixes the message and increases monotonically for each new message in the flow of control. Dewey decimal numbering, e.g., 1.1 for the first message nested in message number 1, is used to indicate nesting in messages.

UML 2.0 refines interaction diagrams. It defines that, depending on their purpose, interaction diagrams can be displayed in several different types of diagrams that include: sequence diagrams, interaction overview diagrams, and communication diagrams. Timing diagrams as well as interaction tables are defined as optional diagrams. Sequence diagrams specify the message interchange between lifelines. Communication diagrams (a term used instead of collaboration diagrams) show, as stated in [OMG, 2003b], interactions through an architectural view. The arcs between the communicating lifelines are annotated with the descriptions of the passed messages and their sequencing. Interaction overview diagrams are a variant of activity diagrams. They focus on the overview of the flow of control where the nodes are interactions. Lifelines and messages do not appear at this overview level, providing a higher level, of abstraction and a way of organizing sequence diagrams in terms of higher level functionality.

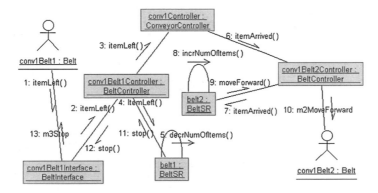

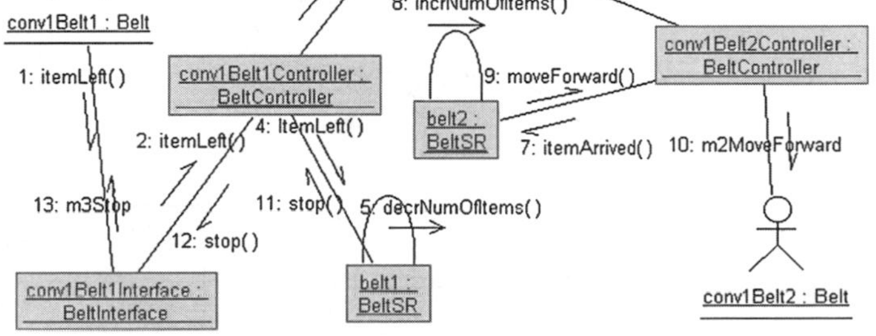

FIGURE 16.27 A simple collaboration diagram from the TBCS.

The above diagrams as well as a number of new constructs introduced by UML 2.0 dramatically reduce the number of interaction diagrams required for the modeling of the system. It allows the expression of parallelism, alternatives, and iterations in the same sequence diagram. It also supports decomposition of a lifeline into a more detailed sequence diagram to allow for different abstraction levels to be applied in the development of sequence diagrams.

The major changes introduced by UML 2.0 in interaction diagrams should allow CASE tool vendors to incorporate in their tools more facilities for model animation, system testing, verification, and validation that are very important, especially for embedded systems' development.

Component Diagram

The constructs of component diagrams are: the component, the relationship, and the interface. A component is a physical and replaceable part of a system that conforms to and provides the realization of a set of interfaces [Booch et al., 1999]. According to this definition, components have many similarities with classes. However, components: (a) are at a different level of abstraction than classes, (b) are physical implementations of a set of other logical elements, such as classes and collaborations, (c) represent physical things while classes represent logical things, and (d) in general have no attributes but only operations that are reachable through their interfaces. A practical way to decide how to model a thing is the following: if the thing you are going to represent lives directly on a node, model it as a component, otherwise use a class [Booch et al., 1999]. Component names usually include extensions such as c, cpp, java, and dll, depending on the target run-time platform. Interfaces represent the functionality of components that is directly available to the users of the component. Relationships are used to represent in the model existing conceptual relationships between the real-world components.

Component diagrams can be used to model: (a) executables and libraries, (b) tables, files, and documents, (c) APIs, and (d) source code. An example of modeling source code is given in Figure 16.28, which presents a component diagram showing a possible organization of the source code of the reverse polish calculator program given by Kernighan in the corresponding C language book [Kernighan and Ritchie, 1988]. The component diagram is composed of components representing. c files as well as the .h files and their relationships that represent the dependencies between these files. These dependencies represent compilation dependencies.

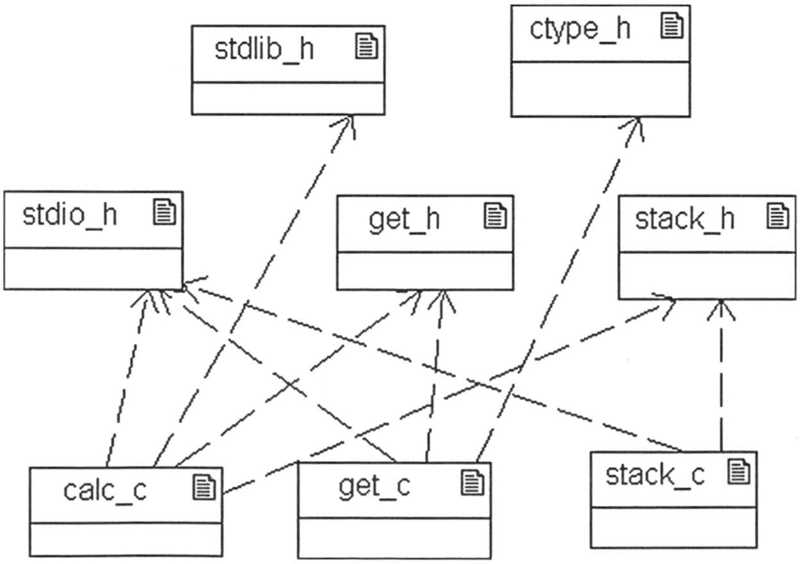

FIGURE 16.28 Component diagram for the organization of source code.

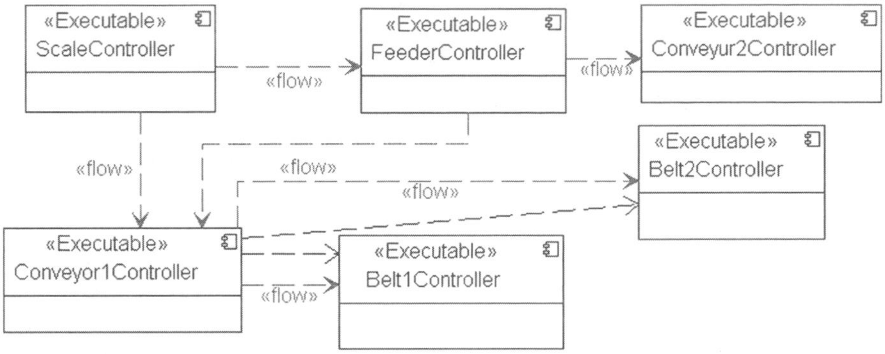

FIGURE 16.29 Component diagram showing a possible configuration of the executable release of the TBCS.

Component diagrams may also be used to represent the model of the executable release of the system as well as the run-time model of the system. The component diagram used for the modeling of executable code includes, except for the system components, all other components required for the system to operate. Different configurations of the system can be shown using run-time component diagrams. Figure 16.29 presents such a possible configuration for the TBCS.

Each component of the system may execute on its own physical node, or they may all execute on the same node. Hybrid configurations are the most common situation in today's embedded systems. Communication between components is by means of messages. Each component must be able to operate independently of the other components for a significant period of time, even when other components are temporarily unavailable. Components of type controller are usually executed on dedicated nodes to obtain the required performance, otherwise performance analysis is required to obtain predictability.

UML 2.0 expands the concept of component to make it more applicable throughout the modeling life cycle since the focus of UML 1.x was only for the implementation phase [OMG, 2003b]. UML 2.0 defines the concept of component, which is used to represent an executable element of the system, as a specialized class that has an external specification in the form of one or more provided and required interfaces, and an internal representation consisting of one or more classifiers that realize its behavior. This definition addresses the requirements of component-based development and component-based system structuring where the reuse of previously constructed components is of primary interest. The component, which can be considered as an autonomous unit within the system: (a) is defined early in the development process and is successively refined into deployment and run time, and (b) is considered as a substitutable unit that can be replaced at design time or run time by another component of the same type. UML 2.0 introduces the concepts of delegation connector, assembly connector, and the concept of realization and results in more complicated component diagrams. These diagrams however are not yet supported by the existing CASE tools.

A system composed of components can be extended by adding new component types that add new functionality. This reconfiguration can be supported for real-time systems even during run time with a suitable execution environment. It is possible to upgrade components in the TBCS even during run time or to expand it to include more belts or conveyors without modifying the already existing components of the system.

Deployment Diagram

Deployment diagrams are composed of nodes and connections. A node represents a computational resource, which generally has at least some memory and, often, processing capability. A node is required for the execution of a system's components, which are defined in component diagrams. Each node may have its own attributes and operations to model the structure and behavior of the physical element. A deployment is the allocation of a component, type, or instance to a node. Deployment diagrams along with component diagrams are used to define the physical perspective of the system [Booch et al., 1999]. Components in component diagrams are used for the physical packaging of the logical things defined in the logical perspective, i.e., classes, interfaces, collaborations, interactions, and statecharts. Nodes in deployment diagrams are used to represent the hardware on which these components are deployed and executed. Dependency, generalization, and association relationships can be defined among nodes. Nodes can be organized by grouping them in packages in the same way that classes are organized into packages. An association represents a physical connection among nodes such as a serial or parallel line, an Ethernet connection, or even a satellite link.

Deployment diagrams are used, according to [Booch et al., 1999], to model processors and devices as well as the distribution of components. Diagrams of the first category are used to model the topology of the embedded distributed system, while diagrams of the second category are used to visualize or specify the physical distribution of the embedded systems components across the processors and devices that constitute the run-time execution environment of the system. Dynamic migration of components from node to node, required by an agent-based or high-reliability system, is also supported. Figure 16.30 presents a deployment diagram that specifies the distribution of the TBCS's components on the run-time infrastructure of the system.

UML 2.0 introduces many changes to the deployment diagram and upgrades them to first-class diagrams for the development of embedded systems. It defines new constructs and refines existing ones to create a set of constructs that can be used to define the execution architecture of the system that represents the assignment of software artifacts to nodes. It extends the definition of deployment to mean the allocation of an artifact or artifact instance to a deployment target. Nodes may be used to represent not only hardware devices, but also software execution environments. The term CommunicationPath is used to denote an association between two nodes through which they are able to exchange signals or messages. Artifacts represent concrete elements in the physical world that are the result of the development process. This means that artifacts in UML 2.0 can manifest any PackageableElement and not just Components as in UML 1.x. Artifacts may have composition associations to other artifacts that are nested within it. This allows the deployment descriptor artifact of a component to be contained within the artifact that implements the component, and in this way the implementation and deployment descriptor can be deployed to a node instance as one artifact instance [OMG, 2003b].

UML 2.0 defines new constructs such as: Device, ExecutionEnvironment, and DeploymentSpecification. A device is a physical computational resource with processing capability upon which artifacts may be deployed

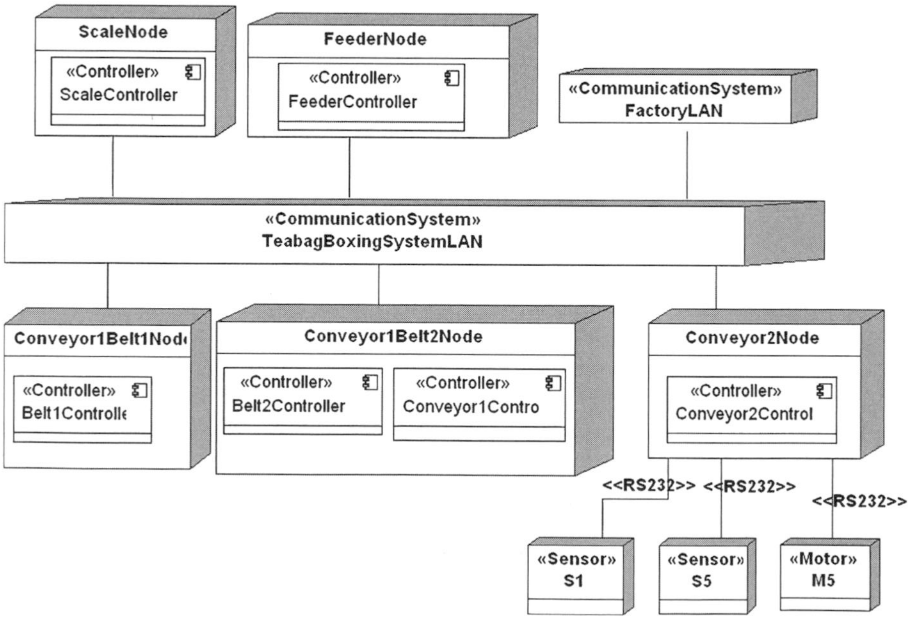

FIGURE 16.30 Deployment diagram showing a possible distribution of TBCS's components.

for execution. A complex device consists of other devices. The ExecutionEnvironment is a node that offers an execution environment for specific types of components that are deployed on it in the form of executable artifacts. A DeploymentSpecification is a set of properties that determines execution parameters of a component artifact that is deployed on a node.

UML 2.0 also allows for extensions through profiles or meta-models in order to satisfy the needs for modeling specific hardware and software environments that require more elaborated deployment models. However, it is estimated that a lot of time and effort is required for these concepts to be mature enough and incorporated into the development process of embedded systems.

Extending UML

UML 1.4 is currently used for modeling of just about any type of application, running on any type and combination of hardware, operating system, programming language, and network. The language's flexibility allows the modeling of distributed applications on any middleware on the market. It is used at different application domains and different levels of abstraction in the system's development process. Examples of application domains where UML has achieved growing acceptability during the last years are embedded systems [Svarstad et al., 2001], real-time systems [Selic, 2000], and industrial automation and control systems [Thramboulidis, 2001; Young et al., 2001; Heck et al., 2003; Thramboulidis and Tranoris, 2004].

UML's extensibility mechanisms, namely stereotype, constraints, and required tags, are used to create profiles for specific application domains. These profiles are primarily intended to tailor UML towards a specific domain by giving to modelers: (a) access to common model elements, and (b) terminology from the specific domain. A profile is a way of packaging UML specializations. Examples of such profiles are: UML for real-time modeling, UML for data modeling, UML for Web modeling, UML for business modeling, UML profile for schedulability, performance and time [OMG, 2002], and UML profile for CORBA (common object request broker architecture).

A number of other profiles have been proposed. Kobryn [2000], for example, explores the synergies between object modeling and component modeling, by focusing on UML component modeling capabilities to examine the possibilities of using the language to support the leading enterprise component architecture standards EJB and COM +. He claims that standard UML profiles should be defined for these specific component technologies. Moreover, he believes that as components enter further into mainstream business computing, UML will evolve from object-oriented to component-based modeling language, as is the case for UML 2.0. Yacoub and Ammar [2001] describe an approach that utilizes UML modeling capabilities to construct applications using patterns. They define techniques and composition mechanisms by which patterns can be integrated and deployed in the design of software applications. Design patterns promise early reuse benefits at the design stage [Gamma et al., 1995]. As Larsen [1999] states,

> Designing systems using components and proven solutions elevates the abstractions at which engineers work. Productivity and quality are the two main drivers this approach brings. *Productivity* will be positively affected by using abstractions for analysis, design and development. *Quality* will be positively affected by reusing known, proven solutions and the components that implement them.

Larsen [1999] discusses and illustrates the use of UML for designing component-based frameworks using patterns. A good interrelationship between patterns, frameworks, and components can also be found in his article. To enhance the UML's usefulness as a graphical programming language, Bjorkander [2000] proposes a merging of UML with the specification and description language (SDL) [Ellsberger et al., 1997]. The objective for this combination is to permit the expressive power of UML to coalesce with SDL's strengths of coherence and semantics. Medvidovic et al. [2002] assess the UML's expressive power for modeling software architectures in the manner in which traditional software architectural languages (ADLs) model architectures. They present two strategies for supporting architectural concerns within UML. One of these strategies uses UML without any extension, while the other incorporates to UML, using its extension mechanisms, useful features of traditional ADLs.

Martin [2002] argues that the development of the UML and a number of extension proposals in the real-time domain holds promise for the development of new design flows that move beyond static and traditional partitions of hardware and software. In the same paper Martin surveys the requirements for system-level design of embedded systems and gives an overview of the extensions required to UML to address these requirements. He also discusses how the notions of platform-based design intersect with a UML-based development approach.

The UML-RT Notation

The use of the object-oriented paradigm in real-time systems has caught on more slowly than in general-purpose applications [Bihari and Gopinath, 1992]. However, the problems that caused such a delay have already been confronted, and the OO paradigm is becoming the model of choice for the majority of real-time systems. A number of research teams have been exploring real-time object-oriented languages, usually as extensions to existing OO languages. Even more OO analysis and design methods were considered as a means to confront the complexity of the new generation of real-time embedded systems. Selic et al. [1994] describes the real-time object-oriented modeling (ROOM) that brings together the power of object-oriented concepts tailored specifically for real-time systems, with an iterative and incremental process that is based on the use of executable models. ROOM is presented as a systems-development methodology composed of the following three elements: (a) a modeling language, (b) modeling heuristics, and (c) a framework for organizing and performing development work. Douglas [1999] describes an embedded systems programming methodology for the development of real-time systems. This methodology uses: (a) best practices from object technology, and (b) the industry standard UML. Gomaa [2000] presents COMET (concurrent object modeling and architectural design method), an object-oriented methodology that covers requirements modeling and analysis and design of distributed and real-time applications. A number of other approaches that address the analysis and design of real-time systems using the object-oriented paradigm have also evolved during recent years.

One of the most commonly used UML variant for modeling of real-time systems is the UML-RT notation that combines ROOM with the basic notation of UML. UML-RT utilizes the concepts that have been originally defined in the ROOM methodology and used for years for the development of real-time systems using the object-oriented programming paradigm. UML-RT represents these concepts using the extension mechanisms of UML to provide a UML profile for the real-time domain. Cheng and Garlan [2001] argue that UML-RT provides a particularly natural home for architectural modeling, since it provides considerable support for modeling the run-time structure and behavior of complex systems. UML-RT adopts the concept of connector between components as a protocol that is very closely aligned with current thinking in the architecture description language (ADL) research community.

UML-RT focuses on the category of real-time systems that are characterized as complex, event driven, and usually distributed. Selic and Rumbaugh [1998] state that:

> Automatic control applications belong to this category of systems along with telecommunications, aerospace and defense systems. For the development of such applications the architecture plays the primary role. A well-defined architecture must not only support the construction of the initial system but must easily accommodate the system's evolution that is forced by new system requirements. Real-Time UML incorporates among other things the concepts of ROOM that constitute a domain specific architectural definition language. These concepts have proven their effectiveness across hundreds of diverse large-scale industrial projects.

The remainder of this subsection is divided into two parts. The first describes the constructs used for modeling the structure of the system, while the second describes the constructs used for modeling behavior. The description is mainly based on [Selic and Rumbaugh, 1998], [Selic, 1999] and [Gullekson, 2000], where a detailed presentation of the UML-RT constructs and the way they are used for the modeling of real-time systems can be found.

Modeling the Structure of the System

Class diagrams and collaboration diagrams are used to capture the logical structure of the system. Entities that constitute the system, as well as the relationships between them such as communication and containment relationships, are identified. Class diagrams capture universal relationships among classes, while collaboration diagrams capture relationships that exist only within a particular context, as for example within the context of a use case. The following three constructs are defined to model the structure of the system:

- Capsules
- Ports
- Connectors

The *capsule* is defined using the stereotype extension mechanism of UML as a specialization of the general UML concept of class. The capsule is used to represent an entity of the system that has its own thread of control, i.e., an active object. Figure 16.31 shows the `BeltController` capsule of the TBCS.

A capsule interacts with other capsules or the environment through one or more boundary objects called ports. There are no public operations or other public parts to support the interaction of the capsule with the environment. Figure 16.32 shows the `conv1Controller` capsule to interact with the `conv-1Belt1Controller` capsule.

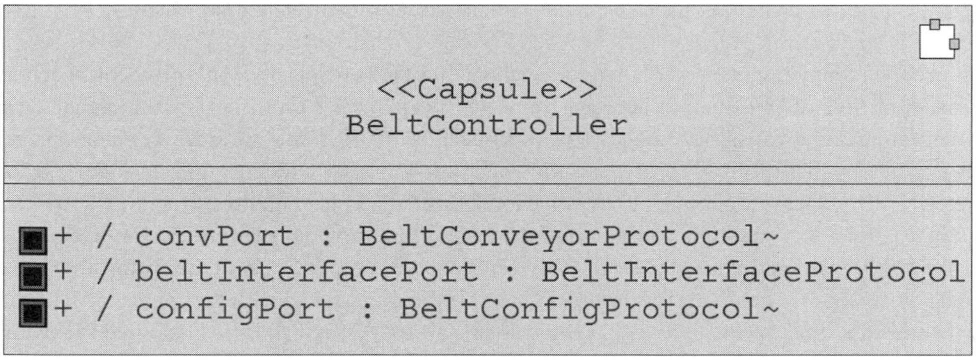

FIGURE 16.31 The `BeltController` capsule.

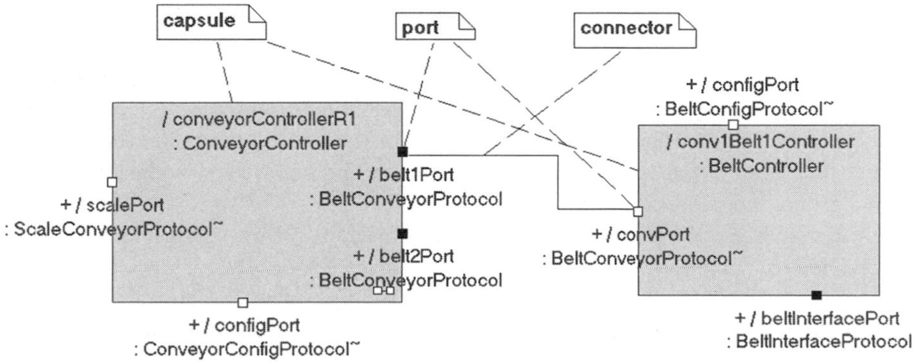

FIGURE 16.32 Capsules interact with each other through one or more boundary objects called ports.

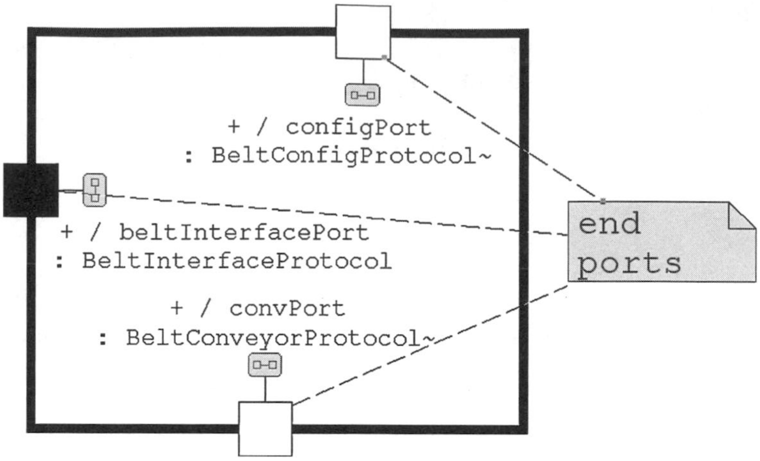

FIGURE 16.33 Ports of the `BeltController` capsule.

Port is an object that implements a specific interface of the capsule and plays a particular role in the collaboration that the capsule has with other objects. Figure 16.33 shows the ports that the `BeltController` capsule uses to collaborate with its environment. These are `convPort`, `beltInterfacePort`, and `configPort`.

To capture the complex semantics of these interactions, ports are associated with protocols. Each protocol defines the valid flow of information between connected ports, i.e., it captures the contractual agreements between communicating capsules. Each port plays a specific role in some protocol, i.e., it implements the behavior specified by that protocol role. This is why each port has its own identity and state. Ports do not map directly to UML interfaces, which are purely behavioral without implementation structure.

A *connector* is used to represent a communication channel that provides the transmission facilities that are required for the implementation of a connection between capsule ports. Connectors can only be used to interconnect ports that play complementary roles in the protocol associated with the connector. The difference between connectors and protocols is that protocols are abstract specifications of desired behavior, while connectors are physical objects whose action is to convey signals between the connected ports. A connector may have physical properties and is usually an object with state and behavior. The actual class that is used to implement a connector is an implementation construct. For the representation of connectors, there is no need for extension mechanisms. A binary protocol connector is represented with a line that is drawn from one port to the complementary one, as the one shown in Figure 16.32 between the `conv-1Controller` and `conv-1Belt-1Controller` capsules.

A capsule can have a state machine that represents its dynamic behavior. The state machine has control of all the internal structure of the capsule and interacts with the capsule's environment, sending and receiving signals through ports. Figure 16.34 shows the state machine of the `BeltController` capsule. According to it, the initial state is `idle`. The event `itemArrived` that is coming from the `ConvController` fires the transition to the `moveForward` state. Guards and actions for the transitions are not shown in the diagram, even though they have been imported in the diagram. However, they are utilized by the CASE tool during the subsequent code generation process.

The capsule's structure is represented by a collaboration diagram, which specifies the capsule's ports, its sub-capsules and its connectors. The collaboration diagram of Figure 16.35 shows the structure of the `TBController` capsule, which represents the TBCS in a higher layer of abstraction. This capsule is composed of the `ScaleController`, `FeederController`, and conveyor and belt controllers. The same information is represented with the class diagram of Figure 16.36, in which some of the protocols are also shown. Ports, subclasses, and connectors are strongly owned by the capsule and cannot exist independently of it. This is why the composition instead of the aggregation relationship was selected.

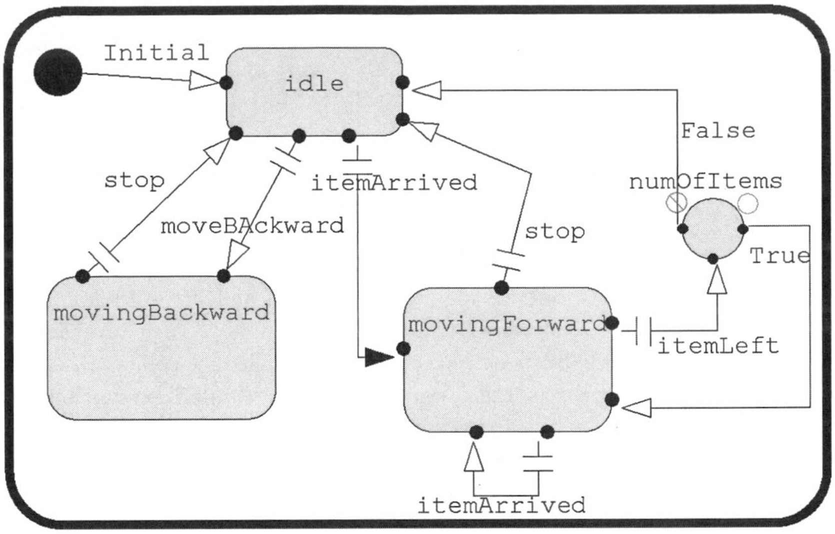

FIGURE 16.34 State machine of the `BeltController` capsule.

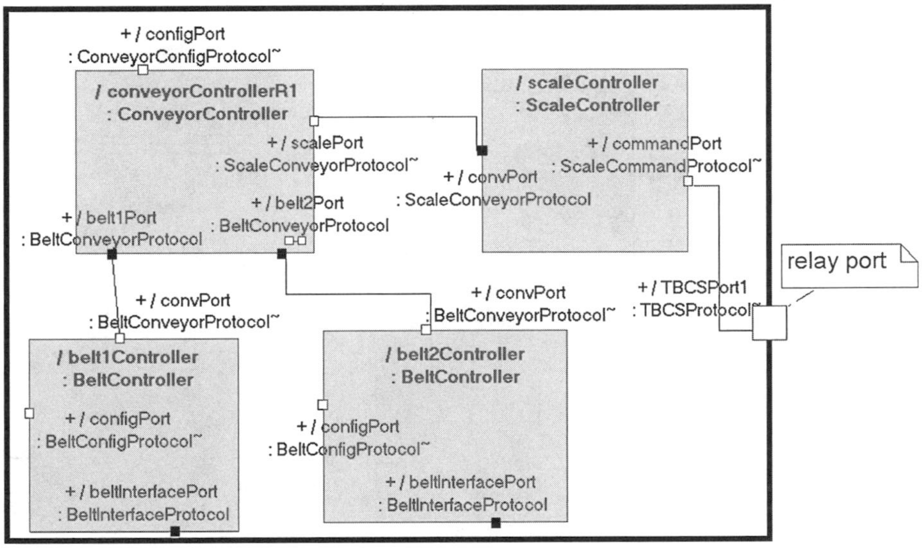

FIGURE 16.35 Part of the collaboration diagram of the `TBController` capsule.

Ports are of two kinds: *relay ports* and *end ports.* Relay ports are used to allow subcapsules to communicate with the outside of the capsule, so they are connected to subcapsules as for example the port `TBCSPort1` of the `TBController` capsule shown in Figure 16.35. End ports, which are connected to the capsule's state machine, are the source or sinks of all messages sent by the capsule. All the ports of the `BeltController` capsule in Figure 16.33, i.e., `convPort`, `beltInterfacePort`, and `configPort`, are all end ports.

Modeling the Behavior

The basic constructs used for modeling the behavior of the system are protocols and state machines. The modeling of time, with timing services, is of great interest for the modeling of behavior as well.

A *protocol* is a special type of UML collaboration that is used for the specification of behavior that exists between communicating capsules. This behavior is specified by defining the set of valid message exchange sequences between two or more active objects. Protocols are defined independently of the specific objects that communicate using this protocol. This makes protocols independent of the specific context, so their reusability is of high degree. Inheritance can also be used to specify new protocols as specializations of already defined protocols. Protocols are defined in terms of protocol roles. Each protocol comprises a set of participants, each of which plays a specific role in the protocol. Each protocol role is specified by a unique name and by the set of valid incoming and outgoing signals that are received and sent by that role. It also defines the set of valid message orderings. Protocol roles are represented by the <<protocolRole>> stereotype that has two compartments: one for the incoming signals and the other for the outgoing ones. Binary protocols, i.e., protocols with two participants, are the most common. Their advantage is that only one role, which is called the base role, has to be specified. The other role, called the conjugate, is defined by simply inverting incoming and outgoing signal sets; this inversion operation is called conjugation.

Figure 16.37 shows the `BeltConveyorProtocol` protocol that is used for the specification of behavior that exists between the communicating capsules `ConveyorController` and `BeltController` of Figure 16.32. The icon shown in the right upper corner of the `BeltConveyorProtocol` is used to denote

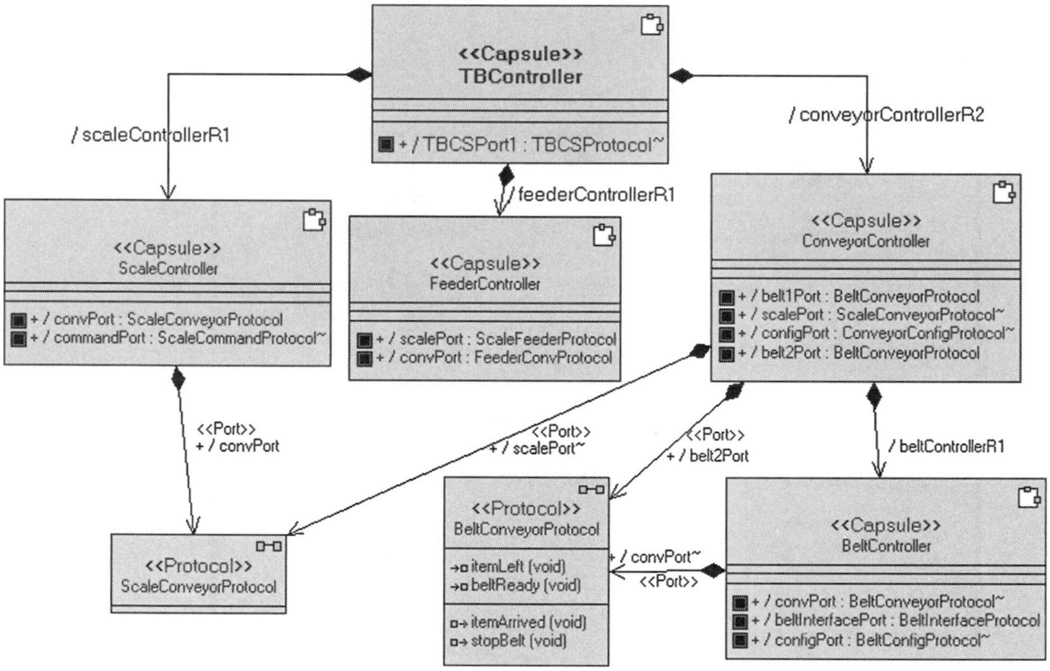

FIGURE 16.36 Class diagram for the `TBController` capsule.

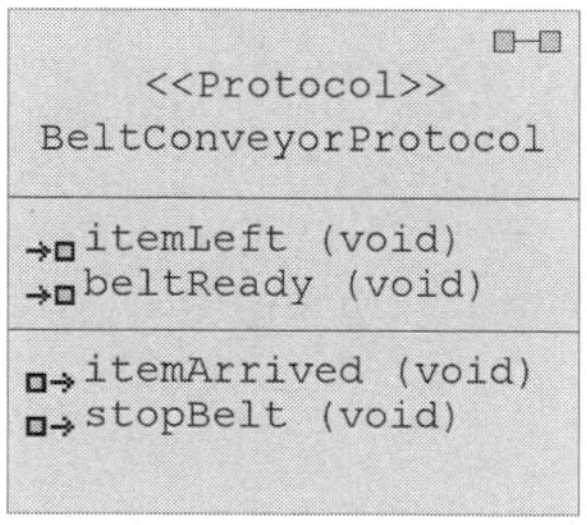

FIGURE 16.37 The binary protocol BeltConveyorProtocol.

binary protocols. `itemLeft` and `beltReady` are the incoming signals for the base role of this binary protocol, while `itemArrived` and `stopBelt` are its outgoing signals. In this case, since the protocol is binary, the role-state machine and the protocol-state machine are the same. The state machine and the interaction diagrams of a protocol role are represented using the standard UML notation.

The *handling of time* is the most important issue in most real-time systems. UML-RT defines the following two forms of time-based situations: (a) the ability to trigger activities at a particular time of day, and (b) the ability to trigger activities after a certain interval has expired from a given point in time. UML-RT defines the time-service facility that can be accessed through a standard port and converts time into events. These events are then handled in the same way as traditional signal-based events. A state machine that uses the time service can be notified with a "timeout" event when a particular time of day has been reached or when a particular interval has expired. The time service is provided by protected end ports, which are of type "TimeServiceSAP," where SAP stands for service access point.

UML-RT has wide acceptance in academia, and many researchers are working on different aspects of it. For example, Selic [1999] demonstrates the feasibility of UML to model recourses and thus predict crucial system properties before fully implementing a system. He proposes the incorporation of a generic QoS framework into UML to provide a standard way of producing quantifiable and, hence, precisely analyzable software models. He and Goddard [2000] argue that even though UML-RT is a widely accepted technique for the development of embedded systems, it fails to extract, represent, and analyze temporal parameters such as release time and deadlines, which are the essential attributes of mission-critical real-time applications. Therefore even though UML-RT models allow for the validation of functional or behavioral requirements, it is difficult to validate for temporal correctness the real-time embedded system during early development stages. They present a method for capturing the temporal parameters of a real-time application in UML-RT models so that schedulability analysis can be performed early in the design phase. The proposed method enriches UML-RT and enhances the modeling of real-time systems by capturing more temporal parameters and making them first-class attributes in the model. It also ensures that a real-time system modeled with UML-RT is schedulable and that this analysis can begin early in the design phase.

UML Profile for Schedulability, Performance, and Time

In recent years, a large number of time-critical and resource-critical systems have been developed utilizing UML. During this time, even though UML was proved to be a useful tool for modeling of such systems, a consensus has emerged that the language is lacking quantifiable notions of time and resource that will allow its broader use in the real-time and embedded domain [OMG, 2003d]. The UML profile for schedulability, performance, and time (SP&T) specification is an attempt in this direction. The remainder of this section is mainly based on [OMG, 2003d] that is the version 1.0 of the adopted specification of the Object Management Group Inc. for the UML profile for schedulability, performance, and time.

The lightweight extensibility mechanisms of UML were utilized to define the lacking quantifiable notions of time and resource, in the form of a common framework that is general enough to be adopted and specialized by the many different modeling and analysis techniques, as well as design paradigms that have emerged within the real-time software community. The main objective of this framework is the modeling of time and time-related aspects that constitute the key characteristics of timeliness, performance, and schedulability. This allows the analysis of software models in the early development phases to make predictions on these characteristics and detect problems that can be removed at a much lower cost and with less rework. The modeler is able to construct UML models of its system and utilize the different types of model-analysis techniques in order to determine whether these models meet their performance and schedulability requirements, without requiring a deep understanding of the inner working of those techniques.

What makes this UML profile so flexible is that the approach taken is not to predefine a canonical library of real-time modeling concepts, but instead to leave modelers to utilize the full power of UML to represent their real-time solutions in the way that is most suitable to their needs. However, even though this approach is characterized by its flexibility and powerfulness, it may be proved to be a balk towards the profile's wide adoption, especially until the proper tools appear in the market.

The key concept of this UML profile is the definition of a unifying framework that captures the common elements of different real-time specific-model analysis methods and the essential patterns that are used in deriving time-based analysis models from application models. The concept of the quality of service (QoS) is adopted to provide a uniform basis for attaching quantitative information to UML models. QoS information represents, either directly or indirectly, the physical properties of the hardware and software environments of the application represented by the model.

The core of the framework of the UML profile is a common model that is called the general resource model (GRM). The GRM includes the modeling of all kinds of resources and their QoS attributes. It provides a set of basic concepts that can be used by every model analysis method as the basis. These basic concepts should then be specialized according to the needs of the specific domain to produce the conceptual domain model. The purpose of the conceptual domain model according to [OMG, 2003d] is: (a) to define and explain the key concepts and relationships of the domain as well as their relationship to the general resource modeling framework, and (b) to serve as a guideline for defining the UML corresponding extensions in terms of stereotypes, tagged values, and constraints. The abstract notion of "QoS characteristic" is very useful as an abstraction in this framework, but it will only appear in analysis models in its concrete forms, i.e., delay, throughput, capacity, etc.

Modeling Resources

The modeling of resources is one of the most important objectives of the UML profile for SP & T. Since the quantification of constraints that apply to real-time systems is of great concern, QoS characteristics are an integral part of the resource model. Resources provide services of specific quality, in terms of offered QoS, while the client, i.e., the application model, specifies demands on quality of services in terms of required QoS. So the model analysis problem is reduced to comparing that required by the application QoS against that offered by the resources QoS. Two ways of looking at the resource model are defined: the peer interpretation and the layered interpretation. According to the *peer interpretation*, a client and its used resource coexist at the same computational level, and the quality of service is compared using associations. The client and the resource instances that it uses are peers in the sense that they are collaborating instances at the same level of abstraction, as for example a software task (the client) that uses a semaphore (the resource) to access shared data. According to the *layered interpretation*, the client and the resource are not really coexisting, but they are two complementary perspectives of the same modeling construct. The client represents, for example, elements of a software model, and the resource instances represent elements of the hardware/software environment on which the software elements are deployed. The first perspective is known as the logical model, while the second is known as the engineering model, where the engineering model realizes the logical model [Selic, 2000]. The term "layered" is used to highlight the concept that the engineering model may next be considered in the development process as a logical model and in turn be realized by another engineering model.

GRM is envisaged as the foundation required for any quantitative analysis of UML models. It comprises two closely related viewpoints: (a) the domain viewpoint and (b) the UML viewpoint.

The *domain viewpoint*, which is defined independently of the UML metamodel, captures, in a generic way, the common structural and behavioral concepts and patterns that characterize real-time systems as well as real-time systems analysis methods. The *UML viewpoint* provides a specification of how the elements of the domain model are realized in UML. It consists of a set of UML extensions, i.e., stereotypes, tagged values and constraints, and specifications of the mapping of the domain concepts to those extensions.

The central concept of the GRM is the notion of a resource instance, which represents a run-time entity that offers one or more services for which we need to express a measure of effectiveness or quality of service (QoS). This measure, for example, might be the execution time of the service or its precision. QoS values can vary in complexity from simple numbers to complex structured values as, for example, probability distributions. Resource usage is another core concept in GRM. It can be either static or dynamic and represents a pattern that describes how a set of clients uses a set of resources and their services. According to the static-resource usage, where the relationship between the clients and resources can be viewed as static, the required QoS values associated with a client can be matched against the offered QoS values of the resource to determine if the requirements can be met, or to determine the necessary characteristics to support the client in the usage.

The dynamic-usage model applies in situations where the order and time of occurrence of the loads on resources is relevant to the model analysis.

Resources are classified based on:

- Purpose, into processor and communication resources and devices
- Activeness, into active and passive resources
- Protection, into protected and unprotected resources

Modeling Time

Time is the most important notion of real-time systems. The profile distinguishes between physical time and simulated time. In the second case the time does not necessarily increase monotonically. The profile defines concepts for modeling time and time values, events in time and time-related stimuli, timing mechanisms such as clocks and timers, and timing services such as those provided by real-time operating systems. Timing patterns are essential for modeling schedulability and performance. They can be used to model whether something is periodic or not. In the case of periodic, the modeling of period, distribution function, and jitter is of great interest.

Modeling Schedulability

This part of the model describes a minimal set of common scheduling annotations that are used specifically in schedulability analysis. These annotations can be used to annotate the model in ways that allow a wide variety of schedulability techniques to be applied. The schedulability model of the system allows the designer to determine whether or not a model is schedulable, i.e., whether it will be able to meet its guaranteed response times. Even though specific model analysis methods such as Rate Monotonic Analysis, Deadline Monotonic Analysis, and Earliest Deadline First were utilized to capture the basic abstractions, the profile is flexible enough to handle a variety of other techniques.

Typical tools supporting schedulability analysis should provide, according to [OMG, 2003d] two important functions. The first is to calculate the schedulability of the systems, i.e., the ability of the system to meet all of the deadlines defined for the individual scheduling jobs. The second is to assist with determining how the system can be improved, as for example suggestions for making an entity schedulable or optimizing system usage for a more balanced system.

Modeling Performance

The main objective of performance analysis for a system is to determine the rate at which the system can perform its function given that it has finite resources with finite QoS characteristics. GRM is utilized to model resources and their performance-related characteristics, such as delay and throughput. A minimal profile is defined that is sufficient for basic performance analysis of complex systems. For a deeper performance analysis, it is expected that further specialized profiles will be defined by tool vendors or specialists, by specializing the basic concepts defined by this profile.

UML-Based Notations and Methodologies for Embedded Applications and Systems

A number of papers describe specific notations and methodologies for the development of embedded real-time systems based on the UML notation. However, despite the claims of the authors, there has been considerable debate over whether these notations and methodologies are mature enough for commercial use. To our knowledge, there is no single best notation and methodology that is ideal for embedded-systems development. The remainder of this section briefly presents the best known UML-based notations and methodologies for ESs development.

The COMET/UML Method

COMET, which adopts a highly iterative software life-cycle model, is a UML-based concurrent object modeling and architectural design method for the development of distributed real-time applications. COMET utilizes use-case models to build the requirements model of the system. A brief description of the methodology is given in [Gomaa, 2001], while a detailed description with case studies can be found in [Gomaa, 2000].

For the analysis-modeling phase, static and dynamic models are developed, with the static model showing structural relationships among problem domain classes, and the dynamic model showing: (a) the objects that participate in each use case, (b) the way they collaborate, and (c) the statecharts for state-dependent objects. In the design modeling phase, a component-based approach is adopted, and special emphasis is given to the division of responsibility between clients and servers, the centralization vs. distribution of data and control, and the design of messages communication interfaces. In the so-created architectural design model, each subsystem is designed as a distributed self-contained component. Each concurrent subsystem is then designed in terms of active and passive objects. Finally, an approach based on rate monotonic analysis (SEI 93) is utilized to estimate the performance of the created model.

The HASoC Methodology

Green and Edwards [2002] present the HASoC methodology for the development of complete embedded systems, i.e., hardware, software and platform architectures that are targeted primarily, but not necessarily, at SoC implementations. HASoC explicitly separates the behavior of a system from its hardware and software implementation technologies. An executable model of the system is incrementally developed and validated. This model is then partitioned into hardware and software to create a committed model, which is mapped onto a system platform.

HASoC was developed utilizing the UML notation and the experience gained using the MOOSE method [Green et al., 1994]. Green and Edwards [2002] considered UML-RT to be applicable to hardware as well as to software, and they argue that further extensions are required to make the notation more suitable for mixed-technology embedded systems. The same is also reported by Martin et al. [2001].

The most important modification to the UML-RT model is the definition of the data-driven capsule. To address the requirements of data-driven applications, the UML-RT model was extended to obtain a data-driven capsule whose ports can receive streams of data and process them in the capsule's thread of execution, a mechanism that is based on the "time-continuous flow" communication model of MOOSE.

The UML and DORIS Combination

Hull et al. [2004] argue that the UML potentials of this method for modeling real-time and embedded systems can be greatly enhanced with the use of DORIS, a method extensively used in the aerospace industry. The authors of this methodology propose an integration of UML and DORIS to model complex embedded systems in industrial, commercial, and defense applications. They mainly consider systems whose critical safety aspects are of primary importance, i.e., "systems that have to respond to multiple streams of events with unpredictable arrival rates, in a manner subject to timing constraints that have been captured by the requirements." The authors propose a way to model the structures of the DORIS notation with UML in order to allow the UML-based models to adequately represent nonfunctional aspects such as performance, throughput, and dependability.

The Embedded UML Profile

Embedded UML presented by Martin et al. [2001] is based on a set of ideas that best allow specification and analysis of mixed HW-SW systems. These ideas have been captured from the most important real-time UML proposals and integrated with concepts from the codesign world.

Embedded UML specifies the system as a collection of reusable communicating objects. It defines the concepts of "reactive" object to be used instead of the active object to better match embedded-systems requirements. It defines the concepts of block that is an extension of the concept of capsule and defines communications explicitly via interfaces and channels, which are extensions of ports and signals. It supports multiple means of communications to favor refinement. Customization of communication protocols at the specification level will allow for better exploitation of the many different implementation artifacts such as finite buffers, interrupts, semaphores, shared memory, etc. Use-case and sequence diagrams are utilized for test cases and test scenarios. Netlists, an extension of collaboration diagrams, are used for functional decomposition. State diagrams with specified action semantics are utilized for code generation, optimization, and synthesis.

Refinement plays a key role in the development process to move from nonexecutable specifications to formal executable specifications and finally to implementation model. Concepts from co-design include:

1. A rigorously defined platform model in both HW and SW for the implementation architecture. This architecture is considered as a collection of resources offering services that constitute the system platform API.
2. The concept of "mapping," which is considered as the platform refinement paradigm and is utilized for performance analysis, communication synthesis, and optimized code synthesis/generation.

Mapping diagrams, which are a rigorous extension of deployment diagrams, are utilized for performance analysis and optimized implementation-model generation.

The Real-Time Perspective Process

The real-time perspective is the ARTiSAN's approach for the development of real-time systems. It is described by Moore and Cooling [2000] and supported by the real-time studio professional CASE tool of ARTiSAN. It is, according to Moore and Cooling, an incremental process and a set of models (requirements, object, software, and system) that are applicable to all real-time projects. It is described in four parts: (a) a requirements architecture that supports real-time requirements, (b) a three-tier solution architecture that supports multitasking and multiprocessor systems, (c) the object-oriented modeling techniques, and (d) an incremental development process. The process extensions include: performance engineering, requirements and traceability, reuse, and verification. From these extensions, performance engineering, which is very important for real-time embedded systems, addresses the capture of performance requirements and the mechanisms that can be used by the performance engineer to meet and verify these requirements.

The Systems Modeling Language

The systems modeling language (SysML) is the proposal of the SysML group (www.sysml.org) to the RFP issued by OMG soliciting submissions that should specify a customization of UML for systems engineering and especially support the phases of analysis, specification, design, and verification of complex systems. SysML proposes a number of modifications to UML 2.0 as well as some new diagrams to support modeling of a broad range of systems, which may include hardware, software, data, personnel, procedures, and facilities [Hause et al., 2004]. The proposed modifications to UML 2.0 are minor and are restricted only to three diagrams, namely the activity, structured class, and state diagrams, while the new diagrams are the requirements diagram and the parametric diagram. It is clear that for the definition of SysML, the extension mechanisms of UML were not enough for specifying the new extension.

The UML 2.0 Specification

UML 2.0, which is now in its final acceptance phase, is composed of two main documents:

1. *UML 2.0: Infrastructure*: Serves as the architectural specification of the language and defines the foundational language constructs required for UML 2.0. It describes the structure of the language, the formal approach used for its specification, the infrastructure library that specifies a flexible metamodel library, and the UML::Classes::Kernel, which is a package that reuses the infrastructure library. The document is already a final adopted specification and is available online at http://www.omg.org/cgi-bin/doc?ptc/2003–09–15.
2. *UML 2.0: Superstructure*: Defines the user-level constructs that are built on top of the infrastructural constructs specified in UML 2.0: Infrastructure. It is organized into two main parts: (a) the "structure" part, which defines the static structural constructs (classes, components, nodes, etc.) used in various structural diagrams such as class diagrams, components, and deployment diagrams, and (b) the "behavior" part, which specifies the dynamic behavioral constructs (activities, interactions, state machines) used in various behavioral diagrams (activity, sequence, and state-machine diagrams). This document is also a final adopted specification and is available online at http://www.omg.org/cgi-bin/doc?ptc/2003-08-02.

The above specifications completely replace the current version of UML 1.4.1 and UML 1.5 with action semantics. The most important feature of the specification is the support provided to the evolving conceptual architecture of the OMG that is called MDA and supports a model-driven approach to software development. MDA, as stated in [OMG, 2003b], is not itself a technology specification, but "it represents an important approach and a plan to achieve a cohesive set of model-driven technology specification." In the context of MDA it is assumed that "extensions of the UML language will be defined and standardized for specific purposes and that many of these will be designed specifically for use in MDA."

According to [OMG, 2003a], UML 2.0 supports the most prominent concepts that are required in the evolving MDA vision. More specifically, UML 2.0 actually defines a family of languages. It defines only the general concepts and refines the language's profile mechanism to make it more robust, flexible, and significantly easier to implement and apply. This will allow its customization for a wide variety of domains, platforms, and methods. UML 2.0 also specifies the system independently of the platform that supports it, and it is intended to be used with a wide range of software methodologies. By significantly enhancing the constructs for specifying component architectures, UML 2.0 also allows modelers to fully specify target implementation environments. Finally, it specifies various relationships such as "realization," "refine," and "trace," that can be used to specify the transformation of the platform-independent model to the platform-specific model that is a key concept in MDA.

UML 2.0 is specified, as stated in [OMG, 2003a], "using a metamodeling approach that adapts formal specification techniques." These formal specification techniques were used to increase the precision and correctness of the specification. The so-produced specification gives UML a great advantage compared with formal methods, whose great disadvantage is that the executable cannot be automatically constructed from specifications [Wang and Shin, 2000]. In this case the manually constructed implementation usually introduces additional errors, even in the case where the specification has been proved to be correct. UML 2.0 utilizing: (a) formal specification techniques, to the extent that they do not add significant complexity to the language, and (b) the model-driven approach, allows for the creation of formal specifications for the system and the subsequent automatic construction of the implementation model. The goals of the specification techniques used to define UML 2.0 are, according to [OMG, 2003a]: correctness, precision, conciseness, consistency, and understandability. The specification uses a subset of UML, an object-constraint language, and precise natural language to provide a self-contained description of the abstract syntax and the static and dynamic semantics of UML.

UML-Based CASE Tools for Embedded Systems

In recent years the number of CASE tools that support the UML notation has increased dramatically. A list of UML CASE tools can be found at http://dmoz.org/Computers/Programming/Methodologies/Modeling_Languages/Unified_Modeling_Language/Tools/ as well as at http://www.qucis.queensu.ca/Software-Engineering/tools.html.

The software developer has a lot of choices, but it is quite difficult to make the right selection. The large number of unsuccessful attempts to introduce CASE tools into the development process forced the software community to give considerable attention to the establishment of processes for the formal assessment of case tools [Mosley, 1992; Le Blanc and Korn, 1994; Kitchenham and Pickard, 1998]. Working groups created by authorities such as the Software Engineering Institute (SEI) and the IEEE, which sought to advance the state of practice in software development through the use of good tools for system and software engineering, resulted in the IEEE 1462 standard, which defines guidelines for the evaluation and selection of CASE tools [IEEE Std. 1462, 1998]. However, a formal assessment following the standards is a very costly and time-consuming task. Further, some weaknesses in standards have been identified and reported so far [Lundel and Lings, 2002].

A less formal assessment is usually adopted. In such an assessment, issues such as ease of use, robustness, functionality, round-trip engineering, model navigation, repository support, support for reusability, ease of insertion, and quality of support must be considered very carefully. A good list of evaluation criteria can be found at http://www.objectsbydesign.com/tools/modeling_tools.html. A lot of articles report evaluation processes and results that may be used appropriately in the CASE tool selection process [Budgen and Thomson, 2002; Frankel, 2000; Yphise, 2000; Eichelberger, 2002; Post and Kagan, 2000; Artim, 2003]. We next make a brief reference to four commercially available CASE tools that provide considerable support for ESs:

a. *Rhapsody by I-Logix.* Rhapsody provides, according to I-Logix, a collaborative model-driven development (MDD) environment for system and software engineers to: (a) work in an iterative fashion, (b) analyze and make rapid system and software design changes, (c) generate the application, and (d) test it quickly and efficiently. The development of real-time embedded applications is supported by a special tool, which is based on executable models [I-Logix, 2003]. The tool for embedded systems is very flexible. A real-time framework is provided in source-code form with the tool, so users can easily adapt the generated code for any other operating system, including proprietary in-house systems. There is a free developer's trial for 30 days provided by Rhapsody. An overview of the tool can be found online at: http://www.ilogix.com/products/rhapsody/index.cfm.
b. *The IBM Rational Rose RT.* Rational Rose RT, which has recently been renamed to IBM Rational Rose Technical Developer, is a complete UML development environment that supports, according to IBM, the challenges of real-time embedded-software development. It provides a single interface for model-based development and supports all phases for development from initial requirements to test and deployment. Among its features, we discriminate the validation with automated test generation and the highly scalable multiuser team environment. To support the distribution of a single Rose RT model over several processors, it provides Connexis, which provides a full distributed communication layer. Connexis, which supports model visualization of distributed applications, is highly optimized for event-driven asynchronous systems and improves the time to market by eliminating the need to develop and test a custom IPC mechanism [Rational, 2003]. More information can be found at http://www-306.ibm.com/software/awdtools/developer/technical/.
c. *ARTiSAN Real-time Studio (RtS) Professional.* RtS supports a UML-based object modeling with numerous real-time extensions. It utilizes the basic UML diagrams and proposes new diagrams such as the constraint diagram, the system architecture diagram, and the concurrency diagram to complement the development process. RtS has a powerful system-validation functionality to support the engineers to: (a) build and simulate advanced state models for system behavior, (b) automatically generate test harnesses for behavior verification, (c) build front-panel simulation, and (d) integrate the state-model simulation. RtS's powerful code synchronizer, according to Artisan, enables software engineers to generate and reverse engineer C, C++, Java, Ada, and SPARK, keeping source code consistent with the design. More information can be found at: http://www.artisansw.com/products/products.asp.
d. *Tau Generation2 by Telelogic.* Tau Generation is a family of role-based and model-centric tools for real-time software and systems development. Tau/Developer™ is a stand-alone tool, based on UML, for the design, analysis, and development of advanced real-time applications. According to Telelogic, the tool forces the engineer to make a shift from the traditional labor-intensive code-centric programming to the highly efficient model-driven development. The tool, according to Telelogic, supports executable UML models with behavioral specifications. The tool's philosophy is that the specifications must be validated at an early stage for timely and cost-effective error correction. Information is available on-line at: http://www.taug2.com/.

There are also some free CASE tools that, even if they lack special support for embedded systems, can be used for practicing UML. Between them we discriminate:

- *Visual Paradigm for UML.*
  A visual development platform. http://www.geocities.com/millermax2001/casetool.html
- *Argo/UML.*
  A free open source UML tool.http://argouml.tigris.org/
- *Poseidon for UML* (community edition).
  http://www.gentleware.com/adgate.php4?in= odb&banner= poseidonCE&url=/products/poseidonCE.php3

Summary

UML is the de facto standard for software development. A large number of practitioners and researchers have adopted UML in many different application domains at a rate exceeding even OMG's most optimistic predictions. Even though UML has emerged as a language for modeling OO systems, it is now evolving to a component-based modeling language. UML 2.0 is in the context of the OMG's model-driven architecture initiative, which considers as primary artifacts of software development not programs, but models created by modeling languages. This ensures that UML will drive the software development industry for the next decades.

However, UML is a very large language. A long time is required to learn the language and efficiently use it in the software development process. In this paper, we have presented a subset of the UML language that includes the most commonly used diagrams to model the most important aspects of software systems, as well as those constructs that are more important in the domain of distributed embedded systems. In particular, we have presented the basic concepts regarding use-case diagrams, class diagrams, statecharts, and sequence and collaboration diagrams. These diagrams have specific constructs and semantics to enable the software developer to model the most important views of the system under development. Component and deployment diagrams play an important role during the final implementation phase and the deployment of the executable model.

One of the most important features of UML is its ability to be extended and specialized in different application domains. Even though UML's flexibility allows modeling in different application domains and at different levels of abstraction during the system's development process, the powerful set of extensibility mechanisms, which are provided by the language, allows the definition of profiles for any specific application domain. A profile is primarily intended to tailor UML towards a specific domain by giving modelers access to common model elements and terminology from the specific domain. This greatly enhances productivity and quality in the development process. We have briefly presented UML-RT that focuses on the category of real-time systems that are characterized as complex, event driven, and usually distributed. We have described the most important constructs of the profile, including capsules, ports, connectors, protocols, etc. that are utilized to create executable models for the system under development. We have also presented the UML profile for SP & T, which provides a significant step for bringing UML in the ES domain.

As was expected, UML has attracted the interest of researchers from the embedded-systems application domain. It is believed that component-based architectures and the object-oriented approach are very promising technologies for this application domain, and the best way to exploit these technologies is through the use of UML. However it is not yet clear how the UML will be utilized in this domain. We have described a number of promising methodologies and notations to this direction that have already been adopted by researchers in this area. Finally, we have made a reference to the family of CASE tools that support UML in the ES domain. The great number of commercially available and free-to-download CASE tools makes the selection a very costly and time-consuming task. Standards may be used to simplify this task, but a less formal assessment is usually adopted to select the tool that addresses the specific developer's requirements.

References

I. Alexander, "Misuse cases: use cases with hostile intent," *IEEE Software*, vol. 20, no. 1, 2003.

B. Anda and D. Sjoberg, "Towards an inspection technique for use case models," *SEKE '02*, Ischia, Italy, July 2002.

M.J. Artim, "UML Tools," 2003, http://www.primaryview.org/UML/Tools.html.

E. Berard, *Essays on Object-Oriented Software Engineering*, vol. 1, Englewood Cliffs, NJ: Prentice Hall, 1993.

T. Bihari and P. Gopinath, "Object-oriented real-time systems: concepts and examples," *IEEE Comput.*, vol. 25, no. 12, pp. 25–32, 1992.

M. Bjorkander, "Graphical programing using UML and SDL," *IEEE Comput.*, vol. 33, no. 12, 2000.

G. Booch, *Object-Oriented Analysis and Design with Applications*, 2nd ed., Reading, MA: Benjamin/Cummings Publishing Company, 1994.

G. Booch, J. Rumbaugh, and I. Jacobson, *The Unified Modeling Language User Guide*, Reading, MA: Addison Wesley Longman, 1999.

D. Budgen and M. Thomson, "CASE tool evaluation: experiences from an empirical study," *Journal of Systems and Software*, 67, 2002.

L. Capretz, "A brief history of the object-oriented approach," *Software Eng. Notes*, vol. 28, no. 2, ACM SIGSOFT, 2003.

S-W. Cheng and D. Garlan, " Mapping architectural concepts to UML-RT," in *Proc. 2001 Int. Conf. Parallel Distrib. Process. Tech. Appl. (PDPTA'2001)*, Nevada, June 2001.

D. Coleman, F. Hayes, and S. Bear, "Introducing objectcharts or how to use statecharts in object-oriented design," *IEEE Trans. Software Eng.*, vol. 18, no 1, 1992.

D. Crawford, "Editorial pointers," *Commun. ACM*, vol. 45, no. 11, 2002.

T. Demarco, *Structured Analysis and System Specification*, Yourdon Press Computing Series, Englewood Cliffs, NJ: Prentice Hall, 1979.

D. Dori, "Why significant UML change is unlikely," *Commun. ACM*, vol. 45, no. 11, 2002.

B. Douglas, *Doing Hard Time: Developing Real-Time Systems with UML, Objects, Frameworks, and Patterns*, Reading, MA: Addison Wesley, 1999.

L. Edward, "What's ahead for embedded software," *IEEE Comput.*, September vol. 39, no. 9, pp. 18–26, 2000.

H. Eichelberger, "Evaluation—report on the layout facilities of UML tools," report no. 298, July 2002, http://www-info2.informatik.uni-wuerzburg.de/mitarbeiter/eichelberger/reports/eval2002Abstract.html.

J. Ellsberger, D. Hogrefe, and A. Sarma, *SDL: Formal Object-Oriented Languages for Communicating Systems*, Englewood Cliffs, NJ: Prentice Hall, 1997.

E. Erikson and M. Penker, *UML Toolkit*, New York: Wiley, 1998.

D. Frankel, "Alternative UML modeling tools," *Software Mag.*, October 2000, http://www.softwaremag.com/L.cfm?Doc = archive/2000oct/Lab.html.

E. Gamma, R. Helm, R. Johnson, and J. Vlissides, *Design Patterns: Elements of Reusable Object-Oriented Software*, Reading, MA: Addison Wesley, 1995.

H. Gomaa, *Designing Concurrent, Distributed, and Real-Time Applications with UML*, Reading, MA: Addison Wesley, 2000.

H. Gomaa, "Designing real-time and embedded systems with the COMET/UML method," *Dedicated Syst. Mag.*, Q1, pp. 44–49, 2001, http://www.omimo.be/magazine/01q1/2001q1_p044.pdf.

P. Green and M. Edwards, "The modelling of embedded systems using HASoC," in *Proc. 2002 Des. Autom. Test Eur. Conf. Exhibition (DATE'02)*, New York: IEEE Press, 2002.

P. Green, M. Edwards, and S. Essa, "HASoC — towards a new method for system-on-a-chip development," *Des. Autom. Embed. Syst.*, vol. 6, no. 4, pp. 333–353, 2002.

P. Green, P. Rushton, and S. Beggs, "An example of applying the codesign method MOOSE," in *Proc. Third Int. Workshop Hardware/Software Codesign*, New York: IEEE Press, 1994.

G. Gullekson, "Designing for concurrency and distribution with rational rose realtime," Rational Software White Paper, 2000, http://www.-128.ibm.com/developerworks/rational/library/269.html.

D. Harel, "Statecharts: a visual formalism for complex systems," *Sci. Comput. Program.*, 1987.

D. Harel and E. Gery, "Executable object modeling with statecharts," *IEEE Comput.*, vol. 30, no. 7, 1997.

M. Hause, F. Thom, and A. Moore, The systems modelling language (SysML), 2004, http://www.artisansw.-com/whitepapers/home.asp.

W. He and S. Goddard, "Capturing an application's temporal properties with UML for real-time," *Fifth IEEE Int. Symp. High Assur. Syst. Eng., HASE*, November 2000, pp. 15–17.

B. Heck, L. Wills, and G. Vachtevanos, "Software technology for implementing reusable, distributed control systems," *IEEE Contr. Syst. Mag.*, vol. 23, no. 1, pp. 21–35, 2003.

J. Holt, *UML for Systems Engineering*, United Kingdom: The Institution of Electrical Engineers, 2001.

M. Hull, S. Ewart, and J. Hanna, "Modelling complex real-time and embedded systems — the UML and DORIS combination," *Real-Time Syst.*, vol. 26, no. 2, pp. 135–159, 2004.

IEEE Std. 1462, Information technology — guideline for the evaluation and selection of CASE tools, IEEE STD 1998.

I-Logix, "Model Driven Development for Real-Time Embedded Applications," 2003, http://www.ilogix.com.

I. Jacobson, *Object-Oriented Software Engineering: A Use Case Driven Approach*, New York/Reading, MA: ACM Press/Addison Wesley, 1992.

I. Jacobson, G. Booch, and J. Rumbaugh, "The unified process," *IEEE Software*, vol. 16, no. 3, 1999a.

I. Jacobson, G. Booch, and J. Rumbaugh, *The Unified Software Development Process*, Reading, MA: Addison Wesley, 1999b.

G. Karsai, J. Stripanovits, A. Ledeczi, and T. Bapty, "Model-integrated development of embedded software," *Proc. IEEE*, vol. 91, no. 1, 2003.

B. Kernighan and D. Ritchie, *The C Programming Language*, 2nd ed., Englewood Cliffs, NJ: Prentice Hall, 1988.

B. Kitchenham and L. Pickard, "Evaluating software engineering methods and tools: part 9: quantitative case study methodology" *Software Eng. Notes*, ACM SIGSOFT, vol. 23 no. 1, 1998.

C. Kobryn, "UML 2001: a standardization odyssey," *Commun. ACM*, vol. 42, no. 10, 1999.

C. Kobryn, "Modeling components and frameworks with UML," *Commun. ACM*, vol. 43, no. 10, 2000.

C. Kobryn, "Will UML 2.0 be agile or awkward?," *Commun. ACM*, vol. 45, no. 1, 2002.

G. Larsen, "Designing component-based frameworks using patterns in the UML," *Commun. ACM*, vol. 42, no. 10, 1999.

C. Le Blanc and L.W. Korn, "A phased approach to the evaluation and selection of CASE tools," *Inf. Software Technol.*, vol. 36, no. 5, 1994.

C. Le Blanc and E. Stiller, UML for Undergraduate Software Engineering, JCSC 15, 5, May 2000, Consortium for Computing in Small Colleges.

D. Lee, "Evaluating real-time software specification languages," *Computer Standards & Interfaces*, vol. 24, pp. 395–409, 2002.

J. Lee and N. Xue, "Analyzing user requirements by use cases: a goal-driven approach," *IEEE Software*, vol. 16, no. 4, 1999.

B. Lundel and B. Lings, "Comments on ISO 14102: the standard for CASE-tool evaluation," *Computer Standards & Interfaces*, vol. 24, no. 5, pp. 381–388, 2002.

G. Martin, "UML for embedded systems specification and design: motivation and overview," in *Proc. 2002 Des. Autom. Test Eur. Conf. Exhibition (DATE'02)*, Los Alamitos, CA: IEEE Computer Society Press, 2002.

G. Martin, L. Lavagno, and J. Louis-Gyerin, "Embedded UML: a merger of real-time UML and co-design," in *Proc. CODES2001*, Copenhagen, April 2001, pp. 23–28.

N. Medvidovic, D. Rosenblum, D. Redmiles, and J. Robbins, "Modeling software architectures in the unified modeling language," *ACM Trans. Software Eng. Methodol.*, vol. 11, no. 1, 2002.

S. Mellor, "Make models be assets," *Commun. ACM*, vol. 45, no. 11, 2002.

J. Miller and J. Mukerji, Model Driven Architecture (MDA), document no. ormsc/2001-07-01, Architecture Board ORMSC, July 9, 2001. http://www.omg.org/docs/ormsc/01-07-01.pdf.

J. Miller, "What UML should be," *Commun. ACM*, vol. 45, no. 11, 2002.

A. Moore and N. Cooling, *Real-Time Perspective: Overview*, Version 1.3, 2000, http://www.artisansw.com/whitepapers/home.asp.

V. Mosley, "How to assess tools efficiently and quantitatively," *IEEE Software*, vol. 9, no. 3, pp. 29–32, 1992.

OMG, *UML Infrastructure 2.0 Final Adopted Specification*, 2003a, http://www.omg.org/cgi-bin/doc?ptc/2003-08-02.

OMG, *UML Superstructure 2.0 Final Adopted Specification*, 2003b, http://www.omg.org/cgi-bin/doc?ptc/2003-09-15.

OMG, *Introduction to OMG's Unified Modeling Language*, 2003c, http://www.omg.org/gettingstarted/what_is_uml.htm.

OMG, *UML Profile for Schedulability, Performance and Time Specification*, 2003d, http://www.omg.org/docs/ptc/02-03-02.pdf.

M. Page-Jones and S. Weiss, "Synthesis: an object-oriented analysis and design method," *Am. Programmer*, vol. 2, no. 7–8, 1989.

N.C. Paulk, "Extreme programming from a CMM perspective," *IEEE Software*, vol. 18 no. 6, 2001.

G. Post and A. Kagan, "OO-CASE tools: an evaluation of Rose," *Inf. Software Technol.*, vol. 42, 2000.

R. Pressman and D. Ince, "Software engineering: a practitioner's approach," *European Adaptation*, 5th ed., New York: McGraw-Hill, 2000.

T. Quatrani, Introduction to the Unified Modeling Language, Rational Rose, white paper, 2001, http://www.rational.com/media/uml/intro_rdn.pdf?SMSESSION=NO

Rational Software Corporation, *IBM Rational Rose Real Time Connexis User's Guide*, Version: 2003.06.00, 2003, http://publibfp.boulder.ibm.com/epubs/pdf/12653660.pdf.

J. Rumbaugh, M.R. Blaha, W. Premerlani, F. Eddy, and W. Lorensen, *Object-Oriented Modeling and Design*, Englewood Cliffs, NJ: Prentice Hall International, 1991.

J. Rumbaugh, I. Jacobson, and G. Booch, *The Unified Modeling Language — Reference Manual*, Reading, MA: Addison Wesley Longman, 1999.

SEI, Carnegie Mellon University Software Engineering Institute, *A Practitioner's Handbook for Real-Time Analysis — Guide to Rate Monotonic Analysis for Real-Time Systems*, Boston: Kluwer Academic Publishers, 1993.

B. Selic, "Turning clockwise: using UML in the real-time domain," *Commun. ACM*, vol. 42, no. 10, 1999.

B. Selic, "A generic framework for modeling resources with UML," *IEEE Comput.*, vol. 33, no. 6, pp. 64–69, 2000.

B. Selic, G. Ramackers, and C. Kobryn, "Evolution not revolution," *Commun. ACM*, vol. 45, no. 11, 2002.

B. Selic and J. Runbaugh, Using UML for Modeling Complex Real-Time Systems, March 1998, http://www.rational.com/products/whitepapers/UML-rt.pdf?SMSESSION=NO.

B. Selic, G. Gullekson, and P. Ward, *Real-Time Object-Oriented Modeling*, New York: Wiley, 1994.

K.G. Svarstad, A. Nicolescu, and Jerraya, "A model for designing communication between aggregate objects in the specification and design of embedded systems," *Proc. Conference on Design, Automation and Test in Europe*, Munich, Germany, pp. 77–85, 2001.

K. Thramboulidis, "Using UML for the development of distributed industrial process measurement and control systems," *IEEE Conference on Control Applications (CCA)*, Mexico, September 2001.

K. Thramboulidis, "A sequence of assignments to teach object-oriented programming: a constructivism design-first approach," *Inform. Educ.*, vol. 2, no. 1, pp. 103–122, 2003.

K. Thramboulidis, "Using UML in control and automation: a model driven approach," *Second IEEE International Conference on Industrial Informatics*, Berlin, Germany, 24–26 June, (INDINÇ04).

K. Thramboulidis, "Model integrated mechatronics — towards a new paradigm in the development of manufacturing systems, *IEEE Trans. Indust. Informatics*, vol. 1, no. 1, pp. 54–61, 2005.

K. Thramboulidis and C. Tranoris, "Developing a CASE tool for distributed control applications," *Int. J. Adv. Manuf. Technol.*, vol. 24, no. 1–2, pp. 24–31, 2004.

S. Wang and K. Shin, "An architecture for embedded software integration using reusable components," in *Proc. CASE'00*, San Jose CA, November 2000.

R. Wieringa, "A survey of structured and object-oriented software specification methods and techniques," *ACM Comput. Surv.*, vol. 30, no. 4, 1998.

S. Yacoub and H. Ammar, *UML Support for Designing Software Systems as a Composition of Design Patterns, LNCS 2185*, Berlin: Springer, 2001, pp. 149–165.

K. Young, R. Piggin, and P. Rachitrangsan, "An object-oriented approach to an agile manufacturing control system design," *International Journal of Advanced Manufacturing Technology*, vol. 17, Berlin: Springer, 2001.

E. Yourdon and L. Constantine, *Structured Design*, Yourdon Press Computing Series, Englewood Cliffs, NJ: Prentice Hall, 1979.

E. Yourdon, *Modern Structured Analysis*, Englewood Cliffs, NJ: Prentice Hall, 1989.

Yphise, Software Evaluation Process, January 2000, http://www.rational.com/media/products/rose/yphise.pdf?SMSESSION=NO.

17

Welding and Bonding

George E. Cook
Vanderbilt University

Reginald Crawford
Vanderbilt University

David R. DeLapp
Vanderbilt University

Alvin M. Strauss
Vanderbilt University

17.1 Automated Welding and Joining

Most welding processes require the application of heat or pressure, or both, to produce a bond between the parts being joined. The welding control system must include means for controlling the applied heat, pressure, and filler material, if used, to achieve the desired weld microstructure and mechanical properties.

Welding usually involves the application or development of localized heat near the intended joint. Welding processes that use an electric arc are the most widely used in industry. Other externally applied heat sources of importance include electron beams, lasers, and exothermic reactions (oxyfuel gas and thermit). With these processes, fusion of the two parts being joined occurs as a result of the localized applied heat. The high-energy-density heat source is normally applied to the prepared edges or surfaces of the members to be joined and is moved along the path of the intended joint. The power and energy density of the heat source must be sufficient to accomplish melting of intended joint locally.

Resistance spot welding is another widely used joining process, particularly in the automotive industry. With this process, the pieces to be joined are typically sandwiched between two electrodes through which a large electric current is allowed to flow for a controlled time duration during which pressure is applied. Heat is developed in the area of the spot from I^2R (resistance) heating, and the applied pressure serves to forge the two materials together. Another relatively new resistance welding process that relies on mechanically produced heat is the *friction stir welding* (FSW) process [Thomas et al., 1991]. With this process, the welding tool is comprised of a shoulder in contact with the workpiece and a pin, which in butt welds extends almost completely through the thickness of the workpiece. As the tool rotates, frictional heat is produced at the contacting shoulder and pin surfaces, causing the workpiece material to soften into the material's plastic state and flow around the tool pin from the front to rear, where it consolidates and forges under the confining pressure of the tool shoulder, and regains strength upon cooling to form the weld. Unlike fusion welding processes, e.g., *arc welding*, *electron beam welding*, and *laser welding*, the FSW process takes place in the solid phase below the melting point of the materials being joined.

Control System Requirements

Insight into the control system requirements of the different fusion welding processes can be obtained by consideration of the power density, effective spot size of the heat source, and interaction time of the heat source on the material.

A heat source power density of approximately 10^3 W/cm^2 is required to melt most metals [Eagar, 1986]. Below this power density the solid metal can be expected to conduct away the heat faster than it is being introduced. On the other hand, a heat source power density of 10^6 or 10^7 W/cm^2 will cause vaporization of most metals within a few microseconds, so for higher power densities no fusion can occur [Lancaster, 1986]. Thus, it can be concluded that the heat sources for all fusion welding processes lie between approximately 10^3 and 10^6 W/cm^2 heat intensity. Examples of welding processes that are characteristic of the low end of this range include *oxyacetylene welding*, *electroslag welding*, and *thermit welding*. The high end of the power density range of welding is occupied by *laser beam welding* and *electron beam welding*. The midrange of heat source power densities is filled in by the various arc welding processes.

For pulsed welding, the interaction time of the heat source on the material is determined by the pulse duration, whereas for continuous welding the interaction time is proportional to the spot diameter divided by the travel speed. The minimum interaction time required to produce melting can be estimated from the relation for a planar heat source given by [Eagar, 1986]

$$t_m = (K/p_d)^2$$

where p_d is the heat source density (watts per square centimeter) and K is a function of the thermal conductivity, and thermal diffusivity of the material. For steel, Eagar gives K equal to 5000 W/cm^2/s. Using this value for K, one sees that the minimum interaction time to produce melting for the low power density processes, such as oxyacetylene welding with a power density on the order of 10^3 W/cm^2, is 25 s, while for the high energy density processes, such as laser beam welding with a power density on the order of 10^6 W/cm^2, is 25 μs. Interaction times for arc welding processes lie somewhere between these extremes.

An example of practical process parameters for a continuous *gas tungsten arc weld* (GTAW) are 100 A, 12 V, and travel speed 10 ipm (4.2 mm/s). The peak power density of a 100-A, 12-V gas tungsten arc with argon shielding gas, 2.4-mm diameter electrode, and 50° tip angle has been found to be approximately 8×10^3 W/cm^2. A typical argon-shielded gas tungsten arc is shown in Figure 17.1. Assuming an estimated spot diameter of 4 mm, the interaction time (taken here as the spot diameter divided by the travel speed) is 0.95 s. At the other extreme, 0.2 mm (0.008 in.) material has been laser welded at 3000 in./min (1270 mm/s) at 6 kW average power. Assuming a spot diameter of 0.5 mm, the interaction time is 3.94×10^{-4} s.

FIGURE 17.1 Gas tungsten arc.

Spot diameters for the high-density processes vary typically from 0.2 mm to 1 mm, while the spot diameters for arc welding processes vary from roughly 3 mm to 10 mm or more. Assuming a rule of thumb of 1/10 the spot diameter for positioning accuracy, we conclude that typical positioning accuracy requirements for the high power density processes is on the order of 0.1 mm and for the arc welding processes is on the order of 1 mm. The required control system response time should be on the order of the interaction time and, hence, may vary from seconds to microseconds, depending on the process chosen. With these requirements it can be concluded that the required accuracy and response speed of control systems designed for welding increases as the power density of the process increases. Furthermore, it is clear that the high power density processes *must* be automated because of the human's inability to react quickly and accurately enough.

Weld Process

The arc welding process may be depicted as shown in Figure 17.2. Inputs to the process may include current, voltage, wire speed, travel speed, etc., which may be varied in real time during the weld, and parameters such as electrode tip geometry, shielding gas type, and wire diameter are usually held constant during the weld. Outputs of the process include geometrical features of the weld, microstructural characteristics, and mechanical properties.

An objective of automated welding control is to choose a set of input parameters that yields a desired set of output characteristics and that maintains those characteristics throughout the weld for any number of parts to be welded. To accomplish this, various output features are sensed, and multivariable feedback control methods are employed to exercise the level of control desired. From the multivariable controls standpoint, a welding process or process variant that offers the greatest decoupling of control parameters is desired [Cook, 1999]. This greatly simplifies the multivariable control systems design and increases the possible range of control over the output features.

Successful implementation of multivariable weld process control involves (1) sensing, (2) modeling, and (3) control. Issues dealing with each of these will be discussed in the following sections.

Sensing

Optical Sensing

Optical sensing technology has been developed and used for a number of applications, including joint tracking and fill control, sensing of molten pool width, sensing of weld bead profile, arc length sensing and control, sensing and control of electrode extension in *gas metal arc welding* (GMAW), and sensing of weld depth or penetration.

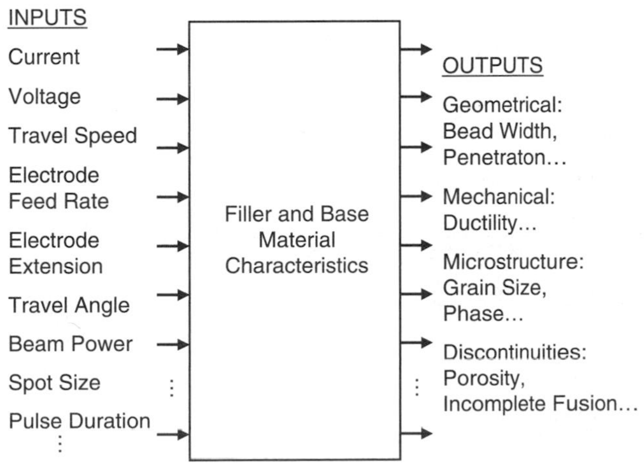

FIGURE 17.2 Input and output variables of welding process.

Hartman [1999] used a nonintrusive, noncontact, topside sensor to collect the arc light reflected from induced oscillations of a GTA weld pool surface. Oscillations were induced in the weld pool by pulsing or modulating the weld current. A software-based, dynamically reconfigurable *phase-locked loop* (PLL) technique enabled synchronization of the weld current pulsations with the fundamental mode of oscillation of the molten weld pool. Improved locking and tracking characteristics in the presence of noise, signal distortions, and harmonic conditions were achieved through the use of intelligent signal monitoring algorithms, which coexisted in parallel with the PLL. Dynamic reconfiguration of the PLL's digital filter coefficients provided superior tracking performance over a wide range of resonant frequency conditions. Assessment of the joint geometry was performed using Rayleigh's [1896] suspended droplet analogy in which the mass, and hence the volume, of the droplet was related to the natural resonant frequency of the molten region. By changing the welding current in a feedback control system to maintain the pool frequency constant, the weld depth, or penetration, was likewise maintained constant for constant pool width. References to other work dealing with weld pool vibration sensing and analysis may be found in Hartman [1999].

Many welds produced using automated welding apparatus exhibit weld surface ripples, as shown in Figure 17.3. DeLapp et al. [1999] used a macrocamera viewing system with a high-speed digital-frame-grabber imaging system to obtain magnified dynamic images of the solidification interface. The macrocamera imaging system is shown schematically in Figure 17.4. Using this equipment, DeLapp et al. [1999] found that weld surface ripples have a distinct tangential formation velocity component relative to the liquid-solid interface.

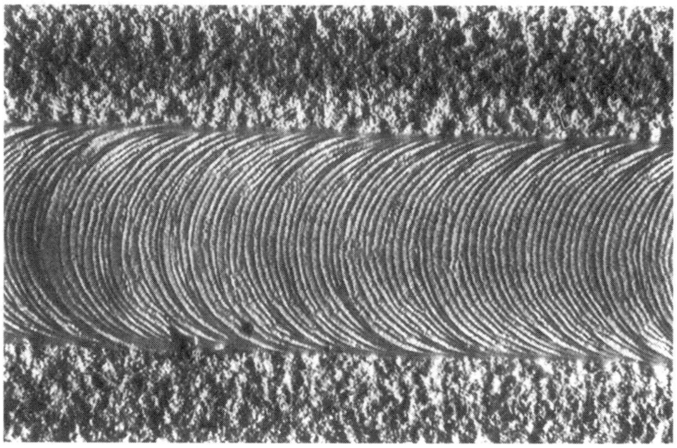

FIGURE 17.3 Typical autogenous GTA weld solidification ripple pattern.

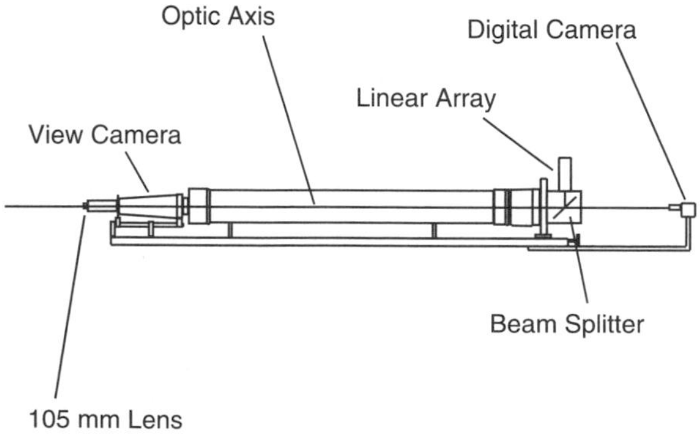

FIGURE 17.4 Macrocamera imaging system showing linear array and digital frame grabber camera in position for measurements.

FIGURE 17.5 Four-head automatic hot-wire GTA pipe welding system.

The formation was found to proceed from the center toward the outside, or from the outside toward the center of the weld pool. As weld pools became larger, the mode of solidification changed from that of ripple formation to a more continuous one, in keeping with established theory. As travel speed was increased with arc parameters held constant, the spacing between weld surface ripples was seen to decrease. This suggests that control systems could be designed to sense and control the surface ripple features. Hall et al. [2001] have shown surface ripples can be related to internal microstructure. Thus, sensing and control of the weld surface ripples may provide a means of indirectly exercising control over certain internal microstructural features.

Arc Sensing

Arc sensing (or through-the-arc sensing) has many applications, some, such as automatic voltage control, dating back 40 years or more. The obvious advantage of arc sensing is that use of the arc itself as a sensor means there is not any need for external sensors, with the associated concern for their reliability in the harsh environment of the welding arc.

Arc sensing is one of the most widely employed means of joint tracking for robotic arc welding [Cook, 1983]. For this application, the sensing method is based on the changes in current and/or voltage when the arc is weaved back and forth across the joint. Control schemes have included both classical PID control and fuzzy logic control [Bingul et al., 2000]. Applications include both consumable and nonconsumable arc welding processes and range from pipe welding (Figure 17.5) to turbine blade repair (Figure 17.6), and from automotive applications to aerospace applications. For *submerged arc welding* (SAW), for example, current variations of approximately 10% at the sidewalls have been observed while welding in a joint consisting of a 45° included angle with a 5-mm root opening. With a nominal current of 580 A at the center of the joint, the current at the sidewalls is approximately 640 A. Variations of this magnitude may be used to implement robust control algorithms for joint tracking and width control.

Andersen et al. [1989] have reported the use of arc signal parameters as a potential control means for GMAW, short-circuiting transfer. Digital signal processing was used to extract from the electrical signals various features, including average and peak values of voltage and current, short-circuiting frequency, arc period, shorting period, and the ratio of the arcing to shorting period. Additionally, a joule-heating model was derived that accurately predicted the melt-back distance during each short. The ratio of the arc period to short period was found to be a good indicator for monitoring and control of stable arc conditions. Any change in the arcing voltage for a given power condition leads to corresponding changes in the arcing/shorting time ratio. Such changes in arcing voltage may occur with change in the shielding gas, in the surface condition (in the form of contaminates) of the electrode wire and work, and in their composition. It is shown that if the average arc current may be assumed to be nominally constant because of constant electrode feed, then the arcing/shorting time ratio serves as a sensitive index of the operation of the GMAW short-circuiting system. The arcing/shorting time ratio can be used to control the short-circuiting gas metal arc in a feedback loop by adjusting the open-circuit voltage to compensate for variations in the arcing voltage.

Infrared Sensing

Infrared sensing has inherent appeal for weld sensing. Applications include cooling-rate measurements, discontinuity sensing, penetration estimation, seam tracking, and weld pool geometry measurements [Nagarajan et al., 1983].

Acoustical Sensing

The acoustical signals generated by the welding arc are a principal source of feedback for manual welders. Acoustical signals have been used as a sensing means for automated welding as well.

Sound generated by the electric arc of a GTAW has been used for arc-length control. With this system the current is pulsed at small amount at an audible rate to generate an audible tone at the arc. The intensity of the arc-generated tone has been shown to be proportional to the arc length and can be suitably processed to provide a feedback signal for arc length control.

Acoustical signals generated by a gas metal arc have been correlated with the detachment of individual droplets from the filler wire. Research has demonstrated the ability to detect the detachment of individual droplets and to distinguish among transfer modes: globular, spray, and streaming transfer. This may lead to a means of closed-loop control of the heat and mass input during pulsed and nonpulsed GMAW.

Acoustical signals have also been reported as a means of plasma monitoring in laser beam welding. Specifically, experiments have been conducted to characterize the interaction between the incident laser light, the plasma formation, and the target material during pulse welding with an Nd/YAG laser. In the experiments, the acoustical signal, picked up by a microphone, was used to signal plasma initiation and propagation. A correlation was observed between the number of plasmas generated and the weld-pool penetration in a target.

Ultrasonic Sensing

The use of ultrasonics for weld process sensing has the potential to detect weld-pool geometry and discontinuities in real time. Most realistic production applications, however, require noncontacting sensors for injecting and receiving the ultrasound. Lasers have been proposed as a sound source, and *electromagnetic acoustic transducers* (EMATs) have been proposed for ultrasound reception [Johnson et al., 1987]. With this approach, the pulsed laser is directed to impinge on the molten pool, setting up stress waves that are transmitted through the workpiece and picked up by the EMAT receiver.

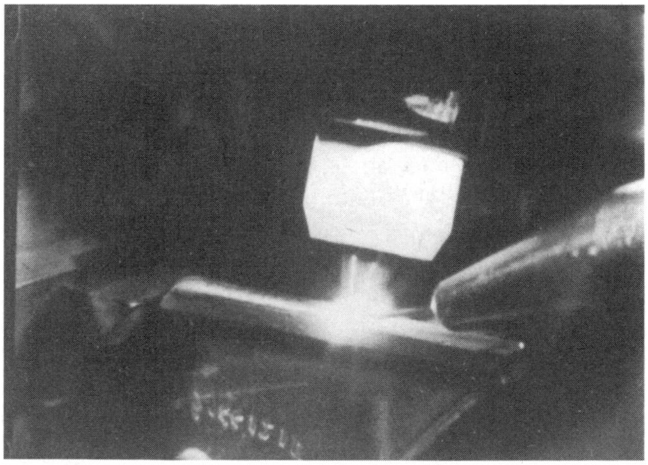

FIGURE 17.6 Turbine blade repair with plasma arc welding process.

Modeling

Weld process models intended for control purposes are characterized by the need to be computable in real time. This rules out many of the more exact numerical models that have been developed for finite element and finite difference methods. However, these computationally intensive numerical models may be quite useful in developing simpler models that can be used in the control of multivariable weld feedback control systems. Another important aspect of process models used for control purposes is that they generally need to provide both static and dynamic information.

Analytical Models

Since the 1940s, considerable research has been focused on developing steady-state thermal models that would predict thermal profiles in the near and far field of the heat source, given a set of input parameters. Easily computed analytical thermal models, based solely on conductive heat transfer, are reasonably accurate but primarily are of value in establishing approximate relationships. Improvements to these early analytical thermal models have been proposed that permit obtaining a better match to actual conditions that may be calibrated in real time. However, accuracy remains limited in the absence of modeling extensions that require computationally intensive numerical solution.

In addition to analytical thermal models, other steady-state models have been developed for characterizing specific aspects of various welding processes. For example, Lesnewich [1958] presented a relationship between electrode burn-off rate, electrode extension, and current density for the GMAW process, based on extensive experimental results. Halmoy [1979] later showed that the relationship could be derived from basic principles. Shepard [1991] extended these steady-state relationships to a dynamic model of the self-regulation process for constant-potential GMAW.

Bingul et al. [2001] derived a dynamic model of the electrode-melting rate in the GMAW process. The components of this model are the electrode melting rate, the temperature-dependent resistivity of the electrode, and the arc voltage. The differential equations describing the dynamic behavior of the electrode extension were derived from the mass continuity and energy relations. The temperature of the electrode extension was determined by analyzing conductive heat transfer and Joule heating effects. One-dimensional solutions for the temperature and heat content were used to obtain the dynamic melting rate equation. Bingul et al. [2001] developed their dynamic model of electrode melting rate for use in adaptive arc-length control of pulsed GMAW with a constant-current power source.

Empirical and Statistical Models

Other approaches taken to developing steady-state weld process models include empirically derived relationships between the input and output parameters, with coefficients chosen to match experimental data, and statistically derived relationships. Both of these approaches have proven to possess only a limited range of applicability, and they do not lend themselves to real-time "tuning" in a multivariable control system application.

Artificial Neural Network Models

Artificial neural networks (ANN) have been studied for weld modeling and found to be accurate and computationally fast in the application mode. Furthermore, the ANN can be refined at any time with the addition of new training data and thus promises a method of continuously adapting to the actual welding conditions.

Andersen [1992] has reported the application of an ANN to mapping among the input parameters, arc current, travel speed, arc length, and plate thickness, and the output parameters, bead width and penetration, for GTAW. A back-propagation network, using ten nodes in a single hidden layer (Figure 17.7) was used for the modeling. A variety of different network configurations were initially evaluated for this purpose. Generally, it was found that one hidden layer was sufficient for weld modeling, and the best training rate was obtained with on the order of 5 to 20 nodes in the hidden layer. The same plate material was assumed throughout the experiment, which eliminated the need for specifying any of the material parameters. Otherwise, additional input parameters might have included thermal conductivity, diffusivity, etc.

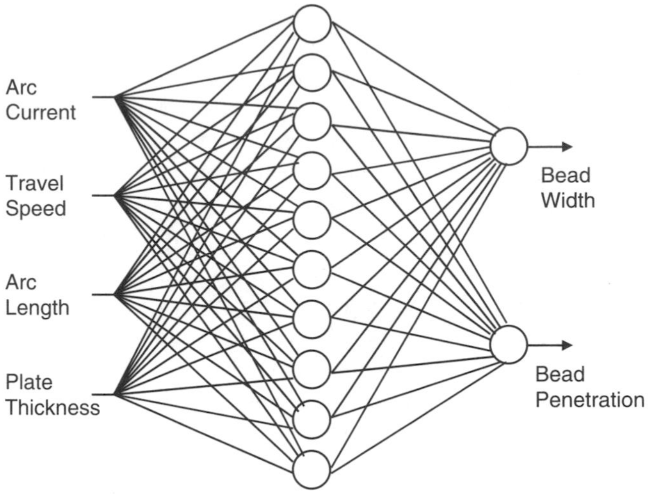

FIGURE 17.7 A neural network used for weld modeling.

A total of 72 welds, produced on two material thicknesses of 3.175 and 6.350 mm, were used for the purpose of training and testing the network for modeling purposes. Weld current values of 80, 100, 120, and 140 A, travel speeds of 2.12, 2.75, and 3.39 mm/s, and arc lengths of 1.52, 2.03, and 2.54 mm were used. Eight of the welds, which were randomly selected, were not used in the training phase but were reserved for testing the model. With a learning rate parameter of 0.6 and a momentum term of 0.9, the network was trained for 200,000 iterations.

Once the network had been trained with the 64 training welds, the remaining eight welds were applied to test the modeling network. The root mean square (RMS) values of the errors were calculated separately for the bead width and penetration, resulting in about 5% and 18% errors, respectively. These results agree with other similar experiments reported by Andersen, in that modeling accuracy is typically on the order of 10 to 20%.

When compared to other control modeling methodologies, neural networks have certain drawbacks as well as advantages. Of the drawbacks, the most notable is the lack of comprehension of the physics of the process. Relating the qualitative effects of the network structure or parameters to the process parameters is usually impossible. On the other hand, other control modeling methods resort to substantial simplifications of either the physical process or more exact numerical models and therefore also trade computability for comprehensibility. The advantages of neural models include relative accuracy and generality. If the training data for a neural network is general enough, spanning the entire ranges of process parameters, the resulting model will capture the complexion of the process, including nonlinearities and parameter cross couplings, over the same ranges. Model development is much simpler than for most other models. Instead of theoretical analysis and development for a new model, the neural network tailors itself to the training data. The network can be refined at any time with addition of new training data. Finally, the neural network can calculate its results relatively quickly, as the input data are only propagated once through the network in the application mode.

The reader is referred to Andersen [1992] for a more thorough discussion of the neural network approach to weld process modeling. Andersen also presents a detailed comparison of neural network modeling to two analytical models and a statistically based multidimensional parameter interpolation approach.

Control

Decoupling of Variables for Improved Control

The easiest approach to controlling multiple weld process parameters can be realized if input variables can be found that affect only a single output quantity. If the output variable is affected by another input variable as well, then one may be the primary variable while the other may constitute a secondary feedback loop, which is capable of controlling the output quantity by a relatively small amount with respect to the basic level set by the primary variable. For example, high-frequency pulsation of the current in GTAW may provide a means of controlling the depth of penetration over a small range without affecting the width of the weld bead. In this case the heat input, as determined by the voltage, current, and travel speed, would be the primary input variable controlling the width and penetration, while the high-frequency pulsation would be the secondary variable capable of producing small corrections to the basic penetration depth.

As a means of decoupling process variables, Doumanidis and Hardt [1990] and Doumanidis and Fourligkas [1996] have employed a multitorch system or a torch scanning system to obtain greater control of the temperature field and hence weld bead geometry. The hot-wire process variant [Manz, 1975] permits essentially independent control of the heat input delivered to the base material and the deposition rate. Pulsed GMAW tends to decouple the average current requirement for stable spray transfer from the deposition rate [Allum, 1983]. This permits welding at low average currents that would otherwise result in globular transfer [Kim and Eagar, 1993]. The pulsed current permits stable spray metal transfer and formation of a uniform bead shape with shallow controlled penetration. This permits tapering the weld profile at the start and stop by sloping the wire-feed speed (and corresponding pulsed current parameters) to achieve smooth starts absent of excessive buildup and crater filling at the end of the weld to avoid cracks [Amin and Naseer-Ahmed, 1987].

Control schemes that rely on arc signals, such as through-the-arc sensing for seam tracking [Cook, 1983] and penetration control based on weld pool vibrations [Hartman, 1999], are also best implemented with process variants offering the greatest decoupling of parameters. For example, the inherent self-regulation characteristics [Shepard, 1991] of conventional constant-potential GMAW serves as a detriment to seam tracking with through-the-arc sensing, which is based on sensing variations in contact tube-to-work distance. Pulsed GMAW without the inherent self-regulation of conventional GMAW offers the possibility of significantly improved arc sensing for seam tracking purposes provided the sensing is properly synchronized with the pulsing and any arc length control employed in the power source [Damrongsak, 1997].

It should be noted that while decoupling of process variables is desirable for automated weld control, such decoupling may not be advantageous from the standpoint of manual welding [Cook, 1999]. For example, the inherent self-regulation characteristic of conventional constant-potential GMAW makes the process easy to use for manual welding by providing built-in compensation for minor variations in *contact-tube-to-work distance* (CTWD). This same self-regulation characteristic of conventional GMAW combined with a relatively small range of acceptable average current for stable spray transfer renders it less than optimum for automated welding control, however.

Adaptive Control

Even for single-variable weld process control, nonlinearities in the process may call for an adaptive system to automatically adjust the parameters of the controller when the process parameters and disturbances are unknown or change with time. For example, Bjorgvinsson [1992] shows that a simple *automatic voltage control* (AVC) system may be unstable over a wide range of current settings because of the variation of the arc sensitivity (voltage change per unit change of arc length) with current. A simplified schematic of an AVC system is shown in Figure 17.8. The arc voltage (proportional to the arc length) is compared with a reference voltage in a simple position servo. If an error exists between the reference voltage and the arc voltage, the servomotor moves the welding torch up or down to reduce the error to zero. If K_a is the gain of the AVC motor-drive system and K_s is the arc sensitivity ($K_s = dV_{arc}/DL_{arc}$), then the overall loop gain K is given by $K = K_a K_s$. The closed-loop stability of the position control system is dependent on the loop gain and will obviously vary from its design setting if K_s changes. Bjorgvinsson shows that for helium shielding gas, the arc sensitivity may vary by approximately a ratio of 5:1 over a current range of 15 to 150 A. In this case, for a standard proportional controller, the overshoot to a step input at 15 A is approximately 40% if the controller gain K_a is fixed and set for optimum response at 150 A. Bjorgvinsson proposes a gain-scheduling adaptive controller (see Figure 17.9) to vary the controller gain in such a manner as to compensate for the changing arc sensitivity for all levels of welding current. Knowing the arc current, the adaptive controller uses information stored in a look-up table or computed from a mathematical model of the arc to adjust K_a in response to changes in K_s such that the product $K_a K_s = K$ is maintained constant independent of the current. The result is

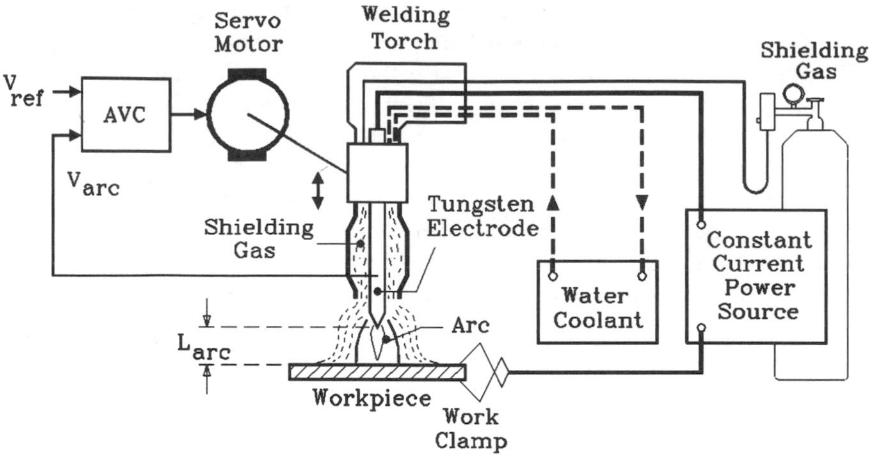

FIGURE 17.8 GTAW system with AVC.

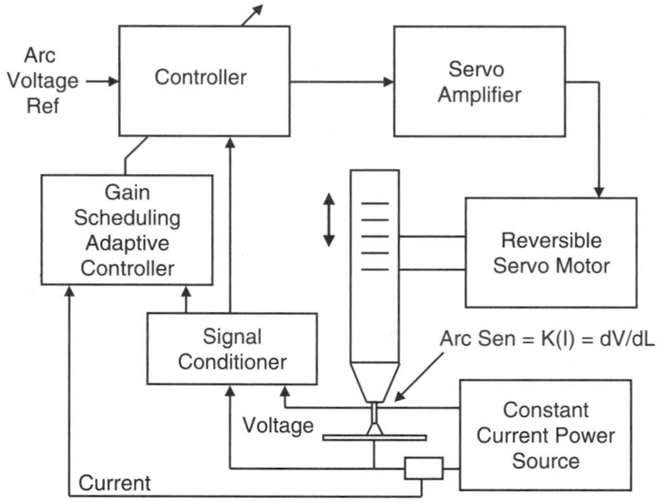

FIGURE 17.9 Gain-scheduling adaptive automatic voltage control.

uniform closed-loop stability characteristics of the AVC system throughout the complete weld. This includes the up-slope period, when the current is varied from the low arc-initiation value to the nominal welding current, which is maintained until the down-slope period, when the current is brought back to a low value for termination of the arc.

While the gain-scheduling adaptive controller employed by Bjorgvinsson [1992] was demonstrated to work well under controlled conditions, the functional relationship between arc sensitivity and current is dependent on a number of parameters that are difficult to preprogram in the look-up table or mathematical relation he used. To circumvent this problem, Koseeyaporn et al. [2000] and Smithmaitrie et al. [2000] used adaptive schemes that did not require any *a priori* knowledge of the current sensitivity relationship.

Koseeyaporn et al. [2000] used the partitioned control method employed by Craig [1988] for adaptive control of robot manipulators. With this approach, the controller is partitioned into a nonlinear part chosen to cancel nonlinearities of the system and a servo part. The nonlinear part is chosen such that the servo part sees a simple unit-mass system. The adaptive estimator, chosen to ensure Lyapunov stability, provides estimates of the unknown parameters in the nonlinear part of the partitioned controller. A block diagram of the adaptive controller is shown in Figure 17.10.

Smithmaitrie et al. [2000] used a fuzzy logic controller with two inputs (voltage error and derivative of voltage error) and one output (servomotor voltage) for the basic inner-loop automatic voltage control. The min/max inference and center of gravity defuzzification method [Passino and Yurkovich, 1998] was used in implementing the fuzzy logic controller. The outer adaptive loop consisted of an identical inverse fuzzy logic controller with two inputs, as shown in Figure 17.11. The difference between the arc voltage and the model arc voltage, $y_e(t)$, and its derivative are the inputs to the fuzzy inverse model. For given information (from the inverse model) about the necessary change in the input needed to make $y_e(t)$ approximately zero, the knowledge-base modifier changes the knowledge base of the fuzzy controller so that the previously applied control action will be modified by the amount specified by the inverse model output. To modify the knowledge base, the knowledge-base modifier shifts the centers of the output membership functions for the rules that were active during the previous control action.

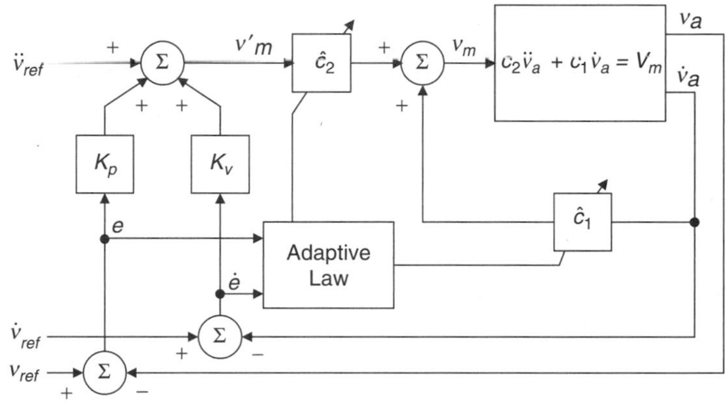

FIGURE 17.10 The welding system with a partitioned adaptive controller.

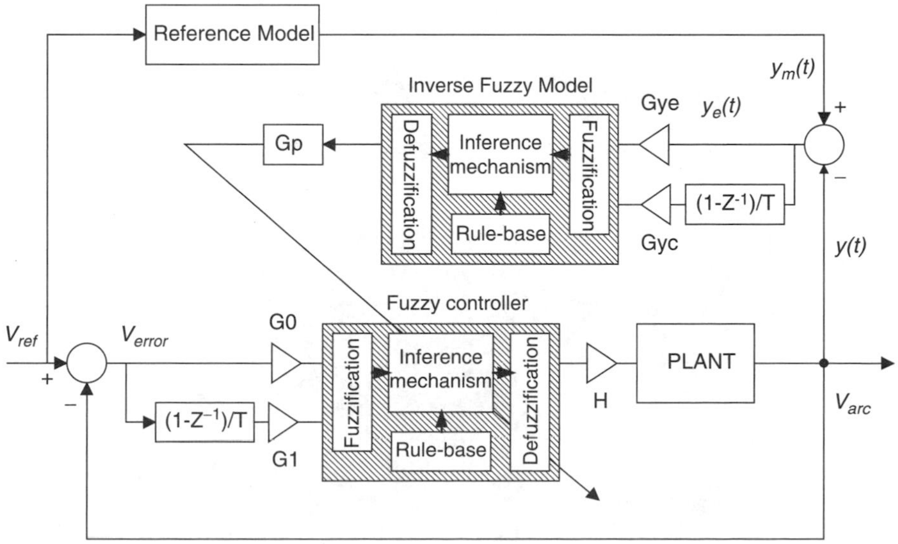

FIGURE 17.11 Adaptive fuzzy control for arc voltage control in GTAW.

Both the adaptive controller of Koseeyaporn et al. [2000] and that of Smithmaitrie et al. [2000] permitted AVC over the full operating range of welding current from low starting values to normal welding values to low termination values. Both approaches proved more robust than that of Bjorgvinsson [1992] based on a preprogrammed relationship between current and arc sensitivity.

Force Control

FSW is depicted in Figure 17.12. The rotating welding tool is comprised of a shoulder in contact with the workpiece and a concentric pin, which extends almost completely through the workpiece thickness. As the tool rotates, frictional heat is produced at the contacting shoulder and pin surfaces, causing the workpiece material to soften into the materials' plastic state and flow around the tool from the front to rear, where it consolidates and forges under the confining pressure of the tool shoulder and regains strength upon cooling to form the weld. No fusion occurs. Most applications currently involve the welding of aluminum alloys, although FSW of steel, titanium, and dissimilar material combinations has been demonstrated.

The FSW tool pin may be smooth, threaded, or take on a number of other configurations [Thomas and Dolby, 2003] designed to enhance material flow vertically and provide improved consolidation of the material at the bottom of the weld. The tool shoulder, whose diameter is relatively large compared to the pin, acts to prevent material from being expelled from the weld zone. This minimizes the formation of voids in the weld zone. A tool has also been patented [Campbell et al., 2001] that utilizes two independent actuated shoulders at the top and bottom of the weld. With its independently actuated shoulders, this tool is able to accommodate varying material thickness.

The rotational speed of the FSW tool may vary, with application, from a few hundred revolutions per minute to several thousand revolutions per minute. The axial force required to counteract the pressure formed in the flowing plasticized weld zone may vary from about 1 to 15 kN. The mechanical power input to the rotating tool is typically of the order of 1.5 to 4 kW. The translational force in the direction of travel may vary from a very low value (or negative value when the rotating tool actually pulls itself along by the friction generated along the returning side) to about 1 kN, and the transverse force, acting toward the advancing side of the pin at an angle of 90° to the direction of travel, varies typically from about 0.2 to 2 kN.

Because of the large forces and torques associated with FSW, workpieces must be rigidly clamped, and, for single-sided tools, a backing bar is required. The large forces and torques also call for machine-tool-type equipment for implementation. This is an advantage in which conventional machine tools, such as milling machines, can be used to apply the process. However, this also leads to a current drawback, which is that most applications are presently constrained to two-dimensional planes by such machine tools.

FIGURE 17.12 Friction stir weld.

FSW applications can be expected to expand substantially when industrial robots are used more in place of currently employed heavy-duty machine-tool equipment. This will permit welding three-dimensional contours, with the robots offering the advantages of greater flexibility and availability, and relatively low cost. However, with a robotic implementation, one cannot assume the rigidity and precision of heavy-duty machine tools. At the forces required, there is usually enough compliance in the manipulator arm that force feedback becomes a necessity.

A force-control scheme that has been successfully used for this application is shown in Figure 17.13. An "outer" force control loop is closed around the "inner" position control loop of the robot manipulator [De Schutter and Van Brussel, 1988]. The programmed z-axis position (with respect to the wrist frame) of the robot is modified as required to maintain the desired axial force set by the outer control loop. This approach is attractive because it does not require access to the basic position control loop of the robot. Stability of this scheme will depend largely on the force/position characteristic of the rotating tool as it acts against the plasticized weld-zone material. Most force-control schemes assume a linear elastic environment. However, in FSW the tool/workpiece environment is nonlinear, nonelastic, and a function of the welding parameters, e.g., tool rotation speed and travel speed. This has not been found to be a major problem, provided that the force control loop is made inactive during the start and stop portions of the weld.

The control scheme shown in Figure 17.13 assumes an independent force sensing means at the weld head. An approach that does not require direct sensing of the axial force is shown in Figure 17.14. Here, use is made of the Jacobian relationship between actuator torques and force given by Craig [1989],

$$F = J\tau$$

where F is a 6×1 Cartesian force-moment vector acting at the end-effector, J is the 6×6 manipulator Jacobian, and τ is a 6×1 vector of actuator torques. This approach has been used [Smith, 2000] and demonstrated to

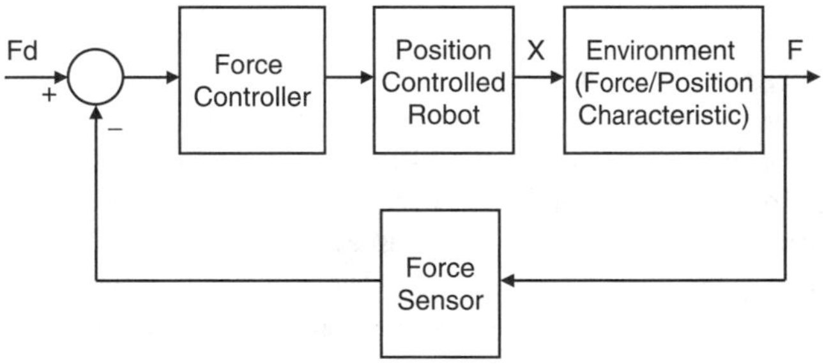

FIGURE 17.13 Force feedback control in FSW implemented as the outer force control loop around the ordinary position control system of the robot manipulator.

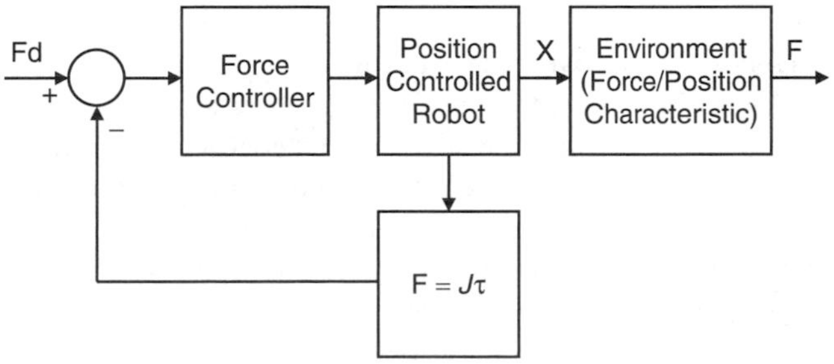

FIGURE 17.14 Force feedback control in FSW with the force derived from the actuator torques of the manipulator arm.

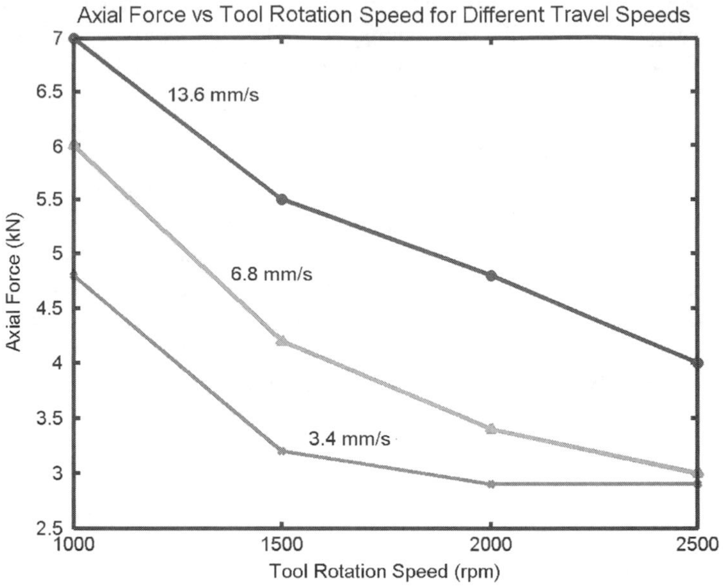

FIGURE 17.15 Axial force vs. tool rotation speed in friction stir welding.

work adequately at rotation speeds of 1500 rpm or greater. However, the update time was limited to 2 Hz because of the computational burden of computing the manipulator Jacobian.

The axial force requirements imposed on the robot can be reduced by operating at high tool rotation speed and lower travel speed, as shown in Figure 17.15 [Cook et al., 2004]. Articulated manipulator arms and arms with a parallel design may be used provided that they are structurally capable of providing the necessary force.

As this technology is perfected, it might become desirable to control the force throughout the starting and ending periods. During the plunge of the pin into the material at weld start (assuming a predrilled starting hole is not used), the axial force may rise initially to three to five times the weld value, depending on the specific weld parameters, tool shape, and the plunge rate [Cook et al., 2003]. With such a wide variation in the tool/workpiece force/position characteristics, stability issues may arise in the force-feedback control, if force control is attempted during the weld start-up. As previously discussed for arc welding, this might call for more advanced control methods including adaptive control to accommodate the widely varying "environment."

Intelligent Control

Practical weld process control implementation, particularly with multivariable and adaptive control, involves a substantial body of heuristic knowledge concerning the weld process and the numerous constraints that are involved in its control. The role that intelligent control concepts can play is to provide a systematic approach to dealing with these constraints.

For example, for a given set of material parameters, one may wish to control several geometrical parameters plus cooling rate for the GMAW process, while maintaining operation in the spray transfer mode of the process. Because of the close coupling among the equipment, material, and geometric parameters, and because of the small latitude of permissible variation of one parameter once the others are specified, tight constraints on the control system will be necessary to achieve the desired process quality.

It will be desirable to specify degrees of control permitted over the various parameters in terms of a hierarchy of parameter importance. For example, while the wire feed rate has an influence on bead width in the GTAW process, it would not be desirable to allow the wire feed rate to be varied excessively as a means of controlling bead width. Further, the allowable variation of a given parameter or parameters may not be symmetrical about the desired set point. Again, for the GTAW process, an increase in current may be partially offset by an increase in travel speed, whereas a reduction in both parameters would tend to more rapidly force the geometrical parameters outside the desired range.

Consideration of the process dynamics is also necessary, particularly for successful control during the initiation and termination phases of the overall welding operation. In addition to the hierarchical considerations referred to above, the time sequence, and rate of change of each parameter should be considered. Intelligent control concepts may be used to handle these practical control issues in a formal and logical manner.

Conclusions

Advances in welding automation and control over the last two decades can be traced in most instances to two major technological innovations — embedded computers and high-performance solid-state power electronics. With these innovations, the welding arc in arc welding, or the energy beam in laser and electron beam welding, can be controlled with high precision, and robotic manipulators with joint sensing technology can maintain the heat source at exactly the desired position along the joint as the weld is made. Additionally, great strides have been made in sensing and controlling the weld geometry. These same advances are available for recently introduced welding processes such as hybrid laser/arc welding systems and friction stir welding. In the case of robotic friction stir welding, for example, force feedback control is required to maintain the desired axial force on the pin tool as the weld progresses.

Today computers are embedded in almost all major welding equipment components. This includes the power supply, wire feeder, automatic voltage/current controllers, tracking systems, penetration control systems, robotic manipulators, positioners, sensors, data acquisition systems, and gas monitoring systems. These embedded computers are designed to communicate with each other, as well as with other computers in the overall manufacturing facility. With this distributed computing capability, very advanced signal analysis techniques and control methods can be easily implemented and maintained.

Advances in solid-state power electronics have shrunk the size of welding power supplies and motor controllers, and they have given us the capability to dynamically sense and respond in microseconds to changes in the electrical signals of the arc. This has given us the ability, for example, to precisely control the droplet transfer in pulsed GMAW over wide variations in the wire feed speed. It has also given us the ability to deliver short-circuiting transfer with virtually zero splatter [Stava, 1993]. Indeed, with high-performance inverter power sources available today, we can literally treat the power supply as a "black box" that can deliver on demand any waveform or voltage–current characteristic we desire at speeds well above actual requirements at the arc.

In the next 20 years, computers will continue to get smaller, and substantial advances in the software technology of real-time systems and embedded computing can be expected. These advances will find their way into welding equipment and control technology, resulting in continued improvements in realizations of truly intelligent welding systems.

Since most material processing for commercial applications is designed to develop specific microstructures that are known to give the final product its desired properties, we can expect the development of process-control algorithms and nondestructive testing procedures to be focused on developing techniques and sensors that provide real-time information on the microstructure of the weld.

Sensors that provide information about the microstructure will rely, in part, on the same advances in computing and signal analysis that will lead to advances in other aspects of the welding system. For example, a critical examination of the material characterization potential of ultrasonics shows that the effect of microstructure on the velocity of sound is often rather small and requires measurement techniques that are reliable to precisions equal to or better than 0.1%. By combining the advantages of EMATs for signal coupling with embedded computers for advanced signal processing and analysis, this may be possible. Already, for example, we are able to predict the hardness or yield strength of a steel from empirical correlations with the velocity of certain ultrasonic waves.

Advances in sensor technology may come from completely new innovations or, as in the case of EMATs and ultrasonics, it may come from further developments of established methods. For example, high-speed macroimaging techniques have recently been shown to be capable of delivering vivid dynamic images of the surface solidification at the trailing edge of a weld pool [DeLapp et al., 1999] If these dynamic surface features can be related to the hidden weld microstructure, then feedback-control methods could be developed to exercise a level of control over the weld properties.

Defining Terms

Arc welding: A group of welding processes that produces coalescence of metals by heating them with an arc, with or without the application of pressure, and with or without the use of filler metal.

Artificial neural networks (ANN): A mathematical model for information processing based on a connectionist approach to computation. The original inspiration for the technique was from examination of bioelectrical networks in the brain formed by neurons and their synapses.

Automatic voltage control (AVC): A servo system designed to sense and control arc voltage in the GTAW process by means of feedback.

Contact-tube-to-work-distance (CTWD): Distance measured between the bottom of the contact tube in GMAW and the workpiece.

Electromagnetic acoustic transducers (EMATS): A noncontacting electromagnetic transducer that is able to induce ultrasound into a material and receive ultrasound signals.

Electron beam welding: A welding process that produces coalescence of metals with the heat obtained from a concentrated beam composed primarily of high-velocity electrons impinging on the surface to be joined.

Electroslag welding: A welding process that produces coalescence of metals with molten slag that melts the filler metal and the surfaces of the parts to be joined.

Gas metal arc welding (GMAW): A welding process that produces coalescence of metals by heating them with an arc between a consumable filler metal electrode and the parts to be joined. The process is used with shielding gas and without the application of pressure.

Gas tungsten arc welding (GTAW): A welding process that produces coalescence of metals by heating them with an arc between a nonconsumable tungsten electrode and the parts to be joined. The process is used with shielding gas and without the application of pressure. Filler metal may or may not be used.

Laser beam welding (LBW): A welding process that produces coalescence of materials, with the heat obtained from the application of a concentrated coherent light beam impinging on the surface to be joined.

Oxyacetylene welding: An oxyfuel gas welding process that produces coalescence of metals by heating them with a gas flame obtained from the combustion of acetylene with oxygen. The process may be used with or without the application of pressure and with or without the use of filler metal.

Phase-locked loop (PLL): An electronic circuit that controls an oscillator so that it maintains a constant phase angle relative to a reference signal.

Resistance spot welding: A resistance welding process that produces coalescence at the faying surfaces of a joint by the heat obtained from resistance to the flow of welding current through the workpieces from electrodes that serve to concentrate the welding current and pressure at the weld area.

Submerged arc welding (SAW): An arc welding process that produces coalescence of metals by heating them with an arc or arcs between a bare metal electrode or electrodes and the workpieces. The arc and molten metal are shielded by a blanket of granular, fusible material on the workpieces. Pressure is not used, and filler metal is obtained from the electrode and sometimes from a supplemental source.

Thermit welding: A welding process that produces coalescence of metals by heating them with superheated liquid metal from a chemical reaction between a metal oxide and aluminum, with or without the application of pressure.

References

C.J. Allum, "MIG welding: time for reassessment," *Met. Constr.*, vol. 15, no. 6, pp. 347–353, 1983.

M. Amin and N. Ahmed, "Synergic control in MIG welding — part 1: parametric relationships for steady DC open arc and short circuiting arc operation," *Met. Constr.*, no. 1, pp. 22–28, 1987.

K. Andersen, Studies and Implementation of Stationary Models of the Gas Tungsten Arc Welding Process, M.S. thesis, Vanderbilt University, Nashville, 1992.

K. Andersen, G.E. Cook, Y. Liu, D.S. Mathews, and M.D. Randall, "Modeling and control parameters for GMAW, short circuiting transfer," in *Advances in Manufacturing Systems Integration and Processes*, D.A. Dornfeld, Ed., Dearborn, MI: Society of Manufacturing Engineers, 1989.

Z. Bingul, G.E. Cook, and A.M. Strauss, "Application of fuzzy logic to spatial thermal control in fusion welding," *IEEE Trans. Ind. Appl.*, vol. 36, no. 6, pp. 1523–1530, 2000.

Z. Bingul, G.E. Cook, and A.M. Strauss, "Dynamic model for electrode melting rate in gas metal arc welding process," *Sci. Technol. Weld. Joining*, vol. 6, no. 1, pp. 41–50, 2001.

J.B. Bjorgvinsson, Adaptive Voltage Control in Gas Tungsten Arc Welding, M.S. thesis, Vanderbilt University, Nashville, 1992.

C.L. Campbell, M.S. Fullen, and M.J. Skinner, U.S. Patent No. 6,199,745, 2001.

G.E. Cook, "Robotic arc welding: research in sensory feedback control," *IEEE Trans. Ind. Electron.*, vol. IE-30, no. 3, pp. 252–268, 1983.

G.E. Cook, "Decoupling of weld variables for improved automatic control," in *Trends Weld. Res. Proc. Fifth Int. Conf.*, J.M. Vitek, S.A. David, J.A. Johnson, H.B. Smartt, and T. DebRoy, Eds., Materials Park, OH: ASM International, 1999, pp. 1007–1015.

G.E. Cook, R. Crawford, and A.M. Strauss, "Robotic friction stir welding," *Ind. Robot*, vol. 31, no. 1, pp. 55–63, 2004.

G.E. Cook, H.B. Smartt, J.E. Mitchell, A.M. Strauss, and R. Crawford, "Controlling robotic friction stir welding," *Weld. J.*, vol. 82, no. 6, pp. 28–34, 2003.

J. Craig, *Adaptive Control of Mechanical Manipulators*, Reading, MA.: Addison Wesley, 1988.

J.J. Craig, *Introduction to Robotics: Mechanics and Control*, 2nd ed., Reading, MA: Addison Wesley, 1989.

D. Damrongsak, Through-the-Arc Sensing and Control in Pulsed Gas Metal Arc Welding, M.S. thesis, Vanderbilt University, Nashville, 1997.

D.R. DeLapp, D.A. Hartman, W.H. Hofmeister, G.E. Cook, and A.M. Strauss, "An investigation into the local solidification rate of the GTA weld pool," in *Trends Weld. Res., Proc. Fifth Int. Conf.*, J.M. Vitek, S.A. David, J.A. Johnson, H.B. Smartt, and T. DebRoy, Eds., Materials Park, OH: ASM International, 1999, pp. 400–404.

D.R. DeLapp, D.A. Hartman, R.J. Barnett, G.E. Cook, and A.M. Strauss, "The development of a GTAW observation system," in *Trends Weld. Res., Proc. Fifth Int. Conf.*, J.M. Vitek, S.A. David, J.A. Johnson, H.B. Smartt, and T. DebRoy, Eds., Materials Park, OH: ASM International, 1999, pp. 405–409.

J. De Schutter and H. Van Brussel, "Compliant robot motion, parts I–II," *Int. J. Rob. Res.*, vol. 7, no. 4, pp. 3–33, 1988.

C. Doumanidis and D. Hardt, "Simultaneous in-process control of heat-affected zone and cooling rate during arc welding," *Weld. J. Res. Suppl.*, vol. 69, no. 5, pp. 186s–196s, 1990.

C. Doumanidis and N. Fourligkas, "Distributed-parameter control of the heat source trajectory in thermal materials processing," *ASME J. Manuf. Sci. Eng.*, vol. 118, no. 4, pp. 571–578, 1996.

T.W. Eagar, "The physics and chemistry of welding processes," in *Advances in Welding Science and Technology*, S.A. David, Ed., Metals Park, OH: ASM International, 1986, pp. 291–298.

A.C. Hall, G.A. Knorovsky, C.V. Robino, J. Brooks, D.O. Maccallum, M. Reece, and G. Poulter, "Characterizing the microstructure of a GTA weld in-process using high-speed, high-magnification, digital imaging," in *Proc. 11th Int. Conf. Comput. Technol. Weld.*, Columbus, OH, 2001, pp. 5–6.

E. Halmoy, "Wire melting rate, droplet temperature, and effective anode melting potential," *Int. Conf. Arc Phys. Weld Pool Behav.*, Abingdon, MA: The Welding Institute, May 1979.

D. A. Hartman, Modal Analysis of GTA Weld Pools for Penetration Control, Ph.D. dissertation, Vanderbilt University, Nashville, 1999.

J.A. Johnson, H.B. Smartt, R.T. Allemeier, N.M. Carlson, C.J. Einerson, and A.D. Watkins, "Sensing and modeling of gas metal arc welding," in *Proc. Fifth Symp. Energy Eng. Sci.*, Argonne, IL: Argonne National Laboratory, June 1987, pp. 168–175.

Y.-S. Kim and T.W. Eagar, "Analyses of metal transfer in gas metal arc welding," *Welding J., Res. Suppl.*, no. 6, pp. 269s–278s, 1993.

P. Koseeyaporn, G.E. Cook, and A.M. Strauss, "Adaptive voltage control in fusion arc welding," *IEEE Trans. Ind. Appl.*, vol. 36, no. 5, pp. 1300–1307, 2000.

J.F. Lancaster, *The Physics of Welding*, Oxford, England: Pergamon Press, 1986.

A. Lesnewich, "Control of melting rate and metal transfer in gas-shielded metal-arc welding, Part 1, control of electrode melting rate," *Weld. J.*, vol. 37, no. 8, pp. 343s–353s, 1958.

A.F. Manz, "Hot wire welding and surfacing technique," *WRC Bulletin 223*, New York: Welding Research Council, 1975.

S. Nagarajan, W.H. Chen, and B.A. Chin, "Infrared thermography for sensing the arc welding process," *Weld. J.*, vol. 62, no. 9, pp. 227s–234s, 1983.

K.M. Passino and S. Yurkovich, *Fuzzy Control*, Menlo Park, CA: Addison Wesley Longman, 1998.

B.J.W.S. Rayleigh, *The Theory of Sound*, vol. II, 2nd ed., New York: McMillan and Co., Ltd, 1896.

M.E. Shepard, Modeling of Self-Regulation in Gas-Metal Arc Welding, Ph.D. dissertation, Vanderbilt University, Nashville, 1991.

C.B. Smith, "Robotic friction stir welding using a standard industrial robot," in *Second Friction Stir Welding International Symposium*, CD-ROM, 2000.

P. Smithmaitrie, P. Koseeyaporn, G.E. Cook, and A.M. Strauss, "Adaptive fuzzy voltage control in GTAW," in *Proc. Seventh Mechatron. Forum Int. Conf.*, Amsterdam: Elsevier, 2000.

E.K. Stava, "The surface-tension-transfer power source: a new low-spatter arc welding machine," *Weld. J.*, vol. 72, no. 1, pp. 25–29, 1993.

W.M. Thomas and R.E. Dolby, "Friction stir welding developments," in *Trends Weld. Res.: Proc. Sixth Int. Conf.*, S.A. David, T. DebRoy, J.C. Lippold, H.B. Smartt, and J.M. Vitek, Eds., Materials Park, OH: ASM International, 2003, pp. 203–211.

W.M. Thomas, E.D. Nicholas, J.C. Needham, M.G. Church, P. Templesmith, and C.J. Dawes, International Patent Application No. PCT/GB92/02203 and GB Patent Application No. 9125978.9, 1991.

Further Information

Other recommended reading on welding technology, welding processes, and welding automation and control includes *Welding Handbook, Volume 1 – Welding Technology* (American Welding Society, Miami, 1987), *Welding Handbook, Volume 2 – Welding Processes* (American Welding Society, Miami, 1987), *Advances in Welding Science and Technology* (edited by S.A. David, ASM International, Metals Park, Ohio, 1986), *Recent Trends in Welding Science and Technology* (edited by S.A. David and J.M. Vitek, ASM International, Metals Park, Ohio, 1990), *International Trends in Welding Science and Technology* (edited by S.A. David and J.M. Vitek, ASM International, Metals Park, Ohio, 1993), *Trends in Welding Research:Proceedings of the 5th International Conference* (edited by J.M. Vitek, S.A. David, J.A. Johnson, H.B. Smartt, and T. DebRoy, ASM International, Metals Park, Ohio, 1999), *Trends in Welding Research: Proceedings of the 6th International Conference* (edited by S.A. David, T. DebRoy, J.C. Lippold, H.B. Smartt, and J.M. Vitek, ASM International, Metals Park, Ohio, 2002), *Developments in Mechanised and Robotic Welding* (edited by G.R. Salter, The Welding Institute, Cambridge, England, 1980), *Developments in Automated and Robotic Welding* (edited by D.N. Waller, The Welding Institute, Cambridge, England, 1987), *Developments and Innovations for Improved Welding Production* (The Welding Institute, Cambridge, England, 1983), *Automated Welding Systems in Manufacturing* (Abington Publishing, Cambridge, England, 1991), *Robotic Welding* (edited by J. Lane, IFS Publications Ltd., Bedford, England, 1987), and *Laser Welding* (Walter W. Duley, John Wiley & Sons, Inc., New York, 1999).

18

Human–Computer Interaction[1]

Evelyn P. Rozanski
Rochester Institute of Technology

Anne R. Haake
Rochester Institute of Technology

18.1 Introduction

Human–computer interaction (HCI) is a relatively young discipline, tracing its origins to a 1982 conference, "Human Factors in Computer Systems" [10]. It draws its theories and methodologies from many complementary disciplines such as cognitive and behavioral psychology, human factors and ergonomics, anthropology, sociology, computer science, graphic design, as well as engineering (see Figure 18.1). The discipline of industrial engineering has long based its curricular study of HCI in the physiological areas of human factors and ergonomics. Interestingly, the early time and motion work analyses of Taylor and Gilbreth, considered as the foundation for the contemporary field of industrial engineering [15], are also forerunners of today's task analysis approaches that are widely employed in HCI.

Another engineering discipline, software engineering, has embraced HCI principles and concepts in the development of complex software products.

So, why should all engineers know about HCI concepts? HCI considers the usability of all products, whether they are hardware, software, or a blend, as an important consideration in their design. A central tenet of HCI is to design with the end-user in mind. There are many critical problems with which engineers are concerned. For example, issues such as failed products, financial losses, and human injuries [2] impact not only the usability of products, but also productivity and revenues. Additionally, there are many different layers of complexity [2], such as easy to learn interfaces and appropriate feedback to the changing working relationships when using such environments as groupware. Effective and efficient use of engineering application tools, such as computerized models and simulations to develop products, also contributes to the

[1]The text of this chapter has been adapted from a paper written by the authors. Please see Ref. 32 for a complete citation.

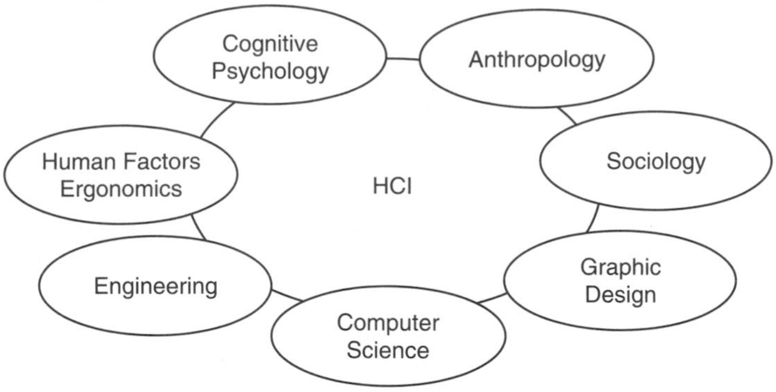

FIGURE 18.1 Multidisciplinary aspects of HCI.

successful design of the end product. In addition, more and more elements of products require some type of human interaction, whether it is with a software interface or with physical components such as buttons and knobs. It follows that HCI should become a key concern for all engineering professionals.

Indeed, software engineers are concerned with good component and architecture design, although in the area of testing the primary emphasis is on software quality, not usability. All engineers are already familiar with the product development life cycle, but today's professionals need to be able to work in all phases of a user-centered development life cycle, as well as in teams with other usability professionals. The engineering professional can benefit from an understanding of the traditional usability and reliability concepts of HCI as well as the implications of the technological developments that are impacting the computing field, the business concerns such as the return on investment, and the overall user experience. Engineering practitioners with an HCI approach will want to understand how such procedures are operationalized into best practices to help guide them to the analytical structures needed for diverse contexts [20]. Thus, HCI is more that just human factors and ergonomics but also about understanding usability and end-user needs.

18.2 Brief History of HCI

As may be expected for a multidisciplinary field, it is difficult to pinpoint the exact beginnings of HCI, although its roots can be traced to research in the computing sciences and the applied social and behavioral sciences for at least 25 years and perhaps more than 50 years. According to Carroll, in his book *Human Computer Interaction in the New Millennium,* the foundations of HCI were primarily laid in the 1960s and 1970s along four major threads of technological development. These are: prototyping and iterative development from the field of software engineering; psychology and human factors in computing systems; user-interface software and computer graphics; and models, theories, and frameworks from cognitive science. Preece extends these influences, citing the ACM SIGCHI (Special Interest Group in Computer Human Interaction) model, when describing the major topics of concern in HCI as the design and development process, the human, computer hardware, graphics and input-output devices, and use and context [29].

Several key individuals and landmark systems illustrate the evolutionary path of HCI. Early visions of personal, desktop access to information, and the basic ideas behind a connected web of information can be traced back to 1945 in the work of Vannevar Bush [9]. The graphical and gestural interactions of the Sketchpad system developed by Sutherland [35] and synchronous collaboration through direct pointing and shared windows [15] have been cited as historically significant [10]. Other early influences came from Licklider [19], who visualized a symbiotic relationship between humans and computers [29].

Technological advances in workstations and displays next led the progress in the areas that now fall under the HCI umbrella. At Xerox, PARC the Dynabook was an early representation of a book-sized personal computer with links to a worldwide computer network. In the late 1970s the same group developed the desk-sized Star with several innovations, such as a high-resolution display, high-quality graphics, and point-and-click capability with a mouse. Then in the early 1980s, Apple capitalized on Xerox's developments with the Apple Lisa, which incorporated the graphical metaphor and marked the beginning of the graphical user interface (GUI). Usenet groups and e-mail sparked an interest in the social and psychosocial consequences of computing. Cognitive science also matured as a discipline in the 1970s, and HCI became one of its original domains, with the vision of bringing cognitive science methods and theories to software development [10].

The convergence of these areas in the 1980s may mark the true beginning of HCI as a discrete field. Along with the initial 1982 conference, groups such as ACM's SIGCHI also were formed during this period.

Not surprisingly, the balance of interest in the topics of HCI has shifted over time, as technologies and scientific theory advanced. In the 1970s and 1980s the emphasis was on the psychology of human information processing. The PC explosion in the early 1980s drove a new focus on usability of single-user computer systems, followed in the next decade by a shift to multiuser workstations, multimedia, hypertext, virtual reality, and a recognition of the importance of group work, integration, and use in home and society [29].

Social and cultural issues have been influential forces in HCI. Early HCI in the United States was primarily concerned with how computers could enrich our lives and make them easier, i.e., a tool facilitating creativity and problem solving. The emphasis was on building models of the interface, empirical evaluation, and psychology of programming. In Europe, the early focus was on hardware and keyboard design, followed in the 1980s by theories and methods of design, and formalization of usability [29]. A postwar social movement in Scandinavia brought about the practice of participatory design that stressed that social objectives are sought in design [18]. The understanding that systems are used in a social context has been taken to another level by the Internet, which has motivated research into the concept of "community" and social software.

Practical experience has also influenced HCI research directions. For example, an early focus on summative evaluation in usability engineering has been replaced with a formative process that emphasizes prototyping and iterative development. The crucial role of requirements analysis in software and product development has been recognized along with the need for cross-functional understanding of requirements for complex systems and diverse domains. As a result, the usability engineering process is now heavily influenced by the methodologies of sociologists and anthropologists who are skilled at the study of application domains and work practices. There is a growing emphasis on design methods and on cost-benefit tradeoffs, with efforts on discount usability engineering methods, such as low-cost inspection and walkthrough methods [10].

18.3 Key Concepts

HCI is more than just the usability of a product but, rather, we can broaden it to be the entire user experience. SIGCHI, the leading professional organization for this area, defines HCI as a discipline "concerned with the design, evaluation, and implementation of interactive computing systems for human use and with the study of major phenomena surrounding them" [1]. Other definitions, such as that defined by John Carroll, specifically mention usability:

> HCI is the study and practice of usability. It is about understanding and creating software and other technology that people will want to use, will be able to use, and will find effective when used. The concept of usability and the methods and tools to encourage it, achieve it, and measure it are now touchstones in the culture of computing [10].

These definitions provide a sound basis for the traditional view of HCI, but do not address the full breadth of the field. HCI is more than the traditional concepts of usability and interface design; it has several other facets.

Traditional HCI Concepts

When the engineering community thinks of HCI, they think primarily of human factors and ergonomics. In this vein, "human factors discover and apply information about human behavior, abilities, limitations, and other characteristics to the design of tools, machines, systems, tasks, jobs, and environments for productive, safe, comfortable, and effective human use" [12]. "Ergonomics is more centered on the physiological aspects of fitting the design of machines and workspaces to accommodate the physiological dimensions of the humans who are attempting to accomplish work" [17]. However, these two disciplines are only some of the many components of the broader discipline of HCI.

Usability is the most traditional concept in HCI. A leading usability proponent, Jakob Nielsen, indicates that it is more narrowly concerned with system acceptability and more broadly concerned with how users can use a function. Does a system satisfy all the needs and tasks of the users of the system? The utility of a system is "whether the functionality of the system in principle can do what is needed" [26]. The implication is that a system not only works, but also does what it was intended to do. This "applies to all aspects of a system that the user interacts with, including the installation and maintenance phases" [26].

There are multiple attributes of usability that deal with performance and preference. These are learnability, how easy the system is to learn; efficiency, how quickly users can complete their tasks; memorability, how easy the system is to remember; control of errors, including prevention and recovery; and satisfaction, how much users like the system [26]. Usability is "a measurable characteristic of a product's user interface that is present to a greater or lesser degree" [23].

Adams and Jansen proposed several key usability design principles that ought to become part of every engineer's toolbox [2]. Some of these usability concepts include appropriate error messages and error checking so as to prevent the user from making irrecoverable mistakes; providing timely feedback and cues to help the user know what is going on; use of appropriate interaction styles such as menus, sliders, or buttons; efficient information organization to aid in navigation through an interface; and, usability testing in addition to typical black-box testing of component parts [2].

Designing for the User Experience

It has become clear that "two factors central to building usability into applications are interaction design and usability testing. Both practices seek to ensure that the user's experience with the software is consistent with expectations; that the use of the software is intuitive; and that there's no needless obstacle to successful completion of the transaction" [6]. As Alan Cooper points out "interaction design" means much more than just "interface design." Interaction design incorporates the end user and the objectives of the end user (and the organization) into the design process, whereas interface design, a subset of interaction design, refers to the look of the software and is not concerned with the overall user experience and satisfaction [13].

As usability has become the defining standard for software quality, it also has been a rapidly changing and expanding specialty [6]. In describing three waves of the usability evolution, Rubin [33] has emphasized the importance of the "user experience" in designs today. According to Rubin, the first wave of usability, started during World War II, was motivated by poor design of airplane cockpits. The major focus of this phase was on human physiology, information processing, and performance under stress. The second wave came with the introduction of the computer. The focus shifted to consideration of the hardware and software with the intent to improve usability so that users could perform tasks easily and efficiently. Rubin refers to this phase as usability with a small "u," because the emphasis was on the product or service rather than the broader issues that affect user perception. With the emergence of the Internet there has been additional pressure for the design to be intuitive because often there is no opportunity to train customers to use the software, and users easily leave a Web site for another if dissatisfied. Usability must extend beyond the issues of ease of use, ease of learning, and navigation. Rubin refers to the current phase as usability with a capital "U," which is more concerned with the user's experience as a whole rather than the user's interaction with the product. "In the broadest sense, usability involves every touch, every interaction, every phone call between an organization and user" [33]. This requires organizations to put the user experience at the forefront and may require new organizational practices.

18.4 Usability Engineering Process

The need for a focus on the user experience has major implications for businesses engaged in computer hardware and software development and for the management of their processes. First, the practice of separating responsibility for customer support, marketing, user interaction, and other concerns into separate departments creates a "silo effect" and a fractured user experience, according to Rubin[33,34] Breaking down such barriers may be necessary in many companies involved in software development and Web design. User-centered design, an organizing theme for software development, may be practiced through a process known as usability engineering. This is "a discipline that provides structured methods for achieving usability in user interface design during product development" [23]. A key feature of user-centered design is the involvement of end users and multidisciplinary design teams, which typically include expertise in usability, human factors, marketing, graphic design, technology engineering, quality assurance, and performance support. Deborah Mayhew, a well-known practitioner in the HCI field, who developed the "Usability Engineering Lifecycle," emphasizes the importance of involving the user throughout the entire process. See Figure 18.2, which shows the relationships of the key steps in the Usability Engineering Lifecycle. The three major phases of the life cycle are: requirements analysis; iterative design/development/testing; and installation. Each stage requires an appropriate level of feedback from the user.

The requirements-analysis phase consists of user and task analysis, consideration of platform capabilities and constraints, and integration of design principles, and the process culminates, unlike a traditional software development process, in the establishment of usability goals [23]. During user analysis, users are studied and profiles and personas are developed for representative users. Alan Cooper, a leading proponent of personas, believes that software should be developed with specific users in mind [13], an approach that lends itself to a more focused product. In task analysis, scenarios are developed that will be used to understand what the users do and how they do it. Personas and scenarios are integrated and used in design, as well as later during the usability-testing phase.

There are a host of design approaches, each of which needs to be evaluated with regard to its appropriateness for the type of application being developed. These design methods are either a variation of the

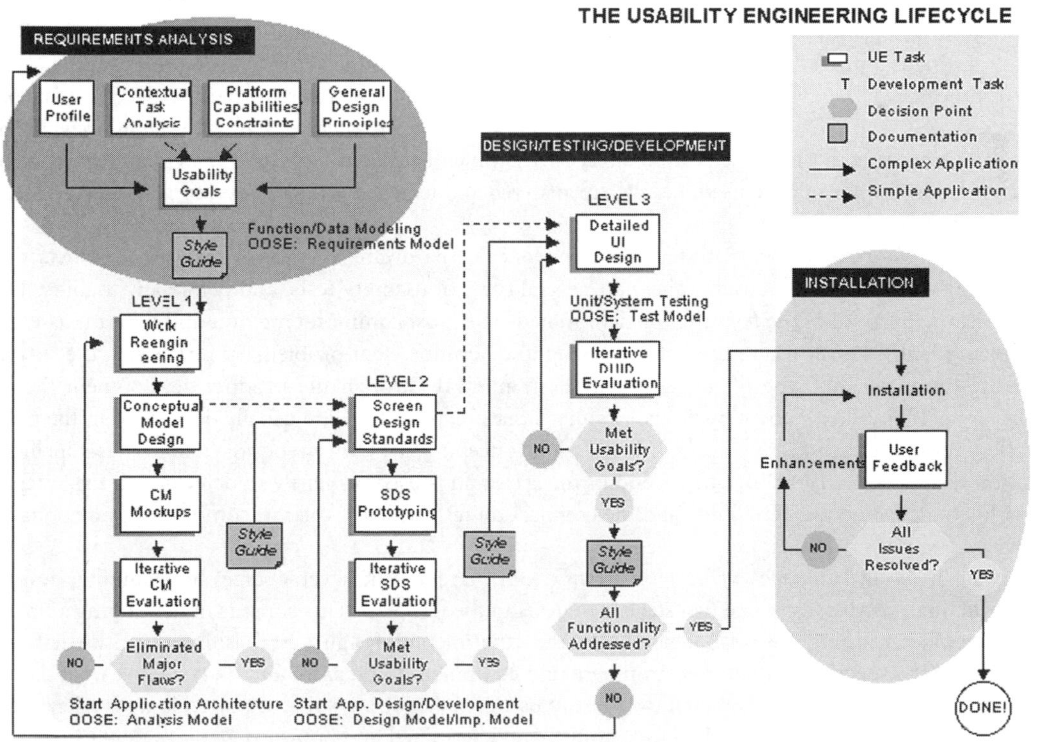

FIGURE 18.2 The usability engineering life cycle [Mayhew, 1999].

basic user-centered design approach or consider the user from a different perspective. Gloria Gery, in her groundbreaking book *Electronic Performance Support Systems,* promoted the design of technology-based systems to improve human performance. This approach considered the support necessary for the user to perform a task. It included the evaluation of the number of steps in the task, what information was needed in its performance, and the identification of the actions required in the performance of the task [16]. Contextual design is a customer-centered approach to designing products. It is based upon customer data gathered in the field and models of how a customer works [4]. Learning-centered design (LCD) is an extension of user-centered design. LCD considers the user as a learner of a new domain [30]. The use of reusable interaction design patterns is a methodology that has gained a lot of momentum over the past few years and fits well into the usability engineering life cycle. A design pattern is "a proven solution to a recurring design problem" (e.g., advanced search), and is expressed as part of a design language [7]. Lastly, scenario-based design is based upon the development of text stories or descriptions of how users interact in particular situations or in doing particular tasks [31]. This approach carries the use of scenarios throughout the entire software development life cycle.

During the iterative prototyping phase, the interface is developed for usability as well as for functionality. Since most graphical user interfaces depend primarily on visual-based interaction, the developer needs to understand any potential impact of the users' cognitive processes. Cognitive psychology plays a very important role in HCI — it addresses the elements of human attention and perception as well as memory, learning, problem solving, and decision making [23]. An understanding of the abilities and limitations of human cognitive processes and the implications for interface design is essential knowledge and contributes to the growing field of cognitive engineering.

18.5 Usability Testing

Product development life cycles include a testing step. However, in a user-centered life cycle, the focus is on usability testing. Usability testing does not "debug" the software or use black-box testing strategies. Rather, usability testing is a process for determining what problems the users may have in using the system [34]. This iterative process involves varying levels and types of users at different times during the development of the prototypes. For example, early in the process development, team members or colleagues may be used, although testing of target end-users, if possible, is recommended; later, various numbers and types of users such as novice or expert may be used. Results are analyzed and fed back into the development of the prototype for the next cycle.

Formal and informal usability testing may be done on hardware as well as software products during different stages of their development. There are several types of usability tests, and before any usability testing takes place, the purpose of the test must be determined. The most common type of usability test is assessment or problem identification, where the tester is trying to determine what problems are impeding the successful use of the product. This type of test is usually done midway through the product development life cycle. Another type of usability study may be exploratory in nature; these tests are usually done early in the product life cycle to evaluate the effectiveness of initial design concepts [34]. Other types of usability tests include the validation test, which is run closer to the end of the cycle and is used primarily to determine if the acceptance criteria for the product are being met, and the comparison test, which is used to compare the functionality of two or more similar products [34].

Rubin in his book *Handbook of Usability Testing* [34] discusses the several stages for planning, designing, and conducting effective tests. The first stage is to develop the test plan. This consists of developing a profile of typical users, determining the methodology (i.e., the experimental design), establishing the tasks that will be used in the study, specifying the test environment and equipment to be used, and determining what data is to be collected, and the metrics to be used to measure usability. Acquiring suitable participants is a key element for success of the test. In many studies, user groups are determined and appropriate users might be recruited by agencies specializing in market research studies. The next stage is to prepare the test materials, some of which include screening, background, and pretest questionnaires; nondisclosure forms; test scenarios;

post-test questionnaires; and debriefing questions. The third stage is conducting the test. This may be done in a local, specialized usability lab, which would include an observation room as well as a private testing room with video and audio recording equipment. Another option is to run the test remotely, which would allow for access to a broader range of users. This option would require specialized logging software as well as communication of the test monitor and participant over telephone. An experienced test monitor, who is able to be impartial in his/her observations and attitudes, is the key person interacting with the user, administering the usability test and recording the data. The final stage, and one of the most important, is data analysis, interpretation, and presentation of the results. A key challenge in usability testing is presenting results to stakeholders, including developers, in a consolidated, prioritized form so that usability issues can be best addressed within the existing development process.

HCI and Business Concerns

Management often has balked at incorporation of usability engineering because of a lack of familiarity with available data regarding savings that can be achieved with attention to usability. Most software and Web development managers view usability costs as added effort and expense, but the opposite is true. Many internal and external benefits of the practice of usability, such as return on investment (ROI), have been identified in a study by Bias and Mayhew [5]. Marcus has provided many examples from the literature and cited statistics to provide evidence for ROI when usability is applied. General categories of benefits include reduced development costs, increased revenues from sales, and improved effectiveness. Recognizing that the first 10% of design process can determine 90% of a product's cost and performance, applying usability techniques early can help keep the product aligned with a company goals and reduce those costs (Smith and Reinersten, 1991 in Ref. 21). For example, "the rule of thumb in many usability-aware organizations is that the cost-benefit ratio for usability is $1:$10:$100. Once a system is in development, correcting a problem costs ten times as much as fixing the same problem in design. If the system has been released, it costs 100 times as much relative to fixing in design" (Gilb, 1988 in Ref. 21). There are also less tangible benefits of usability. Usability increases customer satisfaction and productivity, leads to customer trust and loyalty, and inevitably results in tangible cost savings and profitability. Moreover, usability affects the public's perception and thus brand value and market share. Even court cases have resulted from poor usability [5, 21]. The observation that software developers rarely use the recommended usability engineering methods because they are seen as too costly, too time consuming, and too complex led Nielsen [28] to develop a simplified approach referred to as "discount usability engineering." The discount usability engineering methods include heuristic evaluations, simplified thinking aloud, and scenarios [26]. Nielsen has also addressed another business issue, convincing clients to pay for usability. He suggests that "ultimately, the real answer to getting clients to pay for user testing and other user-centered design methods is to point out usability's astounding return on investment" [27].

18.6 Supportive Technologies

Andrew Dillon proposes that HCI can augment and enhance the design of technologies by supporting the user in the performance of tasks that would otherwise be problematic. Another contribution of HCI to technology advancement would be as a predictive tool rather than only an evaluative one [14].

There is a wide gamut of supportive technology applications. In the information technology realm, digital libraries, a mechanism for bringing massive amounts of information to the user, already has had a profound effect on all users. The design challenges include addressing interface and interaction issues, understanding the right mix of multimedia and instructional technology components, and technical issues such as copyright and bandwidth.

In the engineering realm, there are several developments that will be combined and integrated into new product design. While wireless networking, mobile computing, and telecommunication technologies are allowing the user freedom from their desks to perform a variety of tasks (some of which are not yet envisioned), other technologies such as collaborative software and ubiquitous computing offer another look at how HCI perspective will impact users and their work environments.

CSCW and Groupware

Weiser and Brown's paper, "The Coming of Age of Calm Technology," discusses how the Internet and distributed computing have acted as a transition between the PC (one computer, one person) and ubiquitous computing ("many computers sharing each of us") [38].

In distributed computing, many computers are used in a client-server environment allowing for the access or sharing of information or activities. Some of these applications focus on elements of working in groups. Groupware, which is any type of software designed to facilitate group work, provides for collaboration, cooperation, and communication of groups in a particular environment [11]. CSCW or "computer-supported cooperative work" is a complementary area that studies how people work together using computer technology [36]. In general, groupware applications fall into two categories. Asynchronous groupware, where communication is at different times, includes such applications as e-mail, newsgroups, workflow systems, hypertext, group calendaring, and collaborative writing systems. Synchronous groupware, where communication is real time, includes shared whiteboards, video communications, instant messaging (IM) and chat systems, decision support systems, and multiplayer games [8]. Overall, groupware systems have dramatically impacted the interactions of groups in business, academic, or social environments. In these types of computing environments, not only does the design of the interface need to take into account the tasks to be performed, but also how the user will react and interact with them. For example, in the business environment, workflow applications have changed what tasks are assigned to people, how work is reengineered, and how work is supervised [11].

The major contribution of HCI to the areas of CSCW and groupware for such a wide variety of distributed applications is the understanding of how users interact differently in a new environment (e.g., the development of a new language in the IM environment), the type of support and needs that should be provided, the appropriate visualization and awareness of the information within the interface, and the representation of the presence of the individual within the environment [32].

Ubiquitous Computing

The computer is becoming an invisible entity by slowly disappearing into the background. In 1991, Mark Weiser's *Scientific American* article, "The Computer of the 21st Century," first proposed the notion of the ubiquitous computer. His vision, in which each person would share "thousands of highly distributed, interconnected, often invisible computers, blended into the natural environment, operating without engaging peoples' conscious senses or attention," is beginning to come to fruition [32, 37].

The current computing environment consists of an interface and a "hard" device that is either stationary, in the user's home or office, or portable. Ubiquitous computing (ubicomp), on the other hand, provides for the seamless integration and invisibility of the computer into our society. Computing is becoming second nature. Most users are not conscious of processors (i.e., computers) in their cars, microwaves, VCRs, and a host of other common consumer devices. Unfortunately, the designers of these so-called intelligent devices and appliances have still not solved the HCI problems of ease of use and learnability.

Pervasive computing, a concept that is similar to ubiquitous computing, is computing not bound to the desktop; it "goes beyond the realm of personal computer. It is the idea that almost any device, from clothing to tools to appliances to cars to homes to the human body to your coffee mug, can be embedded with chips to connect the device to an infinite network of other devices. The goal of pervasive computing, which combines current network technologies with wireless computing, voice recognition, Internet capability and artificial intelligence, is to create an environment where the connectivity of devices is embedded in such a way that the connectivity is unobtrusive and always available" [3].

Ubiquitous and pervasive computing can be distinguished in the following manner. "Ubiquitous means everywhere. Pervasive means 'diffused throughout every part of.' ... Pervasive computing involves devices like handhelds ... [e.g.,] Web-enabled phones. ... Ubiquitous computing, though, eschews our having to use computers at all. Instead, it's computing in the background, with technology embedded in the things we already use. That might be a car navigation system that, by accessing satellite pictures, alerts us to a traffic jam ahead, or an oven that shuts off when our food is cooked" [24].

In the future, computing will be a device (i.e., the computer) that will be easily portable, comes in many sizes and forms, and is suited to particular tasks. It will not be uncommon to see a variety of output forms such as *post-it* notes, smart sheets of paper (similar to the tablet PC), and larger devices much like bulletin boards [37]. In an educational environment, students will download instructors' notes real-time in class and there would be no need for hard-copy handouts. Already we see Internet-connected devices such as personal digital assistants (PDAs), pagers, refrigerators, and multifunctional telephones. We are now connected anytime and anywhere.

Wearable computing, that is a computer worn or implanted in an article of clothing, will be the next wave of fashion. These devices, while not as intrusive as an implanted chip, are always on and context-aware with the ability to track and retrieve various types of data about the user's environment. Examples of such devices include jewelry, eyeglasses, company badges, watches, and even the fabric of clothing.

18.7 Future Trends

Usability challenges come with new technological developments. The convergence of multifunctional devices into devices that are smaller and more portable (e.g., smart phones that can take pictures, connect to the Internet, and read e-mail) have presented designers and developers with the challenge to think about users' physical limitations and the tasks and environments they are performed in. New metaphors will need to be developed and the ever-tempting danger of featuritis (i.e., including too many features and functions) will need to be minimized. Limited input and output options, necessitated by the small-form factor of these devices, will be scrutinized for usability issues. Research for defining prescribed solutions or design guidelines for small (and large) display devices will become as important as those for "normal" display devices. In addition, several studies to understand the user performance involving small display devices have shown that different user groups often have conflicting preferences [22].

Interoperability provides yet another challenge to designers and HCI professionals. With a plethora of devices being networked together, physically and wirelessly, these devices will need to easily communicate with one another and to share information that is represented in many different formats. An example of this includes the display of Web sites on a variety of devices such as PDAs, phones, and large display devices.

As computers become invisible, interfaces as we know them will drastically change – from the visual GUI (graphical user interface) that many of us have become accustomed to, to a more physical representation (i.e., as discussed in the "ubiquitous computing" section). Engineering professionals will need to think "out of the box" to develop new designs and metaphors; creativity will need to be encouraged and supported [25]. Also in the case of CSCW and groupware, an understanding of social implications and the impact on on-line communities for many new applications will need to be better understood and embraced.

18.8 Conclusion

HCI has become an important and fundamental part of the product development life cycle. Application of HCI principles will greatly impact success as users interact with a wider range of consumer products that are technically more complex. The paradigm shift to new types of user interfaces will also provide challenges to users as they interact with and learn the new computing media. Engineering professionals will need to understand all facets of the user-centered product development environment. Understanding users' behaviors and tasks as well as the broader impact on the user's environment will enable the design of more usable and cost-effective products.

References

1. ACM SIGCHI, www.sigchi.org/cdg/cdg2.html.
2. W.J. Adams and B.J. Jansen, "Integrating usability design principles into an existing engineering curriculum," *American Society for Engineering Education National Conference*, Milwaukee, WI, 1997.

3. Anonymous, "Pervasive Computing," isp.webopedia.com/TERM/P/pervasive_computing.html.
4. H. Beyer and K. Holtzblatt, *Contextual Design: Defining Customer-Centered Systems*, Los Altos, CA: Morgan Kaufmann, 1997.
5. R.G. Bias and D.J. Mayhew, Eds., *Cost-Justifying Usability*, Boston, MA: Academic Press, 1994.
6. A. Binstock, "New mantra: usability," *Information Week*, 751, 1A–3A, 1999.
7. J.O. Borchers, "A pattern approach to interface design," *Proc. Conf. Des. Interact. Sys. Processes Practices Methods and Tech.*, New York, 2000, pp. 369–378.
8. T. Brinck, Groupware: Applications, http://www.usabilityfirst.com/groupware/applications.txl.
9. V. Bush, "As we may think," *Atl. Mon.*, July, 101–108, 1945.
10. J. Carroll, Ed. *Human-Computer Interaction in the New Millennium*, New York: ACM Press, 2002.
11. D. Chaffey, *Groupware, Workflow and Intranets: Reengineering the Enterprise with Collaborative Software*, Boston, MA: Digital Press, 1998.
12. A. Chapanis, "Some reflections on progress," *Proc. Human Factors Soc. 20th Meet.*, Santa Monica, CA: Human Factors Society, 1985, pp. 1–8.
13. A. Cooper, *The Inmates Are Running the Asylum*, Indianapolis IN: SAMS Publishing, 1999.
14. A. Dillon, "Technologies of Information: HCI and the Digital Library," in *Human-Computer Interaction in the New Millennium*, J. Carroll Ed., New York: ACM Press, 2002.
15. D.C. Engelbart and W.K. English, "A research center for augmenting human intellect," *AFIPS Proc. Fall Joint Comput. Conf.*, vol. 33, pp. 395–410.
16. G. Gery, *Electronic Performance Support Systems*, Cambridge, MA: Ziff Institute, 1991.
17. J.A. Jacko, "HCI education and its role in industrial engineering," *ACM SIGCHI Bull.*, vol. 30, no. 1, pp. 5–6.
18. D. Levinger, "Participatory design history," in *Computer Professionals for Social Responsibility*, PDC98, www.cpsr.org.
19. J.C.R. Licklider, "Man–computer symbiosis," *IRE Trans. Human Factors Electron.*, 1, 4–11.
20. D.S. McCrickard, C.M. Chewar, and J. Somervell, "Design, science, and engineering topics? Teaching HCI with a unified method," *ACM SIGCSE '04*, March, 3–7, 2004.
21. A. Marcus, "Return on investment for usable UI design," *User Exper.*, vol. 1, no. 3, 2002.
22. A. Marcus et al., "Baby faces: User-interface design for small displays," *Proc. CHI '98*, New York: ACM Press, 1998, pp. 96–97.
23. D. Mayhew, *The Usability Engineering Lifecycle — a Practitioner's Handbook for User Interface Design*, Los Altos, CA: Morgan Kaufmann, 1999.
24. A. McCrory, "Ubiquitous? pervasive? sorry, they don't compute," *Computerworld*, 2000, www.computerworld.com/news/2000/story/0,11280,41901,00.html.
25. W. Mitchell, A. Inouye, and M. Blumenthal, *Beyond Productivity: Information, Technology, Innovation, and Creativity*, Committee on Information Technology and Creativity, National Research Council, 2003.
26. J. Nielsen, *Usability Engineering*, San Diego, CA: Academic Press Professional, 1993.
27. J. Nielsen, "Convincing clients to pay for usability," www.useit.com/alertbox/20030519.html.
28. J. Nielsen, "Guerrilla HCI: using discount usability engineering to penetrate the intimidation barrier," in *Cost-Justifying Usability*, R.G. Bias and D.J. Mayhew, Eds., San Diego, CA: Academic Press Professional, 1994.
29. J. Preece et al., *Human-Computer Interaction*, Reading, MA: Addison Wesley, 1994.
30. C. Quintana, J. Krajcik, and E. Soloway, "Exploring a structured definition for learner-centered design," in *Fourth International Conference of the Learning Sciences*, B. Fishman and S. O'Connor-Divelbiss Eds., pp. 256–263.
31. M. Rosson and J. Carroll, *Usability Engineering Scenario-Based Development of Human Computer Interaction*, New York: Academic Press, 2002.
32. E.P. Rozanski and A.R. Haake, "The many facets of HCI," *Proc. 4th Conf. Inf. Technol. Educ.*, Lafayette, IN, 2003, pp. 180–185.
33. J. Rubin, "What business are you in?" *User Exper.*, vol. 1, no. 1, 2002.

34. J. Rubin, *Handbook of Usability Testing — How to Plan, Design, and Conduct Effective Tests*, New York: Wiley, 1994.
35. I. Sutherland, "Sketchpad, a man-machine graphical communications system," *Proc. Spring Joint Comput. Conf.*, pp. 329–346.
36. Usability First, http://www.usabilityfirst.com/groupware/cscw.txl.
37. M. Weiser, "The computer for the 21st century," *Sci. Am.*, September, 1991.
38. M. Weiser and J. Brown, "The coming age of calm technology," Xerox PARC, 1996, www.ubiq.com/hypertext/weiser/acmfuture2endnote.htm.

19

Decision Diagram Technique

S.N. Yanushkevich
University of Calgary, Alberta

V.P. Shmerko
University of Calgary, Alberta

19.1 Introduction

Binary decision diagrams (BDDs) are a state-of-the-art method representing switching functions that are efficient in using time and space. Many algorithms of logic design translate to BDDs and significantly improve their effectiveness. BDDs are useful not only for VLSI CAD but also in electrical engineering.

19.2 Basics of Decision Diagram Technique

Decision diagram techniques in engineering applications include:

- Tree-like structures for software applications
- Decision trees in decision-making and optimization tasks
- Data structures for representing switching functions and manipulation
- Treelike topology of hardware

In electrical engineering, treelike topology and data structures are widely used. We will focus on these topics.

BDD is derived from the binary decision tree that is generated from the truth table for a switching function. The main idea is to represent the data by a treelike structure and use decision-making or other tree-based techniques (including topological and relevcant information theory). Decision-making variables come from the switching function, or are algebraic equations. In other words, structures in the form of algebraic equations are transformed to graphical structures. In contrast to conventional descriptions based on *computation rules*, BDDs are based on a *decision process*. The manipulation of switching functions is replaced by a manipulation of tree attributes (nodes and links). A quick search is based on hash tables and linked-list data structures. The result can be easily transformed from graphical representation to algebraic notation.

BDDs can represent logic functions in more compact ways than with other representations. We will distinguish between *bit-level* and *word-level* BDDs.

BDDs are canonical representations of switching functions using a fixed order of input variables. The permutation of variable order can yield different BDDs for the same function. The effect of variable ordering depends on the nature of the handled functions.

Data Structures for Switching Functions

Representative types of switching functions include:

Boolean formulas, or algebraic models. Boolean formulas are arbitrary expressions that function with Boolean algebra. Among them, the most practical formulas for the logical design of electronic circuits are normal forms. *Normal* forms involve:

Monomials (e.g., literal terms such as sum of literals, $x_1 \vee \bar{x}_3 \vee x_3$, or product of literals, $x_1 x_2 \bar{x}_3$)
Polynomials such as *sum-of-products* (SOP), *product-of-sums* (POS), and *exclusive-OR sum-of-products* (ESOP).

Truth tables. A tabular listing of the function arguments with the corresponding function values. They are of exponential complexity, yet *cube* sets can be generated as a compact representation of truth tables.

Graph-based models. Such as:

Boolean network. A graphical model representing a technology independent (before technology mapping) of multilevel circuit structure, or a circuit netlist. A Boolean network is a directed acyclic graph (DAG).
Decision trees and decision diagrams. Directed graphs where each node is associated with one classifying test.

Boolean formulas allow for the computation of logic functions, while decision trees and diagrams allow for the evaluation of logic functions.

Decision Tree

A decision tree is a rooted acyclic graph consisting of a finite vertex set and a set of edges between vertices. Each vertex v is characterized by in-degree (the number of edges leading to v) and out-degree (the number of edges starting in v). A graph is *rooted* if there is exactly one node with an in-degree of 0, the root.

A *tree* is a rooted acyclic graph where each node but the root has an in-degree of 1. This implies that in a tree, for every vertex v there exists a unique path from the root to v. The length of this path is called the *depth* or *level* of v. The *height* of a tree is equal to the largest depth of any node in the tree. A node with no children is a *terminal* node or *leaf*. A nonleaf node is called an *internal* node. The other parameters of a tree are:

- The *size*. The number of nodes
- The *depth*. The number of levels
- The *width*. The maximum number of nodes for a level
- The *area*. *Depth* × *Width*

A *complete* n-level p-tree is a tree with p^k nodes on level k for $k = 0, \ldots, n-1$. A p^n-leaf complete tree has a level hierarchy (levels 0,1,. . .,n). The root is associated with level zero and its p children are on level 1. This edge model describes the data transmission from the child to the parent, so that data are sent in one direction at a time, either up or down.

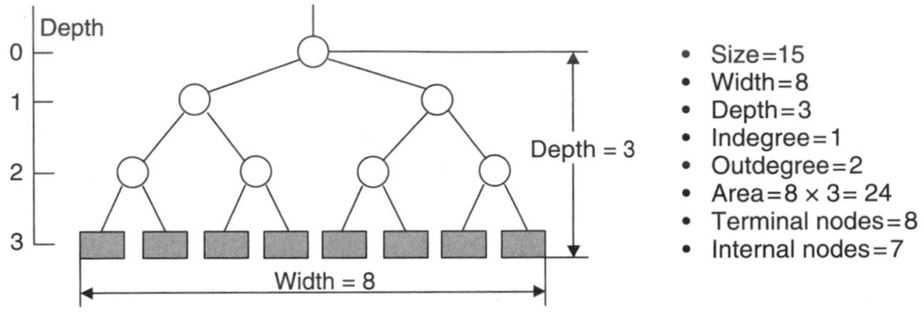

FIGURE 19.1 The complete binary tree (Example 19.1)

Example 19.1

In Figure 19.1, the complete binary ($p = 2$) 3-level ($n = 3$) tree is given. The root corresponds to the level (depth) 0. Its two children are associated with level 1 ($2^1 = 2$). Level 2 includes $2^2 = 4$ nodes. Finally, there are $2^3 = 8$ terminal nodes.

19.3 Shannon Decision Trees and Diagrams

A binary decision tree (BDT) generates the Shannon decision diagram that is also called a BDD.

Binary Decision Tree

A BDT associates with a canonical sum-of-products form of an n-variable switching function. It is a decision tree such that:

- The root and the internal nodes have exactly two outgoing edges, 0-edge and 1-edge.
- The root and internal nodes are labeled with a variable x_i.
- The *width is*, the maximum number of nodes for a level.
- The *area is Depth* × *Width.*

The structural properties of the binary decision tree are as follows:

- The nodes implement the Shannon expansion.
- Each assignment to the variables $x_1, \ldots, x_n$ defines a unique path from the root to the terminal node corresponding to the value of the function given this assignment; this path corresponds to a minterm in the sum-of-products representation of the function; the minterm is determined as the product of labels at the path edges.
- Terminal nodes are assigned with constants 0 and 1.
- The values of terminal nodes are assigned with the truth values $\mathbf{F} = [f_{000}, f_{001} \cdots f_{111}]$ of a switching function f, where "0-node" corresponds to the value $f_0 = f|_{x_i=0}$., and "1-node" corresponds to the value $f_1 = f|_{x_i=1}$..

Example 19.2

An arbitrary switching function f of three variables can be represented by the Shannon decision tree, shown in Figure 19.2: three levels ($n = 3$), seven nodes, eight terminal nodes, 2^k nodes at the k-th level, $k = 0, 1, 2$. To design this tree, Shannon expansion is used as follows:

- With respect to variable x_1:

$$f = \bar{x}_1 f_0 \vee x_1 f_1$$

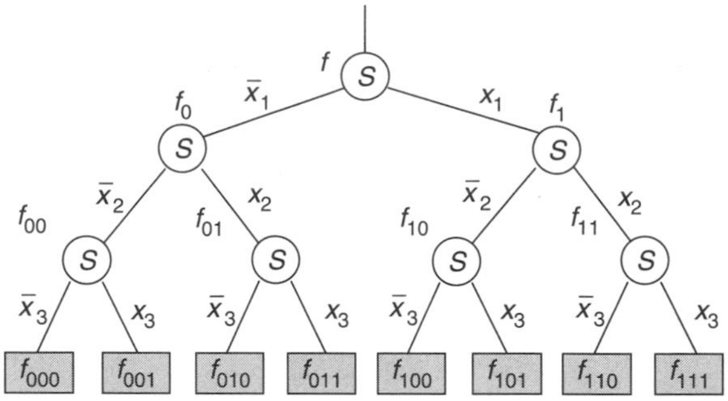

There are 8 paths from f to the terminal nodes

Path 1: $m_1 = \bar{x}_1\bar{x}_2\bar{x}_3$

Path 2: $m_2 = \bar{x}_1\bar{x}_2x_3$

Path 3: $m_3 = \bar{x}_1x_2\bar{x}_3$

Path 4: $m_4 = \bar{x}_1x_2x_3$

Path 5: $m_5 = x_1\bar{x}_2\bar{x}_3$

Path 6: $m_6 = x_1\bar{x}_2x_3$

Path 7: $m_7 = x_1x_2\bar{x}_3$

Path 8: $m_8 = x_1x_2x_3$

Sum-of-products

$f = m_1 \vee m_2 \vee \ldots \vee m_8$

FIGURE 19.2 The Shannon decision tree for a three-variable sum-of-products representation of a switching function (Example 19.2).

- With respect to variable x_2:

$$f_0 = \bar{x}_2 f_{00} \vee x_2 f_{01} \quad \text{and}$$
$$f_1 = \bar{x}_2 f_{10} \vee x_2 f_{11},$$

- With respect to variable x_3:

$$f_{00} = \bar{x}_3 f_{000} \vee x_3 f_{001}, \qquad f_{01} = \bar{x}_3 f_{010} \vee x_3 f_{011},$$
$$f_{10} = \bar{x}_3 f_{100} \vee x_3 f_{101}, \qquad f_{11} = \bar{x}_3 f_{110} \vee x_3 f_{111}.$$

The Shannon decision tree represents a switching function f in the form of the sum-of-products $f = f_{000}\bar{x}_1\bar{x}_2\bar{x}_3 \vee f_{001}\bar{x}_1\bar{x}_2x_3 \vee f_{010}\bar{x}_1x_2\bar{x}_3 \vee f_{011}\bar{x}_1x_2x_3 \vee f_{100}x_1\bar{x}_2\bar{x}_3 \vee f_{101}x_1\bar{x}_2x_3 \vee f_{110}x_1x_2\bar{x}_3 \vee f_{111}x_1x_2x_3$.

Because of the first property, the BDTs are called *Shannon decision trees.*

Formal synthesis. A node in a Shannon decision tree of a switching function f corresponds to the Shannon decomposition of the function with respect to a variable x_i

$$f = \bar{x}_i f_0 \vee x_i f_1, \tag{19.1}$$

where $f_0 = f|_{x_i=0}$ and $f_1 = f|_{x_i=1}$. Here, $f = f|_{x_i=a}$ denotes the cofactor of f after assigning the constant a to the variable x_i. Shannon decomposition Equation (19.1) is labeled by S in Figure 19.3.

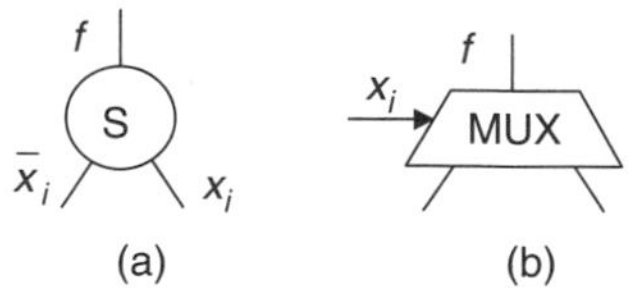

$$f = \bar{x}_i f_0 \vee x_i f_1$$
$$f_0 = f|_{x_i=0}$$
$$f_1 = f|_{x_i=1}$$

(c)

$$f = [\ \bar{x}_i \ \ x_i\ ] \begin{bmatrix} 1 & 0 \\ 0 & 1 \end{bmatrix} \begin{bmatrix} f_0 \\ f_1 \end{bmatrix}$$

(d)

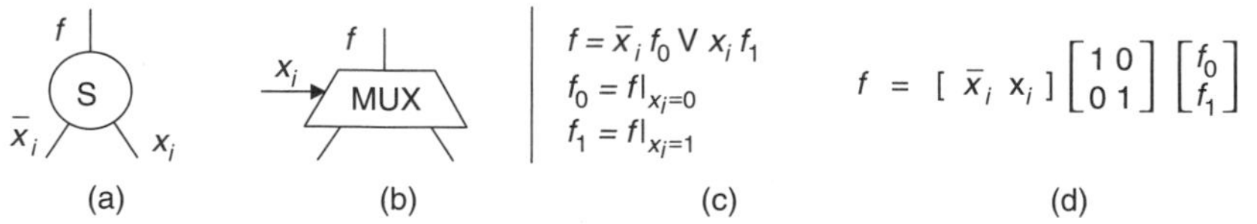

FIGURE 19.3 (a) The node of the Shannon decision tree, (b) its implementation by multiplexer (MUX), (c) algebraic, and (d) matrix descriptions.

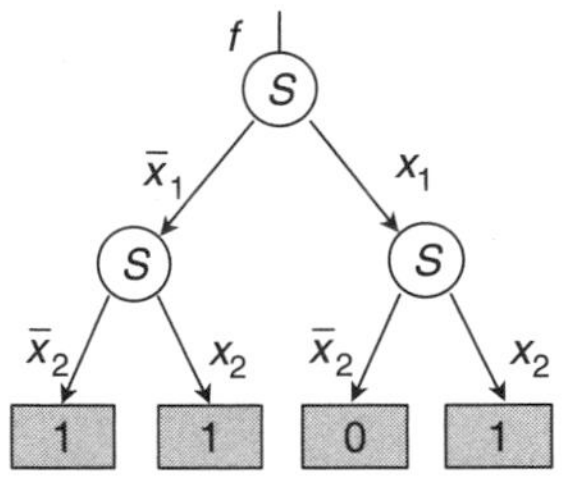

Structure of sum-of-products expression

$$\hat{\mathbf{X}} = [\bar{x}_1 \; x_1] \otimes [\bar{x}_2 \; x_2] = [\bar{x}_1 \bar{x}_2, \; \bar{x}_1 x_2, \; x_1\bar{x}_2, \; x_1x_2]$$

Transform matrix

$$\mathbf{S}_{2^2} = \mathbf{S}_2 \otimes \mathbf{S}_2 = \begin{bmatrix} 1 & 0 \\ 0 & 1 \end{bmatrix} \otimes \begin{bmatrix} 1 & 0 \\ 0 & 1 \end{bmatrix} = \left[\begin{array}{cc|cc} 1 & & & \\ & 1 & & \\ \hline & & 1 & \\ & & & 1 \end{array}\right]$$

Sum-of-products

$$f = \hat{\mathbf{X}}\mathbf{S}_{2^2}\mathbf{F} = \hat{\mathbf{X}} \left[\begin{array}{cc|cc} 1 & & & \\ & 1 & & \\ \hline & & 1 & \\ & & & 1 \end{array}\right] \begin{bmatrix} 1 \\ 1 \\ 0 \\ 1 \end{bmatrix} = \hat{\mathbf{X}} \begin{bmatrix} 1 \\ 1 \\ 0 \\ 1 \end{bmatrix}$$

$$= \bar{x}_1\bar{x}_2 \vee \bar{x}_1 x_2 \vee x_1 x_2$$

FIGURE 19.4 The Shannon decision tree and matrix notation for the switching function $f = \bar{x}_1 \vee x_2$ (Example 19.3).

In matrix notation, the transformation in a node of the Shannon decision tree for a function of a single variable x_i, given by the truth-vector $\mathbf{F} = [f_0 f_1]^T$, is given below

$$f = [\bar{x}_i \, x_i] \begin{bmatrix} 1 & 0 \\ 0 & 1 \end{bmatrix} \begin{bmatrix} f_0 \\ f_1 \end{bmatrix} = [\bar{x}_i x_i] \begin{bmatrix} f_0 \\ f_1 \end{bmatrix} = \bar{x}_i f_0 \vee x_i f_1$$

A recursive application of the Shannon expansion to f, given by truth-vector $\mathbf{F} = [f(0) f(1) \ldots f(2^n - 1)]^T$, is described in the matrix notation as

$$f = \hat{\mathbf{X}} \, \mathbf{S}_{2^n} \, \mathbf{F}, \qquad (19.2)$$

where

$$\hat{\mathbf{X}} = \bigotimes_{i=1}^{n} [\bar{x}_i x_i] \qquad \mathbf{S}_{2^n} = \bigotimes_{i=1}^{n} \mathbf{S}_2 \qquad \mathbf{S}_2 = \begin{bmatrix} 1 & 0 \\ 0 & 1 \end{bmatrix},$$

and $\otimes$ denotes the Kronecker product.

Example 19.3

We will generate the Shannon decision tree for the switching function $f = \bar{x}_1 \vee x_2$, given the truth vector $\mathbf{F} = [1 \;\; 1 \;\; 0 \;\; 1]^T$. Equation (19.2) is applied to obtain the solution in Figure 19.4. The minterms are generated by the Kronecker product $\hat{\mathbf{X}}$. The 4×4 transform matrix S is formed by the Kronecker product of the basic matrix $\mathbf{S}_{2^1}$. The result, in sum-of-products form, is mapped into the Shannon decision tree.

Binary Decision Diagrams

Reduced BDTs normally are called BDDs. A BDD is a directed acyclic graph with one root, whose sinks are labeled by the constants, 0 and 1, and whose internal nodes are labeled by a Boolean variable x_i. BDDs have two outgoing edges, a 0-edge and 1-edge.

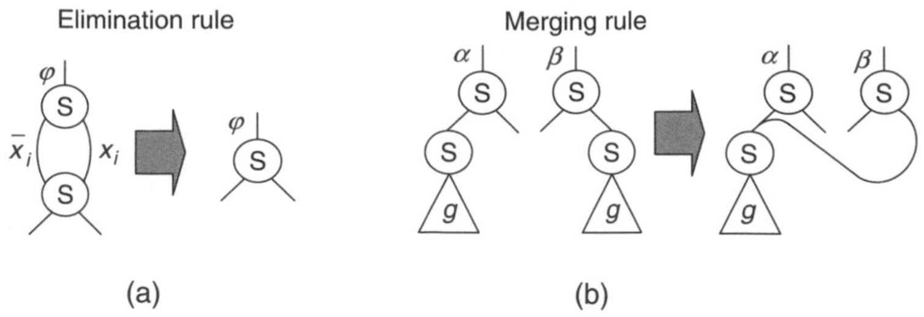

FIGURE 19.5 Reduction of the Shannon decision tree: (a) elimination rule and (b) merging rule.

Decision Tree Reduction

The Shannon decision diagram for a given function f is derived from the Shannon decision tree for f by deleting redundant nodes and by sharing equivalent subgraphs. The reduction rules are as follows (Figure 19.5):

Elimination rule: If two descendent nodes are identical, then delete the node connecting its incoming edges to the corresponding successor.

Merging rule: Share equivalent subgraphs.

In a reduced-decision diagram, edges longer than 1, such as connecting nodes at nonsuccessive levels, can appear. For example, the length of an edge connecting a node at $(i-1)$-th level with a node at $(i+1)$-th level is 2.

Example 19.4

Application of the reduction rules to a decision tree of a three-input NOR function is demonstrated in Figure 19.6.

Order of Variables. An *ordered* BDD (OBDD) is a rooted directed acyclic graph that represents a switching function. A linear variable order is placed on the input variables. The variables' occurrences on each path of this diagram must be consistent with this order. An OBDD is called *reduced* if it does not contain any vertex v such that the 0-edge and 1-edge of v lead to the same node and it does not contain any distinct vertices v and v' such that the subgraphs rooted in v and v' are isomorphic.

Efficiency. A decision diagram is characterized, similarly to a decision tree, by the *size*, *depth*, *width*, *area*, and *efficiency of reduction* of the decision tree of size Size_1 to a BDD of Size_2:

$$100 \times \frac{\text{Size}_1}{\text{Size}_2}\,\%$$

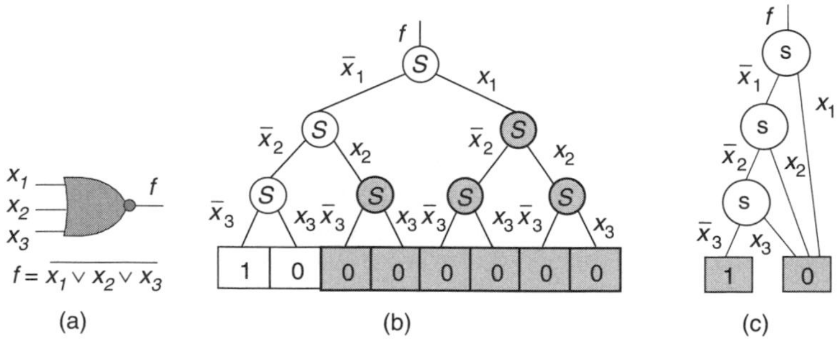

FIGURE 19.6 (a) The three-input NOR function, (b) Shannon decision tree, and (c) decision diagram with the lexicographical order of variables (Example 19.4).

Implementation

In conventional logic design, BDDs are mapped into a circuit netlist, so that each BDD node is associated with a universal element, such as a multiplexer. A multiplexer normally consists of two AND and one OR gate in a gate-level design (Figure 19.7a). Alternatively, pass-transistor logic can be used to implement a multiplexer with fewer transistors (Figure 19.7b), although the level restoration problem causes the transistor count to increase.

In nanoelectronics, the most feasible candidates for logic circuits are single-carrier electronics (at low temperature) and CMOS-molecular electronics (at room temperature). Though single-electron transistor-based AND, EXOR, NOT, and other gate designs have been proposed, such designs suffer from serious difficulties caused by the bilateral nature of a simple tunnel junction. Separation between input and output is poor. It is difficult to construct large logic circuits using such nonunilateral circuits. The latency of the BDD corresponds to the implicit input-output presence in single-electron transistor (SET) based logic devices. In this device, quantum dots are technologically formed using wrap-gate (WRG) structures. Those can be built on Shottky wrap gate device (WPG) or MOS WPG. The core of WPGs is the Shottky or silicon gate wrapped around a quantum wire. This branch-gate quantum-effect device is a universal multiplexer. The quantum dot and three tunnel junctions, two of them controlled by the wrap-gate voltage, are shown schematically in Figure 19.7c. This can be considered a nanomodel of a decision-tree node. In a BDD-based nanodevice, 0-terminal means a connection to the ground, and a 1-terminal is connected to the source of negative voltage that we apply to inject electrons into the nanowire.

Library of BDDs

The efficiency of decision diagram-based techniques has been improved and extended with the various requirements of circuit design and specific applications. In practice, it is necessary to choose the type of decision diagram to solve the problem. Below, the main features of different types of decision diagrams are listed:

Bit-level Decision diagrams are graphical representations of AND-OR or AND-EXOR expressions.

OBDD (ordered BDD) is a BDD where the input variables appear in a fixed order on the paths of the graph and no variable appears more than once in a path.

ROBDD (reduced OBDD) is a result of the application of reduction and merging rules to OBDD; ROBDD is a canonical form of a switching function.

Shared BDDs (multirooted BDDs) represent multi-output switching functions. A set of BDDs can be united into a graph consisting of BDDs sharing subgraphs with each other. This idea saves time and space by avoiding isomorphic subtree duplication.

FDD (functional decision diagrams) include spectral decision diagrams such as Davio decision diagrams.

Word-level decision diagrams are integer counterparts of some bit-level diagrams. In word-level decision diagrams, the constant nodes are elements of finite (Galois) field, integers, or complex numbers.

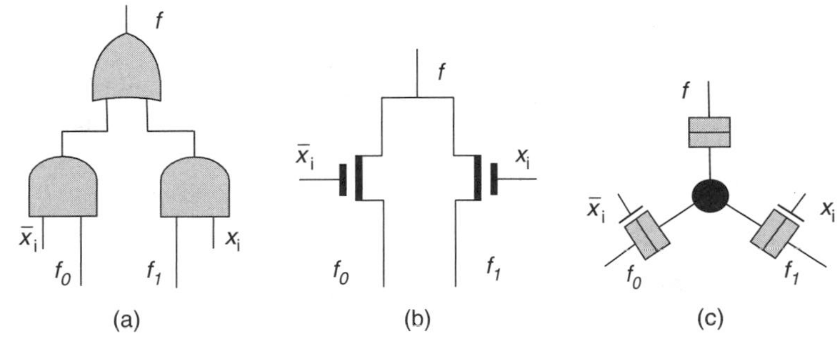

FIGURE 19.7 Implementation of Shannon node by (a) logic gates, (b) pass-gate logic, and (c) single-electron wrap-gate device.

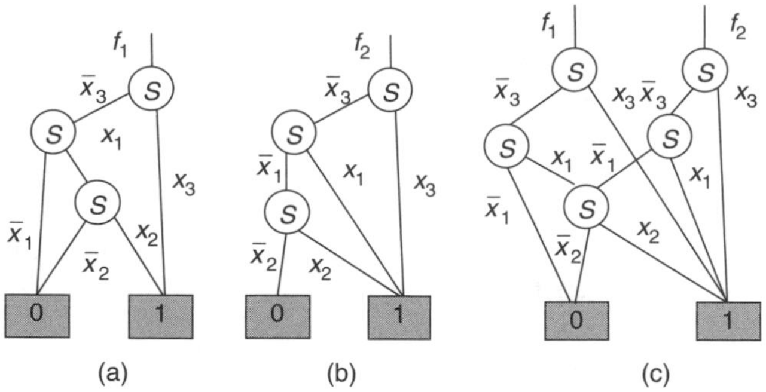

FIGURE 19.8 The ROBDD of two functions, (a) f_1, (b) f_2, and (c) their shared ROBDD (Example 19.5).

EVBDD (edge-valued BDD) is a decision diagram where weighting coefficients are assigned to the edges in the decision diagram for compact representation.

BMD (binary-moment diagram) is an FDD in an integer-valued domain; the FDD nodes are assigned with an arithmetic analog of Davio expansion.

ZBDD (zero-suppressed BDD) is defined as an OBDD, while the reduction rules are similar to ones for FDD (the nodes whose 1-edge points to the 0-terminal node are eliminated). ZBDD represents sets of combinations more efficiently than BDD; the switching function expresses whether a combination specified by the input variables is included in the set.

Representation of Multi-Output Switching Functions

A multi-output switching function is represented by a multirooted decision diagram, called a *shared* decision diagram. This is not word-level representation.

Example 19.5

Given the two-output function

$$f_1 = x_1 x_2 \vee x_3,$$
$$f_2 = x_1 \vee x_2 \vee x_3,$$

two ROBDDs can be derived (Figure 19.8a,b). By inspection, we can find that they have an isomorphic part. Thus, the shared ROBDD can be derived as shown in Figure 19.8c.

The properties of the shared BDDs are as follows:

- Shared BDDs take advantage of sharing isomorphic subgraphs that, combined with BDD reduction rules, contribute to efficient optimization of memory for storage and computations.
- Equivalence of the involved functions can be checked easily by inspection of the root nodes for these functions.

19.4 Functional Decision Trees and Diagrams

Reed-Muller algebra is a universal system that includes constant 1, EXOR, AND, and NOT operations on Boolean variables. Reed–Muller expressions are classified as fixed- and mixed-polarity expressions. In this section, fixed-polarity Reed–Muller expressions are introduced.

Given a switching function f of n variables, the Reed–Muller expression is specified by

$$f = \bigoplus_{i=0}^{2^n-1} r_i \underbrace{(x_1^{i_1} \cdots x_n^{i_n})}_{i\text{-th product}}, \tag{19.3}$$

where $r_i \in \{0, 1\}$ is a coefficient, i_j is the j-th bit $j = 1, 2, \ldots, n$ in the binary representation of the index $i = i_1 i_2 \ldots i_n$, and $x_j^{i_j}$ is defined as

$$x_j^{i_j} = \begin{cases} 1, & \text{if } \ i_j = 0; \\ x_j, & \text{if } \ i_j = 1. \end{cases} \tag{19.4}$$

Example 19.6

An arbitrary switching function of two variables is represented by the Reed–Muller expression using Equation (19.3) and Equation (19.4)

$$\begin{aligned} f &= r_0\left(x_1^0 x_2^0\right) \oplus r_1\left(x_1^0 x_2^1\right) \oplus r_2\left(x_1^1 x_2^0\right) \oplus r_3\left(x_1^1 x_2^1\right) \\ &= r_0 \oplus r_1 x_2 \oplus r_2 x_1 \oplus r_3 x_1 x_2. \end{aligned}$$

Computing the Reed–Muller Coefficients

Given a truth vector **F** of a function f, the vector of Reed–Muller coefficients $\mathbf{R} = [r_0 r_1 \ldots r_{2^n-1}]^{\mathrm{T}}$ is derived by the matrix equation with AND and EXOR operations

$$\mathbf{R} = \mathbf{R}_{2^n}\ \mathbf{F} \,(\text{mod}\ 2) \tag{19.5}$$

the $2^n \times 2^n$ matrix $\mathbf{R}_{2^n}$ is formed by the Kronecker product

$$\mathbf{R}_{2^n} = \bigotimes_{j=1}^{n} \mathbf{R}_{2^j}, \qquad \mathbf{R}_{2^1} = \begin{bmatrix} 1 & 0 \\ 1 & 1 \end{bmatrix}. \tag{19.6}$$

Example 19.7

Computing the Reed–Muller coefficients by Equations (19.5) and (19.6) for the function $f = x_1 \vee \bar{x}_2$ is illustrated in Figure 19.9.

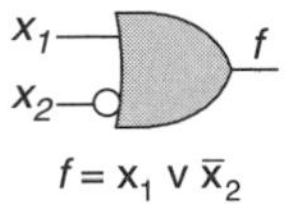

Vector of coefficients

$$\mathbf{R} = \mathbf{R}_{2^2} \cdot \mathbf{F} = \left[\begin{array}{cc|cc} 1 & 0 & 0 & 0 \\ 1 & 1 & 0 & 0 \\ \hline 1 & 0 & 1 & 0 \\ 1 & 1 & 1 & 1 \end{array}\right] \begin{bmatrix} 1 \\ 0 \\ 1 \\ 1 \end{bmatrix} = \begin{bmatrix} 1 \\ 1 \\ 0 \\ 1 \end{bmatrix} (\text{mod}\ 2)$$

Reed–Muller expression

$$f = 1 \oplus x_2 \oplus x_1 x_2$$

FIGURE 19.9 Computing the Reed–Muller expression for the two-input OR (Example 19.7).

Restoration of Sum-of-Products from Reed–Muller Expression

The following matrix equation with AND and EXOR operations restores the truth-vector **F** from the coefficient vector **R**

$$\mathbf{F} = \mathbf{R}_{2^n}^{-1}\mathbf{R} \ (\bmod 2), \tag{19.7}$$

where $\mathbf{R}_{2^n}^{-1} = \mathbf{R}_{2^n}$. Notice that the matrix $\mathbf{R}_{2^n}$ is a self-inverse matrix over Galois Field GF(2), on the set of logic operations AND and EXOR.

Example 19.8

Restore by Equation (19.7) the truth-vector F of a function f given by the vector of Reed–Muller coefficients $\mathbf{R} = [1 \quad 1 \quad 0 \quad 1]^T$:

$$\mathbf{F} = \mathbf{R}_{2^3}^{-1}\mathbf{R} = \left[\begin{array}{cc|cc} 1 & 0 & 0 & 0 \\ 1 & 1 & 0 & 0 \\ \hline 1 & 0 & 1 & 0 \\ 1 & 1 & 1 & 1 \end{array}\right] \begin{bmatrix} 1 \\ 1 \\ 0 \\ 1 \end{bmatrix} = \begin{bmatrix} 1 \\ 0 \\ 1 \\ 1 \end{bmatrix} \quad (\bmod 2).$$

Polarity

The ESOP is a mixed-polarity form where variable entries occur both in complemented and uncomplemented form. The term *fixed polarity* describes Reed–Muller expansions where each variable appears either uncomplemented or complemented and never in both forms. The polarity of variable x is 0 (uncomplemented variable, $x^0 = x$) or 1 (complemented variable $x^1 = \bar{x}$). In a function of n variables, a variable x^i can appear as $\bar{x}_i$ or x^i. With respect to the Reed–Muller domain, a fixed polarity of expansion, c, can vary from 0 to 2^n–1 and means a vector of polarities of variable x_i, $c = (c_1c_2, \ldots, c_n)$, where c_i is a fixed polarity of variable x_i, $x_i^{c_i} = \bar{x}_i$ if $c_i = 0$ and $x_i^{c_i} = x_i$ if $c_i = 1$.

Example 19.9

Let $f = x \vee y$, then four fixed-polarity Reed–Muller (FPRM) expressions can be derived as shown in Figure 19.10.

Davio Trees and Diagrams

A BDT that corresponds to the Reed–Muller canonical representation of a switching function is called a *Davio tree*. This is a functional decision diagram, since it represents function (not Shannon) expansion. A Davio diagram can be derived from a Davio tree.

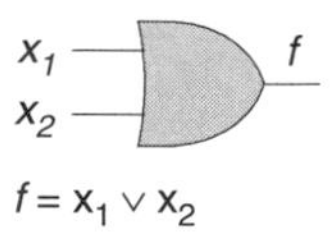

$$\begin{array}{lll} 0-\textit{polarity} & c = 0,\ c_1c_2 = 00: & f = x_1 \oplus x_2 \oplus x_1x_2 \\ 1-\textit{polarity} & c = 1,\ c_1c_2 = 01: & f = 1 \oplus \bar{x}_2 \oplus x_1\bar{x}_2 \\ 2-\textit{polarity} & c = 2,\ c_1c_2 = 10: & f = 1 \oplus \bar{x}_1 \oplus \bar{x}_1x_2 \\ 3-\textit{polarity} & c = 3,\ c_1c_2 = 11: & f = 1 \oplus \bar{x}_1\bar{x}_2 \end{array}$$

FIGURE 19.10 Representation of two-input OR gate by Reed–Muller forms of $2^2 = 4$ polarities (Example 19.9).

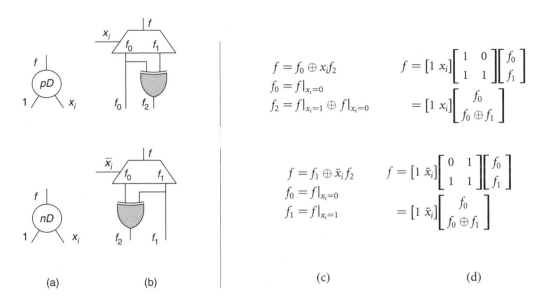

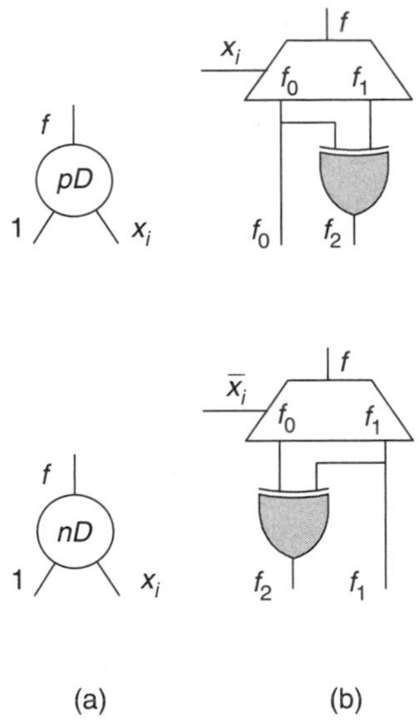

(c)

$$f = f_0 \oplus x_i f_2$$
$$f_0 = f|_{x_i=0}$$
$$f_2 = f|_{x_i=1} \oplus f|_{x_i=0}$$

$$f = f_1 \oplus \bar{x}_i f_2$$
$$f_0 = f|_{x_i=0}$$
$$f_1 = f|_{x_i=1}$$

(d)

$$f = [1\ x_i]\begin{bmatrix} 1 & 0 \\ 1 & 1 \end{bmatrix}\begin{bmatrix} f_0 \\ f_1 \end{bmatrix} = [1\ x_i]\begin{bmatrix} f_0 \\ f_0 \oplus f_1 \end{bmatrix}$$

$$f = [1\ \bar{x}_i]\begin{bmatrix} 0 & 1 \\ 1 & 1 \end{bmatrix}\begin{bmatrix} f_0 \\ f_1 \end{bmatrix} = [1\ \bar{x}_i]\begin{bmatrix} f_0 \\ f_0 \oplus f_1 \end{bmatrix}$$

FIGURE 19.11 (a) The node of a Davio tree, (b) realization, (c) algebraic, and (d) matrix descriptions.

Formal design of Davio trees: A node in a Davio tree of a switching function f corresponds to the Davio decomposition of the function with respect to variable x_i. There exist:

The positive Davio expansion

$$f = f_0 \oplus x_i f_2, \tag{19.8}$$

where $f_0 = f|_{x_i=0}$ and $f_2 = f|_{x_i=1} \oplus f|_{x_i=0}$, and

The negative Davio expansion

$$f = \bar{x}_i f_2 \oplus f_1, \tag{19.9}$$

where $f_1 = f|_{x_i=1}$.

Positive Equation (19.8) and negative Equation (19.9) Davio decompositions are labeled as pD and nD respectively (Figure 19.11).

In matrix notation, the function f of the node is a function of a single variable, x_i, given by truth-vector $\mathbf{F} = [f_0 f_1]^{\mathrm{T}}$ and defined as

$$f = [\bar{x}_i\, x_i]\begin{bmatrix} 1 & 0 \\ 0 & 1 \end{bmatrix}\begin{bmatrix} f_0 \\ f_1 \end{bmatrix} = [\bar{x}_i\, x_i]\begin{bmatrix} f_0 \\ f_1 \end{bmatrix}$$
$$= \bar{x}_i f_0 \oplus x_i f_1 = (1 \oplus x_i) f_0 \oplus x_i f_1 = f_0 \oplus x_i f_2$$

Recursive application of the positive Davio expansion to the function f, given by the truth-vector $\mathbf{F} = [f(0) f(1) \ldots f(2^n - 1)]^{\mathrm{T}}$, can be expressed in matrix notation

$$f = \hat{\mathbf{X}}\, \mathbf{R}_{2^n}\, \mathbf{F}, \tag{19.10}$$

where

$$\hat{\mathbf{X}} = \bigotimes_{i=1}^{n} [1 x_i], \quad \mathbf{R}_{2^n} = \bigotimes_{i=1}^{n} \mathbf{R}_2, \quad \mathbf{R}_2 = \begin{bmatrix} 1 & 0 \\ 1 & 1 \end{bmatrix},$$

and $\otimes$ denotes the Kronecker product.

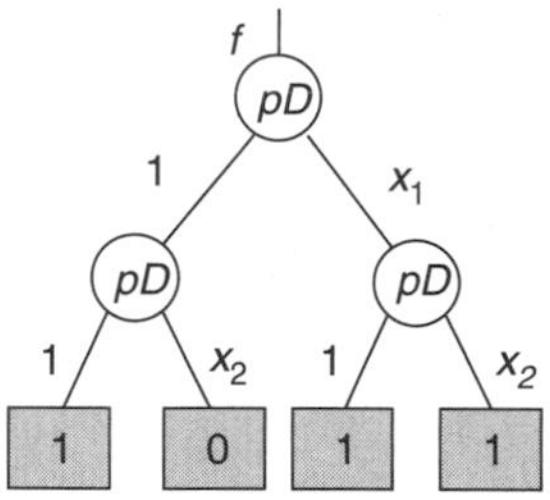

Structure of Reed–Muller expression

$$\hat{\mathbf{X}} = [1 \quad x_1] \otimes [1 \quad x_2]$$
$$= [1, x_2, x_1, x_1x_2]$$

Transform matrix

$$\mathbf{R}_{2^2} = \mathbf{R}_2 \otimes \mathbf{R}_2 = \begin{bmatrix} 1 & 0 \\ 0 & 1 \end{bmatrix} \otimes \begin{bmatrix} 1 & 0 \\ 0 & 1 \end{bmatrix} = \left[\begin{array}{cc|cc} 1 & & & \\ 1 & 1 & & \\ \hline 1 & & 1 & \\ 1 & 1 & 1 & 1 \end{array}\right]$$

Reed–Muller expression

$$f = \hat{\mathbf{X}}\mathbf{R}_{2^2}\,\mathbf{F} = \hat{\mathbf{X}} \left[\begin{array}{cc|cc} 1 & & & \\ 1 & 1 & & \\ \hline 1 & & 1 & \\ 1 & 1 & 1 & 1 \end{array}\right] \begin{bmatrix} 1 \\ 1 \\ 0 \\ 1 \end{bmatrix} = \hat{\mathbf{X}} \begin{bmatrix} 1 \\ 0 \\ 1 \\ 1 \end{bmatrix}$$
$$= 1 \oplus x_1 \oplus x_1x_2$$

FIGURE 19.12 The Davio tree for the switching function $f = \bar{x}_1 \vee x_2$ and calculations in matrix form (Example 19.10).

Example 19.10

We will derive the Davio tree for the switching function $f = \bar{x}_1 \vee x_2$, given its truth-vector $\mathbf{F} = [1\ 1\ 0\ 1]^T$. We apply Equation (19.10) to calculate the solution shown in Figure 19.12. The product terms are generated by the Kronecker product $\hat{\mathbf{X}}$. The 4×4 transform matrix $\mathbf{R}$ is generated by the Kronecker product of the basic matrix $\mathbf{R}_{2^1}$. The final result, the Reed–Muller expression, is mapped to the complete positive Davio tree.

Structural properties. The most important structural properties of the positive Davio tree are described below:

- The Davio expansion is in the nodes of the decision tree.
- n-Variable switching function is represented by an n-level Davio tree.
- The i-th level, $i = 1, \ldots, n$, includes 2^{i-1} nodes.
- Nodes at the n-th level are connected to 2^n terminal nodes, taking values of 0 or 1.
- The nodes, corresponding to the i-th variable, form the i-th level in the Davio tree.
- In every path from the root node to a terminal node, the variables appear in a fixed order. This means that this tree is ordered.
- The values of constant nodes are the values of the positive-polarity Reed–Muller expression for the represented functions. They are elements of the Reed–Muller coefficient vector $\mathbf{R} = [f_{000}\, f_{002}\, f_{020}\, f_{022}\, f_{200}\, f_{202}\, f_{220}\, f_{222}]$, where "0" corresponds to the value $f_0 = f|_{x_i=0}$, and "2" corresponds to the value $f_2 = f|_{x_i=1} \oplus f|_{x_i=0}$.

Example 19.11

An arbitrary switching function f of three variables can be represented by the Davio tree shown in Figure 19.13 (three levels, seven nodes, eight terminal nodes). To design this tree, the positive Davio expansion Equation (19.11) is used as follows:

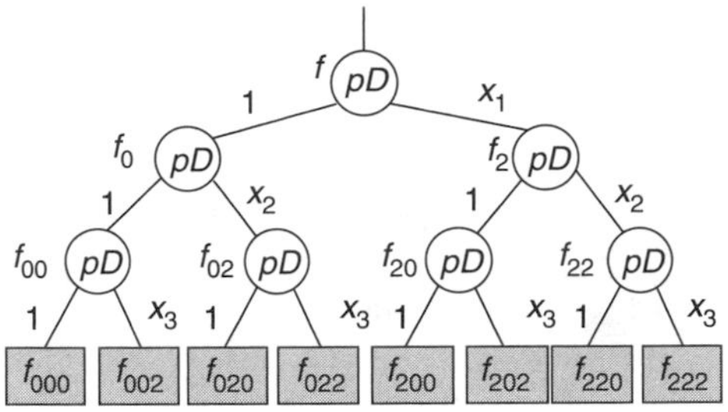

There are 8 pathes from f to the terminal nodes

$Path\,1: t_1 = 1$

$Path\,2: t_2 = x_3$

$Path\,3: t_3 = x_2$

$Path\,4: t_4 = x_2x_3$

$Path\,5: t_5 = x_1$

$Path\,6: t_6 = x_1x_3$

$Path\,7: t_7 = x_1x_2$

$Path\,8: t_8 = x_1x_2x_3$

Reed–Muller expression

$f = f_{000}t_1 \oplus f_{002}t_2 \oplus \ldots \oplus f_{222}t_8$

FIGURE 19.13 Reed–Muller representation of a switching function of three variables by the Davio tree (positive polarity) (Example 19.11).

- With respect to variable x_1:

$$f = f_0 \oplus x_1 f_2$$

- With respect to variable x_2:

$$f_0 = f_{00} \oplus x_2 f_{02}$$
$$f_1 = f_{10} \oplus x_2 f_{22}$$

- With respect to variable x_3:

$$f_{00} = f_{000} \oplus x_3 f_{002} \quad f_{02} = f_{020} \oplus x_3 f_{022},$$

$$f_{20} = f_{200} \oplus x_3 f_{202} \quad f_{22} = f_{220} \oplus x_3 f_{222}.$$

The Davio tree represents switching function f in the form of the Reed–Muller expression $f = f_{000} \oplus f_{002}x_3 \oplus f_{020}x_2 \oplus f_{022}x_2x_3 \oplus f_{200}x_1 \oplus f_{202}x_1x_3 \oplus f_{220}x_1x_2 \oplus f_{222}x_1x_2x_3$.

Davio tree reduction: The Davio diagram is derived from the Davio tree by deleting redundant nodes and by sharing equivalent subgraphs. The rules below produce the reduced Davio diagram (Figure 19.14):

Elimination rule: If the outgoing edge of a node labeled with x_i and $\bar{x}_i$ points to the constant zero, delete the node and connect the edge directly to the other subgraph.

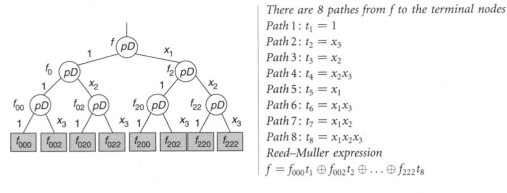

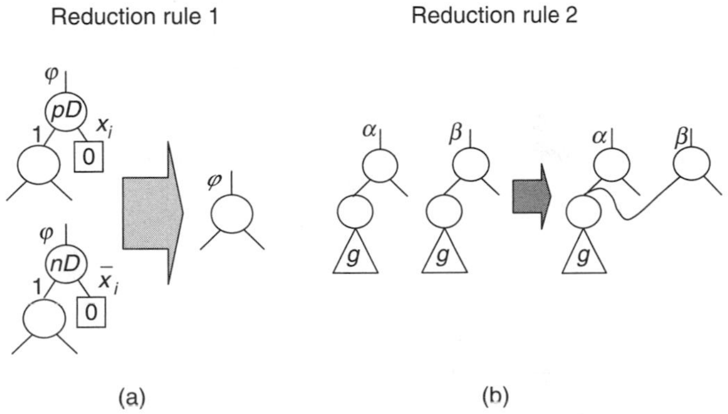

FIGURE 19.14 Reduction of Davio tree using (a) Rule 1 and (b) Rule 2.

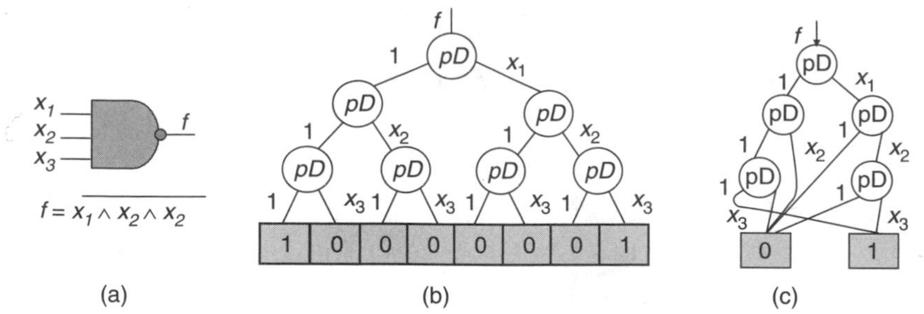

FIGURE 19.15 (a) The three-variable NAND function, (b) its Davio decision tree, and (c) reduced Davio decision diagram (Example 19.12).

Merging rule: Sharing equivalent subgraphs.

In a reduced decision diagram, edges longer than 1, or those that connect nodes at nonsuccessive levels, can appear. For example, the length of an edge connecting a node at $(i-1)$-th level with a node at $(i+1)$-th level is 2.

Example 19.12

Application of reduction rules to the three-variable NAND function is demonstrated in Figure 19.15.

19.5 Word-Level Decision Diagrams

Shared BDD considered above describes m-output logic functions, represented by the cost of increasing the BDD size (m-rooted BDD) and exploiting the isomorphism of the involved functions. The shared BDD still is bit-level. When the edges or terminal nodes are assigned to a weight that requires several bits for representation, or *word*, the decision diagram becomes *word-level*. The variables remain Boolean, so the integer-valued function, describing the multi-output function, sometimes is called *pseudo-Boolean*. This class of decision diagrams includes binary moment diagrams (BMDs), and arithmetic analogs of Davio diagrams.

Binary Moment Diagram

A BMD is characterized as follows:

- A BMD node implements the arithmetic analog of positive Davio expansion:

$$f = f_0 + x_i(-f_0 + f_1), \qquad (19.11)$$

  where $f_0 = f|_{x_i=0}$ and $f_1 = f|_{x_i=1}$;

- Terminal nodes are integer coefficients of arithmetical expansion for the switching function.

In Equation (19.11), expression $-f_0+f_1$ is called the linear moment, and this type of diagram is called BMD.

Word-Level Diagrams for Multi-Output Functions

BMD can be generated to represent multi-output functions. The truth vectors of the involved functions can be treated as word-level forms, so that each word represents the values of these function and is given a variable assignment.

Given a truth vector $\mathbf{F} = [f(0)f(1)\ldots f(2^n-1)]^T$, the vector of coefficients $\mathbf{D} = [d_0d_1\ldots d_{2^n-1}]^T$ is derived by the matrix equation with AND and arithmetic sum operations

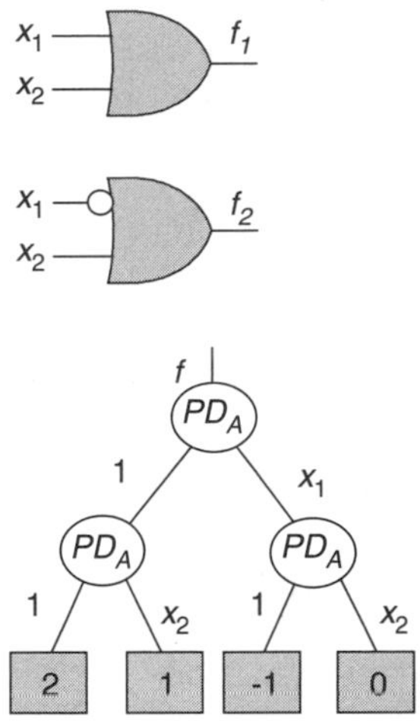

Direct arithmetic transform

$$\mathbf{D} = \mathbf{P}_{2^2} \cdot \mathbf{F} = \left[\begin{array}{cc|cc} 1 & 0 & 0 & 0 \\ -1 & 1 & 0 & 0 \\ \hline -1 & 0 & 1 & 0 \\ 1 & -1 & -1 & 1 \end{array}\right] \begin{bmatrix} 2 \\ 3 \\ 1 \\ 3 \end{bmatrix} = \begin{bmatrix} 2 \\ 1 \\ -1 \\ 1 \end{bmatrix}$$

Arithmetic expression

$$f = 2 + x_2 - x_1 + x_1 x_2$$

The equation is equivalent to

$$\mathbf{F} = 2^1\mathbf{F}_2 + 2^0\mathbf{F}_1 = 2^1 \begin{bmatrix} 1 \\ 1 \\ 0 \\ 1 \end{bmatrix} + 2^0 \begin{bmatrix} 0 \\ 1 \\ 1 \\ 1 \end{bmatrix} = \begin{bmatrix} 2 \\ 3 \\ 1 \\ 3 \end{bmatrix}$$

Applying transformation (Equation (19.12)) implies:

$$\mathbf{D} = \mathbf{P}_{2^2}\mathbf{F} = [2 \quad 1 \quad -1 \quad 1]^{\mathrm{T}},$$

i.e., $f = 2 + x_2 - x_1 + x_1 x_2$

FIGURE 19.16 Computing the arithmetic word-level expression for a two-output circuit (Example 19.13).

$$\mathbf{D} = \mathbf{P}_{2^n}\mathbf{F}, \tag{19.12}$$

where the $2^n \times 2^n$ matrix $\mathbf{P}_{2^n}$ is formed by the Kronecker product

$$\mathbf{P}_{2^n} = \bigotimes_{j=1}^{n} \mathbf{P}_{2^j}, \qquad \mathbf{P}_{2^1} = \begin{bmatrix} 1 & 0 \\ -1 & 1 \end{bmatrix}, \tag{19.13}$$

Example 19.13

Computing the coefficients by Equation (19.12) and Equation (19.13) for the two-output function $f_1 = x_1 \vee x_2$, $f_2 = \overline{x}_1 \vee x_2$ with truth-vector $\mathbf{F}_1$ and $\mathbf{F}_2$ is illustrated in Figure 19.16. The nodes of the diagram correspond to the arithmetic analog of positive Davio expansion denoted as pD_A.

The same result can be obtained by the conversion of binary words representing f_1 and f_2 to the integer domain given

$$f_1 = x_1 \vee x_2 = x_2 + x_1 - x_1 x_2,$$
$$f_2 = \overline{x}_1 \vee x_2 = 1 - x_1 + x_1 x_2,$$
$$\text{i.e.,}\quad f = 2^1 f_2 + 2^0 f_1 = 2^1 \underbrace{(1 - x_1 + x_1 x_2)}_{\text{Output } f_2} + 2^0 \underbrace{(x_2 + x_1 - x_1 x_2)}_{\text{Output } f_1}$$
$$= 2 + x_2 - x_1 + x_1 x_2.$$

BMDs are most efficient at representing linear arithmetic expressions, including the word-level representations of multi-output functions.

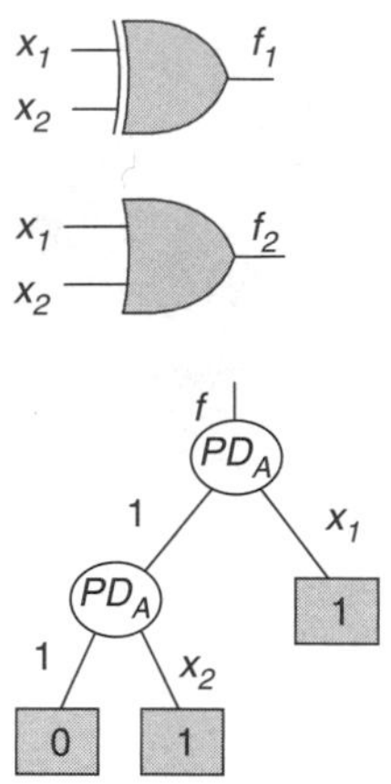

Direct arithmetic transform

$$\mathbf{D} = \mathbf{P}_{2^2} \cdot \mathbf{F} = \left[\begin{array}{cc|cc} 1 & 0 & 0 & 0 \\ -1 & 1 & 0 & 0 \\ \hline -1 & 0 & 1 & 0 \\ 1 & -1 & -1 & 1 \end{array}\right] \begin{bmatrix} 0 \\ 1 \\ 1 \\ 2 \end{bmatrix} = \begin{bmatrix} 0 \\ 1 \\ 1 \\ 0 \end{bmatrix}$$

Arithmetic expression

$$f = 2 + x_1 + x_2$$

The equation is equivalent to

$$\mathbf{F} = 2^1\mathbf{F}_2 + 2^0\mathbf{F}_1 = 2^1 \begin{bmatrix} 0 \\ 0 \\ 0 \\ 1 \end{bmatrix} + 2^0 \begin{bmatrix} 0 \\ 1 \\ 1 \\ 0 \end{bmatrix} = \begin{bmatrix} 0 \\ 1 \\ 1 \\ 2 \end{bmatrix}$$

Applying transformation (Equation (19.12)) implies:

$$\mathbf{D} = \mathbf{P}_{2^2}\mathbf{F} = [0 \quad 1 \quad 1 \quad 0]^T,$$

i.e., $f = x_1 + x_2$

FIGURE 19.17 Computing the linear arithmetic word-level expression for a two-output circuit (Example 19.14).

Example 19.14

Computing the coefficients for the two-output function $f_1 = x_1 \oplus x_2, f_2 = x_1 \wedge x_2$, with truth-vector $\mathbf{F}_1$ and $\mathbf{F}_2$ is illustrated by Figure 19.17. The corresponding decision diagram is linear.

Other Word-Level Decision Diagrams

The direct generation of an integer-valued-output binary-input function from its truth table can be implemented on the BDD with integer-valued terminal nodes. These are called multiterminal BDDs (MTBDDs) (see section on Further Reading).

Yet more word-level decision diagrams are *edge-valued decision diagrams* (EVBDDs). A further improvement, multiplicative BMDs, is denoted by *BMD and combines the properties of BMDs and EVBDDs. They can be used for addition and multiplication and for verifying circuits such as multipliers.

20.6 Available Software and Programming of BDD

Various packages are available, in particular,

- CUDD (Colorado University Decision Diagram) [Somenzi] package that implements a large collection of algorithms for improving the order of variables in various decision diagrams
- BuDDy by Jörn Lind-Nielsen of the IT University of Copenhagen, used mostly to verify finite-state machine systems with up to 1400 concurrent machines working in parallel
- A BDD library with extensions for sequential verification developed by E. Clarke's group at Carnegie-Melon University. They also developed a series of symbolic model checkers, including a word-level checker.

A special Web portal was created in 1999 (www.bdd-portal.org).

References

1. S.B. Akers, "Binary decision diagrams," *IEEE Trans. Comput.*, C-27, no. 6, 509–516, 1978.
2. N. Asahi, M. Akazawa, and Y. Amemiya, "Single-electron logic device based on the binary decision diagram," *IEEE Trans. Electron Devices*, vol. 44, no. 7, pp. 1109–1116, 1997.
3. B. Becker, "Testing with decision diagrams," *INTEGRATION, VLSI J.*, 26, 5–20, 1998.
4. R.E. Bryant, "Graph-based algorithms for Boolean function manipulation," *IEEE Trans. Comput.*, vol. C-35, no. 6, pp. 677–691, 1986.
5. R.E. Bryant and Y-A. Chen, "Verification of arithmetic circuits with binary moment diagrams," in *Proc. 32nd ACM/IEEE Design Autom. Conf.*, June 1995, pp. 535–541.
6. R.E. Bryant and C. Meinel, "Ordered binary decision diagrams," in *Logic Synthesis Verification*, S. Hassoun and T. Sasao, Eds., R.K. Brayton, consulting Ed., Hingham, MA: Kluwer Academic Publishers, 2002, pp. 285–307.
7. P. Buch, A. Narayan, A.R. Newton, and A.L. Sangiovanni-Vincentelli, "Logic synthesis for large pass transistor circuits," in *Proc. Int. Conf. Comput.-Aided Des.*, 1997, pp. 663–670.
8. J.R. Burch, E.M. Clarke, K.L. McMillan, and D.L. Dill, "Sequential circuit verification using symbolic model checking," in *Proc. 27th ACM/IEEE Des. Autom. Conf.*, June 1990, pp. 46–51.
9. V. Cheushev, V. Shmerko, D. Simovici, and S. Yanushkevich, "Functional entropy and decision trees" in *Proc. IEEE 28th Int. Symp. Multiple-Valued Logic*, Japan, 1998, pp. 357–362.
10. H. Cho, G.D. Hachtel, S.-W. Jeong, B. Plessier, E. Schwarz, and F. Somenzi, "ATPG aspects of FSM verification," in *Proc. IEEE/ACM Int. Conf. Comput.-Aided Des.*, November 1990, pp. 134–137.
11. O. Coudert and J.C. Madre, "A new graph based prime computation technique," in, *New Trends in Logic Synthesis*, T. Sasao, Ed., chap. 2, pp. 33–57, Hingham, MA: Kluwer, 1992.
12. R. Drechsler and B. Becker, *Binary Decision Diagrams: Theory and Implementation*, Boston, MA: Kluwer, 1999.
13. P. Dziurzanski, V. Malyugin, S. Shmerko, and S. Yanushkevich, "Linear models of circuits based on the multivalued components," *Autom. Remote Control*, vol. 63, no. 6, pp. 960–980, 2002.
14. S.J. Friedman and K.J. Supowit, "Finding the optimal variable ordering for binary decision diagrams," *IEEE Trans. Comput.*, vol. 39, no. 5, pp. 710–713, 1990.
15. R. Jacobi, A Study of the Application of Binary Decision Diagrams on Multi-Level Logic Synthesis doctoral thesis, Universitat Catholique de Louvain, Belgium, 1993.
16. Y.-T. Lai and M. Pedram, "BDD based decomposition of logic function with application to FPGA Synthesis" in *Proc. 30th ACM/IEEE Des. Autom. Conf.*, 1993, pp. 642–647.
17. Y. Lai, M. Pedram, and S. Vrudhula, "EVBDD-based algorithms for integer linear programming, spectral transformation, and function decomposition" *IEEE Trans. Comput. Aided Des. Integr. Circuits Sys.*, vol. 13, no. 8, pp. 959–975, 1994.
18. P. Lindgren, M. Kerttu, M. Thornton, and R. Drechsler, "Low-power optimization techniques for BDD mapped circuits," in *Proc. Asia-Pacific Des. Autom. Conf.*, January 2001, pp. 615–621.
19. C. Meinel and T. Theobald, *Algorithms and Data Structures in VLSI Design*, Berlin: Springer, 1998.
20. S. Minato, *Binary Decision Diagrams and Applications for VLSI Design*, Hingham, MA: Kluwer, 1996.
21. I. Pomeranz and S.M. Reddy, "Design-for-testability for path delay faults in combinational circuits using test points" *IEEE Trans. Comput. Aided Des. Integr. Circuits Sys.*, 17, 333–343, 1998.
22. R. Rudell, "Dynamic variable ordering for ordered binary decision diagrams," in *Proc. Int. Conf. Comput.-Aided Des.*, Santa Clara, CA, November 1993, pp. 42–47.
23. T. Sasao, *Switching Theory for Logic Synthesis*, Hingham, MA: Kluwer, 1999.
24. V. Shmerko, D.V. Popel, R.S. Stankovic, V.A. Cheushev, and S. Yanushkevich, "Information theoretical approach to minimization of AND/EXOR expressions of switching functions," in *Proc. IEEE Int. Conf. Telecommun.*, Yugoslavia, 1999, pp. 444–451.
25. F. Somenzi, *CUDD Decision Diagram Package*, http://bessie.colorado.edu/~fabio/CUDD.
26. R.S. Stankovic and J.T. Astola, *Spectral Interpretation of Decision Diagrams*, Berlin: Springer, 2003.

27. N. Takahashi, N. Ishiura, and S. Yajima., "Fault simulation for multiple-faults using BDD representation of fault sets," in *Proc. IEEE/ACM Int. Conf. Comput.-Aided Des.*, November 1991, pp. 550–553.
28. T. Yamada, Y. Kinoshita, S. Kasai, H. Hasegawa, and Y. Amemiya, "Quantum-dot logic circuits based on shared binary decision diagram," *J. Appl. Phys.*, vol. 40, no. 7, pp. 4485–4488, 2001.
29. S.N. Yanushkevich, V.P. Shmerko, and S.E. Lyshevski, *Logic Design of NanoICs*, Boca Raton, FL: CRC Press, 2005.

Further Reading

The basic concept of BDD has been introduced by Akers [1]. The introduction of OBDDs in 1986 [4] was a major step forward in the search for suitable data structures for switching functions in circuit design.

The **basics** of decision diagram techniques presented in textbook form can be found in Minato [20] and Sasao [23].

The issue of **complexity** and algorithms for optimizing **variable order** in BDDs has been studied in Meinel and Theobald [19].

Information theoretical measures on decision trees and diagrams aimed at heuristic variable ordering have been studied recently in Cheushev et al. [9] and Shmerko et al. [24].

Shared BDDs, or multirooted BDDs, have been considered in Minato [20].

Multiterminal BDDs (MTBDDs), or arithmetic decision diagrams (ADDs), have been introduced in Bryant and Chen [5].

Yet another word-level representation, **edge-valued decision diagrams** (EVBDDs), have been proposed in Pedram and Vrudhula [17]. An improved diagram is found in multiplicative BMDs, denoted by *BMDs, combining the properties of BMDs and EVBDDs" [5]. This is useful for representing addition and multiplication and for verifying circuits such as multipliers.

Word-level diagrams for the representation of linear arithmetic expressions have been studied in Dziurzanski et al. [13] and Yanushkevich et al. [29].

Spectral interpretation of decision diagrams is systematically presented in Stankovic and Astola [26].

Trends of decision-diagram technique are analyzed in Bryant and Meinel [6] and Drechsler and Becker [12].

Techniques of applying decision diagrams to digital designs include:

Circuit optimization. Reduction of BDDs is relevant, implicitly, to the minimization of area for logic networks [19, 20]. Choosing a variable ordering that significantly reduces the BDD size can be managed by dynamic programming [14, 22] and sifting [19].

Pass-transistor logic circuit optimization. An approach for a pass-transistor logic synthesis, based on a mapping of BDDs, was first considered in Buch et al. [7]. The advantage of this method is that pass-transistor logic circuits originating from BDDs are path-free. There is no assignment to the inputs that produces a conducting path from power supply to ground. The prime implicants in the computation of minimized switching functions, using BDD for representing the cube sets, has been proposed in Coudert and Madre [11].

Testing. FDDs such as Davio decision diagrams involve calculation of Boolean differences that are known for their application to the problem of test-pattern generation. The advantage of this is taken into account in Cho et al. [10].

Fault simulation. It has been proposed to simulate multiple stuck-at faults by representing sets of fault combinations with BDDs [27]. In Minato [20], an improved method based on ZBDDs was developed.

Time-driven analysis. The testability of circuits synthesized by mapping a BDD to a multiplexor circuit was investigated in Becker [3]. It is based on the path-delay fault model (PDFM). The PDFM allows detection of static and dynamic faults. Its major drawback is the large number of paths that must be tested. Solutions based on test points have been proposed in Pomeranz and Reddy [21], though test-point insertion requires additional hardware.

Estimation of power dissipation. In Lindgren et al. [18], power-dissipation reductions are considered for a minimum-size BDD. To evaluate power dissipation in a logic network, the circuit is realized by mapping BDD nodes to multiplexer circuits implemented using CMOS transmission gates and static inverters. The power dissipation is a function of switching activity and its fanout (corresponding to the capacitive load).

The capacitive load for a mapped node is modeled as the number of incoming edges. The heuristic optimization targets the cost expressed in terms of the probability of a signal. Note that the variable order of the underlying BDD heavily affects the number of nodes (or the size) and the switching activity for each node.

Decomposition. Logic decomposition methods, based in the past on algebraic techniques, have been replaced with new BDD-based decomposition techniques [1, 6, 20]. These decomposition methods can be divided into direct and path-oriented decomposition. The direct decomposition extracts shared subgraphs and replaces them with new variables in the original BDD and is aimed at circuit-size optimization [16]. Path-oriented methods are based on algebraic decomposition performed on BDDs [15].

The verification problem is formulated as follows: given two circuits C_1 and C_2 with the same number of inputs and outputs, verify that C_1 and C_2 produce for each input assignment the same output sequence. There are two approaches: complete ordered-decision trees can be generated from the circuit and compared for equivalency; alternatively, ROBDDs can be generated and compared to the decision trees. This approach is possible because both complete ordered-decision trees and ROBDD are canonical forms. The second approach is more compact: the space of global properties of circuits C_1 and C_2 is reduced to local properties [5, 20].

Nanodevices design. BDD can be considered as a candidate for topological models for nanoelectronic architecture. Nanoelectronic demands are of switch-type but have gate-level logic (similar to pass-gate logic). BDD nanowire wrap-gate networks have been proposed in Asahi et al. [2] and developed for quantum-dot networks in Yamada et al. [29]. Hypercube topology of nanostructures based on embedded BDDs has been proposed in Yanushkevich et al. [29].

20

Vehicular Systems

Linda Sue Boehmer
LSB Technology

20.1 Introduction

Vehicular systems have evolved and advanced from many other fields of technology over the past two decades. Instrumentation and controls for the various transportation modes (aircraft, marine vessels, rail vehicles, buses, trucks, and cars) resemble each other more every year. Technology from one mode of transportation used to be of little interest to practitioners of any other mode. A technology historian might notice similarities among the functions of airport beacons, lighthouses, railroad signals, and traffic lights, but the specialists in each field had little to say to each other. This is no longer the case. Computers, microprocessor controls, electronics, GPS, and advanced networking and radio technologies are being applied in all forms of passenger and freight transportation, through the air and over water and land. The vehicles are now considered in context with an entire system, which is increasingly viewed as part of an overall transportation environment encompassing more than one mode.

Although "multimodal" is a term that was coined by policy makers in the 1990s to earmark funding for various transportation modes and to facilitate interfaces among them, it applies equally well to the supporting technologies of the original modes. All modes now utilize microprocessor controls in their subsystems. With microprocessor control has come additional diagnostic capability and the use of system-level intelligence linking all intelligent subsystems, analog sensors, and controls. Propulsion technology for vehicles now varies by mode less than it ever did. Because of microprocessors, propulsion can be controlled more precisely, allowing vehicles to use nontraditional energy sources and to switch from one source to another easily, even automatically. We do not yet have the ideal multimodal vehicle, capable of navigating, either automatically or by a driver/operator, through air, in water, on rails, and along roads; but technology is no longer a limiting factor. We have not yet achieved the best balance among modes so that the most appropriate mode is utilized for passenger or freight transportation, but the tools exist to make those decisions possible. (This section deals primarily with land transportation. See the index for aviation and maritime applications.)

20.2 Design Considerations

If design could begin with a clean slate, the first step would be to decide which mode of transportation and which power source are best suited for the application, based on geography, priority (and mass) of the passenger(s) or cargo to be transported, energy efficiency, safety, cost per mile, and other factors. However, this is not really possible because the factors relating to funding sources and existing infrastructure often outweigh any technical considerations.

Planes, boats, trains, and cars each have their own operating environments, although their design is increasingly considered part of an integrated transportation system rather than simply as independent vehicles. Many of the underlying technologies are utilized across modal boundaries, but the details of design remain highly dependent on mode.

Electrical and electronic systems are extensively used in automobiles today. Electronic systems are used to control the engine, transmission, steering, antilock braking, active suspensions, and traction and offer a wide variety of mobile entertainment options. Most automobiless incorporate an integrated computer system for controlling the functions mentioned and for displaying information regarding them. Sensors and electronic controls are increasingly used to enhance safety, such as in airbag-inflation units that activate automatically within milliseconds after a collision with the appropriate degree of inflation.

The computing power in place in individual vehicles facilitates ITS (intelligent transportation systems) applications. The term refers to a varied assortment of electronics that provide real-time information on accidents, congestion, routing, and roadside services to drivers and traffic managers. ITS technologies such as collision-avoidance systems (similar to those already used in aviation) and lane-tracking devices are designed to make vehicles safer and more autonomous by alerting drivers to impending disaster or allowing a car to drive itself. ITS also includes interfaces between personal traffic and mass transportation, particularly when rail traffic mixes with cars and at rail crossings, whether or not such crossings are protected by gates. Mass-transportation vehicles have most of the same internal subsystems as cars do, and also additional subsystems, for which cars increasingly have analogous functions — thanks to the synergistic effects of ITS.

Major rail vehicle subsystems include propulsion, braking, power conditioning, communication, passenger information (audio and visual), heating/air conditioning, door control, speed control, and monitoring and diagnostics. All of these subsystems interact with each other, and many also interact with systems external to the vehicle.

Some of the initial design considerations for vehicles include:

- Will the vehicle interface with an existing fleet?
- Will the vehicle operate independently or as part of a **consist**, or both?
- What are the physical requirements, including dimensions and number of passengers (seated vs. standing)? Is the system (or portions of it) elevated, in tunnels, at grade, or underground? Will the vehicles mix with or cross other types of traffic? Are ambient conditions exceptionally hot, cold, or dangerous?
- How closely must one vehicle (or consist) follow another, and how fast must they be capable of traveling? Where, how often, and for how long will they stop?
- To what degree will the vehicles be automated; how much control will a human operator (and other crew) have and under what conditions?
- What kind of power is available, how will it be collected, and can energy generated during electric braking be returned to the power system?
- How will system and subsystem failures be handled, what kind of failures are acceptable, and to what degree (or for how long) will they be tolerated? How often and with what degree of expertise will the vehicles be serviced?

20.3 Land Transportation Classifications

Among land transportation vehicles, there are more subdivisions than the lay person might expect. In general, distinctions are based on vehicle size, weight, speed, and passenger or cargo capacity and whether they are designed to operate primarily on roads or rails. Cars, trucks, minivans, vans, sport utility vehicles, etc. are familiar terms. There are some variations among them with regards to the electronics embedded in the systems and the options available to the driver. The same is true for railroads, rail transit, and mass transit, although the distinctions among classifications are important to the manufacturers, regulators, and public operators of the equipment. The classifications include railroad, commuter rail, heavy rail, light rail, street car, trolley bus, bus, paratransit, and "people mover" or monorail.

20.4 Propulsion

In the early 1900s, electric automobiles became commonplace but were gradually replaced by gasoline-fueled automobiles with their higher top speeds and longer ranges. Since the mid-1970s, when the electric vehicle reemerged as an appealing transportation option, many have recognized the potential of electric power for fleet vans and commuter cars. An electric vehicle uses electric energy storage, electronic controls, and electric propulsion devices. Vehicles using batteries to drive electric motors are well suited to the short routes and regular schedules followed by vans in a company fleet or by commuters. Because the vehicles can be recharged regularly at night, they offer electric utilities a desirable off-peak demand. At the same time, electric power is environmentally superior to internal combustion, especially in overcrowded urban settings. As advanced batteries and recharging technologies extend their range and reduce their initial and operating costs, the market for electric vehicles can be expected to expand. A variety of electric and hybrid electric cars continue to be introduced as concept cars and for production, but none have yet enjoyed general use (defined by sales per model year) in spite of increasing gasoline prices. A few modest success stories (General Motors' Griffon fleet vans, Takara's Choro Q, and Honda's Civic Hybrid) indicate that the promise of these technologies will expand.

As products of combustion were acknowledged as major air pollutants and fuels have been acknowledged as nonrenewable resources, alternatives have been sought with increasing diligence. Many alternatives and hybrid technologies are now technically feasible and have been demonstrated or are in use in a limited way.

Today, electric motors are used more widely for rail vehicles than for cars, vans, or buses, although this is slowly changing. Electric motors as a backup mode for buses are increasingly common in certain areas where air pollution is considered a serious problem and in portions of systems, such as North American tunnels, where fumes from internal combustion can be hazardous.

Power distribution for rail vehicles ranges from three-phase ac, at various voltages (usually collected from overhead wires), to several different dc voltages (usually collected from a "third rail"). Until recently (within the past decade in the U.S.), most electric traction motors utilized dc power. Today, most traction motors in new vehicles use ac power. Various techniques are used to cool the motors, depending on the operating environment. Collected power is conditioned continuously to meet motor requirements and the power requirements of other vehicle systems, and also is stored to power critical onboard systems if power is lost.

Traction motors also are used for braking, which generates power that can be reconditioned and recycled to the power system to power other vehicles or returned to the power grid. Power that cannot be recycled or used elsewhere in the system is converted into heat by banks of large braking resistors. Electrical braking is supplemented by mechanical braking systems, which can be actuated pneumatically, hydraulically, and/or electrically. Coordination of propulsion and braking efforts, especially when traction surfaces are slippery, is an important design point, made safer and more reliable by microprocessors.

Microprocessor controls have allowed optimization of automotive internal combustion processes to economize on fuel and minimize air pollution. Alternative fuels, such as natural gas, are becoming more common because the combustion process can be managed more uniformly, responsively, and safely than ever before.

Diesel electric locomotives have become the propulsion vehicle of choice for long-haul freight railroads. Microprocessors control the combustion process that produces electricity to power electric motors when ac power is not available, such as on long sections of rail that are not yet electrified.

Dual-mode (or hybrid) buses utilize internal combustion when operating on the streets, but switch to electric power in tunnels. Dual-mode rail vehicles or streetcars collect power from an overhead catenary or third rail, but can switch to battery power when they are not operating in electrified areas.

Improvements in battery technology (cost, life, power density, weight, and maintenance requirements) and charging technologies and other alternative power sources will encourage the acceptance of alternatives to internal combustion and direct electric power.

20.5 Microprocessor Controls

All major vehicle subsystems and most minor ones are now microprocessor-controlled. Embedded microprocessors replaced banks of relays and mechanical switches to perform functions on the vehicle and to control functions that did not exist prior to the advent of microprocessors. Some major vehicle subsystems have several microprocessors handling different functions and coordinating analog and digital input/output signals.

Intelligent subsystems exchange information within a vehicle, among vehicles in a consist, and between the vehicle and its external environment. This information is exchanged through increasingly sophisticated networks, which may or may not use traditional wiring. Separate networks and layers of error-checking software handle safety-critical data.

A human vehicle operator typically has status indicators, alarms, and controls. These have changed dramatically with the advances in microprocessor-controlled subsystems. The "glass cockpit" and "fly-by-wire" techniques originally developed for aviation transition to cars and trains as they are service-proven and as their cost decreases. A driver or train operator once had a few lights, a gauge or two, a throttle of some sort, and a brake handle or pedal. For vehicles that are not automated but are partially automated or allow manual operation at times, a human operator today is confronted by an array of dials, buttons, data and CCTV screens, LCD panels, microphones, and annunciators (audio and visual) of various types. In some cases there is more information than an operator needs.

There was initial resistance to technology advances, based on the perception that electronics might not be as reliable as electromechanical devices (such as relays) and on a concern that they could be more complicated to troubleshoot and maintain than electromechanical and analog equipment. This resulted in a transition during which there was duplication between the old, analog indicators and controls and new, digital indicators and "soft" controls accessed by voice or through a touch screen. Experience with advanced technologies has proven those initial concerns to be unfounded.

20.6 Monitoring and Diagnostics

A great advantage of microprocessor-controlled systems is the degree to which they can be self-diagnosing. Each intelligent subsystem has internal self-diagnostics, which include routines to perform initial tests on power-up, update checks for "hot" startup, and continuous checking to assure that inputs and outputs are within expected ranges. Internal self-diagnostics are also capable of performing self-tests on request. When fault conditions are noted, typically there are internal resets to allow for inaccurate data, with faults being logged after a certain number of occurrences or duration of a fault condition.

The basis of any monitoring and diagnostics system is the underlying maintenance philosophy. Microprocessor controls make it possible to capture any combination of information from the intelligent subsystems. Information can then be processed and presented in a variety of ways, along with data from analog sensors that are not part of any intelligent system.

Information of interest includes operating status and existing or historical fault information. Data collected may include faults and other expected actions that may or may not be considered as faults. Historical information can include faults and the status of parameters of interest associated with the fault. The amount of information that can be captured is limited only by the amount of memory provided and the speed at which it can be transferred. Typically, some (or all) of the memory is in a form that will allow the fault data to be preserved after power is lost or vehicles are shut down between operating periods.

Decisions on what information to capture, how often to sample it, and how many samples to preserve are ideally based on what will be most valuable for troubleshooting existing faults and predicting future failures.

A thorough understanding of each intelligent subsystem is needed in addition to an understanding of the environment in which it operates, including the other subsystems with which it interacts. It is also important to know what level of skill will be applied for interpreting the saved information. If the level of skill will be low, some degree of artificial intelligence can be designed into the diagnostics to guide a maintenance technician through the troubleshooting process.

During design, the information to be collected is influenced by the target audience. The most detailed internal subsystem information is primarily used by engineers or specialized technicians to troubleshoot detailed failures or fine-tune operation. A subset of less detailed information is used by maintenance staff to determine which components or submodules need replacement or repair. A further subset of that information is used by a general troubleshooting staff to determine which subsystem is malfunctioning or which higher level modules to replace. An even smaller subset of operating status information and only a few major faults are needed by the operator, with a selection of that data being useful to a central-control or maintenance-dispatching facility if real-time links are available. A variety of techniques are used to present this information to the target audience(s) in ways and at times that are most appropriate.

"Event recorders" similar to the "black boxes" required on passenger airliners, capturing selected parameters in hardened storage for post-accident review, are now required by the Fedearl Railroad Administration for railroads and have been standardized by the rail transit industry. The automotive industry is considering similar standards.

Defining Terms

Active suspension: An electronically controlled suspension system for maintaining level suspension of a vehicle.

Consist: Two or more vehicles coupled together in a train. The vehicles may be identical or they may each lack a major subsystem (such as propulsion), whose functions may be handled by another vehicle in the consist.

Dual mode (or hybrid): Vehicles that are designed to switch manually or automatically from one type of propulsion to another, for instance from internal combustion to electric.

Electric braking: Use of traction motor to slow the vehicle.

Electric vehicle: Vehicle using electric energy storage, electric controls, and electric propulsion devices.

Traction motor: Electric motor that provides motive power to move vehicles.

References

IEEE Standard 1482.1, IEEE Standard for Rail Transit Vehicle Event Recorders, 1999.

J. Voelcker, "Top 10 tech cars," *IEEE Spectrum*, vol. 41, no. 3 (NA), pp. 28–35, 2004.

Further Information

The following organizations can provide industry perspectives, applicable standards, guidelines, and technical information:

American Public Transportation Association (APTA), Washington, DC, www.APTA.com.

Federal Transit Administration (FTA), Federal Highway Administration (FHA), Federal Railroad Administration (FRA), Department of Transportation, Washington, DC.

ITS America, Washington, DC, www.ITSA.org

Society of Automotive Engineers, Warrandale, PA.

Vehicular Technology Society, Institute of Electrical and Electronics Engineers, New York, 10017–2394, 1–800–678-IEEE.

III

Mathematics, Symbols, and Physical Constants

Ronald J. Tallarida
Temple University

THE GREAT ACHIEVEMENTS in engineering deeply affect the lives of all of us and also serve to remind us of the importance of mathematics. Interest in mathematics has grown steadily with these engineering achievements and with concomitant advances in pure physical science. Whereas scholars in nonscientific fields, and even in such fields as botany, medicine, geology, etc., can communicate most of the problems and results in nonmathematical language, this is virtually impossible in present-day engineering and physics. Yet it is interesting to note that until the beginning of the twentieth century, engineers regarded calculus as something of a mystery. Modern students of engineering now study calculus, as well as differential equations, complex variables, vector analysis, orthogonal functions, and a variety of other topics in applied analysis. The study of systems has ushered in matrix algebra and, indeed, most engineering students now take linear algebra as a core topic early in their mathematical education.

This section contains concise summaries of relevant topics in applied engineering mathematics and certain key formulas, that is, those formulas that are most often needed in the formulation and solution of engineering problems. Whereas even inexpensive electronic calculators contain tabular material (e.g., tables of trigonometric and logarithmic functions) that used to be needed in this kind of handbook, most calculators do not give symbolic results. Hence, we have included formulas along with brief summaries that guide their use. In many cases we have added numerical examples, as in the discussions of matrices, their inverses, and their use in the solutions of linear systems. A table of derivatives is included, as well as key applications of the derivative in the solution of problems in maxima and minima, related rates, analysis of curvature, and finding approximate roots by numerical methods. A list of infinite series, along with the interval of convergence of each, is also included.

Of the two branches of calculus, integral calculus is richer in its applications, as well as in its theoretical content. Though the theory is not emphasized here, important applications such as finding areas, lengths, volumes, centroids, and the work done by a nonconstant force are included. Both cylindrical and spherical polar coordinates are discussed, and a table of integrals is included. Vector analysis is summarized in a separate section and includes a summary of the algebraic formulas involving dot and cross multiplication, frequently needed in the study of fields, as well as the important theorems of Stokes and Gauss. The part on special functions includes the gamma function, hyperbolic functions, Fourier series, orthogonal functions, and both Laplace and z-transforms. The Laplace transform provides a basis for the solution of differential equations and is fundamental to all concepts and definitions underlying analytical tools for describing feedback control systems. The z-transform, not discussed in most applied mathematics books, is most useful in the analysis of discrete signals as, for example, when a computer receives data sampled at some prespecified time interval. The Bessel functions, also called cylindrical functions, arise in many physical applications, such as the heat transfer in a "long" cylinder, whereas the other orthogonal functions discussed—Legendre, Hermite, and Laguerre polynomials—are needed in quantum mechanics and many other subjects (e.g., solid-state electronics) that use concepts of modern physics.

The world of mathematics, even applied mathematics, is vast. Even the best mathematicians cannot keep up with more than a small piece of this world. The topics included in this section, however, have withstood the test of time and, thus, are truly *core* for the modern engineer.

This section also incorporates tables of physical constants and symbols widely used by engineers. While not exhaustive, the constants, conversion factors, and symbols provided will enable the reader to accommodate a majority of the needs that arise in design, test, and manufacturing functions.

Mathematics, Symbols, and Physical Constants

Greek Alphabet

Greek Letter		Greek Name	English Equivalent	Greek Letter		Greek Name	English Equivalent
Α	α	Alpha	a	Ν	ν	Nu	n
Β	β	Beta	b	Ξ	ξ	Xi	x
Γ	γ	Gamma	g	Ο	ο	Omicron	ŏ
Δ	δ	Delta	d	Π	π	Pi	P
Ε	ε	Epsilon	ĕ	Ρ	ρ	Rho	r
Ζ	ζ	Zeta	z	Σ	σ	Sigma	s
Η	η	Eta	ē	Τ	τ	Tau	t
Θ	θ ϑ	Theta	th	Υ	υ	Upsilon	u
Ι	ι	Iota	i	Φ	ϕ φ	Phi	ph
Κ	κ	Kappa	k	Χ	χ	Chi	ch
Λ	λ	Lambda	l	Ψ	ψ	Psi	ps
Μ	μ	Mu	m	Ω	ω	Omega	ō

International System of Units (SI)

The International System of units (SI) was adopted by the 11th General Conference on Weights and Measures (CGPM) in 1960. It is a coherent system of units built form seven *SI base units,* one for each of the seven dimensionally independent base quantities: they are the meter, kilogram, second, ampere, kelvin, mole, and candela, for the dimensions length, mass, time, electric current, thermodynamic temperature, amount of substance, and luminous intensity, respectively. The definitions of the SI base units are given below. The *SI derived units* are expressed as products of powers of the base units, analogous to the corresponding relations between physical quantities but with numerical factors equal to unity.

In the International System there is only one SI unit for each physical quantity. This is either the appropriate SI base unit itself or the appropriate SI derived unit. However, any of the approved decimal prefixes, called *SI prefixes,* may be used to construct decimal multiples or submultiples of SI units.

It is recommended that only SI units be used in science and technology (with SI prefixes where appropriate). Where there are special reasons for making an exception to this rule, it is recommended always to define the units used in terms of SI units. This section is based on information supplied by IUPAC.

Definitions of SI Base Units

Meter: The meter is the length of path traveled by light in vacuum during a time interval of 1/299,792,458 of a second (17th CGPM, 1983).

Kilogram: The kilogram is the unit of mass; it is equal to the mass of the international prototype of the kilogram (3rd CGPM, 1901).

Second: The second is the duration of 9,192,631,770 periods of the radiation corresponding to the transition between the two hyperfine levels of the ground state of the cesium-133 atom (13th CGPM, 1967).

Ampere: The ampere is that constant current which, if maintained in two straight parallel conductors of infinite length, of negligible circular cross-section, and placed 1 m apart in vacuum, would produce between these conductors a force equal to 2×10^{-7} newton per meter of length (9th CGPM, 1948).

Kelvin: The kelvin, unit of thermodynamic temperature, is the fraction 1/273.16 of the thermodynamic temperature of the triple point of water (13th CGPM, 1967).

Mole: The mole is the amount of substance of a system which contains as many elementary entities as there are atoms in 0.012 kg of carbon-12. When the mole is used, the elementary entities must be specified and may be atoms, molecules, ions, electrons, or other particles or specified groups of such particles (14th CGPM, 1971).

Examples of the use of the mole:

1 mol of H_2 contains about 6.022×10^{23} H_2 molecules, or 12.044×10^{23} H atoms.
1 mol of HgCl has a mass of 236.04 g.
1 mol of Hg_2Cl_2 has a mass of 472.08 g.
1 mol of $Hg_2{}^{2+}$ has a mass of 401.18 g and a charge of 192.97 kC.
1 mol of $Fe_{0.91}S$ has a mass of 82.88 g.
1 mol of e^- has a mass of 548.60 μg and a charge of −96.49 kC.
1 mol of photons whose frequency is 10^{14} Hz has energy of about 39.90 kJ.

Candela: The candela is the luminous intensity in a given direction of a source that emits monochromatic radiation of frequency 540×10^{12} hertz and that has a radiant intensity in that direction of (1/683) watt per steradian (16th CGPM, 1979).

Names and Symbols for the SI Base Units

Physical Quantity	Name of SI Unit	Symbol for SI Unit
Length	meter	m
Mass	kilogram	kg
Time	second	s
Electric current	ampere	A
Thermodynamic temperature	kelvin	K
Amount of substance	mole	mol
Luminous intensity	candela	cd

SI Derived Units with Special Names and Symbols

Physical Quantity	Name of SI Unit	Symbol for SI Unit	Expression in Terms of SI Base Units	
Frequency[1]	hertz	Hz	s^{-1}	
Force	newton	N	$m\ kg\ s^{-2}$	
Pressure, stress	pascal	Pa	$N\ m^{-2}$	$= m^{-1}\ kg\ s^{-2}$
Energy, work, heat	joule	J	N m	$= m^2\ kg\ s^{-2}$
Power, radiant flux	watt	W	$J\ s^{-1}$	$= m^2\ kg\ s^{-3}$
Electric charge	coulomb	C	A s	
Electric potential, electromotive force	volt	V	$J\ C^{-1}$	$= m^2\ kg\ s^{-3}\ A^{-1}$
Electric resistance	ohm	Ω	$V\ A^{-1}$	$= m^2\ kg\ s^{-3}\ A^{-2}$
Electric conductance	siemens	S	Ω^{-1}	$= m^{-2}\ kg^{-1}\ s^3\ A^2$
Electric capacitance	farad	F	$C\ V^{-1}$	$= m^{-2}\ kg^{-1}\ s^4\ A^2$
Magnetic flux density	tesla	T	$V\ s\ m^{-2}$	$= kg\ s^{-2}\ A^{-1}$
Magnetic flux	weber	Wb	V s	$= m^2\ kg\ s^{-2}\ A^{-1}$
Inductance	henry	H	$V\ A^{-1}\ s$	$= m^2\ kg\ s^{-2}\ A^{-2}$
Celsius temperature[2]	degree Celsius	°C	K	

(continued)

SI Derived Units with Special Names and Symbols (continued)

Physical Quantity	Name of SI Unit	Symbol for SI Unit	Expression in Terms of SI Base Units	
Luminous flux	lumen	lm	cd sr	
Illuminance	lux	lx	cd sr m^{-2}	
Activity (radioactive)	becquerel	Bq	s^{-1}	
Absorbed dose (of radiation)	gray	Gy	J kg^{-1}	$= m^2\ s^{-2}$
Dose equivalent (dose equivalent index)	sievert	Sv	J kg^{-1}	$= m^2\ s^{-2}$
Plane angle	radian	rad	1	$= m\ m^{-1}$
Solid angle	steradian	sr	1	$= m^2\ m^{-2}$

[1]For radial (circular) frequency and for angular velocity the unit rad s^{-1}, or simply s^{-1}, should be used, and this may not be simplified to Hz. The unit Hz should be used only for frequency in the sense of cycles per second.

[2]The Celsius temperature θ is defined by the equation:

$$\theta/°C = T/K - 273.15$$

The SI unit of Celsius temperature interval is the degree Celsius, °C, which is equal to the kelvin, K. °C should be treated as a single symbol, with no space between the ° sign and the letter C. (The symbol °K and the symbol ° should no longer be used.)

Units in Use Together with the SI

These units are not part of the SI, but it is recognized that they will continue to be used in appropriate contexts. SI prefixes may be attached to some of these units, such as milliliter, ml; millibar, mbar; megaelectronvolt, MeV; kilotonne, ktonne.

Physical Quantity	Name of Unit	Symbol for Unit	Value in SI Units
Time	minute	min	60 s
Time	hour	h	3600 s
Time	day	d	86,400 s
Plane angle	degree	°	(π/180) rad
Plane angle	minute	′	(π/10,800) rad
Plane angle	second	″	(π/648,000) rad
Length	ångstrom[1]	Å	10^{-10} m
Area	barn	b	10^{-28} m^2
Volume	liter	l, L	$dm^3 = 10^{-3}\ m^3$
Mass	tonne	t	$Mg = 10^3$ kg
Pressure	bar[1]	bar	10^5 Pa $= 10^5$ N m^{-2}
Energy	electronvolt[2]	eV ($= e \times V$)	$\approx 1.60218 \times 10^{-19}$ J
Mass	unified atomic mass unit[2,3]	u ($= m_a(^{12}C)/12$)	$\approx 1.66054 \times 10^{-27}$ kg

[1]The ångstrom and the bar are approved by CIPM for "temporary use with SI units," until CIPM makes a further recommendation. However, they should not be introduced where they are not used at present.

[2]The values of these units in terms of the corresponding SI units are not exact, since they depend on the values of the physical constants e (for the electronvolt) and N_a (for the unified atomic mass unit), which are determined by experiment.

[3]The unified atomic mass unit is also sometimes called the dalton, with symbol Da, although the name and symbol have not been approved by CGPM.

Conversion Constants and Multipliers

Recommended Decimal Multiples and Submultiples

Multiples and Submultiples	Prefixes	Symbols	Multiples and Submultiples	Prefixes	Symbols
10^{18}	exa	E	10^{-1}	deci	d
10^{15}	peta	P	10^{-2}	centi	c
10^{12}	tera	T	10^{-3}	milli	m
10^{9}	giga	G	10^{-6}	micro	μ (Greek mu)
10^{6}	mega	M	10^{-9}	nano	n
10^{3}	kilo	k	10^{-12}	pico	p
10^{2}	hecto	h	10^{-15}	femto	f
10	deca	da	10^{-18}	atto	a

Conversion Factors—Metric to English

To Obtain	Multiply	By
Inches	centimeters	0.3937007874
Feet	meters	3.280839895
Yards	meters	1.093613298
Miles	kilometers	0.6213711922
Ounces	grams	$3.527396195 \times 10^{-2}$
Pounds	kilogram	2.204622622
Gallons (U.S. liquid)	liters	0.2641720524
Fluid ounces	milliliters (cc)	$3.381402270 \times 10^{-2}$
Square inches	square centimeters	0.155003100
Square feet	square meters	10.76391042
Square yards	square meters	1.195990046
Cubic inches	milliliters (cc)	$6.102374409 \times 10^{-2}$
Cubic feet	cubic meters	35.31466672
Cubic yards	cubic meters	1.307950619

Conversion Factors—English to Metric*

To Obtain	Multiply	By
Microns	mils	**25.4**
Centimeters	inches	**2.54**
Meters	feet	**0.3048**
Meters	yards	**0.9144**
Kilometers	miles	**1.609344**
Grams	ounces	28.34952313
Kilograms	pounds	**0.45359237**
Liters	gallons (U.S. liquid)	**3.785411784**
Millimeters (cc)	fluid ounces	29.57352956
Square centimeters	square inches	**6.4516**
Square meters	square feet	**0.09290304**
Square meters	square yards	**0.83612736**
Milliliters (cc)	cubic inches	**16.387064**
Cubic meters	cubic feet	$2.831684659 \times 10^{-2}$
Cubic meters	cubic yards	0.764554858

*Boldface numbers are exact; others are given to ten significant figures where so indicated by the multiplier factor.

Conversion Factors—General*

To Obtain	Multiply	By
Atmospheres	feet of water @ 4°C	2.950×10^{-2}
Atmospheres	inches of mercury @ 0°C	3.342×10^{-2}
Atmospheres	pounds per square inch	6.804×10^{-2}
BTU	foot-pounds	1.285×10^{-3}
BTU	joules	9.480×10^{-4}
Cubic feet	cords	**128**
Degree (angle)	radians	57.2958
Ergs	foot-pounds	1.356×10^{7}
Feet	miles	**5280**
Feet of water @ 4°C	atmospheres	33.90
Foot-pounds	horsepower-hours	1.98×10^{6}
Foot-pounds	kilowatt-hours	2.655×10^{6}
Foot-pounds per min	horsepower	3.3×10^{4}
Horsepower	foot-pounds per sec	1.818×10^{-3}
Inches of mercury @ 0°C	pounds per square inch	2.036
Joules	BTU	1054.8
Joules	foot-pounds	1.35582
Kilowatts	BTU per min	1.758×10^{-2}
Kilowatts	foot-pounds per min	2.26×10^{-5}
Kilowatts	horsepower	0.745712
Knots	miles per hour	0.86897624
Miles	feet	1.894×10^{-4}
Nautical miles	miles	0.86897624
Radians	degrees	1.745×10^{-2}
Square feet	acres	**43,560**
Watts	BTU per min	17.5796

*Boldface numbers are exact; others are given to ten significant figures where so indicated by the multiplier factor.

Temperature Factors

$$°F = 9/5\ (°C) + 32$$

$$\text{Fahrenheit temperature} = 1.8\ (\text{temperature in kelvins}) - 459.67$$

$$°C = 5/9\ [(°F) - 32)]$$

$$\text{Celsius temperature} = \text{temperature in kelvins} - 273.15$$

$$\text{Fahrenheit temperature} = 1.8\ (\text{Celsius temperature}) + 32$$

Conversion of Temperatures

From	To	
°Celsius	°Fahrenheit	$t_F = (t_C \times 1.8) + 32$
	Kelvin	$T_K = t_C + 273.15$
	°Rankine	$T_R = (t_C + 273.15) \times 18$
°Fahrenheit	°Celsius	$t_C = \frac{t_F - 32}{1.8}$
	Kelvin	$T_k = \frac{t_F - 32}{1.8} + 273.15$
	°Rankine	$T_R = t_F + 459.67$
Kelvin	°Celsius	$t_C = T_K - 273.15$
	°Rankine	$T_R = T_K \times 1.8$
°Rankine	Kelvin	$T_K = \frac{T_R}{1.8}$
	°Fahrenheit	$t_F = T_R - 459.67$

Physical Constants

General

Equatorial radius of the Earth = 6378.388 km = 3963.34 miles (statute)
Polar radius of the Earth, 6356.912 km = 3949.99 miles (statute)
1 degree of latitude at 40° = 69 miles
1 international nautical mile = 1.15078 miles (statute) = 1852 m = 6076.115 ft
Mean density of the earth = 5.522 g/cm^3 = 344.7 lb/ft^3
Constant of gravitation $(6.673 \pm 0.003) \times 10^{-8}$ cm^3 gm^{-1} s^{-2}
Acceleration due to gravity at sea level, latitude 45° = 980.6194 cm/s^2 = 32.1726 ft/s^2
Length of seconds pendulum at sea level, latitude 45° = 99.3575 cm = 39.1171 in.
1 knot (international) = 101.269 ft/min = 1.6878 ft/s = 1.1508 miles (statute)/h
1 micron = 10^{-4} cm
1 ångstrom = 10^{-8} cm
Mass of hydrogen atom = $(1.67339 \pm 0.0031) \times 10^{-24}$ g
Density of mercury at 0°C = 13.5955 g/ml
Density of water at 3.98°C = 1.000000 g/ml
Density, maximum, of water, at 3.98°C = 0.999973 g/cm^3
Density of dry air at 0°C, 760 mm = 1.2929 g/l
Velocity of sound in dry air at 0°C = 331.36 m/s – 1087.1 ft/s
Velocity of light in vacuum = $(2.997925 \pm 0.000002) \times 10^{10}$ cm/s
Heat of fusion of water 0°C = 79.71 cal/g
Heat of vaporization of water 100°C = 539.55 cal/g
Electrochemical equivalent of silver 0.001118 g/s international amp
Absolute wavelength of red cadmium light in air at 15° C, 760 mm pressure = 6438.4696 Å
Wavelength of orange-red line of krypton 86 = 6057.802 Å

π Constants

π = 3.14159 26535 89793 23846 26433 83279 50288 41971 69399 37511
$1/\pi$ = 0.31830 98861 83790 67153 77675 26745 02872 40689 19291 48091
π^2 = 9.8690 44010 89358 61883 44909 99876 15113 53136 99407 24079
$\log_e \pi$ = 1.14472 98858 49400 17414 34273 51353 05871 16472 94812 91531
$\log_{10} \pi$ = 0.49714 98726 94133 85435 12682 88290 89887 36516 78324 38044
$\log_{10} \sqrt{2\pi}$ = 0.39908 99341 79057 52478 25035 91507 69595 02099 34102 92128

Constants Involving *e*

e = 2.71828 18284 59045 23536 02874 71352 66249 77572 47093 69996
$1/e$ = 0.36787 94411 71442 32159 55237 70161 46086 74458 11131 03177
e^2 = 7.38905 60989 30650 22723 04274 60575 00781 31803 15570 55185
$M = \log_{10} e$ = 0.43429 44819 03251 82765 11289 18916 60508 22943 97005 80367
$1/M \cdot = \log_e 10$ = 2.30258 50929 94045 68401 79914 54684 36420 67011 01488 62877
$\log_{10} M$ = 9.63778 43113 00536 78912 29674 98645 –10

Numerical Constants

$\sqrt{2}$ = 1.41421 35623 73095 04880 16887 24209 69807 85696 71875 37695
$3\sqrt{2}$ = 1.25992 10498 94873 16476 72106 07278 22835 05702 51464 70151
$\log_e 2$ = 0.69314 71805 59945 30941 72321 21458 17656 80755 00134 36026
$\log_{10} 2$ = 0.30102 99956 63981 19521 37388 94724 49302 67881 89881 46211

$\sqrt{3}$ = 1.73205 08075 68877 29352 74463 41505 87236 69428 05253 81039
$\sqrt[3]{3}$ = 1.44224 95703 07408 38232 16383 10780 10958 83918 69253 49935
$\log_e 3$ = 1.09861 22886 68109 69139 52452 36922 52570 46474 90557 82275
$\log_{10} 3$ = 0.47712 12547 19662 43729 50279 03255 11530 92001 28864 19070

Symbols and Terminology for Physical and Chemical Quantities

Name	Symbol	Definition	SI Unit
		Classical Mechanics	
Mass	m		kg
Reduced mass	μ	$\mu = m_1 m_2/(m_1 + m_2)$	kg
Density, mass density	ρ	$\rho = M/V$	kg m^{-3}
Relative density	d	$d = \rho/\rho^\theta$	1
Surface density	ρ_A, ρ_S	$\rho_A = m/A$	kg m^{-2}
Momentum	p	$p = mv$	kg m s^{-1}
Angular momentum, action	L	$l = r$ ¥ p	J s
Moment of inertia	I, J	$I = \Sigma m_i r_i^2$	kg m^2
Force	F	$F = d\mathbf{p}/dt = ma$	N
Torque, moment of a force	T, (M)	$T = r \times \mathbf{F}$	N m
Energy	E		J
Potential energy	E_p, V, Φ	E_p=Fds	J
Kinetic energy	E_k, T, K	$e_k = (1/2)mv^2$	J
Work	W, w	$w = F$ds	J
Hamilton function	H	$H(q, p) = T(q, p) + V(q)$	J
Lagrange function	L	$L(q, \dot{q})T(q, \dot{q}) - V(q)$	J
Pressure	p, P	$p = F/A$	Pa, N m^{-2}
Surface tension	γ, σ	$\gamma = dW/dA$	N m^{-1}, J m^{-2}
Weight	G, (W, P)	$G = mg$	N
Gravitational constant	G	$F = Gm_1m_2/r^2$	N m^2 kg^{-2}
Normal stress	σ	$\sigma = F/A$	Pa
Shear stress	τ	$\tau = F/A$	Pa
Linear strain, relative elongation	ε, e	$\varepsilon = \Delta l/l$	1
Modulus of elasticity, Young's modulus	E	$E = \sigma/\varepsilon$	Pa
Shear strain	γ	$\gamma = \Delta x/d$	1
Shear modulus	G	$G = \tau/\gamma$	Pa
Volume strain, bulk strain	θ	$\theta = \Delta V/V_0$	1
Bulk modulus,	K	$K = -V_0(dp/dV)$	Pa
compression modulus	η, μ	$\tau_{x,z} = \eta(dv_x/dz)$	Pa s
Viscosity, dynamic viscosity			
Fluidity	ϕ	$\phi = 1/\eta$	m kg^{-1} s
Kinematic viscosity	v	$v = \eta/\rho$	m^2 s^{-1}
Friction coefficient	μ, (f)	$F_{frict} = \mu F_{norm}$	1
Power	P	$P = dW/dt$	W
Sound energy flux	P, P_a	$P = dE/dt$	W
Acoustic factors			
Reflection factor	ρ	$\rho = P_t/P_0$	1
Acoustic absorption factor	α_a, (α)	$\alpha_a = 1 - \rho$	1
Transmission factor	τ	$\tau = P_{tr}/P_0$	1
Dissipation factor	δ	$\delta = \alpha_a - \tau$	1

(continued)

Symbols and Terminology for Physical and Chemical Quantities (continued)

Name	Symbol	Definition	SI Unit
Electricity and Magnetism			
Quantity of electricity, electric charge	Q		C
Charge density	ρ	$\rho = Q/V$	$\mathrm{C\ m^{-3}}$
Surface charge density	σ	$\sigma = Q/A$	$\mathrm{C\ m^{-2}}$
Electric potential	V, ϕ	$V = \mathrm{d}W/\mathrm{d}Q$	$\mathrm{V, J\ C^{-1}}$
Electric potential difference	$U, \Delta V, \Delta\phi$	$U = V_2 - V_1$	V
Electromotive force	E	$E = (F/Q)\mathrm{d}s$	V
Electric field strength	$\mathbf{E}$	$\mathbf{E} = \mathbf{F}/Q = -\mathrm{grad}\ V$	$\mathrm{V\ m^{-1}}$
Electric flux	Ψ	$\Psi = \mathbf{D}\mathrm{d}\mathbf{A}$	C
Electric displacement	$\mathbf{D}$	$\mathbf{D} = \varepsilon\mathbf{E}$	$\mathrm{C\ m^{-2}}$
Capacitance	C	$C = Q/U$	$\mathrm{F, C\ V^{-1}}$
Permittivity	ε	$D = \varepsilon E$	$\mathrm{F\ m^{-1}}$
Permittivity of vacuum	ε_0	$\varepsilon_0 = \mu_0^{-1} c_0^{-2}$	$\mathrm{F\ m^{-1}}$
Relative permittivity	ε_r	$\varepsilon_r = \varepsilon/\varepsilon_0$	1
Dielectric polarization (dipole moment per volume)	$\mathbf{P}$	$\mathbf{P} = \mathbf{D} - \varepsilon_0\mathbf{E}$	$\mathrm{C\ m^{-2}}$
Electric susceptibility	χ_e	$\chi_e = \varepsilon_r - 1$	1
Electric dipole moment	$\mathbf{p}, \mu$	$\mathbf{p} = Q\mathbf{r}$	C m
Electric current	I	$I = \mathrm{d}Q/\mathrm{d}t$	A
Electric current density	$\mathbf{j}, \mathbf{J}$	$I = \mathbf{j}\mathrm{d}x\mathbf{A}$	$\mathrm{A\ m^{-2}}$
Magnetic flux density, magnetic induction	$\mathbf{B}$	$\mathbf{F} = Qv \times \mathbf{B}$	T
Magnetic flux	Φ	$\Phi = \mathbf{B}\mathrm{d}A$	Wb
Magnetic field strength	$\mathbf{H}$	$\mathbf{B} = \mu\mathbf{H}$	$\mathrm{A\ M^{-1}}$
Permeability	μ	$\mathbf{B} = \mu\mathbf{H}$	$\mathrm{N\ A^{-2}, H\ m^{-1}}$
Permeability of vacuum	μ_0		$\mathrm{H\ m^{-1}}$
Relative permeability	μ_r	$\mu_r = \mu/\mu_0$	1
Magnetization (magnetic dipole moment per volume)	$\mathbf{M}$	$\mathbf{M} = \mathbf{B}/\mu_0 - \mathbf{H}$	$\mathrm{A\ m^{-1}}$
Magnetic susceptibility	$\chi, \kappa, (\chi_m)$	$\chi = \mu_r - 1$	1
Molar magnetic susceptibility	χ_m	$\chi_m = V_m\chi$	$\mathrm{m^3\ mol^{-1}}$
Magnetic dipole moment	$\mathbf{m}, \mu$	$E_p = -\mathbf{m}\cdot\mathbf{B}$	$\mathrm{A\ m^2, J\ T^{-1}}$
Electrical resistance	R	$\mathbf{P}=\mathbf{Y}/\mathbf{I}$	Ω
Conductance	G	$G = 1/R$	S
Loss angle	δ	$\delta = (\pi/2) + \phi_I - \phi_U$	1, rad
Reactance	X	$X = (U/I)\sin\delta$	Ω
Impedance (complex impedance)	Z	$Z = R + \mathrm{i}X$	Ω
Admittance (complex admittance)	Y	$Y = 1/Z$	S
Susceptance	B	$Y = G + iB$	S
Resistivity	ρ	$\rho = E/j$	Ω m
Conductivity	κ, γ, σ	$\kappa = 1/\rho$	$\mathrm{S\ m^{-1}}$
Self-inductance	L	$E = -L(\mathrm{d}I/\mathrm{d}t)$	H
Mutual inductance	M, L_{12}	$E_1 = L_{12}(\mathrm{D}i_2/\mathrm{d}t)$	H
Magnetic vector potential	$\mathbf{A}$	$\mathbf{B} = \mathbf{\nabla} \times \mathbf{A}$	$\mathrm{Wb\ m^{-1}}$
Poynting vector	$\mathbf{S}$	$\mathbf{S} = \mathbf{E} \times \mathbf{H}$	$\mathrm{W\ m^{-2}}$
Electromagnetic Radiation			
Wavelength	λ		m
Speed of light			$\mathrm{m\ s^{-1}}$
in vacuum	c_0		
in a medium	c	$c = c_0/n$	

(continued)

Symbols and Terminology for Physical and Chemical Quantities (continued)

Name	*Symbol*	*Definition*	*SI Unit*
		Electromagnetic Radiation	
Wavenumber in vacuum	V	$V = V/c_0 = 1/n\lambda$	m^{-1}
Wavenumber (in a medium)	σ	$\sigma = 1/\lambda$	m^{-1}
Frequency	v	$v = c/\lambda$	Hz
Circular frequency, pulsatance	ω	$\omega = 2\pi v$	s^{-1}, rad s^{-1}
Refractive index	n	$n = c_0/c$	1
Planck constant	h		J s
Planck constant/2π	$\hbar$	$\hbar = h/2\pi$	J s
Radiant energy	Q, W		J
Radiant energy density	ρ, w	$\rho = Q/V$	J m^{-3}
Spectral radiant energy density			
in terms of frequency	ρ_v, w_v	$\rho_v = \delta\rho/dv$	J m^{-3} Hz^{-1}
in terms of wavenumber	$\rho_{\bar{v}}$, $w_{\bar{v}}$	$\rho_{\bar{v}} = d\rho/d\bar{v}$	J m^{-2}
in terms of wavelength	ρ_λ, w_λ	$\rho_\lambda = \delta\rho/d\lambda$	J m^{-4}
Einstein transition probabilities			
Spontaneous emission	A_{nm}	$dN_n/dt = -A_{nm}N_n$	s^{-1}
Stimulated emission	B_{nm}	$dn_n/dt = -\rho\bar{v}(\bar{V}_{nm}) \times B_{nm}N_n$	s kg^{-1}
Radiant power, radiant energy per time	Φ, P	$\Phi = dQ/dt$	W
Radiant intensity	I	$I = d\Phi/d\Omega$	W sr^{-1}
Radiant exitance (emitted radiant flux)	M	$M = d\Phi/dA_{source}$	W m^{-2}
Irradiance (radiant flux received)	E, (I)	$E = d\Phi/\delta A$	W m^{-2}
Emittance	ε	$\varepsilon = M/M_{bb}$	1
Stefan–Boltzmann constant	σ	$M_{bb} = \sigma T^4$	W m^{-2} K^{-4}
First radiation constant	c_1	$c_1 = 2\pi h c_0{}^2$	W m^2
Second radiation constant	c_2	$c_2 = hc_0/k$	K m
Transmittance, transmission factor	τ, T	$\tau = \Phi_{tr}/\Phi_0$	1
Absorptance, absorption factor	α	$\alpha = \phi_{abs}/\phi_0$	1
Reflectance, reflection factor	ρ	$\rho = \phi_{refl}/\Phi_0$	1
(Decadic) absorbance	A	$A = \lg(1 - \alpha_i)$	1
Napierian absorbance	B	$B = \ln(1 - \alpha_i)$	1
Absorption coefficient			
(Linear) decadic	a, K	$a = A/l$	m^{-1}
(Linear) napierian	α	$\alpha = B/l$	m^{-1}
Molar (decadic)	ε	$\varepsilon = a/c = A/cl$	m^2 mol^{-1}
Molar napierian	κ	$\kappa = \alpha/c = B/cl$	m^2 mol^{-1}
Absorption index	k	$k = \alpha/4\pi\bar{v}$	1
Complex refractive index	$\hat{n}$	$\hat{n} = n + ik$	1
Molar refraction	R, R_m	$R = \frac{(n^2-1)}{(n^2+2)} V_m$	m^3 mol^{-1}
Angle of optical rotation	α		1, rad
		Solid State	
Lattice vector	$\mathbf{R}$, $\mathbf{R}_0$		m
Fundamental translation vectors for the crystal lattice	$\mathbf{a}_1$; $\mathbf{a}_2$; $\mathbf{a}_3$, $\mathbf{a}$; $\mathbf{b}$; $\mathbf{c}$	$\mathrm{R} = \mathrm{n}_1\mathbf{a}_1 + \mathrm{n}_2\mathbf{a}_2 + \mathrm{n}_3\mathbf{a}_3$	m
(Circular) reciprocal lattice vector	$\mathbf{G}$	$\mathrm{G} \cdot \mathrm{R} = 2\pi m$	m^{-1}

(continued)

Symbols and Terminology for Physical and Chemical Quantities (continued)

Name	Symbol	Definition	SI Unit
Solid State			
(Circular) fundamental translation vectors for the reciprocal lattice	$\mathbf{b}_1$; $\mathbf{b}_2$; $\mathbf{b}_3$, $\mathbf{a}^*$; $\mathbf{b}^*$; $\mathbf{c}^*$	$\mathbf{a}_i \cdot \mathbf{b}_k = 2\pi\delta_{ik}$	m^{-1}
Lattice plane spacing	d		m
Bragg angle	θ	$n\lambda = 2d \sin \theta$	1, rad
Order of reflection	n		1
Order parameters			
Short range	σ		1
Long range	s		1
Burgers vector	b		m
Particle position vector	r, R_j		m
Equilibrium position vector of an ion	R_o		m
Displacement vector of an ion	$\mathbf{u}$	$\mathbf{u} = \mathbf{R} - \mathbf{R}_0$	m
Debye–Waller factor	B, D		1
Debye circular wavenumber	q_D		m^{-1}
Debye circular frequency	ω_D		s^{-1}
Grüneisen parameter	γ, Γ	$\gamma = \alpha V/\kappa C_V$	1
Madelung constant	α, $\mathcal{M}$	$E_{coul} = \frac{\alpha N_A z_+ z_- e^2}{4\pi\varepsilon_0 R_0}$	1
Density of states	N_E	$N_E = dN(E)/dE$	$J^{-1}\ m^{-3}$
(Spectral) density of vibrational modes	N_ω, g	$N_\omega = dN(\omega)/d\omega$	$s\ m^{-3}$
Resistivity tensor	ρ_{ik}	$E = \rho \cdot j$	Ω m
Conductivity tensor	σ_{ik}	$\sigma = \rho^{-1}$	$S\ m^{-1}$
Thermal conductivity tensor	λ_{ik}	$J_q = -\lambda \cdot \text{grad}\ T$	$W\ m^{-1}\ K^{-1}$
Residual resistivity	ρ_R		Ω m
Relaxation time	τ	$\tau = l/v_F$	s
Lorenz coefficient	L	$L = \lambda/\sigma T$	$V^2\ K^{-2}$
Hall coefficient	A_H, R_H	$\mathbf{E} = \rho \cdot \mathbf{j} + R_H(\mathbf{B} \times \mathbf{j})$	$m^3\ C^{-1}$
Thermoelectric force	E		V
Peltier coefficient	Π		V
Thomson coefficient	μ,(τ)		$V\ K^{-1}$
Work function	Φ	$\Phi = E_\infty - E_F$	J
Number density, number concentration	n, (p)		m^{-3}
Gap energy	E_γ		J
Donor ionization energy	E_δ		J
Acceptor ionization energy	E_α		J
Fermi energy	E_Φ, ε_F		J
Circular wave vector, propagation vector	$\mathbf{k}$, $\mathbf{q}$	$k = 2\pi/\lambda$	m^{-1}
Bloch function	$u_k(\mathbf{r})$	$\psi(\mathbf{r}) = u_k(\mathbf{r}) \exp(i\mathbf{k} \cdot \mathbf{r})$	$m^{-3/2}$
Charge density of electrons	ρ	$\rho(\mathbf{r}) = -e\psi^*(\mathbf{r})\psi(\mathbf{r})$	$C\ m^{-3}$
Effective mass	m^*		kg
Mobility	μ	$\mu = v_{drift}/E$	$m^2\ V^{-1}\ s^{-1}$
Mobility ratio	b	$b = \mu_n/\mu_p$	1
Diffusion coefficient	D	$dN/dt = -DA(dn/dx)$	$m^2\ s^{-1}$
Diffusion length	L	$L = \sqrt{D\tau}$	m
Characteristic (Weiss) temperature	ϕ, ϕ_W		K
Curie temperature	T_C		K
Néel temperature	T_N		K

Credits

Material in Section III was reprinted from the following sources:

D. R. Lide, Ed., *CRC Handbook of Chemistry and Physics,* 76th ed., Boca Raton, FL: CRC Press, 1992: International System of Units (SI), conversion constants and multipliers (conversion of temperatures), symbols and terminology for physical and chemical quantities, fundamental physical constants, classification of electromagnetic radiation.

D. Zwillinger, Ed., *CRC Standard Mathematical Tables and Formulae,* 30th ed., Boca Raton, FL: CRC Press, 1996: Greek alphabet, conversion constants and multipliers (recommended decimal multiples and submultiples, metric to English, English to metric, general, temperature factors), physical constants, series expansion.

Probability for Electrical and Computer Engineers

Charles W. Therrien

The Algebra of Events

The study of probability is based upon experiments that have uncertain outcomes. Collections of these outcomes comprise *events* and the collection of all possible outcomes of the experiment comprise what is called the *sample space*, denoted by S. Outcomes are members of the sample space and events of interest are represented as *sets* of outcomes (see Figure III.1).

The algebra $\mathcal{A}$ that deals with representing events is the usual set algebra. If A is an event, then A^c (the *complement* of A) represents the event that "A did not occur." The complement of the sample space is the *null event*, $\varnothing = S^c$. The event that *both* event A_1 and event A_2 have occurred is the intersection, written as "$A_1 \cdot A_2$" or "A_1A_2" while the event that *either* A_1 or A_2 *or both* have occurred is the union, written as "$A_1 + A_2$."[1]

Table III.1 lists the two postulates that define the algebra $\mathcal{A}$, while Table III.2 lists seven axioms that define properties of its operations. Together these tables can be used to show all of the properties of the algebra of events. Table III.3 lists some additional useful relations that can be derived from the axioms and the postulates.

Since the events "$A_1 + A_2$" and "A_1A_2" are included in the algebra, it follows by induction that for any finite number of events $A_1 + A_2 + \cdots + A_N$ and $A_1 \cdot A_2 \cdots A_N$ are also included in the algebra. Since problems often involve the union or intersection of an *infinite* number of events, however, the algebra of events must be defined to include these infinite intersections and unions. This extension to infinite unions and intersections is known as a sigma algebra.

A set of events that satisfies the two conditions:

1. $A_iA_j = \varnothing \neq$ for $\neq$ $i \neq j$
2. $A_1 + A_2 + A_3 + \cdots = S$

is known as a *partition* and is important for the solution of problems in probability. The events of a partition are said to be *mutually exclusive* and *collectively exhaustive*. The most fundamental partition is the set outcomes defining the random experiment, which comprise the sample space by definition.

Probability

Probability measures the likelihood of occurrence of events represented on a scale of 0 to 1. We often estimate probability by measuring the *relative frequency* of an event, which is defined as

$$\text{relative frequency} = \frac{\text{number of occurrences of the event}}{\text{number of repetitions of the experiment}}$$

(for a large number of repetitions). Probability can be defined formally by the following axioms:

(I) The probability of any event is nonnegative:

$$\Pr[A] \geqslant 0 \tag{III.1}$$

(II) The probability of the universal event (i.e., the entire sample space) is 1:

$$\Pr[S] = 1 \tag{III.2}$$

[1]Some authors use $\cap$ and $\cup$ rather than $\cdot$ and $+$, respectively.

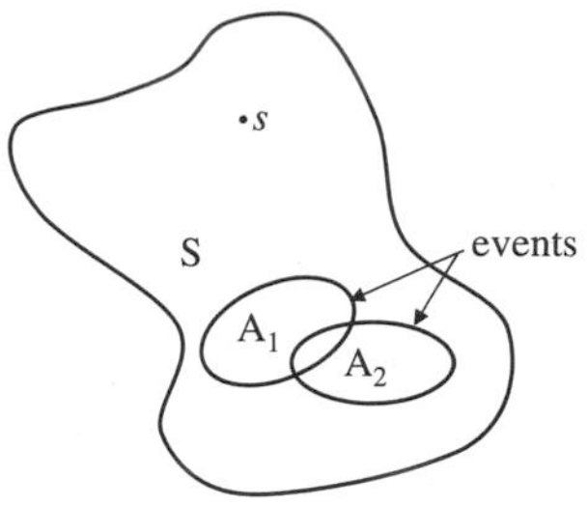

FIGURE III.1 Abstract representation of the sample space S with outcome s and sets A_1 and A_2 representing events.

(III) If A_1 and A_2 are mutually exclusive, i.e., $A_1A_2 = \varnothing$, then

$$\Pr[A_1 + A_2] = \Pr[A_1] + \Pr[A_2] \tag{III.3}$$

(IV) If $\{A_i\}$ represent a countably infinite set of mutually exclusive events, then

$$\Pr[A_1 + A_2 + A_3 + \cdot\cdot\cdot] = \sum_{i=1}^{\infty} \Pr[A_i] \quad (\text{ if } A_iA_j = \varnothing \quad i \neq j) \tag{III.4}$$

Note that although the additivity of probability for any finite set of disjoint events follows from (III), the property has to be stated explicitly for an infinite set in (IV). These axioms and the algebra of events can be used to show a number of other important properties which are summarized in Table III.4. The last item in the table is an especially important formula since it uses probabilistic information about

TABLE III.1 Postulates for an Algebra of Events

1.	*If* $A \in \mathcal{A}$ *then* $A^c \in \mathcal{A}$
2.	*If* $A_1 \in \mathcal{A}$ *and* $A_2 \in \mathcal{A}$ *then* $A_1 + A_2 \in \mathcal{A}$

TABLE III.2 Axioms of Operations on Events

$A_1A_1^c = \varnothing$	Mutual exclusion
$A_1S = A_1$	Inclusion
$(A_1^c)^c = A_1$	Double complement
$A_1 + A_2 = A_2 + A_1$	Commutative law
$A_1 + (A_2 + A_3) = (A_1 + A_2) + A_3$	Associative law
$A_1(A_2 + A_3) = A_1A_2 + A_1A_3$	Distributive law
$(A_1A_2)^c = A_1^c + A_2^c$	DeMorgan's law

TABLE III.3 Additional Identities in the Algebra of Events

$S^c = \varnothing$	
$A_1 + \varnothing = A_1$	Inclusion
$A_1A_2 = A_2A_1$	Commutative law
$A_1(A_2A_3) = (A_1A_2)A_3$	Associative law
$A_1 + (A_2A_3) = (A_1 + A_2)(A_1 + A_3)$	Distributive law
$(A_1 + A_2)^c = A_1{}^cA_2{}^c$	DeMorgan's law

TABLE III.4 Some Corollaries Derived from the Axioms of Probability

$\Pr[A^c] = 1 - \Pr[A]$
$0 \leqslant \Pr[A] \leqslant 1$
If $A_1 \subseteq A_2$ then $\Pr[A_1] \leqslant \Pr[A_2]$
$\Pr[\varnothing] = 0$
If $A_1A_2 = \varnothing$ – then = $\Pr[A_1A_2] = 0$
$\Pr[A_1 + A_2] = \Pr[A_1] + \Pr[A_2] - \Pr[A_1A_2]$

individual events to compute the probability of the union of two events. The term $\Pr[A_1A_2]$ is referred to as the *joint probability* of the two events. This last equation shows that the probabilities of two events add as in Equation (III.3) only if their joint probability is 0. The joint probability is 0 when the two events have no intersection ($A_1A_2 = \varnothing$).

Two events are said to be statistically *independent* if and only if

$$\Pr[A_1A_2] = \Pr[A_1]\cdot\Pr[A_2] \quad \text{(independent events)} \tag{III.5}$$

This definition is not derived from the earlier properties of probability. An argument to give this definition intuitive meaning can be found in Ref. [1]. Independence occurs in problems where two events are not influenced by one another and Equation (III.5) simplifies such problems considerably.

A final important result deals with partitions. *A partition* is a finite or countably infinite set of events $A_1, A_2, A_3, \ldots$ that satisfy the two conditions:

$$A_iA_j = \varnothing \text{ for } i \neq j$$

$$A_1 + A_2 + A_3 + \cdots = S$$

The events in a partition satisfy the relation:

$$\sum_i \Pr[A_i] = 1 \tag{III.6}$$

Further, if B is *any* other event, then

$$\Pr[B] = \sum_i \Pr[A_iB] \tag{III.7}$$

The latter result is referred to as the *principle of total probability* and is frequently used in solving problems. The principle is illustrated by a Venn diagram in Figure III.2. The rectangle represents the sample space and other events are defined therein. The event B is seen to be comprised of all of the pieces

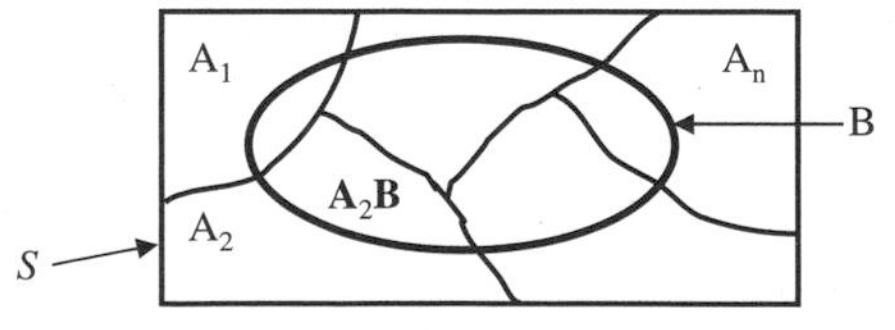

FIGURE III.2 Venn diagram illustrating the principle of total probability.

that represent intersections or overlap of event B with the events A_i. This is the graphical interpretation of Equation (III.7).

An Example

Simon's Surplus Warehouse has large barrels of mixed electronic components (parts) that you can buy by the handful or by the pound. You are not allowed to select parts individually. Based on your previous experience, you have determined that in one barrel, 29% of the parts are bad (faulted), 3% are bad resistors, 12% are good resistors, 5% are bad capacitors, and 32% are diodes. You decide to assign probabilities based on these percentages. Let us define the following events:

Event	Symbol
Bad (faulted) component	B
Good component	G
Resistor	R
Capacitor	C
Diode	D

A Venn diagram representing this situation is shown below along with probabilities of various events as given:

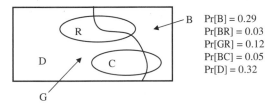

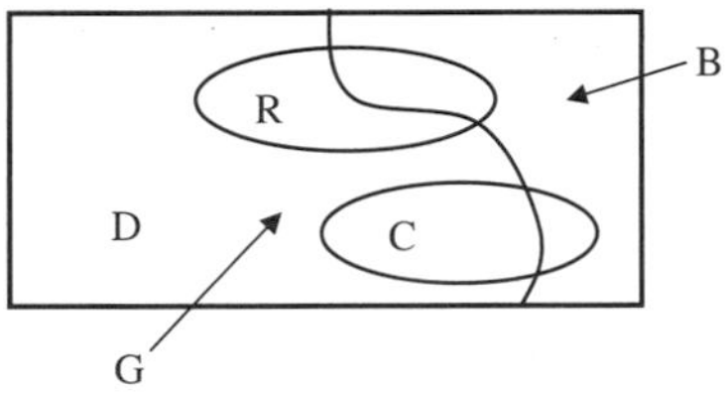

Pr[B] = 0.29
Pr[BR] = 0.03
Pr[GR] = 0.12
Pr[BC] = 0.05
Pr[D] = 0.32

Note that since any component must be a resistor, capacitor, or diode, the region labeled D in the diagram represents everything in the sample space which is not included in R or C.

We can answer a number of questions.

1. What is the probability that a component is a resistor (either good *or* bad)?
   Since the events B and G form a partition of the sample space, we can use the principle of total probability Equation (III.7) to write:

$$\Pr[R] = \Pr[GR] + \Pr[BR] = 0.12 + 0.03 = 0.15$$

2. Are bad parts and resistors independent?
   We know that $\Pr[BR] = 0.03$ and we can compute:

$$\Pr[B] \cdot \Pr[R] = (0.29)(0.15) = 0.0435$$

   Since $\Pr[BR] \neq \Pr[B] \cdot \Pr[R]$, the events are *not* independent.
3. You have no use for either bad parts or resistors. What is the probability that a part is either bad and/or a resistor?

Using the formula from Table III.4 and the previous result we can write:

$$\Pr[B + R] = \Pr[B] + \Pr[R] - \Pr[BR] = 0.29 + 0.15 - 0.03 = 0.41$$

4. What is the probability that a part is useful to you?
   Let U represent the event that the part is useful. Then (see Table III.4):

$$\Pr[U] = 1 - \Pr[U^c] = 1 - 0.41 = 0.59$$

5. What is the probability of a bad diode?
   Observe that the events R, C, and D form a partition, since a component has to be one and only one type of part. Then using Equation (III.7) we write:

$$\Pr[B] = \Pr[BR] + \Pr[BC] + \Pr[BD]$$

   Substituting the known numerical values and solving yields

$$0.29 = 0.03 + 0.05 + \Pr[BD] \text{ or } \Pr[BD] = 0.21$$

Conditional Probability and Bayes' Rule

The *conditional* probability of an event A_1 given that an event A_2 has occurred is defined by

$$\Pr[A_1|A_2] = \frac{\Pr[A_1A_2]}{\Pr[A_2]} \tag{III.8}$$

($\Pr[A_1|A_2]$ is read "probability of A_1 *given* A_2.") As an illustration, let us compute the probability that a component in the previous example is bad given that it is a resistor:

$$\Pr[B|R] = \frac{\Pr[BR]}{\Pr[R]} = \frac{0.03}{0.15} = 0.2$$

(The value for $\Pr[R]$ was computed in question 1 of the example.) Frequently the statement of a problem is in terms of conditional probability rather than joint probability, so Equation (III.8) is used in the form:

$$\Pr[A_1A_2] = \Pr[A_1|A_2] \cdot \Pr[A_2] = \Pr[A_2|A_1] \cdot \Pr[A_1] \tag{III.9}$$

(The last expression follows because $\Pr[A_1A_2]$ and $\Pr[A_2A_1]$ are the same thing.) Using this result, the principle of total probability Equation (III.7) can be rewritten as

$$\Pr[B] = \sum_j \Pr[B|A_j] \Pr[A_j] \tag{III.10}$$

where B is any event and $\{A_j\}$ is a set of events that forms a partition.

Now, consider any one of the events A_i in the partition. It follows from Equation (III.9) that

$$\Pr[A_i|B] = \frac{\Pr[B|A_i] \cdot \Pr[A_i]}{\Pr[B]}$$

Then substituting in Equation (III.10) yields:

$$\Pr[A_i|B] = \frac{\Pr[B|A_i] \cdot \Pr[A_i]}{\sum_j \Pr[B|A_j] \Pr[A_j]} \tag{III.11}$$

This result is known as *Bayes' theorem* or *Bayes' rule.* It is used in a number of problems that commonly arise in electrical engineering. We illustrate and end this section with an example from the field of communications.

Communication Example

The transmission of bits over a binary communication channel is represented in the drawing below:

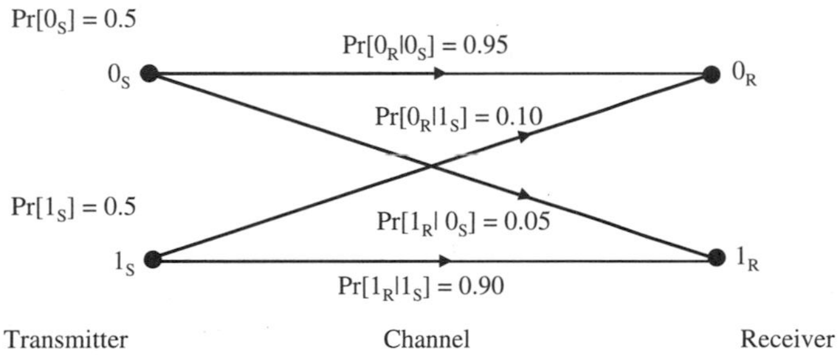

where we use notation like 0_S, $0_R \ldots$ to denote events "0 sent," "0 received," etc. When a 0 is transmitted, it is correctly received with probability 0.95 or incorrectly received with probability 0.05. That is, $\Pr[0_R|0_S] = 0.95$ and $\Pr[1_R|0_S] = 0.05$. When a 1 is transmitted, it is correctly received with probability 0.90 and incorrectly received with probability 0.10. The probabilities of sending a 0 or a 1 are denoted by $\Pr[0_S]$ and $\Pr[1_S]$. It is desired to compute the *probability of error* for the system.

This is an application of the principle of total probability. The two events 0_S and 1_S are mutually exclusive and collectively exhaustive and thus form a partition. Take the event B to be the event that an error occurs. It follows from Equation (III.10) that

$$\begin{aligned}\Pr[\text{error}] &= \Pr[\text{error}|0_S]\Pr[0_S] + \Pr[\text{error}|1_S]\Pr[1_S] \\ &= \Pr[1_R|0_S]\Pr[0_S] + \Pr[0_R|1_S]\Pr[1_S] \\ &= (0.05)(0.5) + (0.10)(0.5) = 0.075\end{aligned}$$

Next, given that an error has occurred, let us compute the probability that a 1 was sent or a 0 was sent. This is an application of Bayes' rule. For a 1, Equation (III.11) becomes

$$\Pr[1_S|\text{error}] = \frac{\Pr[\text{error}|1_S]\Pr[1_S]}{\Pr[\text{error}|1_S]\Pr[1_S] + \Pr[\text{error}|0_S]\Pr[0_S]}$$

Substituting the numerical values then yields:

$$\Pr[1_S|\text{error}] = \frac{(0.10)(0.5)}{(0.10)(0.5) + (0.05)(0.5)} \approx 0.667$$

For a 0, a similar analysis applies:

$$\Pr[0_S|\text{error}] = \frac{\Pr[\text{error}|0_S]\Pr[0_S]}{\Pr[\text{error}|1_S]\Pr[1_S] + \Pr[\text{error}|0_S]\Pr[0_S]}$$
$$= \frac{(0.05)(0.5)}{(0.10)(0.5) + (0.05)(0.5)} \approx 0.333$$

The two resulting probabilities sum to 1 because 0_S and 1_S form a partition for the experiment.

Reference

1. C. W. Therrien and M. Tummala, *Probability for Electrical and Computer Engineers*. Boca Raton, FL: CRC Press, 2004.

Indexes

Author Index

A

B

C

D

E

F

G

H

J

K

L

M

N

O

P

R

S

T

U

V

W

Y

Subject Index

A

Page on which a term is defined is indicated in bold.

F

G

N

O

P

S